CALCULUS AND ITS APPLICATIONS

7TH EDITION

CALCULUS AND ITS APPLICATIONS

Larry J. Goldstein
Goldstein Educational Technologies

David C. Lay
University of Maryland

David I. Schneider
University of Maryland

Prentice Hall, Upper Saddle River, New Jersey 07458

Library of Congress Cataloging-in-Publication Data

GOLDSTEIN, LARRY JOEL.
   Calculus and its applications / Larry J. Goldstein, David C. Lay,
David I. Schneider. — 7th ed.
     p.   cm.
   Includes index.
   ISBN 0-13-321449-4
   1. Calculus.   I. Lay, David C. II. Schneider, David I.
III. Title.
QA303.G625   1996
515—dc20   95-10727
                           CIP

Acquisition editor: George Lobell
Editorial/production supervision: Judy Winthrop
Interior design: Christine Gehring Wolf
Cover design: Jeanette Jacobs
Design director: Amy Rosen
Manufacturing buyer: Alan Fischer

© 1996, 1993, 1990, 1987, 1984, 1980, 1977 by Prentice-Hall, Inc.
Simon & Schuster / A Viacom Company
Upper Saddle River, New Jersey 07458

Printed in the United States of America

10 9 8 7 6 5 4 3

ISBN 0-13-321449-4

Prentice-Hall International (UK) Limited, *London*
Prentice-Hall of Australia Pty. Limited, *Sydney*
Prentice-Hall Canada Inc., *Toronto*
Prentice-Hall Hispanoamericana, S. A., *Mexico*
Prentice-Hall of India Private Limited, *New Delhi*
Prentice-Hall of Japan, Inc., *Tokyo*
Simon & Schuster Asia Pte. Ltd., *Singapore*
Editora Prentice-Hall do Brasil, Ltda., *Rio de Janeiro*

CONTENTS

3 TECHNIQUES OF DIFFERENTIATION, 249

4 THE EXPONENTIAL AND NATURAL LOGARITHM FUNCTIONS, 287

5 APPLICATIONS OF THE EXPONENTIAL AND NATURAL LOGARITHM FUNCTIONS, 325

6 THE DEFINITE INTEGRAL, 373

7 FUNCTIONS OF SEVERAL VARIABLES, 441

8 THE TRIGONOMETRIC FUNCTIONS, 509

9 TECHNIQUES OF INTEGRATION, 547

10 DIFFERENTIAL EQUATIONS, 601

PREFACE

We have been very pleased with the enthusiastic response to the sixth edition of *Calculus and Its Applications* by teachers and students alike. The present work incorporates many of the suggestions they have put forward.

Although there are many changes, we have preserved the approach and the flavor. Our goals remain the same: to begin the calculus as soon as possible; to present calculus in an intuitive yet intellectually satisfying way; and to illustrate the many applications of calculus to the biological, social, and management sciences. We have tried to achieve these goals while paying close attention to students' real and potential problems in learning calculus. Our main concern, as always, is: Will it work for the students? Listed on the following pages are some of the features that illustrate various aspects of this student-oriented approach.

APPLICATIONS

We provide realistic applications that illustrate the uses of calculus in other disciplines. The reader may survey the variety of applications by turning to the Index of Applications on the inside cover. Wherever possible, we have attempted to use applications to motivate the mathematics.

EXAMPLES

We have included many more worked examples than is customary. Furthermore, we have included computational details to enhance readability by students whose basic skills are weak.

EXERCISES

The exercises comprise about one-quarter of the text—the most important part of the text in our opinion. The exercises at the ends of the sections are usually arranged in the order in which the text proceeds, so that the homework assignments may easily be made after only part of a section is discussed. Interesting applications and more challenging problems tend to be located near the ends of the exercise sets. Supplementary exercises at the end of each chapter expand the other exercise sets and provide cumulative exercises that require skills from earlier chapters.

PRACTICE PROBLEMS

The practice problems introduced in the second edition have proved to be a popular and useful feature and are included in the present edition. The practice problems are carefully selected exercises that are located at the end of each section, just before the exercise set. Complete solutions are given following the exercise set. The practice problems often focus on points that are potentially confusing or are likely to be overlooked. We recommend that the reader seriously attempt the practice problems and study their solutions before moving on to the exercises. In effect, the practice problems constitute a built-in workbook.

MINIMAL PREREQUISITES

In Chapter 0, we review those facts that the reader needs to study calculus. A few important topics, such as the laws of exponents, are reviewed again when they are used in a later chapter. A reader familiar with the content of Chapter 0 should begin with Chapter 1 and use Chapter 0 as a reference, whenever needed.

NUMERICAL METHODS

With the common availability of microcomputers, numerical methods assume more significance than ever. We have included many discussions of numerical methods, including the differential in one variable (Section 1.8) and several variables (Section 7.5), numerical integration (Section 9.4), Euler's method for approximating the solutions to linear differential equations (Section 10.3), Taylor polynomials (Section 12.1), the Newton-Raphson algorithm (Section 12.2), and infinite series (Sections 12.3–12.6).

 ## NEW IN THIS EDITION

Among the many changes in this edition, the following are the most significant.

1. *Technology Discussions.*   Throughout the text, we have added discussions of how calculations and mathematical ideas can be elucidated using technology, including graphing calculators and symbolic calculation packages.

2. *Technology Projects.* Each chapter includes from three to eight projects that apply various aspects of technology to the chapter material. Some of these projects are sequences of technical exercises to allow students to learn how the technology is used. Others are mathematical models too complicated or "messy" to explore without the benefit of technology. Many of these models make use of data taken from current newspapers and magazines. In all, the technology projects include over 400 separate exercises. For those not technologically oriented, we have structured the technology projects so that they may be skipped in whole or in part without any loss of continuity.

3. *Additional Examples and Exercises.* We have added new examples and exercises, many based on real-world data. Graphs and tables based on real data are indicated throughout by a shadowed border.

4. *Treatment of Compound Interest.* We have moved the introduction of compound interest to Chapter 0, so that it may be treated with the discussion of exponents. This makes for a more lucid comparison of compound interest and continuous interest in Chapter 5.

5. *Incorporation of Reviewer and User Suggestions.* Throughout, we have made use of the many generous suggestions of colleagues, users, and reviewers.

This edition contains more material than can be covered in most two-semester courses. In addition, the level of theoretical material may be adjusted to the needs of the students. For instance, only the first two pages of Section 1.4 are required in order to introduce the limit notation.

Answers to the odd-numbered exercises are included at the back of the book. Answers to all the exercises are contained in the Instructor's Edition. A Study Guide for students is available that contains detailed explanations and solutions of every sixth exercise. The Study Guide also includes helpful hints and strategies for studying that will help students improve their performance in the course.

An Instructor's Solutions Manual contains worked solutions to every exercise. A Test Item File provides nearly 1000 suggested test questions, keyed to chapter and section. These Test Items are also available on an IBM-compatible disk with Prentice Hall's Custom Test program.

The technology discussions and projects are not calculator- or program-specific. However, instructors wishing to use a particular graphing calculator or symbolic program can make use of the various *Technology Guides* which particularize the technology discussions and projects to Texas Instruments Hewlett Packard graphing calculator, and *Maple*. Computer software for IBM-compatible computers is available at a nominal cost through your book store. *Visual Calculus* by David Schneider contains over twenty routines that provide additional insight into the topics discussed in the text. Although this software has much of the computing power of standard calculus software packages, it is primarily a teaching tool that focuses on understanding mathematical concepts, rather than on computing. These routines incorporate graphics whenever possible to illustrate topics such as secant lines, tangent lines, velocity, optimization, the relationship between the graphs of f, f', and f'', and the various approaches to approximating definite integrals. All the routines in this software package are menu-driven and very easy to use. The software is accompanied by a manual with instructions and additional exercises for the student. Hardware requirements are an IBM-compatible computer with at least 384 K of memory and a graphics adapter: CGA, EGA, VGA, or Hercules.

 ACKNOWLEDGMENTS

While writing this book, we have received assistance from many persons. And our heartfelt thanks goes out to them all. Especially, we should like to thank the following reviewers, who took the time and energy to share their ideas, preferences, and often their enthusiasm with us.

Reviewers of the first edition: Russell Lee, Allan Hancock College; Donald Hight, Kansas State College of Pittsburg; Ronald Rose, American River College; W. R. Wilson, Central Piedmont Community College; Bruce Swenson, Foothill College; Samuel Jasper, Ohio University; Carl David Minda, University of Cincinnati; H. Keith Stumpff, Central Missouri State University; Claude Schochet, Wayne State University; James E. Huneycutt, North Carolina University.

Reviewers of the second edition: Charles Himmelberg, University of Kansas; James A. Huckaba, University of Missouri; Joyce Longman, Villanova University; T. Y. Lam, University of California, Berkeley; W. T. Kyner, University of New Mexico; Shirley A. Goldman, University of California, Davis; Dennis White, University of Minnesota; Dennis Bertholf, Oklahoma State University; Wallace A. Wood, Bryant College; James L. Heitsch, University of Illinois, Chicago Circle; John H. Mathews, California State University, Fullerton; Arthur J. Schwartz, University of Michigan; Gordon Lukesh, University of Texas, Austin.

Reviewers of the third edition: William McCord, University of Missouri; W. E. Conway, University of Arizona; David W. Penico, Virginia Commonwealth University; Howard Frisinger, Colorado State University; Robert Brown, University of California, Los Angeles; Carla Wofsky, University of New Mexico; Heath K. Riggs, University of Vermont; James Kaplan, Boston University; Larry Gerstein, University of California, Santa Barbara; Donald E. Myers, University of Arizona, Tempe; Frank Warner, University of Pennsylvania.

Reviewers of the fourth edition: Edward Spanier, University of California, Berkeley; David Harbater, University of Pennsylvania; Robert Brown, University of Kansas; Bruce Edwards, University of Florida; Ann McGaw, University of Texas, Austin; Michael J. Berman, James Madison University.

Reviewers of the fifth edition: David Harbater, University of Pennsylvania; Fred Brauer, University of Wisconsin; W. E. Conway, University of Arizona; Jack R. Barone, Baruch College, CUNY; James W. Brewer, Florida Atlantic University; Alan Candiotti, Drew University; E. John Hornsby, Jr., University of New Orleans.

Reviewers of the sixth edition: Dennis Brewer, University of Arkansas; Melvin D. Lax, California State University, Long Beach; Lawrence J. Lardy, Syracuse University; Arlene Sherburne, Montgomery College, Rockville; Gabriel Lugo, University of North Carolina, Wilmington; W. R. Hintzman, San Diego State University; George Pyrros, University of Delaware.

Reviewers of the seventh edition: Joan M. Thomas, University of Oregon; Charles Clever, South Dakota State University; James Sochacki, James Madison University; Judy B. Kidd, James Madison University; Jack E. Graves, Syracuse University; Georgia B. Pyrros, University of Delaware; Karabi Datta, Northern Illinois University; James V. Balch, Middle Tennessee State University; H. Suey Quan, Golden West College; Albert G. Fadell, SUNY Buffalo; Murray Schechter, Lehigh University; Betty Fein, Oregon State University.

The authors would like to thank the many people at Prentice Hall who have contributed to the success of our books over the years. We appreciate the tremendous efforts of the production, art, manufacturing, and marketing departments. Our sincere thanks to Judy Winthrop, who has acted as production editor for this edition.

The authors wish to extend a special thank you to Mr. George Lobell, whose ideas, encouragement, and enthusiasm have nurtured us as we prepared this revision.

Larry J. Goldstein
David C. Lay
David I. Schneider

CALCULUS
AND ITS
APPLICATIONS

INTRODUCTION

Often it is possible to give a succinct and revealing description of a situation by drawing a graph. For example, Fig. 1 describes the amount of money in a bank account drawing 5% interest, compounded daily. The graph shows that as time passes, the amount of money in the account grows. In Fig. 2 we have drawn a graph that depicts the weekly sales of a breakfast cereal at various times after advertising has ceased. The graph shows that the longer the time since the last advertisement, the fewer the sales. Figure 3 shows the size of a bacteria culture at various times. The culture grows larger as time passes. But there is a maximum size that the culture cannot exceed. This maximum size reflects the restrictions imposed by food supply, space, and similar factors. The graph in Fig. 4 describes the decay of the radioactive isotope iodine 131. As time passes, less and less of the original radioactive iodine remains.

Each of the graphs in Figs. 1 to 4 describes a change that is taking place. The amount of money in the bank is changing as are the sales of cereal, the size of the bacteria culture, and the amount of the iodine. Calculus provides mathematical tools to study each of these changes in a quantitative way.

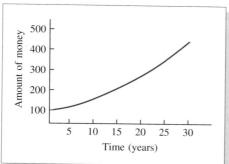

FIGURE 1   Growth of money in a savings account.

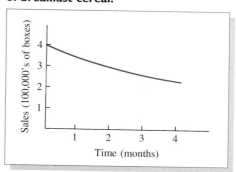

FIGURE 2   Decrease in sales of breakfast cereal.

FIGURE 3   Growth of a bacteria culture.

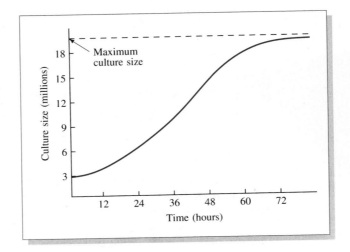

FIGURE 4   Decay of radioactive iodine.

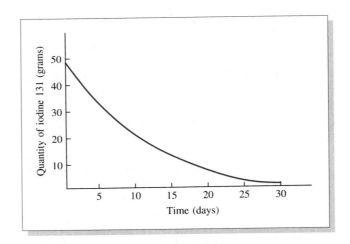

CHAPTER 0

FUNCTIONS

Each of the graphs in Figs. 1 to 4 of the Introduction depicts a relationship between two quantities. For example, Fig. 4 illustrates the relationship between the quantity of iodine (measured in grams) and time (measured in days). The basic quantitative tool for describing such relationships is a *function*. In this preliminary chapter, we develop the concept of a function and review important algebraic operations on functions used later in the text.

0.1 FUNCTIONS AND THEIR GRAPHS

Real Numbers   Most applications of mathematics use real numbers. For purposes of such applications (and the discussions in this text), it suffices to think of a real number as a decimal. A *rational* number is one that may be written as a finite or infinite repeating decimal, such as

$$-\frac{5}{2} = -2.5, \qquad 1, \qquad \frac{13}{3} = 4.333\ldots \qquad \text{(rational numbers)}.$$

An *irrational* number has an infinite decimal representation whose digits form no repeating pattern, such as

$$-\sqrt{2} = -1.414214\ldots, \qquad \pi = 3.14159\ldots \qquad \text{(irrational numbers)}.$$

3

FIGURE I **The real number line.**

The real numbers are described geometrically by a *number line,* as in Fig. 1. Each number corresponds to one point on the line, and each point determines one real number.

We use four types of inequalities to compare real numbers.

$$x < y \qquad x \text{ is less than } y$$

$$x \leq y \qquad x \text{ is less than or equal to } y$$

$$x > y \qquad x \text{ is greater than } y$$

$$x \geq y \qquad x \text{ is greater than or equal to } y$$

The double inequality $a < b < c$ is shorthand for the pair of inequalities $a < b$ and $b < c$. Similar meanings are assigned to other double inequalities, such as $a \leq b < c$. Three numbers in a double inequality, such as $1 < 3 < 4$ or $4 > 3 > 1$, should have the same relative positions as on the number line as in the inequality (when read left to right or right to left). Thus $3 < 4 > 1$ is never written because the numbers are "out of order."

Geometrically, the inequality $x \leq b$ means that either x equals b or x lies to the left of b on the number line. The set of real numbers x that satisfy the double inequality $a \leq x \leq b$ corresponds to the line segment between a and b, including the endpoints. This set is sometimes denoted by $[a, b]$ and is called the *closed*

TABLE I	Intervals on the Number Line	
Inequality	**Geometric Description**	**Interval Notation**
$a \leq x \leq b$	$\bullet_a \quad\quad \bullet_b$	$[a, b]$
$a < x < b$	$\circ_a \quad\quad \circ_b$	(a, b)
$a \leq x < b$	$\bullet_a \quad\quad \circ_b$	$[a, b)$
$a < x \leq b$	$\circ_a \quad\quad \bullet_b$	$(a, b]$
$a \leq x$	$\bullet_a$	$[a, \infty)$
$a < x$	$\circ_a$	(a, ∞)
$x \leq b$	$\bullet_b$	$(-\infty, b]$
$x < b$	$\circ_b$	$(-\infty, b)$

interval from a to b. If a and b are removed from the set, the set is written as (a, b) and is called the *open interval* from a to b. The notation for various line segments is listed in Table 1.

The symbols ∞ ("infinity") and $-\infty$ ("minus infinity") do not represent actual real numbers. Rather, they indicate that the corresponding line segment extends infinitely far to the right or left. An inequality that describes such an infinite interval may be written in two ways. For instance, $a \leq x$ is equivalent to $x \geq a$.

EXAMPLE 1   Describe each of the following intervals.

(a) $(-1, 2)$ (b) $[-2, \pi]$ (c) $(2, \infty)$ (d) $(-\infty, \sqrt{2}]$

Solution   The line segments corresponding to the intervals are shown in Fig. 2(a)–(d).

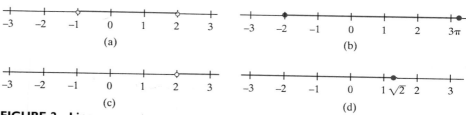

FIGURE 2   Line segments.

EXAMPLE 2   The variable x describes the profit that a company is anticipated to earn in the current fiscal year. Ths business plan calls for a profit of at least 5 million dollars. Describe this aspect of the business plan in the language of intervals.

Solution   The business plan requires that $x \geq 5$ (where the units are millions of dollars). This is equivalent to saying that x lies in the infinite interval $[5, \infty)$. ●

Functions   A *function* of a variable x is a *rule f* that assigns to each value of x a unique number $f(x)$, called *the value of the function at x*. [We read "$f(x)$" as "*f* of *x*."] The variable x is called the *independent variable*. The set of values that the independent variable is allowed to assume is called the *domain* of the function. The domain of a function may be explicitly specified as part of the definition of a function or it may be understood from context. (See the following discussion.) The *range of a function* is the set of values that the function assumes.

The functions we shall meet in this book will usually be defined by algebraic formulas. For example, the domain of the function

$$f(x) = 3x - 1$$

consists of all real numbers x. This function is the rule that takes a number, multiplies it by 3, and then subtracts 1. If we specify a value of x, say $x = 2$, then we find the value of the function at 2 by substituting 2 for x in the formula:

$$f(2) = 3(2) - 1 = 5.$$

EXAMPLE 3   Let f be the function with domain all real numbers x and defined by the formula

$$f(x) = 3x^3 - 4x^2 - 3x + 7.$$

Find $f(2)$ and $f(-2)$.

Solution   To find $f(2)$ we substitute 2 for every occurrence of x in the formula for $f(x)$:

$$f(2) = 3(2)^3 - 4(2)^2 - 3(2) + 7$$
$$= 3(8) - 4(4) - 3(2) + 7$$
$$= 24 - 16 - 6 + 7$$
$$= 9.$$

To find $f(-2)$ we substitute (-2) for each occurrence of x in the formula for $f(x)$. The parentheses ensure that the -2 is substituted correctly. For instance, x^2 must be replaced by $(-2)^2$, not -2^2.

$$f(-2) = 3(-2)^3 - 4(-2)^2 - 3(-2) + 7$$
$$= 3(-8) - 4(4) - 3(-2) + 7$$
$$= -24 - 16 + 6 + 7$$
$$= -27. \ \bullet$$

EXAMPLE 4   If x represents the temperature of an object in degrees Celsius, then the temperature in degrees Fahrenheit is a function of x, given by $f(x) = \frac{9}{5}x + 32$.

(a) Water freezes at 0°C (C = Celsius) and boils at 100°C. What are the corresponding temperatures in degrees Fahrenheit?

(b) Aluminum melts at 660°C. What is its melting point in degrees Fahrenheit?

Solution   (a) $f(0) = \frac{9}{5}(0) + 32 = 32$. Water freezes at 32°F.

$$f(100) = \frac{9}{5}(100) + 32 = 180 + 32 = 212.$$

Water boils at 212°F.

(b) $f(660) = \frac{9}{5}(660) + 32 = 1188 + 32 = 1220$. Aluminum melts at 1220°F.
$\bullet$

EXAMPLE 5   (*A Voting Model*) Let x be the proportion of the total popular vote that a Democratic candidate for president receives in a U.S. national election (so x is a number between 0 and 1). Political scientists have observed that a good estimate of the proportion of seats in the House of Representatives going to Democratic candidates is given by the function

$$f(x) = \frac{x^3}{x^3 + (1 - x)^3}, \qquad 0 \le x \le 1,$$

whose domain is the interval [0, 1]. This formula is called the *cube law*. Compute $f(.6)$ and interpret the result.

Solution   We must substitute .6 for every occurrence of x in $f(x)$:

$$f(.6) = \frac{(.6)^3}{(.6)^3 + (1 - .6)^3} = \frac{(.6)^3}{(.6)^3 + (.4)^3}$$

$$= \frac{.216}{.216 + .064} = \frac{.216}{.280} \approx .77.$$

This calculation shows that the cube law function predicts that if .6 (or 60%) of the total popular vote is for the Democratic candidate for president, then approximately .77 (or 77%) of the seats in the House of Representatives will be won by Democratic candidates; that is, about 335 of the 435 seats will be won by Democrats. ●

In the preceding examples, the functions had domains consisting of all real numbers or an interval. For some functions, the domain may consist of several intervals, with a different formula defining the function on each interval. Here is an illustration of this phenomenon.

EXAMPLE 6   A leading brokerage firm charges a 6% commission on gold purchases in amounts from $50 to $300. For purchases exceeding $300, the firm charges 2% of the amount purchased plus $12.00. Let x denote the amount of gold purchased (in dollars) and let $f(x)$ be the commission charge as a function of x.
(a) Describe $f(x)$.
(b) Find $f(100)$ and $f(500)$.

Solution   (a) The formula for $f(x)$ depends on whether $50 \leq x \leq 300$ or $300 < x$. When $50 \leq x \leq 300$, the charge is $.06x$ dollars. When $300 < x$, the charge is $.02x + 12$. The domain consists of the values x in one of the two intervals [50, 300] and (300, ∞). In each of these intervals, the function is defined by a separate formula:

$$f(x) = \begin{cases} .06x & \text{for } 50 \leq x \leq 300 \\ .02x + 12 & \text{for } 300 < x. \end{cases}$$

Note that an alternate description of the domain is the interval [50, ∞). That is, the value of x may be any real number greater than or equal to 50.
(b) Since $x = 100$ satisfies $50 \leq x \leq 300$, we use the first formula for $f(x)$: $f(100) = .06(100) = 6$. Since $x = 500$ satisfies $300 < x$, we use the second formula for $f(x)$: $f(500) = .02(500) + 12 = 22$. ●

In calculus, it is often necessary to substitute an algebraic expression for x and simplify the result, as illustrated in the following example.

EXAMPLE 7   If $f(x) = (4 - x)/(x^2 + 3)$, what is $f(a)$? $f(a + 1)$?

Solution   Here a represents some number. To find $f(a)$, we substitute a for x wherever x appears in the formula defining $f(x)$:

$$f(a) = \frac{4 - a}{a^2 + 3}.$$

To evaluate $f(a + 1)$, substitute $a + 1$ for each occurrence of x in the formula for $f(x)$:

$$f(a + 1) = \frac{4 - (a + 1)}{(a + 1)^2 + 3}.$$

The expression for $f(a + 1)$ may be simplified, using the fact that $(a + 1)^2 = (a + 1)(a + 1) = a^2 + 2a + 1$:

$$f(a + 1) = \frac{4 - (a + 1)}{(a + 1)^2 + 3} = \frac{4 - a - 1}{a^2 + 2a + 1 + 3} = \frac{3 - a}{a^2 + 2a + 4}. \bullet$$

More About the Domain of a Function   When defining a function, it is necessary to specify the domain of the function, which is the set of acceptable values of the variable. In the preceding examples, we explicitly specified the domains of the functions considered. However, throughout the remainder of the text, we will usually mention functions without specifying domains. In such circumstances, we will understand the intended domain to consist of all numbers for which the defining formula(s) make sense. For example, consider the function

$$f(x) = x^2 - x + 1.$$

The expression on the right may be evaluated for any value of x. So in the absence of any explicit restrictions on x, the domain is understood to consist of all numbers. As a second example consider the function

$$f(x) = \frac{1}{x}.$$

Here x may be any number except zero. (Division by zero is not permissible.) So the domain intended is the set of nonzero numbers. Similarly, when we write

$$f(x) = \sqrt{x},$$

we understand the domain of $f(x)$ to be the set of all nonnegative numbers, since the square root of a number x is defined if and only if $x \geq 0$.

Graphs of Functions   Often it is helpful to describe a function f geometrically, using a rectangular xy-coordinate system. Given any x in the domain of f, we can plot the point $(x, f(x))$. This is the point in the xy-plane whose y-coordinate is the value of the function at x. The set of *all* such points $(x, f(x))$ usually forms a curve in the xy-plane and is called the *graph of the function $f(x)$*.

It is possible to approximate the graph of $f(x)$ by plotting the points $(x, f(x))$ for a representative set of values of x and joining them by a smooth curve. (See Fig. 3.) The more closely spaced the values of x, the closer the approximation.

FIGURE 3

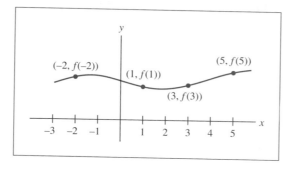

EXAMPLE 8   Sketch the graph of the function $f(x) = x^3$.

Solution   The domain consists of all numbers x. We choose some representative values of x and tabulate the corresponding values of $f(x)$. We then plot the points $(x, f(x))$ and sketch the graph indicated. (See Fig. 4.) ●

FIGURE 4   Graph of
$f(x) = x^3$.

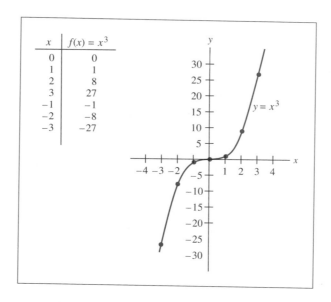

EXAMPLE 9   Sketch the graph of the function $f(x) = 1/x$.

Solution   The domain of the function consists of all numbers except zero. The table in Fig. 5 lists some representative values of x and the corresponding values of $f(x)$. A function often has interesting behavior for x near a number not in the domain. So

FIGURE 5   Graph of $f(x) = \dfrac{1}{x}$.

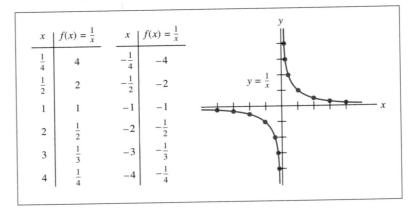

x	$f(x) = \frac{1}{x}$	x	$f(x) = \frac{1}{x}$
$\frac{1}{4}$	4	$-\frac{1}{4}$	-4
$\frac{1}{2}$	2	$-\frac{1}{2}$	-2
1	1	-1	-1
2	$\frac{1}{2}$	-2	$-\frac{1}{2}$
3	$\frac{1}{3}$	-3	$-\frac{1}{3}$
4	$\frac{1}{4}$	-4	$-\frac{1}{4}$

when we chose representative values of x from the domain, we included some values close to zero. The points $(x, f(x))$ are plotted and the graph sketched in Fig. 5. ●

Now that graphing calculators and computer graphing programs are widely available, one seldom needs to sketch graphs by hand-plotting large numbers of points on graph paper. However, to use such a calculator or program effectively, one must know in advance which part of a curve to display. Critical features of a graph may be missed or misinterpreted if, for instance, the scale on the x- or y-axis is inappropriate.

An important use of calculus is to identify key features of a function that should appear in its graph. In many cases, only a few points need be plotted and the general shape of the graph is easy to sketch by hand. For more complicated functions, a graphing program is helpful. Even then, a calculus analysis provides a check that the graph on the computer screen has the correct shape. Algebraic calculations are usually part of the analysis. The appropriate algebraic skills are reviewed in this chapter.

The connection between a function and its graph is explored in this section and in Section 0.6.

EXAMPLE 10   Suppose that f is the function whose graph is given in Fig. 6. Notice that the point $(x, y) = (3, 2)$ is on the graph of f.

FIGURE 6

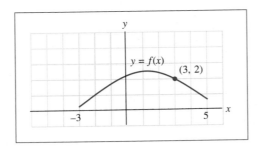

(a) What is the value of the function when $x = 3$?

(b) Find $f(-2)$.

(c) What is the domain of f?

Solution

(a) Since $(3, 2)$ is on the graph of f, the y-coordinate 2 must be the value of f at the x-coordinate 3. That is, $f(3) = 2$.

(b) To find $f(-2)$ we look at the y-coordinate of the point on the graph where $x = -2$. From Fig. 6 we see that $(-2, 1)$ is on the graph of f. Thus $f(-2) = 1$.

(c) The points on the graph of $f(x)$ all have x-coordinates between -3 and 5 inclusive; and for each value of x between -3 and 5 there is a point $(x, f(x))$ on the graph. So the domain consists of those x for which $-3 \leq x \leq 5$. ●

To every x in the domain, a function assigns one and only one value of y, namely, the function value $f(x)$. This implies, among other things, that not every curve is the graph of a function. To see this, refer first to the curve in Fig. 6, which *is* the graph of a function. It has the following important property: For each x between -3 and 5 inclusive there is a *unique* y such that (x, y) is on the curve. The variable y is called the *dependent variable,* since its value depends on the value of the independent variable x. Refer to the curve in Fig. 7. It cannot be the graph of a function because a function f must assign to each x in its domain a *unique* value $f(x)$. However, for the curve of Fig. 7 there corresponds to $x = 3$ (for example) more than one y-value, namely, $y = 1$ and $y = 4$.

FIGURE 7 A curve that is *not* the graph of a function.

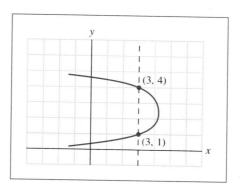

The essential difference between the curves in Figs. 6 and 7 leads us to the following test.

The Vertical Line Test   A curve in the xy-plane is the graph of a function if and only if each vertical line cuts or touches the curve at no more than one point.

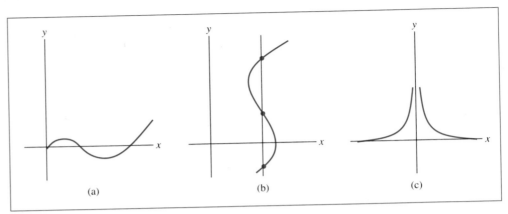

(a)   (b)   (c)

FIGURE 8

EXAMPLE 11   Which of the curves in Fig. 8 are graphs of functions?

Solution   The curve in (a) is the graph of a function. It appears that vertical lines to the left of the y-axis do not touch the curve at all. This simply means that the function represented in (a) is defined only for $x \geq 0$. The curve in (b) is *not* the graph of a function because some vertical lines cut the curve in three places. The curve in (c) is the graph of a function whose domain is all nonzero x. [There is no point on the curve in (c) whose x-coordinate is 0.] ●

There is another notation for functions that we will find useful. Suppose that $f(x)$ is a function. When $f(x)$ is graphed on an xy-coordinate system, the values of $f(x)$ give the y-coordinates of points of the graph. For this reason, the function is often abbreviated by the letter y, and we find it convenient to speak of "the function $y = f(x)$." For example, the function $y = 2x^2 + 1$ refers to the function $f(x)$ for which $f(x) = 2x^2 + 1$. The graph of a function $f(x)$ is often called *the graph of the equation $y = f(x)$.*

Graphing Functions Using Technology   As you might have already inferred from the above examples, sketching the graphs of functions by plotting points can be a tedious procedure. However, it is one that can be carried out using technology. In this text, we will discuss two of the many technological tools currently in use in solving mathematical problems: graphing calculators and symbolic mathematics programs. Either of these technological tools can be used to graph functions. However, we will mostly concern ourselves with graphing calculators, since these are the more available tool for students in this course.*

———————

*However, beginning in Chapter 3, there will be a number of projects that use symbolic packages, such as Maple, MathCAD or Mathematica. These projects are all clearly marked and are optional. Subsequent text discussion and graphing calculator projects do not depend in any way on the symbolic package projects, which are provided to give instructors added flexibility.

We do not give keystrokes for any particular calculator, but discuss the features that are common. For keystrokes for your particular calculator, consult your calculator manual or one of the technology-specific manuals that accompany this text.

To graph a function using a graphing calculator, you must perform the following steps:

1. Enter the equation for the function.
2. Enter the graph parameters.
3. Display the graph.

Let's describe what is involved in each step.

To enter an equation, you must display the equation entry screen, which is usually activated by pressing the key $\boxed{\text{Y=}}$. (See Fig. 9a.) Note that you may enter one or more functions, denoted Y_1, Y_2, Y_3, ... (The exact number of functions will depend on your calculator.)

FIGURE 9a   The equation entry screen of a TI-85 calculator.

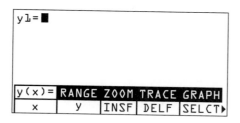

A function must be entered in the form:

$$y = [\text{an expression in } x]$$

Powers are entered using the key $\boxed{\wedge}$. With some calculators, multiplication must be explicitly indicated by $*$. Also, you must use parentheses wherever necessary to clarify the mathematical meaning of an expression.

The following example illustrates how to enter various functions.

EXAMPLE 12   Write the following functions in a form suitable for a graphing calculator.

1. $f(x) = x^2 - 3x$
2. $f(x) = x^{1/3}$
3. $f(x) = \dfrac{x - 1}{x + 2}$
4. $f(x) = \sqrt{x - \dfrac{1}{x}}$

Solution   The variable x is entered by pressing the X key. Note that this is not the same key as $\boxed{\times}$, used for multiplication.

1. $Y_1 = X^2 - 3 * X$.

   Note that the term $3x$ must be entered as $3 * X$.
2. $Y_1 = X^{(1/3)}$.

   Note that it is necessary to surround the exponent by parentheses in order for the calculator to interpret it correctly.
3. $Y_1 = (X - 1)/(X + 2)$.
4. $Y_1 = SQR(X - 1/X)$.

   Here SQR denotes the square root key. ●

The equations for functions are displayed in the equation entry screen. Fig. 9b shows the equation entry screen just after entering the equation for the first function Y_1 in the above example.

FIGURE 9b An equation for Y_1 has been entered.

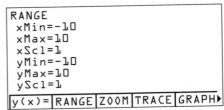

FIGURE 9c The parameters controlling display of a graph.

After you enter the equation of a function, you must then set the parameters that the calculator will use in drawing the graph. There are several parameters you can set. Fig. 9c shows the parameters available on the TI-85 calculator. The most important parameters are:

XMIN = the smallest value of the variable X which is considered

XMAX = the largest value of the variable X which is considered

YMIN = the smallest value of the variable Y which is considered

YMAX = the largest value of the variable Y which is considered

These values are initially set at XMIN = -10, XMAX = 10, YMIN = -10, YMAX = 10. Collectively, these four parameters determine the size of the *display window*. The initial values are called *default values* and are suitable for some graphs. However, one of the most important tasks in using a graphing calculator is to determine the display window which shows the features of the graph you are interested in. To determine these values often requires calculus, as we shall learn in later chapters. For now, we will either use the default settings, specify the settings, or determine appropriate settings by experimentation.

For now, let's leave the parameters set at their default values. To display the graph of Y_1 as defined in Fig. 9b, we press the key $\boxed{\text{GRAPH}}$ and the calculator displays the graph shown in Fig. 9d.

FIGURE 9d   The graph of
$Y_1 = X^2 - 3 * X$
$(-10 \le X \le 10,$
$-10 \le Y \le 10).$

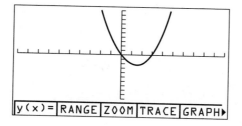

A graphing calculator provides three basic graph manipulations: ZOOM, BOX, and TRACE. Since these manipulations are fundamental for all the technology projects throughout the text, let's discuss these and provide you with some practice in using them.

The ZOOM operation either enlarges or shrinks the display window. There are two options. ZOOM OUT allows you to increase the size of the display window. The exact factor of increase depends on the particular calculator and may be user settable. The center of the display window remains the same. The graph will look as if it has shrunk.

ZOOM IN allows you to shrink the display window in both directions. The center of the display window remains the same. This operation enlarges the central region of the display window. The outer portion of the display window in each direction no longer appears.

ZOOMing expands or shrinks the display window from its current center. A more controlled form of display window alteration is provided by the BOX command, which allows you to change the display window to a box you draw on the screen. For example, Fig. 9f shows a boxed portion of the display window. Fig. 9g shows this region enlarged to fill the entire screen. You should consult your calculator manual for the keystrokes to define the box and to create the new display.

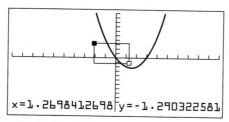

FIGURE 9f   Specifying a boxed region to enlarge.

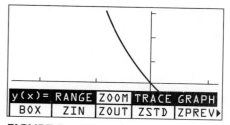

FIGURE 9g   The enlarged portion of the graph defined by the boxed region.

Note that a graphing calculator display does not display numerical coordinates on a graph. In order to read coordinates of points, you use the TRACE operation. When you specify TRACE, a small cross, called the *cursor,* appears on the graph. The coordinates of the cursor are displayed at the bottom of the screen. By using the arrow keys, you can move the cursor along the curve and determine the coordinates of any point. (See Fig. 9h.)

FIGURE 9h   Using the TRACE to determine coordinates of points.

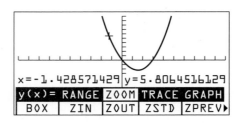

x=-1.428571429 y=5.806451629
y(x)= RANGE ZOOM TRACE GRAPH
BOX | ZIN | ZOUT | ZSTD | ZPREV▸

By combining TRACE and BOX operations, you may determine coordinates of points to as many decimal places as you wish (within the limits of the calculator).

The following project provides some practice in performing basic graphing calculator operations.

TECHNOLOGY PROJECT 0.1
Graphing Functions with a Graphing Calculator

Perform all of these actions:

a. Display the graph with the display window $-10 \leq X \leq 10$, $-10 \leq Y \leq 10$.
b. Zoom out. By reexamining the parameter display screen, determine the new range of parameters.
c. Zoom in twice. By reexamining the parameter display screen, determine the new range of parameters.
d. Zoom out to get back to the original graph. Use BOX to display the portion of the graph corresponding to $-1 \leq X \leq 1$, $-2 \leq Y \leq 2$.
e. Use BOX and ZOOM to display the portion of the graph with $-2 \leq X \leq 2$ and Y range sufficient to display the corresponding points.
f. Use TRACE to determine $f(0)$, $f(.5)$, $f(-1)$.

For each of the following functions:

1. $f(x) = 2.1x - 1.7$ 2. $f(x) = 2x^2 - x + 1$

3. $f(x) = 0.01x^3$ 4. $f(x) = \dfrac{x}{x - 15}$

Graphs of Equations   The equations arising in connection with functions are all of the form

$$y = [\text{an expression in } x].$$

However, not all equations connecting the variables x and y are of this sort. For example, consider these equations:

$$2x + 3y = 5$$

$$x = 3$$

$$x^2 + y^2 = 1.$$

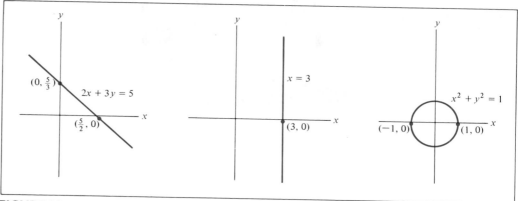

FIGURE 10

It is possible to graph an equation by plotting points just as for functions. The only difference is that the resulting graph may not satisfy the vertical line test. For example, the graphs of the three equations above are shown in Fig. 10. Only the first graph is the graph of a function.

Graphing calculators cannot display the graphs of all equations. They can only display graphs of equations you can solve for y in terms of x. For instance, in order to graph the first equation above, it is necessary to first solve for y in terms of x:

$$2x + 3y = 5$$

$$3y = 5 - 2x$$

$$y = \frac{5}{3} - \frac{2}{3}x$$

The second equation cannot be solved for y in terms of x and so cannot be graphed using a graphing calculator. For the third equation, we may solve for y in terms of x to obtain:

$$x^2 + y^2 = 1$$

$$y^2 = 1 - x^2$$

$$y = \pm\sqrt{1 - x^2}$$

The $\pm$ indicates that the last equation is really two equations, one corresponding to each choice of sign. On a graphing calculator, we must enter each of these equations:

$$y = \sqrt{1 - x^2}, \qquad y = -\sqrt{1 - x^2}$$

(as Y_1 and Y_2, for instance). The calculator will draw the graphs of both equations, which together comprise the graph of the original equation.

Letters other than f can be used to denote functions and letters other than x and y can be used to denote variables. This is especially common in applied problems where letters are chosen to suggest the quantities they depict. For instance, the revenue of a company as a function of time might be written $R(t)$.

PRACTICE PROBLEMS 0.1

1. Is the point $(3, 12)$ on the graph of the function $g(x) = x^2 + 5x - 10$?
2. Sketch the graph of the function $h(t) = t^2 - 2$.

EXERCISES 0.1*

Draw the following intervals on the number line.

1. $[-1, 4]$ **2.** $(4, 3\pi)$ **3.** $[-2, \sqrt{2})$

4. $[1, \frac{3}{2}]$ **5.** $(-\infty, 3)$ **6.** $(4, \infty)$

Use intervals to describe the real numbers satisfying the inequalities in Exercises 7–12.

7. $2 \le x < 3$ **8.** $-1 < x < \frac{3}{2}$

9. $x < 0, x \ge -1$ **10.** $x \ge -1, x < 8$

11. $x < 3$ **12.** $x \ge \sqrt{2}$

13. If $f(x) = x^2 - 3x$, find $f(0), f(5), f(3)$, and $f(-7)$.

14. If $f(x) = 9 - 6x + x^2$, find $f(0), f(2), f(3)$, and $f(-13)$.

15. If $f(x) = x^3 + x^2 - x - 1$, find $f(1), f(-1), f(\frac{1}{2})$, and $f(a)$.

16. If $g(t) = t^3 - 3t^2 + t$, find $g(2), g(-\frac{1}{2}), g(\frac{2}{3})$, and $g(a)$.

17. If $h(s) = s/(1 + s)$, find $h(\frac{1}{2}), h(-\frac{3}{2})$, and $h(a + 1)$.

18. If $f(x) = x^2/(x^2 - 1)$, find $f(\frac{1}{2}), f(-\frac{1}{2})$, and $f(a + 1)$.

19. If $f(x) = x^2 - 2x$, find $f(a + 1)$ and $f(a + 2)$.

20. If $f(x) = x^2 + 4x + 3$, find $f(a - 1)$ and $f(a - 2)$.

21. An office supply firm finds that the number of fax machines sold in year x is given approximately by the function $f(x) = 50 + 4x + \frac{1}{2}x^2$, where $x = 0$ corresponds to 1990.

    (a) What does $f(0)$ represent?

    (b) Find the number of fax machines sold in 1992.

22. When a solution of acetylcholine is introduced into the heart muscle of a frog, it diminishes the force with which the muscle contracts. The data from experiments of A. J. Clark are closely approximated by a function of the form

$$R(x) = \frac{100x}{b + x}, \qquad x \ge 0,$$

where x is the concentration of acetylcholine (in appropriate units), b is a positive constant that depends on the particular frog, and $R(x)$ is the response of the muscle to the acetylcholine, expressed as a percentage of the maximum possible effect of the drug.

    (a) Suppose that $b = 20$. Find the response of the muscle when $x = 60$.

    (b) Determine the value of b if $R(50) = 60$—that is, if a concentration of $x = 50$ units produces a 60% response.

*A complete solution for every third odd-numbered exercise is given in the Study Guide for this text.

In Exercises 23–26, describe the domain of the function.

23. $f(x) = \dfrac{8x}{(x-1)(x-2)}$

24. $f(t) = \dfrac{1}{\sqrt{t}}$

25. $g(x) = \dfrac{1}{\sqrt{3-x}}$

26. $g(x) = \dfrac{4}{x(x+2)}$

In Exercises 27–32, decide which curves are graphs of functions.

27.
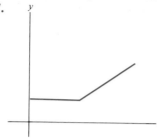

28.
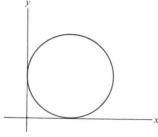

29.
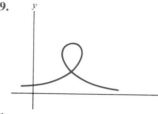

30.
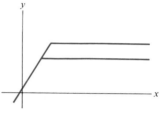

31.
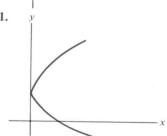

32.
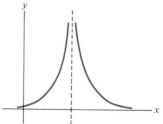

Exercises 33–42 relate to the function whose graph is sketched in Fig. 11.

33. Find $f(0)$.

34. Find $f(7)$.

FIGURE 11

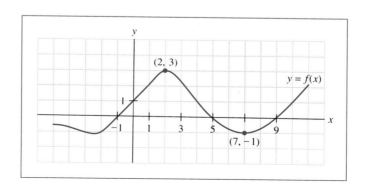

35. Find $f(2)$.

36. Find $f(-1)$.

37. Is $f(4)$ positive or negative?

38. Is $f(6)$ positive or negative?

39. Is $f(-\frac{1}{2})$ positive or negative?

40. Is $f(1)$ greater than $f(6)$?

41. For what values of x is $f(x) = 0$?

42. For what values of x is $f(x) \geq 0$?

Exercises 43–46 relate to Fig. 12. When a drug is injected into a person's muscle tissue, the concentration y of the drug in the blood is a function of the time elapsed since the injection. The graph of a typical time-concentration function f is given in Fig. 12, where $t = 0$ corresponds to the time of the injection.

FIGURE 12 Drug time-concentration curve.

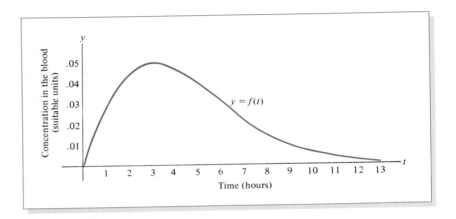

43. What is the concentration of the drug when $t = 1$?

44. What is the value of the time-concentration function f when $t = 6$?

45. Find $f(5)$.

46. At what time does $f(t)$ attain its largest value?

47. Is the point $(3, 12)$ on the graph of the function $f(x) = (x - \frac{1}{2})(x + 2)$?

48. Is the point $(-2, 12)$ on the graph of the function $f(x) = x(5 + x)(4 - x)$?

49. Is the point $(\frac{1}{2}, \frac{2}{5})$ on the graph of the function $g(x) = (3x - 1)/(x^2 + 1)$?

50. Is the point $(\frac{2}{3}, \frac{5}{3})$ on the graph of the function $g(x) = (x^2 + 4)/(x + 2)$?

51. Find the y-coordinate of the point $(a + 1, \blacksquare)$ if this point lies on the graph of the function $f(x) = x^3$.

52. Find the y-coordinate of the point $(2 + h, \blacksquare)$ if this point lies on the graph of the function $f(x) = (5/x) - x$.

In Exercises 53–56, compute $f(1)$, $f(2)$, and $f(3)$.

53. $f(x) = \begin{cases} \sqrt{x} & \text{for } 0 \leq x < 2 \\ 1 + x & \text{for } 2 \leq x \leq 5 \end{cases}$

54. $f(x) = \begin{cases} 1/x & \text{for } 1 \leq x \leq 2 \\ x^2 & \text{for } 2 < x \end{cases}$

55. $f(x) = \begin{cases} \pi x^2 & \text{for } x < 2 \\ 1 + x & \text{for } 2 \leq x \leq 2.5 \\ 4x & \text{for } 2.5 < x \end{cases}$

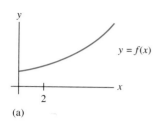

$y = f(x)$

(a)

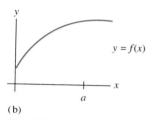

$y = f(x)$

(b)

FIGURE 13

56. $f(x) = \begin{cases} 3/(4 - x) & \text{for } x < 2 \\ 2x & \text{for } 2 \le x < 3 \\ \sqrt{x^2 - 5} & \text{for } 3 \le x \end{cases}$

57. Suppose that the brokerage firm in Example 6 decides to keep the commission charges unchanged for purchases up to and including $600 but to charge only 1.5% plus $15 for gold purchases exceeding $600. Express the brokerage commission as a function of the amount x of gold purchased.

58. Figure 13(a) shows the number 2 on the x-axis and the graph of a function. Let h represent a positive number and label a possible location for the number $2 + h$. Plot the point on the graph whose first coordinate is $2 + h$ and label the point with its coordinates.

59. Figure 13(b) shows the number a on the x-axis and the graph of a function. Let h represent a negative number and label a possible location for the number $a + h$. Plot the point on the graph whose first coordinate is $a + h$ and label the point with its coordinates.

SOLUTIONS TO PRACTICE PROBLEMS 0.1

1. If $(3, 12)$ is on the graph of $g(x) = x^2 + 5x - 10$, then we must have $g(3) = 12$. This is not the case, however, because

$$g(3) = 3^2 + 5(3) - 10$$
$$= 9 + 15 - 10 = 14.$$

Thus $(3, 12)$ is *not* on the graph of $g(x)$.

FIGURE 14 Graph of $h(t) = t^2 - 2$.

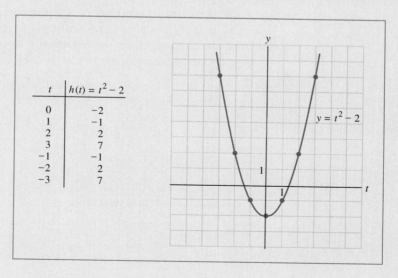

t	$h(t) = t^2 - 2$
0	-2
1	-1
2	2
3	7
-1	-1
-2	2
-3	7

2. Choose some representative values for t, say $t = 0, \pm 1, \pm 2, \pm 3$. For each value of t, calculate $h(t)$ and plot the point $(t, h(t))$. See Fig. 14.

0.2 SOME IMPORTANT FUNCTIONS

In this section we introduce some of the functions that will play a prominent role in our discussion of calculus.

Linear Functions   As we shall see in the next chapter, a knowledge of the algebraic and geometric properties of straight lines is essential for the study of calculus. Every straight line is the graph of a linear equation of the form

$$cx + dy = e,$$

where c, d, and e are given constants, with c and d not both zero. If $d \neq 0$, then we may solve the equation for y to obtain an equation of the form

$$y = mx + b, \tag{1}$$

for appropriate numbers m and b. If $d = 0$, then we may solve the equation for x to obtain an equation of the form

$$x = a, \tag{2}$$

for an appropriate number a. So every straight line is the graph of an equation of type (1) or (2). The graph of an equation of the form (1) is a nonvertical line [Fig. 1(a)], whereas the graph of (2) is a vertical line [Fig. 1(b)].

The straight line of Fig. 1(a) is the graph of the function $f(x) = mx + b$. Such a function, which is defined for all x, is called a *linear function*. Note that the straight line of Fig. 1(b) is not the graph of a function, since the vertical line test is violated.

An important special case of a linear function occurs if the value of m is zero, that is, if $f(x) = b$ for some number b. In this case, $f(x)$ is called a *constant function*, since it assigns the same number b to every value of x. Its graph is the horizontal line whose equation is $y = b$. (See Fig. 2.)

FIGURE I

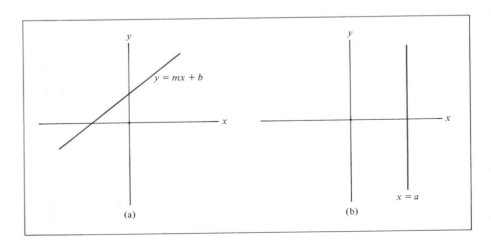

(a)   (b)

FIGURE 2   Graph of the constant function $f(x) = b$.

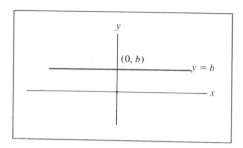

Linear functions often arise in real-life situations, as the first two examples show.

EXAMPLE 1   When the U.S. Environmental Protection Agency found a certain company dumping sulfuric acid into the Mississippi River, it fined the company $125,000, plus $1000 per day until the company complied with federal water pollution regulations. Express the total fine as a function of the number x of days the company continued to violate the federal regulations.

Solution   The variable fine for x days of pollution, at $1000 per day, is $1000x$ dollars. The total fine is therefore given by the function

$$f(x) = 125,000 + 1000x. \bullet$$

Since the graph of a linear function is a line, we may sketch it by locating any two points on the graph and drawing the line through them. For example, to sketch the graph of the function $f(x) = -\frac{1}{2}x + 3$, we may select two convenient values of x, say 0 and 4, and compute $f(0) = -\frac{1}{2}(0) + 3 = 3$ and $f(4) = -\frac{1}{2}(4) + 3 = 1$. The line through the points $(0, 3)$ and $(4, 1)$ is the graph of the function. (See Fig. 3.)

FIGURE 3

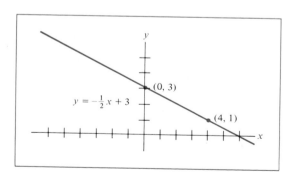

EXAMPLE 2   A simple cost function for a business consists of two parts—the *fixed costs,* such as rent, insurance, and business loans, which must be paid no matter how many items of a product are produced, and the *variable costs,* which depend on the number of items produced.

Suppose a computer software company produces and sells a new spreadsheet program at a cost of $25 per copy, and the company has fixed costs of $10,000 per month. Express the total monthly cost as a function of the number of copies sold, x, and compute the cost when $x = 500$.

Solution   The monthly variable cost is $25x$ dollars. Thus

$$[\text{total cost}] = [\text{fixed costs}] + [\text{variable costs}]$$

$$C(x) = 10{,}000 + 25x.$$

When sales are at 500 copies per month, the cost is

$$C(500) = 10{,}000 + 25(500) = 22{,}500 \quad (\text{dollars}).$$

See Fig. 4. ●

FIGURE 4   A linear cost function.

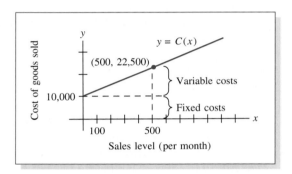

The point at which the graph of a linear function intersects the y-axis is called the *y-intercept* of the graph. The point at which the graph intersects the x-axis is called the *x-intercept*. The next example shows how to determine the intercepts of a linear function.

EXAMPLE 3   Determine the intercepts of the graph of the linear function $f(x) = 2x + 5$.

Solution   Since the y-intercept is on the y-axis, its x-coordinate is 0. The point on the line with x-coordinate zero has y-coordinate

$$f(0) = 2(0) + 5 = 5.$$

So the y-intercept is $(0, 5)$. Since the x-intercept is on the x-axis, its y-coordinate is 0. Since $f(x)$ gives the y-coordinate, we must have

$$2x + 5 = 0$$

$$2x = -5$$

$$x = -\tfrac{5}{2}.$$

So $(-\tfrac{5}{2}, 0)$ is the x-intercept. (See Fig. 5.) ●

FIGURE 5 Graph of
f(x) = 2x + 5.

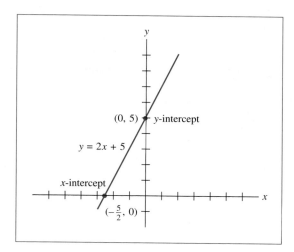

EXAMPLE 4   Sketch the graph of the following function.

$$f(x) = \begin{cases} \frac{5}{2}x - \frac{1}{2} & \text{for } -1 \leq x \leq 1 \\ \frac{1}{2}x - 2 & \text{for } x > 1. \end{cases}$$

Solution   This function is defined for $x \geq -1$. But it is defined by means of two distinct linear functions. We graph the two linear functions $\frac{5}{2}x - \frac{1}{2}$ and $\frac{1}{2}x - 2$. Then the graph of $f(x)$ consists of that part of the graph of $\frac{5}{2}x - \frac{1}{2}$ for which $-1 \leq x \leq 1$ plus that part of the graph of $\frac{1}{2}x - 2$ for which $x > 1$. (See Fig. 6.) ●

FIGURE 6   Graph of a func-
tion specified by two expres-
sions.

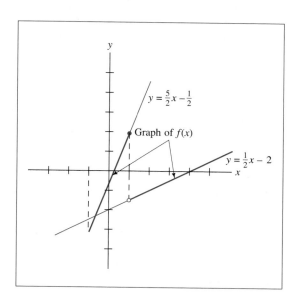

Quadratic Functions   Economists utilize average cost curves that relate the average unit cost of manufacturing a commodity to the number of units to be produced. (See Fig. 7.) Ecologists use curves that relate the net primary production of nutrients in a plant to the surface area of the foliage. (See Fig. 8.) Each of the curves is bowl-shaped, opening either up or down. The simplest functions whose graphs resemble these curves are the quadratic functions.

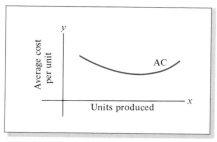

FIGURE 7   Average cost curve.

FIGURE 8   Production of nutrients.

A *quadratic function* is a function of the form

$$f(x) = ax^2 + bx + c,$$

where a, b, and c are constants and $a \neq 0$. The domain of such a function consists of all numbers. The graph of a quadratic function is called a *parabola*. Two typical parabolas are drawn in Figs. 9 and 10. We shall develop techniques for sketching graphs of quadratic functions after we have the properties of the derivative at our disposal.

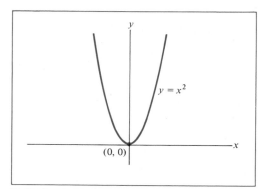

FIGURE 9   Graph of $f(x) = x^2$.

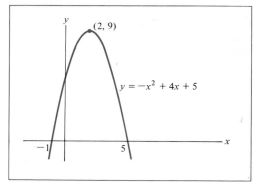

FIGURE 10   Graph of $f(x) = -x^2 + 4x + 5$.

Polynomial and Rational Functions   A *polynomial function $f(x)$* is one of the form

$$f(x) = a_n x^n + a_{n-1} x^{n-1} + \cdots + a_0,$$

where n is a nonnegative integer and $a_0, a_1, \ldots, a_n$ are given numbers. Some examples of polynomial functions are

$$f(x) = 5x^3 - 3x^2 - 2x + 4$$

$$g(x) = x^4 - x + 1.$$

Of course, linear and quadratic functions are special cases of polynomial functions. The domain of a polynomial function consists of all numbers.

A function expressed as the quotient of two polynomials is called a *rational function*. Some examples are

$$h(x) = \frac{x^2 + 1}{x}$$

$$k(x) = \frac{x + 3}{x^2 - 4}.$$

The domain of a rational function excludes all values of x for which the denominator is zero. For example, the domain of $h(x)$ excludes $x = 0$, whereas the domain of $k(x)$ excludes $x = 2$ and $x = -2$. As we shall see, both polynomial and rational functions arise in applications of calculus.

Rational functions are used in environmental studies as *cost-benefit* models. The cost of removing a pollutant from the atmosphere is estimated as a function of the percentage of the pollutant removed. The higher the percentage removed, the greater the "benefit" to the people who breathe that air. The issues here are complex, of course, and the definition of "cost" is debatable. The cost to remove a small percentage of pollutant may be fairly low. But the removal cost of the final 5% of the pollutant, for example, may be terribly expensive.

EXAMPLE 5   Suppose a cost-benefit function is given by

$$f(x) = \frac{50x}{105 - x}, \qquad 0 \le x \le 100,$$

where x is the percentage of some pollutant to be removed and $f(x)$ is the associated cost (in millions of dollars). See Fig. 11. Find the costs to remove 70%, 95%, and 100% of the pollutant.

Solution   The cost to remove 70% is

$$f(70) = \frac{50(70)}{105 - 70} = 100 \quad \text{(million dollars)}.$$

Similar calculations show that

$$f(95) = 475 \quad \text{and} \quad f(100) = 1000.$$

FIGURE 11   A cost-benefit model.

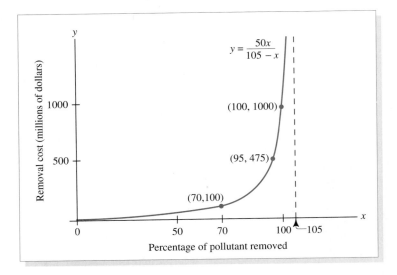

Observe that the cost to remove the last 5% of the pollutant is $f(100) - f(95) = 1000 - 475 = 525$ million dollars. This is more than five times the cost to remove the first 70% of the pollutant! ●

Power Functions   Functions of the form $f(x) = x^r$ are called *power functions*. The meaning of x^r is obvious when r is a positive integer. However, the power function $f(x) = x^r$ may be defined for any number r. Power functions are discussed in Section 0.5.

The Absolute Value Function   The absolute value of a number x is denoted by $|x|$ and is defined by

$$|x| = \begin{cases} x & \text{if } x \text{ is positive or zero,} \\ -x & \text{if } x \text{ is negative.} \end{cases}$$

For example, $|5| = 5$, $|0| = 0$, and $|-3| = -(-3) = 3$.

FIGURE 12   Graph of the absolute value function.

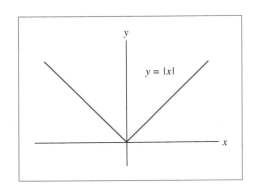

The function defined for all numbers x by

$$f(x) = |x|$$

is called the *absolute value function*. Its graph coincides with the graph of the equation $y = x$ for $x \geq 0$ and with the graph of the equation $y = -x$ for $x < 0$. (See Fig. 12.)

TECHNOLOGY PROJECT 0.2
Graphing Standard Functions with a Graphing Calculator

This project is designed to acquaint you with the graphs of some elementary functions. Graph the following functions.

1. Linear functions: $f(x) = ax + b$ for $a = 1, b = 3; a = -2, b = 4; a = 0, b = 1; a = 4, b = -7$.
2. Quadratic functions: $f(x) = ax^2 + bx$, for $a = 1, b = -1; a = 2, b = 0; a = -1, b = 3; a = 4, b = -2$.
3. Cubic functions: $f(x) = x^3 - ax$ for $a = -2, -1, -.5, 0, .5, 1, 2$.
4. Graph $x + \dfrac{1}{x}$ for $.5 \leq x \leq 10$.
5. Graph $f(x) = \sqrt{2 - x}$. Determine an appropriate display window.

PRACTICE PROBLEMS 0.2

1. A photocopy service has a fixed cost of $2000 per month (for rent, depreciation of equipment, etc.) and variable costs of $.04 for each page it reproduces for customers. Express its total cost as a (linear) function of the number of pages copied per month.
2. Determine the intercepts of the graph of $f(x) = -\frac{3}{8}x + 6$.

EXERCISES 0.2

Graph the following functions.

1. $f(x) = 2x - 1$
2. $f(x) = 3$
3. $f(x) = 3x + 1$
4. $f(x) = -\frac{1}{2}x - 4$
5. $f(x) = -2x + 3$
6. $f(x) = \frac{1}{4}$

Determine the intercepts of the graphs of the following functions.

7. $f(x) = 9x + 3$
8. $f(x) = -\frac{1}{2}x - 1$
9. $f(x) = 5$
10. $f(x) = 14$
11. $f(x) = -\frac{1}{4}x + 3$
12. $f(x) = 6x - 4$

13. In biochemistry, such as in the study of enzyme kinetics, one encounters a linear function of the form $f(x) = (K/V)x + 1/V$, where K and V are constants.
    (a) If $f(x) = .2x + 50$, find K and V so that $f(x)$ may be written in the form $f(x) = (K/V)x + 1/V$.

(b) Find the x-intercept and y-intercept of the line $y = (K/V)x + 1/V$ (in terms of K and V).

14. The constants K and V in Exercise 13 are often determined from experimental data. Suppose that a line is drawn through data points and has x-intercept $(-500, 0)$ and y-intercept $(0, 60)$. Determine K and V so that the line is the graph of the function $f(x) = (K/V)x + 1/V$. [*Hint:* Use Exercise 13(b).]

15. In some cities you can rent a car for $18 per day and $.20 per mile.

    (a) Find the cost of renting the car for one day and driving 200 miles.

    (b) Suppose that the car is to be rented for one day, and express the total rental expense as a function of the number x of miles driven. (Assume that for each fraction of a mile driven, the same fraction of $.20 is charged.)

16. A gas company will pay a property owner $5000 for the right to drill on the land for natural gas and $.10 for each thousand cubic feet of gas extracted from the land. Express the amount of money the landowner will receive as a function of the amount of gas extracted from the land.

17. In 1992, a patient paid $300 per day for a semiprivate hospital room and $1500 for an appendectomy operation. Express the total amount paid for an appendectomy as a function of the number of days of hospital confinement.

18. When a baseball thrown at 85 miles per hour is hit by a bat swung at x miles per hour, the ball travels $6x - 40$ feet.* (This formula assumes that $50 \leq x \leq 90$ and that the bat is 35 inches long, weighs 32 ounces, and strikes a waist-high pitch so that the plane of the swing lies at $35°$ from the horizontal.) How fast must the bat be swung in order for the ball to travel 350 feet?

19. Let $f(x)$ be the cost-benefit function from Example 5. If 70% of the pollutant has been removed, what is the added cost to remove another 5%? How does this compare with the cost to remove the final 5% of the pollutant? (See Example 5.)

20. Suppose that the cost (in millions of dollars) to remove x percent of a certain pollutant is given by the cost-benefit function

$$f(x) = \frac{20x}{102 - x} \qquad \text{for } 0 \leq x \leq 100.$$

    (a) Find the cost to remove 85% of the pollutant.

    (b) Find the cost to remove the final 5% of the pollutant.

Each of the quadratic functions in Exercises 21–26 has the form $y = ax^2 + bx + c$. Identify a, b, and c.

21. $y = 3x^2 - 4x$ 22. $y = \dfrac{x^2 - 6x + 2}{3}$ 23. $y = 3x - 2x^2 + 1$

24. $y = 3 - 2x + 4x^2$ 25. $y = 1 - x^2$ 26. $y = \frac{1}{2}x^2 + \sqrt{3}x - \pi$

Sketch the graphs of the following functions.

27. $f(x) = \begin{cases} 3x & \text{for } 0 \leq x \leq 1 \\ \frac{9}{2} - \frac{3}{2}x & \text{for } x > 1 \end{cases}$ 28. $f(x) = \begin{cases} 1 + x & \text{for } x \leq 3 \\ 4 & \text{for } x > 3 \end{cases}$

29. $f(x) = \begin{cases} 3 & \text{for } x < 2 \\ 2x + 1 & \text{for } x \geq 2 \end{cases}$ 30. $f(x) = \begin{cases} \frac{1}{2}x & \text{for } 0 \leq x < 4 \\ 2x - 3 & \text{for } 4 \leq x \leq 5 \end{cases}$

*Robert K. Adair, *The Physics of Baseball* (New York: Harper & Row, 1990).

31. $f(x) = \begin{cases} 4 - x & \text{for } 0 \le x < 2 \\ 2x - 2 & \text{for } 2 \le x < 3 \\ x + 1 & \text{for } x \ge 3 \end{cases}$ 32. $f(x) = \begin{cases} 4x & \text{for } 0 \le x < 1 \\ 8 - 4x & \text{for } 1 \le x < 2 \\ 2x - 4 & \text{for } x \ge 2 \end{cases}$

Evaluate each of the functions in Exercises 33–38 at the given value of x.

33. $f(x) = x^{100}$, $x = -1$

34. $f(x) = x^5$, $x = \frac{1}{2}$

35. $f(x) = |x|$, $x = 10^{-2}$

36. $f(x) = |x|$, $x = \pi$

37. $f(x) = |x|$, $x = -2.5$

38. $f(x) = |x|$, $x = -\frac{2}{3}$

SOLUTIONS TO PRACTICE PROBLEMS 0.2

1. If x represents the number of pages copied per month, then the variable cost is $.04x$ dollars. Now [total cost] = [fixed cost] + [variable cost]. If we define

$$f(x) = 2000 + .04x,$$

then $f(x)$ gives the total cost per month.

2. To find the y-intercept, evaluate $f(x)$ at $x = 0$.

$$f(0) = -\tfrac{3}{8}(0) + 6 = 0 + 6 = 6.$$

To find the x-intercept set $f(x) = 0$ and solve for x.

$$-\tfrac{3}{8}x + 6 = 0$$

$$\tfrac{3}{8}x = 6$$

$$x = \tfrac{8}{3} \cdot 6 = 16.$$

Therefore, the y-intercept is $(0, 6)$ and the x-intercept is $(16, 0)$.

0.3 THE ALGEBRA OF FUNCTIONS

Many functions we shall encounter later in the text may be viewed as combinations of other functions. For example, let $P(x)$ represent the profit a company makes on the sale of x units of some commodity. If $R(x)$ denotes the revenue received from the sale of x units, and if $C(x)$ is the cost of producing x units, then

$$P(x) = \quad R(x) \quad - \quad C(x)$$

$$[\text{profit}] = [\text{revenue}] - [\text{cost}]$$

Writing the profit function in this way makes it possible to predict the behavior of $P(x)$ from properties of $R(x)$ and $C(x)$. (See Fig. 1.)

The first four examples below review the algebraic techniques needed to combine functions by addition, subtraction, multiplication, and division.

EXAMPLE 1   Let $f(x) = 3x + 4$ and $g(x) = 2x - 6$. Find $f(x) + g(x)$, $f(x) - g(x)$, $\dfrac{f(x)}{g(x)}$, and $f(x)\,g(x)$.

FIGURE I   Profit equals revenue minus cost.

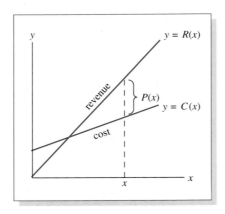

Solution   For $f(x) + g(x)$ and $f(x) - g(x)$ we add or subtract corresponding terms:

$$f(x) + g(x) = (3x + 4) + (2x - 6) = 3x + 4 + 2x - 6 = 5x - 2.$$

$$f(x) - g(x) = (3x + 4) - (2x - 6) = 3x + 4 - 2x + 6 = x + 10.$$

To compute $\dfrac{f(x)}{g(x)}$ and $f(x)g(x)$, we first substitute the formulas for $f(x)$ and $g(x)$.

$$\frac{f(x)}{g(x)} = \frac{3x + 4}{2x - 6}.$$

$$f(x)g(x) = (3x + 4)(2x - 6).$$

The expression for $\dfrac{f(x)}{g(x)}$ is already in simplest form. To simplify the expression for $f(x)g(x)$, we carry out the multiplication indicated in $(3x + 4)(2x - 6)$. We must be careful to multiply each term of $3x + 4$ by each term of $2x - 6$. A common order for multiplying such expressions is: (1) the First terms, (2) the Outer terms, (3) the Inside terms, and (4) the Last terms. (This procedure may be remembered by the word FOIL.)

$$f(x)g(x) = (3x + 4)(2x - 6) = 6x^2 - 18x + 8x - 24$$
$$= 6x^2 - 10x - 24. \;\bullet$$

EXAMPLE 2   Let $g(x) = \dfrac{2}{x}$ and $h(x) = \dfrac{3}{x - 1}$. Express $g(x) + h(x)$ as a rational function.

Solution   First we have

$$g(x) + h(x) = \frac{2}{x} + \frac{3}{x - 1}, \qquad x \neq 0, 1.$$

The restriction $x \neq 0, 1$ comes from the fact that $g(x)$ is defined only for $x \neq 0$ and $h(x)$ is defined only for $x \neq 1$. (A rational function is not defined for values of the variable for which the denominator is 0.) In order for us to add two fractions, their denominators must be the same. A common denominator for $\dfrac{2}{x}$ and $\dfrac{3}{(x-1)}$ is $x(x-1)$. If we multiply $\dfrac{2}{x}$ by $\dfrac{(x-1)}{(x-1)}$, we obtain an equivalent expression whose denominator is $x(x-1)$. Similarly, if we multiply $\dfrac{3}{(x-1)}$ by $\dfrac{x}{x}$, we obtain the equivalent expression whose denominator is $x(x-1)$. Thus

$$\frac{2}{x} + \frac{3}{x-1} = \frac{2}{x} \cdot \frac{x-1}{x-1} + \frac{3}{x-1} \cdot \frac{x}{x}$$

$$= \frac{2(x-1)}{x(x-1)} + \frac{3x}{x(x-1)}$$

$$= \frac{2(x-1) + 3x}{x(x-1)}$$

$$= \frac{5x-2}{x(x-1)}.$$

So

$$g(x) + h(x) = \frac{5x-2}{x(x-1)}. \ \bullet$$

EXAMPLE 3   Find $f(t)g(t)$, where

$$f(t) = \frac{t}{t-1} \quad \text{and} \quad g(t) = \frac{t+2}{t+1}.$$

Solution   To multiply rational functions, multiply numerator by numerator and denominator by denominator:

$$f(t)g(t) = \frac{t}{t-1} \cdot \frac{t+2}{t+1}$$

$$= \frac{t(t+2)}{(t-1)(t+1)}.$$

An alternative way of expressing $f(t)g(t)$ is obtained by carrying out the indicated multiplications:

$$f(t)g(t) = \frac{t^2 + 2t}{t^2 + t - t - 1} = \frac{t^2 + 2t}{t^2 - 1}.$$

The choice of which expression to use for $f(t)g(t)$ depends on the particular application involving $f(t)g(t)$. $\ \bullet$

EXAMPLE 4   Find $\dfrac{f(x)}{g(x)}$, where

$$f(x) = \frac{x}{x-3} \quad \text{and} \quad g(x) = \frac{x+1}{x-5}.$$

Solution   The function $f(x)$ is defined only for $x \neq 3$, and $g(x)$ is defined only for $x \neq 5$. The quotient $f(x)/g(x)$ is therefore not defined for $x = 3, 5$. Moreover, the quotient is not defined for values of x for which $g(x)$ is equal to 0, that is $x = -1$. Thus, the quotient is defined for $x \neq 3, 5, -1$. To divide $f(x)$ by $g(x)$, we multiply $f(x)$ by the reciprocal of $g(x)$:

$$\frac{f(x)}{g(x)} = \frac{x}{x-3} \cdot \frac{x-5}{x+1} = \frac{x(x-5)}{(x-3)(x+1)} = \frac{x^2 - 5x}{x^2 - 2x - 3}, \qquad x \neq 3, -1, 5.$$

●

Composition of Functions   Another important way of combining two functions $f(x)$ and $g(x)$ is to substitute the function $g(x)$ for every occurrence of the variable x in $f(x)$. The resulting function is called the *composition* (or *composite*) of $f(x)$ and $g(x)$ and is denoted by $f(g(x))$.

EXAMPLE 5   Let $f(x) = x^2 + 3x + 1$ and $g(x) = x - 5$. What is $f(g(x))$?

Solution   We substitute $g(x)$ in place of each x in $f(x)$.

$$f(g(x)) = [g(x)]^2 + 3g(x) + 1$$
$$= (x-5)^2 + 3(x-5) + 1$$
$$= (x^2 - 10x + 25) + (3x - 15) + 1$$
$$= x^2 - 7x + 11. \; ●$$

Later in the text we shall need to study expressions of the form $f(x + h)$, where $f(x)$ is a given function and h represents some number. The meaning of $f(x + h)$ is that $x + h$ is to be substituted for each occurrence of x in the formula for $f(x)$. In fact, $f(x + h)$ is just a special case of $f(g(x))$, where $g(x) = x + h$.

EXAMPLE 6   If $f(x) = x^3$, find $f(x + h) - f(x)$.

Solution

$$f(x + h) = (x + h)^3 = x^3 + 3x^2h + 3xh^2 + h^3$$
$$f(x + h) - f(x) = (x^3 + 3x^2h + 3xh^2 + h^3) - x^3$$
$$= 3x^2h + 3xh^2 + h^3. \; ●$$

EXAMPLE 7   In a certain lake, the bass feed primarily on minnows, and the minnows feed on plankton. Suppose that the size of the bass population is a function $f(n)$ of the number n of minnows in the lake, and the number of minnows is a function $g(x)$ of the amount x of plankton in the lake. Express the size of the bass population as a function of the amount of plankton, if $f(n) = 50 + \sqrt{n/150}$ and $g(x) = 4x + 3$.

Solution We have $n = g(x)$. Substituting $g(x)$ for n in $f(n)$, we find that the size of the bass population is given by

$$f(g(x)) = 50 + \sqrt{\frac{g(x)}{150}} = 50 + \sqrt{\frac{4x + 3}{150}}. \quad \bullet$$

TECHNOLOGY PROJECT 0.3
Translating Graphs of Functions

Let $f(x) = x^2$.

1. Graph the functions $f(x + 1), f(x - 1), f(x + 2), f(x - 2)$.
2. Graph the functions $f(x) + 1, f(x) - 1, f(x) + 2, f(x) - 2$.
3. On the basis of the evidence accumulated in problem 1., make a guess about the relationship between the graph of a general function $f(x)$ and the graph of $f(x + h)$.
4. Test your guess in problem 3. using the functions $g(x) = x^3$ and $f(x) = \sqrt{x}$ and various values of h.
5. On the basis of the evidence accumulated in problem 2., make a guess about the relationship between the graph of a general function $f(x)$ and the graph of $f(x) + h$.
6. Test your guess in problem 5. using the functions $g(x) = x^3$ and $f(x) = \sqrt{x}$ and various values of h.
7. Using what you have learned above, sketch the graph of $f(x) = (x - 1)^2 + 2$ without using a graphing calculator. Check your result using a graphing calculator.

PRACTICE PROBLEMS 0.3

1. Let $f(x) = x^5$, $g(x) = x^3 - 4x^2 + x - 8$.
   (a) Find $f(g(x))$.   (b) Find $g(f(x))$.
2. Let $f(x) = x^2$. Calculate $\dfrac{f(1 + h) - f(1)}{h}$ and simplify.

EXERCISES 0.3

Let $f(x) = x^2 + 1$, $g(x) = 9x$, and $h(x) = 5 - 2x^2$. Calculate the following functions.

1. $f(x) + g(x)$ 2. $f(x) - h(x)$ 3. $f(x)g(x)$

4. $g(x)h(x)$ 5. $\dfrac{f(t)}{g(t)}$ 6. $\dfrac{g(t)}{h(t)}$

In Exercises 7–12, express $f(x) + g(x)$ as a rational function. Carry out all multiplications.

7. $f(x) = \dfrac{2}{x - 3}$, $g(x) = \dfrac{1}{x + 2}$

8. $f(x) = \dfrac{3}{x - 6}$, $g(x) = \dfrac{-2}{x - 2}$

9. $f(x) = \dfrac{x}{x - 8}$, $g(x) = \dfrac{-x}{x - 4}$

10. $f(x) = \dfrac{-x}{x + 3}$, $g(x) = \dfrac{x}{x + 5}$

11. $f(x) = \dfrac{x + 5}{x - 10}$, $g(x) = \dfrac{x}{x + 10}$

12. $f(x) = \dfrac{x + 6}{x - 6}$, $g(x) = \dfrac{x - 6}{x + 6}$

Let $f(x) = \dfrac{x}{x - 2}$, $g(x) = \dfrac{5 - x}{5 + x}$, and $h(x) = \dfrac{x + 1}{3x - 1}$. Express the following as rational functions.

13. $f(x) - g(x)$

14. $f(t) - h(t)$

15. $f(x)g(x)$

16. $g(x)h(x)$

17. $\dfrac{f(x)}{g(x)}$

18. $\dfrac{h(s)}{f(s)}$

19. $f(x + 1)g(x + 1)$

20. $f(x + 2) + g(x + 2)$

21. $\dfrac{g(x + 5)}{f(x + 5)}$

22. $f\left(\dfrac{1}{t}\right)$

23. $g\left(\dfrac{1}{u}\right)$

24. $h\left(\dfrac{1}{x^2}\right)$

Let $f(x) = x^6$, $g(x) = \dfrac{x}{1 - x}$, and $h(x) = x^3 - 5x^2 + 1$. Calculate the following functions.

25. $f(g(x))$

26. $h(f(t))$

27. $h(g(x))$

28. $g(f(x))$

29. $g(h(t))$

30. $f(h(x))$

31. If $f(x) = x^2$, find $f(x + h) - f(x)$ and simplify.

32. If $f(x) = 1/x$, find $f(x + h) - f(x)$ and simplify.

33. If $g(t) = 4t - t^2$, find $\dfrac{g(t + h) - g(t)}{h}$ and simplify.

34. If $g(t) = t^3 + 5$, find $\dfrac{g(t + h) - g(t)}{h}$ and simplify.

35. After t hours of operation, an assembly line has assembled $A(t) = 20t - \frac{1}{2}t^2$ power lawn mowers, $0 \le t \le 10$. Suppose that the factory's cost of manufacturing x units is $C(x)$ dollars, where $C(x) = 3000 + 80x$.

    (a) Express the factory's cost as a (composite) function of the number of hours of operation of the assembly line.

    (b) What is the cost of the first 2 hours of operation?

36. During the first $\frac{1}{2}$ hour, the employees of a machine shop prepare the work area for the day's work. After that, they turn out 10 precision machine parts per hour, so that the output after t hours is $f(t)$ machine parts, where $f(t) = 10(t - \frac{1}{2}) = 10t - 5$, $\frac{1}{2} \le t \le 8$. The total cost of producing x machine parts is $C(x)$ dollars, where $C(x) = .1x^2 + 25x + 200$.

    (a) Express the total cost as a (composite) function of t.

    (b) What is the cost of the first 4 hours of operation?

TABLE I	Conversion Table for Men's Hat Sizes							
Britain	$6\frac{1}{2}$	$6\frac{5}{8}$	$6\frac{3}{4}$	$6\frac{7}{8}$	7	$7\frac{1}{8}$	$7\frac{1}{4}$	$7\frac{3}{8}$
France	53	54	55	56	57	58	59	60
United States	$6\frac{5}{8}$	$6\frac{3}{4}$	$6\frac{7}{8}$	7	$7\frac{1}{8}$	$7\frac{1}{4}$	$7\frac{3}{8}$	$7\frac{1}{2}$

37. Table 1 shows a conversion table for men's hat sizes for three countries. The function $g(x) = 8x + 1$ converts from British sizes to French sizes and the function $f(x) = \frac{1}{8}x$ converts from French sizes to U.S. sizes. Determine the function $h(x) = f(g(x))$ and give its interpretation.

SOLUTIONS TO PRACTICE PROBLEMS 0.3

1. (a) $f(g(x)) = [g(x)]^5 = (x^3 - 4x^2 + x - 8)^5$.

   (b) $g(f(x)) = [f(x)]^3 - 4[f(x)]^2 + f(x) - 8$

   $= (x^5)^3 - 4(x^5)^2 + x^5 - 8$

   $= x^{15} - 4x^{10} + x^5 - 8$.

2. $\dfrac{f(1 + h) - f(1)}{h} = \dfrac{(1 + h)^2 - 1^2}{h}$

   $= \dfrac{1 + 2h + h^2 - 1}{h}$

   $= \dfrac{2h + h^2}{h}$

   $= 2 + h$.

0.4 ZEROS OF FUNCTIONS—THE QUADRATIC FORMULA AND FACTORING

A *zero* of a function $f(x)$ is a value of x for which $f(x) = 0$. For instance, the function $f(x)$ whose graph is shown in Fig. 1 has $x = -3$, $x = 3$, and $x = 7$ as zeros. Throughout this book we shall need to determine zeros of functions or, what amounts to the same thing, to solve the equation $f(x) = 0$.

In Section 0.2 we found zeros of linear functions. In this section our emphasis is on zeros of quadratic functions.

The Quadratic Formula   Consider the quadratic function $f(x) = ax^2 + bx + c$, $a \neq 0$. The zeros of this function are precisely the solutions of the quadratic equation

$$ax^2 + bx + c = 0.$$

FIGURE 1   Zeros of a function.

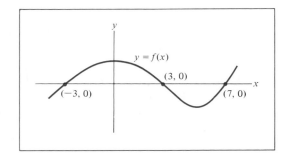

One way of solving such an equation is via the *quadratic formula.*

The solutions of the equation $ax^2 + bx + c = 0$ are

$$x = \frac{-b \pm \sqrt{b^2 - 4ac}}{2a}.$$

The $\pm$ sign tells us to form two expressions, one with $+$ and one with $-$. The quadratic formula is derived at the end of the section.

EXAMPLE 1   Solve the quadratic equation $3x^2 - 6x + 2 = 0$.

Solution   Here $a = 3$, $b = -6$, and $c = 2$. Substituting these values into the quadratic formula, we find that

$$\sqrt{b^2 - 4ac} = \sqrt{(-6)^2 - 4(3)(2)} = \sqrt{36 - 24} = \sqrt{12}$$
$$= \sqrt{4 \cdot 3} = 2\sqrt{3}$$

and

$$x = \frac{-b \pm \sqrt{b^2 - 4ac}}{2a}$$
$$= \frac{-(-6) \pm 2\sqrt{3}}{2(3)}$$
$$= \frac{6 \pm 2\sqrt{3}}{6}$$
$$= 1 \pm \frac{\sqrt{3}}{3}.$$

The solutions of the equation are $1 + \sqrt{3}/3$ and $1 - \sqrt{3}/3$. (See Fig. 2.) ●

EXAMPLE 2   Find the zeros of the following quadratic functions.
   (a) $f(x) = 4x^2 - 4x + 1$ (b) $f(x) = \frac{1}{2}x^2 - 3x + 5$

FIGURE 2

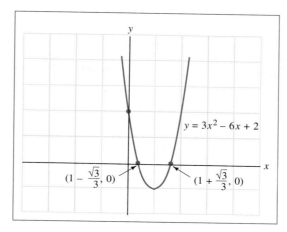

$y = 3x^2 - 6x + 2$

$\left(1 - \frac{\sqrt{3}}{3}, 0\right)$ $\left(1 + \frac{\sqrt{3}}{3}, 0\right)$

Solution   (a) We must solve $4x^2 - 4x + 1 = 0$. Here $a = 4, b = -4$, and $c = 1$, so that
$$\sqrt{b^2 - 4ac} = \sqrt{(-4)^2 - 4(4)(1)} = \sqrt{0} = 0.$$
Thus there is only one zero, namely,
$$x = \frac{-(-4) \pm 0}{2(4)} = \frac{4}{8} = \frac{1}{2}.$$
The graph of $f(x)$ is sketched in Fig. 3.

(b) We must solve $\frac{1}{2}x^2 - 3x + 5 = 0$. Here $a = \frac{1}{2}, b = -3$, and $c = 5$, so that
$$\sqrt{b^2 - 4ac} = \sqrt{(-3)^2 - 4(\tfrac{1}{2})(5)} = \sqrt{9 - 10} = \sqrt{-1}.$$
The square root of a negative number is undefined, so we conclude that $f(x)$ has no zeros. The reason for this is clear from Fig. 4. The graph of $f(x)$ lies entirely above the x-axis and has no x-intercepts. ●

FIGURE 3

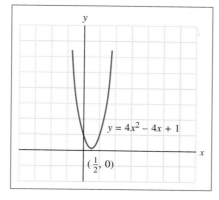

$y = 4x^2 - 4x + 1$

$\left(\frac{1}{2}, 0\right)$

FIGURE 4

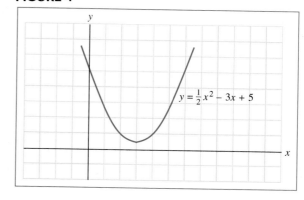

$y = \frac{1}{2}x^2 - 3x + 5$

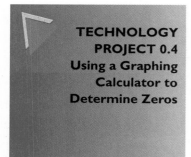

TECHNOLOGY PROJECT 0.4
Using a Graphing Calculator to Determine Zeros

Using the TRACE, you can locate zeros of functions. By successively approximating the zero and then using BOX to expand the region near the zero, you can achieve any degree of accuracy desired. Use this technique to determine the zeros of the following functions in the interval $-10 \le x \le 10$. Determine each zero to within .001.

1. $f(x) = x^2 - x - 2$ 2. $f(x) = x^3 - 3x + 4$

3. $f(x) = \sqrt{x - 2} - x + 2$ 4. $f(x) = \dfrac{x}{x + 1} - x^2 + 1$

The common problem of finding where two curves intersect amounts to finding the zero of a function.

EXAMPLE 3   Find the points of intersection of the graphs of the functions $y = x^2 + 1$ and $y = 4x$. (See Fig. 5.)

Solution   If a point (x, y) is on both graphs, then its coordinates must satisfy both equations. That is, x and y must satisfy $y = x^2 + 1$ and $y = 4x$. Equating the two expressions for y, we have

$$x^2 + 1 = 4x.$$

To use the quadratic formula, we rewrite the equation in the form

$$x^2 - 4x + 1 = 0.$$

FIGURE 5   Points of intersection of two graphs.

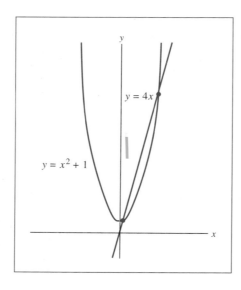

From the quadratic formula,

$$x = \frac{4 \pm \sqrt{16 - 4}}{2} = \frac{4 \pm \sqrt{12}}{2} = \frac{4 \pm 2\sqrt{3}}{2} = 2 \pm \sqrt{3}.$$

Thus the x-coordinates of the points of intersection are $2 + \sqrt{3}$ and $2 - \sqrt{3}$. To find the y-coordinates, we substitute these values of x into either equation, $y = x^2 + 1$ or $y = 4x$. The second equation is simpler. We obtain $y = 4(2 + \sqrt{3}) = 8 + 4\sqrt{3}$ and $y = 4(2 - \sqrt{3}) = 8 - 4\sqrt{3}$. Thus the points of intersection are $(2 + \sqrt{3}, 8 + 4\sqrt{3})$ and $(2 - \sqrt{3}, 8 - 4\sqrt{3})$. ●

EXAMPLE 4　A cable television company estimates that with x thousand subscribers, its monthly revenue and cost (in thousands of dollars) are

$$R(x) = 32x - .21x^2,$$

$$C(x) = 195 + 12x.$$

Determine the company's *break-even points;* that is, find the number of subscribers at which the revenue equals the cost. See Fig. 6.

FIGURE 6　Break-even points.

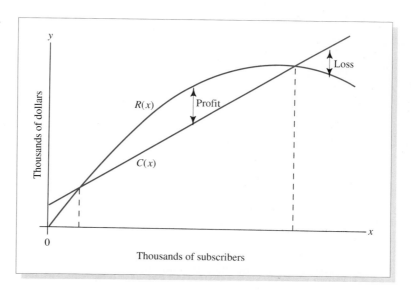

Solution　Let $P(x)$ be the profit function.

$$P(x) = R(x) - C(x)$$

$$= (32x - .21x^2) - (195 + 12x)$$

$$= -.21x^2 + 20x - 195$$

The break-even point occurs where the profit is zero. Thus we must solve

$$-.21x^2 + 20x - 195 = 0.$$

From the quadratic formula,

$$x = \frac{-20 \pm \sqrt{20^2 - 4(-.21)(-195)}}{2(-.21)} = \frac{-20 \pm \sqrt{236.2}}{-.42}$$

$$\approx 47.62 \pm 36.59 = 11.03 \text{ and } 84.21.$$

The break-even points occur where the company has 11,030 or 84,210 subscribers. Between those two levels, the company will be profitable. ●

TECHNOLOGY PROJECT 0.5

Using a Graphing Calculator to Determine Intersection Points of Graphs

The equation entry screen allows you to define more than one function at a time. The calculator can graph all of the currently defined functions on the same screen. (This is usually the default option.) Moreover, by using the vertical arrow keys, you can move the TRACE cursor between the various graphs shown. Consult your calculator manual for details. Combining the TRACE and BOX operations, you can approximate the coordinates of the intersection points of a pair of graphs. This ability can be used, for example, to solve systems of simultaneous equations. Use your graphing calculator to determine the intersection points of the graphs of $f(x)$ and $g(x)$. Approximate the coordinates of each intersection point to within .1.

1. $f(x) = 2x - 1, \quad g(x) = x^2 - 2$

2. $f(x) = -x - 2, \quad g(x) = -4x^2 + x + 1$

3. $f(x) = x^2 - 3x, \quad g(x) = x^3 + 2x^2 - 4x + 1$

4. $f(x) = \dfrac{1}{x}, \quad g(x) = \sqrt{x^2 - 1}$

Factoring If $f(x)$ is a polynomial, we can often write $f(x)$ as a product of linear factors (i.e., factors of the form $ax + b$). If this can be done, then the zeros of $f(x)$ can be determined by setting each of the linear factors equal to zero and solving for x. (The reason is that the product of numbers can be zero only when one of the numbers is zero.)

EXAMPLE 5 Factor the following quadratic functions.

(a) $x^2 + 7x + 12$ (b) $x^2 - 13x + 12$

(c) $x^2 - 4x - 12$ (d) $x^2 + 4x - 12$

Solution Note first that for any numbers c and d,

$$(x + c)(x + d) = x^2 + (c + d)x + cd.$$

In the quadratic on the right, the constant term is the product cd, whereas the coefficient of x is the sum $c + d$.

(a) Think of all integers c and d such that $cd = 12$. Then choose the pair that satisfies $c + d = 7$; that is, take $c = 3$, $d = 4$. Thus

$$x^2 + 7x + 12 = (x + 3)(x + 4).$$

(b) We want $cd = 12$. Since 12 is positive, c and d must be both positive or both negative. We must also have $c + d = -13$. These facts lead us to

$$x^2 - 13x + 12 = (x - 12)(x - 1).$$

(c) We want $cd = -12$. Since -12 is negative, c and d must have opposite signs. Also, they must sum to give -4. We find that

$$x^2 - 4x - 12 = (x - 6)(x + 2).$$

(d) This is almost the same as part (c).

$$x^2 + 4x - 12 = (x + 6)(x - 2). \ \bullet$$

EXAMPLE 6   Factor the following polynomials.
   (a) $x^2 - 6x + 9$ (b) $x^2 - 25$
   (c) $3x^2 - 21x + 30$ (d) $20 + 8x - x^2$

Solution   (a) We look for $cd = 9$ and $c + d = -6$. The solution is $c = d = -3$, and

$$x^2 - 6x + 9 = (x - 3)(x - 3) = (x - 3)^2.$$

In general,

$$x^2 - 2cx + c^2 = (x - c)(x - c) = (x - c)^2.$$

(b) We use the identity

$$x^2 - c^2 = (x + c)(x - c).$$

Hence

$$x^2 - 25 = (x + 5)(x - 5).$$

(c) We first factor out a common factor of 3 and then use the method of Example 5.

$$3x^2 - 21x + 30 = 3(x^2 - 7x + 10)$$
$$= 3(x - 5)(x - 2).$$

(d) We first factor out a -1 in order to make the coefficient of x^2 equal to $+1$.

$$20 + 8x - x^2 = (-1)(x^2 - 8x - 20)$$
$$= (-1)(x - 10)(x + 2). \ \bullet$$

EXAMPLE 7   Factor the following polynomials.
   (a) $x^2 - 8x$ (b) $x^3 + 3x^2 - 18x$ (c) $x^3 - 10x$

Solution   In each case we first factor out a common factor of x.

(a) $x^2 - 8x = x(x - 8)$.

(b) $x^3 + 3x^2 - 18x = x(x^2 + 3x - 18) = x(x + 6)(x - 3)$.

(c) $x^3 - 10x = x(x^2 - 10)$. To factor $x^2 - 10$, we use the identity $x^2 - c^2 = (x + c)(x - c)$, where $c^2 = 10$ and $c = \sqrt{10}$. Thus $x^3 - 10x = x(x^2 - 10) = x(x + \sqrt{10})(x - \sqrt{10})$. ●

The main use of factoring in this text will be to solve equations.

EXAMPLE 8   Solve the following equations.

(a) $x^2 - 2x - 15 = 0$ (b) $x^2 - 20 = x$ (c) $\dfrac{x^2 + 10x + 25}{x + 1} = 0$

Solution   (a) The equation $x^2 - 2x - 15 = 0$ may be written in the form

$$(x - 5)(x + 3) = 0.$$

The product of two numbers is zero only if one or the other of the numbers (or both) is zero. Hence

$$x - 5 = 0 \quad \text{or} \quad x + 3 = 0.$$

That is,

$$x = 5 \quad \text{or} \quad x = -3.$$

(b) First we must rewrite the equation $x^2 - 20 = x$ in the form $ax^2 + bx + c = 0$; that is,

$$x^2 - x - 20 = 0$$
$$(x - 5)(x + 4) = 0.$$

We conclude that

$$x - 5 = 0 \quad \text{or} \quad x + 4 = 0;$$

that is

$$x = 5 \quad \text{or} \quad x = -4.$$

(c) A rational function will be zero only if the numerator is zero. Thus

$$x^2 + 10x + 25 = 0$$
$$(x + 5)^2 = 0$$
$$x + 5 = 0.$$

That is,

$$x = -5.$$

Since the denominator is not 0 at $x = -5$, we conclude that $x = -5$ is the solution. ●

Derivation of the Quadratic Formula

$$ax^2 + bx + c = 0$$

$$ax^2 + bx = -c$$

$$4a^2x^2 + 4abx = -4ac \qquad \text{(both sides multiplied by } 4a)$$

$$4a^2x^2 + 4abx + b^2 = b^2 - 4ac \qquad (b^2 \text{ added to both sides).}$$

Now note that $4a^2x^2 + 4abx + b^2 = (2ax + b)^2$. To check this, simply multiply out the right-hand side. Therefore,

$$(2ax + b)^2 = b^2 - 4ac$$

$$2ax + b = \pm\sqrt{b^2 - 4ac}$$

$$2ax = -b \pm \sqrt{b^2 - 4ac}$$

$$x = \frac{-b \pm \sqrt{b^2 - 4ac}}{2a}.$$

PRACTICE PROBLEMS 0.4

1. Solve the equation $x - 14/x = 5$.
2. Use the quadratic formula to solve $7x^2 - 35x + 35 = 0$.

EXERCISES 0.4

Use the quadratic formula to find the zeros of the functions in Exercises 1–6.

1. $f(x) = 2x^2 - 7x + 6$
2. $f(x) = 3x^2 + 2x - 1$
3. $f(x) = 4x^2 - 12x + 9$
4. $f(x) = \frac{1}{4}x^2 + x + 1$
5. $f(x) = -2x^2 + 3x - 4$
6. $f(x) = 11x^2 - 7x + 1$

Use the quadratic formula to solve the equations in Exercises 7–12.

7. $5x^2 - 4x - 1 = 0$
8. $x^2 - 4x + 5 = 0$
9. $15x^2 - 135x + 300 = 0$
10. $x^2 - \sqrt{2}x - \frac{5}{4} = 0$
11. $\frac{3}{2}x^2 - 6x + 5 = 0$
12. $9x^2 - 12x + 4 = 0$

Factor the polynomials in Exercises 13–24.

13. $x^2 + 8x + 15$
14. $x^2 - 10x + 16$
15. $x^2 - 16$
16. $x^2 - 1$
17. $3x^2 + 12x + 12$
18. $2x^2 - 12x + 18$
19. $30 - 4x - 2x^2$
20. $15 + 12x - 3x^2$
21. $3x - x^2$
22. $4x^2 - 1$
23. $6x - 2x^3$
24. $16x + 6x^2 - x^3$

Find the points of intersection of the pairs of curves in Exercises 25–32.

25. $y = 2x^2 - 5x - 6, \ y = 3x + 4$
26. $y = x^2 - 10x + 9, \ y = x - 9$
27. $y = x^2 - 4x + 4, \ y = 12 + 2x - x^2$
28. $y = 3x^2 + 9, \ y = 2x^2 - 5x + 3$

29. $y = x^3 - 3x^2 + x$, $y = x^2 - 3x$ 30. $y = \frac{1}{2}x^3 - 2x^2$, $y = 2x$

31. $y = \frac{1}{2}x^3 + x^2 + 5$, $y = 3x^2 - \frac{1}{2}x + 5$ 32. $y = 30x^3 - 3x^2$, $y = 16x^3 + 25x^2$

Solve the equations in Exercises 33–38.

33. $\dfrac{21}{x} - x = 4$ 34. $x + \dfrac{2}{x - 6} = 3$ 35. $x + \dfrac{14}{x + 4} = 5$

36. $1 = \dfrac{5}{x} + \dfrac{6}{x^2}$ 37. $\dfrac{x^2 + 14x + 49}{x^2 + 1} = 0$ 38. $\dfrac{x^2 - 8x + 16}{1 + \sqrt{x}} = 0$

39. Suppose the cable television company's cost function in Example 4 changes to $C(x) = 275 + 12x$. Determine the new break-even points.

40. When a car is moving at x miles per hour and the driver decides to slam on the brakes, the car will travel $x + \frac{1}{20}x^2$ feet.* If a car travels 175 feet after the driver decides to stop, how fast was the car moving?

SOLUTION TO PRACTICE PROBLEMS 0.4

1. Multiply both sides of the equation by x. Then

$$x^2 - 14 = 5x.$$

Now, take the term $5x$ to the left side of the equation and solve by factoring.

$$x^2 - 5x - 14 = 0$$
$$(x - 7)(x + 2) = 0$$
$$x = 7 \quad \text{or} \quad x = -2.$$

2. In this case, each coefficient is a multiple of 7. To simplify the arithmetic, we divide both sides of the equation by 7 before using the quadratic formula.

$$x^2 - 5x + 5 = 0$$
$$\sqrt{b^2 - 4ac} = \sqrt{(-5)^2 - 4(1)(5)} = \sqrt{5}$$
$$x = \frac{-b \pm \sqrt{b^2 - 4ac}}{2a} = \frac{5 \pm \sqrt{5}}{2 \cdot 1} = \frac{5}{2} \pm \frac{1}{2}\sqrt{5}.$$

 0.5 EXPONENTS AND POWER FUNCTIONS

In this section we review the operations with exponents that occur frequently throughout the text. We begin with the definition of b^r for various types of numbers b and r.

_____

* The general formula is $f(x) = ax + bx^2$, where the constant a depends on the driver's reaction time and the constant b depends on the weight of the car and the type of tires. This mathematical model was analyzed in D. Burghes, I. Huntley, and J. McDonald, *Applying Mathematics: A Course in Mathematical Modelling* (New York: Halstead Press, 1982), pp. 57–60.

For any nonzero number b and any positive integer n, we have by definition that

$$b^n = \underbrace{b \cdot b \cdot \ \cdots \ \cdot b}_{n \text{ times}},$$

$$b^{-n} = \frac{1}{b^n},$$

and

$$b^0 = 1.$$

For example, $2^4 = 2 \cdot 2 \cdot 2 \cdot 2 = 16$, $2^{-4} = \dfrac{1}{2^4} = \dfrac{1}{16}$, and $2^0 = 1$.

Next, we consider numbers of the form $b^{1/n}$, where n is a positive integer. For instance,

$2^{1/2}$ is the positive number whose square is 2: $\qquad 2^{1/2} = \sqrt{2}$;

$2^{1/3}$ is the positive number whose cube is 2: $\qquad 2^{1/3} = \sqrt[3]{2}$;

$2^{1/4}$ is the positive number whose fourth power is 2: $\quad 2^{1/4} = \sqrt[4]{2}$;

and so on. In general, when b is zero or positive, $b^{1/n}$ is zero or the positive number whose nth power is b.

In the special case when n is odd, we may permit b to be negative as well as positive. For example, $(-8)^{1/3}$ is the number whose cube is -8; that is,

$$(-8)^{1/3} = -2.$$

Thus, when b is negative and n is odd, we again define $b^{1/n}$ to be the number whose nth power is b.

Finally, let us consider numbers of the form $b^{m/n}$ and $b^{-m/n}$, where m and n are positive integers. We may assume that the fraction m/n is in lowest terms (so that m and n have no common factor). Then we define

$$b^{m/n} = (b^{1/n})^m$$

whenever $b^{1/n}$ is defined, and

$$b^{-m/n} = \frac{1}{b^{m/n}}$$

whenever $b^{m/n}$ is defined and is not zero. For example,

$$8^{5/3} = (8^{1/3})^5 = (2)^5 = 32,$$

$$8^{-5/3} = \frac{1}{8^{5/3}} = \frac{1}{32},$$

$$(-8)^{5/3} = [(-8)^{1/3}]^5 = [-2]^5 = -32.$$

Exponents may be manipulated algebraically according to the following rules:

Laws of Exponents

1. $b^r b^s = b^{r+s}$ **4.** $(b^r)^s = b^{rs}$

2. $b^{-r} = \dfrac{1}{b^r}$ **5.** $(ab)^r = a^r b^r$

3. $\dfrac{b^r}{b^s} = b^r \cdot b^{-s} = b^{r-s}$ **6.** $\left(\dfrac{a}{b}\right)^r = \dfrac{a^r}{b^r}$

EXAMPLE 1   Use the laws of exponents to calculate the following quantities.

(a) $2^{1/2} 50^{1/2}$ (b) $(2^{1/2} 2^{1/3})^6$ (c) $\dfrac{5^{3/2}}{\sqrt{5}}$

Solution   (a) $2^{1/2} 50^{1/2} = (2 \cdot 50)^{1/2}$ (Law 5)

$$= \sqrt{100}$$

$$= 10.$$

(b) $(2^{1/2} 2^{1/3})^6 = (2^{(1/2)+(1/3)})^6$ (Law 1)

$$= (2^{5/6})^6$$

$$= 2^{(5/6)6}$$ (Law 4)

$$= 2^5$$

$$= 32.$$

(c) $\dfrac{5^{3/2}}{\sqrt{5}} = \dfrac{5^{3/2}}{5^{1/2}}$

$$= 5^{(3/2)-(1/2)}$$ (Law 3)

$$= 5^1$$

$$= 5. \ \bullet$$

EXAMPLE 2   Simplify the following expressions.

(a) $\dfrac{1}{x^{-4}}$ (b) $\dfrac{x^2}{x^5}$ (c) $\sqrt{x}(x^{3/2} + 3\sqrt{x})$

Solution   (a) $\dfrac{1}{x^{-4}} = x^{-(-4)}$ (Law 2 with $r = -4$)

$$= x^4.$$

(b) $\dfrac{x^2}{x^5} = x^{2-5}$ (Law 3)

$\qquad = x^{-3}.$

It is also correct to write this answer as $\dfrac{1}{x^3}$.

(c) $\sqrt{x}(x^{3/2} + 3\sqrt{x}) = x^{1/2}(x^{3/2} + 3x^{1/2})$

$\qquad = x^{1/2}x^{3/2} + 3x^{1/2}x^{1/2}$

$\qquad = x^{(1/2)+(3/2)} + 3x^{(1/2)+(1/2)}$ (Law 1)

$\qquad = x^2 + 3x.$ ●

A *power function* is a function of the form

$$f(x) = x^r,$$

for some number r.

EXAMPLE 3   Let $f(x)$ and $g(x)$ be the power functions

$$f(x) = x^{-1} \quad \text{and} \quad g(x) = x^{1/2}.$$

Determine the following functions.

(a) $\dfrac{f(x)}{g(x)}$ 　　　　　(b) $f(x)g(x)$ 　　　　　(c) $\dfrac{g(x)}{f(x)}$

Solution   (a) $\dfrac{f(x)}{g(x)} = \dfrac{x^{-1}}{x^{1/2}}$ 　　(b) $f(x)g(x) = x^{-1}x^{1/2}$ 　(c) $\dfrac{g(x)}{f(x)} = \dfrac{x^{1/2}}{x^{-1}}$

$\qquad = x^{-1-(1/2)}$ 　　　　　$= x^{-1+(1/2)}$ 　　　　$= x^{(1/2)-(-1)}$

$\qquad = x^{-3/2}$ 　　　　　　$= x^{-1/2}$ 　　　　　$= x^{3/2}.$ ●

$\qquad = \dfrac{1}{x^{3/2}}.$ 　　　　　$= \dfrac{1}{x^{1/2}}$

$\qquad\qquad\qquad\qquad\qquad = \dfrac{1}{\sqrt{x}}.$

Compound Interest   The subject of compound interest provides a significant application of exponents. Let's introduce this topic at this point with a view toward using it as a source of applied problems throughout the book.

When money is deposited in a savings account, interest is paid at stated intervals. If this interest is added to the account and thereafter earns interest itself, then the interest is called *compound interest*. The original amount deposited is called the *principal amount*. The principal amount plus the compound interest is called the *compound amount*. The interval between interest payments is referred to as the *interest period*. In formulas for compound interest, the interest rate is expressed as a decimal rather than a percent. Thus 6% is written as .06.

If $1000 is deposited at 6% annual interest, compounded annually, the compound amount at the end of the first year will be

$$A_1 = \underset{\text{principal}}{1000} + \underset{\text{interest}}{1000(.06)} = 1000(1 + .06).$$

At the end of the second year the compound amount will be

$$A_2 = \underset{\substack{\text{compound} \\ \text{amount}}}{A_1} + \underset{\text{interest}}{A_1(.06)} = A_1(1 + .06)$$

$$= [1000(1 + .06)](1 + .06) = 1000(1 + .06)^2.$$

At the end of 3 years,

$$A_3 = A_2(.06) = A_2(1 + .06)$$

$$= [1000(1 + .06)^2](1 + .06) = 1000(1 + .06)^3.$$

After n years the compound amount will be

$$A = 1000(1 + .06)^n.$$

In this example the interest period was 1 year. The important point to note, however, is that at the end of each interest period the amount on deposit grew by a factor of $(1 + .06)$. In general, if the interest rate is i instead of .06, the compound amount will grow by a factor of $(1 + i)$ at the end of each interest period.

Suppose that a principal amount P is invested at a compound interest rate i per interest period, for a total of n interest periods. Then the compound amount A at the end of the nth period will be

$$A = P(1 + i)^n. \tag{1}$$

EXAMPLE 4   Suppose that $5000 is invested at 8% per year, with interest compounded annually. What is the compound amount after 3 years?

Solution   Substituting $P = 5000$, $i = .08$, and $n = 3$ into formula (1), we have

$$A = 5000(1 + .08)^3 = 5000(1.08)^3$$

$$= 5000(1.259712) = 6298.56 \text{ dollars.} \bullet$$

It is common practice to state the interest rate as a percent per year ("per annum"), even though each interest period is often shorter than 1 year. If the annual rate is r and if interest is paid and compounded m times per year, then the interest rate i for each period is given by

$$[\text{rate per period}] = i = \frac{r}{m} = \frac{[\text{annual interest rate}]}{[\text{periods per year}]}.$$

Many banks pay interest quarterly. If the stated annual rate is 5%, then $i = .05/4 = .0125$.

If interest is compounded for t years, with m interest periods each year, there will be a total of mt interest periods. If in formula (1) we replace n by mt and replace i by r/m, we obtain the following formula for the compound amount:

$$A = P\left(1 + \frac{r}{m}\right)^{mt},$$

where P = principal amount,
     r = interest rate per annum,   (2)
     m = number of interest periods per year,
     t = number of years.

EXAMPLE 5   Suppose that $1000 is deposited in a savings account that pays 6% per annum, compounded quarterly. If no additional deposits or withdrawals are made, how much will be in the account at the end of 1 year?

Solution   We use (2) with $P = 1000$, $r = .06$, $m = 4$, and $t = 1$.

$$A = 1000\left(1 + \frac{.06}{4}\right)^4 = 1000(1.015)^4$$

$$= 1000(1.06136355) \approx 1061.36 \text{ dollars.} \quad \bullet$$

Note that the $1000 in Example 2 earned a total of $61.36 in (compound) interest. This is 6.136% of $1000. Savings institutions sometimes advertise this rate as the *effective* annual interest rate. That is, the savings institutions mean that *if* they paid interest only once a year, they would have to pay a rate of 6.136% in order to produce the same earnings as their 6% rate compounded quarterly. The stated annual rate of 6% is often called the *nominal rate*.

The effective annual rate can be increased by compounding the interest more often. Some savings institutions compound interest monthly or even daily.

EXAMPLE 6   Suppose that the interest in Example 2 were compounded monthly. How much would be in the account at the end of 1 year? What about the case when 6% annual interest is compounded daily?

Solution   For monthly compounding, $m = 12$. From (2) we have

$$A = 1000\left(1 + \frac{.06}{12}\right)^{12} = 1000(1.005)^{12}$$

$$\approx 1000(1.06167781) \approx 1061.68 \text{ dollars.}$$

The effective rate in this case is 6.168%.

A "bank year" usually consists of 360 days (in order to simplify calculations). So, for daily compounding, we take $m = 360$. Then

$$A = 1000\left(1 + \frac{.06}{360}\right)^{360} \approx 1000(1.00016667)^{360}$$

$$\approx 1000(1.06183133) \approx 1061.83 \text{ dollars.}$$

With daily compounding, the effective rate is 6.183%. ●

EXAMPLE 7   Suppose that a corporation issues a bond costing $200 and paying interest compounded monthly. The interest is accumulated until the bond reaches maturity. (A security of this sort is called a *zero coupon bond*.) Suppose that after 5 years, the bond is worth $500. What is the annual interest rate?

Solution   Let r denote the annual interest rate. The value A of the bond after 5 years = 60 months is given by the compound interest formula:

$$A = 200\left(1 + \frac{r}{12}\right)^{60}$$

We must find r which satisfies:

$$500 = 200\left(1 + \frac{r}{12}\right)^{60}$$

$$2.5 = \left(1 + \frac{r}{12}\right)^{60}$$

Raise both sides to the power $\frac{1}{60}$ and apply the laws of exponents to obtain:

$$(2.5)^{1/60} = \left[\left(1 + \frac{r}{12}\right)^{60}\right]^{\frac{1}{60}} = \left(1 + \frac{r}{12}\right)^{60 \cdot \frac{1}{60}} = 1 + \frac{r}{12}$$

$$r = 12 \cdot ((2.5)^{1/60} - 1)$$

Using a calculator, we see that $r = .18466$. That is, the annual interest rate is 18.466%. (A bond paying a rate of interest this high is generally called a *junk bond*.) ●

PRACTICE PROBLEMS 0.5

1. Compute the following.
   (a) -5^2 (b) $16^{.75}$

2. Simplify the following.
   (a) $(4x^3)^2$ (b) $\dfrac{\sqrt[3]{x}}{x^3}$ (c) $\dfrac{2 \cdot (x + 5)^6}{x^2 + 10x + 25}$

EXERCISES 0.5

In Exercises 1–28, compute the numbers.

1. 3^3

2. $(-2)^3$

3. 1^{100}

4. 0^{25}

5. $(.1)^4$

6. $(100)^4$

7. -4^2

8. $(.01)^3$

9. $(16)^{1/2}$

10. $(27)^{1/3}$

11. $(.000001)^{1/3}$

12. $\left(\dfrac{1}{125}\right)^{1/3}$

13. 6^{-1}

14. $\left(\dfrac{1}{2}\right)^{-1}$

15. $(.01)^{-1}$

16. $(-5)^{-1}$

17. $8^{4/3}$

18. $16^{3/4}$

19. $(25)^{3/2}$

20. $(27)^{2/3}$

21. $(1.8)^0$

22. $9^{1.5}$

23. 16^5

24. $(81)^{.75}$

25. $4^{-1/2}$

26. $\left(\dfrac{1}{8}\right)^{-2/3}$

27. $(.01)^{-1.5}$

28. $1^{-1.2}$

In Exercises 29–40, use the laws of exponents to compute the numbers.

29. $5^{1/3} \cdot 200^{1/3}$

30. $(3^{1/3} \cdot 3^{1/6})^6$

31. $6^{1/3} \cdot 6^{2/3}$

32. $(9^{4/5})^{5/8}$

33. $\dfrac{10^4}{5^4}$

34. $\dfrac{3^{5/2}}{3^{1/2}}$

35. $(2^{1/3} \cdot 3^{2/3})^3$

36. $20^{.5} \cdot 5^{.5}$

37. $\left(\dfrac{8}{27}\right)^{2/3}$

38. $(125 \cdot 27)^{1/3}$

39. $\dfrac{7^{4/3}}{7^{1/3}}$

40. $(6^{1/2})^0$

In Exercises 41–70, use the laws of exponents to simplify the algebraic expressions. Your answer should not involve parentheses or negative exponents.

41. $(xy)^6$

42. $(x^{1/3})^6$

43. $\dfrac{x^4 \cdot y^5}{xy^2}$

44. $\dfrac{1}{x^{-3}}$

45. $x^{-1/2}$

46. $(x^3 \cdot y^6)^{1/3}$

47. $\left(\dfrac{x^4}{y^2}\right)^3$

48. $\left(\dfrac{x}{y}\right)^{-2}$

49. $(x^3 y^5)^4$

50. $\sqrt{1 + x}\,(1 + x)^{3/2}$

51. $x^5 \cdot \left(\dfrac{y^2}{x}\right)^3$

52. $x^{-3} \cdot x^7$

53. $(2x)^4$

54. $\dfrac{-3x}{15x^4}$

55. $\dfrac{-x^3 y}{-xy}$

56. $\dfrac{x^3}{y^{-2}}$

57. $\dfrac{x^{-4}}{x^3}$

58. $(-3x)^3$

59. $\sqrt[3]{x} \cdot \sqrt[3]{x^2}$

60. $(9x)^{-1/2}$

61. $\left(\dfrac{3x^2}{2y}\right)^3$

62. $\dfrac{x^2}{x^5 y}$

63. $\dfrac{2x}{\sqrt{x}}$

64. $\dfrac{1}{yx^{-5}}$

65. $(16x^8)^{-3/4}$

66. $(-8y^9)^{2/3}$

67. $\sqrt{x}\left(\dfrac{1}{4x}\right)^{5/2}$

68. $\dfrac{(25xy)^{3/2}}{x^2 y}$

69. $\dfrac{(-27x^5)^{2/3}}{\sqrt[3]{x}}$

70. $(-32y^{-5})^{3/5}$

The expressions in Exercises 71–74 may be factored as shown. Find the missing factors.

71. $\sqrt{x} - \dfrac{1}{\sqrt{x}} = \dfrac{1}{\sqrt{x}}(\ \ )$

72. $2x^{2/3} - x^{-1/3} = x^{-1/3}(\ \ )$

73. $x^{-1/4} + 6x^{1/4} = x^{-1/4}(\ \ )$

74. $\sqrt{\dfrac{x}{y}} - \sqrt{\dfrac{y}{x}} = \sqrt{xy}\,(\ \ )$

75. Explain why $\sqrt{a} \cdot \sqrt{b} = \sqrt{ab}$.

76. Explain why $\sqrt{a}/\sqrt{b} = \sqrt{a/b}$.

In Exercises 77–84, evaluate $f(4)$.

77. $f(x) = x^2$ **78.** $f(x) = x^3$ **79.** $f(x) = x^{-1}$ **80.** $f(x) = x^{1/2}$

81. $f(x) = x^{3/2}$ **82.** $f(x) = x^{-1/2}$ **83.** $f(x) = x^{-5/2}$ **84.** $f(x) = x^0$

Calculate the compound amount from the given data in Exercises 85–92.

85. principal $= \$500$, compounded annually, 6 years, annual rate $= 6\%$

86. principal $= \$700$, compounded annually, 8 years, annual rate $= 8\%$

87. principal $= \$50,000$, compounded quarterly, 10 years, annual rate $= 9.5\%$

88. principal $= \$20,000$, compounded quarterly, 3 years, annual rate $= 12\%$

89. principal $= \$100$, compounded monthly, 10 years, annual rate $= 5\%$

90. principal $= \$500$, compounded monthly, 1 year, annual rate $= 4.5\%$

91. principal $= \$1500$, compounded daily, 1 year, annual rate $= 6\%$

92. principal $= \$1500$, compounded daily, 3 years, annual rate $= 6\%$

93. Assume that a couple invests $\$1000$ upon the birth of their daughter. Assume that the investment earns 6.8% compounded annually. What will the investment be worth on the daughter's 18th birthday?

94. Assume that a couple invests $\$4000$ each year for four years in an investment that earns 8% compounded annually. What will the value of the investment be 8 years after the first amount is invested?

95. Assume that a $\$500$ investment earns interest compounded quarterly. Express the value of the investment after one year as a polynomial in the annual rate of interest r.

96. Assume that a $\$1000$ investment earns interest compounded semiannually. Express the value of the investment after two years as a polynomial in the annual rate of interest r.

97. When a car's brakes are slammed on at a speed of x miles per hour, the stopping distance is $\frac{1}{20}x^2$ feet. Show that when the speed is doubled, the stopping distance increases fourfold.

**SOLUTIONS TO
PRACTICE
PROBLEMS 0.5**

1. (a) $-5^2 = -25$. [Note that -5^2 is the same as $-(5^2)$. This number is different from $(-5)^2$, which equals 25. Whenever there are no parentheses, apply the exponent first and then apply the other operations.]

(b) Since $.75 = \frac{3}{4}$, $16^{.75} = 16^{3/4} = (\sqrt[4]{16})^3 = 2^3 = 8$.

2. (a) Apply Law 5 with $a = 4$ and $b = x^3$. Then use Law 4.

$$(4x^3)^2 = 4^2 \cdot (x^3)^2 = 16 \cdot x^6.$$

[A common error is to forget to square the 4. If that had been our intent, we would have asked for $4(x^3)^2$.]

(b) $\dfrac{\sqrt[3]{x}}{x^3} = \dfrac{x^{1/3}}{x^3} = x^{(1/3)-3} = x^{-8/3}$. [The answer can also be given as $1/x^{8/3}$.]

When simplifying expressions involving radicals, it is usually a good idea to first convert the radicals to exponents.

(c) $\dfrac{2(x + 5)^6}{x^2 + 10x + 25} = \dfrac{2 \cdot (x + 5)^6}{(x + 5)^2} = 2(x + 5)^{6-2} = 2(x + 5)^4.$ [Here the third law of exponents was applied to $(x + 5)$. The laws of exponents apply to any algebraic expression.]

0.6 FUNCTIONS AND GRAPHS IN APPLICATIONS

The key step in solving many applied problems in this text is to construct appropriate functions or equations. Once this is done, the remaining mathematical steps are usually straightforward. This section focuses on representative applied problems and reviews skills needed to set up and analyze functions, equations, and their graphs.

Geometric Problems   Many examples and exercises in the text involve dimensions, areas, or volumes of objects similar to those in Fig. 1. When a problem involves a plane figure, such as a rectangle or circle, one must distinguish between the *perimeter* and the *area* of the figure. The perimeter of a figure, or "distance around" the figure, is a *length* or a *sum of lengths*. Typical units, if specified, are inches, feet, centimeters, meters, and so on. Area involves the *product of two lengths,* and the units are *square* inches, *square* feet, *square* centimeters, and so on.

EXAMPLE 1   Suppose the longer side of the rectangle in Fig. 1 has twice the length of the shorter side, and let x denote the length of the shorter side.

(a) Express the perimeter of the rectangle as a function of x.

(b) Express the area of the rectangle as a function of x.

(c) Suppose the rectangle represents a kitchen countertop to be constructed of a durable material costing $25 per square foot. Write a function $C(x)$ that expresses the cost of the material as a function of x, where lengths are in feet.

FIGURE 1   Geometric figures.

Rectangle

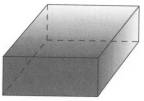

Rectangular box

Cylinder

Solution   (a) The rectangle is shown in Fig. 2. The length of the longer side is $2x$. If the perimeter is denoted by P, then P is the sum of the lengths of the four sides of the rectangle, namely, $x + 2x + x + 2x$. That is, $P = 6x$.

(b) The area A of the rectangle is the product of the lengths of two adjacent sides. That is, $A = x \cdot 2x = 2x^2$.

(c) Here the area is measured in square feet. The basic principle for this part is:

$$\begin{bmatrix} \text{cost of} \\ \text{materials} \end{bmatrix} = \begin{bmatrix} \text{cost per} \\ \text{square foot} \end{bmatrix} \cdot \begin{bmatrix} \text{number of} \\ \text{square feet} \end{bmatrix}$$

$$C(x) = 25 \cdot 2x^2$$

$$= 50x^2 \text{ (dollars)}. \ \bullet$$

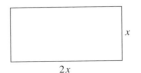

x

$2x$

FIGURE 2

When a problem involves a three-dimensional object, such as a box or cylinder, one must distinguish between the *surface area* of the object and the *volume* of the object. Surface area is an area, of course, so it is measured in *square* units. Typically, the surface area is a *sum of areas* (each area a product of two lengths). The volume of an object is a *product of three lengths* and is measured in *cubic* units.

EXAMPLE 2   A rectangular box has a square copper base, wooden sides, and a wooden top. The copper costs $21 per square foot and the wood costs $2 per square foot.

(a) Write an expression giving the surface area (that is, the sum of the areas of the four sides, the top, and the bottom of the box) in terms of the dimensions of the box. Also, write an expression giving the volume of the box.

(b) Write an expression giving the total cost of the materials used to make the box in terms of the dimensions.

Solution   (a) The first step is to assign letters to the dimensions of the box. Denote the length of one (and therefore every) side of the base by x, and denote the height of the box by h. See Fig. 3.

The top and bottom each have area x^2, and each of the four sides has area xh. Therefore, the surface area is $2x^2 + 4xh$. The volume of the box is the product of the length, width, and height. The volume is x^2h.

(b) When the various surfaces of the box have different costs per square foot, the cost of each surface is computed separately.

$$[\text{cost of bottom}] = [\text{cost per sq. ft.}] \cdot [\text{area of bottom}] = 21x^2$$

$$[\text{cost of top}] = [\text{cost per sq. ft.}] \cdot [\text{area of top}] = 2x^2$$

$$[\text{cost of one side}] = [\text{cost per sq. ft.}] \cdot [\text{area of one side}] = 2xh$$

The total cost is

$$C = [\text{cost of bottom}] + [\text{cost of top}] + 4 \cdot [\text{cost of one side}]$$

$$= 21x^2 + 2x^2 + 4 \cdot 2xh = 23x^2 + 8xh \ \bullet$$

FIGURE 3   **Closed box.**

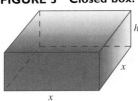

h

x

x

Business Problems   Many applications in the text involve cost, revenue, and profit functions.

EXAMPLE 3   Suppose a toy manufacturer has fixed costs of $3000 (such as rent, insurance, and business loans) that must be paid no matter how many toys are produced. In addition, there are variable costs of $2 per toy. At a production level of x toys, the variable costs are $2 \cdot x$ (dollars) and the total cost is

$$C(x) = 3000 + 2x \quad \text{(dollars)}.$$

(a) Find the cost of producing 2000 toys.

(b) What additional cost is incurred if the production level is raised from 2000 toys to 2200 toys?

(c) To answer the question "How many toys may be produced at a cost of $5000?" should you compute $C(5000)$ or should you solve the equation $C(x) = 5000$?

Solution   (a) $C(2000) = 3000 + 2(2000) = 7000$ (dollars).

(b) The total cost when $x = 2200$ is $C(2200) = 3000 + 2(2200) = 7400$ (dollars). So the *increase* in cost when production is raised from 2000 to 2200 toys is

$$C(2200) - C(2000) = 7400 - 7000 = 400 \quad \text{(dollars)}.$$

(c) This is an important type of question. The phrase "how many toys" implies that the quantity x is unknown. Therefore, the answer is found by solving $C(x) = 5000$ for x.

$$3000 + 2x = 5000$$
$$2x = 2000$$
$$x = 1000.$$

Another way to analyze this problem is to look at the types of units involved. The input x of the cost function is the *quantity* of toys, and the output of the cost function is the *cost*, measured in dollars. Since the question involves 5000 *dollars*, it is the *output* that is specified. The input x is unknown. ●

EXAMPLE 4   Suppose the toys in Example 3 sell for $10 apiece. When x toys are sold, the revenue (amount of money received) $R(x)$ is $10x$ dollars. Given the same cost function, $C(x) = 3000 + 2x$, the profit (or loss) $P(x)$ generated by the x toys will be

$$P(x) = R(x) - C(x)$$
$$= 10x - (3000 + 2x) = 8x - 3000.$$

(a) To determine the revenue generated by 8000 toys, should you compute $R(8000)$ or should you solve the equation $R(x) = 8000$?

(b) If the revenue from the production and sale of some toys is $7000, what is the corresponding profit?

Solution   (a) The revenue is unknown, but the input to the revenue function is known. So compute $R(8000)$ to find the revenue.

(b) The profit is unknown, so we want to compute the value of $P(x)$. Unfortunately, we don't know the value of x. However, the fact that the revenue is $7000 enables us to solve for x. Thus the solution has two steps:

(i) Solve $R(x) = 7000$ to find x.

$$10x = 7000$$

$$x = 700 \quad \text{(toys)}$$

(ii) Compute $P(x)$ when $x = 700$.

$$P(700) = 8(700) - 3000$$

$$= 2600 \quad \text{(dollars)} \bullet$$

Functions and Graphs   When a function arises in an applied problem, the graph of the function provides useful information. Every statement or task involving the function corresponds to a feature or task involving its graph. This "graphical" point of view will broaden your understanding of functions and strengthen your ability to work with them.

Modern graphing calculators and calculus computer software provide excellent tools for thinking geometrically about functions. Most popular graphing calculators and programs provide a *cursor* or *cross hairs* that may be moved to any point on the screen, with the x- and y-coordinates of the cursor displayed somewhere on the screen. The next example shows how geometric calculations with the graph of a function correspond to the more familiar numerical computations. This example is worth reading even if a computer (or calculator) is unavailable.

EXAMPLE 5   To plan for future growth, a company analyzes production costs for one of its products and estimates that the cost (in dollars) of operating at a production level of x units per hour is given by the function

$$C(x) = 150 + 59x - 1.8x^2 + .02x^3.$$

Suppose the graph of this function is available, either displayed on a computer screen or perhaps printed on graph paper in a company report. See Fig. 4.

(a) The point $(16, 715.12)$ is on the graph. What does that say about the cost function $C(x)$?

(b) The equation $C(x) = 900$ may be solved graphically by finding a certain point on the graph and reading its x- and y-coordinates. Describe how to locate the point. How do the coordinates of the point provide the solution to the equation $C(x) = 900$?

FIGURE 4   Graph of a cost function.

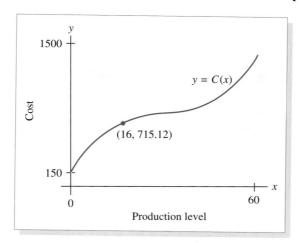

(c) The task "Find $C(45)$" may be completed graphically by finding a point on the graph. Describe how to locate the point. How do the coordinates of the point provide the value of $C(45)$?

Solution   (a) The fact that $(16, 715.12)$ is on the graph of $C(x)$ means that $C(16) = 715.12$. That is, if the production level is 16 units per hour, then the cost is \$715.12.

(b) To solve $C(x) = 900$ graphically, locate 900 on the y-axis and move to the right until you reach the point $(?, 900)$ on the graph of $C(x)$. See Fig. 5. The x-coordinate of the point is the solution of $C(x) = 900$. Estimate x graphically. (On a computer, read the coordinates of the cursor; on graph paper, use a ruler to find x on the x-axis. To two decimal places, $x = 39.04$.)

(c) To find $C(45)$ graphically, locate 45 on the x-axis and move up until you reach the point $(45, ?)$ on the graph of $C(x)$. See Fig. 6. The y-coordinate of the point is the value of $C(45)$. [In fact, $C(45) = 982.50$.] ●

FIGURE 5

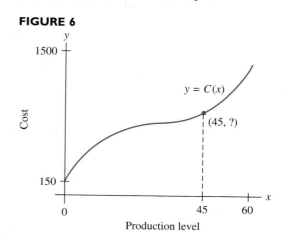

FIGURE 6

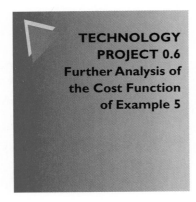

TECHNOLOGY PROJECT 0.6
Further Analysis of the Cost Function of Example 5

Graph the cost function $C(x)$ of Example 5.

1. Graphically determine the y-intercept of the graph. What is the applied significance of the y-intercept?
2. Determine the cost of producing 50 units per hour.
3. By how much does the cost increase if production is increased from 50 to 51 units per hour?
4. Suppose that cash flow restricts operations to spending no more than $15,000 per day. Assuming that production is spread out over 10 hours each day, what will be the number of units produced each hour?

The two final examples illustrate how to extract information about a function by examining its graph.

EXAMPLE 6   Figure 7 is the graph of the function $R(x)$, the revenue obtained from selling x bicycles.

(a) What is the revenue from the sale of 1000 bicycles?
(b) How many bicycles must be sold to achieve a revenue of $102,000?
(c) What is the revenue from the sale of 1100 bicycles?
(d) What additional revenue is derived from the sale of 100 more bicycles if the current sales level is 1000 bicycles?

Solution   (a) Since $(1000, 150,000)$ is on the graph of $R(x)$, the revenue from the sale of 1000 bicycles is $150,000.

(b) The horizontal line at $y = 102,000$ intersects the graph at the point with x-coordinate 600. Therefore, the revenue from the sale of 600 bicycles is $102,000.

FIGURE 7   Graph of a revenue function.

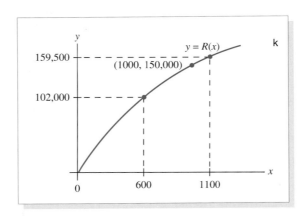

(c) The vertical line at $x = 1100$ intersects the graph at the point with y-coordinate 159,500. Therefore, $R(1100) = 159,500$ and the revenue is $159,500.

(d) When the value of x increases from 1000 to 1100, the revenue increases from 150,000 to 159,500. Therefore, the additional revenue is $9500. ●

EXAMPLE 7   A ball is thrown straight up into the air. The function $h(t)$, the height of the ball (in feet) after t seconds, has the graph shown in Fig. 8. [*Note:* This graph is not a picture of the physical path of the ball; the ball is thrown vertically into the air.]

FIGURE 8   Graph of a height function.

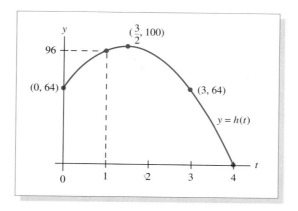

(a) What is the height of the ball after 1 second?
(b) After how many seconds does the ball reach its greatest height, and what is this height?
(c) After how many seconds does the ball hit the ground?
(d) When is the height 64 feet?

Solution   (a) Since the point (1, 96) is on the graph of $h(t)$, $h(1) = 96$. Therefore, the height of the ball after 1 second is 96 feet.

(b) The highest point on the graph of the function has coordinates $(\frac{3}{2}, 100)$. Therefore, after $\frac{3}{2}$ seconds the ball achieves its greatest height, 100 feet.

(c) The ball hits the ground when the height is 0. This occurs after 4 seconds.

(d) The height of 64 feet occurs twice, at times $t = 0$ and $t = 3$ seconds. ●

Table 1 summarizes most of the concepts in Examples 3–7. Although stated here for a profit function, the concepts will arise later for many other types of functions as well. Each statement about the profit is translated into a statement about $f(x)$ and a statement about the graph of $f(x)$. The graph in Fig. 9 illustrates each statement.

TABLE 1   Translating an Applied Problem

Assume that $f(x)$ is the profit in dollars at production level x.

Applied Problem	Function	Graph
When production is at 2 units, the profit is $7.	$f(2) = 7$.	The point $(2, 7)$ is on the graph.
Determine the number of units that generate a profit of $12.	Solve $f(x) = 12$ for x.	Find the x-coordinate(s) of the point(s) on the graph whose y-coordinate is 12.
Determine the profit when the production level is 4 units.	Evaluate $f(4)$.	Find the y-coordinate of the point on the graph whose x-coordinate is 4.
Find the production level that maximizes the profit.	Find x such that $f(x)$ is as large as possible.	Find the x-coordinate of the highest point, M, on the graph.
Determine the maximum profit.	Find the maximum value of $f(x)$.	Find the y-coordinate of the highest point on the graph.
Determine the change in profit when the production level is changed from 6 to 7 units.	Find $f(7) - f(6)$.	Determine the difference in heights of the points with x-coordinates 7 and 6.
The profit decreases when the production level is changed from 6 to 7 units.	The function value decreases when x changes from 6 to 7.	The point on the graph with x-coordinate 6 is higher than the point with x-coordinate 7.

FIGURE 9   Graph of a profit function.

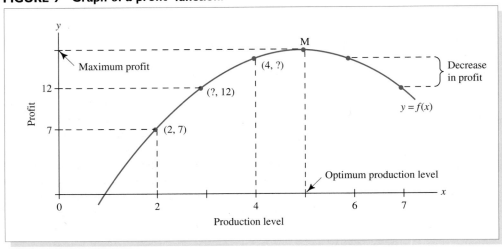

PRACTICE PROBLEMS 0.6

Consider the cylinder shown in Fig. 10.

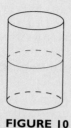

1. Assign letters to the dimensions of the cylinder.
2. The girth of the cylinder is the circumference of the colored circle in the figure. Express the girth in terms of the dimensions of the cylinder.
3. What is the area of the bottom (or top) of the cylinder?
4. What is the surface area of the side of the cylinder? [*Hint*: Imagine cutting the side of the cylinder and unrolling the cylinder to form a rectangle.]

FIGURE 10

EXERCISES 0.6

In Exercises 1–6, assign letters to the dimensions of the geometric object.

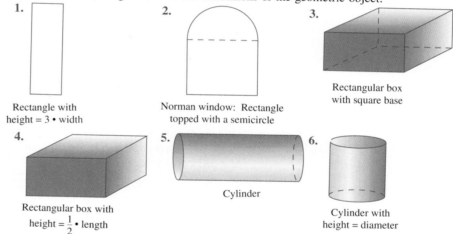

1. Rectangle with height = 3 • width

2. Norman window: Rectangle topped with a semicircle

3. Rectangular box with square base

4. Rectangular box with height = $\frac{1}{2}$ • length

5. Cylinder

6. Cylinder with height = diameter

Exercises 7–14 refer to the letters assigned to the figures in Exercises 1–6.

7. Consider the rectangle in Exercise 1. Write an expression for the perimeter. Suppose the area is 25 square feet, and write this fact as an equation.

8. Consider the rectangle in Exercise 1. Write an expression for the area. Write an equation expressing the fact that the perimeter is 30 centimeters.

9. Consider a circle of radius r. Write an expression for the area. Write an equation expressing the fact that the circumference is 15 centimeters.

10. Consider the Norman window of Exercise 2. Write an expression for the perimeter. Write an equation expressing the fact that the area is 2.5 square meters.

11. Consider the rectangular box in Exercise 3, and suppose that it has no top. Write an expression for the volume. Write an equation expressing the fact that the surface area is 65 square inches.

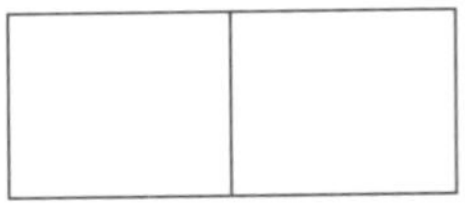

FIGURE 11

FIGURE 12

12. Consider the closed rectangular box in Exercise 4. Write an expression for the surface area. Write an equation expressing the fact that the volume is 10 cubic feet.

13. Consider the cylinder of Exercise 5. Write an equation expressing the fact that the volume is 100 cubic inches. Suppose the material to construct the left end costs $5 per square inch, the material to construct the right end costs $6 per square inch, and the material to construct the side costs $7 per square inch. Write an expression for the total cost of material for the cylinder.

14. Consider the cylinder of Exercise 6. Write an equation expressing the fact that the surface area is 30π square inches. Write an expression for the volume.

15. Consider a rectangular corral with a partition down the middle, as in Fig. 11. Assign letters to the outside dimensions of the corral. Write an equation expressing the fact that 5000 feet of fencing are needed to construct the corral (including the partition). Write an expression for the total area of the corral.

16. Consider a rectangular corral with two partitions, as in Fig. 12. Assign letters to the outside dimensions of the corral. Write an equation expressing the fact that the corral has a total area of 2500 square feet. Write an expression for the amount of fencing needed to construct the corral (including both partitions).

17. Consider the corral of Exercise 16. Suppose the fencing for the boundary of the corral costs $10 per foot and the fencing for the inner partitions costs $8 per foot. Write an expression for the total cost of the fencing.

18. Consider the rectangular box of Exercise 3. Assume the box has no top, the material needed to construct the base costs $5 per square foot, and the material needed to construct the sides costs $4 per square foot. Write an equation expressing the fact that the total cost of materials is $150. (Use the dimensions assigned in Exercise 3.)

19. Suppose the rectangle in Exercise 1 has a perimeter of 40 cm. Find the area of the rectangle.

20. Suppose the cylinder in Exercise 6 has a volume of 54π cubic inches. Find the surface area of the cylinder.

21. A specialty shop prints custom slogans and designs on T-shirts. The shop's total cost at a daily sales level of x T-shirts is $C(x) = 73 + 4x$ dollars.
    (a) At what sales level will the cost be $225?
    (b) If the sales level is at 40 T-shirts, how much will the cost rise if the sales level changes to 50 T-shirts?

22. A college student earns income by typing term papers on a computer, which she leases (along with a printer). The student charges $4 per page for her work, and she estimates that her monthly cost when typing x pages is $C(x) = .10x + 75$ dollars.
    (a) What is the student's profit if she types 100 pages in one month?
    (b) Determine the change in profit when the typing business rises from 100 to 101 pages per month.

23. A frozen yogurt stand makes a profit of $P(x) = .40x - 80$ dollars when selling x scoops of yogurt per day.
    (a) Find the break-even sales level, that is, the level at which $P(x) = 0$.
    (b) What sales level generates a daily profit of $30?
    (c) How many more scoops of yogurt will have to be sold to raise the daily profit from $30 to $40?

24. A cellular telephone company estimates that if it has x thousand subscribers, then its monthly profit is $P(x)$ thousand dollars, where $P(x) = 12x - 200$.

(a) How many subscribers are needed for a monthly profit of 160 thousand dollars?

(b) How many new subscribers would be needed to raise the monthly profit from 160 to 166 thousand dollars?

25. An average sale at a small florist shop is $21, so the shop's weekly revenue function is $R(x) = 21x$, where x is the number of sales in 1 week. The corresponding weekly cost is $C(x) = 9x + 800$ dollars.

(a) What is the florist shop's weekly profit function?

(b) How much profit is made when sales are at 120 per week?

(c) If the profit is $1000 for a week, what is the revenue for the week?

26. A catering company estimates that if it has x customers in a typical week, then its expenses will be approximately $C(x) = 550x + 6500$ dollars, and its revenue will be approximately $R(x) = 1200x$ dollars.

(a) How much profit will the company earn in a week when it has 12 customers?

(b) How much profit is the company making each week if the weekly costs are running at a level of $14,750?

Exercises 27–32 refer to the function $f(r)$, which gives the cost (in cents) of constructing a 100-cubic-inch cylinder of radius r inches. The graph of $f(r)$ is shown in Fig. 13.

FIGURE 13   Cost of a cylinder.

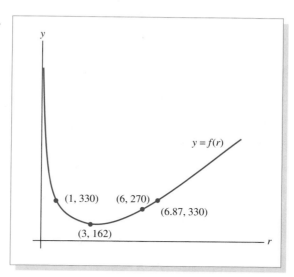

27. What is the cost of constructing a cylinder of radius 6?

28. For what value(s) of r will the cost be 330 cents?

29. Interpret the fact that the point (3, 162) is on the graph of the function.

30. Interpret the fact that the point (3, 162) is the lowest point on the graph of the function. What does this say in terms of cost versus radius?

31. What is the additional cost of increasing the radius from 3 inches to 6 inches?

32. How much is saved by increasing the radius from 1 inch to 3 inches?

FIGURE 14   Cost and revenue functions.

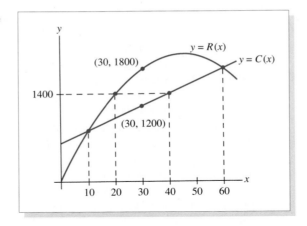

Exercises 33–36 refer to the cost and revenue functions in Fig. 14. The cost of producing x units of goods is $C(x)$ dollars and the revenue from selling x units of goods is $R(x)$ dollars.

33. What are the revenue and cost from the production and sale of 30 units of goods?

34. At what level of production is the revenue $1400?

35. At what level of production is the cost $1400?

36. What is the profit from the manufacture and sale of 30 units of goods?

Exercises 37–40 refer to the cost function in Fig. 15.

37. The point (1000, 4000) is on the graph of the function. Restate this fact in terms of the function $C(x)$.

38. Translate the task "solve $C(x) = 3500$ for x" into a task involving the graph of the function.

39. Translate the task "find $C(400)$" into a task involving the graph.

40. Suppose 500 units of goods are produced. What is the cost of producing 100 more units of goods?

FIGURE 15   A cost function.

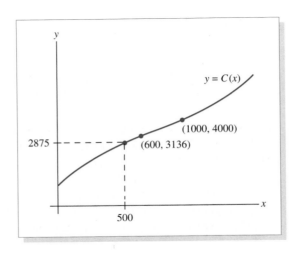

FIGURE 16 A profit function.

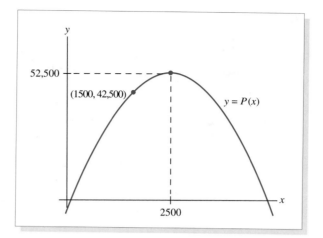

Exercises 41–44 refer to the profit function in Fig. 16.

41. The point (2500, 52,500) is the highest point on the graph of the function. What does this say in terms of profit versus quantity?

42. The point (1500, 42,500) is on the graph of the function. Restate this fact in terms of the function $P(x)$.

43. Translate the task "solve $P(x) = 30,000$" into a task involving the graph of the function.

44. Translate the task "find $P(2000)$" into a task involving the graph.

A ball is thrown straight up into the air. The function $h(t)$ gives the height of the ball (in feet) after t seconds. In Exercises 45–50, translate the task into both a statement involving the function and a statement involving the graph of the function.

45. Find the height of the ball after 3 seconds.

46. Find the time at which the ball attains its greatest height.

47. Find the greatest height attained by the ball.

48. Determine when the ball will hit the ground.

49. Determine when the height is 100 feet.

50. Find the height of the ball when it is first released.

SOLUTIONS TO PRACTICE PROBLEMS 0.6

1. Let r be the radius of circular base and let h be the height of the cylinder.
2. Girth $= 2\pi r$ (the circumference of the circle).
3. Area of the bottom $= \pi r^2$.
4. The cylinder is a rolled-up rectangle of height h and base $2\pi r$ (the circumference of the circle). The area is $2\pi rh$. See Fig. 17.

FIGURE 17 Unrolled side of a cylinder.

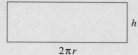

CHAPTER 0
CHECKLIST

✓ Real number
✓ Inequality
✓ Double inequality
✓ Number line
✓ Open interval
✓ Closed interval
✓ Infinite interval
✓ Function
✓ Value of a function at x
✓ Domain of a function
✓ Graph of a function
✓ Vertical line test
✓ Graph of an equation

✓ Linear function
✓ Constant function
✓ x- and y-intercepts
✓ Quadratic function
✓ Polynomial and rational functions
✓ Power function
✓ Absolute value function
✓ Addition, subtraction, multiplication, and division of functions
✓ Composition of functions

✓ Zero of a function
✓ Quadratic formula
✓ Factoring polynomials
✓ Laws of exponents
✓ Compound interest
✓ Dimensions, areas, and volumes of geometric objects
✓ Cost, revenue, and profit functions
✓ Translating an applied problem into mathematical terms

CHAPTER 0   SUPPLEMENTARY EXERCISES

1. Let $f(x) = x^3 + \dfrac{1}{x}$. Evaluate $f(1), f(3), f(-1), f(-\frac{1}{2})$, and $f(\sqrt{2})$.

2. Let $f(x) = 2x + 3x^2$. Evaluate $f(0), f(-\frac{1}{4})$, and $f(1/\sqrt{2})$.

3. Let $f(x) = x^2 - 2$. Evaluate $f(a - 2)$.

4. Let $f(x) = [1/(x + 1)] - x^2$. Evaluate $f(a + 1)$.

Determine the domains of the following functions.

5. $f(x) = \dfrac{1}{x(x + 3)}$

6. $f(x) = \sqrt{x - 1}$

7. $f(x) = \sqrt{x^2 + 1}$

8. $f(x) = \dfrac{1}{\sqrt{3x}}$

9. Is the point $(\frac{1}{2}, -\frac{3}{5})$ on the graph of the function $h(x) = (x^2 - 1)/(x^2 + 1)$?

10. Is the point $(1, -2)$ on the graph of the function $k(x) = x^2 + (2/x)$?

Factor the polynomials in Exercises 11–14.

11. $5x^3 + 15x^2 - 20x$

12. $3x^2 - 3x - 60$

13. $18 + 3x - x^2$

14. $x^5 - x^4 - 2x^3$

15. Find the zeros of the quadratic function $y = 5x^2 - 3x - 2$.

16. Find the zeros of the quadratic function $y = -2x^2 - x + 2$.

17. Find the points of intersection of the curves $y = 5x^2 - 3x - 2$ and $y = 2x - 1$.

18. Find the points of intersection of the curves $y = -x^2 + x + 1$ and $y = x - 5$.

Let $f(x) = x^2 - 2x$, $g(x) = 3x - 1$, and $h(x) = \sqrt{x}$. Find the following functions.

19. $f(x) + g(x)$

20. $f(x) - g(x)$

21. $f(x)h(x)$

22. $f(x)g(x)$

23. $f(x)/h(x)$

24. $g(x)h(x)$

Let $f(x) = x/(x^2 - 1)$, $g(x) = (1 - x)/(1 + x)$, and $h(x) = 2/(3x + 1)$. Express the following as rational functions.

25. $f(x) - g(x)$ **26.** $f(x) - g(x + 1)$ **27.** $g(x) - h(x)$

28. $f(x) + h(x)$ **29.** $g(x) - h(x - 3)$ **30.** $f(x) + g(x)$

Let $f(x) = x^2 - 2x + 4$, $g(x) = 1/x^2$, and $h(x) = 1/(\sqrt{x} - 1)$. Determine the following functions.

31. $f(g(x))$ **32.** $g(f(x))$ **33.** $g(h(x))$

34. $h(g(x))$ **35.** $f(h(x))$ **36.** $h(f(x))$

37. Simplify $(81)^{3/4}$, $8^{5/3}$, and $(.25)^{-1}$.   **38.** Simplify $(100)^{3/2}$ and $(.001)^{1/3}$.

39. The population of a city is estimated to be $750 + 25t + .1t^2$ thousand people t years from the present. Ecologists estimate that the average level of carbon monoxide in the air above the city will be $1 + .4x$ ppm (parts per million) when the population is x thousand people. Express the carbon monoxide level as a function of the time t.

40. The revenue $R(x)$ (in thousands of dollars) a company receives from the sale of x thousand units is given by $R(x) = 5x - x^2$. The sales level x is in turn a function $f(d)$ of the number d of dollars spent on advertising, where

$$f(d) = 6\left(1 - \frac{200}{d + 200}\right).$$

Express the revenue as a function of the amount spent on advertising.

In Exercises 41–44, use the laws of exponents to simplify the algebraic expressions.

41. $(\sqrt{x + 1})^4$ **42.** $\dfrac{xy^3}{x^{-5}y^6}$ **43.** $\dfrac{x^{3/2}}{\sqrt{x}}$ **44.** $\sqrt[3]{x}\,(8x^{2/3})$

CHAPTER **1**

THE DERIVATIVE

T he *derivative* is a mathematical tool used to measure rate of change. To illustrate the sort of change we have in mind, consider the following example. Suppose that a rumor spreads through a town of 10,000 people.* Denote by $N(t)$ the number of people who have heard the rumor after t days. We have tabulated the values for $N(t)$ corresponding to $t = 0, 1, \ldots, 10$ in the chart in Fig. 1. For example, the rumor starts at $t = 0$ with one person $[N(0) = 1]$; after 1 day, the number having heard the rumor increases to 6; after 2 days, to 40; and so forth. Each day the number of people who have heard the rumor increases. We may describe the increase geometrically by plotting the points corresponding to $t = 0, 1, \ldots, 10$ and drawing a smooth curve through them. (See Fig. 1.)

Let us describe the change that is taking place in this example. Initially, the number of *new* people to hear the rumor each day is rather small (5 the

---

* The graph used in this example is derived from a mathematical model used by sociologists. See J. Coleman, *An Introduction to Mathematical Sociology* (London: Collier-Macmillan Ltd., 1964), p. 43.

FIGURE I   Spread of a rumor.

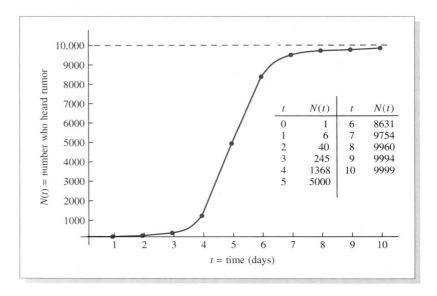

t	$N(t)$	t	$N(t)$
0	1	6	8631
1	6	7	9754
2	40	8	9960
3	245	9	9994
4	1368	10	9999
5	5000		

first day, 34 the second, 205 the third). As we shall see later, this situation is reflected in the fact that the graph is not very steep at $t = 1, 2, 3$. (It is almost flat.) On the fourth day, 1123 additional people hear the rumor; on the fifth day, 3632. The more rapid spread of the rumor on the fourth and fifth days is reflected in the increasing steepness of the graph at $t = 4$ and $t = 5$. After the fifth day the rumor spreads less rapidly, since there are fewer people to whom it can spread. Correspondingly, after $t = 5$, the graph becomes less steep. Finally, at $t = 9$ and $t = 10$, the graph is practically flat, indicating little change in the number who heard the rumor.

In the preceding example, there is a correlation between the rate at which $N(t)$ is changing and the steepness of the graph. This illustrates one of the fundamental ideas of calculus, which may be put roughly as follows: Measure rates of change in terms of steepness of graphs. This chapter is devoted to the *derivative,* which provides a numerical measure of the steepness of a curve at a particular point. By studying the derivative we will be able to deal numerically with the rates of changes which occur in applied problems.

1.1 THE SLOPE OF A STRAIGHT LINE

As we shall see later, the study of straight lines is crucial for the study of the steepness of curves. So this section is devoted to a discussion of the geometric and algebraic properties of straight lines.

A nonvertical line L has an equation of the form $y = mx + b$. The number m is called the *slope* of L, and the point $(0, b)$ is called the y-*intercept*.

If we set $x = 0$, we see that $y = b$, so that $(0, b)$ is on the line L. Thus the y-intercept tells us where the line L crosses the y-axis. The slope measures the steepness of the line. In Fig. 1 we give three examples of lines with slope $m = 2$. In Fig. 2 we give three examples of lines with slope $m = -2$.

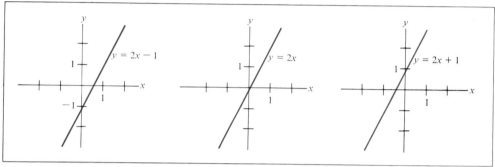

FIGURE 1   Three lines of slope 2.

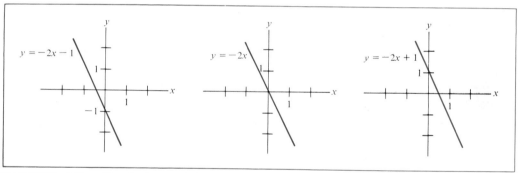

FIGURE 2   Three lines of slope −2.

To conceptualize the meaning of slope, think of walking along a line from left to right. On lines of positive slope we will be walking uphill; the greater the slope, the steeper the ascent. On lines of negative slope we will be walking downhill; the more negative the slope, the steeper the descent. Walking on lines of zero slope corresponds to walking on level ground. In Fig. 3 we have graphed lines with $m = 3, 1, \frac{1}{3}, 0, -\frac{1}{3}, -1, -3$, all having $b = 0$. The reader can readily verify our conceptualization of slope for these lines.

The slope and y-intercept of a straight line often have physical interpretations, as the following three examples illustrate.

FIGURE 3

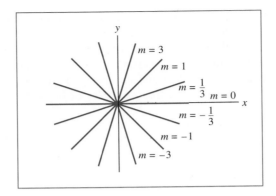

EXAMPLE 1   A manufacturer finds that the total cost of producing x units of a commodity is $2x + 1000$ dollars. What is the economic significance of the y-intercept and the slope of the line $y = 2x + 1000$? (See Fig. 4.)

FIGURE 4   A cost function.

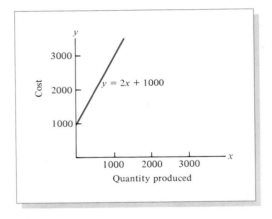

Solution   The y-intercept is $(0, 1000)$. The number 1000 represents the *fixed costs* of the manufacturer—those overhead costs, such as rent and insurance, that must be paid no matter how many items are produced. Thus when $x = 0$ (no units produced), the cost is still $y = 1000$ dollars.

The slope of the line is 2. This number represents the cost of producing each additional unit. To see this, we can calculate some typical costs.

Quantity Produced	Total Cost
$x = 1500$	$y = 2(1500) + 1000 = 4000$
$x = 1501$	$y = 2(1501) + 1000 = 4002$
$x = 1502$	$y = 2(1502) + 1000 = 4004$

Each time x is increased by 1, the value of y increases by 2. The number 2 is called the marginal cost. ●

EXAMPLE 2   An apartment complex has a storage tank to hold its heating oil. The tank was filled on January 1, but no more deliveries of oil will be made until some time in March. Let t denote the number of days after January 1 and let y denote the number of gallons of fuel oil in the tank. Current records of the apartment complex show that y and t are related approximately by the equation

$$y = 30,000 - 400t. \qquad (1)$$

What interpretation can be given to the y-intercept and slope of this line? (See Fig. 5.)

FIGURE 5   Amount of heating oil in a tank.

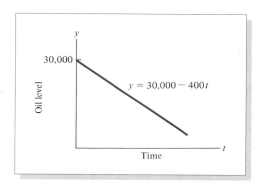

Solution   The y-intercept is (0, 30,000). This value of y corresponds to $t = 0$, so there were 30,000 gallons of oil in the tank on January 1. Let us examine how fast the oil is removed from the tank.

Days After January 1	Gallons of Oil in the Tank
$t = 0$	$y = 30,000 - 400(0) = 30,000$
$t = 1$	$y = 30,000 - 400(1) = 29,600$
$t = 2$	$y = 30,000 - 400(2) = 29,200$
$t = 3$	$y = 30,000 - 400(3) = 28,800$
⋮	⋮

The oil level in the tank drops by 400 gallons each day; that is, the oil is being used at the rate of 400 gallons per day. The slope of the line (1) is -400. Thus the slope gives the rate at which the level of oil in the tank is changing. The negative sign on the -400 indicates that the oil level is decreasing rather than increasing. ●

EXAMPLE 3   For tax purposes, businesses are allowed to regard equipment as decreasing in value (or depreciating) each year. The amount of depreciation may be taken as an income tax deduction.

Suppose that the value y of a piece of equipment x years after its purchase is given by

$$y = 500,000 - 50,000x.$$

Interpret the y-intercept and the slope of the graph.

Solution   The y-intercept is $(0, 500{,}000)$ and corresponds to the value of y when $x = 0$. That is, the y-intercept gives the original value, \$500,000, of the equipment. The slope indicates the rate at which the equipment is changing in value. Thus the value of the equipment is decreasing at the rate of 50,000 dollars per year. ●

Properties of the Slope of a Line   Let us now examine several useful properties of the slope of a straight line. At the end of the section we shall explain why these properties are valid.

Slope Property 1   Suppose that we start at a point on a line of slope m and move one unit to the right. Then we must move m units in the y-direction in order to return to the line.

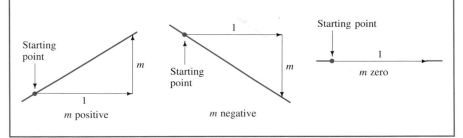

Slope Property 2   We can compute the slope of a line by knowing two points on the line. If (x_1, y_1) and (x_2, y_2) are on the line, then the slope of the line is $\dfrac{y_2 - y_1}{x_2 - x_1}$.

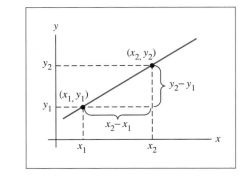

As we move from (x_1, y_1) to (x_2, y_2), the change in the y-coordinates is $y_2 - y_1$ and the change in the x-coordinates is $x_2 - x_1$. Thus the slope of the line is simply the ratio of the change in y to the change in x. This interpretation of the slope is also illustrated by the diagrams given for Slope Property 1. The slope of a line equals the change in y per unit change in x. We say that the slope gives the *rate of change of y with respect to x.*

Slope Property 3   The equation of a line can be obtained if we know the slope and one point on the line. If the slope is m and if (x_1, y_1) is on the line, then the equation of the line is

$$y - y_1 = m(x - x_1).$$

This equation is called the *point-slope form* of the equation of the line.

Slope Property 4   Distinct lines of the same slope are parallel. Conversely, if two lines are parallel, they have the same slope.

Slope Property 5   When two lines are perpendicular, the product of their slopes is -1.

Calculations Involving Slope of a Line

EXAMPLE 4   Find the slope and the y-intercept of the line whose equation is $2x + 3y = 6$.

Solution   We solve for y in terms of x.

$$3y = -2x + 6$$

$$y = -\frac{2}{3}x + 2.$$

The slope is $-\frac{2}{3}$ and the y-intercept is $(0, 2)$. ●

EXAMPLE 5   Sketch the graph of the line

(a) passing through $(2, -1)$ with slope 3,
(b) passing through $(2, 3)$ with slope $-\frac{1}{2}$.

Solution   We use Slope Property 1. (See Fig. 6.) In each case, we begin at the given point, move one unit to the right, and then move m units in the y-direction (upward for positive m, downward for negative m). The new point reached will also be on the line. Draw the straight line through these two points. ●

FIGURE 6

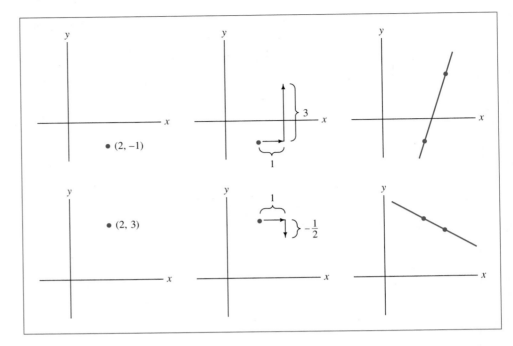

EXAMPLE 6   Find the slope of the line passing through the points $(6, -2)$ and $(9, 4)$.

Solution   We apply Slope Property 2 with $(x_1, y_1) = (6, -2)$ and $(x_2, y_2) = (9, 4)$. Then

$$\frac{y_2 - y_1}{x_2 - x_1} = \frac{4 - (-2)}{9 - 6} = \frac{6}{3} = 2.$$

Thus the slope is 2. [We would have reached the same answer if we had let $(x_1, y_1) = (9, 4)$ and $(x_2, y_2) = (6, -2)$.] The slope is just the difference of the y-coordinates divided by the difference of the x-coordinates, with each difference formed in the same order. ●

EXAMPLE 7   Find an equation of the line passing through $(-1, 2)$ with slope 3.

Solution   We let $(x_1, y_1) = (-1, 2)$ and $m = 3$, and we use Slope Property 3. The equation of the line is

$$y - 2 = 3[x - (-1)]$$

or

$$y - 2 = 3(x + 1).$$

If desired, this equation can be put into the form $y = mx + b$:

$$y - 2 = 3(x + 1) = 3x + 3$$

$$y = 3x + 5. ●$$

EXAMPLE 8   Find an equation of the line passing through the points $(1, -2)$ and $(2, -3)$.

Solution   By Slope Property 2, the slope of the line is

$$\frac{-3 - (-2)}{2 - 1} = \frac{-3 + 2}{1} = -1.$$

Since $(1, -2)$ is on the line, we can use Property 3 to get the equation of the line:

$$y - (-2) = (-1)(x - 1)$$
$$y + 2 = -x + 1$$
$$y = -x - 1. \quad \bullet$$

EXAMPLE 9   Find an equation of the line passing through $(5, 3)$ parallel to the line $2x + 5y = 7$.

Solution   We first find the slope of the line $2x + 5y = 7$.

$$2x + 5y = 7$$
$$5y = 7 - 2x$$
$$y = -\frac{2}{5}x + \frac{7}{5}.$$

The slope of this line is $-\frac{2}{5}$. By Slope Property 4, any line parallel to this line will also have slope $-\frac{2}{5}$. Using the given point $(5, 3)$ and Slope Property 3, we get the desired equation:

$$y - 3 = -\frac{2}{5}(x - 5).$$

This equation can also be written as

$$y = -\frac{2}{5}x + 5. \quad \bullet$$

Verification of the Properties of Slope   It is convenient to verify the properties of slope in the order 2, 3, 1. Verifications of Properties 4 and 5 are outlined in Exercises 54 and 55.

Verification of Property 2   Suppose that the equation of the line is $y = mx + b$. Then, since (x_2, y_2) is on the line, we have $y_2 = mx_2 + b$. Similarly, since (x_1, y_1) is on the line, we have $y_1 = mx_1 + b$. Subtracting, we see that

$$y_2 - y_1 = mx_2 - mx_1$$
$$y_2 - y_1 = m(x_2 - x_1),$$

so that

$$\frac{y_2 - y_1}{x_2 - x_1} = m.$$

This is the formula for the slope m stated in Property 2.

Verification of Property 3 The equation $y - y_1 = m(x - x_1)$ may be put into the form

$$y = mx + \underbrace{(y_1 - mx_1)}_{b}. \tag{2}$$

This is the equation of a line with slope m. Furthermore, the point (x_1, y_1) is on this line because equation (2) remains true when we plug in x_1 for x and y_1 for y. Thus the equation $y - y_1 = m(x - x_1)$ corresponds to the line of slope m passing through (x_1, y_1).

Verification of Property 1 Let $P = (x_1, y_1)$ be on a line l and let y_2 be the y-coordinate of the point on l obtained by moving P one unit to the right and then moving vertically to return to the line. (See Fig. 7.) By Property 2, the slope m of this line l satisfies

$$m = \frac{y_2 - y_1}{(x_1 + 1) - x_1} = \frac{y_2 - y_1}{1} = y_2 - y_1.$$

Thus the difference of the y-coordinates of R and Q is m. This is Property 1.

FIGURE 7

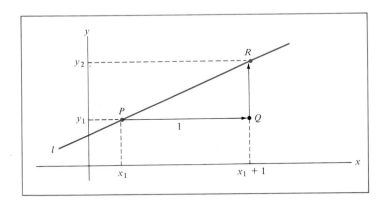

TECHNOLOGY PROJECT 1.1
Application of Slope Using a Graphing Calculator

Let y denote the average amount claimed for itemized deductions on a tax return reporting x dollars of income. According to Internal Revenue Service data, y is a linear function of x. Moreover, in a recent year, income tax returns reporting $20,000 of income averaged $729 in itemized deductions, while returns reporting $50,000 averaged $1380.

1. Determine y as a function of x.
2. Graph this function.
3. Graphically determine the slope of the line.

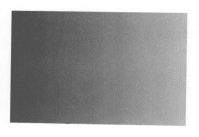

4. Give an interpretation of the slope in applied terms.
5. Determine graphically the average amount of itemized deductions on a return reporting $75,000.
6. Determine graphically the income level at which the average itemized deductions are $5000.
7. Suppose that the income level increases by $15,000. By how much do the average itemized deductions increase?

PRACTICE PROBLEMS 1.1

Find the slopes of the following lines.

1. The line whose equation is $x = 3y - 7$.
2. The line going through the points $(2, 5)$ and $(2, 8)$.

EXERCISES 1.1*

Find the slopes of the following lines.

1. $y = 2 - 5x$ 2. $y = -5x$ 3. $y = 2$

4. $y = \frac{1}{3}(x + 2)$ 5. $y = \dfrac{2x - 1}{7}$ 6. $y = \frac{1}{4}$

7. $2x + 3y = 6$ 8. $x - y = 2$

Find the equations of the following lines.

9. Slope is 3; y-intercept is $(0, -1)$.
10. Slope is $\frac{1}{2}$; y-intercept is $(0, 0)$.
11. Slope is 1; $(1, 2)$ on line.
12. Slope is $-\frac{1}{3}$; $(6, -2)$ on line.
13. Slope is -7; $(5, 0)$ on line.
14. Slope is $\frac{1}{2}$; $(2, -3)$ on line.
15. Slope is 0; $(7, 4)$ on line.
16. Slope is $-\frac{2}{3}$; $(0, 5)$ on line.
17. $(2, 1)$ and $(4, 2)$ on line.
18. $(5, -3)$ and $(-1, 3)$ on line.
19. $(0, 0)$ and $(1, -2)$ on line.
20. $(2, -1)$ and $(3, -1)$ on line.
21. Parallel to $y = -2x + 1$; $(\frac{1}{2}, 5)$ on line.
22. Parallel to $3x + y = 7$; $(-1, -1)$ on line.
23. Parallel to $3x - 6y = 1$; $(1, 0)$ on line.
24. Parallel to $5x + 2y = -4$; $(0, 17)$ on line.

---

*A complete solution of every third odd-numbered exercise is given in the Study Guide for this text.

25. Each of the lines (A), (B), (C), and (D) in Fig. 8 is the graph of one of the equations (a), (b), (c), and (d). Match each equation with its graph.

(a) $x + y = 1$ (b) $x - y = 1$ (c) $x + y = -1$ (d) $x - y = -1$

FIGURE 8

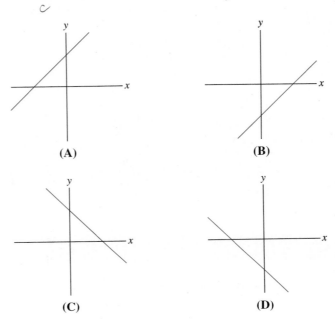

(A)   (B)

(C)   (D)

26. Table 1 gives some points on the line $y = mx + b$. Find m and b.

TABLE I	Points on a Line

x	4.8	4.9	5	5.1	5.2
y	3.6	4.8	6	7.2	8.4

In Exercises 27–30, refer to a line of slope m. Suppose you begin at a point on the line and move h units in the x-direction. How many units must you move in the y-direction to return to the line?

27. $m = \frac{1}{2}, h = 4$ **28.** $m = 2, h = \frac{1}{4}$

29. $m = -3, h = .25$ **30.** $m = .2, h = 5$

In each of Exercises 31–34, we specify a line by giving the slope and one point on the line. We give the first coordinate of some points on the line. Without deriving the equation of the line, find the second coordinate of each of the points.

31. Slope is 2, (1, 3) on line; (2,); (3,); (0,).
32. Slope is -3, (2, 2) on line; (3,); (4,); (1,).
33. Slope is $-\frac{1}{4}$, $(-1, -1)$ on line; (0,); (1,); $(-2,)$.
34. Slope is $\frac{1}{3}$, $(-5, 2)$ on line; $(-4,)$; $(-3,)$; $(-2,)$.

For each pair of lines in the following figures, determine the one with the greater slope.

35.

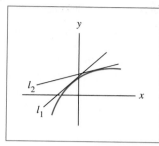

36.

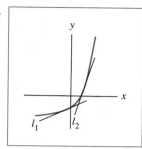

Find the equation and sketch the graph of the following lines.

37. With slope -2 and y-intercept $(0, -1)$.

38. With slope $\frac{1}{3}$ and y-intercept $(0, 1)$.

39. Through $(2, 0)$ with slope $\frac{4}{5}$.

40. Through $(-1, 3)$ with slope 0.

In Exercises 41–46, find the equation of a line with the given property. Each exercise has more than one correct answer.

41. y-intercept is $(0, 5)$.

42. x-intercept is $(9, 0)$.

43. Parallel to the line $4x + 5y = 6$.

44. Horizontal.

45. Slope is -2.

46. Vertical.

Later in the chapter we shall show that the tangent line to the parabola $y = x^2$ passing through the point with coordinates (x, y) has slope $2x$. (See Fig. 9.) Thus the slopes of the tangent lines at the points $(1, 1)$, $(0, 0)$, and $(-\frac{1}{2}, \frac{1}{4})$ are, respectively, 2, 0, and -1. Find the equation of the tangent lines through each of the following points.

FIGURE 9

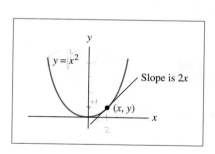

47. $(1, 1)$ **48.** $(0, 0)$ **49.** $(-\frac{1}{2}, \frac{1}{4})$

50. A salesperson's weekly pay depends on the volume of sales. If she sells x units of goods, then her pay is $y = 5x + 60$ dollars. Give an interpretation of the slope and the y-intercept of this straight line.

51. The demand equation for a monopolist is $y = -.02x + 7$, where x is the number of units produced and y is the price. That is, in order to sell x units of goods, the price must be $y = -.02x + 7$ dollars. Interpret the slope and y-intercept of this line.

52. Temperatures of 32°F and 212°F correspond to temperatures of 0°C and 100°C. Suppose the linear equation $y = mx + b$ converts Fahrenheit temperatures to Celsius temperatures. Find m and b. What is the Celsius equivalent of 98.6°F?

53. **(a)** Draw the graph of any function $f(x)$ that passes through the point $(3, 2)$.
    (b) Choose a point to the right of $x = 3$ on the x-axis and label it $3 + h$.
    (c) Draw the straight line through the points $(3, f(3))$ and $(3 + h, f(3 + h))$.
    (d) What is the slope of this straight line (in terms of h)?

54. Prove Property 4 of straight lines. [*Hint:* If $y = mx + b$ and $y = m'x + b'$ are two lines, then they have a point in common if and only if the equation $mx + b = m'x + b'$ has a solution x.]

55. Prove Property 5 of straight lines. [*Hint:* Without loss of generality, assume that both lines pass through the origin. Use Slope Property 1 and the Pythagorean theorem. See Fig. 10.]

FIGURE 10

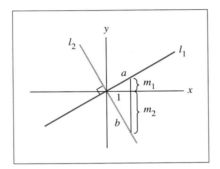

SOLUTIONS TO
PRACTICE
PROBLEMS 1.1

1. We solve for y in terms of x.

$$y = \frac{1}{3}x + \frac{7}{3}.$$

The slope of the line is the coefficient of x, that is, $\frac{1}{3}$.

2. The line passing through these two points is a vertical line; therefore, its slope is undefined.

 1.2 THE SLOPE OF A CURVE AT A POINT

In order to extend the concept of slope from straight lines to more general curves, we must first discuss the notion of the tangent line to a curve at a point.

We have a clear idea of what is meant by the tangent line to a circle at a point P. It is the straight line that touches the circle at just the one point P. Let us focus on the region near P, designated by the dashed rectangle shown in Fig. 1. The enlarged portion of the circle looks almost straight, and the straight line that it resembles is the tangent line. Further enlargements would make the

FIGURE 1   Enlarged portion of a circle.

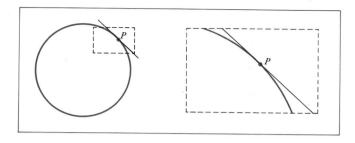

circle near P look even straighter and have an even closer resemblance to the tangent line. In this sense, the tangent line to the circle at the point P is the straight line through P that best approximates the circle near P. In particular, the tangent line at P reflects the steepness of the circle at P. Thus it seems reasonable to define the *slope* of the circle at P to be the slope of the tangent line at P.

Similar reasoning leads us to a suitable definition of slope for an arbitrary curve at a point P. Consider the three curves drawn in Fig. 2. We have drawn an enlarged version of the dashed box around each point P. Notice that the portion of each curve lying in the boxed region looks almost straight. If we further magnify the curve near P, it would appear even straighter. Indeed, if we apply higher and higher magnification, the portion of the curve near P would approach a certain straight line more and more exactly. (See Fig. 3.) This straight line is called the *tangent line to the curve at P*. This line best approximates the curve near P. We define the *slope of a curve at a point P* to be the slope of the tangent line to the curve at P.

FIGURE 2   Enlarged portions of curves.

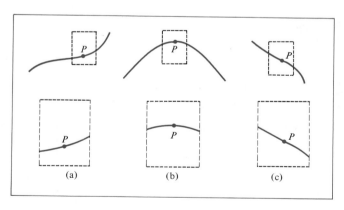

FIGURE 3   Tangent lines to curves.

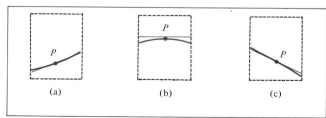

The portion of the curve near P can be, at least within an approximation, replaced by the tangent line at P. Therefore, the slope of the curve at P—that is, the slope of the tangent line at P—measures the rate of increase or decrease of the curve as it passes through P.

EXAMPLE 1   In Example 2 of Section 1.1 the apartment complex used approximately 400 gallons of oil per day. Suppose that we keep a continuous record of the oil level in the storage tank. The graph for a typical 2-day period appears in Fig. 4. What is the physical significance of the slope of the graph at the point P?

FIGURE 4   Oil level in a storage tank.

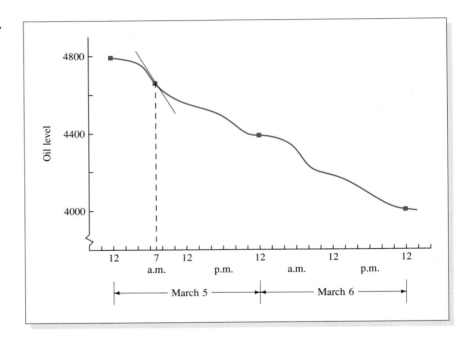

Solution   The curve near P is closely approximated by its tangent line. So think of the curve as replaced by its tangent line near P. Then the slope at P is just the rate of decrease of the oil level at 7 A.M. on March 5. ●

Notice that during the entire day of March 5, the graph in Fig. 4 seems to be the steepest at 7 A.M. That is, the oil level is falling the fastest at that time. This corresponds to the fact that most people awake around 7 A.M., turn up their thermostats, take showers, and so on. Example 1 provides a typical illustration of the manner in which slopes can be interpreted as rates of change. We shall return to this idea in Section 1.8.

In calculus, we can usually compute slopes by using formulas. For instance, in Section 1.3 we shall show that the tangent line to the graph of $y = x^2$ at the point $(1, 1)$ has slope 2; the tangent line at $(3, 9)$ has slope 6; and the tangent line at $\left(-\frac{5}{2}, \frac{25}{4}\right)$ has slope -5. The various tangent lines are shown in Fig.

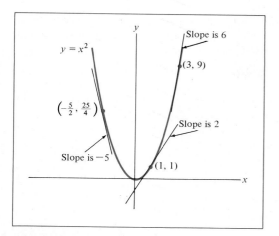

FIGURE 5   **Graph of y = x².**

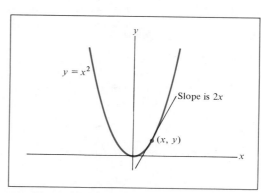

FIGURE 6   **Slope of tangent line to y = x².**

5. Notice that the slope at each point is two times the x-coordinate of the point. This is a general fact. In Section 1.3 we shall derive this simple formula (see Fig. 6.):

$$[\text{slope of the graph of } y = x^2 \text{ at the point } (x, y)] = 2x.$$

EXAMPLE 2   (a) What is the slope of the graph of $y = x^2$ at the point $(\frac{3}{4}, \frac{9}{16})$?

(b) Write the equation of the tangent line to the graph of $y = x^2$ at the point $(\frac{3}{4}, \frac{9}{16})$.

Solution   (a) The x-coordinate of $(\frac{3}{4}, \frac{9}{16})$ is $\frac{3}{4}$, so the slope of $y = x^2$ at this point is $2(\frac{3}{4}) = \frac{3}{2}$.

(b) We shall write the equation of the tangent line in the point-slope form. The point is $(\frac{3}{4}, \frac{9}{16})$, and the slope is $\frac{3}{2}$ by part (a). Hence the equation is

$$y - \frac{9}{16} = \frac{3}{2}\left(x - \frac{3}{4}\right). \quad \bullet$$

TECHNOLOGY PROJECT 1.2
Study of the Approximation of a Curve by Its Tangent Line

Let $f(x) = x^2$.

1. Display the graphs of f and its tangent line at the point $(1, 2)$ on a single screen.

2. Use BOX to closely examine the two graphs in the interval $.5 \le X \le 1.5$. Observe how the portion of the graph of f is almost straight.

3. By how much does the function graph differ from its tangent line over the interval?

4. Repeat 2 and 3 using the interval $.9 \le X \le 1.1$.

5. Repeat 2 and 3 using the interval $.99 \le X \le 1.01$.

**PRACTICE
PROBLEMS 1.2**

1. Refer to Fig. 7.
   (a) What is the slope of the curve at $(3, 4)$?
   (b) What is the equation of the tangent line at the point where $x = 3$?

FIGURE 7

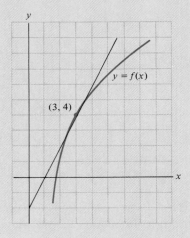

2. What is the equation of the tangent line to the graph of $y = \frac{1}{2}x + 1$ at the point $(4, 3)$?

EXERCISES 1.2

Trace the curves in Exercises 1–6 onto another piece of paper and sketch the tangent line in each case at the designated point P.

1.

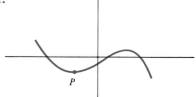

2.

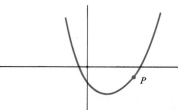

3.

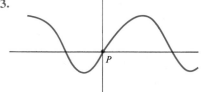

4.

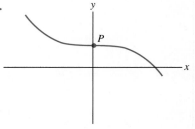

5.

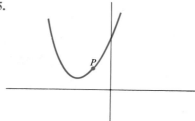

6.

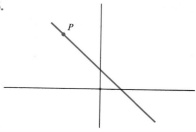

Estimate the slope of each of the following curves at the designated point P.

7.

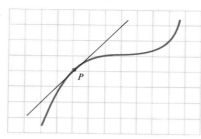

8.

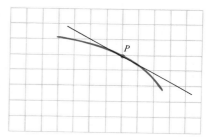

9.

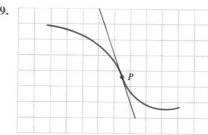

10.

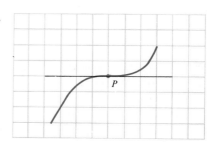

11.

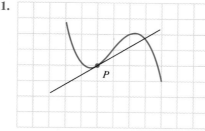

12.

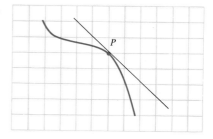

Exercises 13–18 refer to the points in Fig. 8. Assign one of the following descriptors to each point: large positive slope, small positive slope, zero slope, small negative slope, large negative slope.

13. *A*   **14.** *B*   **15.** *C*   **16.** *D*

17. *E*   **18.** *F*

FIGURE 8

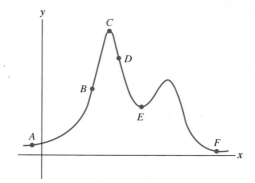

In Exercises 19–21, find the slope of the tangent line to the graph of $y = x^2$ at the point indicated and then write the corresponding equation of the tangent line.

19. $(-2, 4)$ **20.** $(-.4, .16)$ **21.** $(\frac{4}{3}, \frac{16}{9})$

22. Find the slope of the tangent line to the graph of $y = x^2$ at the point where $x = -\frac{1}{2}$.

23. Write the equation of the tangent line to the graph of $y = x^2$ at the point where $x = 1.5$.

24. Write the equation of the tangent line to the graph of $y = x^2$ at the point where $x = .6$.

25. Find the point on the graph of $y = x^2$ where the curve has slope $\frac{5}{3}$.

26. Find the point on the graph of $y = x^2$ where the curve has slope -4.

27. Find the point on the graph of $y = x^2$ where the tangent line is parallel to the line $x + 2y = 4$.

28. Find the point on the graph of $y = x^2$ where the tangent line is parallel to the line $3x - y = 2$.

In the next section we shall see that the tangent line to the graph of $y = x^3$ at the point (x, y) has slope $3x^2$. See Fig. 9. Using this result, find the slope of the curve at the points in Exercises 29–31.

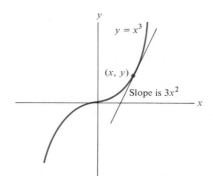

29. $(2, 8)$ **30.** $(\frac{3}{2}, \frac{27}{8})$ **31.** $(-\frac{1}{2}, -\frac{1}{8})$

32. Find the slope of the curve $y = x^3$ at the point where $x = \frac{1}{4}$.

33. Write the equation of the line tangent to the graph of $y = x^3$ at the point where $x = -1$.

34. Write the equation of the line tangent to the graph of $y = x^3$ at the point where $x = \frac{1}{2}$.

35. Let l be the line through the points P and Q in Fig. 10.

    (a) Suppose $P = (2, 4)$ and $Q = (5, 13)$. Find the slope of the line l and the length of the line segment d.

    (b) As the point Q moves toward P, does the slope of the line l increase or decrease?

FIGURE 10   The slope of a secant line.

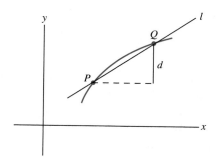

36. In Fig. 11, h represents a positive number, and $3 + h$ is the number h units to the right of 3. Draw line segments on the graph having the following lengths.

    (a) $f(3)$ (b) $f(3 + h)$ (c) $f(3 + h) - f(3)$

    (d) h (e) Draw a line of slope $\dfrac{f(3 + h) - f(3)}{h}$.

FIGURE 11   Geometric representation of values.

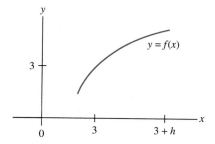

SOLUTIONS TO PRACTICE PROBLEMS 1.2

1. (a) The slope of the curve at the point $(3, 4)$ is, by definition, the slope of the tangent line at $(3, 4)$. Note that the point $(4, 6)$ is also on the line. Therefore, the slope is

$$\frac{6 - 4}{4 - 3} = \frac{2}{1} = 2$$

(b) Use the point-slope formula. The equation of the line passing through the point $(3, 4)$ and having slope 2 is

$$y - 4 = 2(x - 3)$$

or

$$y = 2x - 2.$$

2. The tangent line at (4, 3) is, by definition, the line that best approximates the curve at (4, 3). Since the "curve" in this case is itself a line, the curve and its tangent line at (4, 3) (and at every other point) must be the same. Therefore, the equation is $y = \frac{1}{2}x + 1$.

 1.3 THE DERIVATIVE

Suppose that a curve is the graph of a function $f(x)$. It is usually possible to obtain a formula that gives the slope of the curve $y = f(x)$ at any point. This slope formula is called the *derivative* of $f(x)$ and is written $f'(x)$. For each value of x, $f'(x)$ gives the slope of the curve $y = f(x)$ at the point with first coordinate x.* (See Fig. 1.)

FIGURE 1   Definition of $f'(x)$.

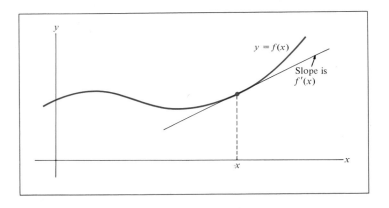

The process of computing $f'(x)$ for a given function $f(x)$ is called *differentiation.*

As we shall see later, the concept of a derivative occurs frequently in applications. In economics, the derivatives are often described by the adjective "marginal." For instance, if $C(x)$ is a cost function (the cost of producing x units of a commodity), then the derivative $C'(x)$ is called the *marginal cost function.* The derivative $P'(x)$ of a *profit function* $P(x)$ is called the *marginal profit function;* the derivative of a revenue function is called the *marginal revenue function;* and so on. We shall discuss the economic meaning of these marginal concepts in Section 1.8.

---

*As we shall see, there are curves that do not have tangent lines at every point. At values of x corresponding to such points, the derivative $f'(x)$ is not defined. For the sake of the current discussion, which is designed to develop an intuitive feeling for the derivative, let us assume that the graph of $f(x)$ has a tangent line for each x in the domain of f.

For the remainder of this section as well as the next few sections we will concentrate on calculating derivatives.

The case of a linear function $f(x) = mx + b$ is particularly simple. The graph of $y = mx + b$ is a straight line L of slope m. The tangent line to L (at any point) is just L itself, and so the slope of the graph is m at every point. (See Fig. 2.) In other words, the value of the derivative $f'(x)$ is always equal to m. We summarize this fact as follows:

If $f(x) = mx + b$, then we have

$$f'(x) = m \qquad (1)$$

FIGURE 2 **Derivative of a linear function.**

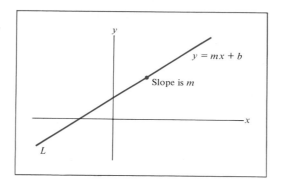

Set $m = 0$ in equation (1). Then the function becomes $f(x) = b$, which has the value b for each value of x. The graph is a horizontal line of slope 0, so $f'(x) = 0$ for all x. (See Fig. 3.) Thus we have:

The derivative of a constant function $f(x) = b$ is zero; that is,

$$f'(x) = 0. \qquad (2)$$

FIGURE 3 **Derivative of a constant function.**

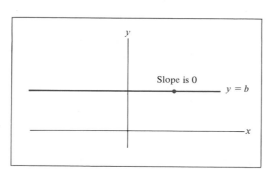

Next, consider the function $f(x) = x^2$. As we stated in Section 1.2 (and will prove at the end of this section), the slope of the graph of $y = x^2$ at the point (x, y) is equal to $2x$. That is, the value of the derivative $f'(x)$ is $2x$:

If $f(x) = x^2$, then its derivative is the function $2x$. That is,

$$f'(x) = 2x. \tag{3}$$

In Exercises 29–34 of Section 1.2 we made use of the fact that the slope of the graph of $y = x^3$ at the point (x, y) is $3x^2$. This can be restated in terms of derivatives as follows:

If $f(x) = x^3$, then the derivative is $3x^2$. That is,

$$f'(x) = 3x^2. \tag{4}$$

We should, at this stage at least, keep the geometric meaning of these formulas clearly in mind. Figure 4 shows the graphs of x^2 and x^3 together with the interpretations of formulas (3) and (4) in terms of slope.

FIGURE 4   Derivatives of x^2 and x^3.

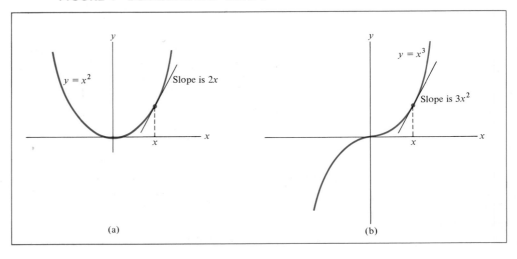

One of the reasons calculus is so useful is that it provides general techniques which can be easily used to determine derivatives. One such general rule, which contains formulas (3) and (4) as special cases, is the so-called power rule.

Power Rule   Let r be any number and let $f(x) = x^r$. Then $f'(x) = rx^{r-1}$.

Indeed, if $r = 2$, then $f(x) = x^2$ and $f'(x) = 2x^{2-1} = 2x$, which is formula (3). If $r = 3$, then $f(x) = x^3$ and $f'(x) = 3x^{3-1} = 3x^2$, which is (4). We shall prove the power rule in Chapter 4. Until then, we shall use it to calculate derivatives.

EXAMPLE 1 Let $f(x) = \sqrt{x}$. What is $f'(x)$?

Solution Recall that $\sqrt{x} = x^{1/2}$. We may apply the power rule with $r = \frac{1}{2}$.

$$f(x) = x^{1/2}$$

$$f'(x) = \frac{1}{2}x^{1/2-1} = \frac{1}{2}x^{-1/2}$$

$$= \frac{1}{2} \cdot \frac{1}{x^{1/2}} = \frac{1}{2\sqrt{x}}. \quad \bullet$$

Another important special case of the power rule occurs for $r = -1$, corresponding to $f(x) = x^{-1}$. In this case, $f'(x) = (-1)x^{-1-1} = -x^{-2}$. However, since $x^{-1} = 1/x$ and $x^{-2} = 1/x^2$, the power rule for $r = -1$ may also be written as follows:*

$$\boxed{\text{If } f(x) = \frac{1}{x}, \text{ then } f'(x) = -\frac{1}{x^2} \quad (x \neq 0).} \qquad (5)$$

EXAMPLE 2 Find the slope of the curve $y = 1/x$ at $(2, \frac{1}{2})$.

Solution Set $f(x) = 1/x$. The point $(2, \frac{1}{2})$ corresponds to $x = 2$, so in order to find the slope at this point, we compute $f'(2)$. From formula (5) we find that

$$f'(2) = -\frac{1}{2^2} = -\frac{1}{4}.$$

Thus the slope of $y = 1/x$ at the point $(2, \frac{1}{2})$ is $-\frac{1}{4}$. (See Fig. 5.) $\bullet$

Warning Do not confuse $f'(2)$, the value of the derivative at 2, with $f(2)$, the value of the y-coordinate at the point on the graph at which $x = 2$. In Example 2 we have $f'(2) = -\frac{1}{4}$, whereas $f(2) = \frac{1}{2}$. The number $f'(2)$ gives the *slope* of the graph at $x = 2$; the number $f(2)$ gives the *height* of the graph at $x = 2$.

Notation The operation of forming a derivative $f'(x)$ from a function $f(x)$ is also indicated by the symbol $\dfrac{d}{dx}$ (read "the derivative with respect to x"). Thus

$$\frac{d}{dx}f(x) = f'(x).$$

---

* The formula gives $f'(x)$ for $x \neq 0$. The derivative of $f(x)$ is not defined at $x = 0$ since $f(x)$ itself is not defined there.

FIGURE 5   Derivative of $\frac{1}{x}$.

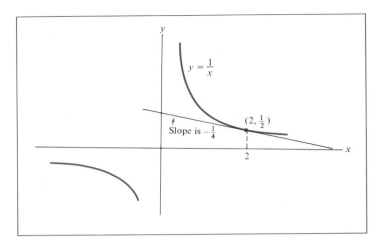

For example,

$$\frac{d}{dx}(x^6) = 6x^5, \qquad \frac{d}{dx}(x^{5/3}) = \frac{5}{3}x^{2/3}, \qquad \frac{d}{dx}\left(\frac{1}{x}\right) = -\frac{1}{x^2}.$$

When working with an equation of the form $y = f(x)$, we often write $\dfrac{dy}{dx}$ as a symbol for the derivative $f'(x)$. For example, if $y = x^6$, we may write

$$\frac{dy}{dx} = 6x^5.$$

The Secant-Line Calculation of the Derivative   So far, we have said nothing about how to derive differentiation formulas such as (3), (4), or (5). Let us remedy that omission now. The derivative gives the slope of the tangent line, so we must describe a procedure for calculating that slope.*

The fundamental idea for calculating the slope of the tangent line at a point P is to approximate the tangent line very closely by *secant lines*. A secant line at P is a straight line passing through P and a nearby point Q on the curve. (See Fig. 6.) By taking Q very close to P, we can make the slope of the secant line approximate the slope of the tangent line to any desired degree of accuracy. Let us see what this amounts to in terms of calculations.

Suppose that the point P is $(x, f(x))$. Suppose also that Q is h horizontal units away from P. Then Q has x-coordinate $x + h$ and y-coordinate $f(x + h)$. The slope of the secant line through the points $P = (x, f(x))$ and $Q = (x + h,$

---

* The following discussion is designed to develop geometric intuition for the derivative. It will also provide the basis for a formal definition of the derivative in terms of limits in Section 1.4.

FIGURE 6 A secant line approximation to a tangent line.

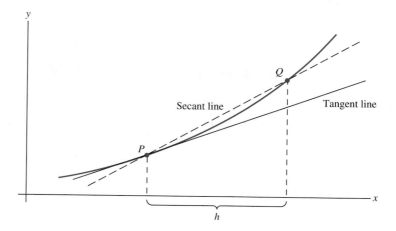

$f(x + h))$ is simply

$$[\text{slope of secant line}] = \frac{f(x + h) - f(x)}{(x + h) - x} = \frac{f(x + h) - f(x)}{h}.$$

(See Fig. 7.)

In order to move Q close to P along the curve, we let h approach zero. Then the secant line approaches the tangent line, and so

$$[\text{slope of secant line}] \quad \text{approaches} \quad [\text{slope of tangent line}];$$

that is,

$$\frac{f(x + h) - f(x)}{h} \quad \text{approaches} \quad f'(x).$$

Since we can make the secant line as close to the tangent line as we wish by taking h sufficiently small, the quantity $[f(x + h) - f(x)]/h$ can be made to approximate $f'(x)$ to any desired degree of accuracy. Thus we arrive at the following method to compute the derivative $f'(x)$.

FIGURE 7 Computing the slope of a secant line.

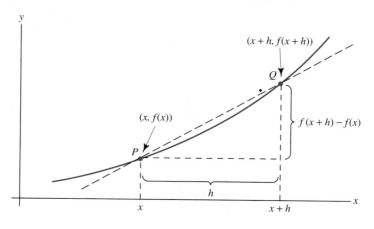

> To calculate $f'(x)$:
>
> 1. First calculate $\dfrac{f(x + h) - f(x)}{h}$ for $h \neq 0$.
> 2. Then let h approach zero.
> 3. The quantity $\dfrac{f(x + h) - f(x)}{h}$ will approach $f'(x)$.

Let us use this method to verify differentiation formulas (3) and (5), which, as we saw, were special cases of the power rule for $r = 2$ and $r = -1$, respectively.

Verification of the Power Rule for $r = 2$

Here $f(x) = x^2$, so that the slope of the secant line is

$$\frac{f(x + h) - f(x)}{h} = \frac{(x + h)^2 - x^2}{h}.$$

However, by multiplying out we have $(x + h)^2 = x^2 + 2xh + h^2$, so that

$$\frac{f(x + h) - f(x)}{h} = \frac{x^2 + 2xh + h^2 - x^2}{h} = \frac{(2x + h)h}{h}$$

$$= 2x + h.$$

As h approaches zero (i.e., as the secant line approaches the tangent line), the quantity $2x + h$ approaches $2x$. Thus we have

$$f'(x) = 2x,$$

which is formula (3).

Verification of the Power Rule for $r = -1$

Here $f(x) = x^{-1} = 1/x$, so that the slope of the secant line is

$$\frac{f(x + h) - f(x)}{h} = \frac{1}{h}\left[\frac{1}{x + h} - \frac{1}{x}\right] = \frac{1}{h}\left[\frac{x - (x + h)}{(x + h)x}\right]$$

$$= \frac{1}{h}\left[\frac{-h}{(x + h)x}\right] = -\frac{1}{x(x + h)}$$

As h approaches zero, $-\dfrac{1}{x(x + h)}$ approaches $-\dfrac{1}{x^2}$. Hence

$$f'(x) = -\frac{1}{x^2},$$

which is formula (5).

Similar arguments can be used to verify the power rule for other values of r. The cases $r = 3$ and $r = \frac{1}{2}$ are outlined in Exercises 55 and 56, respectively.

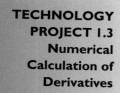

TECHNOLOGY PROJECT 1.3
Numerical Calculation of Derivatives

On a graphing calculator, you may use variables in addition to X by storing data in one of the storage registers available. For example, to use a variable H, you would store the value of H in storage register H. (Consult your calculator manual on how to do this.) You can use these other variables in entering functions in the equation entry screen. For example, to enter the slope of the secant line of the function $f(x) = x^2$ through (X, X^2) and $(X + H, (X + H)^2)$, you would key in:

$$Y_1 = ((X + H)\text{\^{}}2 - (X + H))/H$$

By storing small values for H and using the TRACE, you can approximate the value of the derivative for any value of X. This is a procedure for numerically approximating the value of the derivative of $f(X) = X^2$. A similar procedure works for other functions. Use the above procedure with $H = .001$ to approximate $f'(2)$, where:

1. $f(x) = x^2$

2. $f(x) = x^3$

3. $f(x) = \dfrac{x}{x + 1}$

4. $f(x) = x(2x - 1)^8$

In each case, estimate how good your approximation is by using smaller values of H.

PRACTICE PROBLEMS 1.3

1. Consider the curve $y = f(x)$ in Fig. 8.
   (a) Find $f(5)$.
   (b) Find $f'(5)$.

2. Let $f(x) = 1/x^4$.
   (a) Find its derivative.
   (b) Find $f'(2)$.

FIGURE 8

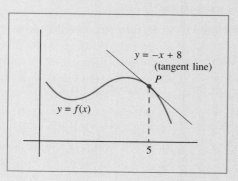

$y = -x + 8$
(tangent line)
P
$y = f(x)$
5

EXERCISES 1.3

Use (1), (2), and the power rule to find the derivatives of the following functions.

1. $f(x) = 2x - 5$ 2. $f(x) = 3 - \frac{1}{2}x$ 3. $f(x) = x^8$

4. $f(x) = x^{75}$ 5. $f(x) = x^{5/2}$ 6. $f(x) = x^{4/3}$

7. $f(x) = \sqrt[3]{x}$ 8. $f(x) = x^{3/4}$ 9. $f(x) = x^{-2}$

10. $f(x) = 5$ 11. $f(x) = x^{-1/4}$ 12. $f(x) = x^{-3}$

13. $f(x) = \frac{3}{4}$ 14. $f(x) = 1/\sqrt[3]{x}$ 15. $f(x) = 1/x^3$

16. $f(x) = 1/x^5$

In Exercises 17–24, find the derivative of $f(x)$ at the designated value of x.

17. $f(x) = x^6$ at $x = -2$ 18. $f(x) = x^3$ at $x = \frac{1}{4}$

19. $f(x) = 1/x$ at $x = 3$ 20. $f(x) = 5x$ at $x = 2$

21. $f(x) = 4 - x$ at $x = 5$ 22. $f(x) = x^{2/3}$ at $x = 1$

23. $f(x) = x^{3/2}$ at $x = 9$ 24. $f(x) = 1/x^2$ at $x = 2$

25. Find the slope of the curve $y = x^4$ at $x = 3$.
26. Find the slope of the curve $y = x^5$ at $x = -2$.
27. Find the slope of the curve $y = \sqrt{x}$ at $x = 9$.
28. Find the slope of the curve $y = x^{-3}$ at $x = 3$.
29. If $f(x) = x^2$, compute $f(-5)$ and $f'(-5)$.
30. If $f(x) = x + 6$, compute $f(3)$ and $f'(3)$.
31. If $f(x) = 1/x^5$, compute $f(2)$ and $f'(2)$.
32. If $f(x) = 1/x^2$, compute $f(5)$ and $f'(5)$.
33. If $f(x) = x^{4/3}$, compute $f(8)$ and $f'(8)$.
34. If $f(x) = x^{3/2}$, compute $f(16)$ and $f'(16)$.
35. Find the slope of the tangent line to the curve $y = x^3$ at the point $(4, 64)$, and write the equation of this line.
36. Find the slope of the tangent line to the curve $y = \sqrt{x}$ at the point $(25, 5)$, and write the equation of this line.

In Exercises 37–44, find the indicated derivative.

37. $\dfrac{d}{dx} (x^8)$ 38. $\dfrac{d}{dx} (x^{-3})$ 39. $\dfrac{d}{dx} (x^{3/4})$

40. $\dfrac{d}{dx} (x^{-1/3})$ 41. $\dfrac{dy}{dx}$ if $y = 1$ 42. $\dfrac{dy}{dx}$ if $y = x^{-4}$

43. $\dfrac{dy}{dx}$ if $y = x^{1/5}$ 44. $\dfrac{dy}{dx}$ if $y = \dfrac{x - 1}{3}$

45. Consider the curve $y = f(x)$ in Fig. 9. Find $f(6)$ and $f'(6)$.
46. Consider the curve $y = f(x)$ in Fig. 10. Find $f(1)$ and $f'(1)$.
47. In Fig. 11 the straight line $y = \frac{1}{4}x + b$ is tangent to the graph of $f(x) = \sqrt{x}$. Find the values of a and b.
48. In Fig. 12 the straight line is tangent to the graph of $f(x) = 1/x$. Find the value of a.
49. Consider the curve $y = f(x)$ in Fig. 13. Find a and $f(a)$. Estimate $f'(a)$.
50. Consider the curve $y = f(x)$ in Fig. 14. Estimate $f'(1)$.

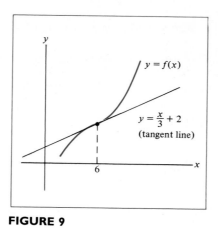

FIGURE 9

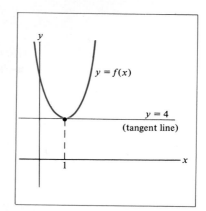

FIGURE 10

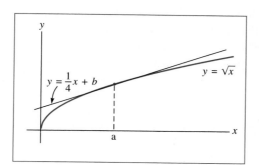

FIGURE 11

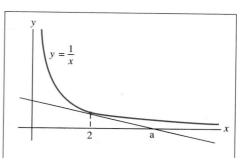

FIGURE 12

FIGURE 13

FIGURE 14

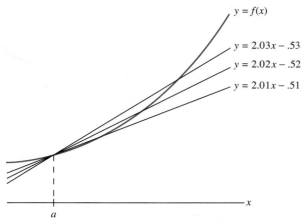

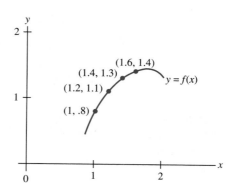

51. As h approaches 0, what value is approached by $\dfrac{(3 + h)^2 - (3)^2}{h}$?

    [*Hint:* This quotient is a secant-line approximation for $f(x) = x^2$.]

52. As h approaches 0, what value is approached by $\dfrac{\sqrt{4 + h} - \sqrt{4}}{h}$?

    [*Hint:* This quotient is a secant-line approximation for $f(x) = \sqrt{x}$.]

53. Given that $f(x + h) - f(x) = 2xh + 5h + h^2$, use the secant-line calculation to find $f'(x)$.

54. Given that $g(x + h) - g(x) = 8xh + h + 4h^2$, use the secant-line calculation to find $g'(x)$.

55. Use an argument like those used to verify formulas (3) and (5) to show that the derivative of $f(x) = x^3$ is $3x^2$. [*Hint:* Recall that $(x + h)^3 = x^3 + 3x^2h + 3xh^2 + h^3$.]

56. Use an argument like those used to verify formulas (3) and (5) to show that the derivative of $f(x) = \sqrt{x}$ is $1/(2\sqrt{x})$. [*Hint:* After forming the expression for the slope of a secant line, eliminate the square roots from the numerator by multiplying both the numerator and denominator by the quantity $\sqrt{x + h} + \sqrt{x}$.]

57. In Fig. 15, find the equation of the tangent line to $f(x)$ at the point A.

58. In Fig. 16, find the equation of the tangent line to $f(x)$ at the point P.

FIGURE 15

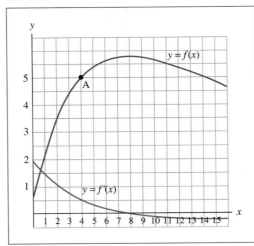

FIGURE 16

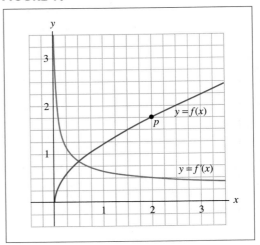

SOLUTIONS TO PRACTICE PROBLEMS 1.3

1. (a) The number $f(5)$ is the y-coordinate of the point P. Since the tangent line passes through P, the coordinates of P satisfy the equation $y = -x + 8$. Since its x-coordinate is 5, its y-coordinate is $-5 + 8 = 3$. Therefore, $f(5) = 3$.

   (b) The number $f'(5)$ is the slope of the tangent line at P, which is readily seen to be -1.

2. (a) The function $1/x^4$ can be written as the power function x^{-4}. Here $r = -4$. Therefore,

$$f'(x) = (-4)x^{(-4)-1} = -4x^{-5} = \frac{-4}{x^5}.$$

(b) $f'(2) = -4/2^5 = -4/32 = -\frac{1}{8}$.

1.4 LIMITS AND THE DERIVATIVE

The notion of a limit is one of the fundamental ideas of calculus. Indeed, any "theoretical" development of calculus rests on an extensive use of the theory of limits. Even in this book, where we have adopted an intuitive viewpoint, limit arguments are used occasionally (although in an informal way). In this section we give a brief introduction to limits and their role in calculus. As we shall see, the limit concept will allow us to define the notion of a derivative independently of our geometric reasoning.

Actually, we have already considered a limit in our discussion of the derivative, although we did not use the term "limit." Using the geometric reasoning of the previous section, we have the following procedure for calculating the derivative of a function $f(x)$ at $x = a$. First calculate the *difference quotient*

$$\frac{f(a + h) - f(a)}{h},$$

where h is a nonzero number. Next, allow h to approach zero by allowing it to assume both positive and negative numbers arbitrarily close to zero but different from zero. In symbols, we write $h \to 0$. The values of the difference quotient then approach the value of the derivative $f'(a)$. We say that the number $f'(a)$ is the *limit* of the difference quotient as h approaches zero, and in symbols we write

$$f'(a) = \lim_{h \to 0} \frac{f(a + h) - f(a)}{h}. \qquad (1)$$

As a numerical example of formula (1), consider the case of the derivative of $f(x) = x^2$ at $x = 2$. The difference quotient in this case has the form

$$\frac{f(2 + h) - f(2)}{h} = \frac{(2 + h)^2 - 2^2}{h}.$$

Table 1 gives some typical values of this difference quotient for progressively smaller values of h, both positive and negative. It is clear that the values of the difference quotient are approaching 4 as $h \to 0$. In other words, 4 is the *limit* of

TABLE I			
h	$\dfrac{(2 + h)^2 - 2^2}{h}$	h	$\dfrac{(2 + h)^2 - 2^2}{h}$
1	$\dfrac{(2 + 1)^2 - 2^2}{1} = 5$	-1	$\dfrac{(2 + (-1))^2 - 2^2}{-1} = 3$
.1	$\dfrac{(2 + .1)^2 - 2^2}{.1} = 4.10$	$-.1$	$\dfrac{(2 + (-.1))^2 - 2^2}{-.1} = 3.9$
.01	$\dfrac{(2 + .01)^2 - 2^2}{.01} = 4.01$	$-.01$	$\dfrac{(2 + (-.01))^2 - 2^2}{-.01} = 3.99$
.001	$\dfrac{(2 + .001)^2 - 2^2}{.001} = 4.001$	$-.001$	$\dfrac{(2 + (-.001))^2 - 2^2}{-.001} = 3.999$
.0001	$\dfrac{(2 + .0001)^2 - 2^2}{.0001} = 4.0001$	$-.0001$	$\dfrac{(2 + (-.0001))^2 - 2^2}{-.0001} = 3.9999$

the difference quotient as $h \to 0$. Thus

$$\lim_{h \to 0} \frac{(2 + h)^2 - 2^2}{h} = 4.$$

Since the values of the difference quotient approach the derivative $f'(2)$, we conclude that $f'(2) = 4$.

Our discussion of the derivative has been based on an intuitive geometric concept of the tangent line. However, the limit on the right in (1) may be considered independently of its geometric interpretation. In fact, we may use (1) to *define* $f'(a)$. We say that f is *differentiable* at $x = a$ if $\dfrac{f(a + h) - f(a)}{h}$ approaches some number as $h \to 0$, and we denote this limiting number by $f'(a)$. If the difference quotient $\dfrac{f(a + h) - f(a)}{h}$ does not approach any specific number as $h \to 0$, we say that f is *nondifferentiable* at $x = a$. Essentially all the functions in this text are differentiable at all points in their domain. A few exceptions are described in Section 1.5.

To better understand the limit concept used to define the derivative, it will be helpful to look at limits in a more general setting. If we let $g(h) = \dfrac{(2 + h)^2 - 2^2}{h}$, then Table 1 gives values of $g(h)$ as $h \to 0$. These values obviously approach 4. We may express this by writing

$$\lim_{h \to 0} g(h) = 4.$$

The preceding discussion suggests the following definition: Let $g(x)$ be a function, a a number. We say that the number L is *the limit of $g(x)$ as x approaches a* provided that, as x gets arbitrarily close (but not equal) to a, the values of $g(x)$ approach L. In this case we write

$$\lim_{x \to 0} g(x) = L.$$

If, as x approaches a, the values $g(x)$ do *not* approach a specific number, then we say that the limit of $g(x)$ as x approaches a *does not exist*. Let us give some further examples of limits.

EXAMPLE 1   Determine $\lim\limits_{x \to 2} (3x - 5)$.

Solution   Let us make a table of values of x approaching 2 and the corresponding values of $3x - 5$:

x	$3x - 5$	x	$3x - 5$
2.1	1.3	1.9	.7
2.01	1.03	1.99	.97
2.001	1.003	1.999	.997
2.0001	1.0003	1.9999	.9997

As x approaches 2, we see that $3x - 5$ approaches 1. In terms of our notation,

$$\lim_{x \to 2} (3x - 5) = 1. \;\bullet$$

EXAMPLE 2   For each of the functions in Fig. 1, determine if $\lim\limits_{x \to 2} g(x)$ exists. (The circles drawn on the graphs are meant to represent breaks in the graph, indicating that the functions under consideration are not defined at $x = 2$.)

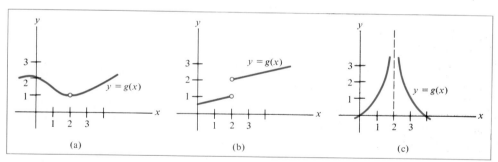

(a)   (b)   (c)

FIGURE 1

Solution   (a) $\lim\limits_{x \to 2} g(x) = 1$. We can see that as x gets closer and closer to 2, the values of $g(x)$ get closer and closer to 1. This is true for values of x to both the right and the left of 2.

(b) $\lim\limits_{x \to 2} g(x)$ does not exist. As x approaches 2 from the right, $g(x)$ approaches 2. However, as x approaches 2 from the left, $g(x)$ approaches 1. In order for a limit to exist, the function must approach the *same* value from each direction.

(c) $\lim\limits_{x \to 2} g(x)$ does not exist. As x approaches 2, the values of $g(x)$ become larger and larger and do not approach a fixed number. $\bullet$

The following limit theorems, which we cite without proof, allow us to reduce the computation of limits for combinations of functions to computations of limits involving the constituent functions.

Limit Theorems   Suppose that $\lim\limits_{x \to a} f(x)$ and $\lim\limits_{x \to a} g(x)$ both exist. Then we have the following results.

(I) If k is a constant, then $\lim\limits_{x \to a} k \cdot f(x) = k \cdot \lim\limits_{x \to a} f(x)$.

(II) If r is a positive constant, then $\lim\limits_{x \to a} [f(x)]^r = \left[\lim\limits_{x \to a} f(x) \right]^r$.

(III) $\lim\limits_{x \to a} [f(x) + g(x)] = \lim\limits_{x \to a} f(x) + \lim\limits_{x \to a} g(x)$.

(IV) $\lim\limits_{x \to a} [f(x) - g(x)] = \lim\limits_{x \to a} f(x) - \lim\limits_{x \to a} g(x)$.

(V) $\lim\limits_{x \to a} [f(x) \cdot g(x)] = \left[\lim\limits_{x \to a} f(x) \right] \cdot \left[\lim\limits_{x \to a} g(x) \right]$.

(VI) If $\lim\limits_{x \to a} g(x) \neq 0$, then $\lim\limits_{x \to a} \dfrac{f(x)}{g(x)} = \dfrac{\lim\limits_{x \to a} f(x)}{\lim\limits_{x \to a} g(x)}$.

EXAMPLE 3   Use the limit theorems to compute the following limits.

(a) $\lim\limits_{x \to 2} x^3$

(b) $\lim\limits_{x \to 2} 5x^3$

(c) $\lim\limits_{x \to 2} (5x^3 - 15)$.

(d) $\lim\limits_{x \to 2} \sqrt{5x^3 - 15}$

(e) $\lim\limits_{x \to 2} (\sqrt{5x^3 - 15}/x^5)$

Solution   (a) Since $\lim\limits_{x \to 2} x = 2$, we have by Limit Theorem II that

$$\lim_{x \to 2} x^3 = \left(\lim_{x \to 2} x \right)^3 = 2^3 = 8.$$

(b) $\lim\limits_{x \to 2} 5x^3 = 5 \lim\limits_{x \to 2} x^3$ (Limit Theorem I with $k = 5$)

$$= 5 \cdot 8 \qquad \text{[by part (a)]}$$

$$= 40.$$

(c) $\lim\limits_{x \to 2} (5x^3 - 15) = \lim\limits_{x \to 2} 5x^3 - \lim\limits_{x \to 2} 15$ (Limit Theorem IV).

Note that $\lim\limits_{x \to 2} 15 = 15$. This is because the constant function $g(x) = 15$ always has the value 15, and so its limit as x approaches *any* number is 15. By part (b), $\lim\limits_{x \to 2} 5x^3 = 40$. Thus

$$\lim_{x \to 2} (5x^3 - 15) = 40 - 15 = 25.$$

(d) $\lim\limits_{x \to 2} \sqrt{5x^3 - 15} = \lim\limits_{x \to 2} (5x^3 - 15)^{1/2}$

$\qquad\qquad\qquad = \left[\lim\limits_{x \to 2} (5x^3 - 15) \right]^{1/2}$ [Limit Theorem II with

$\qquad\qquad\qquad\qquad\qquad\qquad\qquad\qquad\qquad r = \frac{1}{2}, f(x) = 5x^3 - 15]$

$\qquad\qquad\qquad = 25^{1/2}$ [by part (c)]

$\qquad\qquad\qquad = 5.$

(e) The limit of the denominator is $\lim\limits_{x \to 2} x^5$, which is $2^5 = 32$, a nonzero number.

So by Limit Theorem VI we have

$$\lim_{x \to 2} \frac{\sqrt{5x^3 - 15}}{x^5} = \frac{\lim\limits_{x \to 2} \sqrt{5x^3 - 15}}{\lim\limits_{x \to 2} x^5}$$

$$= \frac{5}{32} \qquad \text{[by part (d)].} \; \bullet$$

The following facts, which may be deduced by repeated applications of the various limit theorems, are extremely handy in evaluating limits.

Limit of a Polynomial Function   Let $p(x)$ be a polynomial function, a any number. Then

$$\lim_{x \to a} p(x) = p(a).$$

Limit of a Rational Function   Let $r(x) = p(x)/q(x)$ be a rational function, where $p(x)$ and $q(x)$ are polynomials. Let a be a number such that $q(a) \neq 0$. Then

$$\lim_{x \to a} r(x) = r(a).$$

In other words, to determine a limit of a polynomial or a rational function, simply evaluate the function at $x = a$, provided, of course, that the function is defined at $x = a$. For instance, we can rework the solution to Example 3(c) as follows:

$$\lim_{x \to 2} (5x^3 - 15) = 5(2)^3 - 15 = 25.$$

Many situations require algebraic simplifications before the limit theorems can be applied.

EXAMPLE 4   Compute the following limits.

(a) $\lim\limits_{x \to 3} \dfrac{x^2 - 9}{x - 3}$ (b) $\lim\limits_{x \to 0} \dfrac{\sqrt{x + 4} - 2}{x}$

Solution

(a) The function $\dfrac{x^2 - 9}{x - 3}$ is not defined when $x = 3$, since $\dfrac{3^2 - 9}{3 - 3} = \dfrac{0}{0}$, which is undefined. That causes no difficulty, since the limit as x approaches 3 depends only on the values of x *near* 3 and excludes considerations of the value at $x = 3$ itself. To evaluate the limit, note that $x^2 - 9 = (x - 3)(x + 3)$. So for $x \neq 3$,

$$\frac{x^2 - 9}{x - 3} = \frac{(x - 3)(x + 3)}{x - 3} = x + 3.$$

As x approaches 3, $x + 3$ approaches 6. Therefore,

$$\lim_{x \to 3} \frac{x^2 - 9}{x - 3} = 6.$$

(b) Since the denominator approaches zero when taking the limit, we may not apply Limit Theorem VI directly. However, if we first apply an algebraic trick, the limit may be evaluated. Multiply numerator and denominator by $\sqrt{x + 4} + 2$.

$$\frac{\sqrt{x + 4} - 2}{x} \cdot \frac{\sqrt{x + 4} + 2}{\sqrt{x + 4} + 2} = \frac{(x + 4) - 4}{x(\sqrt{x + 4} + 2)}$$

$$= \frac{x}{x(\sqrt{x + 4} + 2)}$$

$$= \frac{1}{\sqrt{x + 4} + 2}.$$

Thus

$$\lim_{x \to 0} \frac{\sqrt{x + 4} - 2}{x} = \lim_{x \to 0} \frac{1}{\sqrt{x + 4} + 2}$$

$$= \frac{\lim\limits_{x \to 0} 1}{\lim\limits_{x \to 0} (\sqrt{x + 4} + 2)} \qquad \text{(Limit Theorem VI)}$$

$$= \frac{1}{4}. \quad \bullet$$

The basic differentiation rules are obtained from the limit definition of the derivative. The next example (and Exercises 29–40) illustrates this process. There are three main steps.

Using Limits to Calculate a Derivative

1. Write the difference quotient $\dfrac{f(a + h) - f(a)}{h}$.

2. Simplify the difference quotient.

3. Find the limit as $h \to 0$.

EXAMPLE 5   Use limits to compute the derivative $f'(5)$ for the following functions.

(a) $f(x) = 15 - x^2$

(b) $f(x) = \dfrac{1}{2x - 3}$

Solution   In each case, we must calculate $\lim\limits_{h \to 0} \dfrac{f(5 + h) - f(5)}{h}$.

(a) $\dfrac{f(5 + h) - f(5)}{h} = \dfrac{[15 - (5 + h)^2] - (15 - 5^2)}{h}$ **(step 1)**

$= \dfrac{15 - (25 + 10h + h^2) - (15 - 25)}{h}$ **(step 2)**

$= \dfrac{-10h - h^2}{h} = -10 - h.$

Therefore, $f'(5) = \lim\limits_{h \to 0} (-10 - h) = -10.$ **(step 3)**

(b) $\dfrac{f(5 + h) - f(5)}{h} = \dfrac{\dfrac{1}{2(5 + h) - 3} - \dfrac{1}{2(5) - 3}}{h}$ **(step 1)**

$= \dfrac{\dfrac{1}{7 + 2h} - \dfrac{1}{7}}{h} = \dfrac{\dfrac{7 - (7 + 2h)}{(7 + 2h)7}}{h}.$ **(step 2)**

$= \dfrac{-2h}{(7 + 2h)7 \cdot h} = \dfrac{-2}{(7 + 2h)7} = \dfrac{-2}{49 + 14h}.$

$f'(5) = \lim\limits_{h \to 0} \dfrac{-2}{49 + 14h} = -\dfrac{2}{49}.$ **(step 3)**

●

Remark   When computing the limit in Example 5, we considered only values of h near zero (and not $h = 0$ itself). Therefore, we were freely able to divide both numerator and denominator by h.

Infinity and Limits   Consider the function $f(x)$ whose graph is sketched in Fig. 2. As x grows large the value of $f(x)$ approaches 2. In this circumstance, we say that 2 is the *limit of $f(x)$ as x approaches infinity*. Infinity is denoted by the symbol ∞. The preceding limit statement is expressed in the following notation:

$$\lim_{x \to \infty} f(x) = 2.$$

FIGURE 2 Functions with an undefined point.

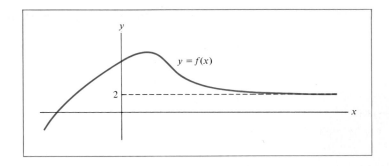

In a similar vein, consider the function whose graph is sketched in Fig. 3. As x grows large in the negative direction, the value of $f(x)$ approaches 0. In this circumstance, we say that 0 is *the limit of $f(x)$ as x approaches minus infinity.* In symbols,

$$\lim_{x \to -\infty} f(x) = 0.$$

FIGURE 3 Function with a limit as x approaches infinity.

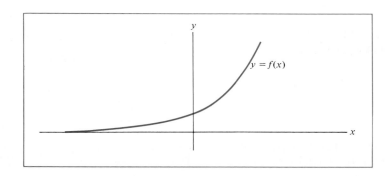

EXAMPLE 6　Calculate the following limits.

(a) $\lim\limits_{x \to \infty} \dfrac{1}{x^2 + 1}$　　　　　　　(b) $\lim\limits_{x \to \infty} \dfrac{x + 1}{x - 1}$

Solution　(a) As x increases without bound, so does $x^2 + 1$. Therefore, $1/(x^2 + 1)$ approaches zero as x approaches ∞.

(b) Both $x + 1$ and $x - 1$ increase without bound as x does. To determine the limit of their quotient, we employ an algebraic trick. Divide both numerator and denominator by x to obtain

$$\lim_{x \to \infty} \frac{x + 1}{x - 1} = \lim_{x \to \infty} \frac{1 + \dfrac{1}{x}}{1 - \dfrac{1}{x}}.$$

As x increases without bound, $1/x$ approaches zero, so that both $1 + (1/x)$ and $1 - (1/x)$ approach 1. Thus the desired limit is $1/1 = 1$. ●

**TECHNOLOGY
PROJECT 1.4
Calculating Limits
Using a Graphing
Calculator**

You can guess the value of limits as $x \to \infty$ or $x \to -\infty$ by examining the behavior of the graph. Use graphical analysis to determine the value of the following limits:

1. $\displaystyle\lim_{x \to \infty} \frac{x^2 - 2x + 3}{2x^2 - 1}$

2. $\displaystyle\lim_{x \to \infty} \left[\sqrt{x} - \sqrt{25 + x} \right]$

3. $\displaystyle\lim_{x \to -\infty} 3 + \frac{1 + x}{x^2}$

4. $\displaystyle\lim_{x \to \infty} \frac{-8x^2 + 1}{x^2 + 1}$

In each case, you will need to examine the graph of the function using a large display window.

**PRACTICE
PROBLEMS 1.4**

Determine which of the following limits exist. Compute the limits that exist.

1. $\displaystyle\lim_{x \to 6} \frac{x^2 - 4x - 12}{x - 6}$

2. $\displaystyle\lim_{x \to 6} \frac{4x + 12}{x - 6}$

EXERCISES 1.4

For each of the following functions $g(x)$, determine whether or not $\displaystyle\lim_{x \to 3} g(x)$ exists. If so, give the limit.

1.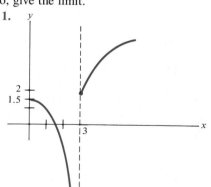

2.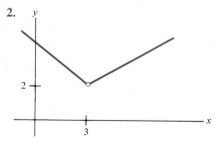

3.

4.

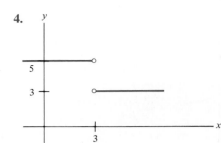

5.

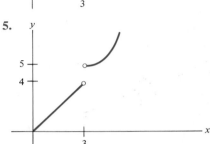

6.

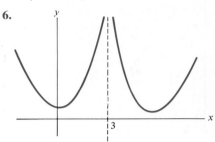

Determine which of the following limits exist. Compute the limits that exist.

7. $\lim\limits_{x \to 1} (1 - 6x)$

8. $\lim\limits_{x \to 2} \dfrac{x}{x - 2}$

9. $\lim\limits_{x \to 3} \sqrt{x^2 + 16}$

10. $\lim\limits_{x \to 4} (x^3 - 7)$

11. $\lim\limits_{x \to 5} \dfrac{x^2 + 1}{5 - x}$

12. $\lim\limits_{x \to 6} \left(\sqrt{6x} + 3x - \dfrac{1}{x} \right)(x^2 - 4)$

13. $\lim\limits_{x \to 7} (x + \sqrt{x - 6})(x^2 - 2x + 1)$

14. $\lim\limits_{x \to 8} \dfrac{\sqrt{5x - 4} - 1}{3x^2 + 2}$

15. $\lim\limits_{x \to 9} \dfrac{\sqrt{x^2 - 5x - 36}}{8 - 3x}$

16. $\lim\limits_{x \to 10} (2x^2 - 15x - 50)^{20}$

17. $\lim\limits_{x \to 0} \dfrac{x^2 + 3x}{x}$

18. $\lim\limits_{x \to 1} \dfrac{x^2 - 1}{x - 1}$

19. $\lim\limits_{x \to 2} \dfrac{-2x^2 + 4x}{x - 2}$

20. $\lim\limits_{x \to 3} \dfrac{x^2 - x - 6}{x - 3}$

21. $\lim\limits_{x \to 4} \dfrac{x^2 - 16}{4 - x}$

22. $\lim\limits_{x \to 5} \dfrac{2x - 10}{x^2 - 25}$

23. $\lim\limits_{x \to 6} \dfrac{x^2 - 6x}{x^2 - 5x - 6}$

24. $\lim\limits_{x \to 7} \dfrac{x^3 - 2x^2 + 3x}{x^2}$

25. $\lim\limits_{x \to 8} \dfrac{x^2 + 64}{x - 8}$

26. $\lim\limits_{x \to 9} \dfrac{1}{(x - 9)^2}$

27. $\lim\limits_{x \to 0} \dfrac{-2}{\sqrt{x + 16} + 7}$

28. $\lim\limits_{x \to 0} \dfrac{4x}{x(x^2 + 3x + 5)}$

Use limits to compute the following derivatives.

29. $f'(3)$ where $f(x) = x^2 + 1$

30. $f'(2)$ where $f(x) = x^3$

31. $f'(0)$ where $f(x) = x^3 + 3x + 1$

32. $f'(0)$ where $f(x) = x^2 + 2x + 2$

33. $f'(3)$ where $f(x) = \dfrac{1}{2x + 5}$

34. $f'(4)$ where $f(x) = \sqrt{2x - 1}$

35. $f'(2)$ where $f(x) = \sqrt{5 - x}$

36. $f'(3)$ where $f(x) = \dfrac{1}{7 - 2x}$

37. $f'(0)$ where $f(x) = \sqrt{1 - x^2}$

38. $f'(2)$ where $f(x) = (5x - 4)^2$

39. $f'(0)$ where $f(x) = (x + 1)^3$

40. $f'(0)$ where $f(x) = \sqrt{x^2 + x + 1}$

Each limit in Exercises 41–46 is a definition of $f'(a)$. Determine the function $f(x)$ and the value of a.

41. $\displaystyle\lim_{h \to 0} \dfrac{(9 + h)^{1/2} - 3}{h}$

42. $\displaystyle\lim_{h \to 0} \dfrac{(2 + h)^3 - 8}{h}$

43. $\displaystyle\lim_{h \to 0} \dfrac{\dfrac{1}{10 + h} - .1}{h}$

44. $\displaystyle\lim_{h \to 0} \dfrac{(64 + h)^{1/3} - 4}{h}$

45. $\displaystyle\lim_{h \to 0} \dfrac{(3(1 + h)^2 + 4) - 7}{h}$

46. $\displaystyle\lim_{h \to 0} \dfrac{(1 + h)^{-1/2} - 1}{h}$

Compute the following limits.

47. $\displaystyle\lim_{x \to \infty} \dfrac{1}{x^2}$

48. $\displaystyle\lim_{x \to -\infty} \dfrac{1}{x^2}$

49. $\displaystyle\lim_{x \to \infty} \dfrac{1}{x - 8}$

50. $\displaystyle\lim_{x \to \infty} \dfrac{1}{3x + 5}$

51. $\displaystyle\lim_{x \to \infty} \dfrac{2x + 1}{x + 2}$

52. $\displaystyle\lim_{x \to \infty} \dfrac{x^2 + x}{x^2 - 1}$

SOLUTIONS TO PRACTICE PROBLEMS 1.4

1. The function under consideration is a rational function. Since the denominator has value 0 at $x = 6$, we cannot immediately determine the limit by just evaluating the function at $x = 6$. Also, $\displaystyle\lim_{x \to 6} (x - 6) = 0$. Since the function in the denominator has limit 0, we cannot apply Limit Theorem VI. However, since the definition of limit considers only values of x different from 6, the quotient can be simplified by factoring and canceling.

$$\frac{x^2 - 4x - 12}{x - 6} = \frac{(x + 2)(x - 6)}{(x - 6)} = x + 2 \qquad \text{for } x \neq 6.$$

Now $\displaystyle\lim_{x \to 6} (x + 2) = 8$. Therefore, $\displaystyle\lim_{x \to 6} \frac{x^2 - 4x - 12}{x - 6} = 8$.

2. No limit exists. It is easily seen that $\displaystyle\lim_{x \to 6} (4x + 12) = 36$ and $\displaystyle\lim_{x \to 6} (x - 6) = 0$. As x approaches 6, the denominator gets very small and the numerator approaches 36. For example, if $x = 6.00001$, then the numerator is 36.00004 and the denominator is .00001. The quotient is 3,600,004. As x approaches 6 even more closely, the quotient gets arbitrarily large and cannot possibly approach a limit.

1.5 DIFFERENTIABILITY AND CONTINUITY

In the preceding section we defined differentiability of $f(x)$ at $x = a$ in terms of a limit. If this limit does not exist, then we say that $f(x)$ is *nondifferentiable* at $x = a$. Geometrically, the nondifferentiability of $f(x)$ at $x = a$ can manifest itself in several different ways. First of all, the graph of $f(x)$ could have no tangent line at $x = a$. Second, the graph could have a vertical tangent line at $x = a$. (Recall that slope is not defined for vertical lines.) Some of the various geometric possibilities are illustrated in Fig. 1.

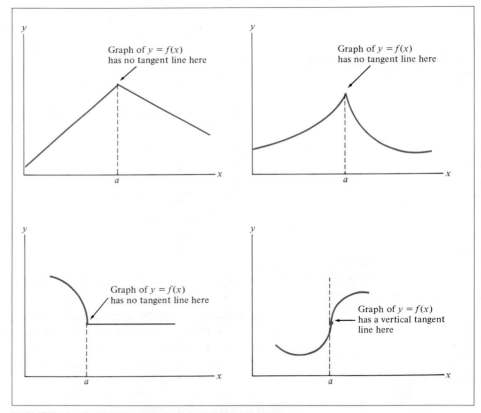

FIGURE 1   Functions that are nondifferentiable at $x = a$.

The following example illustrates how nondifferentiable functions can arise in practice.

EXAMPLE 1   A railroad company charges $10 per mile to haul a boxcar up to 200 miles and $8 per mile for each mile exceeding 200. In addition, the railroad charges a $1000 handling charge per boxcar. Graph the cost of sending a boxcar x miles.

Solution If x is at most 200 miles, then the cost $C(x)$ is given by $C(x) = 1000 + 10x$ dollars. The cost for 200 miles is $C(200) = 1000 + 2000 = 3000$ dollars. If x exceeds 200 miles, then the total cost will be

$$C(x) = \underbrace{3000}_{\substack{\text{cost of first} \\ \text{200 miles}}} + \underbrace{8(x - 200)}_{\substack{\text{cost of miles in} \\ \text{excess of 200}}} = 1400 + 8x.$$

Thus

$$C(x) = \begin{cases} 1000 + 10x, & \text{for} \quad 0 < x \leq 200, \\ 1400 + 8x, & \text{for} \quad x > 200. \end{cases}$$

The graph of $C(x)$ is sketched in Fig. 2. Note that $C(x)$ is not differentiable at $x = 200$. ●

FIGURE 2 Cost of hauling a boxcar.

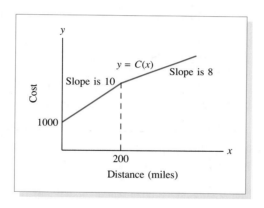

Closely related to the concept of differentiability is that of continuity. We say that a function $f(x)$ is *continuous* at $x = a$ provided that, roughly speaking, its graph has no breaks (or gaps) as it passes through the point $(a, f(a))$. That is, $f(x)$ is continuous at $x = a$ provided that we can draw the graph through $(a, f(a))$ without lifting our pencil from the paper. The functions whose graphs are drawn in Figs. 1 and 2 are continuous for all values of x. By contrast, however, the function whose graph is drawn in Fig. 3(a) is not continuous (we say it is *discontinuous*) at $x = 1$ and $x = 2$, since the graph has breaks there. Similarly, the function whose graph is drawn in Fig. 3(b) is discontinuous at $x = 2$.

Discontinuous functions can occur in applications, as the following example shows.

EXAMPLE 2 Suppose that a manufacturing plant is capable of producing 15,000 units in one shift of 8 hours. For each shift worked, there is a fixed cost of $2000 (for light, heat, etc.). Suppose that the variable cost (the cost of labor and raw materials) is $2 per unit. Graph the cost $C(x)$ of manufacturing x units.

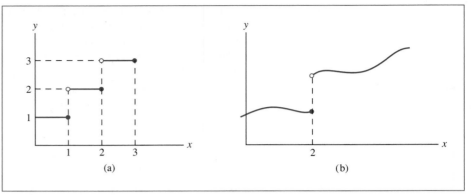

FIGURE 3   Functions with discontinuities.

Solution   If $x \le 15{,}000$, a single shift will suffice, so that

$$C(x) = 2000 + 2x, \qquad 0 \le x \le 15{,}000.$$

If x is between 15,000 and 30,000, one extra shift will be required, and

$$C(x) = 4000 + 2x, \qquad 15{,}000 < x \le 30{,}000.$$

If x is between 30,000 and 45,000, the plant will need to work three shifts, and

$$C(x) = 6000 + 2x, \qquad 30{,}000 < x \le 45{,}000.$$

The graph of $C(x)$ for $0 \le x \le 45{,}000$ is drawn in Fig. 4. Note that the graph has breaks at two points. ●

FIGURE 4   Cost function of a manufacturing plant.

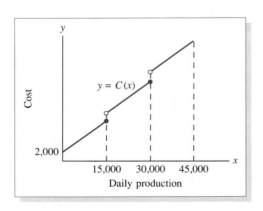

The relationship between differentiability and continuity is this:

Theorem I   If $f(x)$ is differentiable at $x = a$, then $f(x)$ is continuous at $x = a$.

Note, however, that the converse statement is definitely false: A function may be continuous at $x = a$ but still not be differentiable there. The functions whose graphs are drawn in Fig. 1 provide examples of this phenomenon.

Just as with differentiability, the notion of continuity can be phrased in terms of limits. In order for $f(x)$ to be continuous at $x = a$, the values of $f(x)$ for all x near a must be close to $f(a)$ (otherwise, the graph would have a break at $x = a$). In fact, the closer x is to a, the closer $f(x)$ must be to $f(a)$ (again, in order to avoid a break in the graph). In terms of limits, we must therefore have

$$\lim_{x \to a} f(x) = f(a).$$

Conversely, an intuitive argument shows that if the limit relation above holds, then the graph of $y = f(x)$ has no break at $x = a$.

Limit Definition of Continuity   A function $f(x)$ is continuous at $x = a$ provided the following limit relation holds:

$$\lim_{x \to a} f(x) = f(a). \tag{1}$$

In order for (1) to hold, three conditions must be fulfilled.

1. $f(x)$ must be defined at $x = a$.
2. $\lim_{x \to a} f(x)$ must exist.
3. The limit $\lim_{x \to a} f(x)$ must have the value $f(a)$.

A function will fail to be continuous at $x = a$ when any one of these conditions fails to hold. The various possibilities are illustrated in the next example.

EXAMPLE 3   Determine whether the functions whose graphs are drawn in Fig. 5 are continuous at $x = 3$. Use the limit definition.

Solution   (a) Here $\lim_{x \to 3} f(x) = 2$. However, $f(3) = 4$. So

$$\lim_{x \to 3} f(x) \neq f(3)$$

and $f(x)$ is not continuous at $x = 3$. (Geometrically, this is clear. The graph has a break at $x = 3$.)

(b) $\lim_{x \to 3} g(x)$ does not exist, so $g(x)$ is not continuous at $x = 3$.

(c) $\lim_{x \to 3} h(x)$ does not exist, so $h(x)$ is not continuous at $x = 3$.

(d) $f(x)$ is not defined at $x = 3$, so $f(x)$ is not continuous at $x = 3$. ●

Using our result on the limit of a polynomial function (Section 1.4), we see that

$$p(x) = a_0 + a_1 x + \cdots a_n x^n, \qquad a_0, \ldots, a_n \text{ constants,}$$

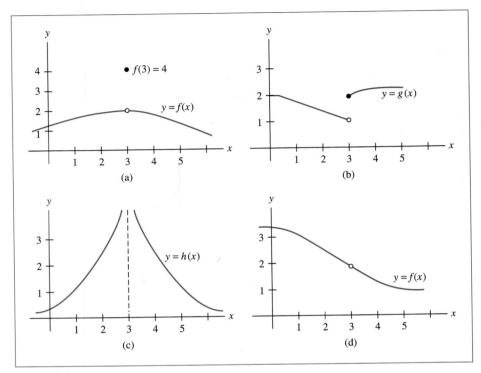

FIGURE 5

is continuous at all x. Similarly, a rational function

$$\frac{p(x)}{q(x)}, \qquad p(x), \, q(x) \text{ polynomials,}$$

is continuous at all x for which $q(x) \neq 0$.

TECHNOLOGY PROJECT 1.5
Graphing Functions Defined by Multiple Expressions

Graphing calculators evaluate inequalities as either true or false. A true inequality is given the value 1 and a false inequality the value 0. For example, if X equals 4, then the inequality $X > 1$ has the value 1 (true). Using this feature, we can enter functions defined by more than one expression. For instance, consider the function:

$$f(x) = \begin{cases} 1 & \text{if } x < 0 \\ -1 & \text{if } x \geq 0 \end{cases}$$

It can be entered into the calculator as:

$$Y_1 = 1 * (X < 0) + (-1) * (X > = 0)$$

The first term on the right is equal to $1 * 1 + (-1) * 0 = 1$ if $X < 0$ and equal to $1 * 0 + (-1) * 1 = -1$ if $X \geq 0$. (Note that the symbol $\geq$ is keyed in as

$> =$.) That is, Y_1 is just the function $f(x)$. In a similar fashion you may key in functions that are defined by multiple expressions. Use this procedure to graph the following functions.

1. $f(x)$ defined above.

2. $f(x) = \begin{cases} x & \text{if } x < 1 \\ 2x - 1 & \text{if } x \le 1 \end{cases}$

3. $f(x) = \begin{cases} x & \text{if } x < 1 \\ x^2 & \text{if } x \ge 1 \end{cases}$

4. A multiple inequality, such as $1 \le X \le 2$ may be keyed in using two inequalities:

$$(1 < = X) * (X < = 2)$$

Use this fact to graph the function:

$$f(x) = \begin{cases} 3 & \text{if } 0 \le x \le 1 \\ 1 & \text{if } x > 1 \end{cases}$$

PRACTICE PROBLEMS 1.5

Let $f(x) = \begin{cases} \dfrac{x^2 - x - 6}{x - 3} & \text{for } x \ne 3. \\ 4 & \text{for } x = 3. \end{cases}$

1. Is $f(x)$ continuous at $x = 3$?
2. Is $f(x)$ differentiable at $x = 3$?

EXERCISES 1.5

Is the function whose graph is drawn in Fig. 6 continuous at the following values of x?

1. $x = 0$ 2. $x = -3$ 3. $x = 3$

4. $x = .001$ 5. $x = -2$ 6. $x = 2$

FIGURE 6

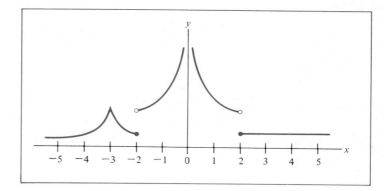

Is the function whose graph is drawn in Fig. 6 differentiable at the following values of x?

7. $x = 0$ 8. $x = -3$ 9. $x = 3$

10. $x = .001$ 11. $x = -2$ 12. $x = 2$

Determine whether each of the following functions is continuous and/or differentiable at $x = 1$.

13. $f(x) = x^2$ 14. $f(x) = \dfrac{1}{x}$

15. $f(x) = \begin{cases} x + 2 & \text{for } -1 \le x \le 1 \\ 3x & \text{for } 1 < x < 5 \end{cases}$ 16. $f(x) = \begin{cases} x & \text{for } 1 \le x \le 2 \\ x^3 & \text{for } 0 \le x < 1 \end{cases}$

17. $f(x) = \begin{cases} 2x - 1 & \text{for } 0 \le x \le 1 \\ 1 & \text{for } 1 < x \end{cases}$ 18. $f(x) = \begin{cases} x & \text{for } x \ne 1 \\ 2 & \text{for } x = 1 \end{cases}$

19. $f(x) = \begin{cases} \dfrac{1}{x - 1} & \text{for } x \ne 1 \\ 0 & \text{for } x = 1 \end{cases}$ 20. $f(x) = \begin{cases} x - 1 & \text{for } 0 \le x < 1 \\ 1 & \text{for } x = 1 \\ 2x - 2 & \text{for } x > 1 \end{cases}$

The functions in Exercises 21–26 are defined for all x except for one value of x. If possible, define $f(x)$ at the exceptional point in a way that makes $f(x)$ continuous for all x.

21. $f(x) = \dfrac{x^2 - 7x + 10}{x - 5}, x \ne 5$ 22. $f(x) = \dfrac{x^2 + x - 12}{x + 4}, x \ne -4$

23. $f(x) = \dfrac{x^3 - 5x^2 + 4}{x^2}, x \ne 0$ 24. $f(x) = \dfrac{x^2 + 25}{x - 5}, x \ne 5$

25. $f(x) = \dfrac{(6 + x)^2 - 36}{x}, x \ne 0$ 26. $f(x) = \dfrac{\sqrt{9 + x} - \sqrt{9}}{x}, x \ne 0$

SOLUTIONS TO PRACTICE PROBLEMS 1.5

1. The function $f(x)$ is defined at $x = 3$, namely, $f(3) = 4$. When computing $\lim_{x \to 3} f(x)$, we exclude consideration of $x = 3$; therefore, we can simplify the expression for $f(x)$ as follows:

Clearly, $f(x) = \dfrac{x^2 - x - 6}{x - 3} = \dfrac{(x - 3)(x + 2)}{x - 3} = x + 2.$

Since $\lim_{x \to 3} f(x) = 5 \ne 4 = f(3), f(x)$ is not continuous at $x = 3$.

2. There is no need to compute any limits in order to answer this question. By Theorem 1, since $f(x)$ is not continuous at $x = 3$, it cannot possibly be differentiable there.

$$\lim_{x \to 3} f(x) = \lim_{x \to 3} (x + 2) = 5.$$

1.6 SOME RULES FOR DIFFERENTIATION

Three additional rules of differentiation greatly extend the number of functions that we can differentiate.

---

1. Constant-Multiple Rule

$$\frac{d}{dx}[k \cdot f(x)] = k \cdot \frac{d}{dx}[f(x)], \quad k \text{ a constant.}$$

2. Sum Rule

$$\frac{d}{dx}[f(x) + g(x)] = \frac{d}{dx}[f(x)] + \frac{d}{dx}[g(x)].$$

3. General Power Rule

$$\frac{d}{dx}([g(x)]^r) = r \cdot [g(x)]^{r-1} \cdot \frac{d}{dx}[g(x)].$$

We shall discuss these rules and then prove the first two.

The Constant-Multiple Rule   Starting with a function $f(x)$, we can multiply it by a constant number k in order to obtain a new function $k \cdot f(x)$. For instance, if $f(x) = x^2 - 4x + 1$ and $k = 2$, then

$$2f(x) = 2(x^2 - 4x + 1) = 2x^2 - 8x + 2.$$

The constant-multiple rule says that the derivative of the new function $k \cdot f(x)$ is just k times the derivative of the original function.* In other words, when faced with the differentiation of a constant times a function, simply carry along the constant and differentiate the function.

EXAMPLE 1   Calculate.

(a) $\dfrac{d}{dx}(2x^5)$ (b) $\dfrac{d}{dx}\left(\dfrac{x^3}{4}\right)$ (c) $\dfrac{d}{dx}\left(-\dfrac{3}{x}\right)$ (d) $\dfrac{d}{dx}(5\sqrt{x})$

Solution   (a) With $k = 2$ and $f(x) = x^5$, we have

$$\frac{d}{dx}(2 \cdot x^5) = 2 \cdot \frac{d}{dx}(x^5) = 2(5x^4) = 10x^4.$$

---

* More precisely, the constant-multiple rule asserts that if $f(x)$ is differentiable at $x = a$, then so is the function $k \cdot f(x)$, and the derivative of $k \cdot f(x)$ at $x = a$ may be computed using the given formula.

(b) Write $\dfrac{x^3}{4}$ in the form $\dfrac{1}{4} \cdot x^3$. Then

$$\frac{d}{dx}\left(\frac{x^3}{4}\right) = \frac{1}{4} \cdot \frac{d}{dx}(x^3) = \frac{1}{4}(3x^2) = \frac{3}{4}x^2.$$

(c) Write $-\dfrac{3}{x}$ in the form $(-3) \cdot \dfrac{1}{x}$. Then

$$\frac{d}{dx}\left(-\frac{3}{x}\right) = (-3) \cdot \frac{d}{dx}\left(\frac{1}{x}\right) = (-3) \cdot \frac{-1}{x^2} = \frac{3}{x^2}.$$

(d) $\dfrac{d}{dx}(5\sqrt{x}) = 5\dfrac{d}{dx}(\sqrt{x}) = 5\dfrac{d}{dx}(x^{1/2}) = \dfrac{5}{2}x^{-1/2}.$

This answer may also be written in the form $\dfrac{5}{2\sqrt{x}}$. ●

The Sum Rule   To differentiate a sum of functions, differentiate each function individually and add the derivatives together.* Another way of saying this is "the derivative of a sum of functions is the sum of the derivatives."

EXAMPLE 2   Find each of the following.

(a) $\dfrac{d}{dx}(x^3 + 5x)$ (b) $\dfrac{d}{dx}\left(x^4 - \dfrac{3}{x^2}\right)$ (c) $\dfrac{d}{dx}(2x^7 - x^5 + 8)$

Solution   (a) Let $f(x) = x^3$ and $g(x) = 5x$. Then

$$\frac{d}{dx}(x^3 + 5x) = \frac{d}{dx}(x^3) + \frac{d}{dx}(5x) = 3x^2 + 5.$$

(b) The sum rule applies to differences as well as sums (see Exercise 46). Indeed, by the sum rule,

$$\frac{d}{dx}\left(x^4 - \frac{3}{x^2}\right) = \frac{d}{dx}(x^4) + \frac{d}{dx}\left(-\frac{3}{x^2}\right) \qquad \text{(sum rule)}$$

$$= \frac{d}{dx}(x^4) - 3\frac{d}{dx}(x^{-2}) \qquad \text{(constant-multiple rule)}$$

$$= 4x^3 - 3(-2x^{-3})$$

$$= 4x^3 + 6x^{-3}.$$

After some practice, one usually omits most or all of the intermediate steps and simply writes

$$\frac{d}{dx}\left(x^4 - \frac{3}{x^2}\right) = 4x^3 + 6x^{-3}.$$

---

* More precisely, the sum rule asserts that if both $f(x)$ and $g(x)$ are differentiable at $x = a$, then so is $f(x) + g(x)$, and the derivative (at $x = a$) of the sum is then the sum of the derivatives (at $x = a$).

(c) We apply the sum rule repeatedly and use the fact that the derivative of a constant function is 0:

$$\frac{d}{dx}(2x^7 - x^5 + 8) = \frac{d}{dx}(2x^7) - \frac{d}{dx}(x^5) + \frac{d}{dx}(8)$$

$$= 2(7x^6) - 5x^4 + 0$$

$$= 14x^6 - 5x^4. \quad \bullet$$

Remark The differentiation of a function *plus* a constant is different from the differentiation of a constant *times* a function. Figure 1 shows the graphs of $f(x)$, $f(x) + 2$, and $2 \cdot f(x)$, where $f(x) = x^3 - \frac{3}{2}x^2$. For each x, the graphs of $f(x)$ and $f(x) + 2$ have the same slope. In contrast, for each x, the slope of the graph of $2 \cdot f(x)$ is twice the slope of the graph of $f(x)$. Upon differentiation, an added constant disappears, whereas a constant that multiplies a function is carried along.

FIGURE 1   Two effects of a constant on the graph of $f(x)$.

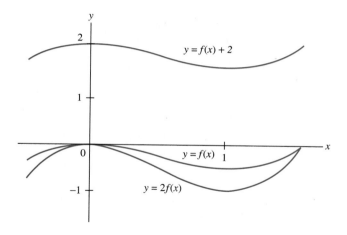

The General Power Rule Frequently, we will encounter expressions of the form $[g(x)]^r$—for instance, $(x^3 + 5)^2$, where $g(x) = x^3 + 5$ and $r = 2$. The general power rule says that, to differentiate $[g(x)]^r$, we must first treat $g(x)$ as if it were simply an x, form $r[g(x)]^{r-1}$, and then multiply it by a "correction factor" $g'(x)$.* Thus

$$\frac{d}{dx}(x^3 + 5)^2 = 2(x^3 + 5)^1 \cdot \frac{d}{dx}(x^3 + 5)$$

$$= 2(x^3 + 5) \cdot (3x^2)$$

$$= 6x^2(x^3 + 5).$$

---

*More precisely, the general power rule asserts that if $g(x)$ is differentiable at $x = a$ and if $g'(x)$ and $[g(x)]^{r-1}$ are both defined at $x = a$, then $[g(x)]^r$ is also differentiable at $x = a$ and its derivative is given by the formula stated.

In this special case it is easy to verify that the general power rule gives the correct answer. We first expand $(x^3 + 5)^2$ and then differentiate.

$$(x^3 + 5)^2 = (x^3 + 5)(x^3 + 5) = x^6 + 10x^3 + 25.$$

From the constant-multiple rule and the sum rule, we have

$$\frac{d}{dx}(x^3 + 5)^2 = \frac{d}{dx}(x^6 + 10x^3 + 25)$$

$$= 6x^5 + 30x^2 + 0$$

$$= 6x^2(x^3 + 5).$$

The two methods give the same answer.

Note that if we set $g(x) = x$ in the general power rule, we recover the power rule. So the general power rule contains the power rule as a special case.

EXAMPLE 3   Differentiate $\sqrt{1 - x^2}$.

Solution

$$\frac{d}{dx}(\sqrt{1 - x^2}) = \frac{d}{dx}[(1 - x^2)^{1/2}] = \frac{1}{2}(1 - x^2)^{-1/2} \cdot \frac{d}{dx}(1 - x^2)$$

$$= \frac{1}{2}(1 - x^2)^{-1/2} \cdot (-2x)$$

$$= \frac{-x}{(1 - x^2)^{1/2}} = \frac{-x}{\sqrt{1 - x^2}}. \quad \bullet$$

EXAMPLE 4   Differentiate $y = \dfrac{1}{x^3 + 4x}$.

Solution

$$y = \frac{1}{x^3 + 4x} = (x^3 + 4x)^{-1}.$$

$$\frac{dy}{dx} = (-1)(x^3 + 4x)^{-2} \cdot \frac{d}{dx}(x^3 + 4x)$$

$$= \frac{-1}{(x^3 + 4x)^2}(3x^2 + 4)$$

$$= -\frac{3x^2 + 4}{(x^3 + 4x)^2}. \quad \bullet$$

Proofs of the Constant-Multiple and Sum Rules   Let us verify both rules when x has the value a. Recall that if $f(x)$ is differentiable at $x = a$, then its derivative is the limit

$$\lim_{h \to 0} \frac{f(a + h) - f(a)}{h}.$$

Constant-Multiple Rule   We assume that $f(x)$ is differentiable at $x = a$.

We must prove that $k \cdot f(x)$ is differentiable at $x = a$ and that its derivative there is $k \cdot f'(a)$. This amounts to showing that the limit

$$\lim_{h \to 0} \frac{k \cdot f(a + h) - k \cdot f(a)}{h}$$

exists and has the value $k \cdot f'(a)$. However,

$$\lim_{h \to 0} \frac{k \cdot f(a + h) - k \cdot f(a)}{h}$$

$$= \lim_{h \to 0} k \left[\frac{f(a + h) - f(a)}{h} \right]$$

$$= k \cdot \lim_{h \to 0} \frac{f(a + h) - f(a)}{h} \qquad \text{(by Limit Theorem I)}$$

$$= k \cdot f'(a) \qquad [\text{since } f(x) \text{ is differentiable at } x = a],$$

which is what we desired to show.

Sum Rule We assume that both $f(x)$ and $g(x)$ are differentiable at $x = a$. We must prove that $f(x) + g(x)$ is differentiable at $x = a$ and that its derivative is $f'(a) + g'(a)$. That is, we must show that the limit

$$\lim_{h \to 0} \frac{[f(a + h) + g(a + h)] - [f(a) + g(a)]}{h}$$

exists and equals $f'(a) + g'(a)$. Using Limit Theorem III and the fact that $f(x)$ and $g(x)$ are differentiable at $x = a$, we have

$$\lim_{h \to 0} \frac{[f(a + h) + g(a + h)] - [f(a) + g(a)]}{h}$$

$$= \lim_{h \to 0} \left[\frac{f(a + h) - f(a)}{h} + \frac{g(a + h) - g(a)}{h} \right]$$

$$= \lim_{h \to 0} \frac{f(a + h) - f(a)}{h} + \lim_{h \to 0} \frac{g(a + h) - g(a)}{h}$$

$$= f'(a) + g'(a).$$

The general power rule will be proven as a special case of the chain rule in Chapter 3.

PRACTICE PROBLEMS 1.6

1. Find the derivative $\dfrac{d}{dx}(x)$.

2. Differentiate the function $y = \dfrac{x + (x^5 + 1)^{10}}{3}$.

EXERCISES 1.6

Differentiate.

1. $y = x^3 + x^2$

2. $y = x^2 + \dfrac{1}{x}$

3. $y = x^2 + 3x - 1$

4. $y = x^3 + 2x + 5$

5. $f(x) = x^5 + \dfrac{1}{x}$

6. $f(x) = x^8 - x$

7. $f(x) = x^4 + x^3 + x$

8. $f(x) = x^5 + x^2 - x$

9. $y = 3x^2$

10. $y = 2x^3$

11. $y = x^3 + 7x^2$

12. $y = -2x$

13. $y = \dfrac{4}{x^2}$

14. $y = 2\sqrt{x}$

15. $y = 3x - \dfrac{1}{x}$

16. $y = -x^2 + 3x + 1$

17. $f(x) = \frac{1}{3}x^3 - \frac{1}{2}x^2$

18. $f(x) = 100x^{100}$

19. $f(x) = -\dfrac{1}{5x^5}$

20. $f(x) = x^2 - \dfrac{1}{x^2}$

21. $f(x) = 1 - \sqrt{x}$

22. $f(x) = -3x^2 + 7$

23. $f(x) = (3x + 1)^{10}$

24. $f(x) = \dfrac{1}{x^2 + x + 1}$

25. $f(x) = 5\sqrt{3x^3 + x}$

26. $y = \dfrac{1}{(x^2 - 7)^5}$

27. $y = (2x^2 - x + 4)^6$

28. $y = \sqrt{-2x + 1}$

29. $y = \dfrac{x}{3} + \dfrac{3}{x}$

30. $y = \dfrac{2x - 1}{5}$

31. $y = \dfrac{2}{1 - 5x}$

32. $y = \dfrac{4}{3\sqrt{x}}$

33. $y = \dfrac{1}{1 - x^4}$

34. $y = \left(x^3 + \dfrac{x}{2} + 1\right)^5$

35. $f(x) = \dfrac{4}{\sqrt{x^2 + x}}$

36. $f(x) = \dfrac{6}{x^2 + 2x + 5}$

37. $f(x) = \left(\dfrac{\sqrt{x}}{2} + 1\right)^{3/2}$

38. $f(x) = \left(4 - \dfrac{2}{x}\right)^3$

Find the slope of the graph of $y = f(x)$ at the designated point.

39. $f(x) = 3x^2 - 2x + 1$, $(1, 2)$

40. $f(x) = x^{10} + 1 + \sqrt{1 - x}$, $(0, 2)$

41. Find the slope of the tangent line to the curve $y = x^3 + 3x - 8$ at $(2, 6)$.

42. Write the equation of the tangent line to the curve $y = x^3 + 3x - 8$ at $(2, 6)$.

43. Find the slope of the tangent line to the curve $y = (x^2 - 15)^6$ at $x = 4$. Then write the equation of this tangent line.

44. Find the equation of the tangent line to the curve $y = \dfrac{8}{x^2 + x + 2}$ at $x = 2$.

45. Differentiate the function $f(x) = (3x^2 + x - 2)^2$ in two ways.
    (a) Use the general power rule.
    (b) Multiply $3x^2 + x - 2$ by itself and then differentiate the resulting polynomial.

46. Using the sum rule and the constant-multiple rule, show that for any functions $f(x)$ and $g(x)$

$$\frac{d}{dx}[f(x) - g(x)] = \frac{d}{dx}f(x) - \frac{d}{dx}g(x).$$

47. Figure 2 contains the curves $y = f(x)$ and $y = g(x)$ and the tangent line to $y = f(x)$ at $x = 1$, with $g(x) = 3 \cdot f(x)$. Find $g(1)$ and $g'(1)$.

48. Figure 3 contains the curves $y = f(x)$, $y = g(x)$, and $y = h(x)$ and the tangent lines to $y = f(x)$ and $y = g(x)$ at $x = 1$, with $h(x) = f(x) + g(x)$. Find $h(1)$ and $h'(1)$.

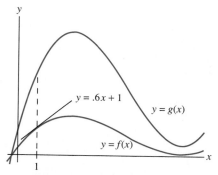

FIGURE 2   Graphs of $f(x)$ and $g(x) = 3f(x)$.

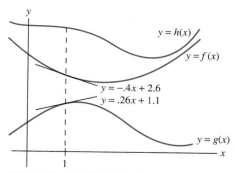

FIGURE 3   Graphs of $f(x)$, $g(x)$, and $h(x) = f(x) + g(x)$.

49. Suppose $f(5) = 2$, $f'(5) = 3$, $g(5) = 4$, and $g'(5) = 1$. Find $h(5)$ and $h'(5)$, where $h(x) = 3f(x) + 2g(x)$.

50. Suppose $g(3) = 2$ and $g'(3) = 4$. Find $f(3)$ and $f'(3)$, where $f(x) = 2 \cdot [g(x)]^3$.

51. Suppose $g(1) = 4$ and $g'(1) = 3$. Find $f(1)$ and $f'(1)$, where $f(x) = 5 \cdot \sqrt{g(x)}$.

52. The tangent line to the curve $y = x^3 - 6x^2 - 34x - 9$ has slope 2 at two points on the curve. Find the two points.

53. The tangent line to the curve $y = \frac{1}{3}x^3 - 4x^2 + 18x + 22$ is parallel to the line $6x - 2y = 1$ at two points on the curve. Find the two points.

54. In Fig. 4 the straight line is tangent to the parabola. Find the value of b.

55. In Fig. 5 the straight line is tangent to the graph of $f(x)$. Find $f(4)$ and $f'(4)$.

FIGURE 4

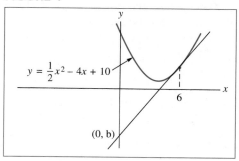

FIGURE 5

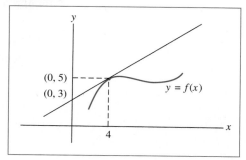

1. The problem asks for the derivative of the function $y = x$, a straight line of slope 1. Therefore, $\dfrac{d}{dx}(x) = 1$. The result can also be obtained from the power rule with $r = 1$. If $f(x) = x^1$, then $\dfrac{d}{dx}(f(x)) = 1 \cdot x^{1-1} = x^0 = 1$. See Fig. 6.

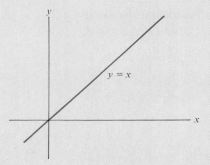

2. All three rules are required to differentiate this function.

$$\frac{dy}{dx} = \frac{d}{dx}\tfrac{1}{3} \cdot [x + (x^5 + 1)^{10}]$$

$$= \frac{1}{3}\frac{d}{dx}[x + (x^5 + 1)^{10}] \qquad \text{(constant-multiple rule)}$$

$$= \frac{1}{3}\left[\frac{d}{dx}(x) + \frac{d}{dx}(x^5 + 1)^{10}\right] \qquad \text{(sum rule)}$$

$$= \tfrac{1}{3}[1 + 10(x^5 + 1)^9 \cdot (5x^4)] \qquad \text{(general power rule)}$$

$$= \tfrac{1}{3}[1 + 50x^4(x^5 + 1)^9].$$

 ## 1.7 MORE ABOUT DERIVATIVES

In many applications it is convenient to use variables other than x and y. One might, for instance, study the function $f(t) = t^2$ instead of writing $f(x) = x^2$. In this case, the notation for the derivative involves t rather than x, but the concept of the derivative as a slope formula is unaffected. (See Fig. 1.) When the independent variable is t instead of x, we write $\dfrac{d}{dt}$ in place of $\dfrac{d}{dx}$. For instance,

$$\frac{d}{dt}(t^3) = 3t^2, \qquad \frac{d}{dt}(2t^2 + 3t) = 4t + 3.$$

Recall that if y is a function of x, say $y = f(x)$, then we may write $\dfrac{dy}{dx}$ in

FIGURE 1 The same function but different variables.

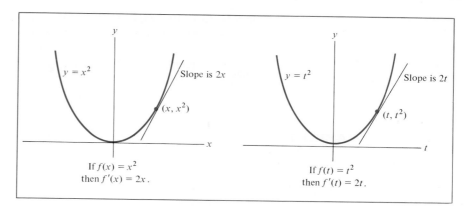

place of $f'(x)$. We sometimes call $\dfrac{dy}{dx}$ "the derivative of y with respect to x."

Similarly, if v is a function of t, then the derivative of v with respect to t is written as $\dfrac{dv}{dt}$. For example, if $v = 4t^2$, then $\dfrac{dv}{dt} = 8t$.

Of course, other letters can be used to denote variables. The formulas

$$\frac{d}{dP}(P^3) = 3P^2, \qquad \frac{d}{ds}(s^3) = 3s^2, \qquad \frac{d}{dz}(z^3) = 3z^2$$

all express the same basic fact that the slope formula for the cubic curve $y = x^3$ is given by $3x^2$.

EXAMPLE 1 Compute.

(a) $\dfrac{ds}{dp}$ if $s = 3(p^2 + 5p + 1)^{10}$ (b) $\dfrac{d}{dt}(at^2 + St^{-1} + S^2)$

Solution (a) $\dfrac{d}{dp} 3(p^2 + 5p + 1)^{10} = 30(p^2 + 5p + 1)^9 \cdot \dfrac{d}{dp}(p^2 + 5p + 1)$

$$= 30(p^2 + 5p + 1)^9(2p + 5).$$

(b) Although the expression $at^2 + St^{-1} + S^2$ contains several letters, the notation $\dfrac{d}{dt}$ indicates that all letters except t are to be considered as constants. Hence

$$\frac{d}{dt}(at^2 + St^{-1} + S^2) = \frac{d}{dt}(at^2) + \frac{d}{dt}(St^{-1}) + \frac{d}{dt}(S^2)$$

$$= a \cdot \frac{d}{dt}(t^2) + S \cdot \frac{d}{dt}(t^{-1}) + 0$$

$$= 2at - St^{-2}.$$

$$\left[\text{The derivative } \frac{d}{dt}(S^2) \text{ is zero because } S^2 \text{ is a constant.} \right] \bullet$$

The Second Derivative   When we differentiate a function $f(x)$, we obtain a new function $f'(x)$ that is a formula for the slope of the curve $y = f(x)$. If we differentiate the function $f'(x)$, we obtain what is called the *second derivative* of $f(x)$, denoted by $f''(x)$. That is,

$$\frac{d}{dx} f'(x) = f''(x).$$

EXAMPLE 2   Find the second derivatives of the following functions.

(a) $f(x) = x^3 + (1/x)$ (b) $f(x) = 2x + 1$ (c) $f(t) = t^{1/2} + t^{-1/2}$

Solution   (a) $f(x) = x^3 + (1/x) = x^3 + x^{-1}$

$f'(x) = 3x^2 - x^{-2}$

$f''(x) = 6x + 2x^{-3}.$

(b) $f(x) = 2x + 1$

$f'(x) = 2$ (a constant function whose value is 2)

$f''(x) = 0.$ (The derivative of a constant function is zero.)

(c) $f(t) = t^{1/2} + t^{-1/2}$

$f'(t) = \frac{1}{2}t^{-1/2} - \frac{1}{2}t^{-3/2}$

$f''(t) = -\frac{1}{4}t^{-3/2} + \frac{3}{4}t^{-5/2}.$ ●

The first derivative of a function $f(x)$ gives the slope of the graph of $f(x)$ at any point. The second derivative of $f(x)$ gives important additional information about the shape of the curve near any point. We shall examine this subject carefully in the next chapter.

Other Notation for Derivatives   Unfortunately, the process of differentiation does not have a standardized notation. Consequently, it is important to become familiar with alternative terminology.

If y is a function of x, say $y = f(x)$, then we may denote the first and second derivatives of this function in several ways.

Prime Notation	$\frac{d}{dx}$ Notation
$f'(x)$	$\frac{d}{dx} f(x)$
y'	$\frac{dy}{dx}$
$f''(x)$	$\frac{d^2}{dx^2} f(x)$
y''	$\frac{d^2 y}{dx^2}$

The notation $\dfrac{d^2}{dx^2}$ is purely symbolic. It reminds us that the second derivative is obtained by differentiating $\dfrac{d}{dx} f(x)$; that is,

$$f'(x) = \frac{d}{dx} f(x),$$

$$f''(x) = \frac{d}{dx} \left[\frac{d}{dx} f(x) \right].$$

If we evaluate the derivative $f'(x)$ at a specific value of x, say $x = a$, we get a number $f'(a)$ that gives the slope of the curve $y = f(x)$ at the point $(a, f(a))$. Another way of writing $f'(a)$ is

$$\frac{dy}{dx} \bigg|_{x=a}$$

If we have a second derivative $f''(x)$, then its value when $x = a$ is written

$$f''(a) \quad \text{or} \quad \frac{d^2y}{dx^2} \bigg|_{x=a}$$

EXAMPLE 3 If $y = x^4 - 5x^3 + 7$, find $\dfrac{d^2y}{dx^2} \bigg|_{x=3}$.

Solution

$$\frac{dy}{dx} = \frac{d}{dx} (x^4 - 5x^3 + 7) = 4x^3 - 15x^2$$

$$\frac{d^2y}{dx^2} = \frac{d}{dx} (4x^3 - 15x^2) = 12x^2 - 30x$$

$$\frac{d^2y}{dx^2} \bigg|_{x=3} = 12(3)^2 - 30(3) = 108 - 90 = 18. \; \bullet$$

EXAMPLE 4 If $s = t^3 - 2t^2 + 3t$, find

$$\frac{ds}{dt} \bigg|_{t=-2} \quad \text{and} \quad \frac{d^2s}{dt^2} \bigg|_{t=-2}$$

Solution

$$\frac{ds}{dt} = \frac{d}{dt} (t^3 - 2t^2 + 3t) = 3t^2 - 4t + 3$$

$$\frac{ds}{dt} \bigg|_{t=-2} = 3(-2)^2 - 4(-2) + 3 = 12 + 8 + 3 = 23.$$

To find the value of the second derivative at $t = -2$, we must first differentiate $\dfrac{ds}{dt}$.

$$\frac{d^2s}{dt^2} = \frac{d}{dt}(3t^2 - 4t + 3) = 6t - 4$$

$$\left.\frac{d^2s}{dt^2}\right|_{t=-2} = 6(-2) - 4 = -12 - 4 = -16. \quad \bullet$$

PRACTICE
PROBLEMS 1.7

1. Let $f(t) = t + (1/t)$. Find $f''(2)$.
2. Differentiate $g(r) = 2\pi rh$.

EXERCISES 1.7

Find the first derivatives.

1. $f(t) = (t^2 + 1)^5$

2. $f(P) = P^4 - P^3 + 4P^2 - P$

3. $v = \sqrt{2t - 1}$

4. $g(z) = (z^3 - z + 1)^2$

5. $y = (T^3 + 5T)^{2/3}$

6. $s = \sqrt{t} + \dfrac{1}{\sqrt{t}}$

7. Find $\dfrac{d}{dP}(3P^2 - \frac{1}{2}P + 1)$.

8. Find $\dfrac{d}{dz}(\sqrt{z^2 - 1})$.

9. Find $\dfrac{d}{dt}(a^2t^2 + b^2t + c^2)$.

10. Find $\dfrac{d}{dx}(x^3 + t^3)$.

Find the first and second derivatives.

11. $f(x) = \frac{1}{2}x^2 - 7x + 2$

12. $y = \dfrac{1}{x^2} + 1$

13. $y = \sqrt{x}$

14. $f(t) = t^{100} + t + 1$

15. $f(r) = \pi hr^2 + 2\pi r$

16. $v = t^{3/2} + t$

17. $g(x) = 2 - 5x$

18. $V(r) = \frac{4}{3}\pi r^3$

19. $f(P) = (3P + 1)^5$

20. $u = \dfrac{t^6}{30} - \dfrac{t^4}{12}$

Compute the following.

21. $\dfrac{d}{dx}(2x^2 - 3)\bigg|_{x=5}$

22. $\dfrac{d}{dt}(1 - 2t - 3t^2)\bigg|_{t=-1}$

23. $\dfrac{d}{dz}(z^2 - 4)^3 \bigg|_{z=1}$

24. $\dfrac{d}{dT}\left(\dfrac{1}{3T + 1}\right)\bigg|_{T=2}$

25. $\dfrac{d^2}{dx^2}(3x^3 - x^2 + 7x - 1)\bigg|_{x=2}$

26. $\dfrac{d}{dt}\left(\dfrac{dv}{dt}\right)$, where $v = 2t^{-3}$

27. $\dfrac{d}{dP}\left(\dfrac{dy}{dP}\right)$, where $y = \dfrac{k}{2P - 1}$

28. $\dfrac{d^2V}{dr^2}\bigg|_{r=2}$, where $V = ar^3$

29. $f'(3)$ and $f''(3)$, when $f(x) = \sqrt{10 - 2x}$

30. $g'(2)$ and $g''(2)$, when $g(T) = (3T - 5)^{10}$

31. Suppose a company finds that the revenue R generated by spending x dollars on advertising is given by $R = 1000 + 80x - .02x^2$, for $0 \le x \le 2000$. Find $\dfrac{dR}{dx}\bigg|_{x=1500}$.

32. A supermarket finds that its average daily volume of business V (in thousands of dollars) and the number of hours t the store is open for business each day are approximately related by the formula

$$V = 20\left(1 - \dfrac{100}{100 + t^2}\right), \qquad 0 \le t \le 24.$$

Find $\dfrac{dV}{dt}\bigg|_{t=10}$.

33. The *third derivative* of a function $f(x)$ is the derivative of the second derivative $f''(x)$ and is denoted by $f'''(x)$. Compute $f'''(x)$ for the following functions.
    (a) $f(x) = x^5 - x^4 + 3x$ (b) $f(x) = 4x^{5/2}$

34. Compute the third derivatives of the following functions.
    (a) $f(t) = t^{10}$ (b) $f(z) = \dfrac{1}{z + 5}$

SOLUTIONS TO PRACTICE PROBLEMS 1.7

1. $f(t) = t + t^{-1}$

$f'(t) = 1 + (-1)t^{(-1)-1} = 1 - t^{-2}$

$f''(t) = -(-2)t^{(-2)-1} = 2t^{-3} = 2t^3$

Therefore, $f''(2) = \dfrac{2}{2^3} = \dfrac{1}{4}$. [*Note:* It is essential first to compute the function $f''(t)$ and *then* to evaluate the function at $t = 2$.]

2. The expression $2\pi rh$ contains two numbers, 2 and π, and two letters, r and h. The notation $g(r)$ tells us that the expression $2\pi rh$ is to be regarded as a function of r. Therefore, h—and hence $2\pi h$—is to be treated as a constant, and differentiation is done with respect to the variable r. That is,

$$g(r) = (2\pi h)r$$

$$g'(r) = 2\pi h.$$

1.8 THE DERIVATIVE AS A RATE OF CHANGE

An important interpretation of the derivative of a function is as a rate of change. In this section, we examine this interpretation and discuss some of the applications in which it proves useful. The first step is to understand what is meant by "average rate of change."

If a function $y = f(x)$ has a derivative at $x = a$, then the difference quotient

$$\frac{f(a + h) - f(a)}{h} \tag{1}$$

approaches $f'(a)$ as $h \to 0$. It will be convenient in this section to rewrite (1) using the Greek letter Δ (delta), a common notation for change in a function or variable. The numerator, $f(a + h) - f(a)$, represents the change in $y = f(x)$ as x goes from a to $a + h$, and we write

$$\Delta y = f(a + h) - f(a).$$

The denominator, h, in (1) gives the change as x goes from a to $a + h$, and we write

$$\Delta x = (a + h) - a = h.$$

Therefore, quotient (1) is

$$\frac{[\text{change in } y = f(x) \text{ from } a \text{ to } a + \Delta x]}{[\text{change in } x \text{ from } a \text{ to } a + \Delta x]} = \frac{\Delta y}{\Delta x}. \tag{2}$$

We call this quotient (or ratio) *the average rate of change of $f(x)$ with respect to x over the interval from a to $a + \Delta x$.* Geometrically, this quotient is the slope of the secant line in Fig. 1.

FIGURE 1   Average rate of change is the slope of the secant line.

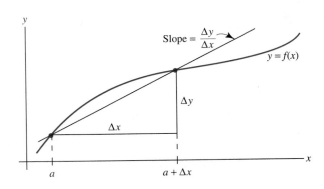

From the secant-line calculation of the derivative, we know that as h approaches 0 in (1)—or, equivalently, as Δx approaches 0 in (2)—the slope of the secant line approaches $f'(a)$. Thus the average rate of change approaches $f'(a)$. For this reason, we may interpret $f'(a)$ as the ("*instantaneous*") rate of change of $y = f(x)$ exactly at the point where $x = a$.

The derivative $f'(a)$ measures the rate of change of $f(x)$ at $x = a$.

As a physical illustration of rate of change, consider the weight of a bear cub as a function of time, say $W = f(t)$. Suppose the bear's weight follows the growth pattern in Fig. 2, and examine the time interval from $t = 3$ to $t = 7$ months. Then

$$\Delta t = 7 - 3 = 4 \text{ months.}$$

According to the graph, the weight changed during this time period from 9 to 33 pounds, so

$$\Delta W = 33 - 9 = 24 \text{ pounds.}$$

A growth of 24 pounds in 4 months amounts to an average gain of 6 pounds per month. That is,

$$\frac{\Delta W}{\Delta t} = \frac{24}{4} = 6 \text{ pounds per month.}$$

This quotient is the slope of the secant line in Fig. 2 through the points $(3, 9)$ and $(7, 33)$.

FIGURE 2   Weight gain of a bear cub.

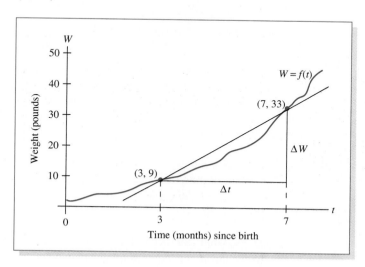

Suppose we consider the bear cub's weight over a small time interval from $t = 3$ to $t = 3 + \Delta t$. As the length of this interval approaches zero, the average rate of weight change, $\Delta W / \Delta t$, approaches $f'(3)$, the derivative of the weight function. Thus it is appropriate to call $f'(3)$ the rate of change of weight *at* $t = 3$.

EXAMPLE 1   Suppose that $f(x) = x^2$.

   (a) Calculate the average rate of change of $f(x)$ over the intervals 1 to 2, 1 to 1.1, and 1 to 1.01.

   (b) Determine the (instantaneous) rate of change of $f(x)$ when $x = 1$.

Solution   (a) The intervals are of the form 1 to $1 + \Delta x$ for $\Delta x = 1$, .1, and .01. The average rate of change is given by the ratio

$$\frac{\Delta y}{\Delta x} = \frac{f(1 + \Delta x) - f(1)}{\Delta x} = \frac{(1 + \Delta x)^2 - 1^2}{\Delta x}.$$

For the three given values of Δx, this expression has the following respective values:

$$\Delta x = 1: \qquad \frac{\Delta y}{\Delta x} = \frac{2^2 - 1^2}{1} = \frac{4 - 1}{1} = \frac{3}{1}.$$

$$\Delta x = .1: \qquad \frac{\Delta y}{\Delta x} = \frac{(1.1)^2 - 1^2}{.1} = \frac{1.21 - 1}{.1} = \frac{.21}{.1} = 2.1.$$

$$\Delta x = .01: \qquad \frac{\Delta y}{\Delta x} = \frac{(1.01)^2 - 1^2}{.01} = \frac{1.0201 - 1}{.01} = \frac{.0201}{.01} = 2.01.$$

Thus, the average rate of change of $f(x)$ over the interval 1 to 2 is 3 units per unit change in x. The average rate of change of $f(x)$ over the interval 1 to 1.1 is 2.1 units per unit change in x. The average rate of change of $f(x)$ over the interval 1 to 1.01 is 2.01 units per unit change in x.

   (b) The instantaneous rate of change of $f(x)$ at $x = 1$ is equal to $f'(1)$. We have

$$f'(x) = 2x$$

$$f'(1) = 2 \cdot 1 = 2.$$

That is, the instantaneous rate of change is 2 units per unit change in x. Note how the average rates of change approach the instantaneous rate of change as the intervals beginning at $x = 1$ shrink. ●

EXAMPLE 2   A flu epidemic hits a midwestern town. Public Health officials estimate that the number of persons sick with the flu at time t (measured in days from the beginning of the epidemic) is approximated by $P(t) = 60t^2 - t^3$, provided that $0 \le t \le 40$.

   (a) At what rate is the flu spreading when $t = 20$?

   (b) When is the flu spreading at the rate of 900 people per day?

Solution   The rate at which the flu spreads is given by the rate of change of $P(t)$, that is, by the derivative

$$P'(t) = 120t - 3t^2 \qquad (0 \le t \le 40).$$

Since $P(t)$ is measured in people and time is measured in days, the rate $P'(t)$ is measured in people per day.

(a) When $t = 20$,

$$P'(20) = 120(20) - 3(20)^2 = 1200.$$

Thus 20 days after the beginning of the epidemic, the flu is spreading at the rate of 1200 people per day.

(b) In this case we are given the rate of change of $P(t)$ and we must find the time corresponding to that rate. We set the expression for $P'(t)$ equal to 900 and solve for t:

$$120t - 3t^2 = 900$$

$$-3t^2 + 120t - 900 = 0.$$

Dividing by -3 and factoring, we have

$$t^2 - 40t + 300 = 0$$

$$(t - 10)(t - 30) = 0.$$

Then $t = 10$ or $t = 30$. At both times the flu is spreading at the rate of 900 people per day. ●

EXAMPLE 3   A common clinical procedure for studying a person's calcium metabolism (the rate at which the body assimilates and uses calcium) is to inject some chemically "labeled" calcium into the bloodstream and then measure how fast this calcium is removed from the blood by the person's bodily processes. Suppose that t days after an injection of calcium, the amount A of the labeled calcium remaining in the blood is $A = t^{-3/2}$ for $t \geq .5$, where A is measured in suitable units.* See Fig. 3. How fast (in units of calcium per day) is the body removing calcium from the blood when $t = 1$ day?

Solution   The rate of change (per day) of calcium in the blood is given by the derivative

$$\frac{dA}{dt} = -\tfrac{3}{2}t^{-5/2}.$$

When $t = 1$, this rate equals

$$\frac{dA}{dt}\bigg|_{t=1} \quad -\tfrac{3}{2}(1)^{-5/2} = -\tfrac{3}{2}.$$

The amount of calcium in the blood is changing at the rate of $-\tfrac{3}{2}$ units per day when $t = 1$. The negative sign indicates that the amount of calcium is decreasing rather than increasing. ●

_____

*For a discussion of this mathematical model, see J. Defares, I. Sneddon, and M. Wise, *An Introduction to the Mathematics of Medicine and Biology* (Chicago: Year Book Publishers, Inc., 1973), pp. 609–619.

FIGURE 3   Labeled calcium in the blood.

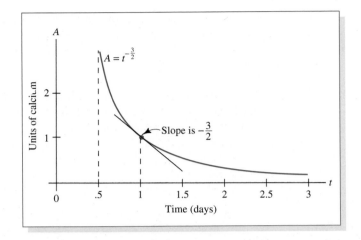

Approximating the Change in a Function   Consider the function $f(x)$ for x near a. As we have just seen, we have the approximation

$$\frac{f(a + h) - f(a)}{h} \approx f'(a).$$

Multiplying both sides of this approximation by h, we have

$$f(a + h) - f(a) \approx f'(a) \cdot h \tag{3}$$

Equivalently, with Δx in place of h,

$$f(a + \Delta x) - f(a) \approx f'(a) \cdot \Delta x \tag{4}$$

If x changes from a to $a + \Delta x$, then the change in the function value is approximately $f'(a)$ times the change Δx in the value of x.

FIGURE 4   Change in y along the tangent line and along the graph of y = f(x).

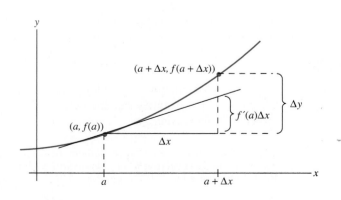

Figure 4 contains a geometric interpretation of (4). Given the small change Δx in x, the quantity $f'(a)\cdot\Delta x$ gives the corresponding change in y *along the tangent line at* $(a, f(a))$. In contrast, the quantity $\Delta y = f(a + \Delta x) - f(a)$ gives the change in y *along the curve* $y = f(x)$.

EXAMPLE 4 Let the production function $p(x)$ give the number of units of goods produced when employing x units of labor. Suppose 5000 units of labor are currently employed, $p(5000) = 300$, and $p'(5000) = 2$. Estimate the number of additional units of goods produced when employing:

(a) One additional unit of labor.
(b) An additional $\frac{1}{2}$ unit of labor.
(c) One less unit of labor.

Solution (a) Here $\Delta x = 1$. By (4), the change in $p(x)$ will be approximately

$$p'(5000)\cdot\Delta x = 2\cdot 1 = 2.$$

About two additional units will be produced.
(b) Here $\Delta x = \frac{1}{2}$. The change in $p(x)$ will be approximately

$$p'(5000)\cdot\tfrac{1}{2} = 2\cdot\tfrac{1}{2} = 1.$$

About one additional unit will be produced.
(c) Here $\Delta x = -1$, since the amount of labor is reduced. The change in $p(x)$ will be approximately

$$p'(5000)\cdot(-1) = 2\cdot(-1) = -2.$$

About two fewer units of goods will be produced. ●

An alternative form of (4) is obtained by adding $f(a)$ to both sides of the approximation

$$f(a + \Delta x) \approx f(a) + f'(a)\cdot\Delta x \tag{5}$$

If $f(a)$ is known, then the value of $f(a + \Delta x)$ may be estimated by adding $f'(a)\cdot\Delta x$ to $f(a)$. The error in this estimate is small if Δx is small enough.

EXAMPLE 5 Use approximation (5) to estimate $\sqrt{4.05}$.

Solution Here $f(x) = \sqrt{x}$. Since 4.05 is very close to 4, we use approximation (5) with $a = 4$, $\Delta x = .05$, and $a + \Delta x = 4.05$.

$$f(4.05) \approx f(4) + f'(4)\cdot(.05).$$

Now, $f(4) = \sqrt{4} = 2$ and $f'(4) = 1/(2\sqrt{4}) = \frac{1}{4}$. Therefore,

$$\sqrt{4.05} = f(4.05) \approx 2 + \tfrac{1}{4}\cdot(.05) = 2.0125. ●$$

The Marginal Concept in Economics   Suppose a company determines that the cost of producing x units of its product is $C(x)$ dollars. Recall from Section 1.3 that $C'(x)$ is called the marginal cost function. The value of the derivative of $C(x)$ at $x = a$—that is, $C'(a)$—is called the *marginal cost at production level a*. Since the marginal cost is just a derivative, its value gives the rate at which costs are increasing with respect to the level of production, assuming that production is at level a. If we apply (3) to the cost function $C(x)$ and take h to be 1 unit, we have

$$C(a + 1) - C(a) \approx C'(a) \cdot 1 = C'(a). \tag{6}$$

The quantity $C(a + 1) - C(a)$ is the amount the cost rises when the production level is increased from a units to $a + 1$ units. See Fig. 5. Economists interpret (6) by saying that the marginal cost is approximately the cost of producing one additional unit.

FIGURE 5   Approximating the change in a cost function.

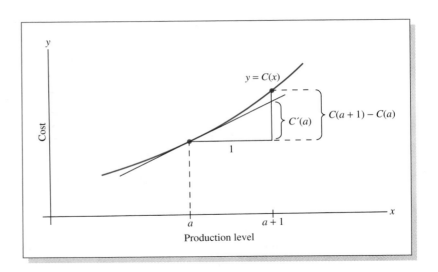

EXAMPLE 6   Suppose that the cost function is $C(x) = .005x^3 - 3x$ and production is proceeding at 1000 units per day.

(a) What is the extra cost of increasing production from 1000 to 1001 units per day?

(b) What is the marginal cost when $x = 1000$?

Solution   (a) The change in cost when production is raised from 1000 to 1001 units per day is $C(1001) - C(1000)$, which equals (rounded to the nearest dollar)

$$[.005(1001)^3 - 3(1001)] - [.005(1000)^3 - 3(1000)]$$

$$= 5,012,012 - 4,997,000$$

$$= 15,012.$$

(b) The marginal cost at production level 1000 is $C'(1000)$.

$$C'(x) = .015x^2 - 3$$

$$C'(1000) = 14,997.$$

Notice that 14,997 is close to the actual cost in (a) of increasing production by one unit. ●

Velocity and Acceleration   An everyday illustration of rate of change is given by the velocity of a moving object. Suppose that we are driving a car along a straight road and at each time t we let $s(t)$ be our position on the road, measured from some convenient reference point. See Fig. 6, where distances are positive to the right of the reference point and negative to the left. We call $s(t)$ the *position function* of the car. For the moment we shall assume that we are proceeding in only the positive direction along the road.

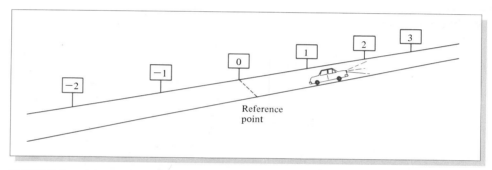

FIGURE 6   Position of a car traveling on a straight road.

At any instant, the car's speedometer tells us how fast we are moving— that is, how fast our position $s(t)$ is changing. To show how the speedometer reading is related to our calculus concept of a derivative, let us examine what is happening at a specific time, say $t = 1$. Consider a short time interval of duration h from $t = 1$ to $t = 1 + h$. Our car will move from position $s(1)$ to position $s(1 + h)$, a distance of $s(1 + h) - s(1)$. Thus the *average velocity from $t = 1$ to $t = 1 + h$* is

$$\frac{[\text{distance traveled}]}{[\text{time elapsed}]} = \frac{s(1 + h) - s(1)}{h}. \tag{7}$$

If the car is traveling at a steady speed during this time period, then the speedometer reading will equal the average velocity in (7).

From our discussion in Section 1.3 the ratio (7) approaches the derivative $s'(1)$ as h approaches zero. For this reason we call $s'(1)$ *the* (instantaneous) *velocity at $t = 1$*. This number will agree with the speedometer reading at $t = 1$ because when h is very small, the car's speed will be nearly steady over the time interval from $t = 1$ to $t = 1 + h$, and so the average velocity over this time interval will be nearly the same as the speedometer reading at $t = 1$.

The reasoning used for $t = 1$ holds for an arbitrary t as well. Thus the following definition makes sense:

> If $s(t)$ denotes the position function of an object moving in a straight line, then the velocity $v(t)$ of the object at time t is given by
> $$v(t) = s'(t).$$

In our discussion we assumed that the car moved in the positive direction. If the car moves in the opposite direction, the ratio (7) and the limiting value $s'(1)$ will be negative. So we interpret negative velocity as movement in the negative direction along the road.

The derivative of the velocity function $v(t)$ is called the *acceleration* function and is often written as $a(t)$:

> $$a(t) = v'(t).$$

Since $v'(t)$ measures the rate of change of the velocity $v(t)$, this use of the word acceleration agrees with our common usage in connection with automobiles. Note that since $v(t) = s'(t)$, the acceleration is actually the second derivative of the position function $s(t)$:

$$a(t) = s''(t).$$

EXAMPLE 7   When a ball is thrown into the air, its position may be measured as the vertical distance from the ground rather than the distance from some reference point. Regard "up" as the positive direction, and let $s(t)$ be the height of the ball in feet after t seconds. Suppose that $s(t) = -16t^2 + 128t + 5$.

(a) What is the velocity after 2 seconds?

(b) What is the acceleration after 2 seconds?

(c) At what time is the velocity -32 feet per second? (The negative sign indicates that the ball's height is decreasing; that is, the ball is falling.)

(d) When is the ball at a height of 117 feet?

Solution   (a) The velocity is the rate of change of the height function, so
$$v(t) = s'(t) = -32t + 128.$$

The velocity when $t = 2$ is $v(2) = -32(2) + 128 = 64$ feet per second.

(b) $a(t) = v'(t) = -32$. The acceleration is -32 feet per second for all t. This constant acceleration is due to the downward (and therefore negative) force of gravity.

(c) Since the velocity is given and the time is unknown, we set $v(t) = -32$ and solve for t:

$$-32t + 128 = -32$$
$$-32t = -160$$
$$t = 5.$$

The velocity is -32 feet per second when t is 5 seconds.

(d) The question here involves the height function, not the velocity. Since the height is given and the time is unknown, we set $s(t) = 117$ and solve for t:

$$-16t^2 + 128t + 5 = 117$$
$$-16(t^2 - 8t + 7) = 0$$
$$-16(t - 1)(t - 7) = 0.$$

The ball is at a height of 117 feet when $t = 1$ and $t = 7$ seconds. ●

TECHNOLOGY PROJECT 1.6
The Rate of Change of Baseball Salaries

Let y denote the average salary of a baseball player (in thousands of dollars) in year x (where years are measured with 1982 corresponding to 0). Throughout the 1980s and early 1990s baseball salaries increased steeply. A mathematical model for these salaries is

$$y = 246 + 64x - 8.9x^2 + 0.95x^3$$

1. Graph y as a function of x.
2. When was the average salary $300,000?
3. What was the average salary in 1990?
4. By how much did the average salary increase from 1990 to 1991?
5. Compute the derivative y'.
6. Use the derivative to estimate the amount that the average salary increased from 1990 to 1991.
7. Compare the answers derived in 4 and 6.

TECHNOLOGY PROJECT 1.7
Analysis of a Falling Ball

A ball thrown straight up into the air has height $s(t) = 102t - 16t^2$ feet after t seconds.

1. Use a graphing calculator to display the graphs of $s(t)$, $s'(t)$, $s''(t)$ for $0 \le t \le 10$. Use these graphs to answer the remaining questions.
2. How high is the ball after 2 seconds?
3. When, during descent, is the height 110 feet?
4. What is the velocity after 6 seconds?
5. When is the velocity 70 feet per second?

TECHNOLOGY PROJECT 1.8
Analysis of a Psychology Experiment

In a psychology experiment*, people improved their ability to recognize common verbal and semantic information with practice. Their judgement time after t days of practice was $f(t) = .36 + .77(t - .5)^{-.36}$.

1. Display the graphs of $f(t), f'(t), f''(t)$ for $.5 \leq t \leq 10$. Use these graphs to answer the following questions.
2. What was the judgement time after 2 days of practice?
3. After how many days of practice was the judgement time about .8 seconds?
4. After 2 days of practice, at what rate was judgement time changing with respect to days of practice?
5. After how many days was judgement time changing at the rate of $-.08$ seconds per day of practice?

PRACTICE PROBLEMS 1.8

Let $f(t)$ be the temperature (in degrees Celsius) of a liquid at time t (in hours). The rate of temperature change at time t is determined from $f'(t)$. Listed next are typical questions about $f(t)$ and $f'(t)$. Match each question with the proper method of solution.

Questions:
1. What is the temperature of the liquid after 6 hours?
2. When is the temperature rising at the rate of 6 degrees per hour?
3. By how many degrees did the temperature rise during the first 6 hours?
4. When is the liquid's temperature only 6°?
5. How fast is the temperature of the liquid changing after 6 hours?
6. What is the average rate of increase in the temperature during the first 6 hours?

Methods of Solutions:
(a) Compute $f(6)$.   (b) Set $f(t) = 6$ and solve for t.
(c) Compute $[f(6) - f(0)]/6$.   (d) Compute $f'(6)$.
(e) Set $f'(t) = 6$ and solve for t.   (f) Compute $f(6) - f(0)$.

EXERCISES 1.8

1. Suppose that $f(x) = 4x^2$.
   (a) What is the average rate of change of $f(x)$ over each of the intervals 0 to 2, 0 to 1, and 0 to .5?
   (b) What is the (instantaneous) rate of change of $f(x)$ when $x = 0$?

---

*John R. Anderson, "Automaticity and the ACT Theory," *American Journal of Psychology*, 105: 2 (Summer 1992), 165–180.

2. Suppose that $f(x) = -6/x$.

   (a) What is the average rate of change of $f(x)$ over each of the intervals 1 to 2, 1 to 1.5, and 1 to 1.2?

   (b) What is the (instantaneous) rate of change of $f(x)$ when $x = 1$?

3. Suppose that $f(t) = t^2 + 3t - 7$.

   (a) What is the average rate of change of $f(t)$ over the interval 5 to 6?

   (b) What is the (instantaneous) rate of change of $f(t)$ when $t = 5$?

4. Suppose that $f(t) = 3t + 2 - \dfrac{12}{t}$.

   (a) What is the average rate of change of $f(t)$ over the interval 2 to 3?

   (b) What is the (instantaneous) rate of change of $f(t)$ when $t = 2$?

5. An analysis of the daily output of a factory assembly line shows that about $60t + t^2 - \frac{1}{12}t^3$ units are produced after t hours of work, $0 \le t \le 8$. What is the instantaneous rate of production (in units per hour) when $t = 2$?

6. Liquid is pouring into a large vat. After t hours, there are $5t - t^{1/2}$ gallons in the vat. At what instantaneous rate is the liquid flowing into the vat (in gallons per hour) when $t = 4$?

7. Suppose that the weight in grams of a cancerous tumor at time t is $W(t) = .1t^2$, where t is measured in weeks.

   (a) What is the instantaneous rate of growth of the tumor (in grams per week) when $t = 5$?

   (b) At what time is the tumor growing at the instantaneous rate of 5 grams per week?

8. After an advertising campaign, the sales of a product often increase and then decrease. Suppose that t days after the end of the advertising, the daily sales are $-3t^2 + 32t + 100$ units.

   (a) At what rate (in units per day) are the sales changing when $t = 2$?

   (b) When will the sales be increasing at the rate of 2 units per day?

9. A sewage treatment plant accidentally discharged untreated sewage into a lake for a few days. This temporarily decreased the amount of dissolved oxygen in the lake. Let $f(t)$ be the amount of oxygen in the lake (measured in suitable units) t days after the sewage started flowing into the lake. See Fig. 7. Experimental data suggest that

FIGURE 7 **Recovery of a lake from pollution.**

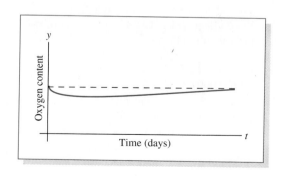

$f(t)$ is given approximately by

$$f(t) = 1 - \frac{10}{t + 10} + \frac{100}{(t + 10)^2}.$$

(a) Find the rate of change (in units per day) of the oxygen content of the lake at $t = 5$ and at $t = 15$.

(b) Is the oxygen content increasing or decreasing when $t = 15$?

10. Flu is spreading through a small school. At time t days after the beginning of the epidemic, there are $P(t)$ students sick, where $P(t) = 20t - t^2$.

(a) At what rate is the flu spreading when $t = 1$?

(b) How many students are sick when the flu is spreading at the rate of 8 students per day?

In Exercises 11–14, decide whether the statement is true or false. [Refer to approximation (4).]

11. If $x = a$ and $f'(a)$ is large and positive, then a small increase in x causes a large increase in the function value.

12. If $x = a$ and $f'(a)$ is small and positive, then a small increase in x causes a very small increase in the function value.

13. If $x = a$ and $f'(a)$ is small and negative, then a small increase in x causes a very small decrease in the function value.

14. If $x = a$ and $f'(a)$ is large and negative, then a small increase in x causes a large decrease in the function value.

15. Suppose $f(100) = 5000$ and $f'(100) = 10$. Estimate each of the following.

(a) $f(101)$ (b) $f(100.5)$ (c) $f(99)$ (d) $f(98)$

(e) $f(99.75)$

16. Suppose $f(25) = 10$ and $f'(25) = -2$. Estimate each of the following.

(a) $f(27)$ (b) $f(26)$ (c) $f(25.25)$ (d) $f(24)$

(e) $f(23.5)$

17. Let $C(x)$ be the cost (in dollars) of manufacturing x radios. Interpret the statements $C(2000) = 50,000$ and $C'(2000) = 10$.

18. Let $P(x)$ be the profit (in dollars) from manufacturing and selling x luxury cars. If $P(100) = 90,000$ and $P'(100) = 1200$, estimate the profit from 99 cars.

19. Let $f(t)$ be the temperature of a cup of coffee t minutes after it is poured. Interpret the statements $f(3) = 170$ and $f'(3) = -5$.

20. Price affects sales. Let $f(p)$ be the number of cars sold when the price is p dollars per car. Interpret the statements $f(10,000) = 200,000$ and $f'(10,000) = -3$.

21. Advertising affects sales. Let $f(x)$ be the number of toys sold when x dollars is spent on advertising. Interpret the statements $f(100,000) = 3,000,000$ and $f'(100,000) = 30$.

22. Suppose $y = f(x)$ and $f'(5) = 3$. Elaborate on the statement "Near $x = 5$, y is changing three times as fast as x."

23. A market finds that if it prices a certain product so as to sell x units each week, then the revenue received will be approximately $2x - .001x^2$ dollars.

(a) Find the marginal revenue at a sales level of 600 units.

(b) At what sales level will the marginal revenue be $1.10 per unit?

24. Suppose the revenue from producing (and selling) x units of a product is given by $R(x) = .01x^2 - 3x$ dollars.

    (a) Find the marginal revenue at a production level of 1800.

    (b) Find the production level where the revenue is $1800.

25. A manufacturer estimates that the hourly cost of producing x units of a product on an assembly line is $.1x^3 - 6x^2 + 136x + 200$ dollars.

    (a) Compute $C(21) - C(20)$, the extra cost of raising the production from 20 to 21 units.

    (b) Find the marginal cost when the production level is 20 units.

26. Suppose that the profit from producing x units of a product is given by $P(x) = .003x^3 + .01x$ dollars.

    (a) Compute the additional profit gained from increasing sales from 100 to 101 units.

    (b) Find the marginal profit at a production level of 100 units.

27. Let $s(t)$ be the height (in feet) after t seconds of a ball thrown straight up into the air. Match each question with the proper solution.

    Questions: A. What will be the velocity of the ball after 3 seconds?

    　　　　　　 B. When will the velocity be 3 feet per second?

    　　　　　　 C. What is the average velocity during the first 3 seconds?

    　　　　　　 D. When will the ball be 3 feet above the ground?

    　　　　　　 E. When will the ball hit the ground?

    　　　　　　 F. How high will the ball be after 3 seconds?

    　　　　　　 G. How far did the ball travel during the first 3 seconds?

    Solutions: a. Set $s(t) = 0$ and solve for t.

    　　　　　　 b. Compute $s'(3)$.

    　　　　　　 c. Compute $s(3)$.

    　　　　　　 d. Set $s(t) = 3$ and solve for t.

    　　　　　　 e. Set $s'(t) = 3$ and solve for t.

    　　　　　　 f. Compute $[s(3) - s(0)]/3$.

    　　　　　　 g. Compute $s(3) - s(0)$.

28. Let $P(x)$ be the profit from producing (and selling) x units of goods. Match each question with the proper solution.

    Questions: A. What is the profit from producing 1000 units of goods?

    　　　　　　 B. For what level of production will the marginal profit be 1000 dollars?

    　　　　　　 C. What is the marginal profit from producing 1000 units of goods?

    　　　　　　 D. For what level of production will the profit be 1000 dollars?

    Solutions: a. Compute $P'(1000)$.

    　　　　　　 b. Set $P'(x) = 1000$ and solve for x.

    　　　　　　 c. Set $P(x) = 1000$ and solve for x.

    　　　　　　 d. Compute $P(1000)$.

29. Table 1 gives a car's trip-meter reading (in miles) at 1 hour into a trip and at several nearby times. What is the average speed during the time interval from 1 to 1.05 hours? Estimate the speed at time 1 hour into the trip.

TABLE 1	Trip-meter Readings at Several Times								
Time	.96	.97	.98	.99	1	1.01	1.02	1.03	1.04	1.05
Trip meter	43.2	43.7	44.2	44.6	45	45.4	45.8	46.3	46.8	47.4

30. A car is traveling from New York to Boston and is halfway between the two cities. Let $s(t)$ be the distance from New York during the next hour. Match each behavior with the corresponding graph of $s(t)$ in Fig. 8.
    (a) The car travels at a steady speed.
    (b) The car is stopped.
    (c) The car is backing up.

FIGURE 8 Possible graphs of $s(t)$.

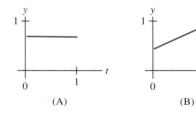 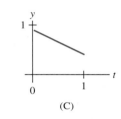

31. A toy rocket fired straight up into the air has height $s(t) = 160t - 16t^2$ feet after t seconds.
    (a) What is the rocket's initial velocity (when $t = 0$)?
    (b) What is the velocity after 2 seconds?
    (c) What is the acceleration when $t = 3$?
    (d) At what time will the rocket hit the ground?
    (e) At what velocity will the rocket be traveling just as it smashes into the ground?
32. Suppose that the position of a car at time t is given by $s(t) = 50t - 7/(t + 1)$, where the position is measured in kilometers. Find the velocity and acceleration of the car at $t = 0$.
33. An object moving in a straight line travels $s(t)$ kilometers in t hours, where $s(t) = \frac{1}{2}t^2 + 4t$.
    (a) What is the object's velocity when $t = 6$?
    (b) How far has the object traveled in 6 hours?
    (c) When is the object traveling at the rate of 6 kilometers per hour?
34. A helicopter is rising straight up in the air. Its distance from the ground t seconds after takeoff is $s(t)$ feet, where $s(t) = t^2 + t$.
    (a) How long will it take for the helicopter to rise 20 feet?
    (b) Find the velocity and the acceleration of the helicopter when it is 20 feet above the ground.
35. The number of people riding the subway daily from Silver Spring to Metro Center is a function $f(x)$ of the fare, x cents. Suppose $f(235) = 4600$ and $f'(235) = -100$.

Approximate the daily number of riders for each of the following costs:

(a) 237 cents   (b) 234 cents   (c) 240 cents   (d) 232 cents

36. The monthly payment on a $10,000, 3-year car loan is a function $f(r)$ of the interest rate, $r\%$. Now, $f(16) = 351.57$ and $f'(16) = 4.94$.

(a) If the interest rate is 16.5%, approximately how large is the monthly payment?

(b) If the interest rate is 15%, approximately how large is the monthly payment?

37. The balance after 10 years in a savings account with a principal of $1000 is a function $f(r)$ of the interest rate $r\%$. Now, $f(6) = 1790.85$ and $f'(6) = 168.95$.

(a) What is the significance of the amount $168.95?

(b) If the interest rate is 6.2%, approximate the balance after 10 years.

(c) If the interest rate is 5.7%, approximate the balance after 10 years.

38. Use approximation (5) with the function $f(x) = x^3$ to estimate $(2.01)^3$.

39. Use approximation (5) with the function $f(x) = \sqrt[3]{x}$ to estimate $\sqrt[3]{8.024}$.

40. Suppose the side of a square is increased from 5 to 5.001 meters. Use the derivative to estimate the increase in area.

41. In an eight-second test run, a vehicle accelerates for four seconds and then decelerates. The function $s(t)$ gives the number of feet traveled after t seconds and is graphed in Fig. 9.

(a) How far has the vehicle traveled after 3.5 seconds?

(b) What is the velocity after 2 seconds?

(c) What is the acceleration after 1 second?

(d) When will the vehicle have traveled 120 feet?

(e) When, during the second half of the test run, will the vehicle be traveling at the rate of 20 feet per second?

(f) What is the greatest velocity? At what time is this greatest velocity reached? How far has the vehicle traveled at this time?

FIGURE 9

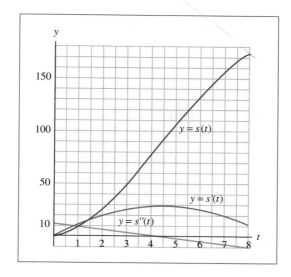

42. National health expenditures (in billions of dollars) from 1960 to 1994 are given by the function $f(t)$ in Fig. 10, where $t = 0$ corresponds to 1960.
    (a) How much money was spent in 1976?
    (b) How fast were expenditures rising in 1980?
    (c) When did expenditures reach 375 billion?
    (d) When were expenditures rising at the rate of $100 billion per year?

FIGURE 10

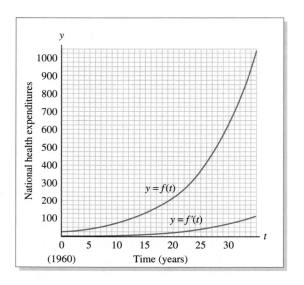

SOLUTIONS TO PRACTICE PROBLEMS 1.8

1. Method (a). The question involves $f(t)$, the temperature at time t. Since the time is given, compute $f(6)$.
2. Method (e). The question involves $f'(t)$, the rate of change of temperature. The interrogative "when" indicates that the time is unknown. Set $f'(t) = 6$ and solve for t.
3. Method (f). The question asks for the change in the value of the function from time 0 to time 6, $f(6) - f(0)$.
4. Method (b). The question involves $f(t)$, and the time is unknown. Set $f(t) = 6$ and solve for t.
5. Method (d). The question involves $f'(t)$, and the time is given. Compute $f'(6)$.
6. Method (c). The question asks for the average rate of change of the function during the time interval 0 to 6, $[f(6) - f(0)]/6$.

CHAPTER 1 CHECKLIST

✓ Slope of a line
✓ y-intercept

✓ Slope Properties 1 to 5
✓ Slope of a curve at a point

✓ The derivative of a constant function is zero

✓ Power rule:

$$\frac{d}{dx}(x^r) = rx^{r-1}$$

✓ Secant-line calculation of the derivative

✓ General power rule:

$$\frac{d}{dx}([g(x)]^r) = r \cdot [g(x)]^{r-1} \cdot \frac{d}{dx}[g(x)]$$

✓ Constant-multiple rule:

$$\frac{d}{dx}[k \cdot f(x)] = k \cdot \frac{d}{dx}[f(x)]$$

✓ Sum rule:

$$\frac{d}{dx}[f(x) + g(x)] = \frac{d}{dx}[f(x)] + \frac{d}{dx}[g(x)]$$

✓ Limit definition of the derivative

✓ Differentiable at $x = a$

✓ Continuous at $x = a$

✓ $\lim_{x \to a} f(x)$

✓ $\lim_{x \to \infty} f(x), \ \lim_{x \to -\infty} f(x)$

✓ Notation for first and second derivatives

✓ $\Delta y = f(a + h) - f(a)$, the change in $y = f(x)$ as x goes from a to $a + h$

✓ $\Delta x = (a + h) - a = h$, the change as x goes from a to $a + h$

✓ $\dfrac{\Delta y}{\Delta x}$, the average rate of change of $f(x)$ with respect to x over the interval from a to $a + \Delta x$

✓ The derivative $f'(a)$ measures the rate of change of $f(x)$ at $x = a$

✓ $f(a + h) - f(a) \approx f'(a) \cdot h$

✓ Marginal concept in economics

✓ Position, velocity, and acceleration functions

CHAPTER 1   SUPPLEMENTARY EXERCISES

Find the equation and sketch the graph of the following lines.

1. With slope -2, y-intercept $(0, 3)$.
2. With slope $\frac{3}{4}$, y-intercept $(0, -1)$.
3. Through $(2, 0)$, with slope 5.
4. Through $(1, 4)$, with slope $-\frac{1}{3}$.
5. Parallel to $y = -2x$, passing through $(3, 5)$.
6. Parallel to $-2x + 3y = 6$, passing through $(0, 1)$.
7. Through $(-1, 4)$, and $(3, 7)$.
8. Through $(2, 1)$ and $(5, 1)$.
9. Perpendicular to $y = 3x + 4$ and passing through $(1, 2)$.

10. Perpendicular to $3x + 4y = 5$ and passing through $(6, 7)$.

11. Horizontal with height 3 units above the x-axis.

12. Vertical and 4 units to the right of the y-axis.

13. The y-axis.

14. The x-axis.

Differentiate.

15. $y = x^7 + x^3$

16. $y = 5x^8$

17. $y = 6\sqrt{x}$

18. $y = x^7 + 3x^5 + 1$

19. $y = \dfrac{3}{x}$

20. $y = x^4 - \dfrac{4}{x}$

21. $y = (3x^2 - 1)^8$

22. $y = \frac{3}{4}x^{4/3} + \frac{4}{3}x^{3/4}$

23. $y = \dfrac{1}{5x - 1}$

24. $y = (x^3 + x^2 + 1)^5$

25. $y = \sqrt{x^2 + 1}$

26. $y = \dfrac{5}{7x^2 + 1}$

27. $f(x) = 1/\sqrt[4]{x}$

28. $f(x) = (2x + 1)^3$

29. $f(x) = 5$

30. $f(x) = \dfrac{5x}{2} - \dfrac{2}{5x}$

31. $f(x) = [x^5 - (x - 1)^5]^{10}$

32. $f(t) = t^{10} - 10t^9$

33. $g(t) = 3\sqrt{t} - \dfrac{3}{\sqrt{t}}$

34. $h(t) = 3\sqrt{2}$

35. $f(t) = \dfrac{2}{t - 3t^3}$

36. $g(P) = 4P^{.7}$

37. $h(x) = \frac{3}{2}x^{3/2} - 6x^{2/3}$

38. $f(x) = \sqrt{x} + \sqrt{x}$

39. If $f(t) = 3t^3 - 2t^2$, find $f'(2)$.

40. If $V(r) = 15\pi r^2$, find $V'(\frac{1}{3})$.

41. If $g(u) = 3u - 1$, find $g(5)$ and $g'(5)$.

42. If $h(x) = -\frac{1}{2}$, find $h(-2)$ and $h'(-2)$.

43. If $f(x) = x^{5/2}$, what is $f''(4)$?

44. If $g(t) = \frac{1}{4}(2t - 7)^4$, what is $g''(3)$?

45. Find the slope of the graph of $y = (3x - 1)^3 - 4(3x - 1)^2$ at $x = 0$.

46. Find the slope of the graph of $y = (4 - x)^5$ at $x = 5$.

Compute.

47. $\dfrac{d}{dx}(x^4 - 2x^2)$

48. $\dfrac{d}{dt}(t^{5/2} + 2t^{3/2} - t^{1/2})$

49. $\dfrac{d}{dP}(\sqrt{1-3P})$

50. $\dfrac{d}{dn}(n^{-5})$

51. $\dfrac{d}{dz}(z^3 - 4z^2 + z - 3)\Big|_{z=-2}$

52. $\dfrac{d}{dx}(4x - 10)^5\Big|_{x=3}$

53. $\dfrac{d^2}{dx^2}(5x + 1)^4$

54. $\dfrac{d^2}{dt^2}(2\sqrt{t})$

55. $\dfrac{d^2}{dt^2}(t^3 + 2t^2 - t)\Big|_{t=-1}$

56. $\dfrac{d^2}{dP^2}(3P + 2)\Big|_{P=4}$

57. $\dfrac{d^2y}{dx^2}$, where $y = 4x^{3/2}$

58. $\dfrac{d}{dt}\left(\dfrac{dy}{dt}\right)$, where $y = \dfrac{1}{3t}$

59. What is the slope of the graph of $f(x) = x^3 - 4x^2 + 6$ at $x = 2$? Write the equation of the line tangent to the graph of $f(x)$ at $x = 2$.

60. What is the slope of the curve $y = 1/(3x - 5)$ at $x = 1$? Write the equation of the line tangent to this curve at $x = 1$.

61. Find the equation of the tangent line to the curve $y = x^2$ at the point $(\frac{3}{2}, \frac{9}{4})$. Sketch the graph of $y = x^2$ and sketch the tangent line at $(\frac{3}{2}, \frac{9}{4})$.

62. Find the equation of the tangent line to the curve $y = x^2$ at the point $(-2, 4)$. Sketch the graph of $y = x^2$ and sketch the tangent line at $(-2, 4)$.

63. Determine the equation of the tangent line to the curve $y = 3x^3 - 5x^2 + x + 3$ at $x = 1$.

64. Determine the equation of the tangent line to the curve $y = (2x^2 - 3x)^3$ at $x = 2$.

65. In Fig. 1 the straight line has slope -1 and is tangent to the graph of $f(x)$. Find $f(2)$ and $f'(2)$.

66. In Fig. 2 the straight line is tangent to the graph of $f(x) = x^3$. Find the value of a.

67. A helicopter is rising at a rate of 32 feet per second. At a height of 128 feet the pilot drops a pair of binoculars. After t seconds, the binoculars have height $s(t) = -16t^2 + 32t + 128$ feet from the ground. How fast will they be falling when they hit the ground?

FIGURE 1

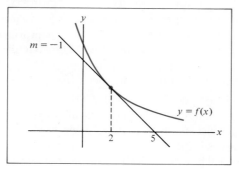

FIGURE 2

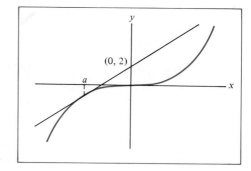

68. Each day, the total output of a coal mine after t hours of operation is approximately $40t + t^2 - \frac{1}{15}t^3$ tons, $0 \le t \le 12$. What is the rate of output (in tons of coal per hour) at $t = 5$ hours?

Exercises 69–72 refer to Fig. 3, where $s(t)$ is the number of feet traveled by a person after t seconds of walking along a straight path.

69. How far has the person traveled after 6 seconds?

70. What is the person's average velocity from time $t = 1$ to $t = 4$?

71. What is the person's velocity at time $t = 3$?

72. Without calculating velocities, determine whether the person is traveling faster at $t = 5$ or at $t = 6$.

FIGURE 3 Walker's progress.

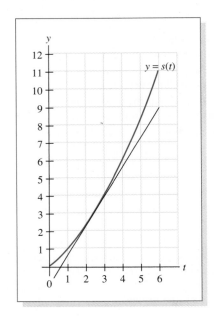

73. Let $f(x)$ be the number of gallons of gas used by a car after it has been driven x miles.
  (a) Is $f'(x)$ positive or negative?
  (b) Would you expect $f'(x)$ to be greater for a subcompact car or a limousine?
  (c) If $f'(25{,}000) = .04$ gallons per mile, approximately how many gallons of gas were required to drive from the 25,000th to the 25,005th mile?

74. Let $f(t)$ be the number of tons of lava spewed by a volcano t hours after its eruption first began. Discuss the significance of each of the following statements.
  (a) $f'(10) = 300$　　　　(b) $f'(10) = 0$　　　　(c) $f'(10) = -10$

75. Let $h(t)$ be a boy's height (in inches) after t years. If $h'(12) = 1.5$, how much will his height increase between ages 12 and $12\frac{1}{2}$?

76. If you deposit $100 into a savings account at the end of each month for 2 years, the balance will be a function $f(r)$ of the interest rate, $r\%$. At 7% interest (compounded monthly), $f(7) = 2568.10$ and $f'(7) = 25.06$. Approximately how much additional money would you earn if the bank paid $7\frac{1}{2}\%$ interest?

Determine whether the following limits exist. If so, compute the limit.

77. $\lim\limits_{x \to 2} \dfrac{x^2 - 4}{x - 2}$

78. $\lim\limits_{x \to 3} \dfrac{1}{x^2 - 4x + 3}$

79. $\lim\limits_{x \to 4} \dfrac{x - 4}{x^2 - 8x + 16}$

80. $\lim\limits_{x \to 5} \dfrac{x - 5}{x^2 - 7x + 2}$

Use limits to compute the following derivatives.

81. $f'(5)$, where $f(x) = 1/(2x)$.

82. $f'(3)$, where $f(x) = x^2 - 2x + 1$.

83. What geometric interpretation may be given to $\dfrac{(3 + h)^2 - 3^2}{h}$ in connection with the graph of $f(x) = x^2$?

84. As h approaches 0, what value is approached by $\dfrac{\dfrac{1}{2 + h} - \dfrac{1}{2}}{h}$?

APPLICATIONS

OF THE

DERIVATIVE

Calculus techniques can be applied to a wide variety of problems in real life. We consider many examples in this chapter. In each case we construct a function as a "mathematical model" of some problem and then analyze the function and its derivatives in order to gain information about the original problem. Our principal method for analyzing a function will be to sketch its graph. For this reason we devote the first part of the chapter to curve sketching.

2.1 DESCRIBING GRAPHS OF FUNCTIONS

Let's examine the graph of a typical function, such as the one shown in Fig. 1, and introduce some terminology to describe its behavior. First observe that the graph is either rising or falling, depending on whether we look at it from left to right or from right to left. To avoid confusion, we shall always follow the accepted practice of reading a graph from left to right.

FIGURE I   An increasing function.

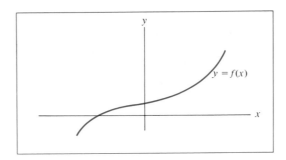

Let's now examine the behavior of a function $f(x)$ in an interval through-out which it is defined. We say that a function $f(x)$ is *increasing in the interval* if the graph continuously rises as x goes from left to right through the interval. That is, whenever x_1 and x_2 are in the interval with $x_1 < x_2$, then we have $f(x_1) < f(x_2)$. We say that $f(x)$ is increasing at $x = c$ provided that $f(x)$ is increasing in some open interval on the x-axis that contains the point c.

We say that a function $f(x)$ is *decreasing in an interval* provided that the graph continuously falls as x goes from left to right through the interval. That is, whenever x_1 and x_2 are in the interval with $x_1 < x_2$, then we have $f(x_1) > f(x_2)$. We say that $f(x)$ is decreasing at $x = c$ provided that $f(x)$ is decreasing in some open interval that contains the point c. Figure 2 shows graphs that are increasing and decreasing at $x = c$. Observe in Fig. 2(d) that when $f(c)$ is negative and $f(x)$ is decreasing, the values of $f(x)$ become *more* negative. When $f(c)$ is negative and $f(x)$ is increasing, as in Fig. 2(e), the values of $f(x)$ become *less* negative.

Extreme Points   A *relative extreme point* of a function is a point at which its graph changes from increasing to decreasing, or vice versa. We distinguish the two possibilities in an obvious way. A *relative maximum point* is a point at which the graph changes from increasing to decreasing; a *relative minimum point* is a point at which the graph changes from decreasing to increasing. (See Fig. 3.) The adjective "relative" in these definitions indicates that a point is maximal or minimal only relative to nearby points on the graph.

The *maximum value* (or *absolute maximum value*) of a function is the largest value that the function assumes on its domain. The *minimum value* (or

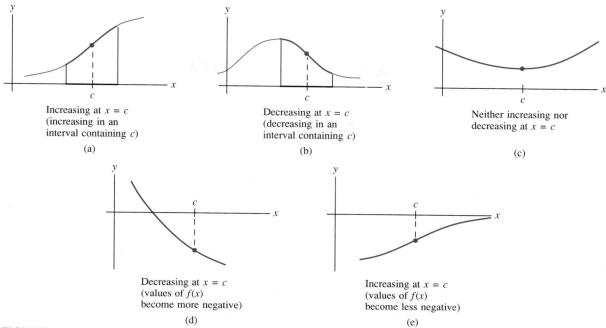

FIGURE 2 **Increasing and decreasing at x = c.**

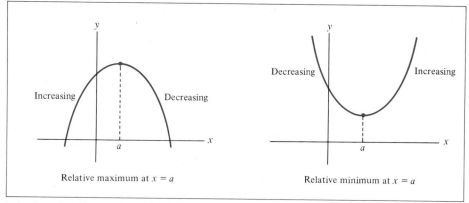

FIGURE 3 **Relative extreme points.**

absolute minumum value) of a function is the smallest value that the function assumes on its domain. Functions may or may not have maximum or minimum values. (See Fig. 4.) However, it can be shown that a continuous function whose domain is an interval of the form $a \le x \le b$ has both a maximum and a minimum value.

Maximum values and minimum values of functions usually occur at relative maximum points and relative minimum points, as in Fig. 4(a). However,

FIGURE 4

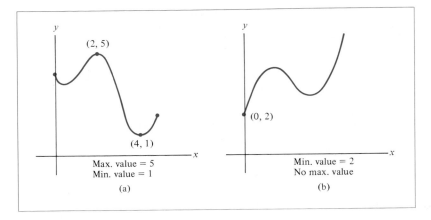

they can occur at endpoints of the domain, as in Fig. 4(b). If so, we say that the function has an *endpoint extreme value* (or *endpoint extremum*).

Relative maximum points and endpoint maximum points are higher than any nearby points. The maximum value of a function is the *y*-coordinate of the highest point on its graph. (The highest point is called the *absolute maximum point*.) Similar considerations apply to minima.

EXAMPLE I   When a drug is injected intramuscularly (into a muscle), the concentration of the drug in the veins has the time-concentration curve shown in Fig. 5. Describe this graph, using the terms introduced above.

FIGURE 5   A drug time-concentration curve.

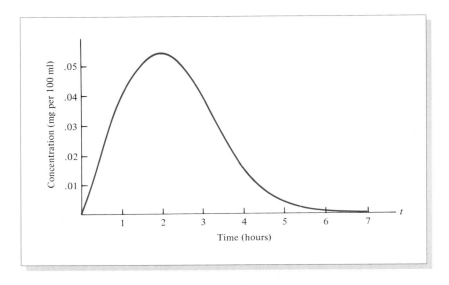

Solution   Initially (when $t = 0$), there is no drug in the veins. When the drug is injected into the muscle, it begins to diffuse into the bloodstream. The concentration in the veins increases until it reaches its maximum value at $t = 2$. After this time the concentration begins to decrease, as the body's metabolic processes remove the drug from the blood. Eventually the drug concentration decreases to a level so small that, for all practical purposes, it is zero. ●

Changing Slope An important but subtle feature of a graph is the way the graph's slope *changes* (as we look from left to right). The graphs in Fig. 6 are both increasing, but there is a fundamental difference in the way they are increasing. Graph I, which describes the U.S. national debt per person, is steeper for 1970 than for 1960. That is, the *slope* of graph I *increases* as we move from left to right. A newspaper description of graph I might read,

> U.S. public debt per capita rose at an increasing rate during the decade from 1960 to 1970.

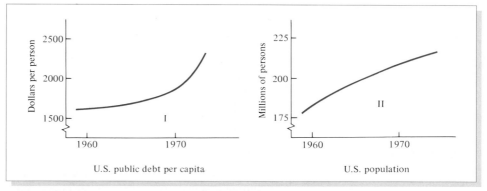

FIGURE 6 **Increasing and decreasing slopes.**

In contrast, the *slope* of graph II *decreases* as we move from left to right. Although the U.S. population is rising each year, the rate of increase declines throughout the decade from 1960 to 1970. That is, the slope becomes less positive. The media might say,

> During the 1960s, U.S. population rose at a decreasing rate.

EXAMPLE 2 Recently, the daily number of hours of sunlight in Washington, D.C., increased from 9.45 hours on December 21 to 12 hours on March 21 and then increased to 14.9 hours on June 21. From December 22 to March 21, the daily increase was greater than the previous daily increase, and from March 22 to June 21, the daily increase was less than the previous daily increase. Draw a possible graph of the number of hours of daylight as a function of time.

Solution Let $f(t)$ be the number of hours of daylight t months after December 21. See Fig. 7. The first part of the graph, December 21 to March 21, is increasing at an increasing rate. The second part of the graph, March 21 to June 21, is increasing at a decreasing rate. ●

**FIGURE 7   Hours of day-
light in Washington, D.C.**

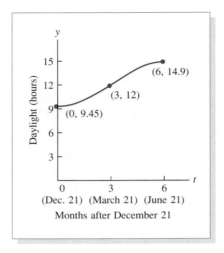

**FIGURE 8   Slope is
decreasing.**

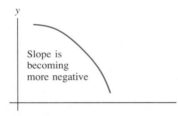

**FIGURE 9   Slope is
increasing.**

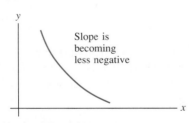

Warning   Recall that when a negative quantity decreases, it becomes more negative. (Think about the temperature outside when it is below zero and the temperature is falling.) So if the slope of a graph is negative and the slope is decreasing, then the slope is becoming more negative, as in Fig. 8. This technical use of the term "decreasing" runs counter to our intuition, because in popular discourse, decrease often means to become smaller in size.

It is true that the curve in Fig. 9 is becoming "less steep" in a nontechnical sense (since steepness, if it were defined, would probably refer to the magnitude of the slope). However, the slope of the curve in Fig. 9 is increasing because it is becoming less negative. The popular press would probably describe the curve in Fig. 9 as decreasing at a decreasing rate, since the rate of fall tends to taper off. Since this terminology is potentially confusing, we shall not use it.

Concavity   The U.S. debt and population graphs in Fig. 6 may also be described in geometric terms: Graph I opens up and lies above its tangent line at each point, whereas graph II opens down and lies below its tangent line at each point (Fig. 10).

We say that a function $f(x)$ is *concave up* at $x = a$ if there is an open interval on the x-axis containing a throughout which the graph of $f(x)$ lies above its tangent line. Equivalently, $f(x)$ is concave up at $x = a$ if the slope of the graph increases as we move from left to right through $(a, f(a))$. Graph I is an example of a function that is concave up at each point.

**FIGURE 10   Relationship
between concavity and
tangent lines.**

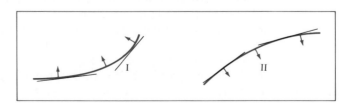

Similarly, we say that a function $f(x)$ is *concave down* at $x = a$ if there is an open interval on the x-axis containing a throughout which the graph of $f(x)$ lies below its tangent line. Equivalently, $f(x)$ is concave down at $x = a$ if the slope of the graph decreases as we move from left to right through $(a, f(a))$. Graph II is concave down at each point.

An *inflection point* is a point on the graph of a function at which the function is continuous and at which the graph changes from concave up to concave down, or vice versa. At such a point, the graph crosses its tangent line (Fig. 11). (The continuity condition means that the graph cannot break at an inflection point.)

FIGURE 11   Inflection points.

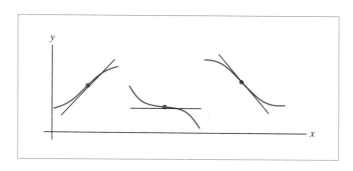

EXAMPLE 3   Use the terms defined earlier to describe the graph shown in Fig. 12.

Solution   (a) For $x < 3$, $f(x)$ is increasing and concave down.
(b) Relative maximum point at $x = 3$.
(c) For $3 < x < 4$, $f(x)$ is decreasing and concave down.
(d) Inflection point at $x = 4$.
(e) For $4 < x < 5$, $f(x)$ is decreasing and concave up.
(f) Relative minimum point at $x = 5$.
(g) For $x > 5$, $f(x)$ is increasing and concave up. ●

FIGURE 12

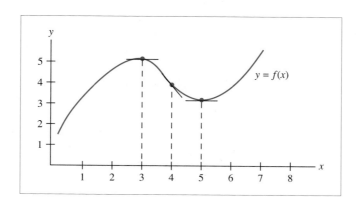

Intercepts, Undefined Points, and Asymptotes   A point at which a graph crosses the *y*- axis is called a *y-intercept*, and a point at which it crosses the *x*- axis is called an *x-intercept*. The *x*- coordinate of an *x*-intercept is sometimes called a "zero" of the function, since the function has the value zero there. (See Fig. 13.)

FIGURE 13   Intercepts of a graph.

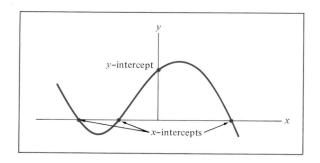

Some functions are not defined for all values of *x*. For instance, $f(x) = 1/x$ is not defined for $x = 0$, and $f(x) = \sqrt{x}$ is not defined for $x < 0$. (See Fig. 14.) Many functions that arise in applications are defined only for $x \geq 0$. A properly drawn graph should leave no doubt as to the values of *x* for which the function is defined.

FIGURE 14   Graphs with undefined points.

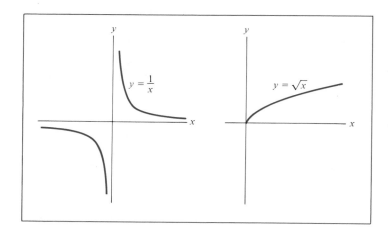

Graphs in applied problems sometimes straighten out and approach some straight line as *x* gets large (Fig. 15). Such a straight line is called an *asymptote* of the curve. The most common asymptotes are horizontal as in (a) and (b) of Fig. 15. In Example 1 the *t*- axis is an asymptote of the drug time-concentration curve.

The horizontal asymptotes of a graph may be determined by calculating the limits

$$\lim_{x \to \infty} f(x) \quad \text{and} \quad \lim_{x \to -\infty} f(x).$$

FIGURE 15 Graphs that approach asymptotes as *x* gets large.

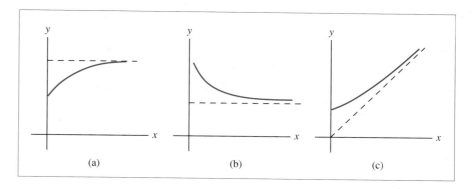

(a)   (b)   (c)

If either limit exists, then the value of the limit determines a horizontal asymptote.

Occasionally, a graph will approach a vertical line as *x* approaches some fixed value, as in Fig. 16. Such a line is a *vertical asymptote*. Most often, we expect a vertical asymptote at a value *x* that would result in division by zero in the definition of $f(x)$. For example, $f(x) = 1/(x - 3)$ has a vertical asymptote $x = 3$.

FIGURE 16 Examples of vertical asymptotes

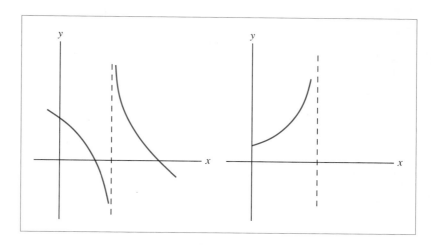

We now have six categories for describing the graph of a function.

1. Intervals in which the function is increasing (resp. decreasing), relative maximum points, relative minimum points
2. Maximum value, minimum value
3. Intervals in which the function is concave up (resp. concave down), inflection points
4. *x*-intercept, *y*-intercept
5. Undefined points
6. Asymptotes

For us, the first three categories will be the most important. However, the last three categories should not be forgotten.

**TECHNOLOGY PROJECT 2.1
Describing Curves Sketched with a Graphing Calculator**

You can use a graphing calculator to determine the intervals in which a graph is increasing or decreasing. Just use the TRACE to traverse the curve from left to right and observe whether the y-coordinate increases or decreases. Use this technique to determine the intervals in which the graphs of the following function increase or decrease:

1. $f(x) = x^2 - 5x - 1$
2. $f(x) = -2x^2 + x + 1$
3. $f(x) = x^3 - 3x + 1$
4. $f(x) = 2x^3 - 7x + 1$

**TECHNOLOGY PROJECT 2.2
Asymptotes of Curves on a Graphing Calculator**

The graph of $f(x) = 1/x$ has a vertical asymptote $x = 0$, corresponding to the point at which the denominator is 0. Graphing calculators do not do well drawing the graphs with asymptotes. Display the graph of this function on your graphing calculator. Note that the asymptote is rendered as a spike and that the graph looks as if the function is defined for $x = 0$. Here's the reason. The calculator computes points on the graph at successive values of x corresponding to screen dots across the x-axis. As it computes each point, the calculator plots it on the display. If the function is not defined at a particular value of x, no point is plotted. However, in order to give the display the appearance of a smooth curve, the calculator draws a line connecting consecutive points on the graph. In this case, the calculator connects a point slightly to the left of $x = 0$ with a point slightly to the right of $x = 0$. No point is plotted for $x = 0$ since the function is undefined there. You must get used to interpreting spikes as asymptotic behavior.

When sketching a graph on a calculator, you can locate the asymptotes by determining the zeros of the denominator. For example, to determine the vertical asymptotes of the function

$$f(x) = \frac{x}{x^3 - 3x + 1}$$

you would approximate the zeros of $x^3 - 3x + 1$ using the TRACE. Use this procedure to determine the vertical asymptotes of the following functions. Use this information to display a graph of the function which shows all the asymptotes.

1. $f(x) = \dfrac{x + 2}{3x - 1}$
2. $f(x) = \dfrac{x^2}{x^3 - x + 1}$
3. $f(x) = \dfrac{2x + 1}{x^4 - 2}$
4. $f(x) = \dfrac{x^3 - 1}{x^4 - 2x^2 - 3}$

1. Does the slope of the curve in Fig. 17 increase or decrease as x increases?
2. At what value of x is the slope of the curve in Fig. 18 minimized?

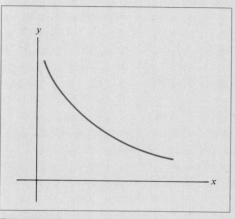

FIGURE 17

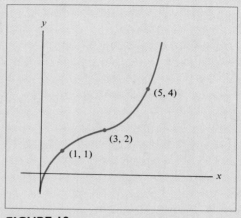

FIGURE 18

EXERCISES 2.1

Exercises 1–4 refer to graphs (a)–(f) in Fig. 19.
1. Which functions are increasing for all x?
2. Which functions are decreasing for all x?

FIGURE 19

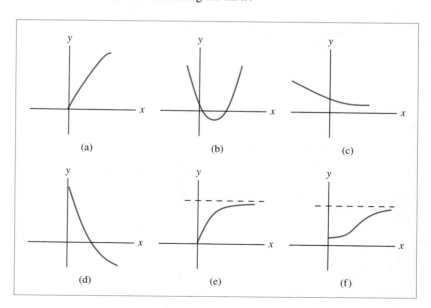

3. Which functions have the property that the slope always increases as x increases?

4. Which functions have the property that the slope always decreases as x increases?

Describe each of the following graphs. Your description should include each of the six categories mentioned above.

5.

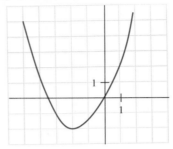

6.

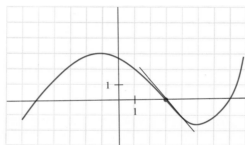

7.

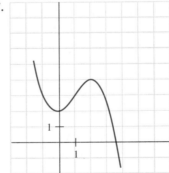

8.

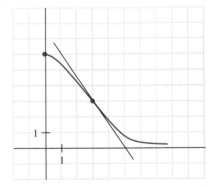

9.

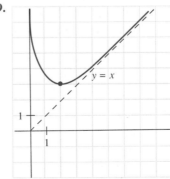

$y = x$

10.

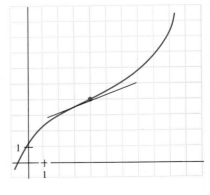

11.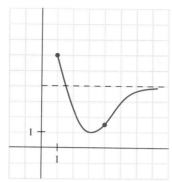

12.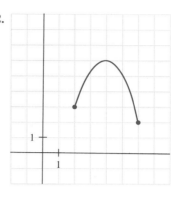

13. Describe the way the *slope* changes as you move along the graph (from left to right) in Exercise 5.

14. Describe the way the *slope* changes on the graph in Exercise 6.

15. Describe the way the *slope* changes on the graph in Exercise 8.

16. Describe the way the *slope* changes on the graph in Exercise 10.

Exercises 17 and 18 refer to the graph in Fig. 20.

17. **(a)** At which labeled points is the function increasing?
    (b) At which labeled points is the graph concave up?
    (c) Which labeled point has the most positive slope?

FIGURE 20

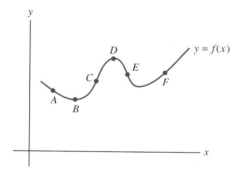

18. **(a)** At which labeled points is the function decreasing?
    (b) At which labeled points is the graph concave down?
    (c) Which labeled point has the most negative slope (that is, negative and with the greatest magnitude)?

In Exercises 19–22, draw the graph of a function $y = f(x)$ with the stated properties.

19. Both the function and the slope increase as x increases.

20. The function increases and the slope decreases as x increases.

21. The function decreases and the slope increases as x increases. [*Note:* The slope is negative but becomes less negative.]

22. Both the function and the slope decrease as x increases. [*Note:* The slope is negative and becomes more negative.]

23. The annual world consumption of oil rises each year. Furthermore, the amount of the annual *increase* in oil consumption is also rising each year. Sketch a graph that could represent the annual world consumption of oil.

24. In certain professions the average annual income has been rising at an increasing rate. Let $f(T)$ denote the average annual income at year T for persons in one of these professions and sketch a graph that could represent $f(T)$.

25. At noon a child's temperature is $101°$ and is rising at an increasing rate. At 1 P.M. the child is given medicine. After 2 P.M. the temperature is still increasing but at a decreasing rate. The temperature reaches a peak of $103°$ at 3 P.M. and decreases to $100°$ by 5 P.M. Draw a possible graph of the function $T(t)$, the child's temperature at time t.

26. The number of parking tickets given out each year in the District of Columbia increased from 114,000 in 1950 to 1,500,000 in 1990. Also, each year's increase was greater than the increase for the previous year. Draw a possible graph of the annual number of parking tickets as a function of time. [*Note:* If $f(t)$ is the yearly rate t years after 1950, then $f(0) = 114$ and $f(40) = 1500$ thousand tickets.]

27. Let $C(x)$ denote the total cost of manufacturing x units of some product. Then $C(x)$ is an increasing function for all x. For small values of x, the rate of increase of $C(x)$ decreases. (This is because of the savings that are possible with "mass production.") Eventually, however, for large values of x, the cost $C(x)$ increases at an increasing rate. (This happens when production facilities are strained and become less efficient.) Sketch a graph that could represent $C(x)$.

28. One method of determining the level of blood flow through the brain requires the person to inhale air containing a fixed concentration of N_2O, nitrous oxide. During the first minute, the concentration of N_2O in the jugular vein grows at an increasing rate to a level of .25%. Thereafter it grows at a decreasing rate and reaches a concentration of about 4% after 10 minutes. Draw a possible graph of the concentration of N_2O in the vein as a function of time.

29. Suppose that some organic waste products are dumped into a lake at time $t = 0$, and suppose that the oxygen content of the lake at time t is given by the graph in Fig. 21. Describe the graph in physical terms. Indicate the significance of the inflection point at $t = b$.

FIGURE 21   A lake's recovery from pollution.

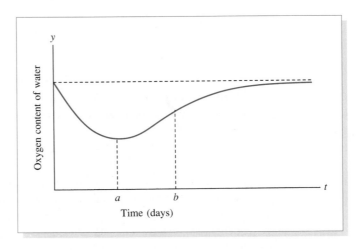

Time (days)

30. Figure 22 gives the U.S. electrical energy production in quadrillion kilowatt-hours from 1910 ($t = 10$) to 1990 ($t = 90$). In what year was the level of production growing at the greatest rate?

FIGURE 22   U.S. electrical energy production.

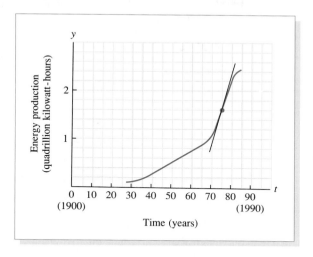

31. Figure 23 gives the number of U.S. farms in millions from 1920 ($t = 20$) to 1990 ($t = 90$). In what year was the number of farms decreasing most rapidly?

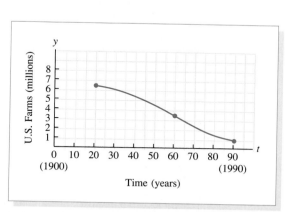

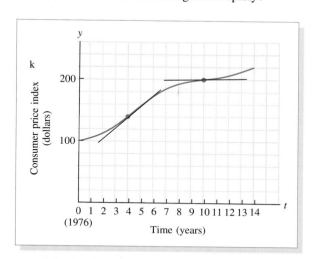

FIGURE 23   Number of U.S. farms.

FIGURE 24   Consumer price index.

32. Figure 24 shows the graph of the consumer price index for the years 1976 ($t = 0$) through 1990 ($t = 14$). This index measures how much a basket of commodities that cost \$100 in the beginning of 1976 would cost at any given time. In what year was the rate of increase of the index greatest? The least?

33. Let $s(t)$ be the distance (in feet) traveled by a parachutist after t seconds from the time of opening the chute, and suppose that $s(t)$ has the line $y = -15t + 10$ as an asymptote. What does this imply about the velocity of the parachutist? [*Note: Distance traveled downward is given a negative value.*]

34. Let $P(t)$ be the population of a bacteria culture after t days and suppose that $P(t)$ has the line $y = 25,000,000$ as an asymptote. What does this imply about the size of the population?

In Exercises 35–38, sketch the graph of a function having the given properties.

35. Defined for $0 \leq x \leq 10$; relative maximum point at $x = 3$; absolute maximum value at $x = 10$.

36. Relative maximum points at $x = 1$ and $x = 5$; relative minimum point at $x = 3$; inflection points at $x = 2$ and $x = 4$.

37. Defined and increasing for all $x \geq 0$; inflection point at $x = 5$; asymptotic to the line $y = (3/4)x + 5$.

38. Defined for $x \geq 0$; absolute minimum value at $x = 0$; relative maximum point at $x = 4$; asymptotic to the line $y = (x/2) + 1$.

39. Consider a smooth curve with no undefined points.

    (a) If it has two relative maximum points, must it have a relative minimum point?

    (b) If it has two relative extreme points, must it have an inflection point?

40. Suppose the function $f(x)$ has a relative minimum at $x = a$ and a relative maximum at $x = b$. Must $f(a)$ be less than $f(b)$?

The difference between the pressure inside the lungs and the pressure surrounding the lungs is called the *transmural pressure* (or transmural pressure gradient). Figure 25 shows for three different persons how the volume of the lungs is related to the transmural pressure, based on static measurements taken while there is no air flowing through the mouth. (The functional reserve capacity mentioned on the vertical axis is the volume of air in the lungs at the end of a normal expiration.) The rate of change in lung volume with respect to transmural pressure is called the *lung compliance.*

FIGURE 25   Lung pressure-volume curves.

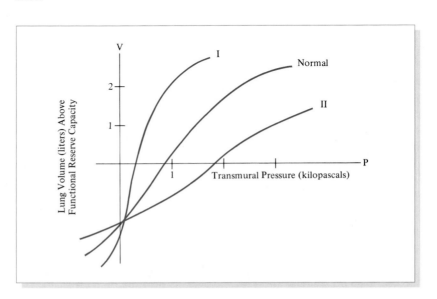

41. If the lungs are less flexible than normal, an increase in pressure will cause a smaller change in lung volume than in a normal lung. In this case, is the compliance relatively high or low?

42. Most lung diseases cause a decrease in lung compliance. However, the compliance of a person with emphysema is higher than normal. Which curve (I or II) in the figure could correspond to a person with emphysema?

SOLUTIONS TO PRACTICE PROBLEMS 2.1

1. The curve is concave up, so the slope increases. Even though the curve itself is decreasing, the slope becomes less negative as we move from left to right.

2. At $x = 3$. We have drawn in tangent lines at various points (Fig. 26). Note that as we move from left to right, the slopes decrease steadily until the point (3, 2), at which time they start to increase. This is consistent with the fact that the graph is concave down (hence, slopes are decreasing) to the left of (3, 2) and concave up (hence, slopes are increasing) to the right of (3, 2). Extreme values of slopes always occur at inflection points.

FIGURE 26

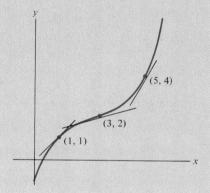

2.2 THE FIRST AND SECOND DERIVATIVE RULES

We shall now show how properties of the graph of a function $f(x)$ are determined by properties of the derivatives, $f'(x)$ and $f''(x)$. These relationships will provide the key to the curve-sketching and optimization problems discussed in the rest of the chapter.*

---

*Throughout this chapter, we shall assume that we are dealing with functions that are not "too badly behaved." More precisely, it suffices to assume that all of our functions have continuous first and second derivatives in the interval(s) (in x) where we are considering their graphs.

We begin with a discussion of the first derivative of a function $f(x)$. Suppose that for some value of x, say $x = a$, the derivative $f'(a)$ is positive. Then the tangent line at $(a, f(a))$ has positive slope and is a rising line (moving from left to right, of course). Since the graph of $f(x)$ near $(a, f(a))$ resembles its tangent line, the function must be increasing at $x = a$. Similarly, when $f'(a) < 0$, the function is decreasing at $x = a$. (See Fig. 1.)

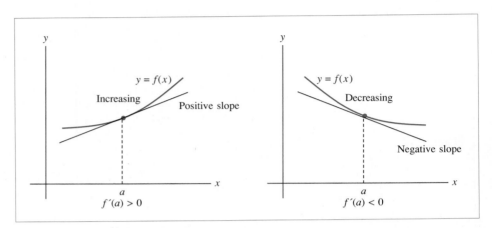

FIGURE I   Illustration of first derivative rule.

Thus we have the following useful result.

First Derivative Rule   If $f'(a) > 0$, then $f(x)$ is increasing at $x = a$. If $f'(a) < 0$, then $f(x)$ is decreasing at $x = a$.

When $f'(a) = 0$, the first derivative rule is not decisive. In this case the function $f(x)$ might be increasing or decreasing or have a relative extreme point at $x = a$.

EXAMPLE I   Sketch the graph of a function $f(x)$ that has all the following properties.

(a) $f(3) = 4$

(b) $f'(x) > 0$ for $x < 3, f'(3) = 0$, and $f'(x) < 0$ for $x > 3$

Solution   The only specific point on the graph is $(3, 4)$ [property (a)]. We plot this point and then use the fact that $f'(3) = 0$ to sketch the tangent line at $x = 3$ (Fig. 2).

From property (b) and the first derivative rule, we know that $f(x)$ must be increasing for x less than 3 and decreasing for x greater than 3. A graph with these properties might look like the curve in Fig. 3. ●

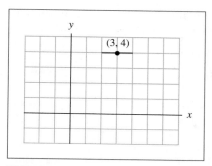

FIGURE 2

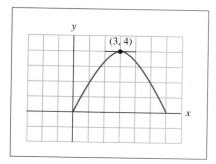

FIGURE 3

The second derivative of a function $f(x)$ gives useful information about the concavity of the graph of $f(x)$. Suppose that $f''(a)$ is negative. Then since $f''(x)$ is the derivative of $f'(x)$, we conclude that $f'(x)$ has a negative derivative at $x = a$. In this case, $f'(x)$ must be a decreasing function at $x = a$; that is, the slope of the graph of $f(x)$ is decreasing as we move from left to right on the graph near $(a, f(a))$. (See Fig. 4.) This means that the graph of $f(x)$ is concave down at $x = a$. A similar analysis shows that if $f''(a)$ is positive, then $f(x)$ is concave up at $x = a$. Thus we have the following rule.

FIGURE 4   Illustration of second derivative rule.

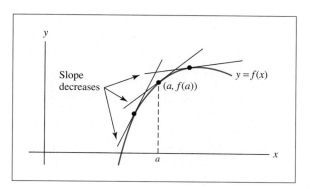

Second Derivative Rule   If $f''(a) > 0$, then $f(x)$ is concave up at $x = a$.
If $f''(a) < 0$, then $f(x)$ is concave down at $x = a$.

When $f''(a) = 0$, the second derivative rule gives no information. In this case, the function might be concave up, concave down, or neither at $x = a$.

The following chart shows how a graph may combine the properties of increasing, decreasing, concave up, and concave down.

Condition on the Derivatives	Description of $f(x)$ at $x = a$	Graph of $y = f(x)$ Near $x = a$
1. $f'(a)$ positive $f''(a)$ positive	$f(x)$ increasing $f(x)$ concave up	
2. $f'(a)$ positive $f''(a)$ negative	$f(x)$ increasing $f(x)$ concave down	
3. $f'(a)$ negative $f''(a)$ positive	$f(x)$ decreasing $f(x)$ concave up	
4. $f'(a)$ negative $f''(a)$ negative	$f(x)$ decreasing $f(x)$ concave down	

EXAMPLE 2   Sketch the graph of a function $f(x)$ with all the following properties.
(a) (2, 3), (4, 5), and (6, 7) are on the graph.
(b) $f'(6) = 0$ and $f'(2) = 0$.
(c) $f''(x) > 0$ for $x < 4$, $f''(4) = 0$, and $f''(x) < 0$ for $x > 4$.

Solution   First we plot the three points from property (a) and then sketch two tangent lines, using the information from property (b). (See Fig. 5). From property (c) and the second derivative rule, we know that $f(x)$ is concave up for $x < 4$. In particular, $f(x)$ is concave up at (2, 3). Also $f(x)$ is concave down for $x > 4$, in particular, at (6, 7). Note that $f(x)$ must have an inflection point at $x = 4$ because the concavity changes there. We now sketch small portions of the curve near (2, 3) and (6, 7). (See Fig. 6.) We can now complete the sketch (Fig. 7), taking care to make the curve concave up for $x < 4$ and concave down for $x > 4$. ●

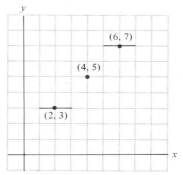

FIGURE 5

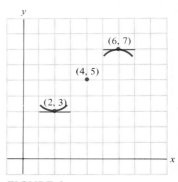

FIGURE 6

FIGURE 7

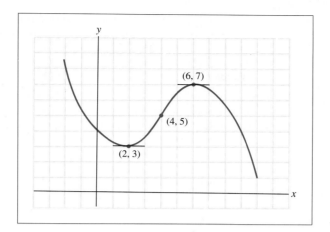

Connections Between the Graphs of $f(x)$ and $f'(x)$ Think of the derivative of $f(x)$ as a "slope-function" for $f(x)$. The "y-values" on the graph of $y = f'(x)$ are the *slopes* of the corresponding points on the original graph $y = f(x)$. This important connection is illustrated in the next three examples.

EXAMPLE 3 The function $f(x) = 8x - x^2$ is graphed in Fig. 8 along with the slope at several points. How is the slope changing on the graph? Compare the slopes on the graph with the y-coordinates of the points on the graph of $f'(x)$ in Fig. 9.

FIGURE 8 Graph of $f(x) = 8x - x^2$.

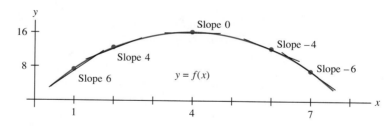

FIGURE 9 Graph of the derivative of the function in Fig. 8.

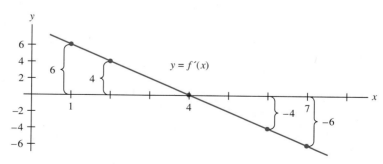

Solution The slopes are decreasing (as we move from left to right). That is, $f'(x)$ is a decreasing function. Observe that the y-values of $f'(x)$ decrease to zero at $x = 4$ and then continue to decrease for x-values greater than 4. ●

The graph in Fig. 8 is the shape of a typical revenue curve for a manufacturer. In this case, the graph of $f'(x)$ in Fig. 9 would be the *marginal revenue curve*. The graph in the next example has the shape of a typical cost curve. Its derivative produces a *marginal cost curve*.

EXAMPLE 4   The function $\frac{1}{3}x^3 - 4x^2 + 18x + 10$ is graphed in Fig. 10. The slope decreases at first and then increases. Use the graph of $f'(x)$ to verify that the slope in Fig. 10 is minimum at the inflection point, where $x = 4$.

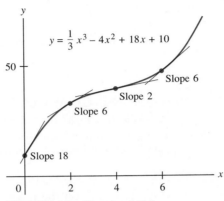

FIGURE 10   Graph of $f(x)$.   **FIGURE 11   Graph of $f'(x)$.**

Solution   Several slopes are marked on the graph of $f(x)$. These values are y-coordinates of points on the graph of the derivative, $f'(x) = x^2 - 8x + 18$. Observe in Fig. 11 that the y-values on the graph of $f'(x)$ decrease at first and then begin to increase. The minimum value of $f'(x)$ occurs at $x = 4$. ●

EXAMPLE 5   Figure 12 shows the graph of $y = f'(x)$, the derivative of a function $f(x)$.

(a) What is the slope of the graph of $f(x)$ when $x = 1$?
(b) For what value of x does the graph of $f(x)$ have a horizontal tangent line?
(c) Describe the shape of the graph of $f(x)$ on the interval $1 \le x \le 2$.

FIGURE 12

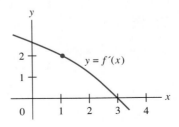

Solution   (a) Since $f'(1)$ is 2, $f(x)$ has slope 2 when $x = 1$.
(b) The graph of $f(x)$ has a horizontal tangent line when the slope is 0—that is, when $f'(x)$ is 0. This occurs at $x = 3$.

(c) On the interval $1 \le x \le 2$, $f'(x)$ is always positive and decreasing as x increases. This means that the *slope* of the graph of $f(x)$ is always positive and is decreasing as x increases. Therefore, the graph of $f(x)$ is increasing and concave down. ●

TECHNOLOGY PROJECT 2.3
Determining Maxima and Minima Using a Graphing Calculator

Use the TRACE on the graph of the derivative to determine all relative extreme points of the following functions. Classify them as relative maximum or relative minimum points.

1. $f(x) = x^3 - 6x^2 + 12x - 6$ 2. $f(x) = 1 - 3x + 3x^2 - x^3$

3. $f(x) = 3 + \dfrac{5}{x}$ 4. $f(x) = \dfrac{1}{x^2 + 1}$

TECHNOLOGY PROJECT 2.4
Determining Concavity and Inflection Points Using A Graphing Calculator

Determine the intervals in which graphs of the following functions are concave up and concave down by displaying the function and its second derivative on the same graph. By using the TRACE on the graph of the second derivative, determine the inflection points of the graph.

1. $f(x) = x^3 - x^2 + 12x - 6$ 2. $f(x) = 1 - 3x + x^2 - x^3$

3. $f(x) = \dfrac{x}{1 - 3x}$ 4. $f(x) = \dfrac{1}{x^2 + 1}$

PRACTICE PROBLEMS 2.2

1. Make a good sketch of the function $f(x)$ near the point where $x = 2$, given that $f(2) = 5$, $f'(2) = 1$, and $f''(2) = -3$.
2. The graph of $f(x) = x^3$ is shown in Fig. 13.
   (a) Is the function increasing at $x = 0$?
   (b) Compute $f'(0)$.
   (c) Reconcile your answers to parts (a) and (b) with the first derivative rule.

FIGURE 13

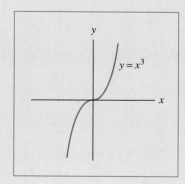

$y = x^3$

3. The graph of $y = f'(x)$ is shown in Fig. 14. Explain why $f(x)$ must have a relative minimum point at $x = 3$.

FIGURE 14

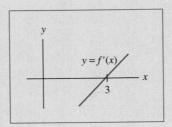

EXERCISES 2.2

Exercises 1–4 refer to the functions whose graphs are given in Fig. 15.

FIGURE 15

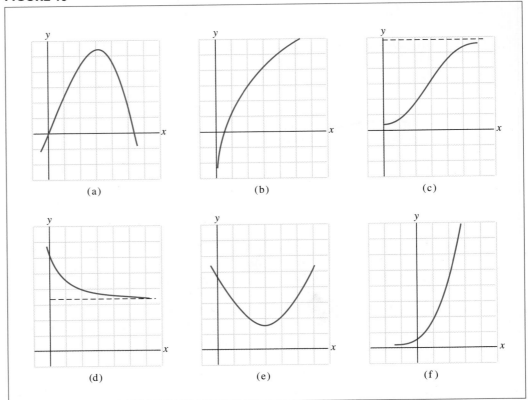

1. Which functions have a positive first derivative for all x?
2. Which functions have a negative first derivative for all x?
3. Which functions have a positive second derivative for all x?
4. Which functions have a negative second derivative for all x?
5. Which one of the graphs in Fig. 16 could represent a function $f(x)$ for which $f(a) > 0, f'(a) = 0$, and $f''(a) < 0$?

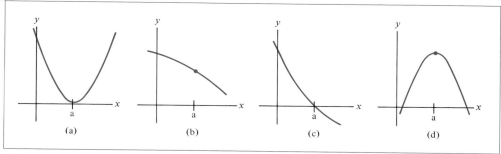

(a)　　　　(b)　　　　(c)　　　　(d)

FIGURE 16

6. Which one of the graphs in Fig. 16 could represent a function $f(x)$ for which $f(a) = 0, f'(a) < 0$, and $f''(a) > 0$?

In Exercises 7–12, sketch the graph of a function that has the properties described.

7. $f(2) = 1; f'(2) = 0$; concave up for all x.

8. $f(-1) = 0; f'(x) < 0$ for $x < -1, f'(-1) = 0$ and $f'(x) > 0$ for $x > -1$.

9. $f(3) = 5; f'(x) > 0$ for $x < 3, f'(3) = 0$ and $f'(x) > 0$ for $x > 3$.

10. $(-2, -1)$ and $(2, 5)$ are on the graph; $f'(-2) = 0$ and $f'(2) = 0; f''(x) > 0$ for $x < 0; f''(0) = 0, f''(x) < 0$ for $x > 0$.

11. $(0, 6), (2, 3)$, and $(4, 0)$ are on the graph; $f'(0) = 0$ and $f'(4) = 0; f''(x) < 0$ for $x < 2; f''(2) = 0, f''(x) > 0$ for $x > 2$.

12. $f(x)$ defined only for $x \geq 0; (0, 0)$ and $(5, 6)$ are on the graph; $f'(x) > 0$ for $x \geq 0$; $f''(x) < 0$ for $x < 5; f''(5) = 0, f''(x) > 0$ for $x > 5$.

In Exercises 13–18, use the given information to make a good sketch of the function $f(x)$ near $x = 3$.

13. $f(3) = 4, f'(3) = -\frac{1}{2}, f''(3) = 5$

14. $f(3) = -2, f'(3) = 0, f''(3) = 1$

15. $f(3) = 1, f'(3) = 0$, inflection point at $x = 3, f'(x) > 0$ for $x > 3$

16. $f(3) = 4, f'(3) = -\frac{3}{2}, f''(3) = -2$

17. $f(3) = -2, f'(3) = 2, f''(3) = 3$

18. $f(3) = 3, f'(3) = 1$, inflection point at $x = 3, f''(x) < 0$ for $x > 3$

19. Refer to the graph in Fig. 17. Fill in each entry of the grid with POS, NEG, or 0.

FIGURE 17

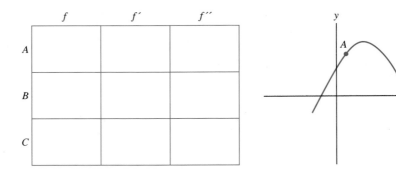

20. The first and second derivatives of the function $f(x)$ have the values given in Table 1.

   (a) Find the x-coordinates of all relative extreme points.
   (b) Find the x-coordinates of all inflection points.

TABLE I	Values of the First Two Derivatives of a Function						
x	$0 \le x < 2$	2	$2 < x < 3$	3	$3 < x < 44$	4	$4 < x \le 6$
$f'(x)$	Positive	0	Negative	Negative	Negative	0	Negative
$f''(x)$	Negative	Negative	Negative	0	Positive	0	Negative

21. The function $s(t)$ in Fig. 18(a) gives the distance traveled by a car after t hours. Is the car going faster at $t = 1$ or $t = 2$?

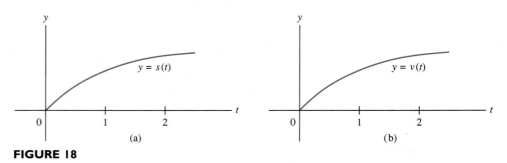

FIGURE 18

22. The function $v(t)$ in Fig. 18(b) gives the velocity of a car after t hours. Is the car going faster at $t = 1$ or $t = 2$?

Exercises 23–34 refer to Fig. 19, which contains the graph of $f'(x)$, the derivative of the function $f(x)$.

FIGURE 19

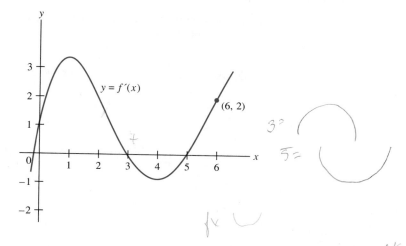

23. Explain why $f(x)$ must be increasing at $x = 6$.
24. Explain why $f(x)$ must be decreasing at $x = 4$.
25. Explain why $f(x)$ has a relative maximum at $x = 3$.
26. Explain why $f(x)$ has a relative minimum at $x = 5$.
27. Explain why $f(x)$ must be concave up at $x = 0$.
28. Explain why $f(x)$ must be concave down at $x = 2$.
29. Explain why $f(x)$ has an inflection point at $x = 1$.
30. Explain why $f(x)$ has an inflection point at $x = 4$.
31. If $f(6) = 3$, what is the equation of the tangent line to the graph of $y = f(x)$ at $x = 6$?
32. If $f(6) = 8$, what is an approximate value of $f(6.5)$?
33. If $f(0) = 3$, what is an approximate value of $f(.25)$?
34. If $f(0) = 3$, what is the equation of the tangent line to the graph of $y = f(x)$ at $x = 0$?
35. Suppose melting snow causes a river to overflow its banks and $h(t)$ is the number of inches of water on Main Street t hours after the melting begins.
    (a) If $h'(100) = \frac{1}{3}$, by approximately how much will the water level change during the next half-hour?
    (b) Which of the following two conditions are the best news?
         (i) $h(100) = 3$, $h'(100) = 2$, $h''(100) = -5$
         (ii) $h(100) = 3$, $h'(100) = -2$, $h''(100) = 5$
36. Suppose $T(t)$ is the temperature on a hot summer day at time t hours.
    (a) If $T'(10) = 4$, by approximately how much will the temperature rise from 10:00 to 10:45?
    (b) Which of the following two conditions are the best news?
         (i) $T(10) = 95$, $T'(10) = 4$, $T''(10) = -3$
         (ii) $T(10) = 95$, $T'(10) = -4$, $T''(10) = 3$

37. By looking at the first derivative, decide which of the curves in Fig. 20 could *not* be the graph of $f(x) = (3x^2 + 1)^4$ for $x \geq 0$.

FIGURE 20

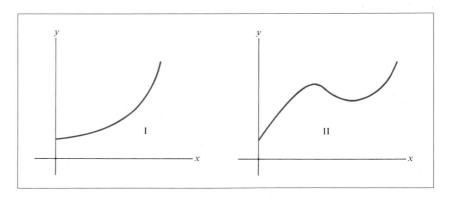

38. By looking at the first derivative, decide which of the curves in Fig. 20 could *not* be the graph of $f(x) = x^3 - 9x^2 + 24x + 1$ for $x \geq 0$. [*Hint:* Factor the formula for $f'(x)$.]

39. By looking at the second derivative, decide which of the curves in Fig. 21 could be the graph of $f(x) = \sqrt{x}$.

FIGURE 21

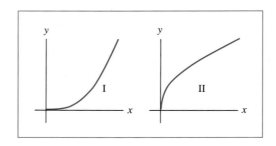

40. By looking at the second derivative, decide which of the curves in Fig. 21 could be the graph of $f(x) = x^{5/2}$.

41. In Fig. 22, the t-axis represents time in minutes.
(a) What is $f(2)$?
(b) Solve $f(t) = 1$.
(c) When does $f(t)$ attain its greatest value?
(d) When does $f(t)$ attain its least value?
(e) What is the rate of change of $f(t)$ at $t = 7.5$?
(f) When is $f(t)$ decreasing at the rate of 1 unit per minute? That is, when is the rate of change equal to -1?
(g) When is $f(t)$ decreasing at the greatest rate?
(h) When is $f(t)$ increasing at the greatest rate?

FIGURE 22

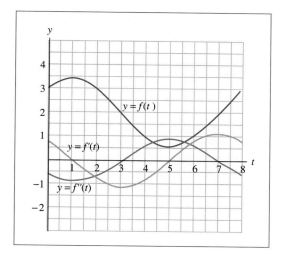

42. In Fig. 23 the t-axis represents time in seconds.
    (a) Solve $f(t) = 10$.
    (b) Find the coordinates of the point on the graph of $f(t)$ at which $f(t)$ has a relative maximum.
    (c) Approximately how fast is the graph of $f(t)$ increasing after 2 seconds?
    (d) For what value of t does $f'(t)$ have a relative minimum?
    (e) Find the coordinates of a point on the graph of $f(t)$ at which the function is increasing at the rate of 3 units per second.
    (f) Find the coordinates of the point on the graph of $f(t)$ at which the function is increasing at the greatest rate.

FIGURE 23

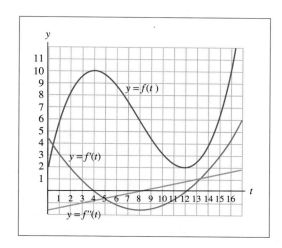

43. Refer to Fig. 24.
    (a) Looking at the graph of $f'(x)$, determine whether $f(x)$ is increasing or decreasing at $x = 9$. Look at the graph of $f(x)$ to confirm your answer.
    (b) Looking at the graph of $f'(x)$, determine the value of x at which $f(x)$ has a relative maximum point. Look at the graph of $f(x)$ to confirm your answer. What are the coordinates of the relative maximum point?
    (c) Repeat part (b) for the relative minimum point of $f(x)$.
    (d) Looking at the graph of $f''(x)$, determine whether $f(x)$ is concave up or concave down at $x = 2$. Look at the graph of $f(x)$ to confirm your answer.
    (e) Looking at the graph of $f''(x)$, determine where $f(x)$ has an inflection point. Look at the graph of $f(x)$ to confirm your answer. What are the coordinates of the inflection point?
    (f) Find the coordinates of the point on the graph of $f(x)$ at which $f(x)$ is increasing at the rate of 6 units per unit change in x.

FIGURE 24

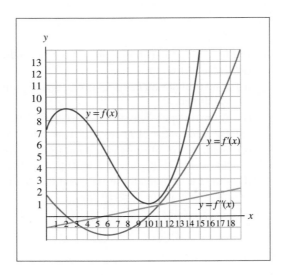

44. In Fig. 25, the t-axis represents time in hours.
    (a) When is $f(t) = 1$?
    (b) Find $f(5)$.
    (c) When is $f(t)$ changing at the rate of $-.08$ units per hour?
    (d) How fast is $f(t)$ changing after 8 hours?
45. The number of farms in the United States t years after 1925 is $f(t)$ million, where f is the function graphed in Fig. 26.
    (a) At what rate was the number of farms declining in 1947?
    (b) In what year were there about 6 million farms?
    (c) Approximately when was the number of farms declining at the rate of 80,000 farms per year?
    (d) When was the number of farms declining fastest?

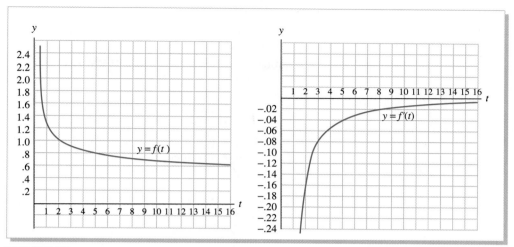

FIGURE 25

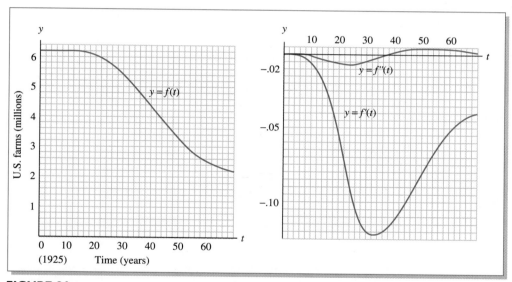

FIGURE 26

46. Flu is spreading on a college campus. After t days, $f(t)$ students will have caught the flu, where f is the function graphed in Fig. 27.

    (a) How many students will be sick after 3 days?

    (b) At what rate will the flu be spreading after 7 days?

    (c) When will 400 students have caught the flu?

    (d) When, during the later stages of the epidemic, will the flu be spreading at the rate of 20 students per day?

    (e) Approximately when will the flu be spreading at the greatest rate? How many students will have contracted the flu at that time?

FIGURE 27

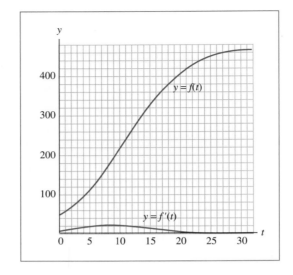

47. U.S. electrical energy production (in trillions of kilowatt-hours) in year t (with 1900 corresponding to $t = 0$) is given by $f(t)$, where f is the function graphed in Fig. 28.
    (a) How much electrical energy was produced in 1960?
    (b) How fast was energy production rising in 1990?
    (c) When did energy production reach 2300 trillion kilowatt-hours?
    (d) When was the level of energy production rising at the rate of 10 trillion kilowatt-hours per year?
    (e) When was energy production growing at the greatest rate? What was the level of production at that time?

FIGURE 28

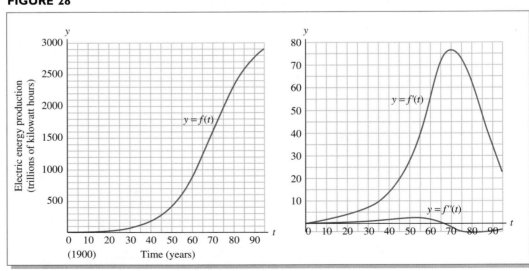

SOLUTIONS TO PRACTICE PROBLEMS 2.2

1. Since $f(2) = 5$, the point $(2, 5)$ is on the graph [Fig. 22(a)]. Since $f'(2) = 1$, the tangent line at the point $(2, 5)$ has slope 1. Draw in the tangent line [Fig. 22(b)]. Near the point $(2, 5)$ the graph looks approximately like the tangent line. Since $f''(2) = -3$, a negative number, the graph is concave down at the point $(2, 5)$. Now we are ready to sketch the graph [Fig. 22(c)].

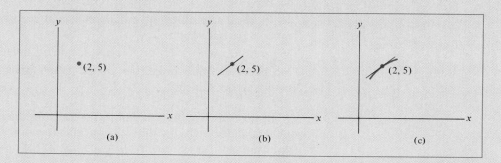

2. (a) Yes. The graph is steadily increasing as we pass through the point $(0, 0)$.

   (b) Since $f'(x) = 3x^2, f'(0) = 3 \cdot 0^2 = 0$.

   (c) There is no contradiction here. The first derivative rule says that if the derivative is positive, the function is increasing. However, it does not say that this is the only condition under which a function is increasing. As we have just seen, sometimes we can have the first derivative zero and the function still increasing.

3. Since $f'(x)$, the derivative of $f(x)$, is negative to the left of $x = 3$ and positive to the right of $x = 3$, $f(x)$ is decreasing to the left of $x = 3$ and increasing to the right of $x = 3$. Therefore, by the definition of a relative minimum point, $f(x)$ has a relative minimum point at $x = 3$.

 # 2.3 CURVE SKETCHING (INTRODUCTION)

In this section and the next we develop our ability to sketch the graphs of functions. There are two important reasons for doing so. First, a geometric "picture" of a function is often easier to comprehend than its abstract formula. Second, the material in this section will provide a foundation for the applications in Sections 2.5 through 2.7.

A "sketch" of the graph of a function $f(x)$ should convey the general shape of the graph—it should show where $f(x)$ is defined and where it is increasing and decreasing and it should indicate, insofar as possible, where $f(x)$ is concave up and concave down. In addition, one or more key points should be accurately located on the graph. These points usually include relative extreme points, inflection points, and x- and y-intercepts. Other features of a graph may be important, too, but we shall discuss them as they arise in examples and applications.

Our general approach to curve sketching will involve four main steps:

1. Starting with $f(x)$, we compute $f'(x)$ and $f''(x)$.
2. Next, we locate all relative maximum and relative minimum points and make a partial sketch.
3. We study the concavity of $f(x)$ and locate all inflection points.
4. We consider other properties of the graph, such as the intercepts, and complete the sketch.

The first step was the main subject of the preceding chapter. We discuss the second and third steps in this section and then present several completely worked examples that combine all four steps in the next.

Locating Relative Extreme Points   The tangent line at a relative maximum or a relative minimum point of a function $f(x)$ has zero slope; that is, the derivative is zero there. Thus we may state the following useful rule.

> Look for possible relative extreme points of $f(x)$ by setting $f'(x) = 0$ and solving for x.   (1)

Suppose that $f'(a) = 0$. Then $x = a$ is a candidate for a relative extreme point of $f(x)$. There are several ways to determine if $f(x)$ has a relative maximum or a relative minimum (or neither) at $x = a$. The method that works in most cases is described in our first two examples.

EXAMPLE 1   The graph of the quadratic function $f(x) = \frac{1}{4}x^2 - x + 2$ is a parabola and so has one relative extreme point. Find it and sketch the graph.

Solution   We begin by computing the first and second derivatives of $f(x)$.

$$f(x) = \frac{1}{4}x^2 - x + 2$$

$$f'(x) = \frac{1}{2}x - 1$$

$$f''(x) = \frac{1}{2}.$$

Setting $f'(x) = 0$, we have $\frac{1}{2}x - 1 = 0$, so that $x = 2$. Thus, $f'(2) = 0$. Geometrically, this means that the graph of $f(x)$ will have a horizontal tangent line

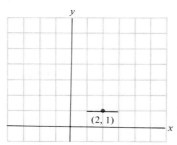

FIGURE 1

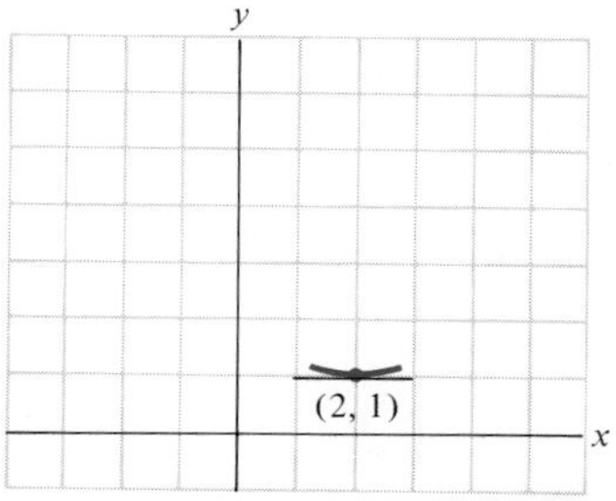

FIGURE 2

at the point where $x = 2$. To plot this point, we substitute the value 2 for x in the original expression for $f(x)$.

$$f(2) = \frac{1}{4}(2)^2 - (2) + 2 = 1.$$

Figure 1 shows the point $(2, 1)$ together with the horizontal tangent line. Is $(2, 1)$ a relative extreme point? In order to decide, we look at $f''(x)$. Since $f''(2) = \frac{1}{2}$, which is positive, the graph of $f(x)$ is concave up at $x = 2$. So a partial sketch of the graph near $(2, 1)$ should look something like Fig. 2.

We see that $(2, 1)$ is a relative minimum point. In fact, it is the only relative extreme point, for there is no other place where the tangent line is horizontal. Since the graph has no other "turning points," it must be decreasing before it gets to $(2, 1)$ and then increasing to the right of $(2, 1)$. Note that since $f''(x)$ is positive (and equal to $\frac{1}{2}$) for all x, the graph is concave upward at each point. A completed sketch is given in Fig. 3. ●

FIGURE 3

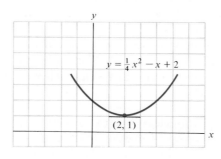

$$y = \frac{1}{4}x^2 - x + 2$$

$(2, 1)$

EXAMPLE 2 Locate all possible relative extreme points on the graph of the function $f(x) = x^3 - 3x^2 + 5$. Check the concavity at these points and use this information to sketch the graph of $f(x)$.

Solution We have

$$f(x) = x^3 - 3x^2 + 5$$
$$f'(x) = 3x^2 - 6x$$
$$f''(x) = 6x - 6.$$

The easiest way to find those values of x for which $f'(x)$ is zero is to factor the expression for $f'(x)$:

$$3x^2 - 6x = 3x(x - 2).$$

From this factorization it is clear that $f'(x)$ will be zero if and only if $x = 0$ or $x = 2$. In other words, the graph will have horizontal tangent lines when $x = 0$ and $x = 2$, and nowhere else.

To plot the points on the graph where $x = 0$ and $x = 2$, we substitute these values back into the original expression for $f(x)$. That is, we compute

$$f(0) = (0)^3 - 3(0)^2 + 5 = 5$$

$$f(2) = (2)^3 - 3(2)^2 + 5 = 1.$$

Figure 4 shows the points $(0, 5)$ and $(2, 1)$, along with the corresponding tangent lines.

Next, we check the concavity of the graph at these points by evaluating $f''(x)$ at $x = 0$ and $x = 2$:

$$f''(0) = 6(0) - 6 = -6$$

$$f''(2) = 6(2) - 6 = +6.$$

Since $f''(0)$ is negative, the graph is concave down at $x = 0$; since $f''(2)$ is positive, the graph is concave up at $x = 2$. A partial sketch of the graph is given in Fig. 5.

It is clear from Fig. 5 that $(0, 5)$ is a relative maximum point and $(2, 1)$ is a relative minimum point. Since they are the only turning points, the graph must be increasing before it gets to $(0, 5)$, decreasing from $(0, 5)$ to $(2, 1)$, and then increasing again to the right of $(2, 1)$. A sketch incorporating these properties appears in Fig. 6. ●

The facts that we used to sketch Fig. 6 could equally well be used to produce the graph in Fig. 7. Which graph really corresponds to $f(x) = x^3 - 3x^2 + 5$? The answer will be clear when we find the inflection points on the graph of $f(x)$.

Locating Inflection Points   An inflection point of a function $f(x)$ can occur only at a value of x for which $f''(x)$ is zero, because the curve is concave up where $f''(x)$ is positive and concave down where $f''(x)$ is negative. Thus we have the following test.

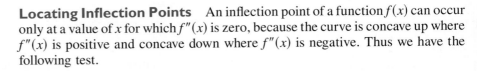

Look for possible inflection points by setting
$f''(x) = 0$ and solving for x.   (2)

Once we have a value of x where the second derivative is zero, say at $x = b$, we must check the concavity of $f(x)$ at nearby points to see if the concavity really changes at $x = b$.

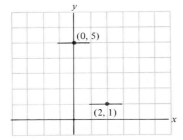

FIGURE 4

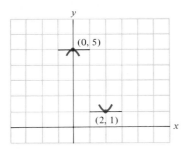

FIGURE 5

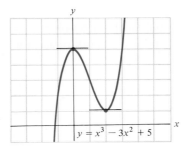

FIGURE 6

EXAMPLE 3 Find the inflection points of the function $f(x) = x^3 - 3x^2 + 5$ and explain why the graph in Fig. 6 has the correct shape.

Solution From Example 2 we have $f''(x) = 6x - 6 = 6(x - 1)$. Clearly, $f''(x) = 0$ if and only if $x = 1$. We will want to plot the corresponding point on the graph, so we compute

$$f(1) = (1)^3 - 3(1)^2 + 5 = 3.$$

Therefore, the only possible inflection point is $(1, 3)$.

Now look back at Fig. 5, where we indicated the concavity of the graph at the relative extreme points. Since $f(x)$ is concave down at $(0, 5)$ and concave up at $(2, 1)$, the concavity must reverse somewhere between these points. Hence $(1, 3)$ must be an inflection point. Furthermore, since the concavity of $f(x)$ reverses nowhere else, the concavity at all points to the left of $(1, 3)$ must be the same (i.e., concave down). Similarly, the concavity of all points to the right of $(1, 3)$ must be the same (i.e., concave up). Thus the graph in Fig. 6 has the correct shape. The graph in Fig. 7 has too many "wiggles," caused by frequent changes in concavity; that is, there are too many inflection points. A correct sketch showing the one inflection point at $(1, 3)$ is given in Fig. 8. ●

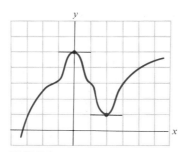

FIGURE 7

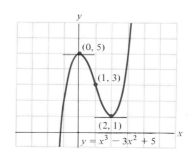

FIGURE 8

EXAMPLE 4 Sketch the graph of $y = -\dfrac{1}{3}x^3 + 3x^2 - 5x$.

Solution Let

$$f(x) = -\frac{1}{3}x^3 + 3x^2 - 5x.$$

Then

$$f'(x) = -x^2 + 6x - 5$$
$$f''(x) = -2x + 6$$

We set $f'(x) = 0$ and solve for x:

$$-(x^2 - 6x + 5) = 0$$
$$-(x - 1)(x - 5) = 0$$
$$x = 1 \quad \text{or} \quad x = 5.$$

Substituting these values of x back into $f(x)$, we find that

$$f(1) = -\frac{1}{3}(1)^3 + 3(1)^2 - 5(1) = -\frac{7}{3}$$

$$f(5) = -\frac{1}{3}(5)^3 + 3(5)^2 - 5(5) = \frac{25}{3}.$$

The information we have so far is given in Fig. 9(a). The sketch in Fig. 9(b) is obtained by computing

$$f''(1) = -2(1) + 6 = 4$$

$$f''(5) = -2(5) + 6 = -4.$$

FIGURE 9

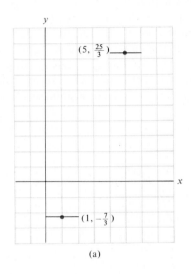

(a)

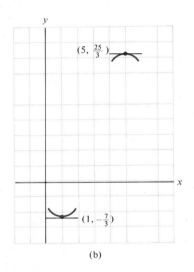

(b)

The curve is concave up at $x = 1$ because $f''(1)$ is positive, and the curve is concave down at $x = 5$ because $f''(5)$ is negative.

Since the concavity reverses somewhere between $x = 0$ and $x = 5$, there must be at least one inflection point. If we set $f''(x) = 0$, we find that

$$-2x + 6 = 0$$

$$x = 3.$$

So the inflection point must occur at $x = 3$. In order to plot the inflection point, we compute

$$f(3) = -\frac{1}{3}(3)^3 + 3(3)^2 - 5(3) = 3.$$

The final sketch of the graph is given in Fig. 10. ●

FIGURE 10

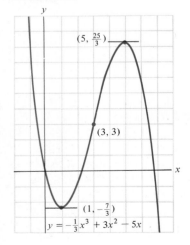

The argument in Example 4 that there must be an inflection point because concavity reverses is valid whenever $f(x)$ is a polynomial. However, it does not always apply to a function whose graph has a break in it. For example, the function $f(x) = 1/x$ is concave down at $x = -1$ and concave up at $x = 1$, but there is no inflection point in between.

A summary of curve-sketching techniques appears at the end of the next section. You may find steps 1, 2, and 3 helpful when working the exercises.

PRACTICE PROBLEMS 2.3

1. Which of the curves in Fig. 11 could possibly be the graph of a function of the form $f(x) = ax^2 + bx + c$, where $a \neq 0$?

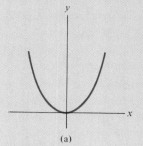
(a)

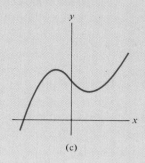
(b)

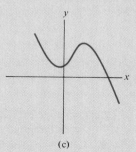
(c)

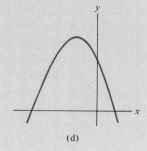
(d)

FIGURE 11

2. Which of the curves in Fig. 12 could possibly be the graph of a function of the form $f(x) = ax^3 + bx^2 + cx + d$, where $a \neq 0$?

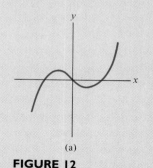
(a)

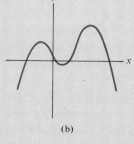
(b)

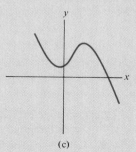
(c)

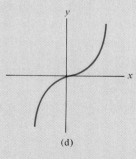
(d)

FIGURE 12

EXERCISES 2.3

Each of the graphs of the functions in Exercises 1–8 has one relative extreme point. Plot this point and check the concavity there. Using only this information, sketch the graph. [As you work the problems, observe that if $f(x) = ax^2 + bx + c$, then $f(x)$ has a relative minimum point when $a > 0$ and a relative maximum point when $a < 0$.]

1. $f(x) = 2x^2 - 8$

2. $f(x) = 3x^2 + 6x - 5$

3. $f(x) = \frac{1}{2}x^2 + x - 4$

4. $f(x) = -\frac{1}{2}x^2 + x - 4$

5. $f(x) = 1 + 6x - x^2$

6. $f(x) = 1 + x + x^2$

7. $f(x) = -x^2 - 8x - 10$

8. $f(x) = -3x^2 + 18x - 20$

Each of the graphs of the functions in Exercises 9–16 has one relative maximum and one relative minimum point. Plot these two points and check the concavity there. Using only this information, sketch the graph.

9. $f(x) = x^3 + 6x^2 + 9x$

10. $f(x) = \frac{1}{9}x^3 - x^2$

11. $f(x) = x^3 - 12x$

12. $f(x) = -\frac{1}{3}x^3 + 9x - 2$

13. $f(x) = -\frac{1}{9}x^3 + x^2 + 9x$

14. $f(x) = 2x^3 - 15x^2 + 36x - 24$

15. $f(x) = -\frac{1}{3}x^3 + 2x^2 - 12$

16. $f(x) = \frac{1}{3}x^3 + 2x^2 - 5x + \frac{8}{3}$

Sketch the following curves, indicating all relative extreme points and inflection points.

17. $y = x^3 - 3x + 2$

18. $y = x^3 - 6x^2 + 9x + 3$

19. $y = 1 + 3x^2 - x^3$

20. $y = -x^3 + 12x - 4$

21. $y = \frac{1}{3}x^3 - x^2 - 3x + 5$

22. $y = x^3 + \frac{3}{2}x^2 - 6x + 4$

23. $y = 2x^3 - 3x^2 - 36x + 20$

24. $y = 11 + 9x - 3x^2 - x^3$

25. Let a, b, c be fixed numbers with $a \neq 0$ and let $f(x) = ax^2 + bx + c$. Is it possible for the graph of $f(x)$ to have an inflection point? Explain your answer.

26. Let a, b, c, d be fixed numbers with $a \neq 0$ and let $f(x) = ax^3 + bx^2 + cx + d$. Is it possible for the graph of $f(x)$ to have more than one inflection point? Explain your answer.

The graph of each function below has one relative extreme point. Find it (giving both x- and y-coordinates) and determine if it is a relative maximum or a relative minimum point. Do not include a sketch of the graph of the function.

27. $f(x) = \frac{1}{4}x^2 - 2x + 7$

28. $f(x) = 5 - 12x - 2x^2$

29. $g(x) = 3 + 4x - 2x^2$

30. $g(x) = x^2 + 10x + 10$

31. $f(x) = 5x^2 + x - 3$

32. $f(x) = 30x^2 - 1800x + 29,000$

In Exercises 33 and 34, determine which function is the derivative of the other.

33.

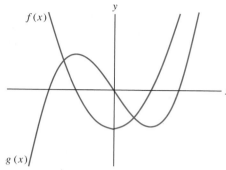

34.

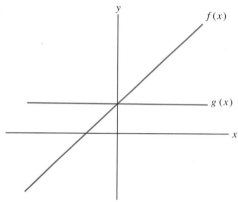

35. Consider the graph of $g(x)$ in Fig. 13.

(a) If $g(x)$ is the first derivative of $f(x)$, what is the nature of $f(x)$ when $x = 2$?

(b) If $g(x)$ is the second derivative of $f(x)$, what is the nature of $f(x)$ when $x = 2$?

FIGURE 13

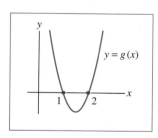

36. The population of the United States (excluding Alaska and Hawaii) t years after 1780 is $f(t)$ million, where $f(t)$ is the function shown in Fig. 14.

(a) What was the population in 1820?

(b) How fast was the population growing in 1870?

(c) Approximately when was the population 90 million?

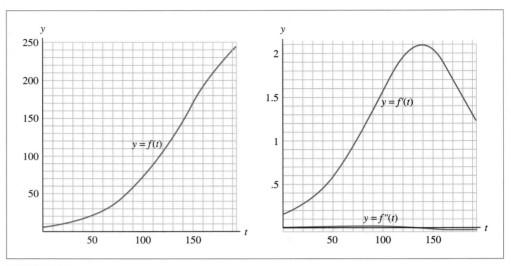

FIGURE 14

(d) Approximately when was the population growing at the rate of 1,100,000 people per year?

(e) When was the population growing at the greatest rate? How large was the population at that time and how fast was it growing?

37. The function $f(t)$ shown in Fig. 15 gives the percentage of homes (with TV sets) that also had a VCR in year t, where year 0 corresponds to 1980.

(a) What percentage of homes had a VCR in 1990?

(b) In what year did just 10% of the homes have a VCR?

(c) At what rate was the percentage of homes with VCRs growing in 1984?

FIGURE 15

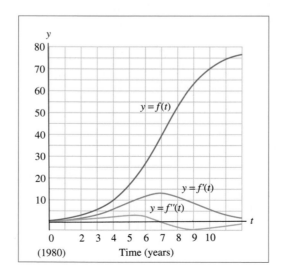

(d) When, in the second half of the 1980s, was the percentage of homes with VCRs growing at the rate of approximately 5 percentage points per year?

(e) When was the percentage of homes with VCRs growing at the greatest rate? Approximately what percentage of homes had VCRs at that time?

SOLUTIONS TO PRACTICE PROBLEMS 2.3

1. Answer: (a) and (d). Curve (b) has the shape of a parabola, but it is not the graph of any function, since vertical lines cross it twice. Curve (c) has two relative extreme points, but the derivative of $f(x)$ is a linear function, which could not be zero for two different values of x.

2. Answer: (a), (c), (d). Curve (b) has three relative extreme points, but the derivative of $f(x)$ is a quadratic function which could not be zero for three different values of x.

2.4 CURVE SKETCHING (CONCLUSION)

In Section 2.3 we discussed the main techniques for curve sketching. Here we add a few finishing touches and examine some slightly more complicated curves.

The more points we plot on a graph, the more accurate the graph becomes. This statement is true even for the simple quadratic and cubic curves in Section 2.3. Of course, the most important points on a curve are the relative extreme points and the inflection points. In addition, the x- and y-intercepts often have some intrinsic interest in an applied problem. The y-intercept is $(0, f(0))$. To find the x-intercepts on the graph of $f(x)$, we must find those values of x for which $f(x) = 0$. Since this can be a difficult (or impossible) problem, we shall find x-intercepts only when they are easy to find or when a problem specifically requires us to find them.

When $f(x)$ is a quadratic function, as in Example 1, we can easily compute the x-intercepts (if they exist) either by factoring the expression for $f(x)$ or by using the quadratic formula.

EXAMPLE 1   Sketch the graph of $y = \frac{1}{2}x^2 - 4x + 7$.

Solution   Let

$$f(x) = \frac{1}{2}x^2 - 4x + 7.$$

Then

$$f'(x) = x - 4$$
$$f''(x) = 1.$$

Since $f'(x) = 0$ only when $x = 4$ and since $f''(4)$ is positive, $f(x)$ must have a relative minimum point at $x = 4$. The relative minimum point is $(4, f(4)) = (4, -1)$.

The y-intercept is $(0, f(0)) = (0, 7)$. To find the x-intercepts, we set $f(x) = 0$ and solve for x:

$$\frac{1}{2}x^2 - 4x + 7 = 0.$$

The expression for $f(x)$ is not easily factored, so we use the quadratic formula to solve the equation.

$$x = \frac{-(-4) \pm \sqrt{(-4)^2 - 4(\frac{1}{2})(7)}}{2(\frac{1}{2})} = 4 \pm \sqrt{2}.$$

The x-intercepts are $(4 - \sqrt{2}, 0)$ and $(4 + \sqrt{2}, 0)$. To plot these points we use the approximation $\sqrt{2} \approx 1.4$. (See Fig. 1.) ●

FIGURE I

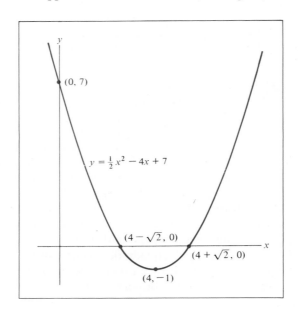

$(0, 7)$

$y = \frac{1}{2}x^2 - 4x + 7$

$(4 - \sqrt{2}, 0)$

$(4 + \sqrt{2}, 0)$

$(4, -1)$

EXAMPLE 2   Sketch the graph of

$$f(x) = \frac{1}{6}x^3 - \frac{3}{2}x^2 + 5x + 1.$$

Solution

$$f(x) = \frac{1}{6}x^3 - \frac{3}{2}x^2 + 5x + 1$$

$$f'(x) = \frac{1}{2}x^2 - 3x + 5$$

$$f''(x) = x - 3.$$

Let us set $f'(x) = 0$ and try to solve for x:

$$\frac{1}{2}x^2 - 3x + 5 = 0.$$

(1)

If we apply the quadratic formula with $a = \frac{1}{2}$, $b = -3$, and $c = 5$, we see that $b^2 - 4ac$ is negative, and so there is no solution to (1). In other words, $f'(x)$ is never zero. Thus the graph cannot have relative extreme points. If we evaluate $f'(x)$ at some x, say $x = 0$, we see that the first derivative is positive, and so $f(x)$ is increasing there. Since the graph of $f(x)$ is a smooth curve with no relative extreme points and no breaks, $f(x)$ must be increasing for all x. (If a function were increasing at $x = a$ and decreasing at $x = b$, then it would have a relative extreme point between a and b.)

Now let us check the concavity.

	$f''(x) = x - 3$	Graph of $f(x)$
$x < 3$	Negative	Concave down
$x = 3$	Zero	Concavity reverses
$x > 3$	Positive	Concave up

The inflection point is $(3, f(3)) = (3, 7)$. The y-intercept is $(0, f(0)) = (0, 1)$. We omit the x-intercept because it is difficult to solve the cubic equation $\frac{1}{6}x^3 - \frac{3}{2}x^2 + 5x + 1 = 0$.

The quality of our sketch of the curve will be improved if we first sketch the tangent line at the inflection point. To do this, we need to know the slope of the graph at $(3, 7)$:

$$f'(3) = \frac{1}{2}(3)^2 - 3(3) + 5 = \frac{1}{2}.$$

We draw a line through $(3, 7)$ with slope $\frac{1}{2}$ and then complete the sketch as shown in Fig. 2. ●

The methods used so far will work for most of the functions we shall study. Occasionally, however, $f'(x)$ and $f''(x)$ are both zero at some value of x, say, $x = a$, and we cannot tell from the second derivative if the function has a relative minimum or a relative maximum at $x = a$ or neither. This exceptional situation can be handled using the following observation: As we move from left to right in the vicinity of a relative maximum, the slope decreases and changes sign from positive to negative. In the vicinity of a relative minimum, the slope increases and changes sign from negative to positive.

The following example shows how this observation may be used in curve sketching.

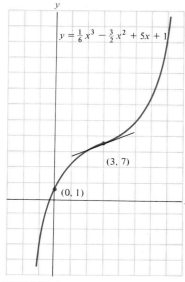

$y = \frac{1}{6}x^3 - \frac{3}{2}x^2 + 5x + 1$

$(3, 7)$

$(0, 1)$

FIGURE 2

EXAMPLE 3   Sketch the graph of $f(x) = (x - 2)^4 - 1$.

Solution

$$f(x) = (x - 2)^4 - 1$$
$$f'(x) = 4(x - 2)^3$$
$$f''(x) = 12(x - 2)^2.$$

Clearly, $f'(x) = 0$ only if $x = 2$. So the curve has a horizontal tangent at $(2, f(2)) = (2, -1)$. Since $f''(2) = 0$, the second derivative rule cannot be used to determine whether the point $(2, -1)$ is a relative maximum, a relative minimum, or neither. However, note that

$$f'(x) = 4(x - 2)^3 \quad \begin{cases} < 0 & \text{if } x < 2 \\ > 0 & \text{if } x > 2 \end{cases}$$

since the cube of a negative number is negative and the cube of a positive number is positive. Therefore, as x goes from left to right in the vicinity of 2, the first derivative goes from negative to positive. By the preceding observation, this means that the point $(2, -1)$ is a relative minimum.

The y-intercept is $(0, f(0)) = (0, 15)$. To find the x-intercepts, we set $f(x) = 0$ and solve for x:

$$(x - 2)^4 - 1 = 0$$
$$(x - 2)^4 = 1$$
$$x - 2 = 1 \quad \text{or} \quad x - 2 = -1$$
$$x = 3 \quad \text{or} \quad x = 1.$$

(See Fig. 3.) ●

A Graph with Asymptotes   Graphs similar to the one in the next example will arise in several applications later in this chapter.

EXAMPLE 4   Sketch the graph of $f(x) = x + (1/x)$, for $x > 0$.

Solution

$$f(x) = x + \frac{1}{x}$$

$$f'(x) = 1 - \frac{1}{x^2}$$

$$f''(x) = \frac{2}{x^3}.$$

We set $f'(x) = 0$ and solve for x:

$$1 - \frac{1}{x^2} = 0$$

$$1 = \frac{1}{x^2}$$

FIGURE 3

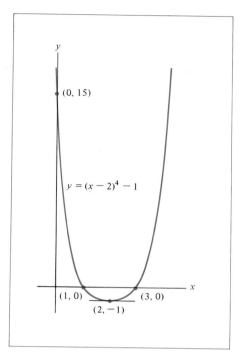

$$x^2 = 1$$

$$x = 1.$$

(We exclude the case $x = -1$ because we are only considering positive values of x.) The graph has a horizontal tangent at $(1, f(1)) = (1, 2)$. Now, $f''(1) = 2 > 0$, and so the graph is concave up at $x = 1$ and $(1, 2)$ is a relative minimum point. In fact, $f''(x) = (2/x^3) > 0$ for all positive x, and therefore the graph is concave up at all points.

Before sketching the graph, notice that as x approaches zero [a point at which $f(x)$ is not defined], the term $1/x$ in the formula for $f(x)$ becomes arbitrarily large. Thus $f(x)$ has the y-axis as an asymptote. For large values of x, $f(x) = x + (1/x)$ is only slightly larger than x; that is, the graph of $f(x)$ is slightly above the graph of $y = x$. As x increases, the graph of $f(x)$ has the line $y = x$ as an asymptote. (See Fig. 4.) ●

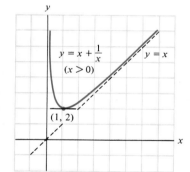

FIGURE 4

Summary of Curve-Sketching Techniques

1. Compute $f'(x)$ and $f''(x)$.
2. Find all relative extreme points.
   (a) Set $f'(x) = 0$ and solve for x. Suppose that $x = a$ is a solution. Substitute $x = a$ into $f(x)$ to find $f(a)$, plot the point $(a, f(a))$, and draw a small horizontal tangent line through the point. Compute $f''(a)$.
      (i) If $f''(a) > 0$, draw a small concave up arc with $(a, f(a))$ as its lowest point. The curve has a relative minimum at $x = a$.

(ii) If $f''(a) < 0$, draw a small concave down arc with $(a, f(a))$ as its peak. The curve has a relative maximum at $x = a$.

(iii) If $f''(a) = 0$, examine $f'(x)$ to the left and right of $x = a$ in order to determine if the function changes from increasing to decreasing, or vice versa. If a relative extreme point is indicated, draw an appropriate arc as in parts (i) and (ii).

(b) Repeat the preceding steps for each of the solutions to $f'(x) = 0$.

3. Find all the inflection points of $f(x)$.

(a) Set $f''(x) = 0$ and solve for x. Suppose that $x = b$ is a solution. Compute $f(b)$ and plot the point $(b, f(b))$.

(b) Test the concavity of $f(x)$ to the right and left of b. If the concavity changes at $x = b$, then $(b, f(b))$ is an inflection point.

4. Consider other properties of the function and complete the sketch.

(a) If $f(x)$ is defined at $x = 0$, the y-intercept is $(0, f(0))$.

(b) Does the partial sketch suggest that there are x-intercepts? If so, they are found by setting $f(x) = 0$ and solving for x. (Solve only in easy cases or when a problem essentially requires you to calculate the x-intercepts.)

(c) Observe where $f(x)$ is defined. Sometimes the function is given only for restricted values of x. Sometimes the formula for $f(x)$ is meaningless for certain values of x.

(d) Look for possible asymptotes.

(i) Examine the formula for $f(x)$. If some terms become insignificant as x gets large and if the rest of the formula gives the equation of a straight line, then that straight line is an asymptote.

(ii) Suppose that there is some point a such that $f(x)$ is defined for x near a but not at a (e.g., $1/x$ at $x = 0$).) If $f(x)$ gets arbitrarily large (in the positive or negative sense) as x approaches a, then the vertical line $x = a$ is an asymptote for the graph.

(e) Complete the sketch.

TECHNOLOGY PROJECT 2.5

Analysis of a Medical Experiment

In a medical experiment, the body weight of a baby rat in the control group after t days was $f(t) = 4.96 + .48t + .17t^2 - .0048t^3$ grams.

1. Graph $f(t)$, $f'(t)$, $f''(t)$ for $0 \leq t \leq 20$.
2. Approximately how much did a rat weight after 7 days?
3. Approximately when did a rat's weight reach 27 grams?
4. Approximately how fast was a rat gaining weight after 4 days?
5. Approximately when was a rat gaining weight at the rate of 2 grams per day?
6. Approximately when was the rat gaining weight at the fastest rate?

*Johnson, Wogenrich, Hsi, Skipper, and Greenberg, "Growth Retardation During the Sucking Period in Expanded Litters of Rats; Observations of Growth Patterns and Protein Turnover," *Growth, Development and Aging,* 55, (1991), 263–273.

TECHNOLOGY PROJECT 2.6
Analysis of a Botanical Study

The height of the tropical bunch-grass elephant millet, t days after mowing, is

$$f(t) = -3.14 + .142t - .0016t^2 + .0000079t^3 - .0000000133t^4$$

meters.*

1. Graph $f(t)$, $f'(t)$, $f''(t)$ for $50 \leq t \leq 200$.
2. How tall was the grass after 160 days?
3. When was the grass 1.75 meters high?
4. How fast was the grass growing after 80 days?
5. When was the grass growing at the rate of .03 meters per day?
6. When was the grass growing at the slowest rate?
7. During the period from the 100th to the 240th day, when was the grass growing at the fastest rate?

TECHNOLOGY PROJECT 2.7
Analysis of a Medical Experiment

The relationship between the area of the pupil of the eye and the intensity of light was analyzed by B. H. Crawford,[†] who concluded that the area of the pupil is

$$f(x) = \frac{160x^{-.4} + 94.8}{4x^{-.4} + 15.8}$$

square millimeters when x units of light are entering the eye per unit time.

1. Graph $f(x)$, $f'(x)$, $f''(x)$ for $0 \leq x \leq 15$.
2. How large is the pupil when 5 units of light are entering the eye per unit time?
3. When 4 units of light are entering the eye per unit time, what is the rate of change of pupil size with respect to a unit change in light intensity?
4. For what light intensity is the pupil size 9 square millimeters?
5. For what light intensity is the rate of change of pupil size with respect to a unit change in light intensity approximately $-.2$?

PRACTICE PROBLEMS 2.4

Determine whether each of the following functions has an asymptote as x gets large. If so, give the equation of the straight line which is the asymptote.

1. $f(x) = \dfrac{3}{x} - 2x + 1$ 2. $f(x) = \sqrt{x} + x$ 3. $f(x) = \dfrac{1}{2x}$

---

* Woodward, K. R., and Prine, G. M., "Crop Quality and Utilization," *Crop Science*, 33 (1993), 818–824.

† B. H. Crawford, "The dependence of pupil size upon the external light stimulus under static and variable conditions," *Proc. Royal Society, Series B*, 121 (1937), 376–395.

EXERCISES 2.4

Find the x-intercepts of the following curves.

1. $y = x^2 - 3x + 1$ **2.** $y = x^2 + 5x + 5$

3. $y = 2x^2 + 5x + 2$ **4.** $y = 4 - 2x - x^2$

5. $y = 4x - 4x^2 - 1$ **6.** $y = 3x^2 + 7x + 2$

7. Show that the function $f(x) = \frac{1}{3}x^3 - 2x^2 + 5x$ has no relative extreme points.

8. Show that the function $f(x) = 5 - 11x + 6x^2 - \frac{4}{3}x^3$ is always decreasing.

Sketch the graphs of the following functions.

9. $f(x) = x^3 - 6x^2 + 12x - 6$ **10.** $f(x) = -x^3$

11. $f(x) = x^3 + 3x + 1$ **12.** $f(x) = 4 - x - x^3$

13. $f(x) = 5 - 13x + 6x^2 - x^3$ **14.** $f(x) = 2x^3 + x - 2$

15. $f(x) = \frac{4}{3}x^3 - 2x^2 + x$ **16.** $f(x) = 1 - 4x - 3x^2 - x^3$

17. $f(x) = 1 - 3x + 3x^2 - x^3$ **18.** $f(x) = x^3 - 6x^2 + 12x - 5$

19. $f(x) = x^4 - 6x^2$ **20.** $f(x) = 1 + 6x^2 - 3x^4$

21. $f(x) = (x - 3)^4$ **22.** $f(x) = (x + 2)^4 - 1$

Sketch the graphs of the following functions for $x > 0$.

23. $y = \dfrac{1}{x} + \dfrac{1}{4}x$ **24.** $y = \dfrac{1}{x} + 9x$

25. $y = \dfrac{9}{x} + x + 1$ **26.** $y = \dfrac{12}{x} + 3x + 1$

27. $y = \dfrac{2}{x} + \dfrac{x}{2} + 2$ **28.** $y = \dfrac{x}{3} + \dfrac{12}{x} - 1$

29. $y = 6\sqrt{x} - x$ **30.** $y = 21x + \dfrac{8400}{x}$

In Exercises 31 and 32, determine which function is the derivative of the other.

31.

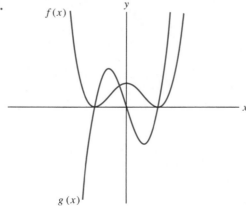

32.

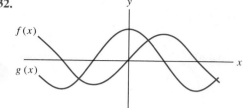

SOLUTIONS TO
PRACTICE
PROBLEMS 2.4

Functions with asymptotes as x gets large have the form $f(x) = g(x) + mx + b$, where $g(x)$ approaches zero as x gets large. The function $g(x)$ often looks like c/x or $c/(ax + d)$. The asymptote will be the straight line $y = mx + b$.

1. Here $g(x)$ is $3/x$ and the asymptote is $y = -2x + 1$.

2. This function has no asymptote as x gets large. Of course, it can be written as $g(x) + mx + b$, where $m = 1$ and $b = 0$. However, $g(x) = \sqrt{x}$ does not approach zero as x gets large.

3. Here $g(x)$ is $\dfrac{1}{2x}$ and the asymptote is $y = 0$. That is, the function has the x-axis as an asymptote.

2.5 OPTIMIZATION PROBLEMS

One of the most important applications of the derivative concept is to "optimization" problems, in which some quantity must be maximized or minimized. Examples of such problems abound in many areas of life. An airline must decide how many daily flights to schedule between two cities in order to maximize its profits. A doctor wants to find the minimum amount of a drug that will produce a desired response in one of her patients. A manufacturer needs to determine how often to replace certain equipment in order to minimize maintenance and replacement costs.

Our purpose in this section is to illustrate how calculus can be used to solve optimization problems. In each example we will find or construct a function that provides a "mathematical model" for the problem. Then, by sketching the graph of this function, we will be able to determine the answer to the original optimization problem by locating the highest or lowest point on the graph. The y-coordinate of this point will be the maximum value or minimum value of the function.

The first two examples are quite simple because the functions to be studied are given explicitly.

EXAMPLE 1   Find the minimum value of the function $f(x) = 2x^3 - 15x^2 + 24x + 19$ for $x \geq 0$.

FIGURE 1

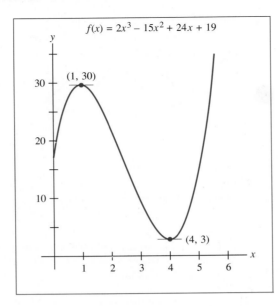

$f(x) = 2x^3 - 15x^2 + 24x + 19$

Solution   Using the curve-sketching techniques from Section 2.3, we obtain the graph in Fig. 1. The lowest point on the graph is $(4, 3)$. The minimum *value* of the function $f(x)$ is the y-coordinate of this point—namely, 3. ●

EXAMPLE 2   Suppose that a ball is thrown straight up into the air and its height after t seconds is $4 + 48t - 16t^2$ feet. Determine how long it will take for the ball to reach its maximum height and determine the maximum height.

Solution   Consider the function $f(t) = 4 + 48t - 16t^2$. For each value of t, $f(t)$ is the height of the ball at time t. We want to find the value of t for which $f(t)$ is the greatest. Using the techniques of Section 2.3, we sketch the graph of $f(t)$. (See Fig. 2.) Note that we may neglect the portions of the graph corresponding to points for which either $t < 0$ or $f(t) < 0$. [A negative value of $f(t)$ would

FIGURE 2

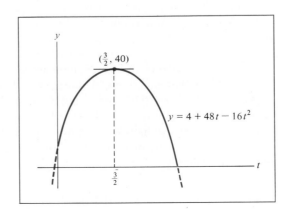

$(\frac{3}{2}, 40)$

$y = 4 + 48t - 16t^2$

$\frac{3}{2}$

correspond to the ball being underneath the ground.] We see that $f(t)$ is greatest when $t = \frac{3}{2}$. At this value of t, the ball attains a height of 40 feet. [Note that the curve in Fig. 2 is the graph of $f(t)$, *not* a picture of the physical path of the ball.]

ANSWER

The ball reaches its maximum height of 40 feet in 1.5 seconds. ●

EXAMPLE 3   A person wants to plant a rectangular garden along one side of a house, with a picket fence on the other three sides of the garden. Find the dimensions of the largest garden that can be enclosed using 40 feet of fencing.

Solution   The first step is to make a simple diagram and assign letters to the quantities that may vary. Let us denote the dimensions of the rectangular garden by w and x (Fig. 3). The phrase "largest garden" indicates that we must maximize the area, A, of the garden. In terms of the variables w and x,

$$A = wx \qquad (1)$$

FIGURE 3

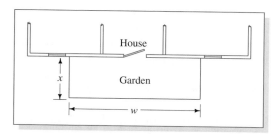

The fencing on three sides must total 40 running feet; that is

$$2x + w = 40. \qquad (2)$$

We now solve equation (2) for w in terms of x:

$$w = 40 - 2x. \qquad (3)$$

Substituting this expression for w into equation (1), we have

$$A = (40 - 2x)x = 40x - 2x^2. \qquad (4)$$

We now have a formula for the area A that depends on just one variable, and so we may graph A as a function of x. From the statement of the problem, the value of $2x$ can be at most 40, so that the domain of the function consists of x in the interval $(0, 20)$.

Using curve-sketching techniques, we obtain the graph in Fig. 4. We see from the graph that the area is maximized when $x = 10$. (The maximum area is 200 square feet, but this fact is not needed for the problem.) From equation (3) we find that when $x = 10$,

$$w = 40 - 2(10) = 20.$$

FIGURE 4

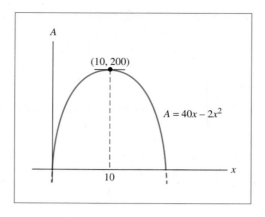

ANSWER

$w = 20$ feet, $x = 10$ feet. ●

Equation (1) in Example 3 is called an *objective equation*. It expresses the quantity to be optimized (the area of the garden) in terms of the variables w and x. Equation (2) is called a *constraint equation* because it places a limit or constraint on the way x and w may vary.

EXAMPLE 4   The manager of a department store wants to build a 600-square-foot rectangular enclosure on the store's parking lot in order to display some equipment. Three sides of the enclosure will be built of redwood fencing, at a cost of $14 per running foot. The fourth side will be built of cement blocks, at a cost of $28 per running foot. Find the dimensions of the enclosure that will minimize the total cost of the building materials.

Solution   Let x be the length of the side built out of cement blocks and let y be the length of an adjacent side, as shown in Fig. 5. The phrase "minimize the total cost" tells us that the objective equation should be a formula giving the total cost of the building materials.

FIGURE 5   Rectangular enclosure.

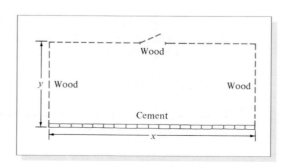

[cost of redwood] = [length of redwood fencing] × [cost per foot]

$$= (x + 2y) \cdot 14 = 14x + 28y.$$

[cost of cement blocks] = [length of cement wall] × [cost per foot]

$$= x \cdot 28.$$

If C denotes the total cost of the materials, then

$$C = (14x + 28y) + 28x$$

$$C = 42x + 28y \qquad \text{(objective equation)}. \qquad (5)$$

Since the area of the enclosure must be 600 square feet, the constraint equation is

$$xy = 600. \qquad (6)$$

We simplify the objective equation by solving (6) for one of the variables, say y, and substituting into (5). Since $y = 600/x$,

$$C = 42x + 28\left(\frac{600}{x}\right) = 42x + \frac{16,800}{x}.$$

We now have C as a function of the single variable x. From the context, we must have $x > 0$, since a length must be positive. However, to any positive value for x, there is a corresponding value for C. So the domain of C consists of all $x > 0$. We may now sketch the graph of C (Fig. 6). (A similar curve was sketched in

FIGURE 6

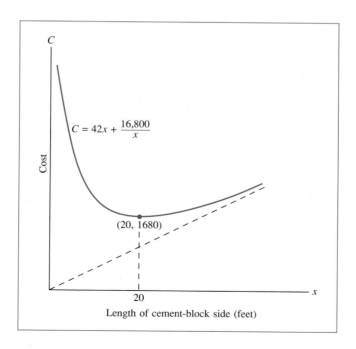

$$C = 42x + \frac{16,800}{x}$$

(20, 1680)

Cost

20

Length of cement-block side (feet)

Example 4 of Section 2.4.) The minimum total cost of $1680 occurs where $x = 20$. From equation (6) we find that the corresponding value of y is $\frac{600}{20} = 30$.

ANSWER

$x = 20$ feet, $y = 30$ feet. ●

EXAMPLE 5　U.S. parcel post regulations state that packages must have length plus girth of no more than 84 inches. Find the dimensions of the cylindrical package of greatest volume that is mailable by parcel post.

Solution　Let l be the length of the package and let r be the radius of the circular end. (See Fig. 7.) The phrase "greatest volume" tells us that the objective equation should express the volume of the package in terms of the dimensions l and r. Let V denote the volume. Then

$$V = [\text{area of base}] \cdot [\text{length}]$$

$$V = \pi r^2 l \qquad \text{(objective equation).} \tag{7}$$

The girth equals the circumference of the end—that is, $2\pi r$. Since we want the package to be as large as possible, we must use the entire 84 inches allowable:

$$\text{length} + \text{girth} = 84$$

$$l + 2\pi r = 84 \text{ (constraint equation).} \tag{8}$$

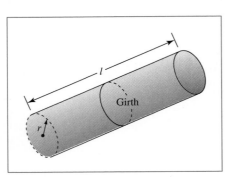

FIGURE 7　Cylindrical mailing package.

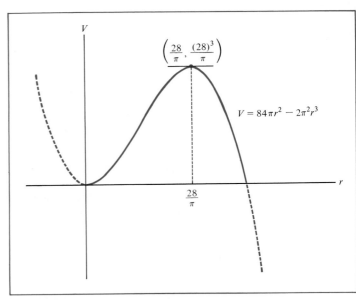

FIGURE 8

We now solve equation (8) for one of the variables, say $l = 84 - 2\pi r$. Substituting this expression into (7), we obtain

$$V = \pi r^2 (84 - 2\pi r) = 84\pi r^2 - 2\pi^2 r^3. \tag{9}$$

Let $f(r) = 84\pi r^2 - 2\pi^2 r^3$. Then, for each value of r, $f(r)$ is the volume of the parcel with end radius r that meets the postal regulations. We want to find that value of r for which $f(r)$ is as large as possible.

Using curve-sketching techniques, we obtain the graph of $f(r)$ in Fig. 8. The domain excludes values of r that are negative and values of r for which the volume $f(r)$ is negative. Points corresponding to values of r not in the domain are shown with a dashed curve. We see that the volume is greatest when $r = 28/\pi$.

From (8) we find that the corresponding value of l is

$$l = 84 - 2\pi r = 84 - 2\pi\left(\frac{28}{\pi}\right)$$

$$= 84 - 56 = 28.$$

The girth when $r = 28/\pi$ is

$$2\pi r = 2\pi\left(\frac{28}{\pi}\right) = 56.$$

ANSWER

$l = 28$ inches, $r = 28/\pi$ inches, girth $= 56$ inches. ●

Suggestions for Solving an Optimization Problem

1. Draw a picture, if possible.
2. Decide what quantity Q is to be maximized or minimized.
3. Assign letters to other quantities that may vary.
4. Determine the "objective equation" that expresses Q as a function of the variables assigned in step 3.
5. Find the "constraint equation" that relates the variables to each other and to any constants that are given in the problem.
6. Use the constraint equation to simplify the objective equation in such a way that Q becomes a function of only one variable. Determine the domain of this function.
7. Sketch the graph of the function obtained in step 6 and use this graph to solve the optimization problem.

Note   Optimization problems often involve geometric formulas. The most common formulas are shown in Fig. 9.

FIGURE 9

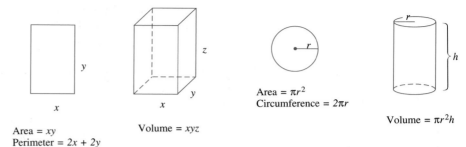

Area = xy
Perimeter = $2x + 2y$

Volume = xyz

Area = πr^2
Circumference = $2\pi r$

Volume = $\pi r^2 h$

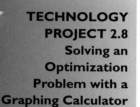

**TECHNOLOGY
PROJECT 2.8
Solving an
Optimization
Problem with a
Graphing Calculator**

A patient's temperature (in degrees Fahrenheit) t hours after contracting an illness is given by

$$T(t) = -.0008t^3 + .0288t^2 + 98.6, \qquad 0 \le t \le 36$$

Determine the time at which the maximum temperature occurs and the value of the temperature at that time.

**PRACTICE
PROBLEMS 2.5**

1. A canvas wind shelter for the beach has a back, two square sides, and a top (Fig. 10). Suppose that 96 square feet of canvas are to be used. Find the dimensions of the shelter for which the space inside the shelter (i.e., the volume) will be maximized.

FIGURE 10   Wind shelter.

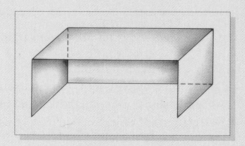

2. In Practice Problem 1, what are the objective equation and the constraint equation?

EXERCISES 2.5

1. For what x does the function $g(x) = 10 + 40x - x^2$ have its maximum value?
2. Find the maximum value of the function $f(x) = 12x - x^2$ and give the value of x where this maximum occurs.

3. Find the minimum value of $f(t) = t^3 - 6t^2 + 40$, $t \geq 0$, and give the value of t where this minimum occurs.

4. For what t does the function $f(t) = t^2 - 24t$ have its minimum value?

5. Three hundred and twenty dollars are available to fence in a rectangular garden. The fencing for the side of the garden facing the road costs $6 per foot and the fencing for the other three sides costs $2 per foot. [See Fig. 11(a).] Consider the problem of finding the dimensions of the largest possible garden.

   (a) Determine the objective and constraint equations.

   (b) Express the quantity to be maximized as a function of x.

   (c) Find the optimal values of x and y.

FIGURE 11

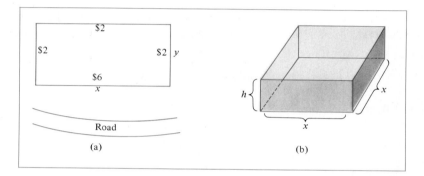

(a)   (b)

6. Figure 11(b) shows an open rectangular box with a square base. Consider the problem of finding the values of x and h for which the volume is 32 cubic feet and the total surface area of the box is minimal. (The surface area is the sum of the areas of the five faces of the box.)

   (a) Determine the objective and constraint equations.

   (b) Express the quantity to be minimized as a function of x.

   (c) Find the optimal values of x and h.

7. Postal requirements specify that parcels must have length plus girth of at most 84 inches. Consider the problem of finding the dimensions of the square-ended rectangular package of greatest volume that is mailable.

   (a) Draw a square-ended rectangular box. Label each edge of the square end with the letter x and label the remaining dimension of the box with the letter h.

   (b) Express the length plus the girth in terms of x and h.

   (c) Determine the objective and constraint equations.

   (d) Express the quantity to be maximized as a function of x.

   (e) Find the optimal values of x and h.

8. Consider the problem of finding the dimensions of the rectangular garden of area 100 square meters for which the amount of fencing needed to surround the garden is as small as possible.

   (a) Draw a picture of a rectangle and select appropriate letters for the dimensions.

   (b) Determine the objective and constraint equations.

   (c) Find the optimal values for the dimensions.

9. A rectangular garden of area 75 square feet is to be surrounded on three sides by a brick wall costing $10 per foot and on one side by a fence costing $5 per foot. Find the dimensions of the garden such that the cost of materials is minimized.

10. A closed rectangular box with square base and a volume of 12 cubic feet is to be constructed using two different types of materials. The top is made of a metal costing $2 per square foot and the remainder of wood costing $1 per square foot. Find the dimensions of the box for which the cost of materials is minimized.

11. Find the dimensions of the closed rectangular box with square base and volume 8000 cubic centimeters that can be constructed with the least amount of material.

12. A canvas wind shelter for the beach has a back, two square sides, and a top. Find the dimensions for which the volume will be 250 cubic feet and that requires the least possible amount of canvas.

13. A farmer has $1500 available to build an E-shaped fence along a straight river so as to create two identical rectangular pastures. (See Fig. 12.) The materials for the side parallel to the river cost $6 per foot and the materials for the three sections perpendicular to the river cost $5 per foot. Find the dimensions for which the total area is as large as possible.

FIGURE 12   Rectangular pastures along a river.

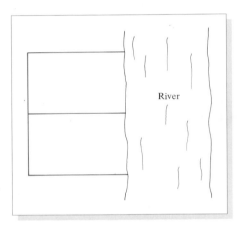

14. Find the dimensions of the rectangular garden of greatest area that can be fenced off (all four sides) with 300 meters of fencing.

15. Find two positive numbers, x and y, whose sum is 100 and whose product is as large as possible.

16. Find two positive numbers, x and y, whose product is 100 and whose sum is as small as possible.

17. Figure 13(a) shows a Norman window, which consists of a rectangle capped by a semicircular region. Find the value of x such that the perimeter of the window will be 14 feet and the area of the window will be as large as possible.

18. A large soup can is to be designed so that the can will hold 16π cubic inches (about 28 ounces) of soup. [See Fig. 13(b).] Find the values of x and h for which the amount of metal needed is as small as possible.

19. In Example 3 one can solve the constraint equation (2) for x instead of w to get $x = 20 - \frac{1}{2}w$. Substituting this for x in (1), one has $A = xw = (20 - \frac{1}{2}w)w$. Sketch

FIGURE 13

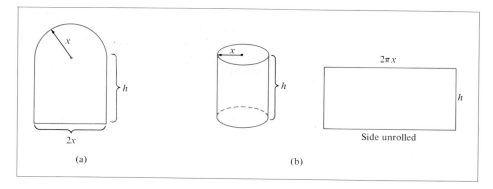

the graph of the equation $A = 20w - \frac{1}{2}w^2$ and show that the maximum occurs when $w = 20$ and $x = 10$.

20. A ship uses $5x^2$ dollars of fuel per hour when traveling at a speed of x miles per hour. The other expenses of operating the ship amount to \$2000 per hour. What speed minimizes the cost of a 500-mile trip? [*Hint:* Express cost in terms of speed and time. The constraint equation is *distance = speed × time.*]

21. Find the point on the graph of $y = \sqrt{x}$ that is closest to the point $(2, 0)$. See Fig. 14. [*Hint:* $\sqrt{(x-2)^2 + y^2}$ has its smallest value when $(x-2)^2 + y^2$ does. Therefore, just minimize the second expression.]

FIGURE 14   Shortest distance from a point to a curve.

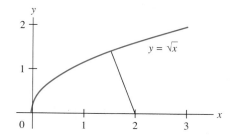

SOLUTIONS TO PRACTICE PROBLEMS 2.5

1. Since the sides of the wind shelter are square, we may let x represent the length of each side of the square. The remaining dimension of the wind shelter can be denoted by the letter h. (See Fig. 15.) The volume of the shelter is $x^2 h$, and this is to be maximized. Since we have learned to maximize only functions of a single variable, we must express h in terms of x. We must use the information that 96 feet of canvas are used—that is, $2x^2 + 2xh = 96$. [*Note:* The roof and the back each have an area xh, and each end has an area x^2.] We now solve this equation for h.

$$2x^2 + 2xh = 96$$

$$2xh = 96 - 2x^2$$

$$h = \frac{96}{2x} - \frac{2x^2}{2x} = \frac{48}{x} - x.$$

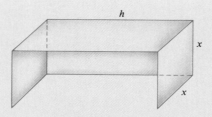

The volume V is

$$x^2h = x^2\left(\frac{48}{x} - x\right) = 48x - x^3.$$

By sketching the graph of $V = 48x - x^3$, we see that V has a maximum value when $x = 4$. Then, $h = \frac{48}{4} - 4 = 12 - 4 = 8$. So each end of the shelter should be a 4-foot by 4-foot square and the top should be 8 feet long.

2. The objective equation is $V = x^2h$, since it expresses the volume (the quantity to be maximized) in terms of the variables. The constraint equation is $2x^2 + 2xh = 96$, for it relates the variables to each other; that is, it can be used to express one of the variables in terms of the other.

 2.6 FURTHER OPTIMIZATION PROBLEMS

In this section we apply the optimization techniques developed in the preceding section to some practical situations.

EXAMPLE 1   Suppose that, on a certain route, an airline carries 8000 passengers per month, each paying $50. The airline wants to increase the fare. However, the market research department estimates that for each $1 increase in fare, the airline will lose 100 passengers. Determine the price that maximizes the airline's revenue.

Solution   Since the problem calls for setting an optimum price, let x be the price per ticket. The other variable is the number of passengers, which we can denote by n. The goal is to maximize revenue.

$$[\text{revenue}] = [\text{number of passengers}] \cdot [\text{price per ticket}]$$

$$= n \cdot x.$$

If R denotes revenue, then the objective equation is

$$R = nx.$$

Since the number of passengers n depends on the price x, the constraint equation can be derived by expressing n in terms of x.

$$\begin{bmatrix} \text{number of} \\ \text{passengers} \end{bmatrix} = \begin{bmatrix} \text{original number} \\ \text{of passengers} \end{bmatrix} - \begin{bmatrix} \text{number of passengers lost} \\ \text{due to fare increase} \end{bmatrix}$$

$$n \quad = \quad 8000 \quad - \quad (x - 50) \cdot 100$$

$$= 13{,}000 - 100x.$$

The number of passengers lost due to the fare increase was obtained by multiplying the number of dollars of fare increase, $x - 50$, by the number of passengers lost for each dollar of fare increase. Therefore,

$$R = nx = (13{,}000 - 100x)x = 13{,}000x - 100x^2.$$

Figure 1 shows that revenue is maximized when the price per ticket is \$65.   ●

FIGURE I   **Revenue function for airline fares.**

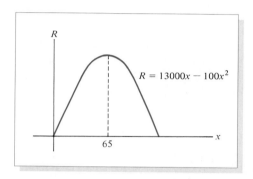

Inventory Control   When a firm regularly orders and stores supplies for later use or resale, it must decide on the size of each order. If it orders enough supplies to last an entire year, the business will incur heavy *carrying costs*. Such costs include insurance, storage costs, and cost of capital that is tied up in inventory. To reduce these carrying costs, the firm could order small quantities of the supplies at frequent intervals. However, such a policy increases the *ordering costs*. These might consist of minimum freight charges, the clerical costs of preparing the orders, and the costs of receiving and checking the orders when they arrive. Clearly, the firm must find an inventory ordering policy that lies between these two extremes.

The following example illustrates the use of calculus to minimize the firm's annual inventory cost, where

$$[\text{inventory cost}] = [\text{ordering cost}] + [\text{carrying cost}].$$

We assume that each order is the same size. The size of the order that minimizes the inventory cost is called the *economic order quantity,* commonly referred to in business as the EOQ.*

---

*See James C. Van Horne, *Financial Management and Policy,* 6th ed. (Englewood Cliffs, N.J.: Prentice Hall, Inc., 1983), pp. 416–420.

EXAMPLE 2   A supermarket manager wants to establish an optimal inventory policy for frozen orange juice. It is estimated that a total of 1200 cases will be sold at a steady rate during the next year. The manager plans to place several orders of the same size spaced equally throughout the year. Use the following data to determine the economic order quantity, that is, the order size that minimizes the total ordering and carrying cost.

1. The ordering cost for each delivery is $75.
2. It costs $8 to carry one case of orange juice in inventory for one year. (Carrying costs should be computed on the average inventory during the order-reorder period.)

Solution   Let x be the order quantity and r the number of orders placed during the year. The number of cases of orange juice in inventory declines steadily from x cases (each time a new order is filled) to 0 cases at the end of each order-reorder period. Figure 2 shows that the average number of cases in storage during the year is $x/2$. Since the carrying cost for one case is $8 per year, the cost for $x/2$ cases is $8 \cdot (x/2)$ dollars. Now

$$[\text{inventory cost}] = [\text{ordering cost}] + [\text{carrying cost}]$$

$$= 75r + 8 \cdot \frac{x}{2}$$

$$= 75r + 4x.$$

If C denotes the inventory cost, then the objective equation is

$$C = 75r + 4x.$$

FIGURE 2   Average inventory level.

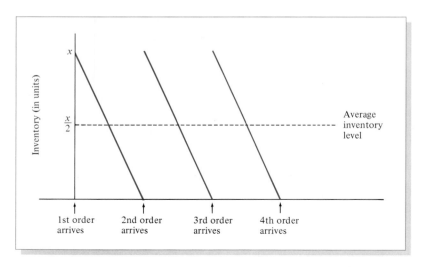

Since there are r orders of x cases each, the total number of cases ordered during the year is $r \cdot x$. Therefore, the constraint equation is

$$r \cdot x = 1200.$$

The constraint equation says that $r = 1200/x$. Substitution into the objective equation yields

$$C = \frac{90{,}000}{x} + 4x.$$

Figure 3 is the graph of C as a function of x, for $x > 0$. The total cost is at a minimum when $x = 150$. Therefore, the optimum inventory policy is to order 150 cases at a time and to place $1200/150 = 8$ orders during the year. ●

FIGURE 3   Cost function for inventory problem.

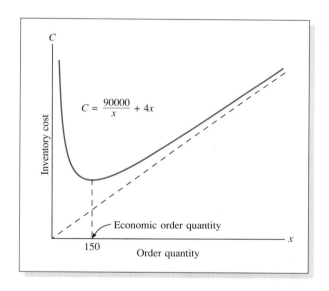

$$C = \frac{90000}{x} + 4x$$

Inventory cost

Economic order quantity

150

Order quantity

EXAMPLE 3   What should the inventory policy of Example 2 be if sales of frozen orange juice increase fourfold (i.e., 4800 cases are sold each year), but all other conditions are the same?

Solution   The only change in our previous solution is in the constraint equation, which now becomes

$$r \cdot x = 4800.$$

The objective equation is, as before,

$$C = 75r + 4x.$$

Since $r = 4800/x$,

$$C = 75 \cdot \frac{4800}{x} + 4x$$

$$= \frac{360{,}000}{x} + 4x.$$

Now

$$C' = -\frac{360{,}000}{x^2} + 4.$$

Setting $C' = 0$ yields

$$\frac{360{,}000}{x^2} = 4$$

$$90{,}000 = x^2$$

$$x = 300.$$

Therefore, the economic order quantity is 300 cases. ●

Notice that although the sales increased by a factor of 4, the economic order quantity increased by only a factor of 2 ($=\sqrt{4}$). In general, a store's inventory of an item should be proportional to the square root of the expected sales. (See Exercise 9 for a derivation of this result.) Many stores tend to keep their average inventories at a fixed percentage of sales. For example, each order may contain enough goods to last for 4 or 5 weeks. This policy is likely to create excessive inventories of high-volume items and uncomfortably low inventories of slower-moving items.

Manufacturers have an inventory-control problem similar to that of retailers. They have the carrying costs of storing finished products and the startup costs of setting up each production run. The size of the production run that minimizes the sum of these two costs is called the *economic lot size*. See Exercises 6 and 7.

EXAMPLE 4   When a person coughs, the trachea (windpipe) contracts. (See Fig. 4.) Let

$r_0 =$ normal radius of the trachea,
$r =$ radius during a cough,
$P =$ increase in air pressure in the trachea during cough,
$v =$ velocity of air through trachea during cough.

Use the following principles of fluid flow to determine how much the trachea should contract in order to create the greatest air velocity—that is, the most effective condition for clearing the lungs and the trachea.

FIGURE 4   Contraction of the windpipe during coughing.

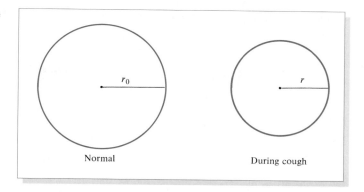

Normal   During cough

1. $r_0 - r = aP$, for some positive constant a. (Experiment has shown that, during coughing, the decrease in the radius of the trachea is nearly proportional to the increase in the air pressure.)
2. $v = b \cdot P \cdot \pi r^2$, for some positive constant b. (The theory of fluid flow requires that the velocity of the air forced through the trachea is proportional to the product of the increase in the air pressure and the area of a cross section of the trachea.)

Solution   In this problem the constraint equation **(1)** and the objective equation **(2)** are given directly. Solving equation **(1)** for P and substituting this result into equation **(2)**, we have

$$v = b\left(\frac{r_0 - r}{a}\right)\pi r^2 = k(r_0 - r)r^2,$$

where $k = b\pi/a$. To find the radius at which the velocity v is a maximum, we first compute the derivatives:

$$v = k(r_0 r^2 - r^3)$$

$$\frac{dv}{dr} = k(2r_0 r - 3r^2) = kr(2r_0 - 3r)$$

$$\frac{d^2 v}{dr^2} = k(2r_0 - 6r).$$

We see that $\dfrac{dv}{dr} = 0$ when $r = 0$ or when $2r_0 - 3r = 0$; that is, when $r = \frac{2}{3}r_0$.

It is easy to see that $\dfrac{d^2 v}{dr^2}$ is positive at $r = 0$ and is negative at $r = \frac{2}{3}r_0$. The graph

FIGURE 5   **Graph for the windpipe problem.**

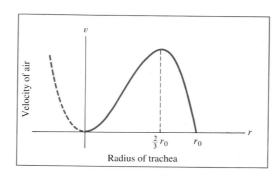

Radius of trachea

of v as a function of r is drawn in Fig. 5. The air velocity is maximized at $r = \frac{2}{3}r_0$. ●

When solving optimization problems, we look for the maximum or minimum point on a graph. From our discussion of curve sketching, we have seen that this point occurs either at a relative extreme point or at an endpoint of the domain of definition. In all our optimization problems so far, the maximum or minimum points were at relative extreme points. In the next example, the optimum point is an endpoint.

EXAMPLE 5   A rancher has 204 meters of fencing from which to build two corrals: one square and the other rectangular with length that is twice the width. Find the dimensions that result in the greatest combined area.

Solution   Let x be the width of the rectangular corral and h be the length of each side of the square corral. (See Fig. 6.) Let A be the combined area. Then

$$A = [\text{area of square}] + [\text{area of rectangle}]$$
$$= h^2 + 2x^2.$$

FIGURE 6   Two corrals.

The constraint equation is

$$204 = [\text{perimeter of square}] + [\text{perimeter of rectangle}] = 4h + 6x.$$

Since the perimeter of the rectangle cannot exceed 204, we must have $0 \leq 6x \leq 204$, or $0 \leq x \leq 34$. Solving the constraint equation for h and substituting into the objective equation leads to the function graphed in Fig. 7. The graph reveals that the area is minimized when $x = 18$. However, the problem asks for

FIGURE 7 Combined area of the corrals.

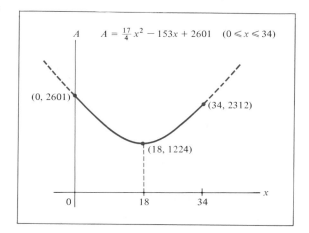

$A = \frac{17}{4}x^2 - 153x + 2601 \quad (0 \leqslant x \leqslant 34)$

(0, 2601)

(34, 2312)

(18, 1224)

the *maximum* possible area. From Fig. 7 we see that this occurs at the endpoint where $x = 0$. Therefore, the rancher should build only the square corral, with $h = 204/4 = 51$ meters. In this example, the objective function has an endpoint extremum; namely, the maximum value occurs at the endpoint $x = 0$. ●

TECHNOLOGY PROJECT 2.9

Solving an Optimization Problem with a Graphing Calculator

Coffee consumption in the United States is greater on a per capita basis than anywhere else in the world. However, due to price fluctuations of coffee beans and worries over the health effects of caffeine, coffee consumption has varied considerably over the years. According to data published in the Wall Street Journal, the number of cups y consumed daily per adult in year x (with 1955 corresponding to $x = 0$) is given by the mathematical model:

$$y = 2.76775 + 0.0847943x - 0.00832058x^2 + 0.000144017x^3$$

1. Graph y as a function of x to show daily coffee consumption from 1955 through 1994.

2. Use the TRACE to determine when coffee consumption was least during this period (to the closest month). What was the daily coffee consumption at that time?

3. Use the TRACE to determine when coffee consumption was greatest during this period (to the closest month). What was the daily coffee consumption at that time?

4. At what rate was coffee consumption changing in 1990?

5. When was the rate of change of coffee consumption the most (to the nearest month)?

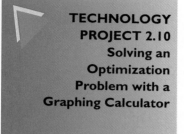

TECHNOLOGY PROJECT 2.10
Solving an Optimization Problem with a Graphing Calculator

The total amount T (in millions of dollars per week) spent on cough and cold remedies x weeks after the beginning of the cough and cold season (September 27) is given by the model

$$T = 11.25 + 0.9597x + 0.5039x^2 - 0.04133x^3 + 0.0007916x^4$$

1. Graph T as a function of x to show amounts spent on cough and cold remedies for the 28-week cough and cold season.
2. Use the TRACE to determine during which week sales are at a maximum. What is the amount of sales for that week?
3. When is the rate of change of sales greatest?
4. When is the rate of change of sales least?

PRACTICE PROBLEMS 2.6

1. An apple orchard produces a profit of $40 a tree when planted with 1000 trees. Because of overcrowding, the profit per tree (for each tree in the orchard) is reduced by 2 cents for each additional tree planted. How many trees should be planted in order to maximize the total profit from the orchard?
2. In the inventory problem of Example 2, suppose that the sales of frozen orange juice increase ninefold; that is, 10,800 cases are sold each year. What is the new economic order quantity?

EXERCISES 2.6

1. An artist is planning to sell signed prints of her latest work. If 50 prints are offered for sale, she can charge $400 each. However, if she makes more than 50 prints, she must lower the price of all the prints by $5 for each print in excess of the 50. How many prints should the artist make in order to maximize her revenue?
2. A swimming club offers memberships at the rate of $200, provided that a minimum of 100 people join. For each member in excess of 100, the membership fee will be reduced $1 per person (for each member). At most, 160 memberships will be sold. How many memberships should the club try to sell in order to maximize its revenue?

3. In the planning of a sidewalk cafe, it is estimated that if there are 12 tables, the daily profit will be $10 per table. Because of overcrowding, for each additional table the profit per table (for every table in the cafe) will be reduced by $.50. How many tables should be provided to maximize the profit from the cafe?
4. A certain toll road averages 36,000 cars per day when charging $1 per car. A survey concludes that increasing the toll will result in 300 fewer cars for each cent of increase. What toll should be charged in order to maximize the revenue?
5. A California distributor of sporting equipment expects to sell 10,000 cases of tennis balls during the coming year at a steady rate. Yearly carrying costs (to be computed

on the average number of cases in stock during the year) are $10 per case, and the cost of placing an order with the manufacturer is $80.

(a) Find the inventory cost incurred if the distributor orders 500 cases at a time during the year.

(b) Determine the economic order quantity, that is, the order quantity that minimizes the inventory cost.

6. The Great American Tire Co. expects to sell 600,000 tires of a particular size and grade during the next year. Sales tend to be roughly the same from month to month. Setting up each production run costs the company $15,000. Carrying costs, based on the average number of tires in storage, amount to $5 per year for one tire.

   (a) Determine the costs incurred if there are 10 production runs during the year.

   (b) Find the economic lot size (i.e., the production run size that minimizes the overall cost of producing the tires).

7. Foggy Optics, Inc. makes laboratory microscopes. Setting up each production run costs $2500. Insurance costs, based on the average number of microscopes in the warehouse, amount to $20 per microscope per year. Storage costs, based on the maximum number of microscopes in the warehouse, amount to $15 per microscope per year. Suppose that the company expects to sell 1600 microscopes at a fairly uniform rate throughout the year. Determine the number of production runs that will minimize the company's overall expenses.

8. A bookstore is attempting to determine the economic order quantity for a popular book. The store sells 8000 copies of this book a year. The store figures that it costs $40 to process each new order for books. The carrying cost (due primarily to interest payments) is $2 per book, to be figured on the maximum inventory during an order-reorder period. How many times a year should orders be placed?

9. A store manager wants to establish an optimal inventory policy for an item. Sales are expected to be at a steady rate and should total Q items sold during the year. Each time an order is placed, a cost of h dollars is incurred. Carrying costs for the year will be s dollars per item, to be figured on the average number of items in storage during the year. Show that the total inventory cost is minimized when each order calls for $\sqrt{2hQ/s}$ items.

10. Refer to the inventory problem of Example 2. Suppose that the distributor offers a discount of $1 per case for orders of 600 or more cases. Should the manager change the quantity ordered?

11. Starting with a 100-foot-long stone wall, a farmer would like to construct a rectangular enclosure by adding 400 feet of fencing as shown in Fig. 8(a). Find the values of x and w that result in the greatest possible area.

FIGURE 8   Rectangular enclosures.

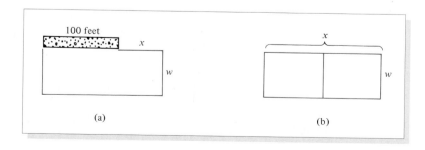

(a)   (b)

12. Rework Exercise 11 for the case where only 200 feet of fencing is added to the stone wall.

13. A rectangular corral of 54 square meters is to be fenced off and then divided by a fence into two sections, as shown in Fig. 8(b). Find the dimensions of the corral so that the amount of fencing required is minimized.

14. Referring to Exercise 13, suppose that the cost of the fencing for the boundary is $5 per meter and the dividing fence costs $2 per meter. Find the dimensions of the corral that minimize the cost of the fencing.

15. A travel agency offers a boat tour of several Caribbean islands for 3 days and 2 nights. For a group of 12 people, the cost per person is $800. For each additional person above the 12-person minimum, the cost per person is reduced by $20 for each person in the group. The maximum tour group size is 25. What tour group size produces the greatest revenue for the travel agency?

16. Design an open rectangular box with square ends, having volume 36 cubic inches, that minimizes the amount of material required for construction.

17. A storage shed is to be built in the shape of a box with a square base. It is to have a volume of 150 cubic feet. The concrete for the base costs $4 per square foot, the material for the roof costs $2 per square foot, and the material for the sides costs $2.50 per square foot. Find the dimensions of the most economical shed.

18. A supermarket is to be designed as a rectangular building with a floor area of 12,000 square feet. The front of the building will be mostly glass and will cost $70 per running foot for materials. The other three walls will be constructed of brick and cement block, at a cost of $50 per running foot. Ignore all other costs (labor, cost of foundation and roof, etc.) and find the dimensions of the base of the building that will minimize the cost of the materials for the four walls of the building.

19. A certain airline requires that rectangular packages carried on an airplane by passengers be such that the sum of the three dimensions is at most 120 centimeters. Find the dimensions of the square-ended rectangular package of greatest volume that meets this requirement.

20. An athletic field [Fig. 9(a)] consists of a rectangular region with a semicircular region at each end. The perimeter will be used for a 440-yard track. Find the value of x for which the area of the rectangular region is as large as possible.

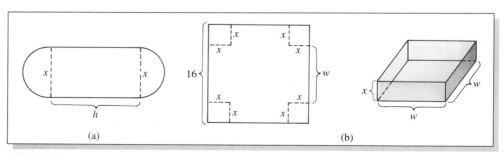

(a)   (b)

FIGURE 9

21. An open rectangular box is to be constructed by cutting square corners out of a 16-by 16-inch piece of cardboard and folding up the flaps. [See Fig. 9(b).] Find the value of x for which the volume of the box will be as large as possible.

22. A closed rectangular box is to be constructed with a base that is twice as long as it is wide. Suppose that the total surface area must be 27 square feet. Find the dimensions of the box that will maximize the volume.

23. Let $f(t)$ be the amount of oxygen (in suitable units) in a lake t days after sewage is dumped into the lake, and suppose that $f(t)$ is given approximately by

$$f(t) = 1 - \frac{10}{t + 10} + \frac{100}{(t + 10)^2}.$$

At what time is the oxygen content increasing the fastest?

24. The daily output of a coal mine after t hours of operation is approximately $40t + t^2 - \frac{1}{15}t^3$ tons, $0 \leq t \leq 12$. Find the maximum rate of output (in tons of coal per hour).

25. Consider a parabolic arch whose shape may be represented by the graph of $y = 9 - x^2$, where the base of the arch lies on the x-axis from $x = -3$ to $x = 3$. Find the dimensions of the rectangular window of maximum area that can be constructed inside the arch.

26. Advertising for a certain product is terminated, and t weeks later the weekly sales are $f(t)$ cases, where $f(t) = 1000(t + 8)^{-1} - 4000(t + 8)^{-2}$. At what time is the weekly sales amount falling the fastest?

27. An open rectangular box of volume 400 cubic inches has a square base and a partition down the middle. See Fig. 10. Find the dimensions of the box for which the amount of material needed to construct the box is as small as possible.

FIGURE 10 Open rectangular box with dividing partition.

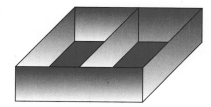

28. Suppose $f(x)$ is defined on the interval $0 \leq x \leq 5$ and $f'(x)$ is negative for all x. For what value of x will $f(x)$ have its greatest value?

SOLUTIONS TO PRACTICE PROBLEMS 2.6

1. Since the question asks for the optimum "number of trees," let x be the number of trees to be planted. The other quantity that varies is "profit per tree." So let p be the profit per tree. The objective is to maximize total profit, call it T. Then

[total profit] = [profit per tree] · [number of trees]

$$T = p \cdot x.$$

Since the profit per tree p depends on the number of trees planted, x, the constraint equation can be derived by expressing p in terms of x.

$$\begin{bmatrix} \text{profit} \\ \text{per tree} \end{bmatrix} = \begin{bmatrix} \text{original profit} \\ \text{per tree} \end{bmatrix} - \begin{bmatrix} \text{loss in profit (per tree)} \\ \text{due to increase} \end{bmatrix}$$

$$p = \qquad 40 \qquad - \qquad (x - 1000)(.02)$$

$$= 60 - .02x.$$

The loss in profit (per tree) due to the increase in the number of trees was obtained by multiplying $x - 1000$, the number of trees in excess of 1000, by the amount of money lost (per tree) for each excess tree. Therefore,

$$T = p \cdot x = (60 - .02x)x = 60x - .02x^2.$$

By computing first and second derivatives and sketching the graph, we easily find that total profit is maximized when $x = 1500$. Therefore, 1500 trees should be planted.

2. This problem can be solved in the same manner that Example 3 was solved. However, the comment made at the end of Example 3 indicates that the economic order quantity should increase by a factor of 3, since $3 = \sqrt{9}$. Therefore, the economic order quantity is $3 \cdot 150 = 450$ cases.

2.7 APPLICATIONS OF CALCULUS TO BUSINESS AND ECONOMICS

In recent years economic decision making has become more and more mathematically oriented. Faced with huge masses of statistical data, depending on hundreds or even thousands of different variables, business analysts and economists have increasingly turned to mathematical methods to help them describe what is happening, predict the effects of various policy alternatives, and choose reasonable courses of action from the myriad possibilities. Among the mathematical methods employed is calculus. In this section we illustrate just a few of the many applications of calculus to business and economics. All our applications will center around what economists call *the theory of the firm*. In other words, we study the activity of a business (or possibly a whole industry) and restrict our analysis to a time period during which background conditions (such as supplies of raw materials, wage rates, taxes) are fairly constant. We then show how calculus can help the management of such a firm make vital production decisions.

Management, whether or not it knows calculus, utilizes many functions of the sort we have been considering. Examples of such functions are

$C(x)$ = cost of producing x units of the product,

$R(x)$ = revenue generated by selling x units of the product,

$P(x) = R(x) - C(x)$ = the profit (or loss) generated by producing and selling x units of the product.

Note that the functions $C(x)$, $R(x)$, $P(x)$ often are defined only for nonnegative integers—that is, for $x = 0, 1, 2, 3, \ldots$. The reason is that it does not make sense to speak about the cost of producing -1 cars or the revenue generated by selling 3.62 refrigerators. Thus each of the functions may give rise to a set of discrete points on a graph, as in Fig. 1(a). In studying these functions, however, economists usually draw a smooth curve through the points and assume that $C(x)$ is actually defined for all positive x. Of course, we must often interpret answers to problems in light of the fact that x is, in most cases, a nonnegative integer.

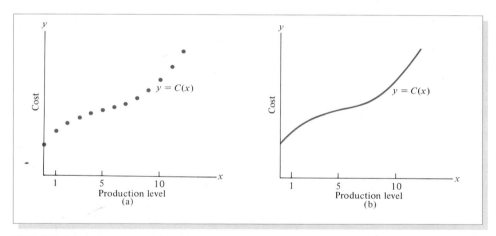

FIGURE I   A cost function.

Cost Functions   If we assume that a cost function $C(x)$ has a smooth graph as in Fig. 1(b), we can use the tools of calculus to study it. A typical cost function is analyzed in Example 1.

EXAMPLE I   Suppose that the cost function for a manufacturer is given by $C(x) = (10^{-6})x^3 - .003x^2 + 5x + 1000$ dollars.

(a) Describe the behavior of the marginal cost.

(b) Sketch the graph of $C(x)$.

Solution   The first two derivatives of $C(x)$ are given by

$$C'(x) = (3 \cdot 10^{-6})x^2 - .006x + 5$$

$$C''(x) = (6 \cdot 10^{-6})x - .006.$$

Let us sketch the marginal cost $C'(x)$ first. From the behavior of $C'(x)$, we will be able to graph $C(x)$. The marginal cost function $y = (3 \cdot 10^{-6})x^2 - .006x + 5$ has as its graph a parabola that opens upward. Since $y' = C''(x) = .000006(x - 1000)$, we see that the parabola has a horizontal tangent at $x = 1000$. So the minimum value of $C'(x)$ occurs at $x = 1000$. The corresponding y-coordinate is

$$(3 \cdot 10^{-6})(1000)^2 - .006 \cdot (1000) + 5 = 3 - 6 + 5 = 2.$$

The graph of $y = C'(x)$ is shown in Fig. 2. Consequently, at first the marginal cost decreases. It reaches a minimum of 2 at production level 1000 and increases thereafter. This answers part (a). Let us now graph $C(x)$. Since the graph shown in Fig. 2 is the graph of the derivative of $C(x)$, we see that $C'(x)$ is never zero, so there are no relative extreme points. Since $C'(x)$ is always positive, $C(x)$ is always increasing (as any cost curve should be). Moreover, since $C'(x)$ decreases for x less than 1000 and increases for x greater than 1000, we see that $C(x)$ is concave down for x less than 1000, is concave up for x greater than 1000, and has an inflection point at $x = 1000$. The graph of $C(x)$ is drawn in Fig. 3. Note that the inflection point of $C(x)$ occurs at the value of x for which marginal cost is a minimum. ●

FIGURE 2   A marginal cost function.

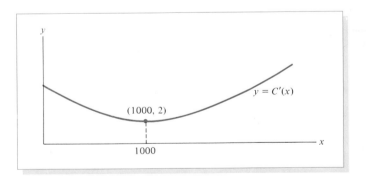

Actually, most marginal cost functions have the same general shape as the marginal cost curve of Example 1. For when x is small, production of additional units is subject to economies of production, which lowers unit costs. Thus, for x small, marginal cost decreases. However, increased production eventually leads to overtime, use of less efficient, older plants, and competition for scarce raw materials. As a result, the cost of additional units will increase for very large x. So we see that $C'(x)$ initially decreases and then increases.

FIGURE 3   **A cost function.**

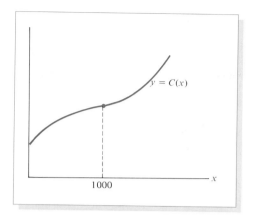

$y = C(x)$

1000

x

Revenue Functions

Revenue Functions   In general, a business is concerned not only with its costs, but also with its revenues. Recall that if $R(x)$ is the revenue received from the sale of x units of some commodity, then the derivative $R'(x)$ is called the *marginal revenue*. Economists use this to measure the rate of increase in revenue per unit increase in sales.

$R'(x)$ marginal revenue

If x units of a product are sold at a price p per unit, then the total revenue $R(x)$ is given by

$$R(x) = x \cdot p.$$

If a firm is small and is in competition with many other companies, its sales have little effect on the market price. Then since the price is constant as far as the one firm is concerned, the marginal revenue $R'(x)$ equals the price p [that is, $R'(x)$ is the amount that the firm receives from the sale of one additional unit]. In this case, the revenue function will have a graph as in Fig. 4.

An interesting problem arises when a single firm is the only supplier of a certain product or service—that is, when the firm has a monopoly. Consumers will buy large amounts of the commodity if the price per unit is low and less if the price is raised. For each quantity x, let $f(x)$ be the highest price per unit that

FIGURE 4   **A revenue curve.**

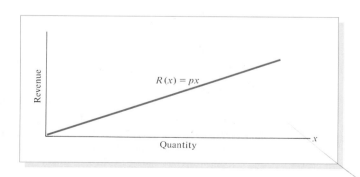

Revenue

$R(x) = px$

Quantity

x

FIGURE 5   A demand curve.

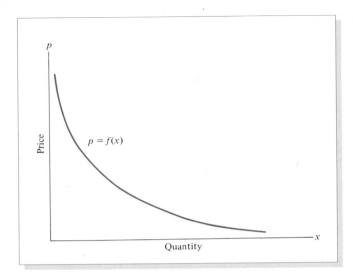

demand equation
$p = f(x)$

can be set in order to sell all x units to consumers. Since selling greater quantities requires a lowering of the price, $f(x)$ will be a decreasing function. Figure 5 shows a typical "demand curve" that relates the quantity demanded, x, to the price $p = f(x)$.

The *demand equation* $p = f(x)$ determines the total revenue function. If the firm wants to sell x units, the highest price it can set is $f(x)$ dollars per unit, and so the total revenue from the sale of x units is

$$R(x) = x \cdot p = x \cdot f(x). \tag{1}$$

The concept of a demand curve applies to an entire industry (with many producers) as well as to a single monopolistic firm. In this case, many producers offer the same product for sale. If x denotes the total output of the industry, then $f(x)$ is the market price per unit of output and $x \cdot f(x)$ is the total revenue earned from the sale of the x units.

EXAMPLE 2   The demand equation for a certain product is $p = 6 - \frac{1}{2}x$. Find the level of production that results in maximum revenue.

Solution   In this case, the revenue function $R(x)$ is

$$R(x) = x \cdot p = x(6 - \tfrac{1}{2}x) = 6x - \tfrac{1}{2}x^2.$$

The marginal revenue is given by

$$R'(x) = 6 - x.$$

The graph of $R(x)$ is a parabola that opens downward (Fig. 6). It has a horizontal tangent precisely at those x for which $R'(x) = 0$—that is, for those x at which

FIGURE 6   Maximizing revenue.

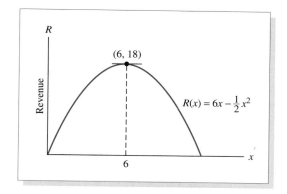

marginal revenue is 0. The only such x is $x = 6$. The corresponding value of revenue is

$$R(6) = 6 \cdot 6 - \frac{1}{2}(6)^2 = 18.$$

Thus the rate of production resulting in maximum revenue is $x = 6$, which results in total revenue of 18. ●

EXAMPLE 3   The WMA Bus Lines offers sightseeing tours of Washington, D.C. One of the tours, priced at \$7 per person, had an average demand of about 1000 customers per week. When the price was lowered to \$6, the weekly demand jumped to about 1200 customers. Assuming that the demand equation is linear, find the tour price that should be charged per person in order to maximize the total revenue each week.

Solution   First we must find the demand equation. Let x be the number of customers per week and let p be the price of a tour ticket. Then $(x, p) = (1000, 7)$ and $(x, p) = (1200, 6)$ are on the demand curve (Fig. 7). Using the point-slope formula for the line through these two points, we have

$$p - 7 = \frac{7 - 6}{1000 - 1200} \cdot (x - 1000)$$

$$= -\frac{1}{200}(x - 1000)$$

$$= -\frac{1}{200}x + 5,$$

so

$$p = 12 - \frac{1}{200}x. \tag{2}$$

FIGURE 7   A demand curve.

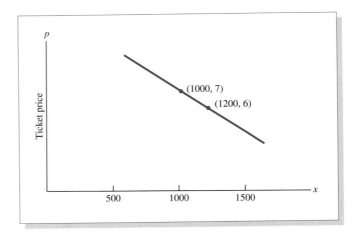

From equation (1) we obtain the revenue function

$$R(x) = x(12 - \frac{1}{200}x) = 12x - \frac{1}{200}x^2.$$

The marginal revenue is

$$R'(x) = 12 - \frac{1}{100}x = -\frac{1}{100}(x - 1200).$$

Using $R(x)$ and $R'(x)$, we can sketch the graph of $R(x)$ (Fig. 8). The maximum revenue occurs when the marginal revenue is zero—that is, when $x = 1200$. The price corresponding to this number of customers is found from the demand equation (2),

$$p = 12 - \frac{1}{200}(1200) = 6.$$

Thus the price of $6 is most likely to bring the greatest revenue per week. ●

FIGURE 8   Maximizing revenue.

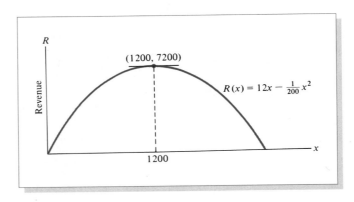

Profit Functions Once we know the cost function $C(x)$ and the revenue function $R(x)$, we can compute the profit function $P(x)$ from

$$P(x) = R(x) - C(x).$$

EXAMPLE 4 Suppose that the demand equation for a monopolist is $p = 100 - .01x$ and the cost function is $C(x) = 50x + 10,000$. Find the value of x that maximizes the profit and determine the corresponding price and total profit for this level of production. (See Fig. 9.)

FIGURE 9 A demand curve.

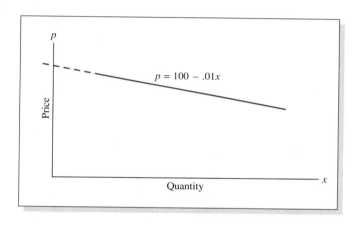

Solution The total revenue function is

$$R(x) = x \cdot p = x(100 - .01x) = 100x - .01x^2.$$

Hence the profit function is

$$P(x) = R(x) - C(x)$$

$$= 100x - .01x^2 - (50x + 10,000)$$

$$= -.01x^2 + 50x - 10,000.$$

[handwritten: $P(x) = x(100 - .01x) - (50x + 10,000)$
$100x - .01x^2 - 50x - 10,000$
$P(x) = -.01x^2 + 50x - 10,000$]

The graph of this function is a parabola that opens downward. (See Fig. 10.) Its highest point will be where the curve has zero slope—that is, where the marginal profit $P'(x)$ is zero. Now

$$P'(x) = -.02x + 50 = -.02(x - 2500).$$

[handwritten: $-.02x + 50 = 0$
$x = 2500$]

So $P'(x) = 0$ when $x = 2500$. The profit for this level of production is

$$P(2500) = -.01(2500)^2 + 50(2500) - 10,000$$

$$= 52,500.$$

[handwritten: $P(2500) = -.01(2500)^2 + 50(2500) - 10,000 = \$52,500$]

FIGURE 10   Maximizing profit.

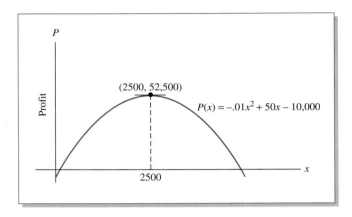

Finally, we return to the demand equation to find the highest price that can be charged per unit to sell all 2500 units:

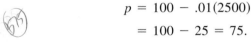

$$p = 100 - .01(2500)$$
$$= 100 - 25 = 75.$$

ANSWER

Produce 2500 units and sell them at $75 per unit. The profit will be $52,500. ●

EXAMPLE 5   Rework Example 4 under the condition that the government imposes an excise tax of $10 per unit.

Solution   For each unit sold, the manufacturer will have to pay $10 to the government. In other words, $10x$ dollars are added to the cost of producing and selling x units. The cost function is now

$$C(x) = (50x + 10,000) + 10x = 60x + 10,000.$$

The demand equation is unchanged by this tax, so the revenue function is still

$$R(x) = 100x - .01x^2.$$

Proceeding as before, we have

$$P(x) = R(x) - C(x)$$
$$= 100x - .01x^2 - (60x + 10,000)$$
$$= -.01x^2 + 40x - 10,000.$$
$$P'(x) = -.02x + 40 = -.02(x - 2000).$$

The graph of $P(x)$ is still a parabola that opens downward, and the highest point is where $P'(x) = 0$—that is, where $x = 2000$. (See Fig. 11.) The corresponding

FIGURE 11   Profit after an excise tax.

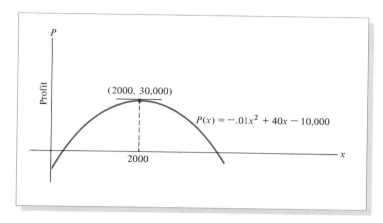

profit is

$$P(2000) = -.01(2000)^2 + 40(2000) - 10,000$$
$$= 30,000.$$

From the demand equation, $p = 100 - .01x$, we find the price that corresponds to $x = 2000$:

$$p = 100 - .01(2000) = 80.$$

ANSWER

Produce 2000 units and sell them at $80 per unit. The profit will be $30,000. ●

Notice in Example 5 that the optimal price is raised from $75 to $80. If the monopolist wishes to maximize profits, he or she should pass only half the $10 tax on to the consumer. The monopolist cannot avoid the fact that profits will be substantially lowered by the imposition of the tax. This is one reason why industries lobby against taxation.

Setting Production Levels   Suppose that a firm has cost function $C(x)$ and revenue function $R(x)$. In a free-enterprise economy the firm will set production x in such a way as to maximize the profit function

$$P(x) = R(x) - C(x).$$

We have seen that if $P(x)$ has a maximum at $x = a$, then $P'(a) = 0$. In other words, since $P'(x) = R'(x) - C'(x)$,

$$R'(a) - C'(a) = 0$$
$$R'(a) = C'(a).$$

FIGURE 12

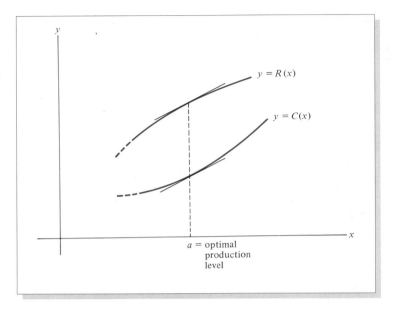

$y = R(x)$

$y = C(x)$

$a = $ optimal
production
level

Thus profit is maximized at a production level for which marginal revenue equals marginal cost. (See Fig. 12.)

PRACTICE PROBLEMS 2.7

1. Rework Example 4 by finding the production level at which marginal revenue equals marginal cost.
2. Rework Example 4 under the condition that the fixed cost is increased from $10,000 to $15,000.

EXERCISES 2.7

1. Given the cost function $C(x) = x^3 - 6x^2 + 13x + 15$, find the minimum marginal cost.
2. Suppose that a total cost function is $C(x) = .0001x^3 - .06x^2 + 12x + 100$. Is the marginal cost increasing, decreasing, or not changing at $x = 100$? Find the minimum marginal cost.
3. The revenue function for a one-product firm is

$$R(x) = 200 - \frac{1600}{x + 8} - x.$$

Find the value of x that results in maximum revenue.
4. The revenue function for a particular product is $R(x) = x(4 - .0001x)$. Find the largest possible revenue.

5. A one-product firm estimates that its daily total cost function (in suitable units) is $C(x) = x^3 - 6x^2 + 13x + 15$ and its total revenue function is $R(x) = 28x$. Find the value of x that maximizes the daily profit.

6. A small tie shop sells ties for \$3.50 each. The daily cost function is estimated to be $C(x)$ dollars, where x is the number of ties sold on a typical day and $C(x) = .0006x^3 - .03x^2 + 2x + 20$. Find the value of x that will maximize the store's daily profit.

7. The demand equation for a certain commodity is $p = \frac{1}{12}x^2 - 10x + 300$, $0 \le x \le 60$. Find the value of x and the corresponding price p that maximize the revenue.

8. The demand equation for a product is $p = 2 - .001x$. Find the value of x and the corresponding price p that maximize the revenue.

9. Some years ago it was estimated that the demand for steel approximately satisfied the equation $p = 256 - 50x$, and the total cost of producing x units of steel was $C(x) = 182 + 56x$. (The quantity x was measured in millions of tons and the price and total cost were measured in millions of dollars.) Determine the level of production and the corresponding price that maximize the profits.

10. Consider a rectangle in the xy-plane, with corners at $(0, 0)$, $(a, 0)$, $(0, b)$, and (a, b). Suppose that (a, b) lies on the graph of the equation $y = 30 - x$. Find a and b such that the area of the rectangle is maximized. What economic interpretation can be given to your answer if the equation $y = 30 - x$ represents a demand curve and y is the price corresponding to the demand x?

11. Until recently hamburgers at the city sports arena cost \$2 each. The food concessionaire sold an average of 10,000 hamburgers on a game night. When the price was raised to \$2.40, hamburger sales dropped off to an average of 8000 per night.

(a) Assuming a linear demand curve, find the price of a hamburger that will maximize the nightly hamburger revenue.

(b) Suppose that the concessionaire has fixed costs of \$1000 per night and the variable cost is \$.60 per hamburger. Find the price of a hamburger that will maximize the nightly hamburger profit.

12. The average ticket price for a concert at the opera house was \$50. The average attendance was 4000. When the ticket price was raised to \$52 attendance declined to an average of 3800 persons per performance. What should the ticket price be in order to maximize the revenue for the opera house? (Assume a linear demand curve.)

13. The monthly demand equation for an electric utility company is estimated to be

$$p = 60 - (10^{-5})x,$$

where p is measured in dollars and x is measured in thousands of kilowatt-hours. The utility has fixed costs of \$7,000,000 per month and variable costs of \$30 per 1000 kilowatt-hours of electricity generated, so that the cost function is

$$C(x) = 7 \cdot 10^6 + 30x.$$

(a) Find the value of x and the corresponding price for 1000 kilowatt-hours that maximize the utility's profit.

(b) Suppose that rising fuel costs increase the utility's variable costs from \$30 to \$40 so that its new cost function is

$$C_1(x) = 7 \cdot 10^6 + 40x.$$

Should the utility pass all this increase of $10 per thousand kilowatt-hours on to consumers? Explain your answer.

14. The demand equation for a monopolist is $p = 200 - 3x$, and the cost function is $C(x) = 75 + 80x - x^2, 0 \leq x \leq 40$.

   (a) Determine the value of x and the corresponding price that maximize the profit.

   (b) Suppose that the government imposes a tax on the monopolist of $4 per unit quantity produced. Determine the new price that maximizes the profit.

   (c) Suppose that the government imposes a tax of T dollars per unit quantity produced so that the new cost function is

$$C(x) = 75 + (80 + T)x - x^2, \qquad 0 \leq x \leq 40.$$

   Determine the new value of x that maximizes the monopolist's profit as a function of T. Assuming that the monopolist cuts back production to this level, express the tax revenues received by the government as a function of T. Finally, determine the value of T that will maximize the tax revenue received by the government.

15. A savings and loan association estimates that the amount of money on deposit will be 1,000,000 times the percentage rate of interest. For instance, a 4% interest rate will generate $4,000,000 in deposits. Suppose the savings and loan can loan all the money it takes in at 10% interest. What interest rate on deposits generates the greatest profit?

16. Let $P(x)$ be the annual profit for a certain product, where x is the amount of money spent on advertising. See Fig. 13.

   (a) Interpret $P(0)$.

   (b) Describe how the marginal profit changes as the amount of money spent on advertising increases.

   (c) Explain the economic significance of the inflection point.

FIGURE 13   Profit as a function of advertising.

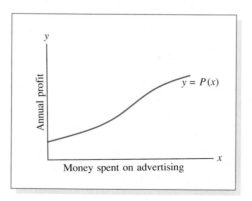

17. The revenue for a manufacturer is $f(x)$ thousand dollars, where x is the number of units of goods produced (and sold) and f is the function given in Fig. 14.

   (a) What is the revenue from producing 30 units of goods?

   (b) What is the marginal revenue when 21 units of goods are produced?

   (c) At what level of production is the revenue $20,000?

FIGURE 14

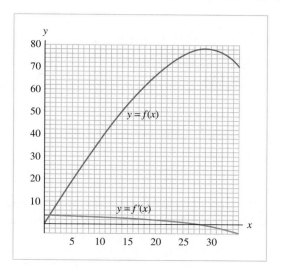

(d) At what level(s) of production is the marginal revenue $-\$2000$?

(e) At what level of production is the revenue greatest?

18. The cost function for a manufacturer is given by $f(x)$, where x is the number of units of goods produced and f is the function given in Fig. 15.

(a) What is the cost of manufacturing 14 units of goods?

(b) What is the marginal cost when 22 units of goods are manufactured?

(c) At what level of production is the cost $\$1300$?

(d) At what level(s) of production is the marginal cost $\$17$?

(e) At what level of production does the marginal cost have the least value? What are the marginal and total costs at this level of production?

FIGURE 15

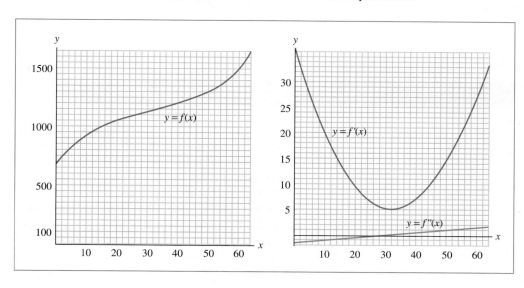

**SOLUTIONS TO
PRACTICE
PROBLEMS 2.7**

1. The revenue function is $R(x) = 100x - .01x^2$, so the marginal revenue function is $R'(x) = 100 - .02x$. The cost function is $C(x) = 50x + 10,000$, so the marginal cost function is $C'(x) = 50$. Let us now equate the two marginal functions and solve for x.

$$R'(x) = C'(x)$$
$$100 - .02x = 50$$
$$-.02x = -50$$
$$x = \frac{-50}{-.02} = \frac{5000}{2} = 2500.$$

Of course, we obtain the same level of production as before.

2. If the fixed cost is increased from $10,000 to $15,000, the new cost function will be $C(x) = 50x + 15,000$ but the marginal cost function will still be $C'(x) = 50$. Therefore, the solution will be the same: 2500 units should be produced and sold at $75 per unit. (Increases in fixed costs should not necessarily be passed on to the consumer if the objective is to maximize the profit.)

**CHAPTER 2
CHECKLIST**

✓ Increasing, decreasing
✓ Relative extreme point, relative maximum point, relative minimum point
✓ Maximum value, minimum value
✓ Concave up: graph above tangent line, opens up, slope increasing
✓ Concave down: graph below tangent line, opens down, slope decreasing
✓ Inflection point
✓ x-intercept, y-intercept
✓ Asymptote

✓ First derivative rule
✓ Second derivative rule
✓ How to find relative extreme points
✓ How to find inflection points
✓ Summary of curve-sketching techniques
✓ Objective equation
✓ Constraint equation
✓ Suggestions for solving an optimization problem

CHAPTER 2   SUPPLEMENTARY EXERCISES

1. The population of a country is growing at an annual rate of 2%. Table 1 shows the population at the beginning of each year for several years. Which of the graphs in Fig. 1 could be the graph of $P(t)$, the population t years after 1989?

2. Table 2 shows the velocity of an accelerating rocket for the first few seconds after lift-off. Which of the graphs in Fig. 2 could be the graph of $v(t)$, the velocity (in feet per second) t seconds after lift-off?

In Exercises 3–6, draw the graph of a function $f(x)$ for which the function and its first derivative have the stated property for all x.

3. $f(x)$ and $f'(x)$ increasing.

4. $f(x)$ and $f'(x)$ decreasing.

5. $f(x)$ increasing and $f'(x)$ decreasing.

6. $f(x)$ decreasing and $f'(x)$ increasing.

TABLE I	Population Growth	
Year	Population	Increase Over Previous Year
1989	50,000,000	
1990	51,000,000	1,000,000
1991	52,020,000	1,020,000
1992	53,060,400	1,040,400

FIGURE I Possible graphs of $P(t)$.

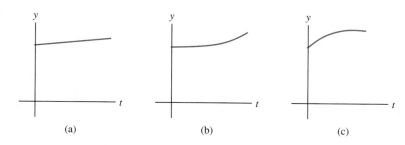

(a)　　　　(b)　　　　(c)

TABLE 2	Velocity of Accelerating Rocket	
Seconds	Velocity	Increase In Velocity
0	0	
1	15	15
2	50	35
3	105	55
4	180	75

FIGURE 2 Velocity of a rocket after lift-off.

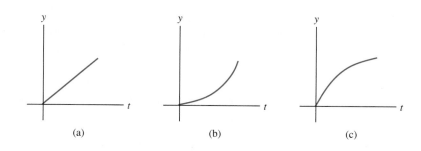

(a)　　　　(b)　　　　(c)

FIGURE 3

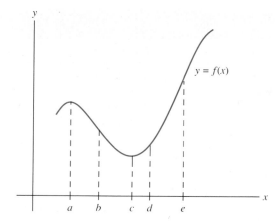

Exercises 7–12 refer to the graph in Fig. 3. List the labeled values of x at which the derivative has the stated property.

7. $f'(x)$ is positive.   8. $f'(x)$ is negative.

9. $f''(x)$ is positive   10. $f''(x)$ is negative.

11. $f'(x)$ is maximized.   12. $f'(x)$ is minimized.

Properties of various functions are described next. In each case draw some conclusion about the graph of the function.

13. $f(1) = 2, f'(1) > 0$ 14. $g(1) = 5, g'(1) = -1$

15. $h'(3) = 4, h''(3) = 1$ 16. $F'(2) = -1, F''(2) < 0$

17. $G(10) = 2, G'(10) = 0, G''(10) > 0$

18. $f(4) = -2, f'(4) > 0, f''(4) = -1$

19. $g(5) = -1, g'(5) = -2, g''(5) = 0$

20. $H(0) = 0, H'(0) = 0, H''(0) = 1$

21. $F(-2) = 0, F'(-2) = 0, F''(-2) = -1$

22. $h(-3) = 4, h'(-3) = 1, h''(-3) = 0$

Sketch the following parabolas. Include their x- and y-intercepts.

23. $y = 3 - x^2$ 24. $y = 7 + 6x - x^2$

25. $y = x^2 + 3x - 10$ 26. $y = 4 + 3x - x^2$

27. $y = -2x^2 + 10x - 10$ 28. $y = x^2 - 9x + 19$

29. $y = x^2 + 3x + 2$ 30. $y = -x^2 + 8x - 13$

31. $y = -x^2 + 20x - 90$ 32. $y = 2x^2 + x - 1$

Sketch the following curves.

33. $y = 2x^3 + 3x^2 + 1$ 34. $y = x^3 - \frac{3}{2}x^2 - 6x$

35. $y = x^3 - 3x^2 + 3x - 2$ 36. $y = 100 + 36x - 6x^2 - x^3$

37. $y = \frac{11}{3} + 3x - x^2 - \frac{1}{3}x^3$

38. $y = x^3 - 3x^2 - 9x + 7$

39. $y = -\frac{1}{3}x^3 - 2x^2 - 5x$

40. $y = x^3 - 6x^2 - 15x + 50$

41. $y = x^4 - 2x^2$

42. $y = x^4 - 4x^3$

43. $y = \frac{x}{5} + \frac{20}{x} + 3$ $(x > 0)$

44. $y = \frac{1}{2x} + 2x + 1$

45. Let $f(x) = (x^2 + 2)^{3/2}$. Show that the graph of $f(x)$ has a possible relative extreme point at $x = 0$.

46. Show that the function $f(x) = (2x^2 + 3)^{3/2}$ is decreasing for $x < 0$ and increasing for $x > 0$.

47. Let $f(x)$ be a function whose *derivative* is

$$f'(x) = \frac{1}{1 + x^2}.$$

Note that $f'(x)$ is always positive. Show that the graph of $f(x)$ definitely has an inflection point at $x = 0$.

48. Let $f(x)$ be a function whose *derivative* is

$$f'(x) = \sqrt{5x^2 + 1}.$$

Show that the graph of $f(x)$ definitely has an inflection point at $x = 0$.

49. Suppose a car is traveling on a straight road and $s(t)$ is the distance traveled after t hours. Match each set of information about $s(t)$ and its derivatives with the corresponding description of the car's motion.

Information:
 A. $s(t)$ is a constant function.
 B. $s'(t)$ is a positive constant function.
 C. $s'(t)$ is positive at $t = a$.
 D. $s'(t)$ is negative at $t = a$.
 E. $s'(t)$ and $s''(t)$ are positive at $t = a$.
 F. $s'(t)$ is positive and $s''(t)$ is negative at $t = a$.

Descriptions:
 a. The car is moving forward and speeding up at time a.
 b. The car is backing up at time a.
 c. The car is standing still.
 d. The car is moving forward but slowing down at time a.
 e. The car is moving forward at a steady rate.
 f. The car is moving forward at time a.

50. The water level in a reservoir varies during the year. Let $h(t)$ be the depth (in feet) of the water at time t days, where $t = 0$ at the beginning of the year. Match each set of information about $h(t)$ and its derivatives with the corresponding description of the reservoir's activity.

Information:
 A. $h(t)$ has the value 50 for $1 \le t \le 2$.
 B. $h'(t)$ has the value .5 for $1 \le t \le 2$.
 C. $h'(t)$ is positive at $t = a$.
 D. $h'(t)$ is negative at $t = a$.
 E. $h'(t)$ and $h''(t)$ are positive at $t = a$.
 F. $h'(t)$ is positive and $h''(t)$ is negative at $t = a$.

Descriptions:
 a. The water level is rising at an increasing rate at time a.
 b. The water level is receding at time a.
 c. The water level stayed at 50 feet on January 2.

    d. At time a the water level is rising, but the rate of increase is slowing down.

    e. On January 2 the water rose steadily at a rate of .5 feet per day.

    f. The water level is rising at time a.

51. Let $f(x)$ be the number of people living within x miles of the center of New York City.

    (a) What does $f(10 + h) - f(10)$ represent?

    (b) Explain why $f'(10)$ cannot be negative.

52. For what x does the function $f(x) = \frac{1}{4}x^2 - x + 2, 0 \le x \le 8$, have its maximum value?

53. Find the maximum value of the function $f(x) = 2 - 6x - x^2, 0 \le x \le 5$, and give the value of x where this maximum occurs.

54. Find the minimum value of the function $g(t) = t^2 - 6t + 9, 1 \le t \le 6$.

55. An open rectangular box is to be 4 feet long and have a volume of 200 cubic feet. Find the dimensions for which the amount of material needed to construct the box is as small as possible.

56. A closed rectangular box with a square base is to be constructed using two different types of wood. The top is made of wood costing \$3 per square foot and the remainder is made of wood costing \$1 per square foot. Suppose that \$48 is available to spend. Find the dimensions of the box of greatest volume that can be constructed.

57. A long rectangular sheet of metal 30 inches wide is to be made into a gutter by turning up strips vertically along the two sides. How many inches should be turned up on each side in order to maximize the amount of water that the gutter can carry?

58. A small orchard yields 25 bushels of fruit per tree when planted with 40 trees. Because of overcrowding, the yield per tree (for each tree in the orchard) is reduced by $\frac{1}{2}$ bushel for each additional tree that is planted. How many trees should be planted in order to maximize the total yield of the orchard?

59. A publishing company sells 400,000 copies of a certain book each year. Ordering the entire amount printed at the beginning of the year ties up valuable storage space and capital. However, running off the copies in several partial runs throughout the year results in added costs for setting up each printing run. Setting up each production run costs \$1000. The carrying costs, figured on the average number of books in storage, are 50¢ per book. Find the economic lot size, that is, the production run size that minimizes the total setting up and carrying costs.

60. A poster is to have an area of 125 square inches. The printed material is to be surrounded by a margin of 3 inches at the top and margins of 2 inches at the bottom and sides. Find the dimensions of the poster that maximize the area of the printed material.

61. Suppose that the demand equation for a monopolist is $p = 150 - .02x$ and the cost function is $C(x) = 10x + 300$. Find the value of x that maximizes the profit.

CHAPTER **3**

TECHNIQUES OF DIFFERENTIATION

We have seen that the derivative is useful in many applications. However, our ability to differentiate functions is somewhat limited. For example, we cannot yet readily differentiate

$$(x^2 - 1)^4(x^2 + 1)^5, \qquad \frac{x^3}{(x^2 + 1)^4}.$$

In this chapter we develop differentiation techniques that apply to functions like those given above. Two new rules are the *product rule* and the *quotient rule*. In Section 3.2 we extend the general power rule into a powerful formula called the *chain rule*.

3.1 THE PRODUCT AND QUOTIENT RULES

We observed in our discussion of the sum rule for derivatives that the derivative of the sum of two differentiable functions is the sum of the derivatives. Unfortunately, however, the derivative of the product $f(x)g(x)$ is *not* the product of the derivatives. Rather, the derivative of a product is determined from the following rule.

Product Rule

$$\frac{d}{dx}[f(x)g(x)] = f(x)g'(x) + g(x)f'(x).$$

The derivative of the product of two functions is the first function times the derivative of the second plus the second function times the derivative of the first. At the end of the section we show why this statement is true.

EXAMPLE 1   Show that the product rule works for the case $f(x) = x^2$, $g(x) = x^3$.

Solution   Since $x^2 \cdot x^3 = x^5$, we know that

$$\frac{d}{dx}[x^2 \cdot x^3] = \frac{d}{dx}[x^5] = 5x^4.$$

On the other hand, using the product rule,

$$\frac{d}{dx}(x^2 \cdot x^3) = x^2 \frac{d}{dx}(x^3) + x^3 \frac{d}{dx}(x^2)$$

$$= x^2(3x^2) + x^3(2x)$$

$$= 3x^4 + 2x^4 = 5x^4.$$

Thus the product rule gives the correct answer. ●

EXAMPLE 2   Differentiate the product $(2x^3 - 5x)(3x + 1)$.

Solution   Let $f(x) = 2x^3 - 5x$ and $g(x) = 3x + 1$. Then

$$\frac{d}{dx}[(2x^3 - 5x)(3x + 1)] = (2x^3 - 5x) \cdot \frac{d}{dx}(3x + 1) + \frac{d}{dx}(2x^3 - 5x)$$

$$= (2x^3 - 5x)(3) + (3x + 1)(6x^2 - 5)$$

$$= 6x^3 - 15x + 18x^3 - 15x + 6x^2 - 5$$

$$= 24x^3 + 6x^2 - 30x - 5. ●$$

EXAMPLE 3   Apply the product rule to $y = g(x) \cdot g(x)$.

Solution

$$\frac{d}{dx}[g(x) \cdot g(x)] = g(x) \cdot g'(x) + g(x) \cdot g'(x)$$

$$= 2g(x)g'(x).$$

This answer is the same as that given by the general power rule:

$$\frac{d}{dx}[g(x) \cdot g(x)] = \frac{d}{dx}[g(x)]^2 = 2g(x)g'(x). ●$$

EXAMPLE 4   Find $\dfrac{dy}{dx}$ where $y = (x^2 - 1)^4(x^2 + 1)^5$.

Solution   Let $f(x) = (x^2 - 1)^4$, $g(x) = (x^2 + 1)^5$, and use the product rule. The general power rule is needed to compute $f'(x)$ and $g'(x)$.

$$\frac{dy}{dx} = (x^2 - 1)^4 \cdot \frac{d}{dx}(x^2 + 1)^5 + (x^2 + 1)^5 \cdot \frac{d}{dx}(x^2 - 1)^4$$

$$= (x^2 - 1)^4 \cdot 5(x^2 + 1)^4(2x) + (x^2 + 1)^5 \cdot 4(x^2 - 1)^3(2x). \qquad (1)$$

This form of $\dfrac{dy}{dx}$ is suitable for some purposes. For example, if we need to compute $\dfrac{dy}{dx}\bigg|_{x=2}$, it is easier just to substitute 2 for x than to simplify and then substitute. However, it is often helpful to simplify the formula for $\dfrac{dy}{dx}$, such as in the case when we need to find x such that $\dfrac{dy}{dx} = 0$.

To simplify the answer in (1), we shall write $\dfrac{dy}{dx}$ as a single product rather than as a sum of two products. The first step is to identify the common factors.

$$\frac{dy}{dx} = (x^2 - 1)^4 \cdot 5(x^2 + 1)^4(2x) + (x^2 + 1)^5 \cdot 4(x^2 - 1)^3(2x).$$

Both terms contain $2x$ and powers of $x^2 - 1$ and $x^2 + 1$. The most we can factor out of each term is $2x(x^2 - 1)^3(x^2 + 1)^4$. We obtain

$$\frac{dy}{dx} = 2x(x^2 - 1)^3(x^2 + 1)^4[5(x^2 - 1) + 4(x^2 + 1)].$$

Simplifying the right most factor in this product, we have

$$\frac{dy}{dx} = 2x(x^2 - 1)^3(x^2 + 1)^4[9x^2 - 1]. \;\bullet \qquad (2)$$

The answers to the exercises for this section appear in two forms, similar to those in (1) and (2) above. The unsimplified answers will allow you to check if you have mastered the differentiation rules. In each case you should make an effort to transform your original answer into the simplified version. Examples 4 and 6 show how to do this.

The Quotient Rule   Another useful formula for differentiating functions is the quotient rule.

$$\boxed{\begin{array}{c} \textit{Quotient Rule} \\[4pt] \dfrac{d}{dx}\left[\dfrac{f(x)}{g(x)}\right] = \dfrac{g(x)f'(x) - f(x)g'(x)}{[g(x)]^2} \end{array}}$$

One must be careful to remember the order of the terms in this formula because of the minus sign in the numerator.

EXAMPLE 5   Differentiate $\dfrac{x}{2x + 3}$.

Solution   Let $f(x) = x$ and $g(x) = 2x + 3$.

$$\frac{d}{dx}\left(\frac{x}{2x + 3}\right) = \frac{(2x + 3)\cdot\dfrac{d}{dx}(x) - (x)\cdot\dfrac{d}{dx}(2x + 3)}{(2x + 3)^2}$$

$$= \frac{(2x + 3)\cdot 1 - x\cdot 2}{(2x + 3)^2} = \frac{3}{(2x + 3)^2}. \quad \bullet$$

EXAMPLE 6   Find $\dfrac{dy}{dx}$ where $y = \dfrac{x^3}{(x^2 + 1)^4}$.

Solution   Let $f(x) = x^3$ and $g(x) = (x^2 + 1)^4$.

$$\frac{d}{dx}\frac{x^3}{(x^2 + 1)^4} = \frac{(x^2 + 1)^4\cdot\dfrac{d}{dx}(x^3) - (x^3)\cdot\dfrac{d}{dx}(x^2 + 1)^4}{[(x^2 + 1)^4]^2}$$

$$= \frac{(x^2 + 1)^4\cdot 3x^2 - x^3\cdot 4(x^2 + 1)^3(2x)}{(x^2 + 1)^8}.$$

If a simplified form of $\dfrac{dy}{dx}$ is desired, we can divide the numerator and the denominator by the common factor $(x^2 + 1)^3$.

$$\frac{dy}{dx} = \frac{(x^2 + 1)\cdot 3x^2 - x^3\cdot 4(2x)}{(x^2 + 1)^5}$$

$$= \frac{3x^4 + 3x^2 - 8x^4}{(x^2 + 1)^5}$$

$$= \frac{3x^2 - 5x^4}{(x^2 + 1)^5} = \frac{x^2(3 - 5x^2)}{(x^2 + 1)^5}. \quad \bullet$$

EXAMPLE 7   Suppose that the total cost of manufacturing x units of a certain product is given by the function $C(x)$. Then the *average cost per unit, AC,* is defined by

$$AC = \frac{C(x)}{x}.$$

Recall that the *marginal cost, MC,* is defined by

$$MC = C'(x).$$

Show that at the level of production where the average cost is at a minimum, the average cost equals the marginal cost.

Solution   In practice, the marginal cost and average cost curves will have the general shapes shown in Fig. 1. So the minimum point on the average cost curve will occur when $\dfrac{d}{dx}(AC) = 0$. To compute the derivative, we need the quotient rule,

$$\frac{d}{dx}(AC) = \frac{d}{dx}\left(\frac{C(x)}{x}\right) = \frac{x \cdot C'(x) - C(x)}{x^2}.$$

FIGURE I   Marginal cost and average cost functions.

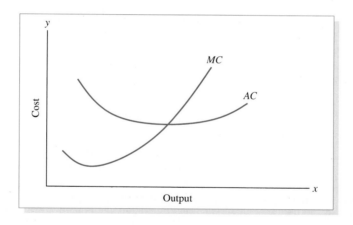

Setting the derivative equal to zero and multiplying by x^2, we obtain

$$0 = x \cdot C'(x) - C(x)$$

$$C(x) = x \cdot C'(x)$$

$$\frac{C(x)}{x} = C'(x)$$

$$AC = MC.$$

Thus when the output x is chosen so that the average cost is minimized, the average cost equals the marginal cost.   ●

Verification of the Product and Quotient Rules

Verification of the Product Rule

From our discussion of limits we compute the derivative of $f(x)g(x)$ at $x = a$ as the limit

$$\frac{d}{dx}[f(x)g(x)]\Big|_{x=a} = \lim_{h \to 0} \frac{f(a + h)g(a + h) - f(a)g(a)}{h}.$$

Let us add and subtract the quantity $f(a)g(a + h)$ in the numerator. After factoring and applying Limit Theorem III, we obtain

$$\lim_{h \to 0} \frac{[f(a + h)g(a + h) - f(a)g(a + h)] + [f(a)g(a + h) - f(a)g(a)]}{h}$$

$$= \lim_{h \to 0} g(a + h) \cdot \frac{f(a + h) - f(a)}{h} + \lim_{h \to 0} f(a) \cdot \frac{g(a + h) - g(a)}{h}.$$

And this expression may be rewritten by Limit Theorem V as

$$\lim_{h \to 0} g(a + h) \cdot \lim_{h \to 0} \frac{f(a + h) - f(a)}{h} + \lim_{h \to 0} f(a) \cdot \lim_{h \to 0} \frac{g(a + h) - g(a)}{h}.$$

Note, however, that since $g(x)$ is differentiable at $x = a$, it is continuous there, so that $\lim_{h \to 0} g(a + h) = g(a)$. Therefore, the expression above equals

$$g(a)f'(a) + f(a)g'(a).$$

That is, we have proved that

$$\frac{d}{dx}[f(x)g(x)]\Big|_{x=a} = g(a)f'(a) + f(a)g'(a),$$

which is the product rule. An alternative verification of the product rule not involving limit arguments is outlined in Exercise 56.

Verification of the Quotient Rule

From the general power rule, we know that

$$\frac{d}{dx}\left[\frac{1}{g(x)}\right] = \frac{d}{dx}[g(x)]^{-1} = (-1)[g(x)]^{-2} \cdot g'(x).$$

We can now derive the quotient rule from the product rule.

$$\frac{d}{dx}\left[\frac{f(x)}{g(x)}\right] = \frac{d}{dx}\left[\frac{1}{g(x)} \cdot f(x)\right]$$

$$= \frac{1}{g(x)} \cdot f'(x) + f(x) \cdot \frac{d}{dx}\left[\frac{1}{g(x)}\right]$$

$$= \frac{g(x)f'(x)}{[g(x)]^2} + f(x) \cdot (-1)[g(x)]^{-2} \cdot g'(x)$$

$$= \frac{g(x)f'(x) - f(x)g'(x)}{[g(x)]^2}.$$

TECHNOLOGY PROJECT 3.1
The Rate of Change of the Election Function

Recall from Chapter 0 the function

$$f(x) = \frac{x^3}{x^3 + (1 - x)^3}, \qquad 0 \le x \le 1$$

which estimates the Democratic proportion of the House of Representatives if x is the proportion of the popular vote for Democrats.

1. Graph this function.
2. Use the TRACE to calculate the effect of an additional 1% Democratic vote from 54% on the composition of the House of Representatives.
3. Calculate the derivative of $f(x)$.
4. Use the derivative to estimate the effect of an additional 1% Democratic vote from 54% on the composition of the House of Representatives.
5. Compare the results of 2 and 4.

TECHNOLOGY PROJECT 3.2
Calculus and Graphing Calculators Complement Each Other

Sketch the graphs of the following functions using both calculus and graphing calculator techniques. Be sure to determine all zeros, asymptotes, relative extreme points, and inflection points. Carry out approximations to one decimal place.

1. $f(x) = \dfrac{x^2 + 1}{x^2 - 1}$

2. $f(x) = \dfrac{x}{x^2 + 1}$

3. $f(x) = \dfrac{1}{(x - 1)(x + 2)}$

4. $f(x) = x^4 + 3x - 1$

TECHNOLOGY PROJECT 3.3
Calculus and Graphing Calculators Complement Each Other, II

Let $f(x) = x^3(1 - x)^4$.

1. Make a table of the values of $f(x)$ for $x = -10, -9, \ldots, 9, 10$.
2. The function values increase rapidly for x away from 0. To get an overall picture of the graph, you need a wide range of y-values. To this end, graph $f(x)$ using the following windows:
   (a) $-2 \le x \le 3$, $-100 \le y \le 100$
   (b) $-1 \le x \le 2$, $-10 \le y \le 10$
   (c) $-1 \le x \le 2$, $-1 \le y \le 1$
   (d) $-1 \le x \le 2$, $-.1 \le y \le .1$
   (e) $-1 \le x \le 2$, $-.01 \le y \le .01$
3. At what range of y-values did the graph give a hint that something might be happening between $x = 0$ and $x = 1$?

4. Describe what appears to be happening in graph 2(e).

5. Determine the derivative $f'(x)$ using the differentiation techniques of this section. Simplify the derivative. Use the formula for the derivative to explain why futher magnification of the y-axis will not reveal additional "bumps" in the graph of $f(x)$.

**TECHNOLOGY
PROJECT 3.4
Symbolic
Differentiation
(Symbolic Package
Required)**

Thus far, our discussions of technology have concentrated on only one tool, the graphing calculator. Another important tool is the **symbolic mathematics program,** which allows you to perform algebraic and calculus operations using variables. Three popular symbolic mathematics programs are Mathematica, Maple, and MathCAD. These programs are all very powerful and are available on a wide range of computer systems, ranging from supercomputers to personal computers. The rest of the technology projects in this chapter assume that you have one of these programs available to you, either on your own computer or on a computer in your mathematics laboratory.

We will use the notation for Mathematica in our discussions. For more information about Mathematica or about the other symbolic programs, you can consult either the user manuals which accompany the programs or one of the program-specific manuals available to accompany this book.

In Mathematica, differentiation can be done using a command of the form:

```
D [function, variable]
```

For example, to differentiate the function X^2, you would use the command:

```
D [X^2, X]
```

The program responds with the answer:

```
2X
```

Note that you must include the name of the variable with the command. This is necessary since your function might include other variables (as in the function AX^2) and you must tell the computer the variable for differentiation.

Use a symbolic mathematics program to determine the following derivatives:

1. $\dfrac{d}{dx}\left[(x^2 - 3)(x^3 + 3x - 1)^4\right]$ 2. $\dfrac{d}{dx}\left[\left(\dfrac{x^2 - 1}{x + 3}\right)^4(3x - 1)\right]$

3. $\dfrac{d}{dx}\left[\dfrac{x^2 + (1 - x)^2}{3x^4 + 4x^3 - 2x^2 + 3}\right]$ 4. $\dfrac{d}{dx}\left[(x - 1)(x^2 + 4)(x - 3)^2\right]$

5. Evaluate $\dfrac{d}{dx}\left[\dfrac{x^2}{x^2 + 1}\right]$ at $x = 3$. [In Mathematica, give the command x := 3 and then request the value of the expression for the derivative. Note that

in order to subsequently use x as a variable, you must clear its value using the command Clear [x].]

6. Evaluate $\dfrac{d}{dx}\left[\dfrac{1}{(x^2 + 1)^5}\right]$ at $x = 0$.

7. Determine the values of x for which $\dfrac{d}{dx}\left[x(1 + x^2)^4\right] = 0$. [In Mathematica, you can solve an equation using the command Solve [rightside == leftside, x], where x is the variable you are solving for.

8. Determine the values of x for which $\dfrac{d}{dx}\left[\dfrac{x}{1 - x^2}\right] = 0$.

9. Determine $\dfrac{d^2}{dx^2}\left[\dfrac{x}{1 + x}\right]$. [In Mathematica, the nth derivative can be calculated using the command D[function, {x, n}].

10. Determine $\dfrac{d^2}{dx^2}\left[\dfrac{x}{(1 + x)^2}\right]$.

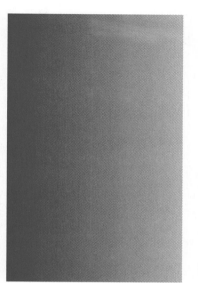

PRACTICE PROBLEMS 3.1

1. Consider the function $y = (\sqrt{x} + 1)x$.
   (a) Differentiate y by the product rule.
   (b) First multiply out the expression for y and then differentiate.

2. Differentiate $y = \dfrac{5}{x^4 - x^3 + 1}$.

EXERCISES 3.1

Differentiate the functions in Exercises 1–30.

1. $(x + 1)(x^3 + 5x + 2)$
2. $(2x - 1)(x^2 - 3)$
3. $(3x^2 - x + 2)(2x^2 - 1)$
4. $(x^4 + 1)(3x + 5)$
5. $(2x - 7)(x - 1)^5$
6. $(x + 1)^2(x - 3)^3$
7. $(x^2 + 3)(x^2 - 3)^{10}$
8. $(x^2 - 4)(x^2 + 3)^6$
9. $\frac{1}{3}(4 - x)^3(4 + x)^3$
10. $\frac{1}{6}(3x + 1)^4(4x - 1)^3$
11. $\dfrac{4 - x}{4 + x}$
12. $\dfrac{x - 2}{x + 2}$
13. $\dfrac{x^2 - 1}{x^2 + 1}$
14. $\dfrac{1 + x^3}{1 - x^3}$
15. $\dfrac{1}{5x^2 + 2x + 5}$
16. $\dfrac{2x - 1}{x}$
17. $\dfrac{x^2 + 2x}{x + 1}$
18. $\dfrac{3x^2 - 2x}{x + 3}$

19. $\dfrac{3x^2 + 5x + 1}{3 - x^2}$

20. $\dfrac{1}{x^2 + 1}$

21. $\dfrac{x}{(x^2 + 1)^2}$

22. $\dfrac{x - 1}{(2x + 1)^2}$

23. $\dfrac{(x - 1)^4}{(x + 2)^2}$

24. $\dfrac{(x + 3)^3}{(x + 4)^2}$

25. $\dfrac{x^4 - 4x^2 + 3}{x}$

26. $(x^2 + 9)\left(x - \dfrac{3}{x}\right)$

27. $2\sqrt{x}\,(3x^2 - 1)^3$

28. $4\sqrt{x}\,(5x - 1)^4$

29. $(x + 3)\sqrt{2x - 3}$

30. $(2x - 5)\sqrt{4x - 1}$

31. Find the equation of the tangent line to the curve $y = (x - 2)^5(x + 1)^2$ at the point $(3, 16)$.

32. Find the equation of the tangent line to the curve $y = (x + 1)/(x - 1)$ at the point $(2, 3)$.

33. Find all x such that $\dfrac{dy}{dx} = 0$, where $y = (x^2 - 4)^3(2x^2 + 5)^5$.

34. Find all x such that $\dfrac{dy}{dx} = 0$, where $y = (3x - 8)^4(2x - 3)^3$.

35. Find all x-coordinates of points (x, y) on the curve $y = (x - 2)^5/(x - 4)^3$ where the tangent line is horizontal.

36. Find all x-coordinates of points (x, y) on the curve $y = x^4/(x - 1)$ where the tangent line is horizontal.

37. Find the point(s) on the graph of $y = (x^2 + 3x - 1)/x$ where the slope is 5.

38. Find the point(s) on the graph of $y = (2x^4 + 1)(x - 5)$ where the slope is 1.

39. An open rectangular box is 3 feet long and has a surface area of 16 square feet. Find the dimensions of the box for which the volume is as large as possible.

40. A closed rectangular box is to be constructed with one side 1 meter long. The material for the top costs $20 per square meter, and the material for the sides and bottom costs $10 per square meter. Find the dimensions of the box with the largest possible volume that can be built at a cost of $240 for materials.

41. A sugar refinery can produce x tons of sugar per week at a weekly cost of $.1x^2 + 5x + 2250$ dollars. Find the level of production for which the average cost is at a minimum and show that the average cost equals the marginal cost at that level of production.

42. A cigar manufacturer produces x cases of cigars per day at a daily cost of $50x(x + 200)/(x + 100)$ dollars. Show that his cost increases and his average cost decreases as the output x increases.

43. Let $R(x)$ be the revenue received from the sale of x units of a product. The *average revenue per unit* is defined by $AR = R(x)/x$. Show that at the level of production where the average revenue is maximized, the average revenue equals the marginal revenue.

44. Let $s(t)$ be the number of miles a car travels in t hours. Then the average velocity during the first t hours is $\bar{v}(t) = s(t)/t$ miles per hour. Suppose that the average velocity is maximized at time t_0. Show that at this time the average velocity $\bar{v}(t_0)$ equals the instantaneous velocity $s'(t_0)$. [*Hint:* Compute the derivative of $\bar{v}(t)$.]

45. The width of a rectangle is increasing at a rate of 3 inches per second and its length is increasing at the rate of 4 inches per second. At what rate is the area of the rectangle increasing when its width is 5 inches and its length is 6 inches? [*Hint:* Let $W(t)$ and $L(t)$ be the widths and lengths, respectively, at time t.]

46. A manufacturer plans to decrease the amount of sulfur dioxide escaping from its smokestacks. The estimated cost-benefit function is

$$f(x) = \frac{3x}{105 - x}, \qquad 0 \le x \le 100$$

where $f(x)$ is the cost in millions of dollars for eliminating x percent of the total sulfur dioxide. (See Fig. 2.) Find the value of x at which the rate of increase of the cost-benefit function is 1.4 million dollars per unit. (Each unit is 1 percentage point increase in pollutant removed.)

FIGURE 2   A cost-benefit function.

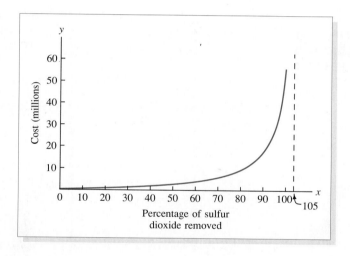

47. Show that the derivative of $\dfrac{x^4}{x^2 + 1}$ is not $\dfrac{4x^3}{2x}$.

48. Show that the derivative of $(5x^3)(2x^4)$ is not $(15x^2)(8x^3)$.

49. Suppose that $f(x)$ is a function whose derivative is $f'(x) = \dfrac{1}{1 + x^2}$. Find the derivative of $\dfrac{f(x)}{1 + x^2}$.

50. Suppose that $f(x)$ and $g(x)$ are differentiable functions such that $f(1) = 2$, $f'(1) = 3$, $g(1) = 4$, and $g'(1) = 5$. Find $\dfrac{d}{dx}\left[f(x)g(x)\right]\Big|_{x=1}$.

51. Consider the functions of Exercise 50. Find $\dfrac{d}{dx}\left[\dfrac{f(x)}{g(x)}\right]\Big|_{x=1}$.

52. Suppose that $f(x)$ is a function whose derivative is $f'(x) = 1/x$. Find the derivative of $xf(x) - x$.

53. Let $f(x) = 1/x$ and $g(x) = x^3$.

    (a) Show that the product rule yields the correct derivative of $(1/x)x^3 = x^2$.

    (b) Compute the product $f'(x)g'(x)$ and note that it is *not* the derivative of $f(x)g(x)$.

54. The derivative of $(x^3 - 4x)/x$ is obviously $2x$ for $x \neq 0$, because $(x^3 - 4x)/x = x^2 - 4$ for $x \neq 0$. Verify that the quotient rule gives the same derivative.

55. Let $f(x)$, $g(x)$, and $h(x)$ be differentiable functions. Find a formula for the derivative of $f(x)g(x)h(x)$. [*Hint:* First differentiate $f(x)[g(x)h(x)]$.]

56. (*Alternative Verification of the Product Rule*) Apply the special case of the general power rule $\dfrac{d}{dx}[h(x)]^2 = 2h(x)h'(x)$ and the identity $fg = \frac{1}{4}[(f + g)^2 - (f - g)^2]$ to prove the product rule.

SOLUTIONS TO PRACTICE PROBLEMS 3.1

1. (a) Apply the product rule to $y = (\sqrt{x} + 1)x$ with

$$f(x) = \sqrt{x} + 1 = x^{1/2} + 1$$

$$g(x) = x$$

$$\frac{dy}{dx} = (x^{1/2} + 1) \cdot 1 + x \cdot \tfrac{1}{2}x^{-1/2}$$

$$= x^{1/2} + 1 + \tfrac{1}{2}x^{1/2}$$

$$= \tfrac{3}{2}\sqrt{x} + 1.$$

(b) $$y = (\sqrt{x} + 1)x = (x^{1/2} + 1)x = x^{3/2} + x$$

$$\frac{dy}{dx} = \tfrac{3}{2}x^{1/2} + 1.$$

Comparing parts (a) and (b), we note that sometimes it is helpful to simplify the function before differentiating.

2. We may apply the quotient rule to $y = \dfrac{5}{x^4 - x^3 + 1}$.

$$\frac{dy}{dx} = \frac{(x^4 - x^3 + 1) \cdot 0 - 5 \cdot (4x^3 - 3x^2)}{(x^4 - x^3 + 1)^2}$$

$$= \frac{-5x^2(4x - 3)}{(x^4 - x^3 + 1)^2}.$$

However, it is slightly faster to use the general power rule, since $y = 5(x^4 - x^3 + 1)^{-1}$. Thus

$$\frac{dy}{dx} = -5(x^4 - x^3 + 1)^{-2}(4x^3 - 3x^2)$$

$$= -5x^2(x^4 - x^3 + 1)^{-2}(4x - 3).$$

3.2 THE CHAIN RULE AND THE GENERAL POWER RULE

In this section we show that the general power rule is a special case of a powerful differentiation technique called the chain rule. Applications of the chain rule appear throughout the text.

A useful way of combining functions $f(x)$ and $g(x)$ is to replace each occurrence of the variable x in $f(x)$ by the function $g(x)$. The resulting function is called the *composition* (or *composite*) of $f(x)$ and $g(x)$ and is denoted by $f(g(x))$.*

EXAMPLE 1 Let $f(x) = \dfrac{x-1}{x+1}$, $g(x) = x^3$. What is $f(g(x))$?

Solution Replace each occurrence of x in $f(x)$ by $g(x)$ to obtain

$$f(g(x)) = \frac{g(x) - 1}{g(x) + 1}$$

$$= \frac{x^3 - 1}{x^3 + 1}. \quad \bullet$$

Given a composite function $f(g(x))$, we may think of $f(x)$ as the "outside" function that acts on the values of the "inside" function $g(x)$. This point of view often helps us to recognize a composite function.

EXAMPLE 2 Write the following functions as composites of simpler functions.
(a) $h(x) = (x^5 + 9x + 3)^8$
(b) $k(x) = \sqrt{4x^2 + 1}$

Solution (a) $h(x) = f(g(x))$, where the outside function is the power function, $(\ldots)^8$, that is, $f(x) = x^8$. Inside this power function is $g(x) = x^5 + 9x + 3$.
(b) $k(x) = f(g(x))$, where the outside function is the square root function, $f(x) = \sqrt{x}$, and the inside function is $g(x) = 4x^2 + 1$. $\bullet$

A function of the form $[g(x)]^r$ is a composite $f(g(x))$, where the outside function is $f(x) = x^r$. We have already given a rule for differentiating this function—namely,

$$\frac{d}{dx} [g(x)]^r = r[g(x)]^{r-1} g'(x).$$

---

*See Section 0.3 for additional information on function composition.

The *chain rule* has the same form, except that the outside function $f(x)$ can be *any* differentiable function.

The Chain Rule   To differentiate $f(g(x))$, first differentiate the outside function $f(x)$ and substitute $g(x)$ for x in the result. Then multiply by the derivative of the inside function $g(x)$. Symbolically,

$$\frac{d}{dx} f(g(x)) = f'(g(x))g'(x).$$

EXAMPLE 3   Use the chain rule to compute the derivative of $f(g(x))$, where $f(x) = x^8$ and $g(x) = x^5 + 9x + 3$.

Solution
$$f'(x) = 8x^7, \qquad g'(x) = 5x^4 + 9$$
$$f'(g(x)) = 8(x^5 + 9x + 3)^7.$$

Finally, by the chain rule,

$$\frac{d}{dx} f(g(x)) = f'(g(x))g'(x)$$
$$= 8(x^5 + 9x + 3)^7(5x^4 + 9).$$

Since in this example the outside function is a power function, the calculations are the same as when the general power rule is used to compute the derivative of $(x^5 + 9x + 3)^8$. However, the organization of the solution here emphasizes the notation of the chain rule. ●

There is another way to write the chain rule. Given the function $y = f(g(x))$, set $u = g(x)$ so that $y = f(u)$. Then y may be regarded either as a function of u or, indirectly through u, as a function of x. With this notation, we have $\dfrac{du}{dx} = g'(x)$ and $\dfrac{dy}{du} = f'(u) = f'(g(x))$. Thus the chain rule says that

---

$$\frac{dy}{dx} = \frac{dy}{du}\frac{du}{dx}. \tag{1}$$

Although the derivative symbols in (1) are not really fractions, the apparent cancellation of the "*du*" symbols provides a mnemonic device for remembering this form of the chain rule.

Here is an illustration that makes (1) a plausible formula. Suppose that y, u, and x are three varying quantities, with y varying three times as fast as u and u varying two times as fast as x. It seems reasonable that y should vary six times as fast as x. That is, $\dfrac{dy}{dx} = \dfrac{dy}{du}\dfrac{du}{dx} = 3 \cdot 2 = 6$.

EXAMPLE 4   Find $\dfrac{dy}{dx}$ if $y = u^5 - 2u^3 + 8$ and $u = x^2 + 1$.

Solution   Since

$$\frac{dy}{du} = 5u^4 - 6u^2 \quad \text{and} \quad \frac{du}{dx} = 2x,$$

we have

$$\frac{dy}{dx} = (5u^4 - 6u^2) \cdot 2x.$$

It is usually desirable to express $\dfrac{dy}{dx}$ as a function of x alone, so we substitute $x^2 + 1$ for u to obtain

$$\frac{dy}{dx} = [5(x^2 + 1)^4 - 6(x^2 + 1)^2] \cdot 2x. \quad \bullet$$

You may have noticed another entirely mechanical way to work Example 4. Namely, first substitute $u = x^2 + 1$ into the original formula for y and obtain $y = (x^2 + 1)^5 - 2(x^2 + 1)^3 + 8$. Then $\dfrac{dy}{dx}$ is easily calculated by the sum rule, the general power rule, and the constant-multiple rule. The solution in Example 4, however, lays the foundation for applications of the chain rule in Section 3.3 and elsewhere.

In many situations involving composition of functions, the basic variable is time, t. It may happen that x is a function of t, say $x = g(t)$, and some other variable such as R is a function of x, say $R = f(x)$. Then $R = f(g(t))$, and the chain rule says that

$$\frac{dR}{dt} = \frac{dR}{dx} \frac{dx}{dt}.$$

EXAMPLE 5   A store sells ties for $12 apiece. Let x be the number of ties sold in one day and let R be the revenue received from the sale of x ties, so that $R = 12x$. Suppose that daily sales are rising at the rate of four ties per day. How fast is the revenue rising?

Solution   Clearly, revenue is rising at the rate of $48 per day, because each of the additional four ties brings in $12. This intuitive conclusion also follows from the chain rule,

$$\frac{dR}{dt} = \frac{dR}{dx} \cdot \frac{dx}{dt}$$

$$\begin{bmatrix} \text{rate of change} \\ \text{of revenue with} \\ \text{respect to time} \end{bmatrix} = \begin{bmatrix} \text{rate of change} \\ \text{of revenue with} \\ \text{respect to sales} \end{bmatrix} \cdot \begin{bmatrix} \text{rate of change} \\ \text{of sales with} \\ \text{respect to time} \end{bmatrix}$$

$$\begin{bmatrix} \$48 \text{ increase} \\ \text{per day} \end{bmatrix} = \begin{bmatrix} \$12 \text{ increase} \\ \text{per add'l tie} \end{bmatrix} \cdot \begin{bmatrix} \text{four additional} \\ \text{ties per day} \end{bmatrix}.$$

Notice that $\dfrac{dR}{dx}$ is actually the marginal revenue, studied earlier. This example shows that the time rate of change of revenue, $\dfrac{dR}{dt}$, is the marginal revenue multiplied by the time rate of change of sales. ●

Verification of the Chain Rule

Suppose that $f(x)$ and $g(x)$ are differentiable, and let $x = a$ be a number in the domain of $f(g(x))$. Since every differentiable function is continuous, we have

$$\lim_{h \to 0} g(a + h) = g(a),$$

which implies that

$$\lim_{h \to 0} [g(a + h) - g(a)] = 0. \tag{2}$$

Now $g(a)$ is a number in the domain of f, and the limit definition of the derivative gives us

$$f'(g(a)) = \lim_{k \to 0} \frac{f(g(a) + k) - f(g(a))}{k}. \tag{3}$$

Let $k = g(a + h) - g(a)$. By equation (2), k approaches zero as h approaches zero. Also, $g(a + h) = g(a) + k$. Therefore, (3) may be rewritten in the form

$$f'(g(a)) = \lim_{h \to 0} \frac{f(g(a + h)) - f(g(a))}{g(a + h) - g(a)}. \tag{4}$$

[Strictly speaking, we must assume that the denominator in (4) is never zero. This assumption may be avoided by a somewhat different and more technical argument which we omit.] Finally, we show that the function $f(g(x))$ has a derivative at $x = a$. We use the limit definition of the derivative, Limit Theorem V, and (4) above.

$$
\begin{aligned}
\frac{d}{dx} [f(g(x))]\bigg|_{x=a} &= \lim_{h \to 0} \frac{f(g(a + h)) - f(g(a))}{h} \\
&= \lim_{h \to 0} \left[\frac{f(g(a + h)) - f(g(a))}{g(a + h) - g(a)} \cdot \frac{g(a + h) - g(a)}{h} \right] \\
&= \lim_{h \to 0} \frac{f(g(a + h)) - f(g(a))}{g(a + h) - g(a)} \cdot \lim_{h \to 0} \frac{g(a + h) - g(a)}{h} \\
&= f'(g(a)) \cdot g'(a).
\end{aligned}
$$

TECHNOLOGY PROJECT 3.5
Symbolic Differentiation and the Chain Rule
(Symbolic Package Required)

Symbolic math programs "know" about the chain rule. If $u = g(x)$, then the derivative of $f(u)$ with respect to x can be specified in Mathematica as:

`D[f, x]`

The program responds with the derivative in terms of x. Calculate $\dfrac{df}{dx}$ where:

1. $f(u) = \sqrt{u^2 + 1}$, $u = \dfrac{1}{x^2 + 1}$

2. $f(u) = \left(u^3 + 3u + \dfrac{4}{u^2}\right)$, $u = \dfrac{x}{x + 1}$

3. $f(u) = (u^3 + 1)^4 (u^2 - 1)^3$, $u = \sqrt{x^2 + 1}$

4. $f(u) = \dfrac{u^2 - 1}{u^2 + 1}$, $u = \sqrt{x + \dfrac{1}{x}}$

PRACTICE PROBLEMS 3.2

Consider the function $h(x) = (2x^3 - 5)^5 + (2x^3 - 5)^4$.

1. Write $h(x)$ as a composite function, $f(g(x))$.
2. Compute $f'(x)$ and $f'(g(x))$.
3. Use the chain rule to differentiate $h(x)$.

EXERCISES 3.2

Compute $f(g(x))$, where $f(x)$ and $g(x)$ are the following.

1. $f(x) = \dfrac{x}{x + 1}$, $g(x) = x^3$

2. $f(x) = x\sqrt{x + 1}$, $g(x) = x^4$

3. $f(x) = x^5 + 3x$, $g(x) = x^2 + 4$

4. $f(x) = \dfrac{x}{(x + 1)^4}$, $g(x) = 5 - 3x$

Each function below may be viewed as a composite function $f(g(x))$. Find $f(x)$ and $g(x)$.

5. $(x^3 + 8x - 2)^5$

6. $(9x^2 + 2x - 5)^7$

7. $\sqrt{4 - x^2}$

8. $(5x^2 + 1)^{-1/2}$

9. $\dfrac{1}{x^3 - 5x^2 + 1}$

10. $(4x - 3)^3 + \dfrac{1}{4x - 3}$

Differentiate the functions in Exercises 11–22 using one or more of the differentiation rules discussed thus far.

11. $(x^2 + 5)^{15}$

12. $(x^4 + x^2)^{10}$

13. $6x^2(x - 1)^3$

14. $5x^3(2 - x)^4$

15. $2(x^3 - 1)(3x^2 + 1)^4$

16. $2(2x - 1)^{5/4}(2x + 1)^{3/4}$

17. $\left(\dfrac{4}{1 - x}\right)^3$

18. $\dfrac{4x^2 + x}{\sqrt{x}}$

19. $\left(\dfrac{4x - 1}{3x + 1}\right)^3$

20. $\left(\dfrac{x}{x^2 + 1}\right)^2$

21. $\left(\dfrac{4 - x}{x^2}\right)^3$ **22.** $\left(\dfrac{1 - x^2}{x}\right)^3$

23. Sketch the graph of $y = 4x/(x + 1)^2$, $x > -1$.

24. Sketch the graph of $y = 2/(1 + x^2)$.

Compute $\dfrac{d}{dx} f(g(x))$, where $f(x)$ and $g(x)$ are the following.

25. $f(x) = x^5$, $g(x) = 6x - 1$ **26.** $f(x) = x^2$, $g(x) = x^3 + 1$

27. $f(x) = \dfrac{1}{x}$, $g(x) = 1 - x^2$ **28.** $f(x) = x^{10}$, $g(x) = x^2 + 3x$

29. $f(x) = x^4 - x^2$, $g(x) = x^2 - 4$ **30.** $f(x) = \dfrac{4}{x} + x^2$, $g(x) = 1 - x^4$

31. $f(x) = (x^3 + 1)^2$, $g(x) = x^2 + 5$

32. $f(x) = x(x - 2)^4$, $g(x) = x^3$

Compute $\dfrac{dy}{dx}$ using the chain rule in formula (1).

33. $y = u^{3/2}$, $u = 4x + 1$ **34.** $y = u^5$, $u = 2 + \sqrt{x}$

35. $y = \dfrac{4}{u^2} + \dfrac{u^2}{4}$, $u = x - 3x^2$ **36.** $y = \frac{1}{2}u^2 + 2u^{1/2}$, $u = 1 - 3x$

37. $y = u(u + 1)^5$, $u = x^2 + x$ **38.** $y = (u^2 + 1)^3$, $u = x - x^2$

39. $y = \dfrac{u - 1}{u + 1}$, $u = 2 + \sqrt{x}$ **40.** $y = \dfrac{u}{1 + u^2}$, $u = 5x + 1$

41. Find the equation of the line tangent to the graph of $y = 2x(x - 4)^6$ at the point $(5, 10)$.

42. Find the equation of the line tangent to the graph of $y = \dfrac{x}{\sqrt{2 - x^2}}$ at the point $(1, 1)$.

43. Find the x-coordinates of all points on the curve $y = (-x^2 + 4x - 3)^3$ with a horizontal tangent line.

44. The function $f(x) = \sqrt{x^2 - 6x + 10}$ has one relative minimum point for $x \geq 0$. Find it.

45. The length, x, of the edge of a cube is increasing.

    (a) Write the chain rule for $\dfrac{dV}{dt}$, the time rate of change of the volume of the cube.

    (b) For what value of x is $\dfrac{dV}{dt}$ equal to 12 times the rate of increase of x?

46. The weight of a male hognose snake is approximately $446x^3$ grams, where x is its length in meters.* Suppose a snake has length .4 meters and is growing at the rate of .2 meters per year. At what rate is the snake gaining weight?

---

    * D. R. Platt, "Natural History of the Hognose Snakes *Heterodon platyrhinos* and *Heterodon nasicus*," University of Kansas Publications, Museum of Natural History **18**, no. 4 (1969), pp. 253–420.

47. Suppose that P, y, and t are variables, where P is a function of y and y is a function of t.

(a) Write the derivative symbols for the following quantities: the rate of change of y with respect to t, the rate of change of P with respect to y, and the rate of change of P with respect to t. Select your answers from the following: $\dfrac{dP}{dy}$, $\dfrac{dy}{dP}$, $\dfrac{dy}{dt}$, $\dfrac{dP}{dt}$, $\dfrac{dt}{dP}$, and $\dfrac{dt}{dy}$.

(b) Write the chain rule for $\dfrac{dP}{dt}$.

48. Suppose that Q, x, and y are variables, where Q is a function of x and x is a function of y. (Read this carefully.)

(a) Write the derivative symbols for the following quantities: the rate of change of x with respect to y, the rate of change of Q with respect to y, and the rate of change of Q with respect to x. Select your answers from the following: $\dfrac{dy}{dx}$, $\dfrac{dx}{dy}$, $\dfrac{dQ}{dx}$, $\dfrac{dx}{dQ}$, $\dfrac{dQ}{dy}$, and $\dfrac{dy}{dQ}$.

(b) Write the chain rule for $\dfrac{dQ}{dy}$.

49. When a company produces and sells x thousand units per week its total weekly profit is P thousand dollars, where $P = \dfrac{200x}{100 + x^2}$. The production level at t weeks from the present is $x = 4 + 2t$.

(a) Find the marginal profit, $\dfrac{dP}{dx}$.

(b) Find the time rate of change of profit, $\dfrac{dP}{dt}$.

(c) How fast (with respect to time) are profits changing when $t = 8$?

50. The cost of manufacturing x cases of cereal is C dollars, where $C = 3x + 4\sqrt{x} + 2$. Weekly production at t weeks from the present is estimated to be $x = 6200 + 100t$ cases.

(a) Find the marginal cost, $\dfrac{dC}{dx}$.

(b) Find the time rate of change of cost, $\dfrac{dC}{dt}$.

(c) How fast (with respect to time) are costs rising when $t = 2$?

51. Ecologists estimate that when the population of a certain city is x persons, the average level L of carbon monoxide in the air above the city will be L ppm (parts per million), where $L = 10 + .4x + .0001x^2$. The population of the city is estimated to be $x = 752 + 23t + .5t^2$ thousand persons t years from the present.

(a) Find the rate of change of carbon monoxide with respect to the population of the city.

(b) Find the time rate of change of the population when $t = 2$.

(c) How fast (with respect to time) is the carbon monoxide level changing at time $t = 2$?

52. A manufacturer of microcomputers estimates that t months from now it will sell x thousand units of its main line of microcomputers per month, where

$x = .05t^2 + 2t + 5$. Because of economies of scale, the profit P from manufacturing and selling x thousand units is estimated to be $P = .001x^2 + .1x - .25$ million dollars. Calculate the rate at which the profit will be increasing 5 months from now.

53. Suppose that $f(x)$ and $g(x)$ are differentiable functions. Find $g(x)$ if you know that

$$\frac{d}{dx} f(g(x)) = 3x^2 \cdot f'(x^3 + 1).$$

54. Suppose that $f(x)$ and $g(x)$ are differentiable functions. Find $g(x)$ if you know that $f'(x) = 1/x$ and

$$\frac{d}{dx} f(g(x)) = \frac{2x + 5}{x^2 + 5x - 4}.$$

55. Suppose that $f(x)$ and $g(x)$ are differentiable functions such that $f(1) = 2$, $f'(1) = 3$, $f'(5) = 4$, $g(1) = 5$, $g'(1) = 6$, $g'(2) = 7$, and $g'(5) = 8$. Find

$$\frac{d}{dx} [f(g(x))]\Big|_{x=1}.$$

56. Consider the functions of Exercise 55. Find $\dfrac{d}{dx} [g(f(x))]\Big|_{x=1}$.

3.3 IMPLICIT DIFFERENTIATION AND RELATED RATES

This section presents two different applications of the chain rule. In each case we will need to differentiate one or more composite functions where the "inside" functions are not known explicitly.

Implicit Differentiation   In some applications, the variables are related by an equation rather than a function. In these cases, we can still determine the rate of change of one variable with respect to the other by the technique of implicit differentiation. As an illustration, consider the equation

$$x^2 + y^2 = 4. \tag{1}$$

The graph of this equation is the circle in Fig. 1. This graph obviously is not the graph of a function since, for instance, there are two points on the graph whose x-coordinate is 1. (Functions must satisfy the vertical line test. See Section 0.1.)

FIGURE 1   **Graph of**
$x^2 + y^2 = 4.$

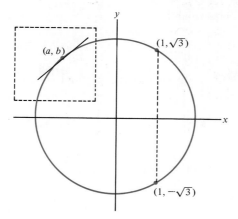

We denote the slope of the curve at the point $(1, \sqrt{3})$ by

$$\left.\frac{dy}{dx}\right|_{\substack{x=1 \\ y=\sqrt{3}}}$$

In general, the slope at the point (a, b) is denoted by

$$\left.\frac{dy}{dx}\right|_{\substack{x=a \\ y=b}}$$

In a small vicinity of the point (a, b), the curve looks like the graph of a function. [That is, on this part of the curve, $y = g(x)$ for some function $g(x)$.] We say that this function is defined *implicitly* by the equation.* We obtain a formula for $\dfrac{dy}{dx}$ by differentiating both sides of the equation with respect to x while treating y as a function of x.

EXAMPLE 1   Consider the graph of the equation $x^2 + y^2 = 4$.

(a) Use implicit differentiation to compute $\dfrac{dy}{dx}$.

(b) Find the slope of the graph at the points $(1, \sqrt{3})$ and $(1, -\sqrt{3})$.

Solution   (a) The first term x^2 has derivative $2x$, as usual. We think of the second term y^2 as having the form $[g(x)]^2$. To differentiate we use the chain rule (specifically, the general power rule):

$$\frac{d}{dx}[g(x)]^2 = 2[g(x)]g'(x)$$

---

*Of course, the tangent line at the point (a, b) must not be vertical. In this section, we assume that the given equations implicitly determine differentiable functions.

or, equivalently,

$$\frac{d}{dx}\,y^2 = 2y\,\frac{dy}{dx}.$$

On the right side of the original equation, the derivative of the constant function 4 is zero. Thus implicit differentiation of $x^2 + y^2 = 4$ yields

$$2x + 2y\,\frac{dy}{dx} = 0.$$

Solving for $\dfrac{dy}{dx}$, we have

$$2y\,\frac{dy}{dx} = -2x.$$

If $y \neq 0$, then

$$\frac{dy}{dx} = \frac{-2x}{2y} = -\frac{x}{y}.$$

Notice that this slope formula involves y as well as x. This reflects the fact that the slope of the circle at a point depends on the y-coordinate of the point as well as the x-coordinate.

(b) At the point $(1,\ \sqrt{3})$ the slope is

$$\left.\frac{dy}{dx}\right|_{\substack{x=1\\y=\sqrt{3}}} = \left.-\frac{x}{y}\right|_{\substack{x=1\\y=\sqrt{3}}} = -\frac{1}{\sqrt{3}}.$$

At the point $(1,\ -\sqrt{3})$ the slope is

$$\left.\frac{dy}{dx}\right|_{\substack{x=1\\y=-\sqrt{3}}} = \left.-\frac{x}{y}\right|_{\substack{x=1\\y=-\sqrt{3}}} = -\frac{1}{-\sqrt{3}} = \frac{1}{\sqrt{3}}.$$

(See Fig. 2.) The formula for $\dfrac{dy}{dx}$ gives the slope at every point on the graph of $x^2 + y^2 = 4$ except $(-2, 0)$, and $(2, 0)$. At these two points the tangent line is vertical and the slope of the curve is undefined. ●

FIGURE 2 **Slope of the tangent line.**

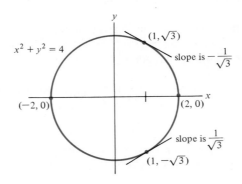

$x^2 + y^2 = 4$

$(1, \sqrt{3})$

slope is $-\dfrac{1}{\sqrt{3}}$

$(-2, 0)$

$(2, 0)$

slope is $\dfrac{1}{\sqrt{3}}$

$(1, -\sqrt{3})$

The difficult step in Example 1(a) was to differentiate y^2 correctly. The derivative of y^2 *with respect to y* would be $2y$, by the ordinary power rule. But the derivative of y^2 *with respect to x* must be computed by the general power rule. In general,

$$\frac{d}{dx} y^r = ry^{r-1} \frac{dy}{dx}. \tag{1}$$

This rule is used to compute slope formulas in the next two examples.

EXAMPLE 2 Use implicit differentiation to calculate $\dfrac{dy}{dx}$ for the equation $x^2y^6 = 1$.

Solution Differentiate each side of the equation $x^2y^6 = 1$ with respect to x. On the left side of the equation use the product rule and treat y as a function of x.

$$x^2 \frac{d}{dx} (y^6) + y^6 \frac{d}{dx} (x^2) = \frac{d}{dx} (1)$$

$$x^2 \cdot 6y^5 \frac{dy}{dx} + y^6 \cdot 2x = 0.$$

Solve for $\dfrac{dy}{dx}$ by moving the term not involving $\dfrac{dy}{dx}$ to the right side and dividing by the factor that multiplies $\dfrac{dy}{dx}$.

$$6x^2y^5 \frac{dy}{dx} = -2xy^6$$

$$\frac{dy}{dx} = \frac{-2xy^6}{6x^2y^5} = -\frac{y}{3x}. \quad \bullet$$

EXAMPLE 3 Use implicit differentiation to calculate $\dfrac{dy}{dx}$ when y is related to x by the equation $x^2y + xy^3 - 3x = 5$.

Solution Differentiate the equation term by term, taking care to differentiate x^2y and xy^3 by the product rule.

$$x^2 \frac{d}{dx} (y) + y \frac{d}{dx} (x^2) + x \frac{d}{dx} (y^3) + y^3 \frac{d}{dx} (x) - 3 = 0$$

$$x^2 \frac{dy}{dx} + y \cdot 2x + x \cdot 3y^2 \frac{dy}{dx} + y^3 \cdot 1 - 3 = 0.$$

Solve for $\dfrac{dy}{dx}$ in terms of x and y.

$$x^2 \frac{dy}{dx} + 3xy^2 \frac{dy}{dx} = 3 - y^3 - 2xy$$

$$(x^2 + 3xy^2) \frac{dy}{dx} = 3 - y^3 - 2xy$$

$$\frac{dy}{dx} = \frac{3 - y^3 - 2xy}{x^2 + 3xy^2} . \quad \bullet$$

Reminder   When a power of y is differentiated with respect to x, the result must include the factor $\frac{dy}{dx}$. When a power of x is differentiated, there is no factor $\frac{dy}{dx}$.

Here is the general procedure for implicit differentiation.

Finding $\frac{dy}{dx}$ by Implicit Differentiation

Differentiate each term of the equation *with respect to x*, treating y as a function of x.

Move all terms involving $\frac{dy}{dx}$ to the left side of the equation and move the other terms to the right side.

Factor out $\frac{dy}{dx}$ on the left side of the equation.

Divide both sides of the equation by the factor that multiplies $\frac{dy}{dx}$.

Equations that implicitly define functions arise frequently in economic models. The economic background for the equation in the next example is discussed in Section 7.1.

EXAMPLE 4   Suppose that x and y represent the amounts of two basic inputs into a production process and the equation
$$60x^{3/4}y^{1/4} = 3240$$

describes all input amounts (x, y) for which the output of the process is 3240 units. (The graph of this equation is called a *production isoquant* or *constant product curve*.) See Fig. 3. Use implicit differentiation to calculate the slope of the graph at the point on the curve where $x = 81$, $y = 16$.

Solution   We use the product rule and treat y as a function of x.

$$60x^{3/4} \frac{d}{dx} (y^{1/4}) + y^{1/4} \frac{d}{dx} (60x^{3/4}) = 0$$

$$60x^{3/4} \cdot \left(\frac{1}{4}\right) y^{-3/4} \frac{dy}{dx} + y^{1/4} \cdot 60\left(\frac{3}{4}\right) x^{-1/4} = 0$$

FIGURE 3   A production isoquant.

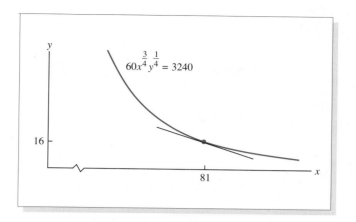

$$15x^{3/4}y^{-3/4}\frac{dy}{dx} = -45x^{-1/4}y^{1/4}$$

$$\frac{dy}{dx} = \frac{-45x^{-1/4}y^{1/4}}{15x^{3/4}y^{-3/4}} = \frac{-3y}{x}.$$

When $x = 81$, $y = 16$, we have

$$\frac{dy}{dx}\bigg|_{\substack{x=81 \\ y=16}} = \frac{-3(16)}{81} = -\frac{16}{27}.$$

The number $-\frac{16}{27}$ is the slope of the production isoquant at the point $(81, 16)$. If the first input (corresponding to x) is increased by one small unit, then the second input (corresponding to y) must decrease by approximately $\frac{16}{27}$ unit in order to keep the production output unchanged [i.e., to keep (x, y) on the curve]. In economic terminology, the absolute value of $\frac{dy}{dx}$ is called the *marginal rate of substitution* of the first input for the second input. ●

TECHNOLOGY PROJECT 3.6

Plotting Implicitly Defined Functions by Computer (Symbolic Package Required)

As we mentioned in Chapter 0, graphing calculators can only plot equations written in the form $y = $ [an expression in x]. In order to plot equations not in this form generally requires the power of a computer, since more sophisticated and laborious calculations are required to determine the points to plot. Most symbolic mathematics programs have commands to plot general equations in two variables. In *Mathematica,* the command to do this is called `ImplicitPlot`. Note that this command may not be automatically loaded in your installation of the program and may need to be loaded as a separate package. (Consult the program documentation for details.) For example, to plot the equation

$$x^2 + x - y^3 = 1$$

with the variables in the ranges $-1 \le x \le 1$, $-1 \le y \le 1$, you would use the command:

```
ImplicitPlot [x^2+x-y^3==1, {x, -1, 1}, {y, -1, 1}]
```

Plot the following implicitly defined functions:

1. $x^2 + y^2 = 4$, $-2 \le x \le 2$, $-2 \le y \le 2$
2. $x^2 + \dfrac{y^2}{4} = 1$, $-2 \le x \le 2$, $-2 \le y \le 2$
3. $x^4 + 2x^2y^2 + y^4 = 4x^2 - 4y^2$, $-2 \le x \le 2$, $-2 \le y \le 2$
4. $xy + y^2 = 14$, $-5 \le x \le 5$, $-5 \le y \le 5$

Related Rates   In implicit differentiation we differentiate an equation involving x and y, with y treated as a function of x. However, in some applications where x and y are related by an equation, both variables are functions of a third variable t (which may represent time). Often the formulas for x and y as functions of t are not known. When we differentiate such an equation with respect to t, we derive a relationship between the rates of change $\dfrac{dy}{dt}$ and $\dfrac{dx}{dt}$. We say that these derivatives are *related rates*. The equation relating the rates may be used to find one of the rates when the other is known.

EXAMPLE 5   Suppose that x and y are both differentiable functions of t and are related by the equation

$$x^2 + 5y^2 = 36. \tag{2}$$

(a) Differentiate each term in the equation with respect to t and solve the resulting equation for $\dfrac{dy}{dt}$.

(b) Calculate $\dfrac{dy}{dt}$ at a time when $x = 4$, $y = 2$, and $\dfrac{dx}{dt} = 5$.

Solution   (a) Since x is a function of t, the general power rule gives

$$\frac{d}{dt}(x^2) = 2x\frac{dx}{dt}.$$

A similar formula holds for the derivative of y^2. Differentiating each term in (2) with respect to t, we obtain

$$\frac{d}{dt}(x^2) + \frac{d}{dt}(5y^2) = \frac{d}{dt}(36)$$

$$2x\frac{dx}{dt} + 5\cdot 2y\frac{dy}{dt} = 0$$

$$10y \frac{dy}{dt} = -2x \frac{dx}{dt}$$

$$\frac{dy}{dt} = -\frac{x}{5y} \frac{dx}{dt}.$$

(b) When $x = 4$, $y = 2$, and $\frac{dx}{dt} = 5$,

$$\frac{dy}{dt} = -\frac{4}{5(2)} \cdot (5) = -2. \ \bullet$$

There is a helpful graphical interpretation of the calculations in Example 5. Imagine a point that is moving clockwise along the graph of the equation $x^2 + 5y^2 = 36$. (See Fig. 4.) Suppose that when the point is at $(4, 2)$, the x-coordinate of the point is changing at the rate of 5 units per minute, so that $\frac{dx}{dt} = 5$. In Example 5(b) we found that $\frac{dy}{dt} = -2$. This means that the y-coordinate of the point is decreasing at the rate of 2 units per minute when the moving point reaches $(4, 2)$.

EXAMPLE 6 Suppose that x thousand units of a commodity can be sold weekly when the price is p dollars per unit, and suppose that x and p satisfy the demand equation

$$p + 2x + xp = 38.$$

See Fig. 5. How fast are weekly sales changing at a time when $x = 4$, $p = 6$, and the price is falling at the rate of \$.40 per week?

Solution Assume that p and x are differentiable functions of t, and differentiate the demand equation with respect to t.

FIGURE 5   A demand curve.

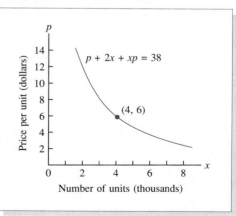

FIGURE 4   Graph of $x^2 + 5y^2 = 36$.

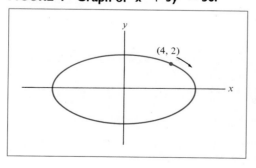

$$\frac{d}{dt}(p) + \frac{d}{dt}(2x) + \frac{d}{dt}(xp) = \frac{d}{dt}(38)$$

$$\frac{dp}{dt} + 2\frac{dx}{dt} + x\frac{dp}{dt} + p\frac{dx}{dt} = 0. \qquad (3)$$

We want to know $\dfrac{dx}{dt}$ at a time when $x = 4$, $p = 6$, and $\dfrac{dp}{dt} = -.40$. (The derivative $\dfrac{dp}{dt}$ is negative because the price is decreasing.) We could solve (3) for $\dfrac{dx}{dt}$ and then substitute the given values, but since we do not need a general formula for $\dfrac{dx}{dt}$, it is easier to substitute first and then solve.

$$-.40 + 2\frac{dx}{dt} + 4(-.40) + 6\frac{dx}{dt} = 0$$

$$8\frac{dx}{dt} = 2$$

$$\frac{dx}{dt} = .25.$$

Thus sales are rising at the rate of .25 thousand units (i.e., 250 units) per week. ●

Suggestions for Solving Related Rate Problems

1. Draw a picture, if possible.
2. Assign letters to quantities that vary, and identify one variable, say t, on which the other variables depend.
3. Find an equation that relates the variables to each other.
4. Differentiate the equation with respect to the independent variable t. Use the chain rule whenever appropriate.
5. Substitute all specified values for the variables and their derivatives.
6. Solve for the derivative that gives the unknown rate.

PRACTICE PROBLEMS 3.3

Suppose that x and y are related by the equation $3y^2 - 3x^2 + y = 1$.

1. Use implicit differentiation to find a formula for the slope of the graph of the equation.
2. Suppose that the x and y in the equation above are both functions of t. Differentiate both sides of the equation with respect to t and find a formula for $\dfrac{dy}{dt}$ in terms of x, y, and $\dfrac{dx}{dt}$.

EXERCISES 3.3

In Exercises 1–18, suppose that x and y are related by the given equation and use implicit differentiation to determine $\dfrac{dy}{dx}$.

1. $x^2 - y^2 = 1$ 2. $x^3 + y^3 - 6 = 0$ 3. $y^5 - 3x^2 = x$

4. $x^4 + (y + 3)^4 = x^2$ 5. $y^4 - x^4 = y^2 - x^2$

6. $x^3 + y^3 = x^2 + y^2$ 7. $2x^3 + y = 2y^3 + x$

8. $x^4 + 4y = x - 4y^3$ 9. $xy = 5$

10. $xy^3 = 2$ 11. $x(y + 2)^5 = 8$ 12. $x^2y^3 = 6$

13. $x^3y^2 - 4x^2 = 1$ 14. $(x + 1)^2(y - 1)^2 = 1$

15. $x^3 + y^3 = x^3y^3$ 16. $x^2 + 4xy + 4y = 1$

17. $x^2y + y^2x = 3$ 18. $x^3y + xy^3 = 4$

Use implicit differentiation of the equations in Exercises 19–24 to determine the slope of the graph at the given point.

19. $4y^3 - x^2 = -5; x = 3, y = 1$ 20. $y^2 = x^3 + 1; x = 2, y = -3$

21. $xy^3 = 2; x = -\frac{1}{4}, y = -2$ 22. $\sqrt{x} + \sqrt{y} = 7; x = 9, y = 16$

23. $xy + y^3 = 14; x = 3, y = 2$ 24. $y^2 = 3xy - 5; x = 2, y = 1$

25. Find the equation of the tangent line to the graph of $x^2y^4 = 1$ at the point $(4, \frac{1}{2})$ and at the point $(4, -\frac{1}{2})$.

26. Find the equation of the tangent line to the graph of $x^4y^2 = 144$ at the point $(2, 3)$ and at the point $(2, -3)$.

27. The graph of $x^4 + 2x^2y^2 + y^4 = 4x^2 - 4y^2$ is the "lemniscate" in Fig. 6.

   (a) Find $\dfrac{dy}{dx}$ by implicit differentiation.

   (b) Find the slope of the tangent line to the lemniscate at $(\sqrt{6}/2, \sqrt{2}/2)$.

FIGURE 6   A lemniscate.

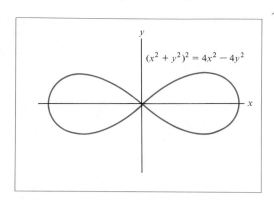

$$(x^2 + y^2)^2 = 4x^2 - 4y^2$$

28. The graph of $x^4 + 2x^2y^2 + y^4 = 9x^2 - 9y^2$ is a lemniscate similar to that in Fig. 6.

(a) Find $\dfrac{dy}{dx}$ by implicit differentiation.

(b) Find the slope of the tangent line to the lemniscate at $(\sqrt{5}, -1)$.

29. Suppose that x and y represent the amounts of two basic inputs for a production process and suppose that the equation

$$30x^{1/3}y^{2/3} = 1080$$

describes all input amounts where the output of the process is 1080 units. Find $\dfrac{dy}{dx}$ when $x = 16$, $y = 54$.

30. Suppose that x and y represent the amounts of two basic inputs for a production process and

$$10x^{1/2}y^{1/2} = 600.$$

Find $\dfrac{dy}{dx}$ when $x = 50$, $y = 72$.

In Exercises 31–36, suppose that x and y are both differentiable functions of t and are related by the given equation. Use implicit differentiation with respect to t to determine $\dfrac{dy}{dt}$ in terms of x, y, and $\dfrac{dx}{dt}$.

31. $x^4 + y^4 = 1$ 32. $y^4 - x^2 = 1$ 33. $3xy - 3x^2 = 4$

34. $y^2 = 8 + xy$ 35. $x^2 + 2xy = y^3$ 36. $x^2y^2 = 2y^3 + 1$

37. A point is moving along the graph of $x^2 - 4y^2 = 9$. When the point is at $(5, -2)$, its x-coordinate is increasing at the rate of 3 units per second. How fast is the y-coordinate changing at that moment?

38. A point is moving along the graph of $x^3y^2 = 200$. When the point is at $(2, 5)$, its x-coordinate is changing at the rate of -4 units per minute. How fast is the y-coordinate changing at that moment?

39. Suppose that the price p (in dollars) and the weekly sales x (in thousands of units) of a certain commodity satisfy the demand equation

$$2p^3 + x^2 = 4500.$$

Determine the rate at which sales are changing at a time when $x = 50$, $p = 10$, and the price is falling at the rate of $\$.50$ per week.

40. Suppose that the price p (in dollars) and the demand x (in thousands of units) of a commodity satisfy the demand equation

$$6p + x + xp = 94.$$

How fast is the demand changing at a time when $x = 4$, $p = 9$, and the price is rising at the rate of $\$2$ per week?

41. The monthly advertising revenue A and the monthly circulation x of a magazine are related approximately by the equation

$$A = 6\sqrt{x^2 - 400}, \qquad x \geq 20,$$

where A is given in thousands of dollars and x is measured in thousands of copies sold. At what rate is the advertising revenue changing if the current circulation is

$x = 25$ thousand copies and the circulation is changing at the rate of 2 thousand copies per month?

$$\left[\textit{Hint}: \text{Use the chain rule } \frac{dA}{dt} = \frac{dA}{dx}\frac{dx}{dt}. \right]$$

42. Suppose that in Boston the wholesale price p of oranges (in dollars per crate) and the daily supply x (in thousands of crates) are related by the equation $px + 7x + 8p = 328$. If there are 4 thousand crates available today at a price of $25 per crate, and if the supply is changing at the rate of $-.3$ thousand crates per day, at what rate is the price changing?

43. Under certain conditions (called adiabatic expansion) the pressure P and volume V of a gas such as oxygen satisfy the equation $P^5V^7 = k$, where k is a constant. Suppose that at some moment the volume of the gas is 4 liters, the pressure is 200 units, and the pressure is increasing at the rate of 5 units per second. Find the (time) rate at which the volume is changing.

44. The volume V of a spherical cancer tumor is given by $V = \pi x^3/6$, where x is the diameter of the tumor. A physician estimates that the diameter is growing at the rate of .4 millimeters per day, at a time when the diameter is already 10 millimeters. How fast is the volume of the tumor changing at that time?

45. Figure 7(a) shows a 10-foot ladder leaning against a wall.
    (a) Use the Pythagorean theorem to find an equation relating x and y.
    (b) Suppose that the foot of the ladder is being pulled along the ground at the rate of 3 feet per second. How fast is the top end of the ladder sliding down the wall at the time when the foot of the ladder is 8 feet from the wall? That is, what is $\frac{dy}{dt}$ at the time when $\frac{dx}{dt} = 3$ and $x = 8$?

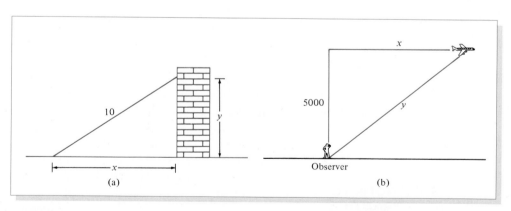

(a)   (b)

FIGURE 7

46. An airplane flying 390 feet per second at an altitude of 5000 feet flew directly over an observer. Figure 7(b) shows the relationship of the airplane to the observer at a later time.
    (a) Find an equation relating x and y.
    (b) Find the value of x when y is 13,000.

(c) How fast is the distance from the observer to the airplane changing at the time when the airplane is 13,000 feet from the observer? That is, what is $\dfrac{dy}{dt}$ at the time when $\dfrac{dx}{dt} = 390$ and $y = 13{,}000$?

47. A baseball diamond is a 90-foot by 90-foot square. See Fig. 8. A player runs from first to second base at the speed of 22 feet per second. How fast is the player's distance from third base changing when he is halfway between first and second base? [*Hint*: If x is the distance from the player to second base and y is his distance from third base, then $x^2 + 90^2 = y^2$.]

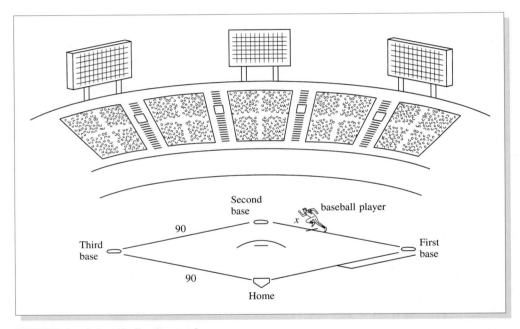

FIGURE 8   A baseball diamond.

| SOLUTIONS TO PRACTICE PROBLEMS 3.3 | **1.** | |

$$\frac{d}{dx}(3y^2) - \frac{d}{dx}(3x^2) + \frac{d}{dx}(y) = \frac{d}{dx}(1)$$

$$6y\,\frac{dy}{dx} - 6x + \frac{dy}{dx} = 0$$

$$(6y + 1)\,\frac{dy}{dx} = 6x$$

$$\frac{dy}{dx} = \frac{6x}{6y + 1}$$

2. First, here is the solution without reference to Practice Problem 1.

$$\frac{d}{dt}(3y^2) - \frac{d}{dt}(3x^2) + \frac{d}{dt}(y) = \frac{d}{dt}(1)$$

$$6y\frac{dy}{dt} - 6x\frac{dx}{dt} + \frac{dy}{dt} = 0$$

$$(6y + 1)\frac{dy}{dt} = 6x\frac{dx}{dt}$$

$$\frac{dy}{dt} = \frac{6x}{6y + 1}\frac{dx}{dt}.$$

Alternatively, we can use the chain rule and the formula for $\dfrac{dy}{dx}$ from Practice Problem 1.

$$\frac{dy}{dt} = \frac{dy}{dx}\frac{dx}{dt} = \frac{6x}{6y + 1}\frac{dx}{dt}.$$

 **CHAPTER 3 CHECKLIST**

✓ $\dfrac{d}{dx}[f(x) \cdot g(x)] = f(x)g'(x) + g(x)f'(x)$

✓ $\dfrac{d}{dx}\left[\dfrac{f(x)}{g(x)}\right] = \dfrac{g(x)f'(x) - f(x)g'(x)}{[g(x)]^2}$

✓ $\dfrac{d}{dx}[f(g(x))] = f'(g(x)) \cdot g'(x)$

✓ Implicit differentiation

✓ $\dfrac{d}{dx}y^r = ry^{r-1}\dfrac{dy}{dx}$

✓ Related rates

CHAPTER 3   SUPPLEMENTARY EXERCISES

Differentiate the following functions.

1. $(4x - 1)(3x + 1)^4$

2. $2(5 - x)^3(6x - 1)$

3. $x(x^5 - 1)^3$

4. $(2x + 1)^{5/2}(4x - 1)^{3/2}$

5. $5(\sqrt{x} - 1)^4(\sqrt{x} - 2)^2$

6. $\dfrac{\sqrt{x}}{\sqrt{x} + 4}$

7. $3(x^2 - 1)^3(x^2 + 1)^5$

8. $\dfrac{1}{(x^2 + 5x + 1)^6}$

9. $\dfrac{x^2 - 6x}{x - 2}$

10. $\dfrac{2x}{2 - 3x}$

11. $\left(\dfrac{3 - x^2}{x^3}\right)^2$

12. $\dfrac{x^3 + x}{x^2 - x}$

13. Let $f(x) = (3x + 1)^4(3 - x)^5$. Find all x such that $f'(x) = 0$.

14. Let $f(x) = (x^2 + 1)/(x^2 + 5)$. Find all x such that $f'(x) = 0$.

15. Find the equation of the line tangent to the graph of $y = (x^3 - 1)(x^2 + 1)^4$ at the point where $x = -1$.

16. Find the equation of the line tangent to the graph of $y = \dfrac{x - 3}{\sqrt{4 + x^2}}$ at the point where $x = 0$.

17. A botanical display is to be constructed as a rectangular region with a river as one side and a sidewalk 2 meters wide along the inside edges of the other three sides. (See Fig. 1.) The area for the plants must be 800 square meters. Find the outside dimensions of the region that minimize the area of the sidewalk (and hence minimize the amount of concrete needed for the sidewalk).

FIGURE I   A botanical display.

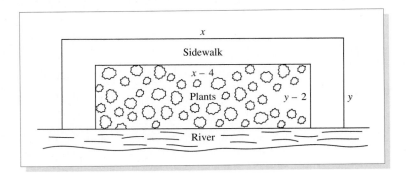

18. Repeat Exercise 17, with the sidewalk on the inside of all four sides. In this case, the 800-square-meter planted region has dimensions $x - 4$ meters by $y - 4$ meters.

19. A store estimates that its cost when selling x lamps per day is C dollars, where $C = 40x + 30$ (i.e., the marginal cost per lamp is $40). Suppose that daily sales are rising at the rate of three lamps per day. How fast are the costs rising? Explain your answer using the chain rule.

20. A company pays y dollars in taxes when its annual profit is P dollars. Suppose that y is some (differentiable) function of P and P is some function of time t. Give a chain rule formula for the time rate of change of taxes $\dfrac{dy}{dt}$.

Exercises 21–26 refer to the graphs of the functions $f(x)$ and $g(x)$ in Fig. 2.

21. Let $h(x) = 2f(x) - 3g(x)$. Determine $h(1)$ and $h'(1)$.

FIGURE 2

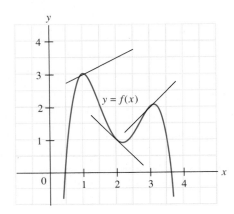

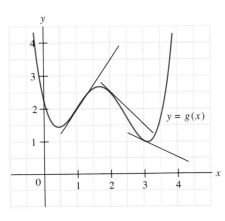

22. Let $h(x) = f(x) \cdot g(x)$. Determine $h(1)$ and $h'(1)$.

23. Let $h(x) = \dfrac{f(x)}{g(x)}$. Determine $h(1)$ and $h'(1)$.

24. Let $h(x) = [f(x)]^2$. Determine $h(1)$ and $h'(1)$.

25. Let $h(x) = f(g(x))$. Determine $h(1)$ and $h'(1)$.

26. Let $h(x) = g(f(x))$. Determine $h(1)$ and $h'(1)$.

In Exercises 27–29, find a formula for $\dfrac{d}{dx} f(g(x))$, where $f(x)$ is a function such that $f'(x) = 1/(x^2 + 1)$.

27. $g(x) = x^3$

28. $g(x) = \dfrac{1}{x}$

29. $g(x) = x^2 + 1$

In Exercises 30–32, find a formula for $\dfrac{d}{dx} f(g(x))$, where $f(x)$ is a function such that $f'(x) = x\sqrt{1 - x^2}$.

30. $g(x) = x^2$

31. $g(x) = \sqrt{x}$

32. $g(x) = x^{3/2}$

In Exercises 33–35, find $\dfrac{dy}{dx}$ where y is a function of u such that $\dfrac{dy}{du} = \dfrac{u}{u^2 + 1}$.

33. $u = x^{3/2}$

34. $u = x^2 + 1$

35. $u = \dfrac{5}{x}$

In Exercises 36–38, find $\dfrac{dy}{dx}$ where y is a function of u such that $\dfrac{dy}{du} = \dfrac{u}{\sqrt{1 + u^4}}$.

36. $u = x^2$

37. $u = \sqrt{x}$

38. $u = \dfrac{2}{x}$

39. The revenue R a company receives is a function of the weekly sales x. Also, the sales level x is a function of the weekly advertising expenditures A, and A in turn is a varying function of time.

   (a) Write the derivative symbols for the following quantities: rate of change of revenue with respect to advertising expenditures, time rate of change of advertising expenditures, marginal revenue, and rate of change of sales with respect to advertising expenditures. Select your answers from the following: $\dfrac{dR}{dx}$, $\dfrac{dR}{dt}$, $\dfrac{dA}{dt}$, $\dfrac{dA}{dR}$, $\dfrac{dA}{dx}$, $\dfrac{dx}{dA}$, and $\dfrac{dR}{dA}$.

   (b) Write a type of chain rule that expresses the time rate of change of revenue, $\dfrac{dR}{dt}$, in terms of three of the derivatives described in part (a).

40. The amount A of anesthetics that a certain hospital uses each week is a function of the number S of surgical operations performed each week. Also, S in turn is a function of the population P of the area served by the hospital, while P is a function of time t.

   (a) Write the derivative symbols for the following quantities: population growth rate, rate of change of anesthetic usage with respect to the population size, rate of change of surgical operations with respect to the population size, and rate of change of anesthetic usage with respect to the number of surgical operations.

Select your answers from the following: $\dfrac{dS}{dP}$, $\dfrac{dS}{dt}$, $\dfrac{dP}{dS}$, $\dfrac{dP}{dt}$, $\dfrac{dA}{dS}$, $\dfrac{dA}{dP}$, and $\dfrac{dS}{dA}$.

(b) Write a type of chain rule that expresses the time rate of change of anesthetic usage, $\dfrac{dA}{dt}$, in terms of three of the derivatives described in part (a).

41. The graph of $x^{2/3} + y^{2/3} = 8$ is the *astroid* in Fig. 3.

 (a) Find $\dfrac{dy}{dx}$ by implicit differentiation.

 (b) Find the slope of the tangent line at $(8, -8)$.

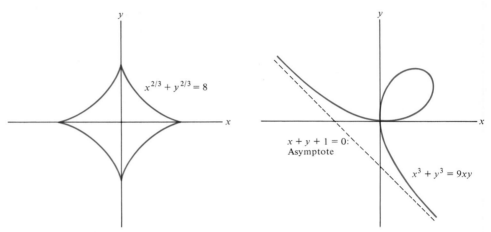

FIGURE 3   Astroid.   **FIGURE 4   Folium of Descartes.**

42. The graph of $x^3 + y^3 = 9xy$ is the folium of Descartes, shown in Fig. 4.

 (a) Find $\dfrac{dy}{dx}$ by implicit differentiation.

 (b) Find the slope of the curve at $(2, 4)$.

In Exercises 43–46, x and y are related by the given equation. Use implicit differentiation to calculate the value of $\dfrac{dy}{dx}$ for the given values of x and y.

43. $x^2y^2 = 9$; $x = 1$, $y = 3$ 44. $xy^4 = 48$; $x = 3$, $y = 2$

45. $x^2 - xy^3 = 20$; $x = 5$, $y = 1$ 46. $xy^2 - x^3 = 10$; $x = 2$, $y = 3$

47. A factory's weekly production costs y and its weekly production quantity x are related by the equation $y^2 - 5x^3 = 4$, where y is in thousands of dollars and x is in thousands of units of output.

 (a) Use implicit differentiation to find a formula for $\dfrac{dy}{dx}$, the marginal cost of production.

 (b) Find the marginal cost of production when $x = 4$ and $y = 18$.

(c) Suppose that the factory begins to vary its weekly production level. Assuming that x and y are differentiable functions of time t, use the method of related rates to find a formula for $\dfrac{dy}{dt}$, the time rate of change of production costs.

(d) Compute $\dfrac{dy}{dt}$ when $x = 4$, $y = 18$, and the production level is rising at the rate of .3 thousand units per week (i.e., when $\dfrac{dx}{dt} = .3$).

48. A town library estimates that when the population is x thousand persons, approximately y thousand books will be checked out of the library during one year, where x and y are related by the equation $y^3 - 8000x^2 = 0$.

    (a) Use implicit differentiation to find a formula for $\dfrac{dy}{dx}$, the rate of change of library circulation with respect to population size.

    (b) Find the value of $\dfrac{dy}{dx}$ when $x = 27$ thousand persons and $y = 180$ thousand books per year.

    (c) Assume that x and y are both differentiable functions of time t, and use the method of related rates to find a formula for $\dfrac{dy}{dt}$, the time rate of change of library circulation.

    (d) Compute $\dfrac{dy}{dt}$ when $x = 27$, $y = 180$, and the population is rising at the rate of 1.8 thousand persons per year $\left(\text{i.e.,} \ \dfrac{dx}{dt} = 1.8 \right)$. Either use part (c), or use part (b) and the chain rule.

49. Suppose that the price p and quantity x of a certain commodity satisfy the demand equation $6p + 5x + xp = 50$, and suppose that p and x are functions of time, t. Determine the rate at which the quantity is changing when $x = 4$, $p = 3$, and $\dfrac{dp}{dt} = -2$.

50. An offshore oil well is leaking oil onto the ocean surface, forming a circular oil slick about .005 meter thick. If the radius of the slick is r meters, then the volume of oil spilled is $V = .005\pi r^2$ cubic meters. Suppose that the oil is leaking at a constant rate of 20 cubic meters per hour, so that $\dfrac{dV}{dt} = 20$. Find the rate at which the radius of the oil slick is increasing, at a time when the radius is 50 meters. $\left[\textit{Hint:} \text{ Find a relation between } \dfrac{dV}{dt} \text{ and } \dfrac{dr}{dt}. \right]$

51. Animal physiologists have determined experimentally that the weight W (in kilograms) and the surface area S (in square meters) of a typical horse are related by the empirical equation $S = 0.1W^{2/3}$. How fast is the surface area of a horse increasing at a time when the horse weighs 350 kg and is gaining weight at the rate of 200 kg per year? [*Hint:* Use the chain rule.]

52. Suppose that a kitchen appliance company's monthly sales and advertising expenses are approximately related by the equation $xy - 6x + 20y = 0$, where x is thousands of dollars spent on advertising and y is thousands of dishwashers sold. Currently, the company is spending 10 thousand dollars on advertising and is selling 2 thousand dishwashers each month. If the company plans to increase monthly advertising expenditures at the rate of $1.5 thousand per month, how fast will sales rise? Use implicit differentiation to answer the question.

CHAPTER 4

THE EXPONENTIAL AND NATURAL LOGARITHM FUNCTIONS

When an investment grows steadily at 15% per year, the rate of growth of the investment at any time is proportional to the value of the investment at that time. When a bacteria culture grows in a laboratory dish, the rate of growth of the culture at any moment is proportional to the total number of bacteria in the dish at that moment. These situations are examples

287

of what is called *exponential growth.* A pile of radioactive uranium ^{235}U decays at a rate that at each moment is proportional to the amount of ^{235}U present. This decay of uranium (and of radioactive elements in general) is called *exponential decay.* Both exponential growth and exponential decay can be described and studied in terms of exponential functions and the natural logarithm function. The properties of these functions are investigated in this chapter. Subsequently, we shall explore a wide range of applications, in fields such as business, biology, archeology, public health, and psychology.

 4.1 EXPONENTIAL FUNCTIONS

Throughout this section b will denote a positive number. The function

$$f(x) = b^x$$

is called an *exponential function,* because the variable x is in the exponent. The number b is called the *base* of the exponential function. In Section 0.5 we reviewed the definition of b^x for various values of b and x (although we used the letter r there instead of x). For instance, if $f(x)$ is the exponential function with base 2,

$$f(x) = 2^x,$$

then

$$f(0) = 2^0 = 1, \qquad f(1) = 2^1 = 2, \qquad f(4) = 2^4 = 2 \cdot 2 \cdot 2 \cdot 2 = 16,$$

and

$$f(-1) = 2^{-1} = \tfrac{1}{2}, \qquad f(\tfrac{1}{2}) = 2^{1/2} = \sqrt{2}, \qquad f(\tfrac{3}{5}) = (2^{1/5})^3 = (\sqrt[5]{2})^3.$$

Actually, in Section 0.5, we defined b^x only for rational (i.e., integer or fractional) values of x. For other values of x (such as $\sqrt{3}$ or π), it is possible to define b^x by first approximating x with rational numbers and then applying a limiting process. We shall omit the details and simply assume henceforth that b^x can be defined for all numbers x in such a way that the usual laws of exponents remain valid.

Let us state the laws of exponents for reference.

(i) $b^x \cdot b^y = b^{x+y}$ (ii) $b^{-x} = \dfrac{1}{b^x}$

(iii) $\dfrac{b^x}{b^y} = b^x \cdot b^{-y} = b^{x-y}$ (iv) $(b^y)^x = b^{xy}$

(v) $a^x b^x = (ab)^x$ (vi) $\dfrac{a^x}{b^x} = \left(\dfrac{a}{b}\right)^x$

Property (iv) may be used to change the appearance of an exponential function. For instance, the function $f(x) = 8^x$ may also be written as $f(x) = (2^3)^x = 2^{3x}$, and $g(x) = (\frac{1}{9})^x$ may be written as $g(x) = (1/3^2)^x = (3^{-2})^x = 3^{-2x}$.

EXAMPLE 1 Use properties of exponents to write the following functions in the form 2^{kx} for a suitable constant k.

(a) $4^{5x/2}$ (b) $(2^{4x} \cdot 2^{-x})^{1/2}$ (c) $8^{x/3} \cdot 16^{3x/4}$ (d) $\dfrac{10^x}{5^x}$

Solution (a) First express the base 4 as a power of 2, and then use Property (iv):

$$4^{5x/2} = (2^2)^{5x/2} = 2^{2(5x/2)} = 2^{5x}.$$

(b) First use Property (i) to simplify the quantity inside the parentheses, and then use Property (iv):

$$(2^{4x} \cdot 2^{-x})^{1/2} = (2^{4x-x})^{1/2} = (2^{3x})^{1/2} = 2^{(3/2)x}.$$

(c) First express the bases 8 and 16 as powers of 2, and then use (iv) and (i):

$$8^{x/3} \cdot 16^{3x/4} = (2^3)^{x/3} \cdot (2^4)^{3x/4} = 2^x \cdot 2^{3x} = 2^{4x}.$$

(d) Use (v) to change the numerator 10^x, and then cancel the common term 5^x:

$$\frac{10^x}{5^x} = \frac{(2 \cdot 5)^x}{5^x} = \frac{2^x \cdot 5^x}{5^x} = 2^x.$$

An alternative method is to use Property (vi):

$$\frac{10^x}{5^x} = \left(\frac{10}{5}\right)^x = 2^x. \quad \bullet$$

Let us now study the graph of the exponential function $y = b^x$ for various values of b. We begin with the special case $b = 2$.

We have tabulated the values of 2^x for $x = 0, \pm 1, \pm 2, \pm 3$ and plotted these values in Fig. 1. Other intermediate values of 2^x for $x = \pm.1, \pm.2, \pm.3, \ldots,$ may be obtained from tables or from a calculator with a y^x key. [See Fig. 2(a).] By passing a smooth curve through these points, we obtain the graph of $y = 2^x$, in Fig. 2(b).

In the same manner, we have sketched the graph of $y = 3^x$ (Fig. 3). The graphs of $y = 2^x$ and $y = 3^x$ have the same basic shape. Also note that they both pass through the point $(0, 1)$ (because $2^0 = 1, 3^0 = 1$).

In Fig. 4 we have sketched the graphs of several more exponential functions. Notice that the graph of $y = 5^x$ has a large slope at $x = 0$, since the graph at $x = 0$ is quite steep; however, the graph of $y = (1.1)^x$ is nearly horizontal at $x = 0$, and hence the slope is close to zero.

There is an important property of the function 3^x that is readily apparent from its graph. Since the graph is always increasing, the function 3^x never assumes the same y-value twice. That is, the only way 3^r can equal 3^s is to have $r = s$. This fact is useful when solving certain equations involving exponentials.

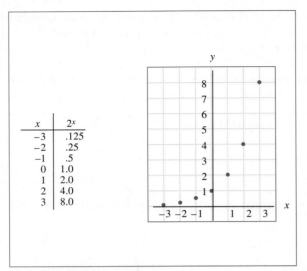

x	2^x
-3	.125
-2	.25
-1	.5
0	1.0
1	2.0
2	4.0
3	8.0

FIGURE 1   Values of 2^x.

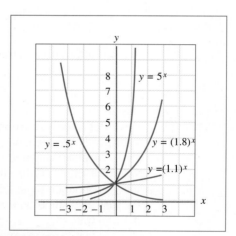

(a)   (b)

FIGURE 2   Graph of $y = 2^x$.

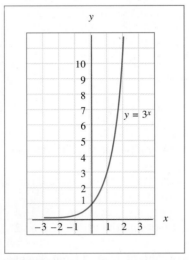

FIGURE 3

FIGURE 4

EXAMPLE 2   Let $f(x) = 3^{5x}$. Determine all x for which $f(x) = 27$.

Solution   Since $27 = 3^3$, we must determine all x for which

$$3^{5x} = 3^3.$$

Equating exponents, we have

$$5x = 3$$

$$x = \frac{3}{5}. \ \bullet$$

In general, for $b > 1$, the equation $b^r = b^s$ implies that $r = s$. This is because the graph of $y = b^x$ has the same basic shape as $y = 2^x$ and $y = 3^x$. Similarly, when $0 < b < 1$, the equation $b^r = b^s$ implies that $r = s$, because the graph of $y = b^x$ resembles the graph of $y = (\frac{1}{3})^x$ and is always decreasing.

There is no need at this point to become familiar with the graphs of the functions b^x. We have shown a few graphs merely to make the reader more comfortable with the concept of an exponential function. The main purpose of this section has been to review properties of exponents in a context that is appropriate for our future work.

PRACTICE PROBLEMS 4.1

1. Can a function such as $f(x) = 5^{3x}$ be written in the form $f(x) = b^x$? If so, what is b?

2. Solve the equation $7 \cdot 2^{6-3x} = 28$.

EXERCISES 4.1

Write each function in Exercises 1–14 in the form 2^{kx} or 3^{kx}, for a suitable constant k.

1. 4^x, $(\sqrt{3})^x$, $(\frac{1}{9})^x$

2. 27^x, $(\sqrt[3]{2})^x$, $(\frac{1}{8})^x$

3. $8^{2x/3}$, $9^{3x/2}$, $16^{-3x/4}$

4. $9^{-x/2}$, $8^{4x/3}$, $27^{-2x/3}$

5. $(\frac{1}{4})^{2x}$, $(\frac{1}{8})^{-3x}$, $(\frac{1}{81})^{x/2}$

6. $(\frac{1}{9})^{2x}$, $(\frac{1}{27})^{x/3}$, $(\frac{1}{16})^{-x/2}$

7. $2^{3x} \cdot 2^{-5x/2}$, $3^{2x} \cdot (\frac{1}{3})^{2x/3}$

8. $2^{5x/4} \cdot (\frac{1}{2})^x$, $3^{-2x} \cdot 3^{5x/2}$

9. $(2^{-3x} \cdot 2^{-2x})^{2/5}$, $(9^{1/2} \cdot 9^4)^{x/9}$

10. $(3^{-x} \cdot 3^{x/5})^5$, $(16^{1/4} \cdot 16^{-3/4})^{3x}$

11. $\dfrac{3^{4x}}{3^{2x}}$, $\dfrac{2^{5x+1}}{2 \cdot 2^{-x}}$, $\dfrac{9^{-x}}{27^{-x/3}}$

12. $\dfrac{2^x}{6^x}$, $\dfrac{3^{-5x}}{3^{-2x}}$, $\dfrac{16^x}{8^{-x}}$

13. $6^x \cdot 3^{-x}$, $\dfrac{15^x}{5^x}$, $\dfrac{12^x}{2^{2x}}$

14. $7^{-x} \cdot 14^x$, $\dfrac{2^x}{6^x}$, $\dfrac{3^{2x}}{18^x}$

From a calculator we have $2^{1/2} \approx 1.414$, $2^{1/10} \approx 1.072$, $(2.7)^{1/2} \approx 1.643$, and $(2.7)^{1/10} \approx 1.104$. (These figures are accurate to three decimal places.) Compute the following numbers, rounding off your answer to two decimal places, if necessary.

15. (a) 2^x for $x = 3$ (b) 2^x for $x = -3$
    (c) 2^x for $x = \frac{5}{2}$ [*Hint:* $\frac{5}{2} = 2 + \frac{1}{2}$.]   (d) 2^x for $x = 4.1$
    (e) 2^x for $x = .2$
    (f) 2^x for $x = .9$ [*Hint:* $.9 = 1 - .1$.]
    (g) 2^x for $x = -2.5$ [*Hint:* $-2.5 = -3 + .5$.]
    (h) 2^x for $x = -3.9$

16. (a) $(2.7)^x$ for $x = .2$ (b) $(2.7)^x$ for $x = 1.5$ (c) $(2.7)^x$ for $x = 0$
    (d) $(2.7)^x$ for $x = -1$ (e) $(2.7)^x$ for $x = 1.1$ (f) $(2.7)^x$ for $x = .6$

Solve the following equations for x.

17. $5^{2x} = 5^2$

18. $10^{-x} = 10^2$

19. $(2.5)^{2x+1} = (2.5)^5$

20. $(3.2)^{x-3} = (3.2)^5$

21. $10^{1-x} = 100$

22. $2^{4-x} = 8$

23. $3(2.7)^{5x} = 8.1$

24. $4(2.7)^{2x-1} = 10.8$

25. $(2^{x+1} \cdot 2^{-3})^2 = 2$

26. $(3^{2x} \cdot 3^2)^4 = 3$

27. $2^{3x} = 4 \cdot 2^{5x}$

28. $3^{5x} \cdot 3^x - 3 = 0$

29. $(1 + x)2^{-x} - 5 \cdot 2^{-x} = 0$

30. $(2 - 3x)5^x + 4 \cdot 5^x = 0$

The expressions in Exercises 31–38 may be factored as shown. Find the missing factors.

31. $2^{3+h} = 2^3 ($ $)$

32. $5^{2+h} = 25 ($ $)$

33. $2^{x+h} - 2^x = 2^x ($ $)$

34. $5^{x+h} + 5^x = 5^x ($ $)$

35. $3^{x/2} + 3^{-x/2} = 3^{-x/2} ($ $)$

36. $5^{7x/2} - 5^{x/2} = \sqrt{5^x} ($ $)$

37. $3^{10x} - 1 = (3^{5x} - 1) ($ $)$

38. $2^{6x} - 2^x = (2^{3x} - 2^{x/2}) ($ $)$

SOLUTIONS TO PRACTICE PROBLEMS 4.1

1. If $5^{3x} = b^x$, then when $x = 1$, $5^{3(1)} = b^1$, which says that $b = 125$. This value of b certainly works, because

$$5^{3x} = (5^3)^x = 125^x.$$

2. Divide both sides of the equation by 7. We then obtain

$$2^{6-3x} = 4.$$

Now 4 can be written as 2^2. So we have

$$2^{6-3x} = 2^2.$$

Equate exponents to obtain

$$6 - 3x = 2$$

$$4 = 3x$$

$$x = \frac{4}{3}.$$

 4.2 THE EXPONENTIAL FUNCTION e^x

Let us begin by examining the graphs of the exponential functions shown in Fig. 1. They all pass through $(0, 1)$, but with different slopes there. Notice that the graph of 5^x is quite steep at $x = 0$, while the graph of $(1.1)^x$ is nearly horizontal at $x = 0$. It turns out that at $x = 0$, the graph of 2^x has a slope of approximately .69, while the graph of 3^x has a slope of approximately 1.1.

FIGURE I   Several exponential functions.

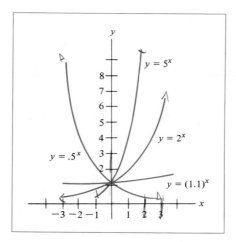

Evidently, there is a particular value of the base b, between 2 and 3, where the graph of b^x has slope *exactly* 1 at $x = 0$. We denote this special value of b by the letter e, and we call

$$f(x) = e^x$$

the exponential function. The number e is an important constant of nature that has been calculated to thousands of decimal places. To 10 significant digits, we have $e = 2.718281828$. For our purposes, it is usually sufficient to think of e as "approximately 2.7."

Our goal in this section is to find a formula for the derivative of e^x. It turns out that the calculations for e^x and 2^x are very similar. Since many people are more comfortable working with 2^x rather than e^x, we shall first analyze the graph of 2^x. Then we shall draw the appropriate conclusions about the graph of e^x.

Before computing the slope of $y = 2^x$ at an arbitrary x, let us consider the special case $x = 0$. Denote the slope at $x = 0$ by m. We shall use the secant-line approximation of the derivative to approximate m. We proceed by constructing the secant line in Fig. 2. The slope of the secant line through $(0, 1)$ and $(h, 2^h)$ is $\dfrac{2^h - 1}{h}$. As h approaches zero, the slope of the secant line approaches the slope of $y = 2^x$ at $x = 0$. That is,

$$m = \lim_{h \to 0} \frac{2^h - 1}{h}. \tag{1}$$

We can estimate the value of m by taking h smaller and smaller. When $h = .1$, we have $2^h \approx 1.072$ (from a calculator), and

$$\frac{2^h - 1}{h} \approx \frac{.072}{.1} = .72.$$

When $h = .01$, we have $2^h \approx 1.00696$, and

$$\frac{2^h - 1}{h} \approx \frac{.00696}{.01} \approx .696.$$

FIGURE 2   A secant line to the graph of y = 2^x.

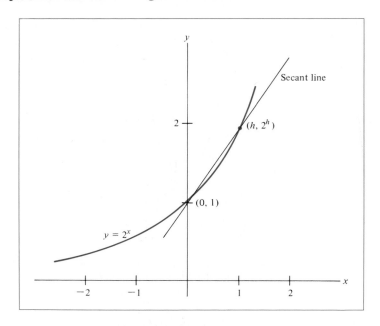

When $h = .001$, $2^h \approx 1.0006934$, so that

$$\frac{2^h - 1}{h} \approx \frac{.0006934}{.001} \approx .693.$$

Thus it is reasonable to conclude that $m \approx .69.$* Since m equals the slope of $y = 2^x$ at $x = 0$, we have

$$m = \frac{d}{dx}(2^x)\Big|_{x=0} \approx .69. \tag{2}$$

Now that we have estimated the slope of $y = 2^x$ at $x = 0$, let us compute the slope for an arbitrary value of x. We construct a secant line through $(x, 2^x)$ and a nearby point $(x + h, 2^{x+h})$ on the graph. The slope of the secant line is

$$\frac{2^{x+h} - 2^x}{h}. \tag{3}$$

By law of exponents (1), we have $2^{x+h} - 2^x = 2^x(2^h - 1)$ so that, by (1), we see that

$$\lim_{h \to 0}\frac{2^{x+h} - 2^x}{h} = \lim_{h \to 0} 2^x\frac{2^h - 1}{h} = 2^x \lim_{h \to 0}\frac{2^h - 1}{h} = m2^x. \tag{4}$$

---

* To 10 decimal places, m is .6931471806.

However, the slope of the secant (3) approaches the derivative of 2^x as h approaches zero. Consequently, we have

$$\frac{d}{dx}(2^x) = m\,2^x, \qquad \text{where } m = \frac{d}{dx}(2^x)\Big|_{x=0} \qquad (5)$$

EXAMPLE 1   Calculate (a) $\dfrac{d}{dx}(2^x)\Big|_{x=3}$ and (b) $\dfrac{d}{dx}(2^x)\Big|_{x=-1}$

Solution   **(a)** $\dfrac{d}{dx}(2^x)\Big|_{x=3} = m\cdot 2^3 = 8m \approx 8(.69) = 5.52.$

         (b) $\dfrac{d}{dx}(2^x)\Big|_{x=-1} = m\cdot 2^{-1} = .5m \approx .5(.69) = .345.$ ●

The calculations just carried out for $y = 2^x$ can be carried out for $y = b^x$, where b is any positive number. Equation (5) will read exactly the same except that 2 will be replaced by b. Thus we have the following formula for the derivative of the function $f(x) = b^x$.

$$\frac{d}{dx}(b^x) = m\,b^x, \qquad \text{where } m = \frac{d}{dx}(b^x)\Big|_{x=0}. \qquad (6)$$

Our calculations showed that if $b = 2$, then $m \approx .69$. If $b = 3$, then it turns out that $m \approx 1.1$. (See Exercise 1.) Obviously, the derivative formula in (6) is simplest when $m = 1$, that is, when the graph of b^x has slope 1 at $x = 0$. As we said earlier, this special value of b is denoted by the letter e. Thus the number e has the property that

$$\frac{d}{dx}(e^x)\Big|_{x=0} = 1 \qquad (7)$$

and

$$\frac{d}{dx}(e^x) = 1\cdot e^x = e^x. \qquad (8)$$

The graphical interpretation of (7) is that the curve $y = e^x$ has slope 1 at $x = 0$. The graphical interpretation of (8) is that the slope of the curve $y = e^x$ at an arbitrary value of x is exactly equal to the value of the function e^x at that point. (See Fig. 3.)

FIGURE 3 Fundamental properties of e^x.

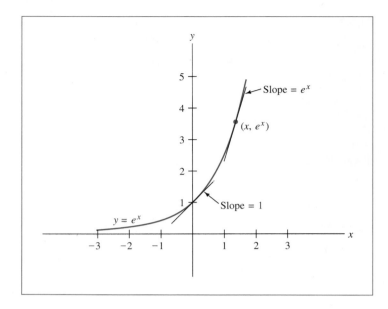

The function e^x is the same type of function as 2^x and 3^x except that taking derivatives of e^x is much easier. For this reason, functions based on e^x are used in almost all applications that require an exponential-type function to describe a physical phenomenon. Extensive tables are available from which to determine the value of e^x for a wide range of values of x. Also, scientific calculators provide for calculating e^x at the push of a button.

TECHNOLOGY PROJECT 4.1
Slopes of Exponential Functions

Use the method of the text to calculate the values of the following derivatives to three significant figures.

1. $\dfrac{d}{dx}[2.5^x]\Big|_{x=0}$

2. $\dfrac{d}{dx}[2.5^x]\Big|_{x=3}$

3. $\dfrac{d}{dx}[0.7^x]\Big|_{x=0}$

4. $\dfrac{d}{dx}[0.7^x]\Big|_{x=-1}$

5. $\dfrac{d}{dx}[10^x]\Big|_{x=0}$

6. $\dfrac{d}{dx}[10^x]\Big|_{x=-2}$

7. $\dfrac{d}{dx}[5.4^x]\Big|_{x=0}$

8. $\dfrac{d}{dx}[5.4^x]\Big|_{x=1}$

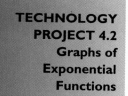

TECHNOLOGY PROJECT 4.2
Graphs of Exponential Functions

Sketch the graphs of the following functions, which are types that will be used in various applications of the exponential function. Experiment to determine a suitable range that shows the interesting properties of the graph. Determine all asymptotes.

1. $f(x) = 1000e^{-.07x}$
2. $f(x) = 50e^{3x}$
3. $f(x) = 300(1 - e^{-3x})$
4. $f(x) = e^{-x} + 4$
5. $f(x) = e^{-x^2}$
6. $f(x) = \dfrac{10}{1 + e^{-x}}$

PRACTICE PROBLEMS 4.2

In the following problems use the number 20 as the (approximate) value of e^3.

1. Find the equation of the tangent line to the graph of $y = e^x$ at $x = 3$.
2. Solve the following equation for x:

$$4e^{6x} = 80.$$

EXERCISES 4.2

1. Use the table below to show that

$$\frac{d}{dx}(3^x)\bigg|_{x=0} \approx 1.1.$$

That is, calculate the slope

$$\frac{3^h - 1}{h}$$

of the secant line passing through the point $(0, 1)$ and $(h, 3^h)$. Take $h = .1, .01$, and $.001$.

x	3ˣ
0	1.00000
.001	1.00110
.010	1.01105
.100	1.11612

2. Use the table below to show that

$$\frac{d}{dx}(2.7)^x\bigg|_{x=0} \approx .99.$$

That is, calculate

$$\frac{(2.7)^h - 1}{h}$$

for $h = .1, .01,$ and $.001$.

x	$(2.7)^x$
0	1.00000
.001	1.00099
.010	1.00998
.100	1.10443

3. Consider the secant line on the graph of e^x passing through $(0, 1)$ and (h, e^h). Its slope is

$$\frac{e^h - 1}{h}.$$

Compute this quantity for $h = .01, .005,$ and $.001$.

x	e^x
0	1.00000
.001	1.00100
.005	1.00501
.010	1.01005

4. Use (8) and a familiar rule for differentiation to find

$$\frac{d}{dx}(5e^x).$$

5. Use (8) and a familiar rule for differentiation to find

$$\frac{d}{dx}(e^x)^{10}.$$

6. Use the fact that $e^{2+x} = e^2 \cdot e^x$ to find

$$\frac{d}{dx}(e^{2+x}).$$

[Remember that e^2 is just a constant—approximately $(2.7)^2$.]

7. Use the fact that $e^{4x} = (e^x)^4$ to find

$$\frac{d}{dx}(e^{4x}).$$

8. Find $\dfrac{d}{dx}(e^x + x^2)$.

Simplify.

9. $e^{2x}(1 + e^{3x})$ 10. $(e^x)^2$ 11. $e^{1-x} \cdot e^{2x}$

12. $\dfrac{5e^{3x}}{e^x}$ 13. $\dfrac{1}{e^{-2x}}$ 14. $e^3 \cdot e^{x+1}$

Use a calculator to determine the following numbers:

15. e^2 16. $e^{-1.30}$ 17. $e^{-.5}$ 18. $e^{3/2}$

Solve the following equations for x.

19. $e^{5x} = e^{20}$ 20. $e^{1-x} = e^2$

21. $e^{x^2 - 2x} = e^8$ 22. $e^{-x} = 1$ e^0 $x = 0$

Differentiate the following functions.

23. xe^x 24. $\dfrac{e^x}{x}$ $4(1+5e^x)^3(5e^x)$ 25. $\dfrac{e^x}{1 + e^x}$ $-3(xe^x - 1)(xe^x)$

26. $(1 + x^2)e^x$ 27. $(1 + 5e^x)^4$ 28. $(xe^x - 1)^{-3}$

SOLUTIONS TO PRACTICE PROBLEMS 4.2

1. When $x = 3$, $y = e^3 \approx 20$. So the point $(3, 20)$ is on the tangent line. Since $\dfrac{d}{dx}(e^x) = e^x$, the slope of the tangent line is e^3 or 20. Therefore, the equation of the tangent line in point-slope form is $y - 20 = 20(x - 3)$.

2. This problem is similar to Practice Problem 2 of Section 4.1. First divide both sides of the equation by 4.

$$e^{6x} = 20.$$

The idea is to express 20 as a power of e and then equate exponents.

$$e^{6x} = e^3$$

$$6x = 3$$

$$x = \frac{1}{2}.$$

4.3 DIFFERENTIATION OF EXPONENTIAL FUNCTIONS

We have shown that $\dfrac{d}{dx}(e^x) = e^x$. Using this fact and the chain rule, we can differentiate functions of the form $e^{g(x)}$, where $g(x)$ is any differentiable function. This is because $e^{g(x)}$ is the composite of two functions. Indeed, if $f(x) = e^x$, then

$$e^{g(x)} = f(g(x)).$$

Thus, by the chain rule, we have

$$\frac{d}{dx}(e^{g(x)}) = f'(g(x))g'(x)$$

$$= f(g(x))g'(x) \quad [\text{since } f'(x) = f(x)]$$

$$= e^{g(x)}g'(x).$$

So we have the following result:

Chain Rule for Exponential Functions   Let $g(x)$ be any differentiable function. Then

$$\frac{d}{dx}(e^{g(x)}) = e^{g(x)}g'(x). \tag{1}$$

EXAMPLE 1   Differentiate e^{x^2+1}.

Solution   Here $g(x) = x^2 + 1$, $g'(x) = 2x$, so

$$\frac{d}{dx}(e^{x^2+1}) = e^{x^2+1} \cdot 2x$$

$$= 2xe^{x^2+1}. \quad \bullet$$

EXAMPLE 2   Differentiate $e^{3x^2-(1/x)}$.

Solution

$$\frac{d}{dx}(e^{3x^2-(1/x)}) = e^{3x^2-(1/x)} \cdot \frac{d}{dx}\left(3x^2 - \frac{1}{x}\right)$$

$$= e^{3x^2-(1/x)}\left(6x + \frac{1}{x^2}\right). \quad \bullet$$

EXAMPLE 3   Differentiate e^{5x}.

Solution

$$\frac{d}{dx}(e^{5x}) = e^{5x} \cdot \frac{d}{dx}(5x) = e^{5x} \cdot 5 = 5e^{5x}. \quad \bullet$$

Using a computation similar to that used in Example 3, we may differentiate e^{kx} for any constant k. (In Example 3 we have $k = 5$.) The result is the following useful formula.

$$\frac{d}{dx}(e^{kx}) = ke^{kx}. \tag{2}$$

Many applications involve exponential functions of the form $y = Ce^{kx}$, where C and k are constants. In the next example we differentiate such functions.

EXAMPLE 4   Differentiate the following exponential functions.
(a) $3e^{5x}$ (b) $3e^{kx}$, where k is a constant
(c) Ce^{kx}, where C and k are constants

Solution   (a) $\dfrac{d}{dx}(3e^{5x}) = 3\dfrac{d}{dx}(e^{5x}) = 3 \cdot 5e^{5x} = 15e^{5x}.$

(b) $\dfrac{d}{dx}(3e^{kx}) = 3\dfrac{d}{dx}(e^{kx})$

$= 3 \cdot ke^{kx}$ [by (2)]

$= 3ke^{kx}.$

(c) $\dfrac{d}{dx}(Ce^{kx}) = C\dfrac{d}{dx}(e^{kx}) = Cke^{kx}.$ ●

The result of part (c) may be summarized in an extremely useful fashion as follows: Suppose that we let $y = Ce^{kx}$. By part (c) we have

$$y' = Cke^{kx}$$
$$= k \cdot (Ce^{kx})$$
$$= ky.$$

In other words, the derivative of the function Ce^{kx} is k times the function itself. Let us record this fact.

Let C, k be any constants and let $y = Ce^{kx}$. Then y satisfies the equation

$$y' = ky.$$

The equation $y' = ky$ expresses a relationship between the function y and its derivative y'. Any equation expressing a relationship between a function y and one or more of its derivatives is called a *differential equation*.

Very often an applied problem will involve a function $y = f(x)$ which satisfies the differential equation $y' = ky$. It can be shown that y must then necessarily be an exponential function of the form Ce^{kx}. That is, we have the following result.

Suppose that $y = f(x)$ satisfies the differential equation

$$y' = ky. \tag{3}$$

Then y is an exponential function of the form

$$y = Ce^{kx}, \qquad C \text{ a constant.}$$

We shall verify this result in Exercise 49.

EXAMPLE 5   Determine all functions $y = f(x)$ such that $y' = -.2y$.

Solution   The equation $y' = -.2y$ has the form $y' = ky$ with $k = -.2$. Therefore, any solution of the equation has the form

$$y = Ce^{-.2x},$$

where C is a constant. ●

EXAMPLE 6   Determine all functions $y = f(x)$ such that $y' = y/2$ and $f(0) = 4$.

Solution   The equation $y' = y/2$ has the form $y' = ky$ with $k = \frac{1}{2}$. Therefore,

$$f(x) = Ce^{(1/2)x}$$

for some constant C. We also require that $f(0) = 4$. That is,

$$4 = f(0) = Ce^{(1/2)\cdot 0} = Ce^0 = C.$$

So $C = 4$ and

$$f(x) = 4e^{(1/2)x}. \quad ●$$

The Functions e^{kx}   Exponential functions of the form e^{kx} occur in many applications. Figure 1 shows the graphs of several functions of this type when k is a positive number. These curves $y = e^{kx}$, k positive, have several properties in common:

1. $(0, 1)$ is on the graph.
2. The graph lies strictly above the x-axis (e^{kx} is never zero).
3. The x-axis is an asymptote as x becomes large negatively.
4. The graph is always increasing and concave up.

When k is negative, the graph of $y = e^{kx}$ is decreasing. (See Fig. 2.) Note the following properties of the curves $y = e^{kx}$, k negative:

FIGURE 1

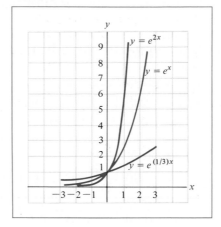

FIGURE 2

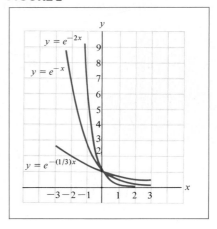

1. $(0, 1)$ is on the graph.
2. The graph lies strictly above the x-axis.
3. The x-axis is an asymptote as x becomes large positively.
4. The graph is always decreasing and concave up.

The Functions b^x If b is a positive number, then the function b^x may be written in the form e^{kx} for some k. For example, take $b = 2$. From Fig. 3 of the preceding section it is clear that there is some value of x such that $e^x = 2$. Call this value k, so that $e^k = 2$. Then

$$2^x = (e^k)^x = e^{kx}$$

for all x. In general, if b is any positive number, there is a value of x, say $x = k$, such that $e^k = b$. In this case, $b^x = (e^k)^x = e^{kx}$. Thus all the curves $y = b^x$ discussed in Section 4.1 can be written in the form $y = e^{kx}$. This is one reason why we have focused on exponential functions with base e instead of studying 2^x, 3^x, and so on.

TECHNOLOGY PROJECT 4.3
Finding Maxima and Minima of Functions Involving Exponential Functions

Use the TRACE to determine the relative extreme points of the following functions. Carry out your calculations to one decimal place accuracy.

1. $f(x) = xe^{-3x}$

2. $f(x) = \dfrac{1}{1 + e^x}$

3. $f(x) = -x^2 + 3x + e^{-x}$

4. $f(x) = e^{-x} - 10e^{-2x} + 3e^{-3x}$

TECHNOLOGY PROJECT 4.4
The Normal Curve

In probability and statistics, it is necessary to consider random variables whose probability distributions are given by the functions

$$f_a(x) = \frac{1}{a\sqrt{2\pi}} e^{-\frac{1}{2}(x/a)^2}, \ a > 0$$

The graphs of these functions are called **normal curves.**

1. Graph the normal curves for $a = 1, 2, 5, 10$.
2. Explain what happens to the normal curves as the value of a increases.
3. Determine the relative extreme points and inflection points of the normal curve with $a = 1$. Carry out your calculations to two decimal places.

PRACTICE PROBLEMS 4.3

1. Differentiate $[e^{-3x}(1 + e^{6x})]^{12}$.
2. Determine all functions $y = f(x)$ such that $y' = -y/20, f(0) = 2$.

Differentiate the following.

1. $y = e^{-x}$

2. $f(x) = e^{10x}$

3. $f(x) = 5e^x$

4. $y = \dfrac{e^x + e^{-x}}{2}$

5. $f(t) = e^{t^2}$

6. $f(t) = e^{-2t}$

7. $f(x) = \dfrac{e^x - e^{-x}}{2}$

8. $f(x) = 2e^{1-x}$

9. $y = e^{-2x} - 2x$

10. $f(x) = \frac{1}{10}e^{-x^2/2}$

11. $g(x) = (e^x + e^{-x})^3$

12. $y = (e^{-x})^2$

13. $y = \frac{1}{3}e^{3 - 2x}$

14. $g(x) = e^{1/x}$

15. $f(t) = e^t(e^{2t} - e^{-t})$

16. $f(t) = \dfrac{e^t + e^{-t}}{e^t}$

[*Hint:* In Exercises 15 and 16, simplify $f(t)$ before differentiating.]

17. $y = e^{x^3 + x - (1/x)}$

18. $y = (e^{x^2} + x^2)^5$

19. $(2x + 1 - e^{2x+1})^4$

20. $e^{1/(3x-7)}$

21. $x^3 e^{x^2}$

22. $x^2 e^{-3x}$

23. e^{x^3}/x

24. $e^{(x-1)/x}$

25. $(x + 1)e^{-x+2}$

26. $x\sqrt{2 + e^x}$

27. $\left(\dfrac{1}{x} + 3\right)e^x$

28. $\dfrac{e^{-3x}}{1 - 3x}$

29. $\dfrac{e^x - 1}{e^x + 1}$

30. $\dfrac{xe^x - 3}{x + 1}$

In Exercises 31–40, find all values of x such that the function has a possible relative maximum or relative minimum point. Use the second derivative test to determine the nature of the function at these points. (Recall that e^x is positive for all x.)

31. $f(x) = (1 + x)e^{-x/2}$

32. $f(x) = (1 - x)e^{-x/2}$

33. $f(x) = \dfrac{3 - 2x}{e^{x/4}}$

34. $f(x) = \dfrac{4x - 3}{e^{x/2}}$

35. $f(x) = (8 - 2x)e^{x+5}$

36. $f(x) = (4x - 1)e^{3x-2}$

37. $f(x) = \dfrac{(x - 1)^2}{e^x}$

38. $f(x) = (x + 3)^2 e^x$

39. $f(x) = (x + 5)^2 e^{2x-1}$

40. $f(x) = \dfrac{(x - 3)^2}{e^{2x}}$

41. Let a and b be positive numbers: A curve whose equation is $y = e^{-ae^{-bx}}$ is called a *Gompertz growth curve*. These curves are used in biology to describe certain types of population growth. Compute the derivative of $y = e^{-2e^{-.01x}}$.

42. The value of a computer t years after purchase is $v(t) = 2000e^{-.35t}$ dollars. At what rate (in dollars per year) is the computer depreciating after 4 years?

43. A painting purchased in 1990 for \$100,000 is estimated to be worth $v(t) = 100,000e^{.2t}$ dollars after t years. At what rate (in dollars per year) will the painting be appreciating in 1995?

44. Find $\dfrac{dy}{dx}$ if $y = e^{-(1/10)e^{-x/2}}$.

45. Determine all solutions of the differential equation $y' = -4y$.

46. Determine all solutions of the differential equation $y' = \frac{1}{3}y$.

47. Determine all functions $y = f(x)$ such that $y' = -.5y$ and $f(0) = 1$.

48. Determine all functions $y = f(x)$ such that $y' = 3y$ and $f(0) = \frac{1}{2}$.

49. Verify the result (3). [*Hint:* Let $g(x) = f(x)e^{-kx}$. Show that $g'(x) = 0$.] You may assume that only a constant function has a zero derivative.

50. Let $f(x)$ be a function with the property that $f'(x) = 1/x$. Let $g(x) = f(e^x)$, and compute $g'(x)$.

Graph the following functions.

51. $y = e^{-x^2}$

52. $y = xe^{-x}$ for $x \geq 0$

53. As h approaches 0, what value is approached by the difference quotient $\dfrac{e^h - 1}{h}$? [*Hint:* $1 = e^0$.]

54. As h approaches 0, what value is approached by $\dfrac{e^{2h} - 1}{h}$? [*Hint:* $1 = e^0$.]

SOLUTIONS TO PRACTICE PROBLEMS 4.3

1. We must use the general power rule. However, this is most easily done if we first use the laws of exponents to simplify the function inside the brackets.

$$e^{-3x}(1 + e^{6x}) = e^{-3x} + e^{-3x} \cdot e^{6x}$$
$$= e^{-3x} + e^{3x}.$$

Now $\dfrac{d}{dx}[e^{-3x} + e^{3x}]^{12} = 12 \cdot [e^{-3x} + e^{3x}]^{11} \cdot (-3e^{-3x} + 3e^{3x})$

$$= 36 \cdot [e^{-3x} + e^{3x}]^{11} \cdot (-e^{-3x} + e^{3x}).$$

2. The differential equation $y' = -y/20$ is of the type $y' = ky$, where $k = -\frac{1}{20}$. Therefore, any solution has the form $f(x) = Ce^{-(1/20)x}$. Now $f(0) = Ce^{-(1/20) \cdot 0} = Ce^0 = C$, so that $C = 2$ when $f(0) = 2$. Therefore, the desired function is $f(x) = 2e^{-(1/20)x}$.

4.4 THE NATURAL LOGARITHM FUNCTION

As a preparation for the definition of the natural logarithm, we shall make a geometrical digression. In Fig. 1 we have plotted several pairs of points. Observe how they are related to the line $y = x$.

FIGURE 1   Reflections of points through the line y = x.

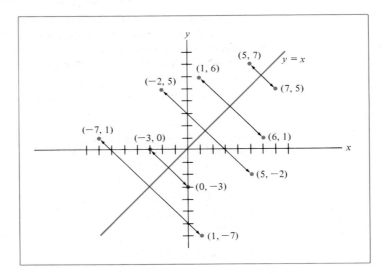

The points $(5, 7)$ and $(7, 5)$, for example, are the same distance from the line $y = x$. If we were to plot the point $(5, 7)$ with wet ink and then fold the page along the line $y = x$, the ink blot would produce a second blot at the point $(7, 5)$. If we think of the line $y = x$ as a mirror, then $(7, 5)$ is the mirror image of $(5, 7)$. We say that $(7, 5)$ is the *reflection* of $(5, 7)$ through the line $y = x$. Similarly, $(5, 7)$ is the reflection of $(7, 5)$ through the line $y = x$.

Now let us consider all points lying on the graph of the exponential function $y = e^x$ [see Fig. 2(a)]. If we reflect each such point through the line

FIGURE 2   Obtaining the graph of ln x as a reflection of e^x.

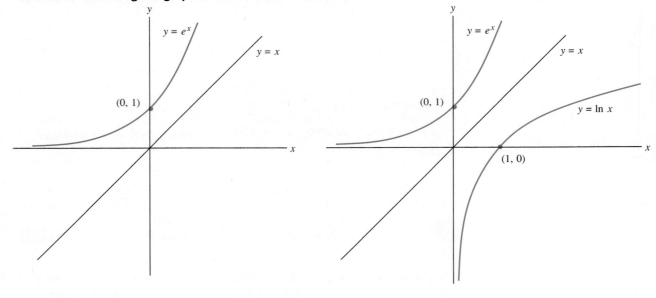

$y = x$, we obtain a new graph [see Fig. 2(b)]. For each positive x, there is exactly one value of y such that (x, y) is on the new graph. We call this value of y the *natural logarithm of x*, denoted $\ln x$. Thus the reflection of the graph of $y = e^x$ through the line $y = x$ is the graph of the natural logarithm function $y = \ln x$.

We may deduce some properties of the natural logarithm function from an inspection of its graph.

1. The point $(1, 0)$ is on the graph of $y = \ln x$ (because $(0, 1)$ is on the graph of $y = e^x$). In other words,

$$\ln 1 = 0. \tag{1}$$

2. $\ln x$ is defined only for positive values of x.
3. $\ln x$ is negative for x between 0 and 1.
4. $\ln x$ is positive for x greater than 1.
5. $\ln x$ is an increasing function.

Let us study the relationship between the natural logarithm and exponential functions more closely. From the way in which the graph of $\ln x$ was obtained we know that (a, b) is on the graph of $\ln x$ if and only if (b, a) is on the graph of e^x. However, a typical point on the graph of $\ln x$ is of the form $(a, \ln a)$, $a > 0$. So for any positive value of a, the point $(\ln a, a)$ is on the graph of e^x. That is,

$$e^{\ln a} = a.$$

Since a was an arbitrary positive number, we have the following important relationship between the natural logarithm and exponential functions.

$$e^{\ln x} = x \qquad \text{for } x > 0. \tag{2}$$

Equation (2) can be put into verbal form.

> For each positive number x, $\ln x$ is that exponent to which we must raise e in order to get x.

If b is any number, then e^b is positive and hence $\ln (e^b)$ makes sense. What is $\ln (e^b)$? Since (b, e^b) is on the graph of e^x, we know that (e^b, b) must be on the graph of $\ln x$. That is, $\ln (e^b) = b$. Thus we have shown that

$$\ln (e^x) = x \qquad \text{for any } x. \tag{3}$$

The identities (2) and (3) express the fact that the natural logarithm is the *inverse* of the exponential function (for $x > 0$). For instance, if we take a number x and compute e^x, then, by (3), we can undo the effect of the exponentiation by taking the natural logarithm; that is, the logarithm of e^x equals the original number x. Similarly, if we take a positive number x and compute $\ln x$, then, by (2), we can undo the effect of the logarithm by raising e to the $\ln x$ power; that is, $e^{\ln x}$ equals the original number x.

Scientific calculators have an "$\ln x$" key that will compute the natural logarithm of a number to as many as ten significant figures. For instance, entering the number 2 into the calculator and pressing the $\ln x$ key, one obtains $\ln 2 = .6931471806$ (to 10 significant figures). If a scientific calculator is unavailable, one may use a table of logarithms.

The relationships (2) and (3) between e^x and $\ln x$ may be used to solve equations, as the next examples show.

EXAMPLE 1   Solve the equation $5e^{x-3} = 4$ for x.

Solution   First divide each side by 5,

$$e^{x-3} = .8.$$

Taking the logarithm of each side and using (3), we have

$$\ln(e^{x-3}) = \ln .8$$

$$x - 3 = \ln .8$$

$$x = 3 + \ln .8.$$

[If desired, the numerical value of x can be obtained by using a scientific calculator or a natural logarithm table, namely, $x = 3 - .22314 = 2.77686$ (to five decimal places).] ●

EXAMPLE 2   Solve the equation $2 \ln x + 7 = 0$ for x.

Solution
$$2 \ln x = -7$$

$$\ln x = -3.5$$

$$e^{\ln x} = e^{-3.5}$$

$$x = e^{-3.5} \qquad [\text{by (2)}]. \ ●$$

Other Exponential and Logarithm Functions   In our discussion of the exponential function, we mentioned that all exponential functions of the form b^x, where b is a fixed positive number, can be expressed in terms of *the* exponential function e^x. Now we can be quite explicit. For since $b = e^{\ln b}$, we see that

$$b^x = (e^{\ln b})^x = e^{(\ln b)x}.$$

Hence we have shown that

$$b^x = e^{kx}, \qquad \text{where } k = \ln b.$$

The natural logarithm function is sometimes called the *logarithm to the base e*, for it is the inverse of the exponential function e^x. If we reflect the graph of the function $y = 2^x$ through the line $y = x$, we obtain the graph of a function called the *logarithm to the base* 2, denoted by $\log_2 x$. Similarly, if we reflect the graph of $y = 10^x$ through the line $y = x$, we obtain the graph of a function called the *logarithm to the base* 10, denoted by $\log_{10} x$. (See Fig. 3.)

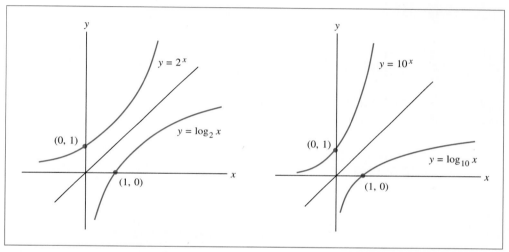

FIGURE 3   Graphs of $\log_2 x$ and $\log_{10} x$ as reflections of 2^x and 10^x.

Logarithms to the base 10 are sometimes called *common* logarithms. Common logarithms are usually introduced into algebra courses for the purpose of simplifying certain arithmetic calculations. However, with the advent of the modern digital computer and the widespread availability of pocket electronic calculators, the need for common logarithms has diminished considerably. It can be shown that

$$\log_{10} x = \frac{1}{\ln 10} \cdot \ln x,$$

so that $\log_{10} x$ is simply a constant multiple of $\ln x$. However, we shall not need this fact.

The natural logarithm function is used in calculus because differentiation and integration formulas are simpler than for $\log_{10} x$ or $\log_2 x$, and so on. (Recall that we prefer the function e^x over the functions 10^x and 2^x for the same reason.) Also, $\ln x$ arises "naturally" in the process of solving certain differential equations that describe various growth processes.

Graph the following functions. Determine their relative extreme points to one decimal place accuracy.

1. $f(x) = x^3 + 10 \ln x$ 2. $f(x) = (\ln x)^2$

3. $f(x) = \dfrac{\ln x}{x^2 + 1}$ 4. $f(x) = x^2 \ln x$

Solve the following equations graphically. Determine the solutions to two decimal places accuracy.

1. $e^{2x} = 3x + 2$ 2. $5e^{-2x} = 2x^2 + 1$

3. $(\ln x)^3 = x^2 - 10$ 4. $e^{-x^2} = \ln x$

**PRACTICE
PROBLEMS 4.4**

1. Find $\ln e$.
2. Solve $e^{-3x} = 2$ using the natural logarithm function.

EXERCISES 4.4

1. Find $\ln (1/e)$ 2. Find $\ln (\sqrt{e})$.
3. If $e^{-x} = 1.7$, write x in terms of the natural logarithm.
4. If $e^x = 3.5$, write x in terms of the natural logarithm.
5. If $\ln x = 2.2$, write x using the exponential function.
6. If $\ln x = -5.7$, write x using the exponential function.

Simplify the following expressions.

7. $\ln e^2$ 8. $e^{\ln 1.37}$ 9. $e^{e \ln 1}$

10. $\ln (e^{.73 \ln e})$ 11. $e^{5 \ln 1}$ 12. $\ln(\ln e)$

Solve the following equations for x.

13. $e^{2x} = 5$ 14. $e^{3x-1} = 4$ 15. $\ln (4 - x) = \tfrac{1}{2}$

16. $\ln 3x = 2$ 17. $\ln x^2 = 6$ 18. $e^{x^2} = 7$

19. $6e^{-.00012x} = 3$ 20. $2 - \ln x = 0$ 21. $\ln 5x = \ln 3$

22. $\ln (x^2 - 3) = 0$ 23. $\ln(\ln 2x) = 0$ 24. $3 \ln x = 8$

25. $2e^{x/3} - 9 = 0$ 26. $4 - 3e^{x+6} = 0$ 27. $300e^{.2x} = 1800$

28. $750e^{-.4x} = 375$ 29. $e^{5x} \cdot e^{\ln 5} = 2$ 30. $e^{x^2 - 5x + 6} = 1$

31. $4e^x \cdot e^{-2x} = 6$ 32. $(e^x)^2 \cdot e^{2-3x} = 4$

In Exercises 33–36, find the coordinates of each relative extreme point of the given function and determine if the point is a relative maximum point or a relative minimum point.

33. $f(x) = e^{-x} + 3x$ 34. $f(x) = 5x - 2e^x$

35. $f(x) = \frac{1}{3}e^{2x} - x + \frac{1}{2}\ln\frac{3}{2}$ 36. $f(x) = 5 - \frac{1}{2}x - e^{-3x}$

37. When a drug or vitamin is administered intramuscularly (into a muscle), the concentration in the blood at time t after injection can be approximated by a function of the form $f(t) = c(e^{-k_1 t} - e^{-k_2 t})$. The graph of $f(t) = 5(e^{-.01t} - e^{-.51t})$, for $t \geq 0$, has the general shape shown in Fig. 12 on page 20. Find the value of t at which this function reaches its maximum value.

38. Under certain geographic conditions, the wind velocity v at a height x centimeters above the ground is given by $v = K\ln(x/x_0)$, where K is a positive constant (depending on the air density, average wind velocity, etc.), and x_0 is a roughness parameter (depending on the roughness of the vegetation on the ground).* Suppose that $x_0 = .7$ centimeter (a value that applies to lawn grass 3 centimeters high) and $K = 300$ centimeters per second.

(a) At what height above the ground is the wind velocity zero?

(b) At what height is the wind velocity 1200 centimeters per second?

39. Use a calculator to estimate $(1.6)^{10}$.

40. Find k such that $2^x = e^{kx}$ for all x.

SOLUTIONS TO PRACTICE PROBLEMS 4.4

1. Answer: 1. The number $\ln e$ is that exponent to which e must be raised in order to obtain e.

2. Take the logarithm of each side and use (3) to simplify the left side:

$$\ln e^{-3x} = \ln 2$$

$$-3x = \ln 2$$

$$x = -\frac{\ln 2}{3}.$$

 4.5 THE DERIVATIVE OF ln x

Let us now compute the derivative of $\ln x$ for $x > 0$. Since $e^{\ln x} = x$, we have

$$\frac{d}{dx}\left(e^{\ln x}\right) = \frac{d}{dx}(x) = 1. \tag{1}$$

---

*G. Cox, B. Collier, A. Johnson, and P. Miller, *Dynamic Ecology* (Englewood Cliffs, N.J.: Prentice-Hall, Inc., 1973), pp. 113–115.

On the other hand, if we differentiate $e^{\ln x}$ by the chain rule, we find that

$$\frac{d}{dx}(e^{\ln x}) = e^{\ln x} \cdot \frac{d}{dx}(\ln x) = x \cdot \frac{d}{dx}(\ln x), \qquad (2)$$

where the last equality used the fact that $e^{\ln x} = x$. By combining equations (1) and (2) we obtain

$$x \cdot \frac{d}{dx}(\ln x) = 1.$$

In other words,

$$\frac{d}{dx}(\ln x) = \frac{1}{x}, \qquad x > 0. \qquad (3)$$

By combining this differentiation formula with the chain rule, product rule, and quotient rule, we can differentiate many functions involving $\ln x$.

EXAMPLE I Differentiate.

(a) $(\ln x)^5$ (b) $x \ln x$ (c) $\ln(x^3 + 5x^2 + 8)$

Solution (a) By the general power rule,

$$\frac{d}{dx}(\ln x)^5 = 5(\ln x)^4 \cdot \frac{d}{dx}(\ln x)$$

$$= 5(\ln x)^4 \cdot \frac{1}{x}$$

$$= \frac{5(\ln x)^4}{x}.$$

(b) By the product rule,

$$\frac{d}{dx}(x \ln x) = x \cdot \frac{d}{dx}(\ln x) + (\ln x) \cdot 1$$

$$= x \cdot \frac{1}{x} + \ln x$$

$$= 1 + \ln x.$$

(c) By the chain rule,

$$\frac{d}{dx}\ln(x^3 + 5x^2 + 8) = \frac{1}{x^3 + 5x^2 + 8} \cdot \frac{d}{dx}(x^3 + 5x^2 + 8)$$

$$= \frac{3x^2 + 10x}{x^3 + 5x^2 + 8}. \quad \bullet$$

Let $g(x)$ be any differentiable function. For any value of x for which $g(x)$ is positive, the function $\ln(g(x))$ is defined. For such a value of x, the derivative is given by the chain rule as

$$\frac{d}{dx}[\ln g(x)] = \frac{1}{g(x)} \cdot \frac{d}{dx} g(x)$$

$$= \frac{g'(x)}{g(x)}.$$

Example 1(c) illustrates a special case of this formula.

EXAMPLE 2   The function $f(x) = (\ln x)/x$ has a relative extreme point for some $x > 0$. Find the point and determine whether it is a relative maximum or a relative minimum point.

Solution   By the quotient rule,

$$f'(x) = \frac{x \cdot \dfrac{1}{x} - \ln x \cdot 1}{x^2} = \frac{1 - \ln x}{x^2}$$

$$f''(x) = \frac{x^2 \cdot \left(-\dfrac{1}{x}\right) - (1 - \ln x)(2x)}{x^4} = \frac{2 \ln x - 3}{x^3}.$$

If we set $f'(x) = 0$, then

$$1 - \ln x = 0$$

$$\ln x = 1$$

$$e^{\ln x} = e^1 = e$$

$$x = e.$$

Therefore, the only possible relative extreme point is at $x = e$. When $x = e$, $f(e) = (\ln e)/e = 1/e$. Furthermore,

$$f''(e) = \frac{2 \ln e - 3}{e^3} = -\frac{1}{e^3} < 0,$$

which implies that the graph of $f(x)$ is concave down at $x = e$. Therefore, $(e, 1/e)$ is a relative maximum point of the graph of $f(x)$. ●

The next example introduces a function that will be needed later when we study integration.

EXAMPLE 3   The function $\ln|x|$ is defined for all nonzero values of x. Its graph is sketched in Fig. 1. Compute the derivative of $\ln|x|$.

FIGURE I   Graph of ln |x|.

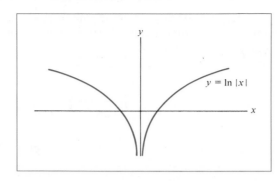

Solution   If x is positive, then $|x| = x$, so

$$\frac{d}{dx} \ln|x| = \frac{d}{dx} \ln x = \frac{1}{x}.$$

If x is negative, then $|x| = -x$; and, by the chain rule,

$$\frac{d}{dx} \ln|x| = \frac{d}{dx} \ln(-x)$$

$$= \frac{1}{-x} \cdot \frac{d}{dx}(-x)$$

$$= \frac{1}{-x} \cdot (-1) = \frac{1}{x}. \quad \bullet$$

Therefore, we have established the following useful fact.

$$\frac{d}{dx} \ln|x| = \frac{1}{x}, \qquad x \neq 0.$$

PRACTICE PROBLEMS 4.5

Differentiate.

1. $f(x) = \dfrac{1}{\ln(x^4 + 5)}$

2. $f(x) = \ln(\ln x)$

EXERCISES 4.5

Differentiate the following functions.

1. $\ln 2x$

2. $\ln x^2$

3. $\ln(x + 5)$

4. $x^2 \ln x$

5. $\dfrac{1}{x}\ln(x + 1)$

6. $\sqrt{\ln x}$

7. $e^{\ln x + x}$

8. $\ln\left(\dfrac{x}{x - 3}\right)$

9. $4 + \ln\left(\dfrac{x}{2}\right)$

10. $\ln \sqrt{x}$ 11. $(\ln x)^2 + \ln x$ 12. $\ln(x^3 + 2x + 1)$

13. $\ln(kx)$, k constant   14. $\dfrac{x}{\ln x}$ 15. $\dfrac{x}{(\ln x)^2}$

16. $(\ln x)e^{-x}$ 17. $e^{2x} \ln x$ 18. $(\ln x + 1)^3$

19. $\ln(e^{5x} + 1)$ 20. $\ln(e^{e^x})$

Find.

21. $\dfrac{d}{dt}(t^2 \ln 4)$ 22. $\dfrac{d^2}{dx^2} \ln(1 + x^2)$ 23. $\dfrac{d^2}{dt^2}(\ln t)^3$

24. Find the slope of the graph of $y = \ln |x|$ at $x = 3$ and $x = -3$.

25. Write the equation of the tangent line to the graph of $y = \ln(x^2 + e)$ at $x = 0$.

26. The function $f(x) = (\ln x + 1)/x$ has a relative extreme point for $x > 0$. Find the coordinates of the point. Is it a relative maximum point?

27. The function $f(x) = (\ln x)/\sqrt{x}$ has a relative extreme point for $x > 0$. Find the coordinates of the point. Is it a relative maximum point?

28. The function $f(x) = x/(\ln x + x)$ has a relative extreme point for $x > 1$. Find the coordinates of the point. Is it a relative minimum point?

29. Sketch the graph of the function $y = \ln x + (1/x) - \frac{1}{2}$.

30. Sketch the graph of $y = 1 + \ln(x^2 - 6x + 10)$.

31. If a cost function is $C(x) = (\ln x)/(40 - 3x)$, find the marginal cost when $x = 10$.

32. Suppose that the demand equation for a certain commodity is $p = 45/(\ln x)$. Determine the marginal revenue function for this commodity, and compute the marginal revenue when $x = 20$.

33. Suppose that the total revenue function for a manufacturer is $R(x) = 300 \ln(x + 1)$, so that the sale of x units of a product brings in about $R(x)$ dollars. Suppose also that the total cost of producing x units is $C(x)$ dollars, where $C(x) = 2x$. Find the value of x at which the profit function $R(x) - C(x)$ will be maximized. Show that the profit function has a relative maximum and not a relative minimum point at this value of x.

34. Evaluate $\lim\limits_{h \to 0} \dfrac{\ln(7 + h) - \ln 7}{h}$.

35. Find the maximum area of a rectangle in the first quadrant with one corner at the origin, two sides on the coordinate axes, and one corner on the graph of $y = -\ln x$.

SOLUTIONS TO PRACTICE PROBLEMS 4.5

1. Here $f(x) = [\ln(x^4 + 5)]^{-1}$. By the chain rule,

$$f'(x) = (-1) \cdot [\ln(x^4 + 5)]^{-2} \cdot \frac{d}{dx} \ln(x^4 + 5)$$

$$= -[\ln(x^4 + 5)]^{-2} \cdot \frac{4x^3}{x^4 + 5}.$$

2. $f'(x) = \dfrac{d}{dx} \ln(\ln x) = \dfrac{1}{\ln x} \cdot \dfrac{d}{dx} \ln x$

$$= \frac{1}{\ln x} \cdot \frac{1}{x} = \frac{1}{x \ln x}.$$

4.6 PROPERTIES OF THE NATURAL LOGARITHM FUNCTION

The natural logarithm function $\ln x$ possesses many of the familiar properties of logarithms to base 10 (or common logarithms) that are encountered in algebra.

Let x and y be positive numbers, b any number.

LI $\ln(xy) = \ln x + \ln y.$

LII $\ln\left(\dfrac{1}{x}\right) = -\ln x.$

LIII $\ln\left(\dfrac{x}{y}\right) = \ln x - \ln y.$

LIV $\ln(x^b) = b \ln x.$

Verification of LI   By equation (2) of Section 4.4 we have $e^{\ln(xy)} = xy$, $e^{\ln x} = x$, and $e^{\ln y} = y$. Therefore,
$$e^{\ln(xy)} = xy = e^{\ln x} \cdot e^{\ln y} = e^{\ln x + \ln y}.$$

By equating exponents, we get LI.

Verification of LII   Since $e^{\ln(1/x)} = 1/x$, we have
$$e^{\ln(1/x)} = \frac{1}{x} = \frac{1}{e^{\ln x}} = e^{-\ln x}.$$

By equating exponents, we get LII.

Verification of LIII   By LI and LII, we have
$$\ln\left(\frac{x}{y}\right) = \ln\left(x \cdot \frac{1}{y}\right)$$
$$= \ln x + \ln\left(\frac{1}{y}\right)$$
$$= \ln x - \ln y.$$

Verification of LIV   Since $e^{\ln(x^b)} = x^b$, we have
$$e^{\ln(x^b)} = x^b = (e^{\ln x})^b = e^{b \ln x}.$$

Equating exponents, we get LIV.

These properties of the natural logarithm should be learned thoroughly. You will find them useful in many calculations involving $\ln x$ and the exponential function.

EXAMPLE 1   Write ln 5 + 2 ln 3 as a single logarithm.

Solution
$$\ln 5 + 2 \ln 3 = \ln 5 + \ln 3^2 \qquad \text{(LIV)}$$
$$= \ln 5 + \ln 9$$
$$= \ln(5 \cdot 9) \qquad \text{(LI)}$$
$$= \ln 45. \quad \bullet$$

EXAMPLE 2   Write $\dfrac{1}{2}\ln(4t) - \ln(t^2 + 1)$ as a single logarithm.

Solution
$$\tfrac{1}{2}\ln(4t) - \ln(t^2 + 1) = \ln[(4t)^{1/2}] - \ln(t^2 + 1) \qquad \text{(LIV)}$$
$$= \ln(2\sqrt{t}) - \ln(t^2 + 1)$$
$$= \ln\!\left(\frac{2\sqrt{t}}{t^2 + 1}\right). \qquad \text{(LIII)}$$
$$\bullet$$

EXAMPLE 3   Simplify ln x + ln 3 + ln y − ln 5.

Solution   Use (LI) twice and (LIII) once.
$$(\ln x + \ln 3) + \ln y - \ln 5 = \ln 3x + \ln y - \ln 5$$
$$= \ln 3xy - \ln 5$$
$$= \ln\!\left(\frac{3xy}{5}\right). \quad \bullet$$

EXAMPLE 4   Differentiate $f(x) = \ln[x(x + 1)(x + 2)]$.

Solution   First rewrite $f(x)$, using (LI).
$$f(x) = \ln[x(x + 1)(x + 2)]$$
$$= \ln x + \ln(x + 1) + \ln(x + 2).$$
Then $f'(x)$ is easily calculated:
$$f'(x) = \frac{1}{x} + \frac{1}{x + 1} + \frac{1}{x + 2}. \quad \bullet$$

The natural logarithm function can be used to simplify the task of differentiating products. Suppose, for example, that we wish to differentiate the function
$$g(x) = x(x + 1)(x + 2).$$
As we showed in Example 4,
$$\frac{d}{dx}\ln g(x) = \frac{1}{x} + \frac{1}{x + 1} + \frac{1}{x + 2}.$$

However,

$$\frac{d}{dx} \ln g(x) = \frac{g'(x)}{g(x)}.$$

Therefore, equating the two expressions for $\dfrac{d}{dx} \ln g(x)$, we have

$$\frac{g'(x)}{g(x)} = \frac{1}{x} + \frac{1}{x + 1} + \frac{1}{x + 2}.$$

Finally, we solve for $g'(x)$:

$$g'(x) = g(x) \cdot \left(\frac{1}{x} + \frac{1}{x + 1} + \frac{1}{x + 2} \right)$$

$$= x(x + 1)(x + 2)\left(\frac{1}{x} + \frac{1}{x + 1} + \frac{1}{x + 2} \right).$$

In a similar way, we differentiate the product of any number of terms by first taking natural logarithms, then differentiating, and finally solving for the desired derivative. This procedure is called *logarithmic differentiation*.

EXAMPLE 5   Differentiate the function $g(x) = (x^2 + 1)(x^3 - 3)(2x + 5)$ using logarithmic differentiation.

Solution   Begin by taking the natural logarithm of both sides of the given equation:

$$\ln g(x) = \ln[(x^2 + 1)(x^3 - 3)(2x + 5)]$$

$$= \ln(x^2 + 1) + \ln(x^3 - 3) + \ln(2x + 5).$$

Now differentiate and solve for $g'(x)$:

$$\frac{g'(x)}{g(x)} = \frac{2x}{x^2 + 1} + \frac{3x^2}{x^3 - 3} + \frac{2}{2x + 5}$$

$$g'(x) = g(x)\left(\frac{2x}{x^2 + 1} + \frac{3x^2}{x^3 - 3} + \frac{2}{2x + 5} \right)$$

$$= (x^2 + 1)(x^3 - 3)(2x + 5)\left(\frac{2x}{x^2 + 1} + \frac{3x^2}{x^3 - 3} + \frac{2}{2x + 5} \right). \; \bullet$$

Let us now use logarithmic differentiation to finally establish the power rule:

$$\frac{d}{dx} (x^r) = rx^{r-1}.$$

Verification of the Power Rule   Let $f(x) = x^r$. Then

$$\ln f(x) = \ln x^r = r \ln x.$$

Differentiation of this equation yields

$$\frac{f'(x)}{f(x)} = r \cdot \frac{1}{x}$$

$$f'(x) = r \cdot \frac{1}{x} \cdot f(x) = r \cdot \frac{1}{x} \cdot x^r = rx^{r-1}.$$

**TECHNOLOGY
PROJECT 4.7
Testing the Laws of
Logarithms**

Here is a graphical approach to testing the laws of logarithms. While not an actual proof, the following arguments can provide strong evidence that the laws of logarithms actually hold.

1. Graph on the same coordinate system the function $f(x) = \ln x + \ln a$ and $g(x) = \ln (ax)$ for at least 10 positive values of a. What do you observe? What conclusion do the data suggest?

2. Graph on the same coordinate system the functions $f(x) = \ln x - \ln a$ and $g(x) = \ln\left(\frac{x}{a}\right)$ for at least 10 positive values of a. What do you observe? What conclusion do the data suggest?

3. Graph on the same coordinate system the functions $f(x) = a \ln x$ and $g(x) = \ln(x^a)$ for at least 10 positive values of a. What do you observe? What conclusion do the data suggest?

**PRACTICE
PROBLEMS 4.6**

1. Differentiate $f(x) = \ln\left[\dfrac{e^x \sqrt{x}}{(x+1)^6}\right]$.

2. Use logarithmic differentiation to differentiate $f(x) = (x+1)^7 (x+2)^8 (x+3)^9$.

EXERCISES 4.6

Simplify the following expressions.

1. $\ln 5 + \ln x$

2. $\ln x^5 - \ln x^3$

3. $\frac{1}{2} \ln 9$

4. $3 \ln \frac{1}{2} + \ln 16$

5. $\ln 4 + \ln 6 - \ln 12$

6. $\ln 2 - \ln x + \ln 3$

7. $e^{2 \ln x}$

8. $\frac{3}{2} \ln 4 - 5 \ln 2$

9. $5 \ln x - \frac{1}{2} \ln y + 3 \ln z$

10. $e^{\ln x^2 + 3 \ln y}$

11. $\ln x - \ln x^2 + \ln x^4$

12. $\frac{1}{2} \ln xy + \frac{3}{2} \ln \frac{x}{y}$

13. Which is larger, $2 \ln 5$ or $3 \ln 3$?

14. Which is larger, $\frac{1}{2} \ln 16$ or $\frac{1}{3} \ln 27$?

15. Which of the following is the same as $4 \ln 2x$?

    (a) $\ln 8x$ (b) $8 \ln x$ (c) $\ln 8 + \ln x$ (d) $\ln 16x^4$

16. Which of the following is the same as $\ln(9x) - \ln(3x)$?

    (a) $\ln 6x$ (b) $\ln(9x)/\ln(3x)$ (c) $6 \cdot \ln(x)$ (d) $\ln 3$

17. Which of the following is the same as $\dfrac{\ln 8x^2}{\ln 2x}$?

    (a) $\ln 4x$ (b) $4x$ (c) $\ln 8x^2 - \ln 2x$ (d) none of these

18. Which of the following is the same as $\ln 9x^2$?

    (a) $2 \cdot \ln 9x$ (b) $3x \cdot \ln 3x$ (c) $2 \cdot \ln 3x$ (d) none of these

Differentiate.

19. $\ln[(x + 5)(2x - 1)(4 - x)]$ **20.** $\ln[x^3(x + 1)^4]$

21. $\ln\left[\dfrac{(x + 1)(3x - 2)}{x + 2}\right]$ **22.** $\ln\left[\dfrac{x^2}{(3 - x)^3}\right]$

23. $\ln\left[\dfrac{\sqrt{x}}{x^2 + 1}\right]$ **24.** $\ln[e^{x^2}(x^4 + x^2 + 1)]$

Use logarithmic differentiation to differentiate the following functions.

25. $f(x) = (x + 1)^3(4x - 1)^2$ **26.** $f(x) = e^x(x - 4)^8$

27. $f(x) = (x - 2)^3(x - 3)^5(x + 2)^{-7}$

28. $f(x) = (x + 1)(2x + 1)(3x + 1)(4x + 1)$

29. $f(x) = x^x$ **30.** $f(x) = x^{1/x}$

31. $f(x) = e^x\sqrt{x^2 - 1}$ **32.** $f(x) = 2^x$

33. $f(x) = x^{\ln x}$ **34.** $f(x) = (2x - 1)^{2x}$?

35. $f(x) = \dfrac{\sqrt{x - 1}(x - 2)}{x^2 - 3}$ **36.** $f(x) = \dfrac{xe^x}{\sqrt{3x^2 + 1}}$

37. There are substantial empirical data to show that if x and y measure the sizes of two organs of a particular animal, then x and y are related by an *allometric equation* of the form

$$\ln y - k \ln x = \ln c,$$

where k and c are positive constants that depend only on the type of parts or organs that are measured, and are constant among animals belonging to the same species.[*] Solve this equation for y in terms of x, k, and c.

38. In the study of epidemics, one finds the equation

$$\ln(1 - y) - \ln y = C - rt,$$

where y is the fraction of the population that has a specific disease at time t. Solve the equation for y in terms of t and the constants C and r.

39. Determine the values of h and k for which the graph of $y = he^{kx}$ passes through the points $(1, 6)$ and $(4, 48)$.

---

[*] E. Batschelet, *Introduction to Mathematics for Life Scientists* (New York: Springer-Verlag, 1971), pp. 305–307.

40. Find values of k and r for which the graph of $y = kx^r$ passes through the points (2, 3) and (4, 15).

SOLUTIONS TO PRACTICE PROBLEMS 4.6

1. Use the properties of the natural logarithm to express $f(x)$ as a sum of simple functions before differentiating.

$$f(x) = \ln\left[\frac{e^x\sqrt{x}}{(x+1)^6}\right]$$

$$f(x) = \ln e^x + \ln \sqrt{x} - \ln(x+1)^6$$

$$= x + \tfrac{1}{2}\ln x - 6\ln(x+1).$$

$$f'(x) = 1 + \frac{1}{2x} - \frac{6}{x+1}.$$

2. $f(x) = (x+1)^7(x+2)^8(x+3)^9.$

$$\ln f(x) = 7\ln(x+1) + 8\ln(x+2) + 9\ln(x+3).$$

Now we differentiate both sides of the equation.

$$\frac{f'(x)}{f(x)} = \frac{7}{x+1} + \frac{8}{x+2} + \frac{9}{x+3}.$$

$$f'(x) = f(x)\left(\frac{7}{x+1} + \frac{8}{x+2} + \frac{9}{x+3}\right)$$

$$= (x+1)^7(x+2)^8(x+3)^9\left(\frac{7}{x+1} + \frac{8}{x+2} + \frac{9}{x+3}\right).$$

✓ **CHAPTER 4 CHECKLIST**

✓ $b^x \cdot b^y = b^{x+y}$
✓ $b^{-x} = 1/b^x$
✓ $b^x/b^y = b^{x-y}$
✓ $(b^y)^x = b^{xy}$
✓ $a^x b^x = (ab)^x$
✓ $a^x/b^x = (a/b)^x$
✓ $b^0 = 1$
✓ definition of e and e^x
✓ $\frac{d}{dx}(e^{kx}) = ke^{kx}$
✓ $\frac{d}{dx}(e^{g(x)}) = e^{g(x)}g'(x)$
✓ graph of e^{kx}
✓ if $y = f(x)$ satisfies $y' = ky$, then $y = ce^{kx}$ for some constant c.
✓ Reflection in the line $y = x$

✓ Definition of $\ln x$
✓ $\ln 1 = 0$
✓ $\ln e = 1$
✓ $e^{\ln x} = x, x > 0$
✓ $\ln e^x = x$
✓ $\frac{d}{dx}(\ln x) = \frac{1}{x}$
✓ $\frac{d}{dx}[\ln g(x)] = \frac{g'(x)}{g(x)}$
✓ $\ln(xy) = \ln x + \ln y$
✓ $\ln\left(\frac{x}{y}\right) = \ln x - \ln y$
✓ $\ln\left(\frac{1}{x}\right) = -\ln x$
✓ $\ln(x^b) = b\ln x$

CHAPTER 4   SUPPLEMENTARY EXERCISES

Calculate the following.

1. $27^{4/3}$ **2.** $4^{1.5}$ **3.** 5^{-2} **4.** $16^{-.25}$

5. $(2^{5/7})^{14/5}$ **6.** $8^{1/2} \cdot 2^{1/2}$ **7.** $\dfrac{9^{5/2}}{9^{3/2}}$ **8.** $4^{.2} \cdot 4^{.3}$

Simplify the following.

9. $(e^{x^2})^3$ **10.** $e^{5x} \cdot e^{2x}$ **11.** $\dfrac{e^{3x}}{e^x}$

12. $2^x \cdot 3^x$ **13.** $(e^{8x} + 7e^{-2x})e^{3x}$ **14.** $\dfrac{e^{5x/2} - e^{3x}}{\sqrt{e^x}}$

Solve the following equations for x.

15. $e^{-3x} = e^{-12}$ **16.** $e^{x^2-x} = e^2$

17. $(e^x \cdot e^2)^3 = e^{-9}$ **18.** $e^{-5x} \cdot e^4 = e$

Differentiate the following functions.

19. $10e^{7x}$ **20.** $e^{\sqrt{x}}$ **21.** xe^{x^2}

22. $\dfrac{e^x + 1}{x - 1}$ **23.** e^{e^x} **24.** $(\sqrt{x} + 1)e^{-2x}$

25. $\dfrac{x^2 - x + 5}{e^{3x} + 3}$ **26.** x^e

27. Determine all solutions of the differential equation $y' = -y$.

28. Determine all functions $y = f(x)$ such that $y' = -1.5y$ and $f(0) = 2000$.

29. Determine all solutions of the differential equation $y' = 1.5y$ and $f(0) = 2$.

30. Determine all solutions of the differential equation $y' = \frac{1}{3}y$.

Graph the following functions.

31. $e^{-x} + x$ **32.** $e^x - x$

33. $e^{-(1/2)x^2}$ **34.** $100(x - 2)e^{-x}$ for $x \geq 2$

Simplify the following expressions.

35. $\dfrac{\ln x^2}{\ln x^3}$ **36.** $e^{2\ln 2}$ **37.** $e^{-5\ln 1}$

38. $[e^{\ln x}]^2$ **39.** $e^{(\ln 5)/2}$ **40.** $e^{\ln(x^2)}$

Solve the following equations for t.

41. $3e^{2t} = 15$ **42.** $3e^{t/2} - 12 = 0$ **43.** $2\ln t = 5$?

44. $2e^{-.3t} = 1$ **45.** $t^{\ln t} = e$ **46.** $\ln(\ln 3t) = 0$

Differentiate the following functions.

47. $\ln(5x - 7)$ **48.** $\ln(9x)$

49. $(\ln x)^2$ **50.** $(x \ln x)^3$

51. $\ln(x^6 + 3x^4 + 1)$

52. $\dfrac{x}{\ln x}$

53. $\ln\left(\dfrac{xe^x}{\sqrt{1 + x}}\right)$

54. $\ln[(x^2 + 3)^5(x^3 + 1)^{-4}]$

55. $\ln(\ln \sqrt{x})$

56. $\dfrac{1}{\ln x}$

57. $x \ln x - x$

58. $e^{2\ln(x+1)}$

59. $e^x \ln x$

60. $\ln(x^2 + e^x)$

Use logarithmic differentiation to differentiate the following functions.

61. $f(x) = (x^2 + 5)^6(x^3 + 7)^8(x^4 + 9)^{10}$

62. $f(x) = x^{1+x}$ **63.** $f(x) = 10^x$ **64.** $f(x) = \sqrt{x^2 + 5}\ e^{x^2}$

Sketch the following curves.

65. $y = x - \ln x$

66. $y = \ln(x^2 + 1)$

67. $y = (\ln x)^2$

68. $y = (\ln x)^3$

CHAPTER **5**

APPLICATIONS OF THE EXPONENTIAL AND NATURAL LOGARITHM FUNCTIONS

E arlier, we introduced the exponential function e^x and the natural logarithm function $\ln x$ and studied their most important properties. From
the way we introduced these functions, it is by no means clear that they have
any substantial connection with the physical world. However, as this chapter
will demonstrate, the exponential and natural logarithm functions intrude
into the study of many physical problems, often in a very curious and unexpected way.

Here the most significant fact that we require is that the exponential
function is uniquely characterized by its differential equation. In other words,
we will constantly make use of the following fact, stated previously.

The function* $y = Ce^{kt}$ satisfies the differential equation
$$y' = ky.$$
Conversely, if $y = f(t)$ satisfies the differential equation, then $y = Ce^{kt}$
for some constant C.

If $f(t) = Ce^{kt}$, then, by setting $t = 0$, we have
$$f(0) = Ce^0 = C.$$
Therefore, C is the value of $f(t)$ at $t = 0$.

 ## 5.1 EXPONENTIAL GROWTH AND DECAY

In biology, chemistry, and economics it is often necessary to study the behavior
of a quantity that is increasing as time passes. If, at every instant, the rate of
increase of the quantity is proportional to the quantity at that instant, then we
say that the quantity is *growing exponentially* or is *exhibiting exponential
growth*. A simple example of exponential growth is exhibited by the growth of
bacteria in a culture. Under ideal laboratory conditions a bacteria culture grows
at a rate proportional to the number of bacteria present. It does so because the
growth of the culture is accounted for by the division of the bacteria. The more
bacteria there are at a given instant, the greater the possibilities for division and
hence the greater is the rate of growth.

Let us study the growth of a bacteria culture as a typical example of
exponential growth. Suppose that $P(t)$ denotes the number of bacteria in a
certain culture at time t. The rate of growth of the culture at time t is $P'(t)$. We

---

* Note that we use the variable t instead of x. The reason is that, in most applications, the variable of our exponential function is time. The variable t will be used
throughout this chapter.

assume that this rate of growth is proportional to the size of the culture at time t, so that

$$P'(t) = kP(t), \tag{1}$$

where k is a positive constant of proportionality. If we let $y = P(t)$, then (1) can be written as

$$y' = ky.$$

Therefore, from our discussion at the beginning of this chapter we see that

$$y = P(t) = P_0 e^{kt}, \tag{2}$$

where P_0 is the number of bacteria in the culture at time $t = 0$. The number k is called the *growth constant.*

EXAMPLE I   Suppose that a certain bacteria culture grows at a rate proportional to its size. At time $t = 0$, approximately 20,000 bacteria are present. In 5 hours there are 400,000 bacteria. Determine a function that expresses the size of the culture as a function of time, measured in hours.

Solution   Let $P(t)$ be the number of bacteria present at time t. By assumption, $P(t)$ satisfies a differential equation of the form $y' = ky$, so $P(t)$ has the form

$$P(t) = P_0 e^{kt},$$

where the constants P_0 and k must be determined. The value of P_0 and k can be obtained from the data that give the population size at two different times. We are told that

$$P(0) = 20,000, \qquad P(5) = 400,000. \tag{3}$$

The first condition immediately implies that $P_0 = 20,000$, so

$$P(t) = 20,000 e^{kt}.$$

Using the second condition in (3), we have

$$20,000 e^{k \cdot 5} = P(5) = 400,000$$

$$e^{5k} = 20$$

$$5k = \ln 20$$

$$k = \frac{\ln 20}{5} \approx .60. \tag{4}$$

So we may take

$$P(t) = 20,000 e^{.6t}.$$

FIGURE I   A model for bacteria growth.

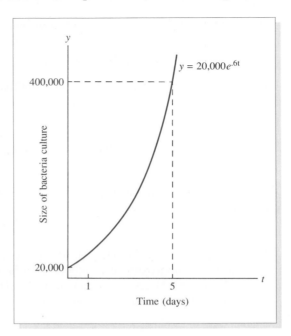

This function is a mathematical model of the growth of the bacteria culture. (See Fig. 1.) ●

EXAMPLE 2   Suppose that a colony of fruit flies is growing according to the exponential law $P(t) = P_0 e^{kt}$ and suppose that the size of the colony doubles in 12 days. Determine the growth constant k.

Solution   We do not know the initial size of the population at $t = 0$. However, we are told that $P(12) = 2P(0)$; that is,

$$P_0 e^{k \cdot 12} = 2P_0$$

$$e^{12k} = 2$$

$$12k = \ln 2$$

$$k = \tfrac{1}{12} \ln 2 \approx .058. \quad ●$$

Notice that the initial size P_0 of the population was not given in Example 2. We were able to determine the growth constant because we were told the amount of time required for the colony to double in size. Thus the growth constant does not depend on the initial size of the population. This property is characteristic of exponential growth.

EXAMPLE 3   Suppose that the initial size of the colony in Example 2 was 300. At what time will the colony contain 1800 fruit flies?

Solution   From Example 2 we have $P(t) = P_0e^{.058t}$. Since $P(0) = 300$, we conclude that

$$P(t) = 300e^{.058t}.$$

Now that we have the explicit formula for the size of the colony, we can set $P(t) = 1800$ and solve for t:

$$300e^{.058t} = 1800$$

$$e^{.058t} = 6$$

$$.058t = \ln 6$$

$$t = \frac{\ln 6}{.058} \approx 31 \text{ days.} \bullet$$

The table below shows the growth of the colony in Example 3. Notice that 1800 is exactly halfway between 1200 ($t = 24$) and 2400 ($t = 36$). It is incorrect to guess that the population will equal 1800 when t is halfway between $t = 24$ and $t = 36$—that is, when $t = 30$. We saw in Example 3 that it takes approximately 31 days for the colony to reach 1800 fruit flies.

Population Size	Day
300	0
600	12
1,200	24
2,400	36
4,800	48
9,600	60
19,200	72
⋮	⋮

Exponential Decay   An example of negative exponential growth, or *exponential decay,* is given by the disintegration of a radioactive element such as uranium 235. It is known that, at any instant, the rate at which a radioactive substance is decaying is proportional to the amount of the substance that has not yet disintegrated. If $P(t)$ is the quantity present at time t, then $P'(t)$ is the rate of decay. Of course, $P'(t)$ must be negative, since $P(t)$ is decreasing. Thus we may write $P'(t) = kP(t)$ for some negative constant k. To emphasize the fact that the constant is negative, k is often replaced by $-\lambda$, where λ is a positive constant.* Then $P(t)$ satisfies the differential equation

$$P'(t) = -\lambda P(t). \tag{5}$$

The general solution of (5) has the form

$$P(t) = P_0e^{-\lambda t}$$

---

*λ is the Greek lowercase letter lambda.

for some positive number P_0. We call such a function an *exponential decay function*. The constant λ is called the *decay constant*.

EXAMPLE 4 The decay constant for strontium 90 is $\lambda = .0244$, where the time is measured in years. How long will it take for a quantity P_0 of strontium 90 to decay to one-half its original mass?

Solution We have

$$P(t) = P_0 e^{-.0244t}.$$

Next, set $P(t)$ equal to $\frac{1}{2} P_0$ and solve for t:

$$P_0 e^{-.0244t} = \tfrac{1}{2} P_0$$

$$e^{-.0244t} = \tfrac{1}{2} = .5$$

$$-.0244t = \ln .5$$

$$t = \frac{\ln .5}{-.0244} \approx 28 \text{ years.} \quad \bullet$$

The *half-life* of a radioactive element is the length of time required for a given quantity of that element to decay to one-half its original mass. Thus strontium 90 has a half-life of about 28 years. (See Fig. 2.) Notice from Example 4 that the half-life does not depend on the initial amount P_0.

FIGURE 2   Half-life of radio-active strontium 90.

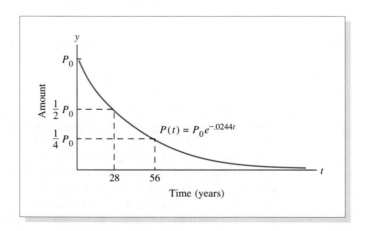

Time (years)

EXAMPLE 5 Radioactive carbon 14 has a half-life of about 5730 years. Find its decay constant.

Solution If P_0 denotes the initial amount of carbon 14, then the amount after t years will be

$$P(t) = P_0 e^{-\lambda t}.$$

After 5730 years, $P(t)$ will equal $\frac{1}{2}P_0$. That is,

$$P_0 e^{-\lambda(5730)} = P(5730) = \tfrac{1}{2}P_0 = .5P_0.$$

Solving for λ gives

$$e^{-5730\lambda} = .5$$

$$-5730\lambda = \ln .5$$

$$\lambda = \frac{\ln .5}{-5730} \approx .00012. \quad \bullet$$

One of the problems connected with aboveground nuclear explosions is the radioactive debris that falls on plants and grass, thereby contaminating the food supply of animals. Strontium 90 is one of the most dangerous components of "fallout" because it has a relatively long half-life and because it is chemically similar to calcium and is absorbed into the bone structure of animals (including humans) who eat contaminated food. Iodine 131 is also produced by nuclear explosions, but it presents less of a hazard because it has a half-life of 8 days.

EXAMPLE 6 If dairy cows eat hay containing too much iodine 131, their milk will be unfit to drink. Suppose that some hay contains 10 times the maximum allowable level of iodine 131. How many days should the hay be stored before it is fed to dairy cows?

Solution Let P_0 be the amount of iodine 131 present in the hay. Then the amount at time t is $P(t) = P_0 e^{-\lambda t}$ (t in days). The half-life of iodine 131 is 8 days, so

$$P_0 e^{-8\lambda} = .5P_0$$

$$e^{-8\lambda} = .5$$

$$-8\lambda = \ln .5$$

$$\lambda = \frac{\ln .5}{-8} \approx .087,$$

and

$$P(t) = P_0 e^{-.087t}.$$

Now that we have the formula for $P(t)$, we want to find t such that $P(t) = \frac{1}{10}P_0$. We have

$$P_0 e^{-.087t} = .1P_0,$$

so

$$e^{-.087t} = .1$$

$$-.087t = \ln .1$$

$$t = \frac{\ln .1}{-.087} \approx 26 \text{ days}. \quad \bullet$$

Radiocarbon Dating Knowledge about radioactive decay is valuable to social scientists who want to estimate the age of objects belonging to ancient civilizations. Several different substances are useful for radioactive-dating techniques; the most common is radiocarbon, ^{14}C. Carbon 14 is produced in the upper atmosphere when cosmic rays react with atmospheric nitrogen. Because the ^{14}C eventually decays, the concentration of ^{14}C cannot rise above certain levels. An equilibrium is reached where ^{14}C is produced at the same rate as it decays. Scientists usually assume that the total amount of ^{14}C in the biosphere has remained constant over the past 50,000 years. Consequently, it is assumed that the *ratio* of ^{14}C to ordinary nonradioactive carbon 12, ^{12}C, has been constant during this same period. (The ratio is about one part ^{14}C to 10^{12} parts of ^{12}C.) Both ^{14}C and ^{12}C are in the atmosphere as constituents of carbon dioxide. All living vegetation and most forms of animal life contain ^{14}C and ^{12}C in the same proportion as the atmosphere. The reason is that plants absorb carbon dioxide through photosynthesis. The ^{14}C and ^{12}C in plants are distributed through the food chain to almost all animal life.

When an organism dies, it stops replacing its carbon, and therefore the amount of ^{14}C begins to decrease through radioactive decay. (The ^{12}C in the dead organism remains constant.) The ratio of ^{14}C to ^{12}C can be later measured in order to determine when the organism died. (See Fig. 3.)

FIGURE 3 $^{14}C-^{12}C$ ratio compared to the ratio in living plants.

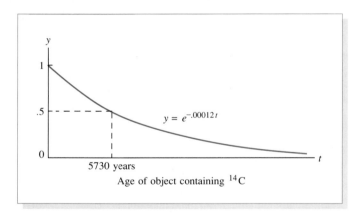

$$y = e^{-.00012t}$$

5730 years

Age of object containing ^{14}C

EXAMPLE 7 A parchment fragment was discovered that had about 80% of the ^{14}C level found today in living matter. Estimate the age of the parchment.

Solution We assume that the original ^{14}C level in the parchment was the same as the level in living organisms today. Consequently, about eight-tenths of the original ^{14}C remains. From Example 5 we obtain the formula for the amount of ^{14}C present t years after the parchment was made from an animal skin:

$$P(t) = P_0 e^{-.00012t},$$

where P_0 = initial amount. We want to find t such that $P(t) = .8P_0$.

$$P_0 e^{-.00012t} = .8P_0$$

$$e^{-.00012t} = .8$$

$$-.00012t = \ln .8$$

$$t = \frac{\ln .8}{-.00012} \approx 1900 \text{ years old.} \bullet$$

TECHNOLOGY PROJECT 5.1
Graphing Calculator Solution of a Decay Problem

Give a graphical solution to the preceding example by using TRACE to determine the solution of the equation:

$$e^{-.00012t} = .8$$

A Sales Decay Curve   Marketing studies* have demonstrated that if advertising and other promotions of a particular product are stopped and if other market conditions remain fairly constant, then, at any time t, the sales of that product will be declining at a rate proportional to the amount of current sales at t. (See Fig. 4.) If S_0 is the number of sales in the last month during which advertising occurred and if $S(t)$ is the number of sales in the tth month following the cessation of promotional effort, then a good mathematical model for $S(t)$ is

$$S(t) = S_0 e^{-\lambda t},$$

where λ is a positive number called the *sales decay constant*. The value of λ depends on many factors, such as the type of product, the number of years of prior advertising, the number of competing products, and other characteristics of the market.

FIGURE 4   Exponential decay of sales.

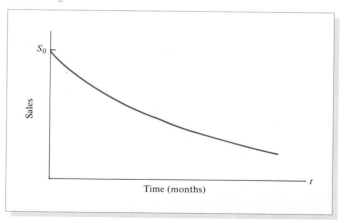

Sales

S_0

Time (months)

t

*M. Vidale and H. Wolfe, "An Operations-Research Study of Sales Response to Advertising," *Operations Research,* 5 (1957), 370–381. Reprinted in F. Bass et al., *Mathematical Models and Methods in Marketing* (Homewood, Ill.: Richard D. Irwin, Inc., 1961).

The Time Constant   Consider an exponential decay function $y = Ce^{-\lambda t}$. In Fig. 5, we have drawn the tangent line to the decay curve when $t = 0$. The slope there is the initial rate of decay. If the decay process were to continue at this rate, the decay curve would follow the tangent line and y would be zero at some time T. This time is called the *time constant* of the decay curve. It can be shown (see Exercise 31) that $T = 1/\lambda$ for the curve $y = Ce^{-\lambda t}$. Thus $\lambda = 1/T$, and the decay curve can be written in the form

$$y = Ce^{-t/T}.$$

If one has experimental data that tend to lie along an exponential decay curve, then the numerical constants for the curve may be obtained from Fig. 5. First, sketch the curve and estimate the y-intercept, C. Then sketch an approximate tangent line and from this estimate the time constant, T. This procedure is sometimes used in biology and medicine.

FIGURE 5   The time constant T in exponential decay: $T = 1/\lambda$,

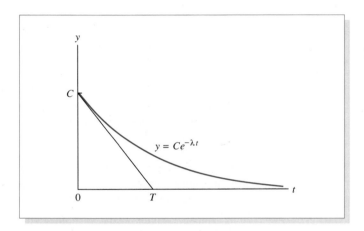

PRACTICE PROBLEMS 5.1

1. (a) Solve the differential equation $P'(t) = -.6P(t)$, $P(0) = 50$.
   (b) Solve the differential equation $P'(t) = kP(t)$, $P(0) = 4000$, where k is some constant.
   (c) Find the value of k in part (b) for which $P(2) = 100P(0)$.
2. Under ideal conditions a colony of *Escherichia coli* bacteria can grow by a factor of 100 every 2 hours. If initially 4000 bacteria are present, how long will it take before there are 1,000,000 bacteria?

EXERCISES 5.1

1. Let $P(t)$ be the size of a certain insect population after t days, and suppose that $P(t)$ satisfies the differential equation

$$P'(t) = .07P(t), \qquad P(0) = 400.$$

Find the formula for $P(t)$.

2. Let $P(t)$ be the number of bacteria present in a culture after t minutes, and suppose that $P(t)$ satisfies the differential equation

$$P'(t) = .55P(t).$$

Find the formula for $P(t)$ if initially there are approximately 10,000 bacteria present.

3. Suppose that after t hours there are $P(t)$ cells present in a culture, where $P(t) = 5000e^{.2t}$.

  (a) How many cells were present initially?

  (b) When will 20,000 cells be present?

4. The size of a certain insect population is given by $P(t) = 300e^{.01t}$, where t is measured in days. At what time will the population equal 600? 1200?

5. Determine the growth constant of a population that is growing at a rate proportional to its size, where the population doubles in size every 40 days.

6. Determine the growth constant of a population that is growing at a rate proportional to its size, where the population triples in size every 5 hours.

7. The world's population was 5.4 billion on January 1, 1991, and was expected to reach 6 billion on January 1, 1995. Assume that at any time the population grows at a rate proportional to the population at that time.

$p(t) = 5.4e^{.0263t}$

  (a) Find the formula for $P(t)$, the world's population t years after January 1, 1991.

  (b) What will the population be on January 1, 2010?

  (c) In what year will the world's population reach 7 billion?

8. Mexico City is expected to become the most heavily populated city in the world by the end of this century. At the beginning of 1990, 20.2 million people lived in the metropolitan area of Mexico City, and the population was growing exponentially with growth constant .032. (Part of the growth is due to immigration.)

  (a) If this trend continues, how large will the population be in the year 2000?

  (b) In what year will the 1990 population have doubled?

9. The growth rate of a certain bacteria culture is proportional to its size. If the bacteria culture doubles in size every 20 minutes, how long will it take for the culture to increase twelve-fold?

10. The growth rate of a certain cell culture is proportional to its size. In 10 hours a population of 1 million cells grew to 9 million. How large will the cell culture be after 15 hours?

11. The population of a certain country is growing exponentially. The total population (in millions) in t years is given by the function $P(t)$. Match each of the following answers with its corresponding question.

    Answers: a. Solve $P(t) = 2$ for t.
                        b. $P(2)$
                        c. $P'(2)$
                        d. Solve $P'(t) = 2$ for t.
                        e. $y' = ky$
                        f. Solve $P(t) = 2P(0)$ for t.
                        g. P_0e^{kt}, $k > 0$
                        h. $P(0)$

Questions: **A.** How fast will the population be growing in 2 years?

B. Give the general form of the function $P(t)$.

C. How long will it take for the current population to double?

D. What will be the size of the population in 2 years?

E. What is the initial size of the population?

F. When will the size of the population be 2 million?

G. When will the population be growing at the rate of 2 million people per year?

H. Give a differential equation satisfied by $P(t)$.

12. A certain bacteria culture grows at a rate proportional to its size, and it doubles every half hour. Suppose that the culture contains 3 million bacteria at time $t = 0$ (with time in hours).

(a) At what time will there be 600 million bacteria present?

(b) At what time will the culture be growing at the rate of 600 million bacteria per hour?

13. The weight in grams after t years, $P(t)$, of a certain radioactive substance satisfies the differential equation

$$P'(t) = -.08P(t), \qquad P(0) = 30.$$

(a) Find the formula for $P(t)$.

(b) What is $P(10)$?

14. Five milligrams of a drug is injected into a patient, and the amount of drug present t hours after the injection satisfies the differential equation

$$P'(t) = -.09P(t).$$

(a) Find the formula for $P(t)$.

(b) Determine the amount of drug present 20 hours after the injection.

15. One hundred grams of a radioactive substance with decay constant .01 is buried in the ground. Assume that time is measured in years.

(a) Give the formula for the amount remaining after t years.

(b) How much will remain after 30 years?

(c) What is the half-life of this radioactive substance?

16. The decay constant for cesium 137 is .023 when time is measured in years. Find the half-life of cesium 137.

17. Radioactive cobalt 60 has a half-life of 5.3 years.

(a) Find the decay constant of cobalt 60.

(b) If the initial amount of cobalt 60 is 10 grams, how much will be present after 2 years?

18. Five grams of a certain radioactive material decays to 3 grams in 1 year. After how many years will just 1 gram remain?

19. A 4500-year-old wooden chest was found in the tomb of the twenty-fifth century B.C. Chaldean king Meskalumdug of Ur. What percentage of the original ^{14}C would you expect to find in the wooden chest? (Recall that the decay constant for ^{14}C is .00012.)

20. In 1947, a cave with beautiful prehistoric wall paintings was discovered in Lascaux, France. Some charcoal found in the cave contained 20% of the ^{14}C expected in living trees. How old are the Lascaux cave paintings?

21. Sandals woven from strands of tree bark were found recently in Fort Rock Creek Cave in Oregon. The bark contained 34% of the level of ^{14}C found in living bark. Approximately how old are the sandals?

22. Many scientists believe there have been four ice ages in the past one million years. Before the technique of carbon dating was known, geologists erroneously believed that the retreat of the Fourth Ice Age began about 25,000 years ago. In 1950, logs from ancient spruce trees were found under glacial debris near Two Creeks, Wisconsin. Geologists determined that these trees had been crushed by the advance of ice during the Fourth Ice Age. Wood from the spruce trees contained 27% of the level of ^{14}C found in living trees. Approximately how long ago did the Fourth Ice Age actually occur?

23. Let $f(t)$ be the value of the dollar t years after January 1, 1990, where the value is in terms of the purchasing power on January 1, 1990, and $f(0) = 1.00$. Suppose that the rate of decrease of $f(t)$ at any time t is proportional to $f(t)$. Give the formula for $f(t)$ if by January 1, 1992, the dollar had lost 15% of its purchasing power.

24. The consumer price index (CPI) gives a measure of the prices of commodities commonly purchased by consumers. For example, an increase of 10% in the CPI corresponds to a 10% average increase in the prices of consumer goods. Let $f(t)$ be the CPI at time t, where time is years since January 1, 1980. Suppose that $f(t)$ satisfies the differential equation

$$f'(t) = .12f(t), \qquad f(0) = 100.$$

(This means that at each time t, the CPI is rising at an annual rate of 12%, and the index is set equal to 100 on January 1, 1980.) How many years did it take for the CPI to double?

25. An island in the Pacific Ocean is contaminated by fallout from a nuclear explosion. If the strontium 90 is 100 times the level that scientists believe is "safe," how many years will it take for the island to once again be "safe" for human habitation? The half-life of strontium 90 is 28 years.

26. A common infection of the urinary tract in humans is caused by the bacterium *E. coli*. The infection is generally noticed when the bacteria colony reaches a population of about 10^8. The colony doubles in size about every 20 minutes. When a full bladder is emptied, about 90% of the bacteria are eliminated. Suppose that at the beginning of a certain time period, a person's bladder and urinary tract contain 10^8 *E. coli* bacteria. During an interval of T minutes the person drinks enough liquid to fill the bladder. Find the value of T such that if the bladder is emptied after T minutes, about 10^8 bacteria will still remain. [*Note:* The average bladder holds about 1 liter of urine. It is seldom possible to eliminate an *E. coli* infection by diuresis without drugs—such as by drinking large amounts of water.]

27. By 1974 the United States had an estimated 80 million gallons of radioactive products from nuclear power plants and other nuclear reactors. These waste products were stored in various sorts of containers (made of such materials as stainless steel and cement), and the containers were buried in the ground and the ocean. Scientists feel that the waste products must be prevented from contaminating the rest of the earth until more than 99.99% of the radioactivity is gone (i.e., until the level is less than 0.0001 times the original level). If a storage cylinder contains waste products whose half-life is 1500 years, how many years must the container survive without leaking? [*Note:* Some of the containers are already leaking.]

28. In 1950 the world's population required 1×10^9 hectares* of arable land for food growth, and in 1980, 2×10^9 hectares were required. Current population trends indicate that if $A(t)$ denotes the amount of land needed t years after 1950, then

$$\frac{dA}{dt} = kA$$

for some constant k.

(a) Derive a formula for $A(t)$.

(b) The total amount of arable land on the earth's surface is estimated at 3.2×10^9 hectares. In what year will the earth exhaust its supply of land for growing food? [Data based on the Club of Rome's report, *The Limits to Growth* by D. H. Meadows, D. L. Meadows, J. Randers, and W. Behrens III (New York: Universe Books, 1972).]

29. A drought in the African veldt causes the death of much of the animal population. A typical herd of wildebeests suffers a death rate proportional to its size. The herd numbers 500 at the onset of the drought and only 200 remain 4 months later.

(a) Find a formula for the herd's population at time t months.

(b) How long will it take for the herd to diminish to one-tenth its original size?

30. In a laboratory experiment, 8 units of a sulfate were injected into a dog. After 50 minutes, only 4 units remained in the dog. Let $f(t)$ be the amount of sulfate present after t minutes. At any time, the rate of change of $f(t)$ is proportional to the value of $f(t)$. Find the formula for $f(t)$.

31. Let T be the time constant of the curve $y = Ce^{-\lambda t}$ as defined in Fig. 5. Show that $T = 1/\lambda$. [*Hint:* Express the slope of the tangent line in Fig. 5 in terms of C and T. Then set this slope equal to the slope of the curve $y = Ce^{-\lambda t}$ at $t = 0$.]

32. Suppose that a person is given an injection of 300 milligrams of penicillin at time $t = 0$, and let $f(t)$ be the amount (in milligrams) of penicillin present in the person's bloodstream t hours after the injection. Then the amount of penicillin present decays exponentially, and a typical formula for $f(t)$ is $f(t) = 300e^{-(2/3)t}$.

(a) What is the initial rate of decay of the penicillin?

(b) What is the time constant for this decay curve $y = f(t)$? (See the discussion accompanying Fig. 5.)

33. Let $M(t)$ be the median price of a home in the United States, where $t = 0$ corresponds to 1986. During the second part of the 1980s, $M(t)$ satisfied the differential equation

$$M'(t) = .05M(t), \qquad M(0) = 80{,}000.$$

Assuming $M(t)$ continues to satisfy this equation in the 1990s, estimate the median price of a home at the turn of the century.

34. According to legend, in the fifth century King Arthur and his knights sat at a huge round table. A round table alleged to have belonged to King Arthur was found at Winchester Castle in England. In 1976 carbon dating revealed the amount of radiocarbon in the table to be 91% of the radiocarbon present in living wood. Could the table possibly have belonged to King Arthur? Why?

35. Suppose you have 80 grams of a certain radioactive material and the amount remaining after t years is given by the function $f(t)$ shown in Fig. 6.

---

*A hectare equals 2.471 acres.

(a) How much will remain after 5 years?
(b) When will 40 grams remain?
(c) At what rate will the radioactive material be distintegrating after 1 year?
(d) After how many years will the radioactive material be disintegrating at the rate of about 5 grams per year?

FIGURE 6

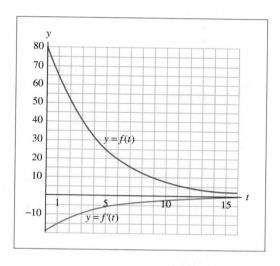

36. The monthly payment on a $100,000 30-year mortgage at $r\%$ interest is $f(r)$ dollars, where f is the function shown in Fig. 7.
    (a) What will the monthly payment be at 7.5% interest?
    (b) For what interest rate will the monthly payment be $600?
    (c) If the interest rate is 2% and interest is increased by 1%, by approximately how much will the monthly payment increase?

FIGURE 7

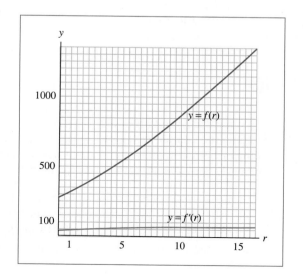

SOLUTIONS TO PRACTICE PROBLEMS 5.1

1. (a) Answer: $P(t) = 50e^{-.6t}$. Differential equations of the type $y' = ky$ have as their solution $P(t) = Ce^{kt}$, where C is $P(0)$.

   (b) Answer: $P(t) = 4000e^{kt}$. This problem is like the previous one except that the constant is not specified. Additional information is needed if one wants to determine a specific value for k.

   (c) Answer: $P(t) = 4000e^{2.3t}$. From the solution to part (b) we know that $P(t) = 4000e^{kt}$. We are given that $P(2) = 100P(0) = 100(4000) = 400,000$. So

$$P(2) = 4000e^{k(2)} = 400,000$$

$$e^{2k} = 100$$

$$2k = \ln 100$$

$$k = \frac{\ln 100}{2} \approx 2.3.$$

2. Let $P(t)$ be the number of bacteria present after t hours. We must first find an expression for $P(t)$ and then determine the value of t for which $P(t) = 1,000,000$. From the discussion at the beginning of the section we know that $P'(t) = k \cdot P(t)$. Also, we are given that $P(2)$ (the population after 2 hours) is $100P(0)$ (100 times the initial population). From Problem 1(c) we have an expression for $P(t)$:

$$P(t) = 4000e^{2.3t}.$$

Now we must solve $P(t) = 1,000,000$ for t.

$$4000e^{2.3t} = 1,000,000$$

$$e^{2.3t} = 250$$

$$2.3t = \ln 250$$

$$t = \frac{\ln 250}{2.3} \approx 2.4.$$

Therefore, after 2.4 hours there will be 1,000,000 bacteria.

5.2 COMPOUND INTEREST

In Section 0.5, we introduced the notion of compound interest. Recall the fundamental formula developed there. If a principal amount P is compounded m times per year at an annual rate of interest r for t years, then the compound amount A, the balance at the end of time t, is given by the formula:

$$A = P\left(1 + \frac{r}{m}\right)^{mt}$$

As an illustration of this formula, we showed in Example 6 of Section 0.5 that if $1,000 were compounded monthly at an annual interest rate of 6%, then the compound amount after a year would be $1061.68. We also showed that if the same principal amount were compounded daily, then the compound amount would be $1061.83. Would the total interest be much more if the interest were compounded every hour? Every minute? To answer these questions, let's connect the concept of compound interest with the exponential function. Start from the above formula for the compound amount and write it in the form:

$$A = P\left(1 + \frac{r}{m}\right)^{mt} = P\left(1 + \frac{r}{m}\right)^{(m/r)\cdot rt}$$

If we set $h = r/m$, then $1/h = m/r$, and

$$A = P(1 + h)^{(1/h)\cdot rt}.$$

As the frequency of compounding is increased, m gets large and $h = r/m$ approaches 0. To determine what happens to the compound amount, we must therefore examine the limit

$$\lim_{h\to 0} P(1 + h)^{(1/h)\cdot rt}.$$

The following remarkable fact is proved in the appendix at the end of this section:

$$\lim_{h\to 0} (1 + h)^{1/h} = e.$$

Using this fact together with two limit theorems, we have

$$\lim_{h\to 0} P(1 + h)^{(1/h)rt} = P\left[\lim_{h\to 0} (1 + h)^{1/h}\right]^{rt} = Pe^{rt}.$$

These calculations show that the compound amount calculated from the formula $P\left(1 + \frac{r}{m}\right)^{mt}$ gets closer to Pe^{rt} as the number m of interest periods per year is increased. When the formula

$$A = Pe^{rt} \tag{1}$$

is used to calculate the compound amount, we say that the interest is *compounded continuously*.

   We can now answer the question posed above: If $1,000 is deposited for 1 year in an account paying 6% per annum, then as the frequency of compounding is increased, the compound amount approaches that reached when interest is compounded continuously. In this case, that amount may be determined from the above formula with $P = 1000$, $r = .06$, $t = 1$ and thus equals:

$$1000e^{.06} \approx 1061.84 \text{ dollars}$$

Daily compounding of interest produces $1,061.83. Consequently, frequent compounding (such as every hour or every second) will produce at most one cent more.

In many computations, it is simpler to use the formula for interest compounded continuously that the formula for ordinary compound interest. In these instances, it is commonplace to use interest compounded continuously as an approximation to ordinary compound interest.

When interest is compounded continuously, the compound amount $A(t)$ is an exponential function of the number of years t that interest is earned, $A(t) = Pe^{rt}$. Hence $A(t)$ satisfies the differential equation

$$\frac{dA}{dt} = rA.$$

The rate of growth of the compound amount is proportional to the amount of money present. Since the growth comes from the interest, we conclude that under continuous compounding, interest is earned continuously at a rate of growth proportional to the amount of money present.

The formula $A = Pe^{rt}$ contains four variables. (Remember that the letter e here represents a specific constant, $e = 2.718. \ldots$.) In a typical problem, we are given values for three of these variables and must solve for the remaining variable.

EXAMPLE 1   How long is required for an investment of $1000 to double if the interest is 10%, compounded continuously?

Solution   Here $P = 1000$ and $r = .10$. For each time t, the value of the investment is $Pe^{rt} = 1000e^{.10t}$. We must find t such that this value is $2000. So we set

$$2000 = 1000e^{.10t}$$

and solve for t. We divide both sides by 1000 and then take logarithms of both sides to obtain

$$2 = e^{.10t}$$

$$\ln 2 = .10t,$$

and

$$t = 10 \ln 2 \approx 6.9 \text{ years.} \quad \bullet$$

Remark   The calculations in Example 4 would be essentially unchanged after the first step if the initial amount of the investment were changed from $1000 to any arbitrary amount P. When this investment doubles, the compound amount will be $2P$. So one sets $2P = Pe^{.10t}$ and solves for t as we did above, to conclude that any amount doubles in about 6.9 years.

If P dollars are invested today, the formula $A = Pe^{rt}$ gives the value of this investment after t years (assuming continuously compounded interest). We say that P is the *present value* of the amount A to be received in t years. If we solve

for P in terms of A, we obtain

$$P = Ae^{-rt}. \tag{2}$$

The concept of the present value of money is an important theoretical tool in business and economics. Problems involving depreciation of equipment, for example, may be analyzed by calculus techniques when the present value of money is computed from (2) using continuously compounded interest.

EXAMPLE 2   Find the present value of $5000 to be received in 2 years if money can be invested at 12% compounded continuously.

Solution   Use (2) with $A = 5000$, $r = .12$, and $t = 2$.

$$P = 5000e^{-(.12)(2)} = 5000e^{-.24}$$
$$\approx 5000(.78663) = 3933.15 \text{ dollars.} \quad \bullet$$

TECHNOLOGY PROJECT 5.2
Internal Rate of Return on an Investment

Suppose that an investment of $2000 yields payments of $2000 in 3 years, $1000 in 4 years and $1000 in 5 years. Thereafter, the investment is worthless. What constant rate of return r would the investment need to produce in order to yield the payments specified? The number r is called the **internal rate of return** on the investment. We can consider the investment as consisting of three parts, one part yielding each payment. The sum of the present values of the three parts must total $2000. This yields the equation:

$$2000 = 2000e^{-3r} + 1000e^{-4r} + 1000e^{-5r}$$

Solve this equation using the TRACE.

 APPENDIX A LIMIT FORMULA FOR e

For $h \neq 0$, we have

$$\ln(1 + h)^{1/h} = (1/h) \ln(1 + h).$$

Taking the exponential of both sides, we find that

$$(1 + h)^{1/h} = e^{(1/h)\ln(1+h)}.$$

Since the exponential function is continuous,

$$\lim_{h \to 0} (1 + h)^{1/h} = e^{\left[\lim_{h \to 0} (1/h)\ln(1+h)\right]}. \tag{3}$$

To examine the limit inside the exponential function, we note that $\ln 1 = 0$, and

hence

$$\lim_{h \to 0} \left(\frac{1}{h}\right) \ln(1 + h) = \lim_{h \to 0} \frac{\ln(1 + h) - \ln 1}{h}.$$

The limit on the right is a difference quotient of the type used to compute a derivative. In fact,

$$\lim_{h \to 0} \frac{\ln(1 + h) - \ln 1}{h} = \frac{d}{dx} \ln x \Big|_{x=1}$$

$$= \frac{1}{x} \Big|_{x=1} = 1.$$

Thus the limit inside the exponential function in (5) is 1. That is,

$$\lim_{h \to 0} (1 + h)^{1/h} = e^{[1]} = e.$$

PRACTICE PROBLEMS 5.2

1. One thousand dollars is to be invested in a bank for 4 years. Would 8% interest compounded semiannually be better than $7\frac{3}{4}\%$ interest compounded continuously?

2. A building was bought for $150,000 and sold 10 years later for $400,000. What interest rate (compounded continuously) was earned on the investment?

EXERCISES 5.2

1. Suppose that $1000 is deposited in a savings account at 10% interest compounded annually. What is the compound amount after 2 years?

2. Five thousand dollars is deposited in a savings account at 6% interest compounded monthly. Give the formula that describes the compound amount after 4 years.

3. Ten thousand dollars is invested at 8% interest compounded quarterly. Give the formula that describes the value of the investment after 3 years.

4. What is the effective annual rate of interest of a savings account paying 8% interest compounded semiannually?

5. One thousand dollars is invested at 14% interest compounded continuously. Compute the value of the investment at the end of 6 years.

6. A painting was purchased in 1983 for $100,000. If it appreciated at 12% compounded continuously, how much was it worth in 1990?

7. Five hundred dollars is deposited in a savings account paying 7% interest compounded daily. *Estimate* the balance in the account at the end of 3 years.

8. Ten thousand dollars is deposited into a money market fund paying 8% interest compounded continuously. How much interest will be earned during the first half-year if this rate of 8% does not change?

9. An office building, built in 1979 at a cost of 10 million dollars, was appraised at 25 million dollars in 1987. At what annual rate of interest (compounded continuously) did the building appreciate?

10. One thousand dollars is deposited in a savings account at 6% interest compounded continuously. How many years are required for the balance in the account to reach $2500?

11. You have ten thousand dollars to be invested in a highly speculative venture for 1 year. Would you rather receive 40% interest compounded semiannually or 39% interest compounded continuously?

12. A lot purchased in 1966 for $5000 was appraised at $60,000 in 1985. If the lot continued to appreciate at the same rate, when was it worth $100,000?

13. Ten thousand dollars is invested at 15% interest compounded continuously. When will the investment be worth $38,000?

14. How many years are required for an investment to double in value if it is appreciating at the rate of 13% compounded continuously?

15. A farm purchased in 1975 for $1,000,000 was valued at $3,000,000 in 1985. If the farm continues to appreciate at the same rate (with continuous compounding), when will it be worth $10,000,000?

16. Find the present value of $1000 payable at the end of 3 years, if money may be invested at 14% with interest compounded continuously.

17. Find the present value of $2000 to be received in 10 years, if money may be invested at 15% with interest compounded continuously.

18. A parcel of land bought in 1985 for $10,000 was worth $16,000 in 1990. If the land continues to appreciate at this rate, in what year will it be worth $45,000?

19. One hundred dollars is deposited in a savings account at 7% interest compounded continuously. What is the effective annual rate of return?

20. In a certain town, property values tripled from 1980 to 1991. If this trend continues, when were property values be at five times their 1980 level? (Use an exponential model for the property value at time t.)

21. How much money must you invest now at 12% interest compounded continuously, in order to have $10,000 at the end of 5 years?

22. Investment A is currently worth $70,200 and is growing at the rate of 13% per year compounded continuously. Investment B is currently worth $60,000 and is growing at the rate of 14% per year compounded continuously. After how many years will the two investments have the same value?

23. Suppose that the present value of $1000 to be received in 2 years is $559.90. What rate of interest, compounded continuously, was used to compute this present value?

24. Two thousand dollars is deposited in a savings account at 5% interest compounded continuously. Let $f(t)$ be the compound amount after t years. Find and interpret $f'(2)$.

25. A small amount of money is deposited in a savings account with interest compounded continuously. Let $A(t)$ be the balance in the account after t years. Match each of the following answers with its corresponding question.

    Answers: **a.** Pe^{rt}
                b. $A(3)$
                c. $A(0)$
                d. $A'(3)$
                e. Solve $A'(t) = 3$ for t.
                f. Solve $A(t) = 3$ for t.
                g. $y' = ky$
                h. Solve $A(t) = 3A(0)$ for t.

Questions: **A.** How fast will the balance be growing in 3 years?
   B. Give the general form of the function $A(t)$.
   C. How long will it take for the initial deposit to triple?
   D. Find the balance after 3 years.
   E. When will the balance be 3 dollars?
   F. When will the balance be growing at the rate of 3 dollars per year?
   G. What was the principal amount?
   H. Give a differential equation satisfied by $A(t)$.

26. When \$1000 is deposited into the bank at $r\%$ interest for ten years, the balance is $f(r)$ dollars, where f is the function shown in Fig. 1.

   (a) What will the balance be at 6% interest?

   (b) For what interest rate will the balance be \$2600?

   (c) If the interest rate is 8% and interest is increased by 1%, by approximately how much will the balance increase?

FIGURE I

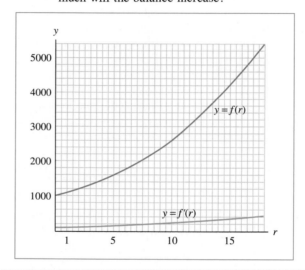

SOLUTIONS TO PRACTICE PROBLEMS 5.2

1. Let us compute the balance after 4 years for each type of interest.

   8% compounded semiannually: Use formula (2). Here $P = 1000$, $r = .08$, $m = 2$ (semiannually means that there are two interest periods per year), and $t = 4$. Therefore,

   $$A = 1000\left(1 + \frac{.08}{2}\right)^{2 \cdot 4}$$

   $$= 1000(1.04)^8 = 1368.57.$$

   $7\frac{3}{4}\%$ *compounded continuously:* Use the formula $A = Pe^{rt}$, where $P = 1000$, $r = .0775$, and $t = 4$. Then

   $$A = 1000e^{(.0775) \cdot 4}$$

   $$= 1000e^{.31} = 1363.43.$$

   Therefore, 8% compounded semiannually is best.

2. If the \$150,000 had been compounded continuously for 10 years at interest rate r, the balance would be $150,000e^{r \cdot 10}$. The question asks: For what value of r will the balance be 400,000? We need just solve an equation for r.

$$150,000e^{r \cdot 10} = 400,000$$

$$e^{r \cdot 10} \approx 2.67$$

$$r \cdot 10 = \ln 2.67$$

$$r = \frac{\ln 2.67}{10} \approx .098.$$

Therefore, the investment earned 9.8% interest per year.

5.3 APPLICATIONS OF THE NATURAL LOGARITHM FUNCTIONS TO ECONOMICS

In this section we consider two applications of the natural logarithm to the field of economics. Our first application is concerned with relative rates of change and the second with elasticity of demand.

Relative Rates of Change   The *logarithmic derivative* of a function $f(t)$ is defined by the equation

$$\frac{d}{dt} \ln f(t) = \frac{f'(t)}{f(t)} \tag{1}$$

The quantity on either side of equation (1) is often called the *relative rate of change of $f(t)$ per unit change of t*. Indeed, this quantity compares the rate of change of $f(t)$ [namely, $f'(t)$] with $f(t)$ itself]. The *percentage rate of change* is the relative rate of change of $f(t)$ expressed as a percentage.

A simple example will illustrate these concepts. Suppose that $f(t)$ denotes the average price per pound of sirloin steak at time t and $g(t)$ denotes the average price of a new car (of a given make and model) at time t, where $f(t)$ and $g(t)$ are given in dollars and time is measured in years. Then the ordinary derivatives $f'(t)$ and $g'(t)$ may be interpreted as the rate of change of the price of a pound of sirloin steak and of a new car, respectively, where both are measured in dollars per year. Suppose that, at a given time t_0, we have $f(t_0) = \$5.25$ and $g(t_0) = \$12,000$. Moreover, suppose that $f'(t_0) = \$.75$ and $g'(t_0) = \$1500$. Then at time t_0 the price per pound of steak is increasing at a rate of \$.75 per year, while the price of a new car is increasing at a rate of \$1500 per year. Which price is increasing more quickly? It is not meaningful to say that the car price is increasing faster simply because \$1500 is larger than \$.75. We must take into account the vast difference between the actual cost of a car and the cost of steak.

The usual basis of comparison of price increases is the percentage rate of increase. In other words, at $t = t_0$, the price of sirloin steak is increasing at the percentage rate

$$\frac{f'(t_0)}{f(t_0)} = \frac{.75}{5.25} \approx .143 = 14.3\%$$

per year, but at the same time the price of a new car is increasing at the percentage rate

$$\frac{g'(t_0)}{g(t_0)} = \frac{1500}{12,000} \approx .125 = 12.5\%$$

per year. Thus the price of sirloin steak is increasing at a faster percentage rate than the price of a new car.

Economists often use relative rates of change (or percentage rates of change) when discussing the growth of various economic quantities, such as national income or national debt, because such rates of change can be meaningfully compared.

EXAMPLE 1 Suppose that a certain school of economists modeled the Gross National Product of the United States at time t (measured in years from January 1, 1990) by the formula

$$f(t) = 3.4 + .04t + .13e^{-t},$$

where the Gross National Product is measured in trillions of dollars. What was the predicted percentage rate of growth (or decline) of the economy at $t = 0$ and $t = 1$?

Solution Since

$$f'(t) = .04 - .13e^{-t},$$

we see that

$$\frac{f'(0)}{f(0)} = \frac{.04 - .13}{3.4 + .13} = -\frac{.09}{3.53} \approx -2.6\%.$$

$$\frac{f'(1)}{f(1)} = \frac{.04 - .13e^{-1}}{3.4 + .04 + .13e^{-1}} = -\frac{.00782}{3.4878} \approx -.2\%.$$

So on January 1, 1990, the economy is predicted to contract at a relative rate of 2.6% per year; on January 1, 1991, the economy is predicted to be still contracting but only at a relative rate of .2% per year. ●

EXAMPLE 2 Suppose that the value in dollars of a certain business investment at time t may be approximated empirically by the function $f(t) = 750,000e^{.6\sqrt{t}}$. Use a logarithmic derivative to describe how fast the value of the investment is increasing when $t = 5$ years.

Solution   We have

$$\frac{f'(t)}{f(t)} = \frac{d}{dt} \ln f(t) = \frac{d}{dt} (\ln 750{,}000 + \ln e^{.6\sqrt{t}})$$

$$= \frac{d}{dt} (\ln 750{,}000 + .6\sqrt{t})$$

$$= (.6)\left(\frac{1}{2}\right)t^{-1/2} = \frac{.3}{\sqrt{t}}.$$

When $t = 5$,

$$\frac{f'(5)}{f(5)} = \frac{.3}{\sqrt{5}} \approx .1345 = 13.4\%.$$

Thus, when $t = 5$ years, the value of the investment is increasing at the relative rate of 13.4% per year. •

In certain mathematical models, it is assumed that for a limited period of time, the percentage rate of change of a particular function is constant. The following example shows that such a function must be an exponential function.

EXAMPLE 3   Suppose that the function $f(t)$ has a constant relative rate of change k. Show that $f(t) = Ce^{kt}$ for some constant C.

Solution   We are given that

$$\frac{d}{dt} \ln f(t) = k.$$

That is,

$$\frac{f'(t)}{f(t)} = k.$$

Hence $f'(t) = kf(t)$. But this is just the differential equation satisfied by the exponential function. Therefore, we must have $f(t) = Ce^{kt}$ for some constant C. •

Elasticity of Demand   In Section 2.7 we considered demand equations for monopolists and for entire industries. Recall that a demand equation expresses, for each quantity q to be produced, the market price that will generate a demand of exactly q. For instance, the demand equation

$$p = 150 - .01x \tag{2}$$

says that in order to sell x units, the price must be set at $150 - .01x$ dollars. To be specific: In order to sell 6000 units, the price must be set at $150 - .01(6000) = \$90$ per unit.

Equation (2) may be solved for x in terms of p to yield

$$x = 100(150 - p). \tag{3}$$

This last equation gives quantity in terms of price. If we let the letter q represent quantity, equation (3) becomes

$$q = 100(150 - p). \tag{3'}$$

This equation is of the form $q = f(p)$, where in this case $f(p)$ is the function $f(p) = 100(150 - p)$. In what follows it will be convenient to always write our demand functions so that the quantity q is expressed as a function $f(p)$ of the price p.

Usually, raising the price of a commodity lowers demand. Therefore, the typical demand function $q = f(p)$ is decreasing and has a negative slope everywhere.

A demand function $q = f(p)$ relates the quantity demanded to the price. Therefore, the derivative $f'(p)$ compares the change in quantity demanded with the change in price. By way of contrast, the concept of elasticity is designed to compare the *relative* rate of change of the quantity demanded with the *relative* rate of change of price.

Let us be more explicit. Consider a particular demand function $q = f(p)$ and a particular price p. Then at this price, the ratio of the relative rates of change of the quantity demanded and the price is given by

$$\frac{[\text{relative rate of change of quantity}]}{[\text{relative rate of change of price}]} = \frac{\dfrac{d}{dp} \ln f(p)}{\dfrac{d}{dp} \ln p}$$

$$= \frac{f'(p)/f(p)}{1/p}$$

$$= \frac{pf'(p)}{f(p)}.$$

Since $f'(p)$ is always negative for a typical demand function, the quantity $pf'(p)/f(p)$ will be negative for all values of p. For convenience, economists prefer to work with positive numbers and therefore the *elasticity of demand* is taken to be this quantity multiplied by -1.

The elasticity of demand $E(p)$ at price p for the demand function $q = f(p)$ is defined to be

$$E(p) = \frac{-pf'(p)}{f(p)}.$$

EXAMPLE 4   Suppose that the demand function for a certain metal is $q = 100 - 2p$, where p is the price per pound and q is the quantity demanded (in millions of pounds).

(a) What quantity can be sold at $30 per pound?

(b) Determine the function $E(p)$.

(c) Determine and interpret the elasticity of demand at $p = 30$.

(d) Determine and interpret the elasticity of demand at $p = 20$.

Solution (a) In this case, $q = f(p)$, where $f(p) = 100 - 2p$. When $p = 30$, we have $q = f(30) = 100 - 2(30) = 40$. Therefore, 40 million pounds of the metal can be sold. We also say that the *demand* is 40 million pounds.

(b)
$$E(p) = \frac{-pf'(p)}{f(p)}$$

$$= \frac{-p(-2)}{100 - 2p}$$

$$= \frac{2p}{100 - 2p}.$$

(c) The elasticity of demand at price $p = 30$ is $E(30)$.

$$E(30) = \frac{2(30)}{100 - 2(30)}$$

$$= \frac{60}{40}$$

$$= \frac{3}{2}.$$

When the price is set at $30 per pound, a small increase in price will result in a relative rate of decrease in quantity demanded of about $\frac{3}{2}$ times the relative rate of increase in price. For example, if the price is increased from $30 by 1%, then the quantity demanded will decrease by about 1.5%.

(d) When $p = 20$, we have

$$E(20) = \frac{2(20)}{100 - 2(20)} = \frac{40}{60} = \frac{2}{3}.$$

When the price is set at $20 per pound, a small increase in price will result in a relative rate of decrease in quantity demanded of only $\frac{2}{3}$ of the relative rate of increase of price. For example, if the price is increased from $20 by 1%, the quantity demanded will decrease by $\frac{2}{3}$ of 1%. ●

Economists say that demand is *elastic* at price p_0 if $E(p_0) > 1$ and *inelastic* at price p_0 if $E(p_0) < 1$. In Example 4, the demand for the metal is elastic at $30 per pound and inelastic at $20 per pound.

The significance of the concept of elasticity may perhaps best be appreciated by studying how revenue, $R(p)$, responds to changes in price. Recall that

$$[\text{revenue}] = [\text{quantity}] \cdot [\text{price per unit}],$$

that is,
$$R(p) = f(p) \cdot p.$$

If we differentiate $R(p)$ using the product rule, we find that

$$R'(p) = \frac{d}{dp}[f(p) \cdot p] = f(p) \cdot 1 + p \cdot f'(p)$$

$$= f(p)\left[1 + \frac{pf'(p)}{f(p)}\right]$$

$$= f(p)[1 - E(p)]. \qquad (4)$$

Now suppose that demand is elastic at some price p_0. Then $E(p_0) > 1$ and $1 - E(p_0)$ is negative. Since $f(p)$ is always positive, we see from (4) that $R'(p_0)$ is negative. Therefore, by the first derivative rule, $R(p)$ is decreasing at p_0. So an increase in price will result in a decrease in revenue, and a decrease in price will result in an increase in revenue. On the other hand, if demand is inelastic at p_0, then $1 - E(p_0)$ will be positive and hence $R'(p_0)$ will be positive. In this case an increase in price will result in an increase in revenue, and a decrease in price will result in a decrease in revenue. We may summarize this as follows:

> The change in revenue is in the opposite direction of the change in price when demand is elastic and in the same direction when demand is inelastic.

TECHNOLOGY PROJECT 5.3
Elasticity of Demand

Suppose that the number of cars using a toll road is described by the following demand function, where p is toll charged:

$$q = 40{,}000e^{-.5p} - 2000p + 5000, \ 0 \le p \le 4$$

1. Graph the demand function.
2. Graph the elasticity of demand $E(p)$.
3. Determine the tolls for which the demand is elastic and the tolls for which the demand is inelastic.
4. Graph the revenue function $R(p)$ on the same coordinate system as $E(p)$.
5. Determine the values of p for which $R(p)$ is increasing and those for which $R(p)$ is decreasing.

PRACTICE PROBLEMS 5.3

The current toll for the use of a certain toll road is \$2.50. A study conducted by the state highway department determined that with a toll of p dollars, q cars will use the road each day, where $q = 60{,}000e^{-.5p}$.

1. Compute the elasticity of demand at $p = 2.5$.
2. Is demand elastic or inelastic at $p = 2.5$?
3. If the state increases the toll slightly, will the revenue increase or decrease?

Determine the percentage rate of change of the functions at the points indicated.

1. $f(t) = t^2$ at $t = 10$ and $t = 50$

2. $f(t) = t^{10}$ at $t = 10$ and $t = 50$

3. $f(x) = e^{3x}$ at $x = 10$ and $x = 20$

4. $f(x) = e^{-.05x}$ at $x = 1$ and $x = 10$

5. $f(t) = e^{.3t}$ at $t = 1$ and $t = 5$

6. $G(s) = e^{-.05s^2}$ at $s = 1$ and $s = 10$

7. $f(p) = 1/(p + 2)$ at $p = 2$ and $p = 8$

8. $g(p) = 5/(2p + 3)$ at $p = 1$ and $p = 11$

9. Suppose that the annual sales S (in dollars) of a company may be approximated empirically by the formula

$$S = 50{,}000\sqrt{e}\sqrt{t},$$

where t is the number of years beyond some fixed reference date. Use a logarithmic derivative to determine the percentage rate of growth of sales at $t = 4$.

10. Suppose that the price of wheat per bushel at time t (in months) is approximated by

$$f(t) = 4 + .001t + .01e^{-t}.$$

What is the percentage rate of change of $f(t)$ at $t = 0$? $t = 1$? $t = 2$?

11. Suppose that an investment grows at a continuous 12% rate per year. In how many years will the value of the investment double?

12. Suppose that the value of a piece of property is growing at a continuous $r\%$ rate per year and that the value doubles in 3 years. Find r.

For each demand function, find $E(p)$ and determine if demand is elastic or inelastic (or neither) at the indicated price.

13. $q = 700 - 5p, p = 80$

14. $q = 600e^{-.2p}, p = 10$

15. $q = 400(116 - p^2), p = 6$

16. $q = (77/p^2) + 3, p = 1$

17. $q = p^2e^{-(p+3)}, p = 4$

18. $q = 700/(p + 5), p = 15$

19. Currently, 1800 people ride a certain commuter train each day and pay $4 for a ticket. The number of people q willing to ride the train at price p is $q = 600(5 - \sqrt{p})$. The railroad would like to increase its revenue.

(a) Is demand elastic or inelastic at $p = 4$?

(b) Should the price of a ticket be raised or lowered?

20. A company can sell $q = 9000/(p + 60) - 50$ radios at a price of p dollars per radio. The current price is $30.

(a) Is demand elastic or inelastic at $p = 30$?

(b) If the price is lowered slightly, will revenue increase or decrease?

21. A movie theater has a seating capacity of 3000 people. The number of people attending a show at price p dollars per ticket is $q = (18{,}000/p) - 1500$. Currently, the price is $6 per ticket.

(a) Is demand elastic or inelastic at $p = 6$?

(b) If the price is lowered, will revenue increase or decrease?

22. A subway charges 65 cents per person and has 10,000 riders each day. The demand function for the subway is $q = 2000\sqrt{90 - p}$.
    (a) Is demand elastic or inelastic at $p = 65$?
    (b) Should the price of a ride be raised or lowered in order to increase the amount of money taken in by the subway?

23. A country which is the major supplier of a certain commodity wishes to improve its balance of trade position by lowering the price of the commodity. The demand function is $q = 1000/p^2$.
    (a) Compute $E(p)$.
    (b) Will the country succeed in raising its revenue?

24. Show that any demand function of the form $q = a/p^m$ has constant elasticity m.

A cost function $C(x)$ gives the total cost of producing x units of a product. The *elasticity of cost at quantity x* is defined to be

$$E_c(x) = \frac{\dfrac{d}{dx} \ln C(x)}{\dfrac{d}{dx} \ln x}.$$

25. Show that $E_c(x) = x \cdot C'(x)/C(x)$.
26. Show that E_c is equal to the marginal cost divided by the average cost.
27. Let $C(x) = (1/10)x^2 + 5x + 300$. Show that $E_c(50) < 1$. (Hence when producing 50 units, a small relative increase in production results in an even smaller relative increase in total cost. Also, the average cost of producing 50 units is greater than the marginal cost at $x = 50$.)
28. Let $C(x) = 1000e^{.02x}$. Determine and simplify the formula for $E_c(x)$. Show that $E_c(60) > 1$ and interpret this result.

SOLUTIONS TO PRACTICE PROBLEMS 5.3

1. The demand function is $f(p) = 60{,}000e^{-.5p}$

$$f'(p) = -30{,}000e^{-.5p}$$

$$E(p) = \frac{-pf'(p)}{f(p)} = \frac{-p(-30{,}000)e^{-.5p}}{60{,}000e^{-.5p}} = \frac{p}{2}$$

$$E(2.5) = \frac{2.5}{2} = 1.25$$

2. The demand is elastic, because $E(2.5) > 1$.
3. Since demand is elastic at \$2.50, a slight change in price causes revenue to change in the *opposite* direction. Hence revenue will decrease.

5.4 FURTHER EXPONENTIAL MODELS

A skydiver, on jumping out of an airplane, falls at an increasing rate. However, the wind rushing past the skydiver's body creates an upward force that begins to counterbalance the downward force of gravity. This air friction finally becomes so great that the skydiver's velocity reaches a limiting speed called the *terminal velocity*. If we let $v(t)$ be the downward velocity of the skydiver after t seconds of free fall, then a good mathematical model for $v(t)$ is given by

$$v(t) = M(1 - e^{-kt}), \tag{1}$$

where M is the terminal velocity and k is some positive constant (Fig. 1). When t is close to zero, e^{-kt} is close to one and the velocity is small. As t increases, e^{-kt} becomes small and so $v(t)$ approaches M.

FIGURE I **Velocity of a sky-diver.**

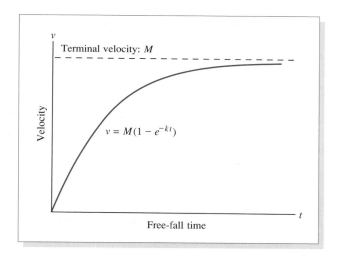

EXAMPLE I   Show that the velocity given in (1) satisfies the differential equation

$$\frac{dv}{dt} = k[M - v(t)], \qquad v(0) = 0. \tag{2}$$

Solution   From (1) we have $v(t) = M - Me^{-kt}$. Then

$$\frac{dv}{dt} = Mke^{-kt}.$$

However,

$$k[M - v(t)] = k[M - (M - Me^{-kt})] = kMe^{-kt},$$

so that the differential equation $\dfrac{dv}{dt} = k[M - v(t)]$ holds. Also,

$$v(0) = M - Me^0 = M - M = 0. \ \bullet$$

The differential equation (2) says that the rate of change in v is proportional to the difference between the terminal velocity M and the actual velocity v. It is not difficult to show that the only solution of (2) is given by the formula in (1).

The two equations (1) and (2) arise as mathematical models in a variety of situations. Some of these applications are described below.

The Learning Curve   Psychologists have found that in many learning situations a person's rate of learning is rapid at first and then slows down. Finally, as the task is mastered, the person's level of performance reaches a level above which it is almost impossible to rise. For example, within reasonable limits, each person seems to have a certain maximum capacity for memorizing a list of nonsense syllables. Suppose that a subject can memorize M syllables in a row if given sufficient time, say an hour, to study the list but cannot memorize $M + 1$ syllables in a row even if allowed several hours of study. By giving the subject different lists of syllables and varying lengths of time to study the lists, the psychologist can determine an empirical relationship between the number of nonsense syllables memorized accurately and the number of minutes of study time. It turns out that a good model for this situation is

$$y = M(1 - e^{-kt}) \tag{3}$$

for some appropriate positive constant k. (See Fig. 2.)

FIGURE 2 Learning curve, $y = M(1 - e^{-kt})$.

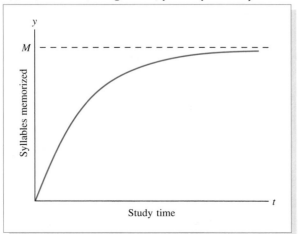

FIGURE 3 Diffusion of information by mass media.

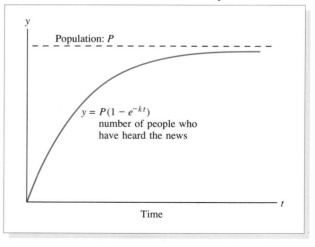

The *slope* of this learning curve at time t is approximately the number of additional syllables that can be memorized if the subject is given one more minute of study time. Thus the slope is a measure of the *rate of learning*. The differential equation satisfied by the function in (3) is

$$y' = k(M - y), \qquad f(0) = 0.$$

This equation says that if the subject is given a list of M nonsense syllables, then the rate of memorization is proportional to the number of syllables remaining to be memorized.

TECHNOLOGY PROJECT 5.4
The Learning Curve

Suppose that the maximum number of nonsense syllables that can be learned is 30 and that 10 can be learned in 1 hour.

1. Use a graphical approach to determine the function $y = f(t)$, the number of nonsense syllables that can be learned in time t hours.
2. Graph the function y.
3. Using TRACE, determine the number of nonsense syllables that can be learned in 3 hours.
4. Using TRACE, determine the length of time it takes to learn 20 nonsense syllables.

Diffusion of Information by Mass Media Sociologists have found that the differential equation (2) provides a good model for the way information is spread (or "diffused") through a population when the information is being propagated constantly by mass media, such as television or magazines.* Given a fixed population P, let $f(t)$ be the number of people who have already heard a certain piece of information by time t. Then $P - f(t)$ is the number who have not yet heard the information. Also, $f'(t)$ is the rate of increase of the number of people who have heard the news (the "rate of diffusion" of the information). If the information is being publicized often by some mass media, then it is likely that the number of *newly informed* people per unit time is proportional to the number of people who have not yet heard the news. Therefore,

$$f'(t) = k[P - f(t)].$$

Assume that $f(0) = 0$ (i.e., there was a time $t = 0$ when nobody had heard the news). Then the remark following Example 1 shows that

$$f(t) = P(1 - e^{-kt}). \tag{4}$$

(See Fig. 3.)

---

* J. Coleman, *Introduction to Mathematical Sociology* (New York: The Free Press, 1964), p. 43.

EXAMPLE 2   Suppose that a certain piece of news (such as the resignation of a public official) is broadcast frequently by radio and television stations. Also suppose that one-half of the residents of a city have heard the news within 4 hours of its initial release. Use the exponential model (4) to estimate when 90% of the residents will have heard the news.

Solution   We must find the value of k in (4). If P is the number of residents, then the number who will have heard the news in the first four hours is given by (4) with $t = 4$. By assumption, this number is half the population. So

$$\tfrac{1}{2}P = P(1 - e^{-k \cdot 4})$$
$$.5 = 1 - e^{-4k}$$
$$e^{-4k} = 1 - .5 = .5.$$

Solving for k, we find that $k \approx .17$. So the model for this particular situation is

$$f(t) = P(1 - e^{-.17t}).$$

Now we want to find t such that $f(t) = .90P$. We solve for t:

$$.90P = P(1 - e^{-.17t})$$
$$.90 = 1 - e^{-.17t}$$
$$e^{-.17t} = 1 - .90 = .10$$
$$-.17t = \ln .10$$
$$t = \frac{\ln .10}{-.17} \approx 14.$$

Therefore, 90% of the residents will hear the news within 14 hours of its initial release. ●

TECHNOLOGY PROJECT 5.5

The Yield Curve

Each day hundreds of billions of dollars worth of bonds are traded on the world's stock exchanges. At any given moment, bonds of different maturities yield different interest rates. Generally speaking, the longer before a bond is to be refunded, the higher the current interest rate yielded by the bond. If you graph the length of the bond maturity vs. its yield, you get a curve called the **yield curve**. Based on data published in Business Week in November 1993, the following equation was derived describing the yield curve: if x is the length of the bond maturity and y is the current interest rate yielded by the bond, then

$$y = 0.062 - 0.031162e^{-0.2188x}$$

1. Graph the yield curve described by the above equation.
2. What is the asymptote on the yield curve? What is the applied significance of the asymptote?
3. What is the yield on a bond that has 8 years left before maturity?
4. How long must the maturity be in order for the yield to be 5.95%?

Intravenous Infusion of Glucose   The human body both manufactures and uses glucose ("blood sugar"). Usually, there is a balance in these two processes, so that the bloodstream has a certain "equilibrium level" of glucose. Suppose that a patient is given a single intravenous injection of glucose and let $A(t)$ be the amount of glucose (in milligrams) above the equilibrium level. Then the body will start using up the excess glucose at a rate proportional to the amount of excess glucose; that is,

$$A'(t) = -\lambda A(t), \tag{5}$$

where λ is a positive constant called the *velocity constant of elimination*. This constant depends on how fast the patient's metabolic processes eliminate the excess glucose from the blood. Equation (5) describes a simple exponential decay process.

   Now suppose that, instead of a single shot, the patient receives a continuous intravenous infusion of glucose. A bottle of glucose solution is suspended above the patient, and a small tube carries the glucose down to a needle that runs into a vein. In this case, there are two influences on the amount of excess glucose in the blood: the glucose being added steadily from the bottle and the glucose being removed from the blood by metabolic processes. Let r be the rate of infusion of glucose (often from 10 to 100 milligrams per minute). If the body did not remove any glucose, the excess glucose would increase at a constant rate of r milligrams per minute; that is,

$$A'(t) = r. \tag{6}$$

Taking into account the two influences on $A'(t)$ described by (5) and (6), we can write

$$A'(t) = r - \lambda A(t). \tag{7}$$

If we let $M = r/\lambda$, then

$$A'(t) = \lambda(M - A(t)).$$

As stated in Example 1, a solution of this differential equation is given by

$$A(t) = M(1 - e^{-\lambda t}) = \frac{r}{\lambda}(1 - e^{-\lambda t}). \tag{8}$$

Reasoning as in Example 1, we conclude that the amount of excess glucose rises until it reaches a stable level. (See Fig. 4).

The Logistic Growth Curve   The model for simple exponential growth discussed in Section 5.1 is adequate for describing the growth of many types of populations, but obviously a population cannot increase exponentially forever. The simple exponential growth model becomes inapplicable when the environment begins to inhibit the growth of the population. The logistic growth curve is an important exponential model that takes into account some of the effects of the environment on a population (Fig. 5). For small values of t, the curve has the same basic shape as an exponential growth curve. Then when the population

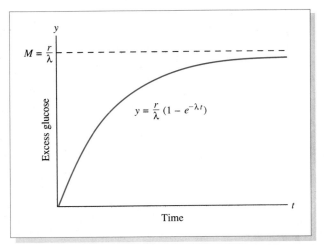

FIGURE 4   Continuous infusion of glucose.

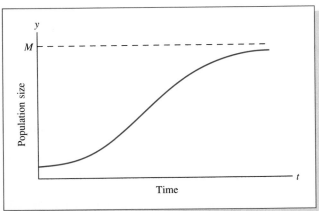

FIGURE 5   Logistic growth.

begins to suffer from overcrowding or lack of food, the growth rate (the slope of the population curve) begins to slow down. Eventually, the growth rate tapers off to zero as the population reaches the maximum size that the environment will support. This latter part of the curve resembles the growth curves studied earlier in this section.

The equation for logistic growth has the general form

$$y = \frac{M}{1 + Be^{-Mkt}}, \tag{9}$$

where B, M, and k are positive constants. We can show that y satisfies the differential equation

$$y' = ky(M - y). \tag{10}$$

The factor y reflects the fact that the growth rate (y') depends in part on the size y of the population. The factor $M - y$ reflects the fact that the growth rate also depends on how close y is to the maximum level M.

The logistic curve is often used to fit experimental data that lie along an "S-shaped" curve. Examples are given by the growth of a fish population in a lake and the growth of a fruit fly population in a laboratory container. Also, certain enzyme reactions in animals follow a logistic law. One of the earliest applications of the logistic curve occurred in about 1840 when the Belgian sociologist P. Verhulst fit a logistic curve to six U.S. census figures, 1790 to 1840, and predicted the U.S. population for 1940. His prediction missed by less than 1 million persons (an error of about 1%).

EXAMPLE 3   Suppose that a lake is stocked with 100 fish. After 3 months there are 250 fish. A study of the ecology of the lake predicts that the lake can support 1000 fish. Find a formula for the number $P(t)$ of fish in the lake t months after it has been stocked.

Solution The limiting population M is 1000. Therefore, we have

$$P(t) = \frac{1000}{1 + Be^{-1000kt}}.$$

At $t = 0$ there are 100 fish, so that

$$100 = P(0) = \frac{1000}{1 + Be^0} = \frac{1000}{1 + B}.$$

Thus $1 + B = 10$, or $B = 9$. Finally, since $P(3) = 250$, we have

$$250 = \frac{1000}{1 + 9e^{-3000k}}$$

$$1 + 9e^{-3000k} = 4$$

$$e^{-3000k} = \frac{1}{3}.$$

$$-3000k = \ln \frac{1}{3}$$

$$k \approx .00037.$$

Therefore,

$$P(t) = \frac{1000}{1 + 9e^{-.37t}}. \quad \bullet$$

Several theoretical justifications can be given for using (9) and (10) in situations where the environment prevents a population from exceeding a certain size. A discussion of this topic may be found in *Mathematical Models and Applications* by D. Maki and M. Thompson (Englewood Cliffs, N.J.: Prentice-Hall, Inc., 1973), pp. 312–317.

An Epidemic Model It will be instructive to actually "build" a mathematical model. Our example concerns the spread of a highly contagious disease. We begin by making several simplifying assumptions:

1. The population is a fixed number P and each member of the population is susceptible to the disease.
2. The duration of the disease is long, so that no cures occur during the time period under study.
3. All infected individuals are contagious and circulate freely among the population.
4. During each unit time period (such as 1 day or 1 week) each infected person makes c contacts, and each contact with an uninfected person results in transmission of the disease.

Consider a short period of time from t to $t + h$. Each infected person makes $c \cdot h$ contacts. How many of these contacts are with uninfected persons? If $f(t)$ is the number of infected persons at time t, then $P - f(t)$ is the number of uninfected persons, and $[P - f(t)]/P$ is the fraction of the population that is uninfected. Thus, of the $c \cdot h$ contacts made,

$$\left[\frac{P - f(t)}{P} \right] \cdot c \cdot h$$

will be with uninfected persons. This is the number of new infections produced by one infected person during the time period of length h. The total number of *new* infections during this period is

$$f(t) \left[\frac{P - f(t)}{P} \right] ch.$$

But this number must equal $f(t + h) - f(t)$, where $f(t + h)$ is the total number of infected persons at time $t + h$. So

$$f(t + h) - f(t) = f(t) \left[\frac{P - f(t)}{P} \right] ch.$$

Dividing by h, the length of the time period, we obtain the average number of new infections per unit time (during the small time period):

$$\frac{f(t + h) - f(t)}{h} = \frac{c}{P} f(t)[P - f(t)].$$

If we let h approach zero and let y stand for $f(t)$, the left-hand side approaches the rate of change in the number of infected persons and we derive the following equation:

$$\frac{dy}{dt} = \frac{c}{P} y(P - y). \tag{11}$$

This is the same type of equation as that used in (10) for logistic growth, although the two situations leading to this model appear to be quite dissimilar.

Comparing (11) with (10), we see that the number of infected individuals at time t is described by a logistic curve with $M = P$ and $k = c/P$. Therefore, by (9), we can write

$$f(t) = \frac{P}{1 + Be^{-ct}}.$$

B and c can be determined from the characteristics of the epidemic. (See Example 4.)

The logistic curve has an inflection point at that value of t for which $f(t) = P/2$. The position of this inflection point has great significance for applications of the logistic curve. From inspecting a graph of the logistic curve, we see that the inflection point is the point at which the curve has greatest slope. In other words, the inflection point corresponds to the instant of fastest growth of the logistic curve. This means, for example, that in the foregoing epidemic model the disease is spreading with the greatest rapidity precisely when half the population is infected. Any attempt at disease control (through immunization, for example) must strive to reduce the incidence of the disease to as low a point as possible, but in any case at least below the inflection point at $P/2$, at which point the epidemic is spreading fastest.

EXAMPLE 4   The Public Health Service monitors the spread of an epidemic of a particularly long-lasting strain of flu in a city of 500,000 people. At the beginning of the first week of monitoring, 200 cases have been reported; during the first week 300 new cases are reported. Estimate the number of infected individuals after 6 weeks.

Solution   Here $P = 500{,}000$. If $f(t)$ denotes the number of cases at the end of t weeks, then

$$f(t) = \frac{P}{1 + Be^{-ct}}$$

$$= \frac{500{,}000}{1 + Be^{-ct}}.$$

Moreover, $f(0) = 200$, so that

$$200 = \frac{500{,}000}{1 + Be^0} = \frac{500{,}000}{1 + B},$$

and $B = 2499$. Consequently, since $f(1) = 300 + 200 = 500$, we have

$$500 = f(1) = \frac{500{,}000}{1 + 2499e^{-c}},$$

so that $e^{-c} \approx .4$ and $c \approx .92$. Finally,

$$f(t) = \frac{500{,}000}{1 + 2499e^{-.92t}}$$

and

$$f(6) = \frac{500{,}000}{1 + 2499e^{-.92(6)}} \approx 45{,}000.$$

After 6 weeks, about 45,000 individuals are infected. ●

      This epidemic model is used by sociologists (who still call it an epidemic model) to describe the spread of a rumor. In economics the model is used to describe the diffusion of knowledge about a product. An "infected person" represents an individual who possesses knowledge of the product. In both cases, it is assumed that the members of the population are themselves primarily responsible for the spread of the rumor or knowledge of the product. This situation is in contrast to the model described earlier where information was spread through a population by external sources, such as radio and television.

There are several limitations to this epidemic model. Each of the four simplifying assumptions made at the outset is unrealistic in varying degrees. More complicated models can be constructed that rectify one or more of these defects, but they require more advanced mathematical tools.

TECHNOLOGY PROJECT 5.6
Spread of an Epidemic

Suppose that an epidemic of flu spreads among students at a certain college according to the logistic equation

$$N = \frac{4000}{1 + 50e^{-t}}$$

where N denotes the number who have been infected by time t days.

1. Graph the function N.
2. What are the asymptotes of the graph?
3. At what rate is the epidemic spreading 5 days after it begins?
4. When is the epidemic spreading fastest? (To answer this, it may be easiest to graph the derivative of N.)

The Exponential Function in Lung Physiology   Let us conclude this section by deriving a useful model for the pressure in a person's lungs when the air is allowed to escape passively from the lungs with no use of the person's muscles. Let V be the volume of air in the lungs and let P be the relative pressure in the lungs when compared with the pressure in the mouth. The *total compliance* (of the respiratory system) is defined to be the derivative $\frac{dV}{dP}$. For normal values of V and P we may assume that the total compliance is a positive constant, say C. That is,

$$\frac{dV}{dP} = C. \tag{12}$$

We shall assume that the airflow during the passive respiration is smooth and not turbulent. Then Poiseuille's law of fluid flow says that the rate of change of volume as a function of time (i.e., the rate of airflow) satisfies

$$\frac{dV}{dt} = -\frac{P}{R}, \tag{13}$$

where R is (positive) constant called the airway resistance. Under these conditions, we may derive a formula for P as a function of time. Since the volume is a function of the pressure, and the pressure is in turn a function of time, we may use the chain rule to write

$$\frac{dV}{dt} = \frac{dV}{dP} \cdot \frac{dP}{dt}.$$

From (12) and (13),

$$-\frac{P}{R} = C \cdot \frac{dP}{dt},$$

so

$$\frac{dP}{dt} = -\frac{1}{RC} \cdot P.$$

From this differential equation we conclude that P must be an exponential function of t. In fact,

$$P = P_0 e^{kt},$$

where P_0 is the initial pressure at time $t = 0$ and $k = -1/RC$. This relation between k and the product RC is useful to lung specialists, because they can experimentally compute k and the compliance C, and then use the formula $k = -1/RC$ to determine the airway resistance R.

TECHNOLOGY PROJECT 5.7

Analysis of the Effectiveness of an Insect Repellent

Human hands covered with cotton fabrics that had been impregnated with the insect repellent DEPA were inserted for five minutes into a test chamber containing 200 female mosquitos.* The function $f(x) = 26.48 - 14.09 \ln x$ gives the number of mosquito bites received when the concentration was x percent.

1. Graph $f(x)$ and $f'(x)$ for $0 \le x \le 4$.
2. How many bites were received when the concentration was 3.25%?
3. What concentration resulted in 15 bites?
4. At what rate is the number of bites changing with respect to concentration of DEPA when $x = 2.75$?
5. For what concentration does the rate of change of bites with respect to concentration equal -10 bites per percentage increase in concentration?

---

* Rao, K. M., Prakash, S., Kumar, S., Suryanarayana, M. V. S., Bhagwat, M. M., Gharia, M. M., and Bhavsar, R. B., "N-diethylphenylacetamide in Treated Fabrics as a Repellent Against Aedes aegypti and Culex quinquefasciatus (Deptera: Culiciae)," *Journal of Medical Entomology*, 28: 1 (January 1991).

TECHNOLOGY PROJECT 5.8
Analysis of the Absorption of a Drug

After a drug is taken orally, the amount of the drug in the bloodstream after t hours is $f(t) = 122(e^{-.2t} - e^{-t})$ units.

1. Graph $f(t), f'(t)$, and $f''(t)$ for $0 \le t \le 15$.
2. How many units of the drug are in the bloodstream after 7 hours?
3. At what rate is the level of drug in the bloodstream increasing after 1 hour?
4. Approximately when (while the level is decreasing) is the level of the drug in the bloodstream 20 units?
5. What is the greatest level of drug in the bloodstream and when is this level reached?
6. When is the level of the drug in the bloodstream decreasing the fastest?

TECHNOLOGY PROJECT 5.9
Analysis of the Growth of a Tumor

A cancerous tumor has volume $f(t) = 1.825^3(1 - 1.6e^{-.4196t})^3$ milliliters after t weeks, with $t > 1$.*

1. Graph $f(t), f'(t)$, and $f''(t)$ for $0 \le t \le 15$.
2. Approximately how large is the tumor after 14 weeks?
3. Appoximately when will the tumor have a volume of 4 milliliters?
4. Approximately how fast is the tumor growing after 5 weeks?
5. Approximately when is the tumor growing at the rate of .5 milliliters per week?
6. Approximately when is the tumor growing at the fastest rate?

TECHNOLOGY PROJECT 5.10
Growth of a Bacteria Culture with Growth Restrictions

A model incorporating growth restrictions for the number of bacteria in a culture after t days is given by:

$$f(t) = 9000(20 + te^{-.04t})$$

1. Graph $f(t), f'(t)$, and $f''(t)$ for $0 \le t \le 100$.
2. How large is the culture after 20 days?
3. How fast is the culture changing after 100 days?
4. Approximately when are there 230,000 bacteria?
5. Approximately when is the culture growing at the rate of 1500 bacteria per day?
6. When is the size of the culture decreasing the fastest?
7. When is the size of the culture greatest?

---

* Baker, Goddard, Clark, and Whimster, "Proportion of Necrosis in Transplanted Murine Adenocarcinomas and Its Relationship to Tumor Growth," *Growth, Development and Aging*, 54 (1990), 85–93.

PRACTICE
PROBLEM 5.4

1. A sociological study* was made to examine the process by which doctors decide to adopt a new drug. The doctors were divided into two groups. The doctors in group A had little interaction with other doctors and so received most of their information through mass media. The doctors in group B had extensive interaction with other doctors and so received most of their information through word of mouth. For each group, let $f(t)$ be the number who have learned about a new drug after t months. Examine the appropriate differential equations to explain why the two graphs were of the types shown in Fig. 6.

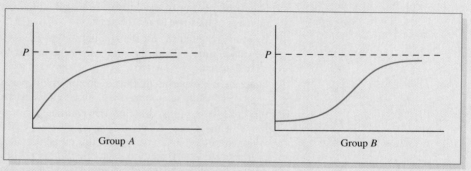

FIGURE 6   **Results of a sociological study**

EXERCISES 5.4

1. Consider the function $f(x) = 5(1 - e^{-2x})$, $x \geq 0$.
   (a) Show that $f(x)$ is increasing and concave down for all $x \geq 0$.
   (b) Explain why $f(x)$ approaches 5 as x gets large.
   (c) Sketch the graph of $f(x)$, $x \geq 0$.
2. Consider the function $g(x) = 10 - 10e^{-.1x}$, $x \geq 0$.
   (a) Show that $g(x)$ is increasing and concave down for $x \geq 0$.
   (b) Explain why $g(x)$ approaches 10 as x gets large.
   (c) Sketch the graph of $g(x)$, $x \geq 0$.
3. Suppose that $y = 2(1 - e^{-x})$. Compute y' and show that $y' = 2 - y$.
4. Suppose that $y = 5(1 - e^{-2x})$. Compute y' and show that $y' = 10 - 2y$.
5. Suppose that $f(x) = 3(1 - e^{-10x})$. Show that $y = f(x)$ satisfies the differential equation

$$y' = 10(3 - y), \qquad f(0) = 0.$$

*James S. Coleman, Eliku Katz, and Herbert Menzel, "The Diffusion of an Innovation Among Physicians," *Sociometry,* 20 (1957), 253–270.

6. (*Ebbinghaus Model for Forgetting*) Suppose that a student learns a certain amount of material for some class. Let $f(t)$ denote the percentage of the material that the student can recall t weeks later. The psychologist Ebbinghaus found that this percent retention can be modeled by a function of the form

$$f(t) = (100 - a)e^{-\lambda t} + a,$$

where λ and a are positive constants and $0 < a < 100$. Sketch the graph of the function $f(t) = 85e^{-.5t} + 15,\ t \geq 0$.

7. When a grand jury indicted the mayor of a certain town for accepting bribes, the newspaper, radio, and television immediately began to publicize the news. Within an hour, one-quarter of the citizens heard about the indictment. Estimate when three-quarters of the town heard the news.

8. Examine formula (8) for the amount $A(t)$ of excess glucose in the bloodstream of a patient at time t. Describe what would happen if the rate r of infusion of glucose were doubled.

9. Describe an experiment that a doctor could perform in order to determine the velocity constant of elimination of glucose for a particular patient.

10. Physiologists usually describe the continuous intravenous infusion of glucose in terms of the excess *concentration* of glucose, $C(t) = A(t)/V$, where V is the total volume of blood in the patient. In this case, the rate of increase in the concentration of glucose due to the continuous injection is r/V. Find a differential equation that gives a model for the rate of change of the excess concentration of glucose.

11. A news item is spread by word of mouth to a potential audience of 10,000 people. After t days,

$$f(t) = \frac{10,000}{1 + 50e^{-.4t}}$$

people will have heard the news. The graph of this function is shown in Fig. 7.
(a) Approximately how many people will have heard the news after 7 days?
(b) At approximately what rate will the news be spreading after 10 days?

FIGURE 7

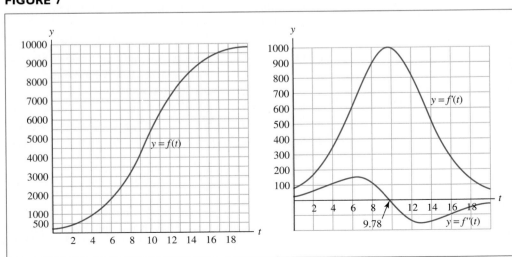

(c) Approximately when will 7000 people have heard the news?

(d) Approximately when will the news be spreading at the rate of 600 people per day?

(e) When will the news be spreading at the greatest rate?

12. The speed of a parachutist during free fall is

$$f(t) = 65(1 - e^{-.165t})$$

meters per second. The graph of this function is shown in Fig. 8.

(a) How fast is the parachutist falling after 25 seconds?

(b) What is the parachutist's acceleration after 10 seconds?

(c) After approximately how many seconds is the parachutist falling at the speed of 56 meters per second?

(d) Approximately when is the parachutist accelerating at 4 meters per sec².

(e) What is the terminal speed, that is, the greatest speed that can be approached by the parachutist?

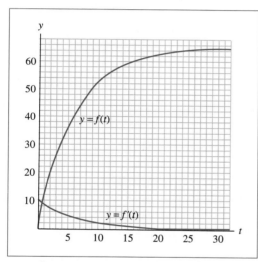

FIGURE 8   **FIGURE 9**

13. The length of a certain weed after t weeks is

$$f(t) = \frac{30}{1 + 25e^{-.5t}}$$

centimeters. The graph of this function is shown in Fig. 9.

(a) What is the length of the weed after 1 week?

(b) How fast is the weed growing after 10 weeks?

(c) Approximately when is the weed 10 centimeters long?

(d) Approximately when is the weed growing at the rate of 2 centimeters per day?

(e) When is the weed growing at the greatest rate? Approximately how long is the weed at that time?

14. A news item is broadcast by mass media to a potential audience of 50,000 people. After t days,

$$f(t) = 50000(1 - e^{-.3t})$$

people will have heard the news. The graph of this function is shown in Fig. 10.

(a) How many people will have heard the news after 10 days?

(b) At what rate is the news spreading initially?

(c) When will 22,500 people have heard the news?

(d) Approximately when will the news be spreading at the rate of 2500 people per day?

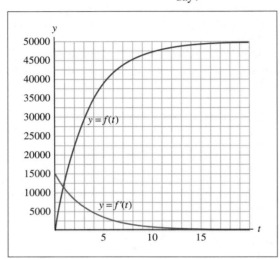

FIGURE 10

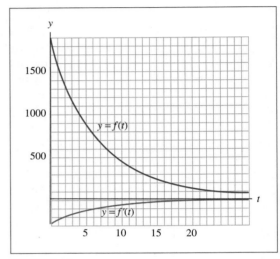

FIGURE 11

15. When a rod of molten steel with a temperature of 1800° F is placed in a large vat of water at temperature 60° F, the temperature of the rod after t seconds is

$$f(t) = 60(1 + 29e^{-.15t})$$

degrees Fahrenheit. The graph of this function is shown in Fig. 11.

(a) What is the temperature of the rod after 11 seconds?

(b) At what rate is the temperature of the rod changing after 6 seconds?

(c) Approximately when is the temperature of the rod 200 degrees?

(d) Approximately when is the rod cooling at the rate of 200 degrees per second?

SOLUTION TO PRACTICE PROBLEM 5.4

1. The difference between transmission of information via mass media and via word of mouth is that in the second case the rate of transmission depends not only on the number of people who have not yet received the information, but also on the number of people who know the information and therefore are capable of spreading it. Therefore, for group A, $f'(t) = k[P - f(t)]$, and for group B, $f'(t) = kf(t)[P - f(t)]$. Note that the spread of information by word of mouth follows the same pattern as the spread of an epidemic.

CHAPTER 5
CHECKLIST

✓ $y' = ky$ has the solution $y = Ce^{kt}$
✓ Exponential growth
✓ Exponential decay
✓ Half-life of a radioactive element
✓ Continuous compounding of interest

✓ Present value of money
✓ Percentage rate of change
✓ Elasticity of demand
✓ $y = M(1 - e^{-kt})$
✓ $y = M/(1 + Be^{-Mkt})$(logistic growth)

CHAPTER 5   SUPPLEMENTARY EXERCISES

1. The atmospheric pressure $P(x)$ (measured in inches of mercury) at height x miles above sea level satisfies the differential equation $P'(x) = -.2P(x)$. Find the formula for $P(x)$ if the atmospheric pressure at sea level is 29.92.

2. The herring gull population in North America has been doubling every 13 years since 1900. Give a differential equation satisfied by $P(t)$, the population t years after 1900.

3. Find the present value of $10,000 payable at the end of 5 years if money can be invested at 12% with interest compounded continuously.

4. One thousand dollars is deposited in a savings account at 10% interest compounded continuously. How many years are required for the balance in the account to reach $3000?

5. The half-life of the radioactive element tritium is 12 years. Find its decay constant.

6. A piece of charcoal found at Stonehenge contained 63% of the level of ^{14}C found in living trees. Approximately how old is the charcoal?

7. From 1980 to 1990, the population of Texas grew from 14.2 million to 17 million.
   (a) Give the formula for the population t years after 1980.
   (b) If this growth rate continues, how large will the population be in 2000?
   (c) In what year will the population reach 25 million?

8. A stock portfolio increased in value from $100,000 to $117,000 in 2 years. What rate of interest, compounded continuously, did this investment earn?

9. An investor initially invests $10,000 in a speculative venture. Suppose that the investment earns 20% interest compounded continuously for 5 years and then 6% interest compounded continuously for 5 years thereafter.
   (a) How much does the $10,000 grow to after 10 years?
   (b) Suppose that the investor has the alternative of an investment paying 14% interest compounded continuously. Which investment is superior over a 10-year period, and by how much?

10. Two different bacteria colonies are growing near a pool of stagnant water. Suppose that the first colony initially has 1000 bacteria and doubles every 21 minutes. The second colony has 710,000 bacteria and doubles every 33 minutes. How much time will elapse before the first colony becomes as large as the second?

11. Find the percentage rate of change of the function $f(t) = 50e^{.2t^2}$ at $t = 10$.

12. Find $E(p)$ for the demand function $q = 4000 - 40p^2$, and determine if demand is elastic or inelastic at $p = 5$.

13. Suppose that for a certain demand function, $E(8) = 1.5$. If the price is increased to $8.16, estimate the percentage decrease in the quantity demanded. Will the revenue increase or decrease?

14. Find the percentage rate of change of the function $f(p) = \dfrac{1}{3p + 1}$ at $p = 1$.

15. A company can sell $q = 1000p^2 e^{-.02(p+5)}$ calculators at a price of p dollars per calculator. The current price is \$200. If the price is decreased, will the revenue increase or decrease?

16. Consider a demand function of the form $q = ae^{-bp}$, where a and b are positive numbers. Find $E(p)$ and show that the elasticity equals 1 when $p = 1/b$.

17. Refer to Practice Problem 5.4. Out of 100 doctors in group A, none knew about the drug at time $t = 0$, but 66 of them were familiar with the drug after 13 months. Find the formula for $f(t)$.

18. The growth of the yellow nutsedge weed is described by a formula $f(t)$ of type (9) in Section 5.4. A typical weed has length 8 centimeters after 9 days, length 48 centimeters after 25 days, and reaches length 55 centimeters at maturity. Find the formula for $f(t)$.

19. The amount (in grams) of a certain radioactive material present after t years is given by the function $P(t)$. Match each of the following answers with its corresponding question.

Answers: **a.** Solve $P(t) = .5P(0)$ for t.
            b. Solve $P(t) = .5$ for t.
            c. $P(.5)$
            d. $P'(.5)$
            e. $P(0)$
            f. Solve $P'(t) = .5$ for t.
            g. $y' = ky$
            h. $P_0 e^{kt},\ k < 0$

Questions: **A.** Give a differential equation satisfied by $P(t)$.
            B. How fast will the radioactive material be disintegrating in $\frac{1}{2}$ year?
            C. Give the general form of the function $P(t)$.
            D. Find the half-life of the radioactive material.
            E. How many grams of the material will remain after $\frac{1}{2}$ year?
            F. When will the radioactive material be disintegrating at the rate of $\frac{1}{2}$ gram per year?
            G. When will there be $\frac{1}{2}$ gram remaining?
            H. How much radioactive material was present initially?

CHAPTER 6

THE DEFINITE INTEGRAL

There are two fundamental problems of calculus. The first is to find the slope of a curve at a point, and the second is to find the area of a region under a curve. These problems are quite simple when the curve is a straight line, as in Fig. 1. Both the slope of the line and the area of the shaded trapezoid can be calculated by geometric principles. When the graph consists of several line segments, as in Fig. 2, the slope of each line segment can be

FIGURE 1

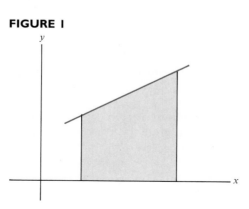

FIGURE 2

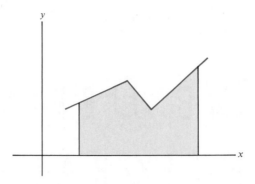

computed separately, and the area of the region can be found by adding the areas of the regions under each line segment.

Calculus is needed when the curves are not straight lines. We have seen that the slope problem is resolved with the derivative of a function. In this chapter, we describe how the area problem is connected with the notion of the "integral" of a function. Both the slope problem and the area problem were studied by the ancient Greeks and solved in special cases, but it was not until the development of calculus in the seventeenth century that the intimate connection between the two problems was discovered. In this chapter we will discuss this connection as stated in the Fundamental Theorem of Calculus.

 6.1 ANTIDIFFERENTIATION

We have developed several techniques for calculating the derivative $F'(x)$ of a function $F(x)$. In many applications, however, it is necessary to proceed in reverse. We are given the derivative $F'(x)$ and must determine the function $F(x)$. The process of determining $F(x)$ from $F'(x)$ is called *antidifferentiation*. The next example gives a typical application involving antidifferentiation.

EXAMPLE 1   During the early 1970s, the annual worldwide rate of oil consumption grew exponentially with a growth constant of about .07. At the beginning of 1970, the rate was about 16.1 billion barrels of oil per year. Let $R(t)$ denote the rate of oil consumption at time t, where t is the number of years since the beginning of 1970. Then a reasonable model for $R(t)$ is given by

$$R(t) = 16.1e^{.07t}. \tag{1}$$

Use this formula for $R(t)$ to determine the total amount of oil that would have been consumed from 1970 to 1980 had this rate of consumption continued throughout the decade.

Solution   Let $T(t)$ be the total amount of oil consumed from time 0 (1970) until time t. We wish to calculate $T(10)$, the amount of oil consumed from 1970 to 1980. We do this by first determining a formula for $T(t)$. Since $T(t)$ is the total oil consumed, the derivative $T'(t)$ is the *rate* of oil consumption, namely, $R(t)$. Thus, although we do not yet have a formula for $T(t)$, we do know that

$$T'(t) = R(t).$$

Thus the problem of determining a formula for $T(t)$ has been reduced to a problem of antidifferentiation: Find a function whose derivative is $R(t)$. We shall solve this particular problem after developing some techniques for solving antidifferentiation problems in general. ●

Suppose that $f(x)$ is a given function and $F(x)$ is a function having $f(x)$ as its derivative—that is, $F'(x) = f(x)$. We call $F(x)$ an *antiderivative* of $f(x)$.

EXAMPLE 2   Find an antiderivative of $f(x) = x^2$.

Solution   One such function is $F(x) = \frac{1}{3}x^3$, since

$$F'(x) = \frac{1}{3} \cdot 3x^2 = x^2.$$

Another antiderivative is $F(x) = \frac{1}{3}x^3 + 2$, since

$$\frac{d}{dx}\left(\frac{1}{3}x^3 + 2\right) = \frac{1}{3} \cdot 3x^2 + 0 = x^2.$$

In fact, if C is any constant, the function $F(x) = \frac{1}{3}x^3 + C$ is also an antiderivative of x^2, since

$$\frac{d}{dx}\left(\frac{1}{3}x^3 + C\right) = \frac{1}{3} \cdot 3x^2 + 0 = x^2.$$

(The derivative of a constant function is zero.) ●

EXAMPLE 3   Find an antiderivative of the function $f(x) = 2x - (1/x^2)$.

Solution   Since

$$\frac{d}{dx}(x^2) = 2x \quad \text{and} \quad \frac{d}{dx}\left(\frac{1}{x}\right) = -\frac{1}{x^2},$$

we see that one antiderivative of $f(x)$ is given by

$$F(x) = x^2 + \frac{1}{x}.$$

However, any function of the form $x^2 + (1/x) + C$, C a constant, will do, since

$$\frac{d}{dx}\left(x^2 + \frac{1}{x} + C\right) = 2x - \frac{1}{x^2} + 0 = 2x - \frac{1}{x^2}. \quad ●$$

Using the same reasoning as in Examples 2 and 3, we see that if $F(x)$ is an antiderivative of $f(x)$, then so is $F(x) + C$, where C is any constant. Thus if we know one antiderivative $F(x)$ of a function $f(x)$, we can write down an infinite number by adding all possible constants C to $F(x)$. It turns out that in this way we obtain all antiderivatives of $f(x)$. That is, we have the following fundamental result.

Theorem I   If $F_1(x)$ and $F_2(x)$ are two antiderivatives of the same function $f(x)$, then $F_1(x)$ and $F_2(x)$ differ by a constant. In other words, there is a constant C such that

$$F_2(x) = F_1(x) + C.$$

Geometrically, the graph of any antiderivative $F_2(x)$ is obtained by shifting the graph of $F_1(x)$ vertically. See Fig. 1.

FIGURE I   Two antiderivatives of the same function.

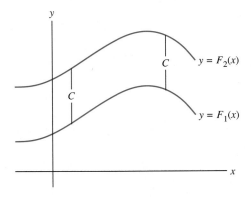

Our verification of Theorem I is based on the following fact, which is important in its own right.

Theorem II   If $F'(x) = 0$ for all x, then $F(x) = C$ for some constant C.

It is easy to see why Theorem II is reasonable. (A formal proof of the theorem requires an important theoretical result called the mean value theorem.) If $F'(x) = 0$ for all x, then the curve $y = F(x)$ has slope equal to zero at every point. Thus the tangent line to $y = F(x)$ at any point is horizontal, which implies that the graph of $y = F(x)$ is a horizontal line. (Try to draw the graph of a function with a horizontal tangent everywhere. There is no choice but to keep your pencil moving on a constant, horizontal line!) If the horizontal line is $y = C$, then $F(x) = C$ for all x.

Verification of Theorem I   If $F_1(x)$ and $F_2(x)$ are two antiderivatives of $f(x)$, then the function $F(x) = F_2(x) - F_1(x)$ has the derivative

$$F'(x) = F_2'(x) - F_1'(x)$$
$$= f(x) - f(x)$$
$$= 0.$$

So, by Theorem II, we know that $F(x) = C$ for some constant C. In other words, $F_2(x) - F_1(x) = C$, so that

$$F_2(x) = F_1(x) + C,$$

which is Theorem I.

Using Theorem I, we can find *all* antiderivatives of a given function once we know one antiderivative. For instance, since one antiderivative of x^2 is $\frac{1}{3}x^3$ (Example 2), all antiderivatives of x^2 have the form $\frac{1}{3}x^3 + C$, where C is a constant.

Suppose that $f(x)$ is a function whose antiderivatives are $F(x) + C$. The standard way to express this fact is to write

$$\int f(x)\, dx = F(x) + C.$$

The symbol $\int$ is called an *integral sign*. The entire notation $\int f(x)\, dx$ is called an *indefinite integral* and stands for antidifferentiation of the function $f(x)$. We always record the variable of interest by prefacing it by the letter d. For example, if the variable of interest is t rather than x, then we write $\int f(t)\, dt$ for the antiderivative of $f(t)$.

EXAMPLE 4 Determine

(a) $\displaystyle\int x^r\, dx$, r a constant $\neq -1$ (b) $\displaystyle\int e^{kx}\, dx$, k a constant $\neq 0$

Solution (a) By the constant-multiple and power rules,

$$\frac{d}{dx}\left(\frac{1}{r+1}x^{r+1}\right) = \frac{1}{r+1}\cdot\frac{d}{dx}x^{r+1} = \frac{1}{r+1}\cdot(r+1)x^r = x^r.$$

Thus $x^{r+1}/(r+1)$ is an antiderivative of x^r. Letting C represent any constant, we have

$$\int x^r\, dx = \frac{1}{r+1}x^{r+1} + C, \qquad r \neq -1. \tag{2}$$

(b) An antiderivative of e^{kx} is e^{kx}/k, since

$$\frac{d}{dx}\left(\frac{1}{k}e^{kx}\right) = \frac{1}{k}\cdot\frac{d}{dx}e^{kx} = \frac{1}{k}(ke^{kx}) = e^{kx}.$$

Hence

$$\int e^{kx}\, dx = \frac{1}{k} e^{kx} + C, \qquad k \neq 0. \tag{3}$$

Formula (2) does not give an antiderivative of x^{-1} because $1/(r+1)$ is undefined for $r = -1$. However, we know that for $x \neq 0$, the derivative of $\ln|x|$ is $1/x$. Hence $\ln|x|$ is an antiderivative of $1/x$, and we have

$$\int \frac{1}{x}\, dx = \ln|x| + C, \qquad x \neq 0. \tag{4}$$

Formulas (2), (3), and (4) are each followed by "reversing" a familiar differentiation rule. In a similar fashion, one may use the sum rule and constant-multiple rule for derivatives to obtain corresponding rules for antiderivatives:

$$\int [f(x) + g(x)]\, dx = \int f(x)\, dx + \int g(x)\, dx \tag{5}$$

$$\int kf(x)\, dx = k \int f(x)\, dx, \quad k \text{ a constant.} \tag{6}$$

In words, (5) says that a sum of functions may be antidifferentiated term by term, and (6) says that a constant multiple may be moved through the integral sign. ●

EXAMPLE 5   Compute:

$$\int \left(x^{-3} + 7e^{5x} + \frac{4}{x} \right) dx.$$

Solution   Using the preceding rules, we have

$$\int \left(x^{-3} + 7e^{5x} + \frac{4}{x} \right) dx = \int x^{-3}\, dx + \int 7e^{5x}\, dx + \int \frac{4}{x}\, dx$$

$$= \int x^{-3}\, dx + 7 \int e^{5x}\, dx + 4 \int \frac{1}{x}\, dx$$

$$= \frac{1}{-2} x^{-2} + 7\left(\frac{1}{5} e^{5x} \right) + 4 \ln|x| + C$$

$$= -\frac{1}{2} x^{-2} + \frac{7}{5} e^{5x} + 4 \ln|x| + C. \, ●$$

After some practice, most of the intermediate steps shown in the solution of Example 5 can be omitted.

A function has infinitely many different antiderivatives, corresponding to the various choices of the constant C. In applications, it is often necessary to satisfy an additional condition, which then determines a specific value of C.

EXAMPLE 6 Find the function $f(x)$ for which $f'(x) = x^2 - 2$ and $f(1) = \frac{4}{3}$.

Solution The unknown function $f(x)$ is an antiderivative of $x^2 - 2$. One antiderivative of $x^2 - 2$ is $\frac{1}{3}x^3 - 2x$. Therefore, by Theorem 1,

$$f(x) = \frac{1}{3}x^3 - 2x + C, \qquad C \text{ a constant.}$$

Figure 2 shows the graphs of $f(x)$ for several choices of C. We want the function whose graph passes through $(1, \frac{4}{3})$. To find the value of C that makes $f(1) = \frac{4}{3}$, we set

$$\frac{4}{3} = f(1) = \frac{1}{3}(1)^3 - 2(1) + C = -\frac{5}{3} + C$$

and find $C = \frac{4}{3} + \frac{5}{3} = 3$. Therefore, $f(x) = \frac{1}{3}x^3 - 2x + 3$. ●

FIGURE 2 Several antiderivatives of $x^2 - 2$.

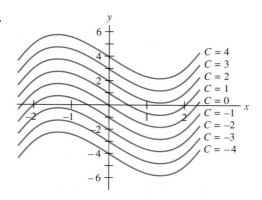

Having introduced the basics of antidifferentiation, let us now solve the oil-consumption problem.

**Solution of Example 1
(Continued)** The rate of oil consumption at time t is $R(t) = 16.1e^{.07t}$ billion barrels per year. Moreover, we observed that the total consumption $T(t)$, from time 0 to time t, is an antiderivative of $R(t)$. Using (3) and (6), we have

$$T(t) = \int 16.1e^{.07t} \, dt = \frac{16.1}{.07}e^{.07t} + C = 230e^{.07t} + C,$$

where C is a constant. However, in our particular example, $T(0) = 0$, since $T(0)$ is the amount of oil used from time 0 to time 0. Therefore, the constant C must satisfy

$$0 = T(0) = 230e^{.07(0)} + C = 230 + C$$

$$C = -230.$$

Therefore,

$$T(t) = 230e^{.07t} - 230 = 230(e^{.07t} - 1).$$

The total amount of oil that would have been consumed from 1970 to 1980 is

$$T(10) = 230(e^{.07(10)} - 1) \approx 233 \text{ billion barrels.} \quad \bullet$$

Antidifferentiation can be used to solve a variety of applied problems, of which the next two examples are typical.

EXAMPLE 7   A rocket is fired vertically into the air. Its velocity at t seconds after lift-off is $v(t) = 6t + .5$ meter per second. Before launch, the top of the rocket is 8 meters above the launch pad. Find the height of the rocket (measured from the top of the rocket to the launch pad) at time t.

Solution   If $s(t)$ denotes the height of the rocket at time t, then $s'(t)$ is the rate at which the height is changing. That is, $s'(t) = v(t)$, and therefore $s(t)$ is an antiderivative of $v(t)$. Thus

$$s(t) = \int v(t) \, dt = \int (6t + .5) \, dt$$

$$= 3t^2 + .5t + C,$$

where C is a constant. When $t = 0$, the rocket's height is 8 meters. That is, $s(0) = 8$ and

$$8 = s(0) = 3(0)^2 + .5(0) + C = C.$$

Thus $C = 8$ and

$$s(t) = 3t^2 + .5t + 8. \quad \bullet$$

EXAMPLE 8   A company's marginal cost function is $.015x^2 - 2x + 80$ dollars, where x denotes the number of units produced in one day. The company has fixed costs of $1000 per day.

(a) Find the cost of producing x units per day.

(b) Suppose the current production level is $x = 30$. Determine the amount costs will rise if the production level is raised to $x = 60$ units.

Solution   (a) Let $C(x)$ be the cost of producing x units in one day. The derivative $C'(x)$ is the marginal cost. In other words, $C(x)$ is an antiderivative of the marginal cost function. Thus

$$C(x) = \int (.015x^2 - 2x + 80) \, dx$$

$$= .005x^3 - x^2 + 80x + C.$$

The \$1000 fixed costs are the costs incurred when producing 0 units. That is, $C(0) = 1000$. So

$$1000 = C(0) = .005(0)^3 - (0)^2 + 80(0) + C.$$

Therefore $C = 1000$, and

$$C(x) = .005x^3 - x^2 + 80x + 1000.$$

(b) The cost when $x = 30$ is $C(30)$, and the cost when $x = 60$ is $C(60)$. So the *increase* in cost when production is raised from $x = 30$ to $x = 60$ is $C(60) - C(30)$. We compute

$$C(60) = .005(60)^3 - (60)^2 + 80(60) + 1000$$

$$= 3280$$

$$C(30) = .005(30)^3 - (30)^2 + 80(30) + 1000$$

$$= 2635.$$

Thus the increase in cost is $3280 - 2635 = \$645$. ●

TECHNOLOGY PROJECT 6.1
Antidifferentiation Using Technology (Symbolic Program Required)

Symbolic mathematics programs typically have a command for antidifferentiation. In *Mathematica*, this command is `Integrate`. For example, to calculate an antiderivative of x^2, you would give the command:

```
Integrate [x^2, x]
```

Note that the second parameter, x, indicates the variable for the integration. In response to this command, the program gives the antiderivative:

```
X^3/3
```

Note that the program does not supply the arbitrary constant. Use a symbolic math program to calculate the following antiderivatives:

1. $\int (3x - 4x^6) \, dx$

2. $\int (5 - 3x + 2x^3 + 9x) \, dx$

3. $\int e^{.003x} \, dx$

4. $\int e^{-40x} \, dx$

5. $\int \dfrac{5}{3 + 2x} \, dx$

6. $\int \dfrac{x}{5x^2 - 1} \, dx$

7. $\int xe^{-4x} \, dx$

8. $\int xe^{-4x^2} \, dx$

9. $\int (4x + 1)e^{-9x-6} \, dx$

10. $\int (x^3 - 4)(x^5 + 1)^4 \, dx$

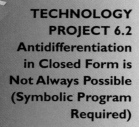

TECHNOLOGY PROJECT 6.2
Antidifferentiation in Closed Form is Not Always Possible (Symbolic Program Required)

Not all simple functions have simple antiderivatives. A number of complications in antidifferentiation can arise. An antiderivative may involve advanced functions with which you are not familiar, which you may have no knowledge of. Or an antiderivative may not be possible in terms of "standard" functions. In the latter case, a symbolic math program gives the original problem as its answer. Another possibility is that an antidifferentiation problem may be too complicated for the program. Antidifferentiation is a complicated task and can often require mathematical ingenuity of a type that is difficult to build into a computer program. The following problems should provide some interesting results from your symbolic math program:

1. $\int \sqrt{1 - x^2}\, dx$

2. $\int e^{-x^2}\, dx$

3. $\int \dfrac{1}{\sqrt{1 - x^2}}\, dx$

4. $\int \dfrac{1}{\sqrt{1 - x^3}}\, dx$

PRACTICE PROBLEMS 6.1

1. Determine each of the following.

(a) $\displaystyle\int t^{7/2}\, dt$

(b) $\displaystyle\int \left(\dfrac{x^3}{3} + \dfrac{3}{x^3} + \dfrac{3}{x}\right) dx$

2. Find the value of k that makes the following antidifferentiation formula true.

$$\int (1 - 2x)^3\, dx = k(1 - 2x)^4 + C$$

EXERCISES 6.1

Find all antiderivatives of each of the following functions.

1. $f(x) = x$ 2. $f(x) = 9x^8$ 3. $f(x) = e^{3x}$

4. $f(x) = e^{-3x}$ 5. $f(x) = 3$ 6. $f(x) = -4x$

In Exercises 7–22, find the value of k that makes the antidifferentiation formula true. [*Note*: You can check your answer without looking in the answer section. How?]

7. $\displaystyle\int x^{-5}\, dx = kx^{-4} + C$ 8. $\displaystyle\int x^{1/3}\, dx = kx^{4/3} + C$

9. $\displaystyle\int \sqrt{x}\, dx = kx^{3/2} + C$ 10. $\displaystyle\int \dfrac{6}{x^3}\, dx = \dfrac{k}{x^2} + C$

11. $\displaystyle\int \dfrac{10}{t^6}\, dt = kt^{-5} + C$ 12. $\displaystyle\int \dfrac{3}{\sqrt{t}}\, dt = k\sqrt{t} + C$

13. $\displaystyle\int 5e^{-2t}\, dt = ke^{-2t} + C$ 14. $\displaystyle\int 3e^{t/10}\, dt = ke^{t/10} + C$

15. $\int 2e^{4x-1} \, dx = ke^{4x-1} + C$

16. $\int \dfrac{4}{e^{3x+1}} \, dx = \dfrac{k}{e^{3x+1}} + C$

17. $\int (x - 7)^{-2} \, dx = k(x - 7)^{-1} + C$

18. $\int \sqrt{x + 1} \, dx = k(x + 1)^{3/2} + C$

19. $\int (x + 4)^{-1} \, dx = k \ln|x + 4| + C$

20. $\int \dfrac{5}{(x - 8)^4} \, dx = \dfrac{k}{(x - 8)^3} + C$

21. $\int (3x + 2)^4 \, dx = k(3x + 2)^5 + C$

22. $\int (2x - 1)^3 \, dx = k(2x - 1)^4 + C$

Determine the following.

23. $\int (x^2 - x - 1) \, dx$

24. $\int (x^3 + 6x^2 - x) \, dx$

25. $\int \left(\dfrac{2}{\sqrt{x}} - 3\sqrt{x} \right) dx$

26. $\int \left[\dfrac{\sqrt{t}}{4} - 4(t - 3)^{-2} \right] dt$

27. $\int \left(4 - 5e^{-5t} + \dfrac{e^{2t}}{3} \right) dt$

28. $\int (e^2 + 3t^2 - 2e^{3t}) \, dt$

Find all functions f(t) with the following property.

29. $f'(t) = t^{3/2}$

30. $f'(t) = \dfrac{4}{6 + t}$

31. $f'(t) = 0$

32. $f'(t) = t^2 - 5t - 7$

Find all functions f(x) with the following properties.

33. $f'(x) = x, f(0) = 3$

34. $f'(x) = 8x^{1/3}, f(1) = 4$

35. $f'(x) = \sqrt{x} + 1, f(4) = 0$

36. $f'(x) = x^2 + \sqrt{x}, f(1) = 3$

37. Figure 3 shows the graphs of several functions $f(x)$ for which $f'(x) = \dfrac{2}{x}$. Find the expression for the function $f(x)$ whose graph passes through $(1, 2)$.

FIGURE 3

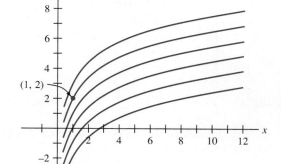

38. Figure 4 shows the graphs of several functions $f(x)$ for which $f'(x) = \frac{1}{3}$. Find the expression for the function $f(x)$ whose graph passes through $(6, 3)$.

FIGURE 4

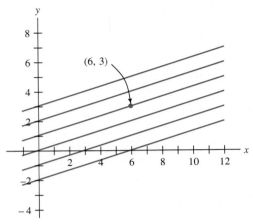

39. Which of the following is $\int \ln x \, dx$?

(a) $\dfrac{1}{x} + C$ (b) $x \cdot \ln x - x + C$ (c) $\dfrac{1}{2} \cdot (\ln x)^2 + C$

40. Which of the following is $\int x\sqrt{x + 1} \, dx$?

(a) $\frac{2}{5}(x + 1)^{5/2} - \frac{2}{3}(x + 1)^{3/2} + C$ (b) $\frac{1}{2}x^2 \cdot \frac{2}{3}(x + 1)^{3/2} + C$

41. Figure 5 contains the graph of a function $F(x)$. On the same coordinate system, draw the graph of the function $G(x)$ having the properties $G(0) = 0$ and $G'(x) = F'(x)$ for each x.

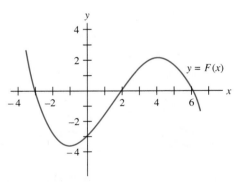

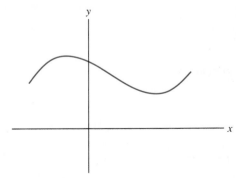

FIGURE 5   **FIGURE 6**

42. Figure 6 contains an antiderivative of the function $f(x)$. Draw the graph of another antiderivative of $f(x)$.

43. The function $g(x)$ in Fig. 7 was obtained by shifting the graph of $f(x)$ up three units. If $f'(5) = \frac{1}{4}$, what is $g'(5)$?

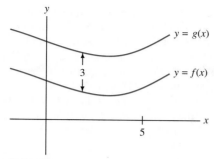

FIGURE 7

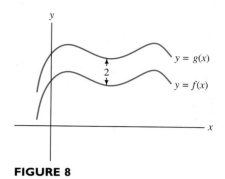

FIGURE 8

44. The function $g(x)$ in Fig. 8 was obtained by shifting the graph of $f(x)$ up two units. What is the derivative of $h(x) = g(x) - f(x)$?

45. A ball is thrown upward from a height of 256 feet above the ground, with an initial velocity of 96 feet per second. From physics it is known that the velocity at time t is $96 - 32t$ feet per second.
    (a) Find $s(t)$, the function giving the height of the ball at time t.
    (b) How long will the ball take to reach the ground?
    (c) How high will the ball go?

46. A rock is dropped from the top of a 400-foot cliff. Its velocity at time t seconds is $v(t) = -32t$ feet per second.
    (a) Find $s(t)$, the height of the rock above the ground at time t.
    (b) How long will the rock take to reach the ground?
    (c) What will be its velocity when it hits the ground?

47. Let $P(t)$ be the total output of a factory assembly line after t hours of work. Suppose that the rate of production at time t is $60 + 2t - \frac{1}{4}t^2$ units per hour. Find the formula for $P(t)$. [*Hint:* The rate of production is $P'(t)$ and $P(0) = 0$.]

48. After t hours of operation a coal mine is producing coal at the rate of $40 + 2t - \frac{1}{5}t^2$ tons of coal per hour. Find a formula for the total output of the coal mine after t hours of operation.

49. A package of frozen strawberries is taken from a freezer at $-5°C$ into a room at $20°C$. At time t the average temperature of the strawberries is increasing at the rate of $10e^{-.4t}$ degrees Celsius per hour. Find the temperature of the strawberries at time t.

50. A flu epidemic hits a town. Let $P(t)$ be the number of persons sick with the flu at time t, where time is measured in days from the beginning of the epidemic and $P(0) = 100$. Suppose that after t days the flu is spreading at the rate of $120t - 3t^2$ people per day. Find the formula for $P(t)$.

51. A small tie shop finds that at a sales level of x ties per day its marginal profit is $MP(x)$ dollars per tie, where $MP(x) = 1.30 + .06x - .0018x^2$. Also, the shop will lose \$95 per day at a sales level of $x = 0$. Find the profit from operating the shop at a sales level of x ties per day.

52. A soap manufacturer estimates that its marginal cost of producing soap powder is $.2x + 1$ hundred dollars per ton at a production level of x tons per day. Fixed costs are \$200 per day. Find the cost of producing x tons of soap powder per day.

53. The United States has been consuming iron ore at the rate of $R(t)$ million metric tons per year at time t, where $t = 0$ corresponds to 1980 and $R(t) = 94e^{.016t}$. Find a formula for the total U.S. consumption of iron ore from 1980 until time t.

54. Since 1987, the rate of production of natural gas in the United States has been approximately $R(t)$ quadrillion British thermal units per year at time t, with $t = 0$ corresponding to 1987 and $R(t) = 17.04e^{.016t}$. Find a formula for the total U.S. production of natural gas from 1987 until time t.

55. Suppose the drilling of an oil well has a fixed cost of $10,000 and a marginal cost of $C'(x) = 1000 + 50x$ dollars per foot, where x is the depth in feet. Find the expression for $C(x)$, the total cost of drilling x feet. [*Note:* $C(0) = 10,000$.]

SOLUTIONS TO PRACTICE PROBLEMS 6.1

1. (a) $\displaystyle\int t^{7/2}\, dt = \frac{1}{\frac{9}{2}}t^{9/2} + C = \frac{2}{9}t^{9/2} + C$

   (b) $\displaystyle\int \left(\frac{x^3}{3} + \frac{3}{x^3} + \frac{3}{x}\right) dx = \int \left(\frac{1}{3}\cdot x^3 + 3x^{-3} + 3\cdot\frac{1}{x}\right) dx$

   $$= \frac{1}{3}\left(\frac{1}{4}x^4\right) + 3\left(-\frac{1}{2}x^{-2}\right) + 3\ \ln|x| + C$$

   $$= \frac{1}{12}x^4 - \frac{3}{2}x^{-2} + 3\ \ln|x| + C$$

2. Since we are told that the antiderivative has the general form $k(1 - 2x)^4$, all we have to do is determine the value of k. Differentiating, we obtain

$$4k(1 - 2x)^3(-2) \quad \text{or} \quad -8k(1 - 2x)^3,$$

which is supposed to equal $(1 - 2x)^3$. Therefore, $-8k = 1$, so $k = -\frac{1}{8}$.

6.2 AREAS AND RIEMANN SUMS

This section and the next reveal the important connection between antiderivatives and areas of regions under curves. Although the full story will have to wait until the next section, we can give a hint now by describing how "area" is related to a problem solved in Section 6.1.

Example 8 in Section 6.1 concerned a company's marginal cost function, $f(x) = .015x^2 - 2x + 80$. A calculation with an antiderivative of $f(x)$ showed that if the production level is increased from $x = 30$ to $x = 60$ units per day, the change in total cost is $645. As we'll see, this change in cost exactly equals the area of the region under the graph of the marginal cost curve in Fig. 1 from $x = 30$ to $x = 60$. First, however, we need to learn how to find areas of such regions.

Area Under a Graph   If $f(x)$ is a continuous nonnegative function on the interval $a \leq x \leq b$, we refer to the area of the region shown in Fig. 2 as the *area under the graph of $f(x)$ from a to b.*

FIGURE 1   Area under a marginal cost curve.

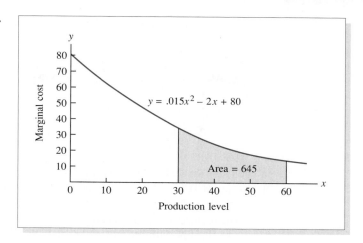

FIGURE 2   Area under a graph.

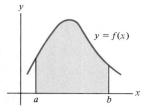

The computation of the area in Fig. 2 is not a trivial matter when the top boundary of the region is curved. However, we can *estimate* the area to any desired degree of accuracy. The basic idea is to construct rectangles whose total area is approximately the same as the area to be computed. The area of each rectangle, of course, is easy to compute.

Figure 3 shows three rectangular approximations to the area under a graph. When the rectangles are thin, the mismatch between the rectangles and the region under the graph is quite small. In general, a rectangular approximation can be made as close as desired to the exact area simply by making the width of the rectangles sufficiently small.

Given a continuous nonnegative function $f(x)$ on the interval $a \le x \le b$, divide the x-axis interval into n equal subintervals, where n represents some positive integer. Such a subdivision is called a *partition* of the interval from a to b. Since the entire interval is of width $b - a$, the width of each of the n subintervals is $(b - a)/n$. For brevity, denote this width by Δx. That is,

$$\Delta x = \frac{b - a}{n} \qquad \text{(width of one subinterval).}$$

FIGURE 3   Approximating a region with rectangles.

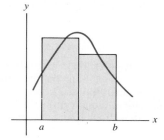 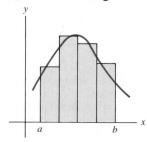 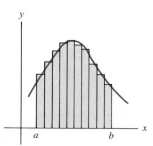

FIGURE 4   Rectangles with heights $f(x_1), \ldots, f(x_n)$.

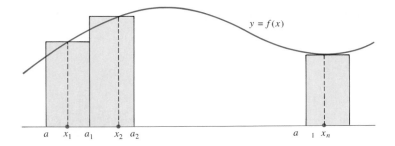

In each subinterval, select a point. (Any point in the subinterval will do.) Let x_1 be the point selected from the first subinterval, x_2 the point from the second subinterval, and so on. These points are used to form the rectangles that approximate the region under the graph of $f(x)$. Construct the first rectangle with height $f(x_1)$ and the first subinterval as base, as in Fig. 4.

The top of the rectangle touches the graph directly above x_1. Notice that

$$[\text{area of first rectangle}] = [\text{height}][\text{width}] = f(x_1)\Delta x.$$

The second rectangle rests on the second subinterval and has height $f(x_2)$. Thus

$$[\text{area of second rectangle}] = [\text{height}][\text{width}] = f(x_2)\Delta x.$$

Continuing in this way, we construct n rectangles with a combined area of

$$f(x_1)\Delta x + f(x_2)\Delta x + \cdots + f(x_n)\Delta x. \tag{1}$$

A sum as in (1) is called a *Riemann sum.* It provides an approximation to the area under the graph of $f(x)$ when $f(x)$ is nonnegative and continuous. In fact, as the number of subintervals increases indefinitely, the Riemann sums (1) approach a limiting value, the area under the graph.*

EXAMPLE 1   Estimate the area under the graph of the marginal cost function $f(x) = .015x^2 - 2x + 80$ from $x = 30$ to $x = 60$. Use partitions of 5, 20, and 100 subintervals. Use the midpoints of the subintervals as $x_1, x_2, \ldots, x_n$ to construct the rectangles. See Fig. 5.

Solution   The partition of $30 \le x \le 60$ with $n = 5$ is shown in Fig. 6. The length of each subinterval is

$$\Delta x = \frac{60 - 30}{5} = 6.$$

---

* Riemann sums are named after the nineteenth-century German mathematician G. B. Riemann (pronounced "Reemahn"), who used them extensively in his work on calculus. The concept of a Riemann sum has several uses: to approximate areas under curves, to construct mathematical models in applied problems, and to give a formal definition of area. In the next section, Riemann sums are used to define the definite integral of a function.

FIGURE 5 **Estimating the area under a marginal cost curve.**

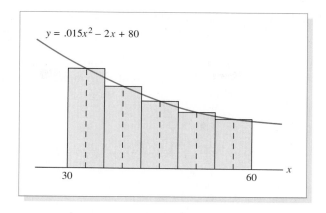

$y = .015x^2 - 2x + 80$

30   60 x

FIGURE 6 **A partition of the interval $30 \le x \le 0$.**

Δx

30   x_1 x_2 x_3 x_4 x_5 60 x

Δx
$= 6$

Observe that the first midpoint is $\Delta x/2$ units from the left endpoint, and the midpoints themselves are Δx units apart. The first midpoint is $x_1 = 30 + \Delta x/2 = 30 + 3 = 33$. Subsequent midpoints are found by successively adding $\Delta x = 6$.

$$\text{midpoints: } 33, 39, 45, 51, 57.$$

The corresponding estimate for the area under the graph of $f(x)$ is

$$f(33)\Delta x + f(39)\Delta x + f(45)\Delta x + f(51)\Delta x + f(57)\Delta x$$
$$= (30.335) \cdot 6 + (24.815) \cdot 6 + (20.375) \cdot 6$$
$$+ (17.015) \cdot 6 + (14.735) \cdot 6$$
$$= 182.01 + 148.89 + 122.25 + 102.09 + 88.41$$
$$= 643.65.$$

A similar calculation with 20 subintervals produces an area estimate of 644.916. With 100 subintervals the estimate is 644.997. ●

The approximations in Example 1 seem to confirm the claim made at the beginning of the section that the area under the marginal cost curve equals the change in total cost, $645. The verification of this fact will be given in the next section.

Although the midpoints of subintervals are often selected as the x_1, $x_2, \ldots, x_n$ in a Riemann sum, left endpoints and right endpoints are also convenient.

EXAMPLE 2   Use a Riemann sum with $n = 4$ to estimate the area under the graph of $f(x) = x^2$ from 1 to 3. Select the right endpoints of the subintervals as x_1, x_2, x_3, x_4.

Solution   Here $\Delta x = (3 - 1)/4 = .5$. The right endpoint of the first subinterval is $1 + \Delta x = 1.5$. Subsequent right endpoints are obtained by successively adding .5, as shown below.

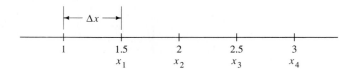

The corresponding Riemann sum is

$$f(x_1)\Delta x + f(x_2)\Delta x + f(x_3)\Delta x + f(x_4)\Delta x$$

$$= (1.5)^2(.5) + (2)^2(.5) + (2.5)^2(.5) + (3)^2(.5)$$

$$= 1.125 + 2 + 3.125 + 4.5 = 10.75.$$

The rectangles used for this Riemann sum are shown in Fig. 7. The right endpoints here give an area estimate that is obviously greater than the exact area. Midpoints would work better. But if the rectangles are sufficiently narrow, even a Riemann sum using right endpoints will be close to the exact area. ●

FIGURE 7   A Riemann sum using right endpoints.

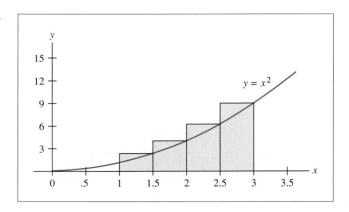

EXAMPLE 3   To estimate the area of a 100-foot-wide waterfront lot, a surveyor measured the distance from the street to the waterline at 20-foot intervals, starting 10 feet from one corner of the lot. Use the data to construct a Riemann sum approximation to the area of the lot. See Fig. 8.

Solution   Treat the street as the x-axis, and consider the waterline along the property as the graph of a function $f(x)$ over the interval from 0 to 100. The five "vertical" distances give $f(x_1), \ldots, f(x_5)$, where $x_1 = 10, \ldots, x_5 = 90$. Since there are five points $x_1, \ldots, x_5$ spread across the interval $0 \le x \le 100$, we partition the

FIGURE 8   Survey of a water-front property.

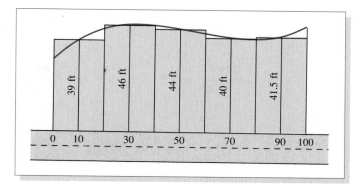

interval into five subintervals, with $\Delta x = \frac{100}{5} = 20$. Fortunately, each subinterval contains one x_i. (In fact, each x_i is the midpoint of a subinterval.) Thus the area of the lot is approximated by the Riemann sum

$$f(x_1)\Delta x + \cdots + f(x_5)\Delta x$$
$$= (39)20 + (46)20 + (44)20 + 40(20) + (41.5)20$$
$$= 4210 \text{ square feet.}$$

For a better estimate of the area, the surveyor will have to make more measurements from the street to the waterline. ●

The final example shows how Riemann sums arise in an application. Since the variable is time, t, we write Δt in place of Δx.

EXAMPLE 4   The velocity of a rocket at time t is $v(t)$ feet per second. Construct a Riemann sum that estimates how far the rocket travels in the first 10 seconds. What happens when the number of subintervals in the partition increases without bound?

Solution   Partition the interval $0 \le t \le 10$ into n subintervals of width Δt, and select points $t_1, t_2, \ldots, t_n$ from these subintervals. Although the rocket's velocity is not constant, it does not change much during a small subinterval of time. So we may use $v(t_1)$ as an approximation of the rocket's velocity during the first subinterval. Over this time subinterval of length $\Delta t = (10 - 0)/n$,

$$[\text{distance traveled}] \approx [\text{velocity}] \cdot [\text{time}]$$
$$= v(t_1)\ \Delta t. \tag{2}$$

The distance traveled during the second time subinterval is approximately $v(t_2)\Delta t$, and so on. Thus an estimate for the total distance traveled is

$$v(t_1)\Delta t + v(t_2)\Delta t + \cdots + v(t_n)\Delta t \tag{3}$$

The sum in (3) is a Riemann sum for the velocity function on the interval $0 \le t \le 10$. As n increases, such a Riemann sum approaches the area under the graph of the velocity function. However, from our derivation of (3), it seems

FIGURE 9   Area under a velocity curve.

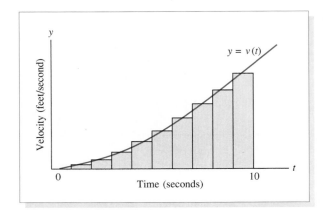

reasonable that as n increases, the sums become a better and better estimate of the total distance traveled. We conclude that

$$\begin{bmatrix} \text{total distance rocket travels} \\ \text{during the first 10 seconds} \end{bmatrix} = \begin{bmatrix} \text{area under graph of} \\ v(t) \text{ over } 0 \le t \le 10 \end{bmatrix}$$

See Fig. 9. ●

It is helpful to look at Example 4 from a slightly different point of view. The height of the rocket is an increasing quantity and $v(t)$ is the *rate* of change of height at time t. The area under the graph of $v(t)$ from time $a = 0$ to time $b = 10$ is the *amount* of increase in the height during the first 10 seconds.

This connection between the rate of change of a function and the amount of increase of the function generalizes to a variety of situations.

> If a quantity is increasing, then the area under the rate of change function from a to b is the amount of increase in the quantity from a to b.

Table 1 shows four instances of this principle. The result is justified in the next section. The first example in the table was discussed in Example 1.

TABLE 1   Interpretation of Areas

Function	a to b	Area Under the Graph from a to b
Marginal cost at production level x	30 to 60	Additional cost when production is increased from 30 to 60 units
Rate of sulfur emissions from a power plant t years after 1990	1 to 3	Amount of sulfur released from 1991 to 1993
Birth rate t years after 1980 (in babies per year)	0 to 10	Number of babies born from 1980 to 1990
Rate of gas consumption t years after 1985	3 to 6	Amount of gas used from 1988 to 1991

TECHNOLOGY
PROJECT 6.3
Estimation of
Definite Integrals
Using Riemann
Sums (Symbolic
Program Required)

As we have just described, it is possible to estimate definite integrals using Riemann sums. For small numbers of subdivisions, this can be done by hand. However, when the number of subdivisions increases, some form of technology is a necessity. Basically, what is needed is a way to add up sums of numbers specified as values of a function. This can be done on a graphing calculator by writing a program. However, we will bypass that approach and, instead, discuss how the computation can be performed using a symbolic math program. Most such programs have commands that allow you to add up values of a function at a sequence of numbers. For instance, in *Mathematica,* you can use the function Sum. For example, to compute the sum

$$5^2 + 6^2 + 7^2 + \cdots + 100^2$$

you would give the command:

```
Sum [n^2, {n, 5, 100}]
```

The program responds with the answer 338,320. In the above sum, the numbers increase by 1. However, *Mathematica* also allows you to specify sequences that increase (or decrease) by an arbitrary amount. For example, you can compute the sum

$$100^2 + 99.5^2 + 99^2 + \cdots + 50^2$$

Using the command:

```
Sum[n^2, {n, 100, 50, - .5}]
```

The program gives the answer 589,588.

The following exercises provide some practice in calculating Riemann sums. Let $f(x) = 4 - x^2$. Calculate the following Riemann sums for the function $f(x)$ over the interval $-2 \le x \le 2$:

1. $n = 10$, right endpoints
2. $n = 50$, right endpoints
3. $n = 100$, right endpoints
4. $n = 500$, right endpoints
5. $n = 1000$, right endpoints
6. $n = 10$, left endpoints
7. $n = 50$, left endpoints
8. $n = 100$, left endpoints
9. $n = 500$, left endpoints
10. $n = 1000$, left endpoints
11. $n = 10$, midpoints
12. $n = 50$, midpoints
13. $n = 100$, midpoints
14. $n = 500$, midpoints
15. $n = 1000$, midpoints
16. On the basis of the above data, what is the approximate area bounded by the graph of y and the x-axis? To how many decimal places does the approximation appear to be accurate?

PRACTICE PROBLEMS 6.2

1. Determine Δx and the midpoints of the subintervals formed by partitioning the interval $-2 \leq x \leq 2$ into five subintervals.

2. The graph in Fig. 10 gives the rate at which new jobs were created (in millions of jobs per year) in the United States, where $t = 0$ corresponds to 1983. For instance, on January 1, 1986, jobs were being created at the rate of 2.4 million new jobs per year. Interpret the area of the shaded region.

FIGURE 10 Rate of job creation.

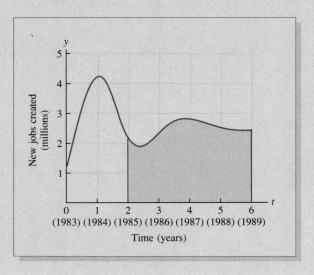

EXERCISES 6.2

Determine Δx and the midpoints of the subintervals formed by partitioning the given interval into n subintervals. [*Hint*: Decimals are sometimes easier to use than fractions.]

1. $0 \leq x \leq 2; n = 4$
2. $0 \leq x \leq 3; n = 6$
3. $1 \leq x \leq 4; n = 5$
4. $3 \leq x \leq 5; n = 5$

In Exercises 5–10, use a Riemann sum to approximate the area under the graph of $f(x)$ on the given interval, with selected points as specified.

5. $f(x) = x^2; 1 \leq x \leq 3, n = 4$, midpoints of subintervals.
6. $f(x) = x^2; -2 \leq x \leq 2, n = 4$, midpoints of subintervals.
7. $f(x) = x^3; 1 \leq x \leq 3, n = 5$, left endpoints.
8. $f(x) = x^3; 0 \leq x \leq 1, n = 5$, right endpoints.
9. $f(x) = e^{-x}; 2 \leq x \leq 3, n = 5$, right endpoints.
10. $f(x) = \ln x; 2 \leq x \leq 4, n = 5$, left endpoints.
11. Use a Riemann sum with $n = 4$ and left endpoints to estimate the area under the graph of $f(x) = 4 - x$ on the interval $1 \leq x \leq 4$. Then repeat with $n = 4$ and

midpoints. Compare the answers with the exact answer, 4.5, which can be computed from the formula for the area of a triangle.

12. Use a Riemann sum with $n = 4$ and right endpoints to estimate the area under the graph of $f(x) = 2x - 4$ on the interval $2 \leq x \leq 3$. Then repeat with $n = 4$ and midpoints. Compare the answers with the exact answer, 1, which can be computed from the formula for the area of a triangle.

13. The graph of the function $f(x) = \sqrt{1 - x^2}$ on the interval $-1 \leq x \leq 1$ is a semicircle. The area under the graph is $\frac{1}{2} \cdot \pi(1)^2 = \pi/2 = 1.57080$, to five decimal places. Use a Riemann sum with $n = 5$ and midpoints to estimate the area under the graph. See Fig. 11(a). Carry out the calculations to five decimal places and compute the error (the difference between the estimate and 1.57080).

FIGURE 11

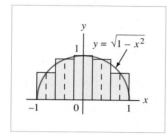 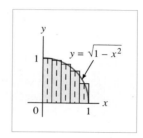

14. Use a Riemann sum with $n = 5$ and midpoints to estimate the area under the graph of $f(x) = \sqrt{1 - x^2}$ on the interval $0 \leq x \leq 1$. The graph is a quarter circle, and the area under the graph is .78540, to five decimal places. See Fig. 11(b). Carry out the calculations to five decimal places and compute the error. If you double the estimate from this exercise, will it be more accurate than the estimate computed in Exercise 13?

15. Lung physiologists measure the velocity of air passing through a patient's throat by having the patient breathe into a pneumotachograph. This machine produces a graph that plots air-flow rate as a function of time. The graph in Fig. 12 shows the flow rate while a patient is breathing out. The area under the graph gives the total

FIGURE 12   Data from a pneumotachograph.

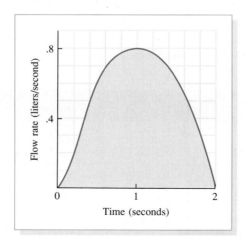

16. volume of air during exhalation. Estimate this volume with a Riemann sum. Use $n = 5$ and midpoints.

Estimate the area (in square feet) of the piece of land shown in Fig. 13.

FIGURE 13 Area of residential property

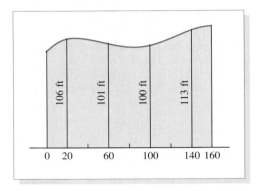

17. The velocity of a car (in feet per second) is recorded from the speedometer every 10 seconds, beginning 5 seconds after the car starts to move. See Table 2. Use a Riemann sum to estimate the distance the car travels during the first 60 seconds.

TABLE 2	A Car's Velocity					
Time	5	15	25	35	45	55
Velocity	20	44	32	39	65	80

18. Table 3 shows the velocity (in feet per second) at the end of each second for a person starting a morning jog. Make three Riemann sum estimates of the total distance jogged during the time from $t = 2$ to $t = 8$.

(a) $n = 6$, left endpoints.

(b) $n = 6$, right endpoints.

(c) $n = 3$, midpoints.

TABLE 3	A Jogger's Velocity								
Time	0	1	2	3	4	5	6	7	8
Velocity	0	2	3	5	5	6	6	7	8

19. Complete the missing entries in Table 4.

TABLE 4	Interpretation of Areas	
Function	*a* to *b*	Area Under the Graph from *a* to *b*
Rate of growth of a population *t* years after 1900 (in millions of people per year)	10 to 50	
	5 to 7	Number of cigarettes smoked from 1990 to 1992
Marginal profit at production level *x*		Additional profit created by increasing production from 20 to 50 units

20. Interpret the area of the shaded region in Fig. 14.

21. Interpret the area of the shaded region in Fig. 15.

22. Interpret the area of the shaded region in Fig. 16.

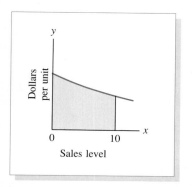

FIGURE 14   Marginal revenue.

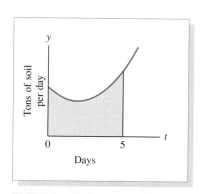

FIGURE 15   Rate of soil erosion.

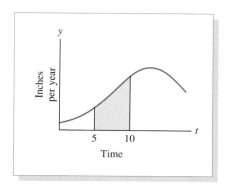

FIGURE 16   A child's growth rate.

23. How does the area under $y = x^2$ from $x = 2$ to $x = 3$ compare with the area under $y = \frac{1}{5}x^2$ from 2 to 3? To answer this, write Riemann sums that approximate each area and compare them. Use $n = 5$ with midpoints. Write a brief paragraph about your conclusions.

24. How are the areas under $y = x^2$ and $y = 2 + x^3$ from $x = 1$ to $x = 2$ related to the area under the graph of $y = x^2 + 2 + x^3$? The answer is not obvious from an examination of the graphs in Fig. 17. Examine Riemann sums for the three areas with $n = 5$ and midpoints, and write a short paragraph about what you find.

25. Let $f(x)$ and $g(x)$ be nonnegative functions on $a \leq x \leq b$. Write a short paragraph about how you think the areas under the graphs of $f(x)$, $g(x)$, and $f(x) + g(x)$ are related. Use a general Riemann sum for $f(x) + g(x)$ in your discussion.

FIGURE 17

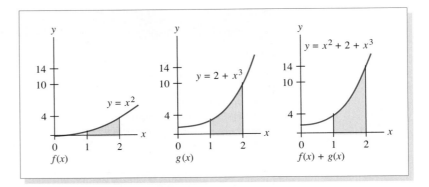

26. Let $f(x)$ be a nonnegative continuous function on $a \leq x \leq b$, and let k be a positive constant. Write a short paragraph about how you think the area under the graph of $k \cdot f(x)$ is related to the area under the graph of $f(x)$. Use a general Riemann sum for $k \cdot f(x)$ in your discussion.

SOLUTIONS TO PRACTICE PROBLEMS 6.2

1. Since $n = 5$, $\Delta x = \dfrac{2 - (-2)}{5} = \dfrac{4}{5} = .8$. The first midpoint is $x_1 = -2 + .8/2 = -1.6$. Subsequent midpoints are found by successively adding .8, to obtain $x_2 = -.8$, $x_3 = 0$, $x_4 = .8$, and $x_5 = 1.6$.

2. The area under the curve from $t = 2$ to $t = 6$ is the number of new jobs created from 1985 to 1989.

6.3 DEFINITE INTEGRALS AND THE FUNDAMENTAL THEOREM

In Section 6.2 we saw that the area under the graph of a continuous nonnegative function $f(x)$ from a to b is the limiting value of Riemann sums of the form

$$f(x_1)\Delta x + f(x_2)\Delta x + \cdots + f(x_n)\Delta x$$

as n increases without bound or, equivalently, as Δx approaches zero. (Recall that $x_1, x_2, \ldots, x_n$ are selected points from a partition of $a \leq x \leq b$ and Δx is the width of each of the n subintervals.) It can be shown that even if $f(x)$ has negative values, the Riemann sums still approach a limiting value as $\Delta x \to 0$. This number is called the *definite integral of $f(x)$ from a to b* and is denoted by

$$\int_a^b f(x)\, dx.$$

That is,

$$\int_a^b f(x)\ dx = \lim_{\Delta x \to 0} \left[f(x_1)\Delta x + f(x_2)\Delta x + \cdots + f(x_n)\Delta x \right]. \quad (1)$$

If $f(x)$ is a nonnegative function, we know from Section 6.2 that the Riemann sum on the right side of (1) approaches the area under the graph of $f(x)$ from a to b. Thus, *the definite integral of a nonnegative function $f(x)$ equals the area under the graph of $f(x)$.*

EXAMPLE 1 Calculate $\int_1^4 (\frac{1}{3}x + \frac{2}{3})\ dx.$

Solution Figure 1 shows the graph of the function $f(x) = \frac{1}{3}x + \frac{2}{3}$. Since $f(x)$ is nonnegative for $1 \le x \le 4$, the definite integral of $f(x)$ equals the area of the shaded region in Fig. 1. The region consists of a rectangle and a triangle. By geometry,

$$[\text{area of rectangle}] = [\text{width}] \cdot [\text{height}] = 3 \cdot 1 = 3,$$

$$[\text{area of triangle}] = \tfrac{1}{2}[\text{width}] \cdot [\text{height}] = \tfrac{1}{2} \cdot 3 \cdot 1 = \tfrac{3}{2}.$$

Thus the area under the graph is $4\frac{1}{2}$, and hence

$$\int_1^4 (\tfrac{1}{3}x + \tfrac{2}{3})\ dx = 4\tfrac{1}{2}. \quad \bullet$$

FIGURE 1   A definite integral as the area under a curve.

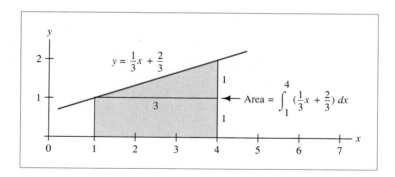

In case $f(x)$ is negative at some points in the interval, we may also give a geometric interpretation of the definite integral. Consider the function $f(x)$ shown in Fig. 2. It shows a rectangular approximation of the region between the graph and the x-axis from a to b. Consider a typical rectangle located above or below the representative point x_i. If $f(x_i)$ is nonnegative, the area of the rectangle equals $f(x_i)\ \Delta x$. In case $f(x_i)$ is negative, the area of the rectangle equals $(-f(x_i))\Delta x$. So the expression $f(x_i)\Delta x$ equals either the area of the correspond-

FIGURE 2

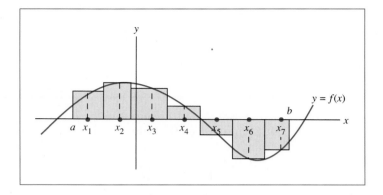

ing rectangle or the negative of the area, according to whether $f(x_i)$ is nonnegative or negative, respectively. In particular, the Riemann sum

$$f(x_1)\Delta x + f(x_2)\Delta x + \cdots + f(x_n)\Delta x$$

is equal to the area of the rectangles above the x-axis minus the area of the rectangles below the x-axis. Now take the limit as Δx approaches 0. On the one hand, the Riemann sum approaches the definite integral. On the other hand, the rectangular approximations approach the area bounded by the graph that is above the x-axis minus the area bounded by the graph that is below the x-axis. This gives us the following geometric interpretation of the definite integral.

Suppose that $f(x)$ is continuous on the interval $a \leq x \leq b$. Then

$$\int_a^b f(x)\ dx$$

is equal to the area above the x-axis bounded by the graph of $y = f(x)$ from $x = a$ to $x = b$ minus the area below the x-axis. Referring to Fig. 3, we have

$$\int_a^b f(x)\ dx = [\text{area of } B \text{ and } D] - [\text{area of } A \text{ and } C]$$

FIGURE 3   Regions above and below the x axis.

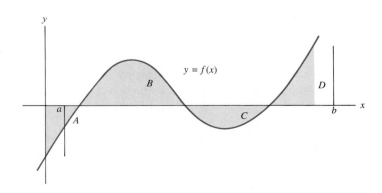

FIGURE 4

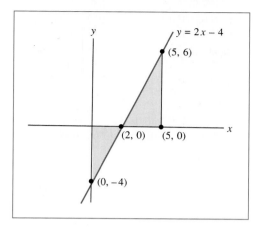

EXAMPLE 2 Calculate $\int_0^5 (2x - 4)\, dx$.

Solution Figure 4 shows the graph of the function $f(x) = 2x - 4$ on the interval $0 \le x \le 5$. The area of the triangle above the x-axis is 9 and the area of the triangle below the x-axis is 4. Therefore, from geometry, we find that

$$\int_0^5 (2x - 4)\, dx = 9 - 4 = 5. \quad \bullet$$

The values of the definite integrals in Examples 1 and 2 follow from simple geometric formulas. For integrals of more complex functions, analogous area formulas are not available. Of course, Riemann sums may always be used to estimate the value of a definite integral to any desired degree of accuracy. However, for most of the integrals in this text, the following fundamental theorem of calculus will rescue us from such calculations.

> *Fundamental Theorem of Calculus* Suppose that $f(x)$ is continuous on the interval $a \le x \le b$, and let $F(x)$ be an antiderivative of $f(x)$. Then
>
> $$\int_a^b f(x)\, dx = F(b) - F(a). \tag{2}$$

This theorem connects the two key concepts of calculus—the integral and the derivative. An explanation of why the theorem is true is given later. First we show how to use the theorem to evaluate definite integrals.

EXAMPLE 3 Use the fundamental theorem of calculus to evaluate the following definite integrals.

(a) $\int_1^4 \left(\tfrac{1}{3}x + \tfrac{2}{3}\right) dx$

(b) $\int_0^5 (2x - 4)\, dx$

Solution   (a) An antiderivative of $\frac{1}{3}x + \frac{2}{3}$ is $F(x) = \frac{1}{6}x^2 + \frac{2}{3}x$. Therefore, by the fundamental theorem,

$$\int_1^4 \left(\tfrac{1}{3}x + \tfrac{2}{3}\right) dx = F(4) - F(1)$$

$$= \left[\tfrac{1}{6}(4)^2 + \tfrac{2}{3}(4)\right] - \left[\tfrac{1}{6}(1)^2 + \tfrac{2}{3}(1)\right]$$

$$= \left[\tfrac{16}{6} + \tfrac{8}{3}\right] - \left[\tfrac{1}{6} + \tfrac{2}{3}\right] = 4\tfrac{1}{2}.$$

This result is the same as in Example 1.

(b) An antiderivative of $2x - 4$ is $F(x) = x^2 - 4x$. Therefore,

$$\int_0^5 (2x - 4) dx = F(5) - F(0)$$

$$= [5^2 - 4(5)] - [0^2 - 4(0)].$$

$$= 5.$$

This result is the same as in Example 2. ●

The next example illustrates the fact that when computing the definite
integral of a function, we may use *any* antiderivative of the function.

EXAMPLE 4   Evaluate $\displaystyle\int_2^5 3x^2 \, dx$.

Solution   An antiderivative of $f(x) = 3x^2$ is $F(x) = x^3 + C$, where C is any constant.
Then

$$\int_2^5 3x^2 \, dx = F(5) - F(2) = [5^3 + C] - [2^3 + C]$$

$$= 5^3 + C - 2^3 - C = 117.$$

Notice how the C in $F(2)$ is subtracted from the C in $F(5)$. Thus the value of the
definite integral does not depend on the choice of the constant C. For convenience, we may take $C = 0$ when evaluating a definite integral. ●

The quantity $F(b) - F(a)$ is called the *net change of* $F(x)$ *from* $x = a$ *to*
$x = b$. It is abbreviated by the symbol $F(x)\big|_a^b$. For instance, the net change of
$F(x) = \frac{1}{3}e^{3x}$ from $x = 0$ to $x = 2$ is written as $\frac{1}{3}e^{3x}\big|_0^2$ and is evaluated as
$F(2) - F(0)$.

EXAMPLE 5   Evaluate $\displaystyle\int_0^2 e^{3x} \, dx$.

Solution   An antiderivative of e^{3x} is $\frac{1}{3}e^{3x}$. Therefore,

$$\int_0^2 e^{3x} \, dx = \tfrac{1}{3}e^{3x}\bigg|_0^2$$

$$= \tfrac{1}{3}e^{3(2)} - \tfrac{1}{3}e^{3(0)} = \tfrac{1}{3}e^6 - \tfrac{1}{3}. \bullet$$

EXAMPLE 6 Compute the area under the curve $y = x^2 - 4x + 5$ from $x = -1$ to $x = 3$.

Solution The graph of $f(x) = x^2 - 4x + 5$ is shown in Fig. 5. Since $f(x)$ is nonnegative for $-1 \le x \le 3$, the area under the curve is given by the definite integral

$$\int_{-1}^{3} (x^2 - 4x + 5) \, dx = \left(\frac{x^3}{3} - 2x^2 + 5x \right)\Big|_{-1}^{3}$$

$$= \left[\frac{(3)^3}{3} - 2(3)^2 + 5(3) \right]$$

$$- \left[\frac{(-1)^3}{3} - 2(-1)^2 + 5(-1) \right]$$

$$= [9 - 18 + 15] - [-\tfrac{1}{3} - 2 - 5]$$

$$= [6] - [-\tfrac{22}{3}] = \tfrac{18}{3} + \tfrac{22}{3} = \tfrac{40}{3}. \; \bullet$$

FIGURE 5

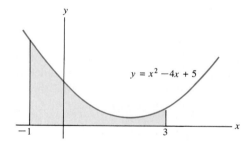

$y = x^2 - 4x + 5$

Study this solution carefully. The calculations show how to use parentheses to avoid errors in arithmetic. It is particularly important to include the outside brackets around the value of the antiderivative at -1.

TECHNOLOGY PROJECT 6.4
Calculating Definite Integrals Using a Symbolic Math Program (Symbolic Program Required)

Symbolic math programs allow you to calculate definite integrals. For example, in *Mathematica* you can calculate the definite integral

$$\int_{-1}^{5} (x^3 - 3x + 1) \, dx$$

using the command

```
Integrate [x^3 - 3x + 1, {x, -1, 5}]
```

The program gives the answer 126. Note that this method of calculating definite integrals requires the program to first calculate an antiderivative. If the program can't calculate an antiderivative, it will refuse to do the calculation and will just display the definite integral unevaluated.

Use your symbolic mathematics program to calculate the values of the following definite integrals:

1. $\int_0^3 (3x + 1)^5 \, dx$

2. $\int_1^{20} \sqrt{5x - 2} \, dx$

3. $\int_3^5 (x + 1)(x - 3)^4 \, dx$

4. $\int_0^1 \dfrac{x + 1}{x + 2} \, dx$

5. $\int_{12}^{15} 51e^{-.02t} \, dt$

6. $\int_0^9 [3 + e^{-4x}]^3 \, dx$

Areas in Applications   At the end of Section 6.2, we described how the area under a graph can represent the amount of change (now called the "net change") in a quantity. This interpretation of area follows immediately from the fundamental theorem of calculus, which we may rewrite as

$$\int_a^b F'(x) \, dx = F(b) - F(a) \tag{3}$$

because the $f(x)$ in (2) is the derivative of $F(x)$. If the derivative is nonnegative, then the integral in (3) is the area under the graph of the "rate of change" function $F'(x)$. Equation (3) says that this area equals the net change in $F(x)$ over the interval from a to b.

EXAMPLE 7   A rocket is fired vertically into the air. Its velocity at t seconds after lift-off is $v(t) = 6t + .5$ meter per second.

(a) Describe a region whose area represents the distance the rocket travels from time $t = 40$ to $t = 100$ seconds.

(b) Compute the distance in (a).

Solution   (a) Let $s(t)$ be the position of the rocket at time t, measured from some reference point. (Example 7 in Section 6.1 measured distances from the launch pad, for example.) Then the required distance is $s(100) - s(40)$, the net change in position over the interval $40 \le t \le 100$. This distance is represented by the area under the velocity curve from 40 to 100. See Fig. 6.

(b) Now that we know the connection between antiderivatives and area, we can easily compute the area in (a) corresponding to the distance traveled by the rocket:

$$s(100) - s(40) = \int_{40}^{100} s'(t) \, dt = \int_{40}^{100} v(t) \, dt$$

$$= \int_{40}^{100} (6t + .5) \, dt = (3t^2 + .5t) \Big|_{40}^{100}$$

$$= [3(100)^2 + .5(100)] - [3(40)^2 + .5(40)]$$

$$= 30{,}050 - 4820 = 25{,}230 \text{ meters} \quad \bullet$$

FIGURE 6   Area under velocity curve is distance traveled.

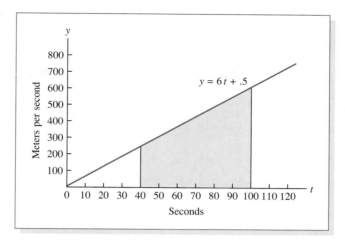

EXAMPLE 8   During the early 1970s, the annual worldwide rate of oil consumption was $R(t) = 16.1e^{.07t}$ billion barrels of oil per year, where t is the number of years since the beginning of 1970.

(a) Determine the amount of oil consumed from 1972 to 1974.

(b) Represent the answer to part (a) as an area.

Solution   (a) We are interested in $T(4) - T(2)$, where $T(t)$ is the total consumption of oil since 1970. This difference is the net change in $T(t)$ over the time period from $t = 2$ (1972) to $t = 4$ (1974). Now, $T(t)$ is an antiderivative of the rate function $R(t)$. Hence

$$T(4) - T(2) = \int_2^4 R(t)\, dt = \int_2^4 16.1e^{.07t}\, dt$$

$$= \frac{16.1}{.07} e^{.07t} \bigg|_2^4 = 230e^{.07(4)} - 230e^{.07(2)}$$

$$\approx 39.76 \text{ billion barrels of oil.}$$

(b) The net change in $T(t)$ is the area of the region under the graph of the rate function $T'(t) = R(t)$ from $t = 2$ to $t = 4$. See Fig. 7. ●

Verification of the Fundamental Theorem of Calculus   We shall give two explanations of why the fundamental theorem of calculus is true. Both convey important ideas about definite integrals and the fundamental theorem.

The first explanation of the fundamental theorem is based directly on the Riemann sum definition of the integral. As we observed earlier, the fundamental theorem may be written in the form

$$\int_a^b F'(x)\, dx = F(b) - F(a), \tag{4}$$

where $F(x)$ is any function with a continuous derivative on $a \le x \le b$. There are three key ideas in an explanation of why (4) is true.

FIGURE 7   Total oil consumed.

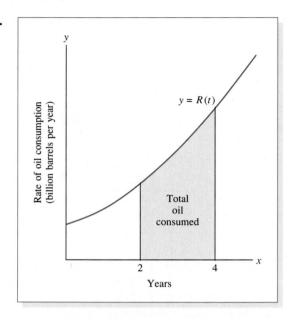

I. If the interval $a \leq x \leq b$ is partitioned into n subintervals of width $\Delta x = (b - a)/n$, then the net change in $F(x)$ over $a \leq x \leq b$ is the sum of the net changes in $F(x)$ over each subinterval.

For example, partition $a \leq x \leq b$ into three subintervals and denote the left endpoints by x_1, x_2, x_3, as shown below.

$$\begin{array}{ccccc} & & & & \\ a = x_1 & & x_2 & x_3 & b \end{array}$$

Then

$$[\text{change in } F(x) \text{ over 1st interval}] = F(x_2) - F(a),$$

$$[\text{change in } F(x) \text{ over 2nd interval}] = F(x_3) - F(x_2),$$

$$[\text{change in } F(x) \text{ over 3rd interval}] = F(b) - F(x_3).$$

When these changes are summed, the intermediate terms cancel:

$$F(b) - F(x_3) + F(x_3) - F(x_2) + F(x_2) - F(a) = F(b) - F(a).$$

II. If Δx is small, then the change in $F(x)$ over the ith subinterval is approximately $F'(x_i)\Delta x$.

This is the approximation of the change in a function discussed in Section 1.8. Here x_i is the left endpoint of the ith subinterval, as shown below.

$$\overleftarrow{\quad \Delta x \quad}\overrightarrow{\quad}$$
$$x_i \qquad x_i + \Delta x$$

III. The sum of the approximations in (II) is a Riemann sum for the definite integral $\int_a^b F'(x)\,dx$.

That is,

$F(b) - F(a)$

$$= \begin{bmatrix} \text{change over} \\ \text{1st subinterval} \end{bmatrix} + \begin{bmatrix} \text{change over} \\ \text{2nd subinterval} \end{bmatrix} + \cdots + \begin{bmatrix} \text{change over} \\ n\text{th subinterval} \end{bmatrix}$$

$$= F'(x_1)\Delta x + F'(x_2)\Delta x + \cdots + F'(x_n)\Delta x.$$

As $\Delta x \to 0$, these approximations improve. Thus, in the limit

$$F(b) - F(a) = \int_a^b F'(x)\,dx.$$

To summarize, the derivative of $F(x)$ determines the approximate change of $F(x)$ on small subintervals. The definite integral sums these approximate changes and in the limit gives the exact change of $F(x)$ over the entire interval $a \le x \le b$.

If you look back at the solution of Example 4 of Section 6.2, you will see essentially the same argument. The distance a rocket travels is the sum of the distances it travels over small intervals of time. On each subinterval, this distance is approximately the velocity (the derivative) times Δt.

An Area Function as an Antiderivative   The second explanation of the fundamental theorem of calculus applies only when $f(x)$ is nonnegative. We begin with the following theorem, which describes the basic relationship between area and antiderivatives. In fact, this theorem is sometimes referred to as an alternative version of the fundamental theorem of calculus.

Theorem III   Let $f(x)$ be a continuous nonnegative function for $a \le x \le b$. Let $A(x)$ be the area of the region under the graph of the function from a to the number x. (See Fig. 8.) Then $A(x)$ is an antiderivative of $f(x)$.

FIGURE 8

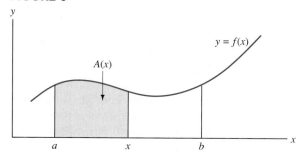

The fundamental theorem of calculus for a nonnegative function follows easily from Theorem III. Let $F(x)$ be an antiderivative of $f(x)$. Since the "area function" $A(x)$ is also an antiderivative of $f(x)$, by Theorem III, we have

$$A(x) = F(x) + C$$

for some constant C. Notice that $A(a)$ is 0 and $A(b)$ equals the area of the region under the graph of $f(x)$ for $a \leq x \leq b$. Therefore,

$$\int_a^b f(x)\,dx = A(b) = A(b) - A(a)$$
$$= [F(b) + C] - [F(a) + C]$$
$$= F(b) - F(a).$$

It is not difficult to explain why Theorem III is reasonable, although we shall not give a formal proof of the theorem. If h is a small positive number, then $A(x + h) - A(x)$ is the area of the shaded region in Fig. 9. This shaded region is approximately a rectangle of width h, height $f(x)$, and area $h \cdot f(x)$. Thus

$$A(x + h) - A(x) \approx h \cdot f(x),$$

FIGURE 9

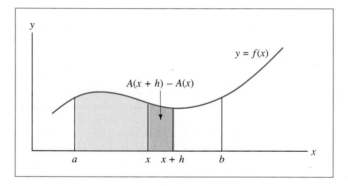

where the approximation becomes better as h approaches zero. Dividing by h, we have

$$\frac{A(x + h) - A(x)}{h} \approx f(x).$$

Since the approximation improves as h approaches zero, the quotient must approach $f(x)$. However, the limit definition of the derivative tells us that the quotient approaches $A'(x)$ as h approaches zero. Therefore, we have $A'(x) = f(x)$. Since x represented any number between a and b, this shows that $A(x)$ is an antiderivative of $f(x)$.

TECHNOLOGY PROJECT 6.5
Using a Graphing Calculator to Explore the Fundamental Theorem of Calculus

Let $f(x) = x^3 - 9x^2 + 15x + 20$. Find an antiderivative $F(x)$ of $f(x)$ such that $F(0) = 0$. (Do this without using technology.)

1. Graph $f(x)$ and $F(x)$ using the window $-2 \le x \le 8$, $-10 \le y \le 100$.
2. Determine an interval $a \le x \le b$ on which $f(x)$ has nonnegative values. Describe what happens to the graph of $F(x)$ on this interval.
3. Use the TRACE to estimate the values of x at which $f(x)$ has a relative extreme point. Describe what happens to the graph of $F(x)$ at these values of x.

TECHNOLOGY PROJECT 6.6
Using a Graphing Calculator to Explore the Fundamental Theorem of Calculus, II

Let $f(x) = x^3 - 7x^2 + 8x + 10$. Find an antiderivative $F(x)$ of $f(x)$ such that $F(0) = 0$. (Do this without using technology.)

1. Graph $f(x)$ and $F(x)$ using the window $-1 \le x \le 10$, $-10 \le y \le 50$.
2. Using TRACE, estimate the nonnegative values of x at which $F(x)$ has a relative extreme point. Describe what happens to the graph of $f(x)$ at these values of x.
3. Use the graph of $f(x)$ to explain why $F(x)$ is decreasing over the interval between the relative extreme points found in problem 2.

PRACTICE PROBLEMS 6.3

1. Find the area under the curve $y = e^{x/2}$ from $x = -3$ to $x = 2$.
2. Let $MR(x)$ be a company's marginal revenue at production level x. Give an economic interpretation of the number $\int_{75}^{80} MR(x)\, dx$.

EXERCISES 6.3

Calculate the following definite integrals.

1. $\displaystyle\int_{-1}^{1} x\, dx$

2. $\displaystyle\int_{4}^{5} e^{2x}\, dx$

3. $\displaystyle\int_{1}^{2} 5\, dx$

4. $\displaystyle\int_{-1}^{-1/2} \frac{1}{x^2}\, dx$

5. $\displaystyle\int_{1}^{2} 8x^3\, dx$

6. $\displaystyle\int_{0}^{1} e^{x/3}\, dx$

7. $\displaystyle\int_{0}^{1} 4e^{-3x}\, dx$

8. $\displaystyle\int_{1}^{3} \frac{5}{x}\, dx$

9. $\displaystyle\int_{1}^{4} 3\sqrt{x}\, dx$

10. $\int_{1}^{8} 2x^{1/3} \, dx$ **11.** $\int_{0}^{5} e^{-2t} \, dt$ **12.** $\int_{0}^{1} \frac{5}{e^{3t}} \, dt$

13. $\int_{3}^{6} x^{-1} \, dx$ **14.** $\int_{1}^{3} (5t - 1)^3 \, dt$ **15.** $\int_{-1}^{1} \frac{4}{(t + 2)^3} \, dt$

16. $\int_{-3}^{0} \sqrt{25 + 3t} \, dt$ **17.** $\int_{2}^{3} (5 - 2t)^4 \, dt$ **18.** $\int_{4}^{9} \frac{3}{t - 2} \, dt$

19. $\int_{0}^{3} (x^3 + x - 7) \, dx$ **20.** $\int_{-5}^{5} (e^{x/10} - x^2 - 1) \, dx$

21. $\int_{2}^{4} \left(x^2 + \frac{2}{x^2} - \frac{1}{x + 5} \right) dx$ **22.** $\int_{1}^{2} (4x^3 + 3x^{-4} - 5) \, dx$

Find the area under each of the given curves.

23. $y = 4x; x = 2$ to $x = 3$ **24.** $y = 3x^2; x = -1$ to $x = 1$

25. $y = e^{x/2}; x = 0$ to $x = 1$ **26.** $y = \sqrt{x}; x = 0$ to $x = 4$

27. $y = (x - 3)^4; x = 1$ to $x = 4$ **28.** $y = e^{3x}; x = -\frac{1}{3}$ to $x = 0$

Find the area under each of the given curves by antidifferentiation and use elementary geometry to check the answer.

29. $y = 5; x = -1$ to $x = 2$ **30.** $y = x + 1; x = 2$ to $x = 4$

31. $y = 2x; x = 0$ to $x = 3$ **32.** $y = 2x + 1; x = 0$ to $x = 3$

33. Use geometry to evaluate $\int_{0}^{3} f(x) \, dx$, where the graph of $f(x)$ is shown in Fig. 10.

34. Determine $\int_{0}^{13} f(x) \, dx$, where the graph of $f(x)$ is shown in Fig. 11. The graph consists of two straight line segments and two quarter-circles of radius 3. [*Hint:* The area of a triangle is $\frac{1}{2}$[base]·[altitude] and the area of a quarter-circle is $\frac{1}{4}\pi r^2$.]

FIGURE 10

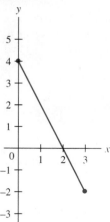

FIGURE 11

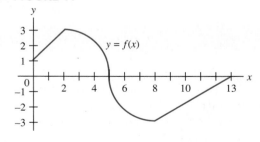

$y = f(x)$

35. Let $f(x)$ be the function pictured in Fig. 12. Determine whether $\int_{0}^{7} f(x) \, dx$ is positive, negative, or zero.

36. Let $g(x)$ be the function pictured in Fig. 13. Determine whether $\int_{0}^{7} g(x) \, dx$ is positive, negative, or zero.

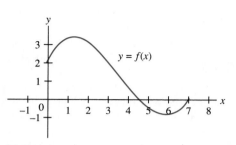

FIGURE 12

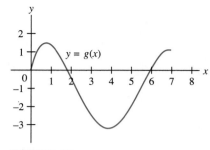

FIGURE 13

37. The worldwide rate of cigarette consumption (in trillions of cigarettes per year) since 1960 is given approximately by the function $c(t) = .1t + 2.4$, where $t = 0$ corresponds to 1960. Determine the number of cigarettes sold from 1980 to 1990.

38. Suppose $p(t)$ is the rate (in tons per year) pollutants are discharged into a lake, where t is the number of years since 1985. Interpret $\int_5^7 p(t)\, dt$.

39. A helicopter is rising straight up in the air. Its velocity at time t is $2t + 1$ feet per second.

    (a) How high does the helicopter rise during the first 5 seconds?

    (b) Represent the answer to part (a) as an area.

40. After t hours of operation, an assembly line is producing power lawn mowers at the rate of $21 - \frac{4}{3}t$ mowers per hour.

    (a) How many mowers are produced during the time from $t = 2$ to $t = 5$ hours?

    (b) Represent the answer to part (a) as an area.

41. Suppose that the marginal cost function of a handbag manufacturer is $\frac{3}{32}x^2 - x + 200$ dollars per unit at production level x (where x is measured in units of 100 handbags).

    (a) Find the total cost of producing 6 additional units if 2 units are currently being produced.

    (b) Describe the answer to part (a) as an area. (Give a written description rather than a sketch.)

42. Suppose that the marginal profit function for a company is $100 + 50x - 3x^2$ at production level x.

    (a) Find the extra profit earned from the sale of 3 additional units if 5 units are currently being produced.

    (b) Describe the answer to part (a) as an area. (Do not make a sketch.)

43. Let $MP(x)$ be a company's marginal profit at production level x. Give an economic interpretation of the number $\int_{44}^{48} MP(x)\, dx$.

44. Let $MC(x)$ be a company's marginal cost at production level x. Give an economic interpretation of the number $\int_0^{100} MC(x)\, dx$. [*Note:* At any production level, the total cost equals the fixed cost plus the total variable cost.]

45. Some food is placed in a freezer. After t hours the temperature of the food is dropping at the rate of $r(t)$ degrees Fahrenheit per hour, where $r(t) = 12 + 4/(t + 3)^2$.

    (a) Compute the area under the graph of $y = r(t)$ over the interval $0 \le t \le 2$.

    (b) What does the area in part (a) represent?

46. Suppose that the velocity of a car at time t is $40 + 8/(t + 1)^2$ kilometers per hour.

(a) Compute the area under the velocity curve from $t = 1$ to $t = 9$.

(b) What does the area in part (a) represent?

47. Deforestation is one of the major problems facing sub-Saharan Africa. Although the clearing of land for farming has been the major cause, the steadily increasing demand for fuelwood has become a significant factor. Figure 14 summarizes projections of the World Bank. The rate of fuelwood consumption (in millions of cubic meters per year) in the Sudan t years after 1980 is given approximately by the function $c(t) = 76.2e^{.03t}$. Determine the amount of fuelwood that will be consumed from 1980 to 2000.

FIGURE 14 Data from "Sudan and Options in the Energy Sector," World Bank/ UNDP, July 1983.

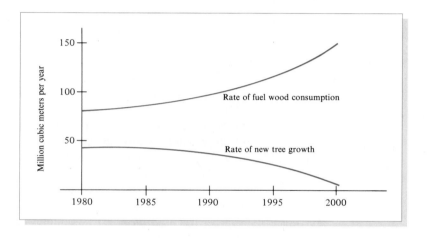

48. (a) Compute $\displaystyle\int_1^b \frac{1}{t}\, dt$, where $b > 1$.

(b) Explain how the logarithm of a number greater than 1 may be interpreted as the area of a region under a curve. (What is the curve?)

49. For each positive number x, let $A(x)$ be the area of the region under the curve $y = x^2 + 1$ from 0 to x. Find $A'(3)$.

50. For each number $x > 2$, let $A(x)$ be the area of the region under the curve $y = x^3$ from 2 to x. Find $A'(6)$.

SOLUTIONS TO PRACTICE PROBLEMS 6.3

1. The desired area is

$$\int_{-3}^2 e^{x/2}\, dx = 2e^{x/2}\Big|_{-3}^2 = 2e - 2e^{-3/2} \approx 4.99.$$

2. The number $\int_{75}^{80} MR(x)\, dx$ is the net change in revenue received when the production level is raised from $x = 75$ to $x = 80$ units.

6.4 AREAS IN THE xy-PLANE

In this section we show how to use the definite integral to compute the area of a region that lies between the graphs of two or more functions. Three simple but important properties of the integral will be used repeatedly.

Let $f(x)$ and $g(x)$ be functions and a, b, and k be any constants. Then

$$\int_a^b f(x)\, dx + \int_a^b g(x)\, dx = \int_a^b [f(x) + g(x)]\, dx \tag{1}$$

$$\int_a^b f(x)\, dx - \int_a^b g(x)\, dx = \int_a^b [f(x) - g(x)]\, dx \tag{2}$$

$$\int_a^b kf(x)\, dx = k\int_a^b f(x)\, dx. \tag{3}$$

To verify (1), let $F(x)$ and $G(x)$ be antiderivatives of $f(x)$ and $g(x)$, respectively. Then $F(x) + G(x)$ is an antiderivative of $f(x) + g(x)$. By the fundamental theorem of calculus,

$$\int_a^b [f(x) + g(x)]\, dx = [F(x) + G(x)]\Big|_a^b$$

$$= [F(b) + G(b)] - [F(a) + G(a)]$$

$$= [F(b) - F(a)] + [G(b) - G(a)]$$

$$= \int_a^b f(x)\, dx + \int_a^b g(x)\, dx.$$

The verifications of (2) and (3) are similar and use the facts that $F(x) - G(x)$ is an antiderivative of $f(x) - g(x)$ and $kF(x)$ is an antiderivative of $kf(x)$.

Let us now consider regions that are bounded both above and below by graphs of functions. Referring to Fig. 1, we would like to find a simple expression for the area of the shaded region under the graph of $y = f(x)$ and above the graph of $y = g(x)$ from $x = a$ to $x = b$. It is the region under the graph of $f(x)$ with the region under the graph of $g(x)$ taken away. Therefore,

[area of shaded region] = [area under $f(x)$] − [area under $g(x)$]

$$= \int_a^b f(x)\, dx - \int_a^b g(x)\, dx$$

$$= \int_a^b [f(x) - g(x)]\, dx \qquad \text{[by property (2)].}$$

FIGURE I

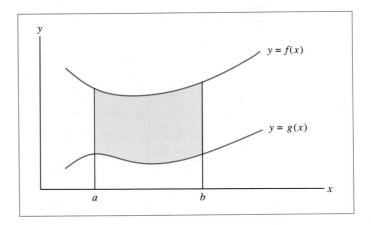

Area Between Two Curves   If $y = f(x)$ lies above $y = g(x)$ from $x = a$ to $x = b$, the area of the region between $f(x)$ and $g(x)$ from $x = a$ to $x = b$ is

$$\int_a^b [f(x) - g(x)]\, dx.$$

EXAMPLE I   Find the area of the region between $y = 2x^2 - 4x + 6$ and $y = -x^2 + 2x + 1$ from $x = 1$ to $x = 2$.

Solution   Upon sketching the two graphs (Fig. 2), we see that $f(x) = 2x^2 - 4x + 6$ lies above $g(x) = -x^2 + 2x + 1$ for $1 \le x \le 2$. Therefore, our formula gives the area of the shaded region as

$$\int_1^2 [(2x^2 - 4x + 6) - (-x^2 + 2x + 1)]\, dx = \int_1^2 (3x^2 - 6x + 5)\, dx$$

$$= (x^3 - 3x^2 + 5x)\Big|_1^2$$

$$= 6 - 3 = 3. \ \bullet$$

EXAMPLE 2   Find the area of the region between $y = x^2$ and $y = (x - 2)^2 = x^2 - 4x + 4$ from $x = 0$ to $x = 3$.

Solution   Upon sketching the graphs (Fig. 3), we see that the two graphs cross; by setting $x^2 = x^2 - 4x + 4$, we find that they cross when $x = 1$. Thus one graph does not always lie above the other from $x = 0$ to $x = 3$, so that we cannot directly apply our rule for finding the area between two curves. However, the difficulty is easily surmounted if we break the region into two parts, namely, the area from $x = 0$ to $x = 1$ and the area from $x = 1$ to $x = 3$. For from $x = 0$ to $x = 1$,

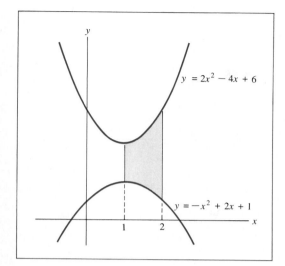

FIGURE 2

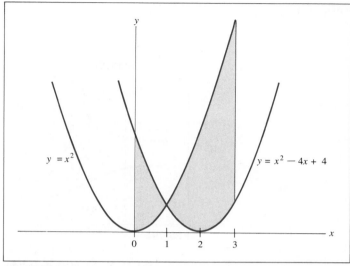

FIGURE 3

$y = x^2 - 4x + 4$ is on top; and from $x = 1$ to $x = 3$, $y = x^2$ is on top. Consequently,

$$[\text{area from } x = 0 \text{ to } x = 1] = \int_0^1 [(x^2 - 4x + 4) - (x^2)]\, dx$$

$$= \int_0^1 (-4x + 4)\, dx$$

$$= (-2x^2 + 4x)\Big|_0^1 = 2 - 0 = 2.$$

$$[\text{area from } x = 1 \text{ to } x = 3] = \int_1^3 [(x^2) - (x^2 - 4x + 4)]\, dx$$

$$= \int_1^3 (4x - 4)\, dx$$

$$= (2x^2 - 4x)\Big|_1^3$$

$$= 6 - (-2) = 8.$$

Thus the total area is $2 + 8 = 10$. ●

In our derivation of the formula for the area between two curves, we examined functions that are nonnegative. However, the statement of the rule does not contain this stipulation, and rightfully so. Consider the case where $f(x)$ and $g(x)$ are not always positive. Let us determine the area of the shaded region in Fig. 4(a). Select some constant c such that the graphs of the functions

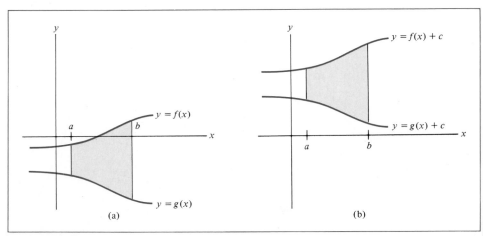

FIGURE 4

$f(x) + c$ and $g(x) + c$ lie completely above the x-axis [Fig. 4(b)]. The region between them will have the same area as the original region. Using the rule as applied to nonnegative functions, we have

$$[\text{area of the region}] = \int_a^b [(f(x) + c) - (g(x) + c)] \, dx$$

$$= \int_a^b [f(x) - g(x)] \, dx.$$

Therefore, we see that our rule is valid for any functions $f(x)$ and $g(x)$ as long as the graph of $f(x)$ lies above the graph of $g(x)$ for all x from $x = a$ to $x = b$.

EXAMPLE 3   Set up the integral that gives the area between the curves $y = x^2 - 2x$ and $y = -e^x$ from $x = -1$ to $x = 2$.

Solution   Since $y = x^2 - 2x$ lies above $y = -e^x$ (Fig. 5), the rule for finding the area between two curves can be applied directly. The area between the curves is

$$\int_{-1}^2 (x^2 - 2x + e^x) \, dx. \; \bullet$$

Sometimes we are asked to find the area between two curves without being given the values of a and b. In these cases there is a region that is completely enclosed by the two curves. As the next examples illustrate, we must first find the points of intersection of the two curves in order to obtain the values of a and b. In such problems careful curve sketching is especially important.

EXAMPLE 4   Set up the integral that gives the area bounded by the curves $y = x^2 + 2x + 3$ and $y = 2x + 4$.

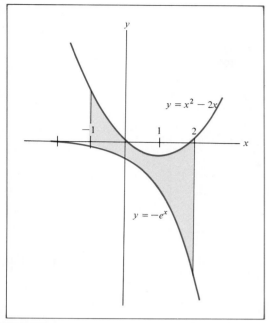

FIGURE 5

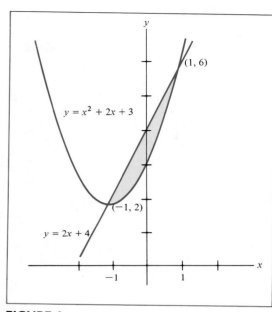

FIGURE 6

Solution The two curves are sketched in Fig. 6, and the region bounded by them is shaded. In order to find the points of intersection, we set $x^2 + 2x + 3 = 2x + 4$ and solve for x. We obtain $x^2 = 1$, or $x = -1$ and $x = +1$. When $x = -1$, $2x + 4 = 2(-1) + 4 = 2$. When $x = 1$, $2x + 4 = 2(1) + 4 = 6$. Thus the curves intersect at the points $(1, 6)$ and $(-1, 2)$.

Since $y = 2x + 4$ lies above $y = x^2 + 2x + 3$ from $x = -1$ to $x = 1$, the area between the curves is given by

$$\int_{-1}^{1} [(2x + 4) - (x^2 + 2x + 3)] \, dx = \int_{-1}^{1} (1 - x^2) \, dx. \quad \bullet$$

EXAMPLE 5 Set up the integral that gives the area bounded by the two curves $y = 2x^2$ and $y = x^3 - 3x$.

Solution First we make a rough sketch of the two curves, as in Fig. 7. The curves intersect where $x^3 - 3x = 2x^2$, or $x^3 - 2x^2 - 3x = 0$. Note that

$$x^3 - 2x^2 - 3x = x(x^2 - 2x - 3) = x(x - 3)(x + 1).$$

So the solutions to $x^3 - 2x^2 - 3x = 0$ are $x = 0, 3, -1$, and the curves intersect at $(-1, 2)$, $(0, 0)$, and $(3, 18)$. From $x = -1$ to $x = 0$, the curve $y =$

FIGURE 7

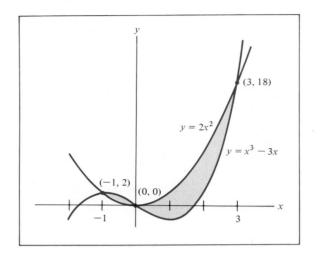

$x^3 - 3x$ lies above $y = 2x^2$. But from $x = 0$ to $x = 3$, the reverse is true. Thus the area between the curves is given by

$$\int_{-1}^{0} (x^3 - 3x - 2x^2) \, dx + \int_{0}^{3} (2x^2 - x^3 + 3x) \, dx. \quad \bullet$$

EXAMPLE 6   Beginning in 1974, with the advent of dramatically higher oil prices, the exponential rate of growth of world oil consumption slowed down from a growth constant of 7% to a growth constant of 4% per year. A fairly good model for the annual rate of oil consumption since 1974 is given by

$$R_1(t) = 21.3e^{.04(t-4)}, \qquad t \geq 4,$$

where $t = 0$ corresponds to 1970. Determine the total amount of oil saved between 1976 and 1980 by not consuming oil at the rate predicted by the model of Example 1, Section 6.1, namely,

$$R(t) = 16.1e^{.07t}, \qquad t \geq 0.$$

Solution   If oil consumption had continued to grow as it did prior to 1974, then the total oil consumed between 1976 and 1980 would have been

$$\int_{6}^{10} R(t) \, dt. \tag{4}$$

However, taking into account the slower increase in the rate of oil consumption since 1974, we find that the total oil consumed between 1976 and 1980 was approximately

$$\int_{6}^{10} R_1(t) \, dt. \tag{5}$$

The integrals in (4) and (5) may be interpreted as the areas under the curves $y = R(t)$ and $y = R_1(t)$, respectively, from $t = 6$ to $t = 10$. (See Fig. 8.) By

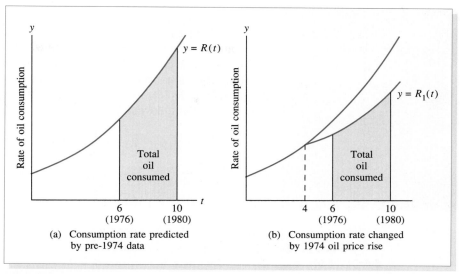

(a) Consumption rate predicted by pre-1974 data

(b) Consumption rate changed by 1974 oil price rise

FIGURE 8

superimposing the two curves we see that the area between them from $t = 6$ to $t = 10$ represents the total oil that was saved by consuming oil at the rate given by $R_1(t)$ instead of $R(t)$. (See Fig. 9.) The area between the two curves equals

$$\int_6^{10} [R(t) - R_1(t)] \, dt = \int_6^{10} [16.1e^{.07t} - 21.3e^{.04(t-4)}] \, dt$$

$$= \left(\frac{16.1}{.07} e^{.07t} - \frac{21.3}{.04} e^{.04(t-4)} \right) \Big|_6^{10}$$

$$\approx 13.02.$$

Thus about 13 billion barrels of oil were saved between 1976 and 1980. •

FIGURE 9

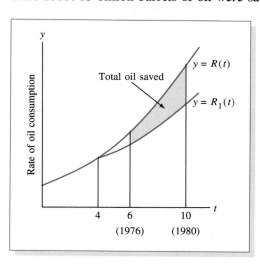

TECHNOLOGY PROJECT 6.7
Calculating the Area Between Curves Using a Graphing Calculator

As you have already seen from the examples above, one of the most tedious tasks in calculating the area bounded by curves is calculating the intersection points of the curves. Of course, this can be done using a graphing calculator to solve the resulting equations and then setting up the necessary integrals and allowing the program to calculate their values. Use this approach to calculate the areas bounded by the following curves:

1. $y = 5x - 3$, $y = 2x^2 + x - 4$ 2. $y = 1 + x$, $y = x^3 - 5x + 1$

3. $y = \dfrac{50}{x + 1}$, $y = 3x + 1$, $y = 0$, $x = 0$

4. $y = x^2 + x + 1$, $y = \sqrt{x} + 3$, $x = 0$

PRACTICE PROBLEMS 6.4

1. Find the area between the curves $y = x + 3$ and $y = \frac{1}{2}x^2 + x - 7$ from $x = -2$ to $x = 1$.

2. A company plans to increase its production from 10 to 15 units per day. The present marginal cost function is $MC_1(x) = x^2 - 20x + 108$. By redesigning the production process and purchasing new equipment, the company can change the marginal cost function to $MC_2(x) = \frac{1}{2}x^2 - 12x + 75$. Determine the area between the graphs of the two marginal cost curves from $x = 10$ to $x = 15$. Interpret this area in economic terms.

EXERCISES 6.4

1. Write down a definite integral or sum of definite integrals that gives the area of the shaded portion of Fig. 10.
2. Write down a definite integral or sum of definite integrals that gives the area of the shaded portion of Fig. 11.

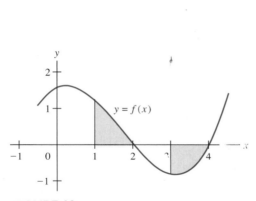

FIGURE 10

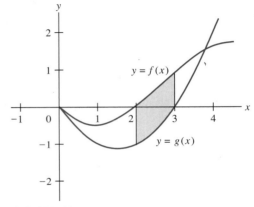

FIGURE 11

3. Shade the portion of Fig. 12 whose area is given by the integral

$$\int_0^2 [f(x) - g(x)] \, dx + \int_2^4 [h(x) - g(x)] \, dx.$$

4. Shade the portion of Fig. 13 whose area is given by the integral

$$\int_0^1 [f(x) - g(x)] \, dx + \int_1^2 [g(x) - f(x)] \, dx.$$

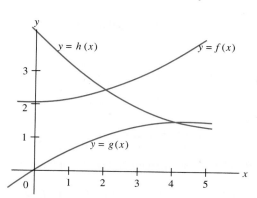

FIGURE 12

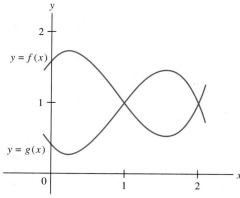

FIGURE 13

Find the area of the region between the curves.

5. $y = 2x^2$ and $y = 8$ (a horizontal line) from $x = -2$ to $x = 2$.

6. $y = 13 - 3x^2$ and $y = 1$ from $x = -2$ to $x = 2$.

7. $y = x^2 - 6x + 12$ and $y = 1$ from $x = 0$ to $x = 4$.

8. $y = x(2 - x)$ and $y = 4$ from $x = 0$ to $x = 2$.

9. $y = 3x^2$ and $y = -3x^2$ from $x = -1$ to $x = 2$.

10. $y = e^{2x}$ and $y = -e^{2x}$ from $x = -1$ to $x = 1$.

Find the area of the region bounded by the curves.

11. $y = x^2 + x$ and $y = 3 - x$.

12. $y = 3x - x^2$ and $y = 4 - 2x$.

13. $y = -x^2 + 6x - 5$ and $y = 2x - 5$.

14. $y = 2x^2 + x - 7$ and $y = x + 1$.

15. $y = x^2$ and $y = 18 - x^2$.

16. $y = 4x^2 - 24x + 20$ and $y = 2 - 2x^2$.

17. $y = x^2 - 6x - 7$ and $y = -2x^2 + 6x + 8$.

18. $y = 2x^2 - 8x + 8$ and $y = -x^2 + 7x + 8$.

19. Find the area of the region between $y = x^2 - 3x$ and the *x*-axis

(a) from $x = 0$ to $x = 3$, (b) from $x = 0$ to $x = 4$,

(c) from $x = -2$ to $x = 3$.

20. Find the area of the region between $y = x^2$ and $y = 1/x^2$

   (a) from $x = 1$ to $x = 4$,   (b) from $x = \frac{1}{2}$ to $x = 4$.

21. Find the area of the region bounded by $y = 1/x^2$, $y = x$, and $y = 8x$, for $x \geq 0$. [See Fig. 14(a).]

FIGURE 14

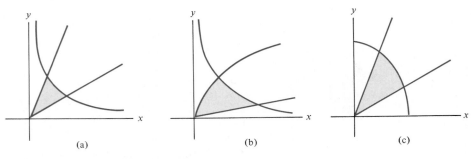

(a)   (b)   (c)

22. Find the area of the region bounded by $y = 1/x$, $y = 4x$, and $y = x/2$, for $x \geq 0$. [The region resembles the shaded region in Fig. 14(a).]

23. Find the area of the shaded region in Fig. 14(b) bounded by $y = 12/x$, $y = \frac{3}{2}\sqrt{x}$, and $y = x/3$.

24. Find the area of the shaded region in Fig. 14(c) bounded by $y = 12 - x^2$, $y = 4x$, and $y = x$.

25. Refer to Exercise 47 of Section 6.3. The rate of new tree growth (in millions of cubic meters per year) in the Sudan t years after 1980 is given approximately by the function $g(t) = 50 - 6.03e^{.09t}$. Set up the definite integral giving the amount of depletion of the forests due to the excess of fuel wood consumption over new growth from 1980 to 2000.

26. Refer to the oil-consumption data in Example 1, Section 6.1. Suppose that in 1970 the growth constant for the annual rate of oil consumption had been held to .04. What effect would this action have had on oil consumption from 1970 to 1974?

27. The marginal profit for a certain company is $MP_1(x) = -x^2 + 14x - 24$. The company expects the daily production level to rise from $x = 6$ to $x = 8$ units. The management is considering a plan that would have the effect of changing the marginal profit to $MP_2(x) = -x^2 + 12x - 20$. Should the company adopt the plan? Determine the area between the graphs of the two marginal profit functions from $x = 6$ to $x = 8$. Interpret this area in economic terms.

FIGURE 15 Velocity functions of two cars.

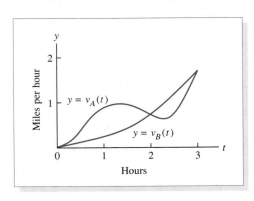

value of $f(x)$ on this interval. To calculate the average of a collection of numbers $y_1, y_2, \ldots, y_n$, we add the numbers and divide by n to obtain

$$\frac{y_1 + y_2 + \cdots + y_n}{n}$$

To determine the average value of $f(x)$, we proceed similarly. Choose n values of x, say $x_1, x_2, \ldots, x_n$, and calculate the corresponding function values $f(x_1)$, $f(x_2), \ldots, f(x_n)$. The average of these values is

$$\frac{f(x_1) + f(x_2) + \cdots + f(x_n)}{n}. \tag{1}$$

Our goal now is to obtain a reasonable definition of the average of all the values of $f(x)$ on the interval $a \le x \le b$. If the points $x_1, x_2, \ldots, x_n$ are spread "evenly" throughout the interval, then the average (1) should be a good approximation to our intuitive concept of the average value of $f(x)$. In fact, as n becomes large, the average (1) should approximate the average value of $f(x)$ to any arbitrary degree of accuracy. To guarantee that the points $x_1, x_2, \ldots, x_n$ are "evenly" spread out from a to b, let us divide the interval from $x = a$ to $x = b$ into n subintervals of equal length $\Delta x = (b - a)/n$. Then choose x_1 from the first subinterval, x_2 from the second, and so forth. The average (1) that corresponds to these points may be arranged in the form of a Riemann sum as follows:

$$\frac{f(x_1) + f(x_2) + \cdots + f(x_n)}{n}$$

$$= f(x_1) \cdot \frac{1}{n} + f(x_2) \cdot \frac{1}{n} + \cdots + f(x_n) \cdot \frac{1}{n}$$

$$= \frac{1}{b - a} \left[f(x_1) \cdot \frac{b - a}{n} + f(x_2) \cdot \frac{b - a}{n} + \cdots + f(x_n) \cdot \frac{b - a}{n} \right]$$

$$= \frac{1}{b - a} [f(x_1)\Delta x + f(x_2)\Delta x + \cdots + f(x_n)\Delta x].$$

The sum inside the brackets is a Riemann sum for the definite integral of $f(x)$. Thus we see that for a large number of points x_i, the average in (1) approaches the quantity

$$\frac{1}{b - a} \int_a^b f(x) \, dx.$$

This argument motivates the following definition.

The *average value* of a continuous function $f(x)$ over the interval $a \le x \le b$ is defined as the quantity

$$\frac{1}{b - a} \int_a^b f(x) \, dx \tag{2}$$

EXAMPLE 2   Compute the average value of $f(x) = \sqrt{x}$ over the interval $0 \le x \le 9$.

Solution   Using (2) with $a = 0$ and $b = 9$, the average value of $f(x) = \sqrt{x}$ over the interval $0 \le x \le 9$ is equal to

$$\frac{1}{9 - 0} \int_0^9 \sqrt{x} \, dx.$$

Since $\sqrt{x} = x^{1/2}$, an antiderivative of $\sqrt{x}$ is $\frac{2}{3}x^{3/2}$. Therefore,

$$\frac{1}{9} \int_0^9 \sqrt{x} \, dx = \frac{1}{9}\left(\frac{2}{3}x^{3/2}\right)\Big|_0^9 = \frac{1}{9}\left(\frac{2}{3} \cdot 9^{3/2} - 0\right) = \frac{1}{9}\left(\frac{2}{3} \cdot 27\right) = 2,$$

so that the average value of $\sqrt{x}$ over the interval $0 \le x \le 9$ is 2. The area of the shaded region is the same as the area of the rectangle pictured in Fig. 1. ●

FIGURE 1   Average value of a function.

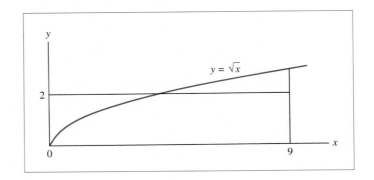

EXAMPLE 3   Suppose that the current world population is 5 billion and the population t years from now is given by the exponential growth law

$$P(t) = 5e^{.023t}.$$

Determine the average population of the earth during the next 30 years. (This average is important in long-range planning for agricultural production and the allocation of goods and services.)

Solution   The average value of the population $P(t)$ from $t = 0$ to $t = 30$ is

$$\frac{1}{30 - 0} \int_0^{30} P(t) \, dt = \frac{1}{30} \int_0^{30} 5e^{.023t} \, dt$$

$$= \frac{1}{30}\left(\frac{5}{.023}e^{.023t}\right)\Big|_0^{30}$$

$$= \frac{5}{.69}(e^{.69} - 1)$$

$$\approx 7.2 \text{ billion.} \quad ●$$

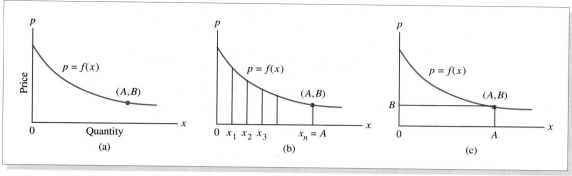

FIGURE 2   Consumers' surplus.

Consumers' Surplus   Using a demand curve from economics, we can derive a formula showing the amount that consumers benefit from an open system that has no price discrimination. Figure 2(a) is a *demand curve* for a commodity. It is determined by complex economic factors and gives a relationship between the quantity sold and the unit price of a commodity. Specifically, it says that, in order to sell x units, the price must be set at $f(x)$ dollars per unit. Since, for most commodities, selling larger quantities requires a lowering of the price, demand functions are usually decreasing. Interactions between supply and demand determine the amount of a quantity available. Let A designate the amount of the commodity currently available and $B = f(A)$ the current selling price.

Divide the interval from 0 to A into n subintervals, each of length $\Delta x = (A - 0)/n$, and take x_i to be the right-hand endpoint of the ith interval. Consider the first subinterval, from 0 to x_1. [See Fig. 2(b).] Suppose that only x_1 units had been available. Then the price per unit could have been set at $f(x_1)$ dollars and these x_1 units sold. Of course, at this price we could not have sold any more units. However, those people who paid $f(x_1)$ dollars had a great demand for the commodity. It was extremely valuable to them, and there was no advantage in substituting another commodity at that price. They were actually paying what the commodity was worth to them. In theory, then, the first x_1 units of the commodity could be sold to these people at $f(x_1)$ dollars per unit, yielding (price per unit) $\cdot$ (number of units) $= f(x_1) \cdot (x_1) = f(x_1) \cdot \Delta x$ dollars.

After selling the first x_1 units, suppose that more units become available, so that now a total of x_2 units have been produced. Setting the price at $f(x_2)$, the remaining $x_2 - x_1 = \Delta x$ units can be sold, yielding $f(x_2) \cdot \Delta x$ dollars. Here, again, the second group of buyers would have paid as much for the commodity as it was worth to them. Continuing this process of price discrimination, the amount of money paid by the consumers would be

$$f(x_1)\Delta x + f(x_2)\Delta x + \cdots + f(x_n)\Delta x.$$

Taking n large, we see that this Riemann sum approaches $\int_0^A f(x)\,dx$. Since $f(x)$ is positive, this integral equals the area under the graph of $f(x)$ from $x = 0$ to $x = A$.

Of course, in an open system, everyone pays the same price, B, so the total amount paid is [price per unit] · [number of units] = BA. Since BA is the area of the rectangle under the graph of the line $p = B$ from $x = 0$ to $x = A$, the amount of money saved by the consumers is the area of the shaded region in Fig. 2(c). That is, the area between the curves $p = f(x)$ and $p = B$ gives a numerical value to one benefit of a modern efficient economy.

The *consumers' surplus* for a commodity having demand curve $p = f(x)$ is

$$\int_0^A [f(x) - B] \, dx,$$

where the quantity demanded is A and the price is $B = f(A)$.

EXAMPLE 4   Find the consumers' surplus for the demand curve $p = 50 - .06x^2$ at the sales level 20.

Solution   Since 20 units are sold, the price must be

$$B = 50 - .06(20)^2$$
$$= 50 - 24 = 26.$$

Therefore, the consumers' surplus is

$$\int_0^{20} [(50 - .06x^2) - 26] \, dx = \int_0^{20} (24 - .06x^2) \, dx$$

$$= 24x - .02x^3 \Big|_0^{20}$$

$$= 24(20) - .02(20)^3$$

$$= 480 - 160 = 320.$$

That is, the consumers' surplus is $320. ●

Future Value of an Income Stream   The next example shows how the definite integral can be used to approximate the sum of a large number of terms.

EXAMPLE 5   Suppose that money is deposited daily into a savings account at an annual rate of $1000. The account pays 6% interest compounded continuously. Approximate the amount of money in the account at the end of 5 years.

Solution   Divide the time interval from 0 to 5 years into daily subintervals. Each subinterval is then of duration $\Delta t = \frac{1}{365}$ years. Let $t_1, t_2, \ldots, t_n$ be points chosen from these subintervals. Since we deposit money at an annual rate of $1000, the amount deposited during one of the subintervals is $1000\Delta t$ dollars. If this amount is deposited at time t_i, the $1000\Delta t$ dollars will earn interest for the remaining $5 - t_i$ years. The total amount resulting from this one deposit at time t_i is then

$$1000\Delta t e^{.06(5-t_i)}.$$

Add the effects of the deposits at times $t_1, t_2, \ldots, t_n$ to arrive at the total balance in the account:

$$A = 1000e^{.06(5-t_1)} \Delta t + 1000e^{.06(5-t_2)} \Delta t + \cdots + 1000e^{.06(5-t_n)} \Delta t.$$

This is a Riemann sum for the function $f(t) = 1000e^{.06(5-t)}$ on the interval $0 \le t \le 5$. Since Δt is very small when compared with the interval, the total amount in the account, A, is approximately

$$\int_0^5 1000e^{.06(5-t)} \, dt = \frac{1000}{-.06} e^{.06(5-t)} \Big|_0^5$$

$$= \frac{1000}{-.06}(1 - e^{.3})$$

$$\approx 5831.$$

That is, the approximate balance in the account at the end of 5 years is \$5831. ●

Note The antiderivative was computed by observing that since the derivative of $e^{.06(5-t)}$ is $e^{.06(5-t)}(-.06)$, we must divide the integrand by $-.06$ to obtain an antiderivative.

In Example 5, money was deposited daily into the account. If the money had been deposited several times a day, the definite integral would have given an even better approximation to the balance. Actually, the more frequently the money is deposited, the better the approximation. Economists consider a hypothetical situation where money is deposited steadily throughout the year. This flow of money is called a *continuous income stream,* and the balance in the account is given exactly by the definite integral.

FIGURE 3

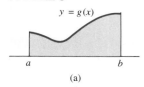

$y = g(x)$

a b

(a)

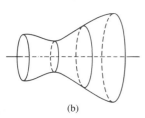

(b)

The *future value of a continuous income stream* of P dollars per year for N years at interest rate r compounded continuously is

$$\int_0^N P e^{r(N-t)} \, dt.$$

Solids of Revolution When the region of Fig. 3(a) is revolved about the x-axis, it sweeps out a solid [Fig. 3(b)]. Riemann sums can be used to derive a formula for this *solid of revolution.* Let us break the x-axis between a and b into a large number n of equal subintervals, each of length $\Delta x = (b - a)/n$. Using each subinterval as a base, we can divide the region into strips (see Fig. 4).

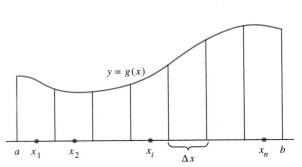

FIGURE 4

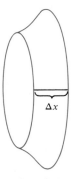
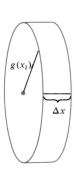

FIGURE 5

Let x_i be a point in the ith subinterval. Then the volume swept out by revolving the ith strip is approximately the same as the volume of the cylinder swept out by revolving the rectangle of height $g(x_i)$ and base Δx around the x-axis (Fig. 5). The volume of the cylinder is

$$[\text{area of circular side}] \cdot [\text{width}] = \pi[g(x_i)]^2 \cdot \Delta x.$$

The total volume swept out by all the strips is approximated by the total volume swept out by the rectangles, which is

$$[\text{volume}] \approx \pi[g(x_1)]^2 \, \Delta x + \pi[g(x_2)]^2 \, \Delta x + \cdots + \pi[g(x_n)]^2 \, \Delta x.$$

As n gets larger and larger, this approximation becomes arbitrarily close to the true volume. The expression on the right is a Riemann sum for the definite integral of $f(x) = \pi[g(x)]^2$. Therefore, the volume of the solid equals the value of the definite integral.

> The volume of the *solid of revolution* obtained from revolving the region below the graph of $y = g(x)$ from $x = a$ to $x = b$ about the x-axis is
>
> $$\int_a^b \pi[g(x)]^2 \, dx.$$

EXAMPLE 6   Find the volume of the solid of revolution obtained by revolving the region of Fig. 6 about the x-axis.

Solution   Here $g(x) = e^{kx}$, and

$$[\text{volume}] = \int_0^1 \pi(e^{kx})^2 \, dx$$

$$= \int_0^1 \pi e^{2kx} \, dx = \frac{\pi}{2k} e^{2kx} \Big|_0^1 = \frac{\pi}{2k}(e^{2k} - 1). \quad \bullet$$

FIGURE 6

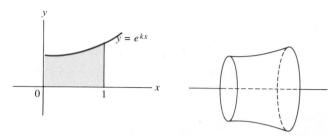

EXAMPLE 7   Find the volume of a right circular cone of radius r and height h.

Solution   The cone [Fig. 7(a)] is the solid of revolution swept out when the shaded region in Fig. 7(b) is revolved about the x-axis. Using the formula developed previously, the volume of the cone is

$$\int_0^h \pi\left(\frac{r}{h}x\right)^2 dx = \frac{\pi r^2}{h^2}\int_0^h x^2\, dx = \frac{\pi r^2}{h^2}\frac{x^3}{3}\Big|_0^h = \frac{\pi r^2 h}{3}. \quad \bullet$$

FIGURE 7

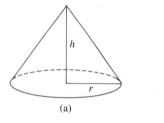   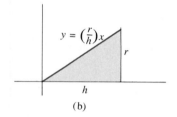

  (a)   (b)

TECHNOLOGY PROJECT 6.8
Estimating the Volume of a Solid of Revolution (Symbolic Program Required)

Use a symbolic math program to calculate the volume of the solids of revolution generated by the following functions. [Note that the number π can be entered as Pi.]

1. $y = \dfrac{x}{x+1}$, $0 \le x \le 5$ 2. $y = \sqrt{5x-1}$, $2 \le x \le 4$

3. $y = (x^2+1)^6$, $0 \le x \le 4$

4. The area bounded by $y = 2x+5$ and $x^2 - 3x + 5$

TECHNOLOGY PROJECT 6.9
Using a Graphing Calculator to Compute

Use your graphing calculator to estimate the intersection point of the following supply and demand curves. Then determine the consumers' surplus.

1. Demand curve $p = 500/(x+10)$; supply curve $p = 11 + .04x^2$

2. Demand curve $p = 225/(x+5)$; supply curve $p = 15 + 2.4x^2$

PRACTICE
PROBLEMS 6.5

1. A rock dropped from a bridge has a velocity of $-32t$ feet per second after t seconds. Find the average velocity of the rock during the first three seconds.

2. An investment yields \$300 per year compounded continuously for 10 years at 10% interest. What is the (future) value of this income stream at the end of 10 years?

EXERCISES 6.5

Determine the average value of $f(x)$ over the interval from $x = a$ to $x = b$, where:

1. $f(x) = x^2; a = 0, b = 3$
2. $f(x) = e^{x/3}; a = 0, b = 3$
3. $f(x) = x^3; a = -1, b = 1$
4. $f(x) = 5; a = 1, b = 10$
5. $f(x) = 1/x^2; a = \frac{1}{4}, b = \frac{1}{2}$
6. $f(x) = 2x - 6; a = 2, b = 4$

7. During a certain 12-hour period the temperature at time t (measured in hours from the start of the period) was $47 + 4t - \frac{1}{3}t^2$ degrees. What was the average temperature during that period?

8. Assuming that a country's population is now 3 million and is growing exponentially with growth constant .02, what will be the average population during the next 50 years?

9. One hundred grams of radioactive radium having a half-life of 1690 years is placed in a concrete vault. What will be the average amount of radium in the vault during the next 1000 years?

10. One hundred dollars are deposited in the bank at 5% interest compounded continuously. What will be the average value of the money in the account during the next 20 years?

Find the consumers' surplus for each of the following demand curves at the given sales level x.

11. $p = 3 - \dfrac{x}{10}; x = 20$
12. $p = \dfrac{x^2}{200} - x + 50; x = 20$

13. $p = \dfrac{500}{x + 10} - 3; x = 40$
14. $p = \sqrt{16 - .02x}; x = 350$

Figure 8 shows a supply curve for a commodity. It gives the relationship between the selling price of the commodity and the quantity that producers will manufacture. At a higher selling price, a greater quantity will be produced. Therefore, the curve is increasing. If (A, B) is a point on the curve, then, in order to stimulate the production of A units of the commodity, the price per unit must be B dollars. Of course, some producers will be willing to produce the commodity even with a lower selling price. Since everyone receives the same price in an open efficient economy, most producers are receiving more than their minimal required price. The excess is called the *producers' surplus*. Using an argument analogous to that of the *consumers' surplus*, one can show that the total producers' surplus when the price is B is the area of the shaded region in Fig. 8. Find the producers' surplus for each of the following supply curves at the given sales level x.

15. $p = .01x + 3; x = 200$
16. $p = \dfrac{x^2}{9} + 1; x = 3$

17. $p = \dfrac{x}{2} + 7; x = 10$
18. $p = 1 + \frac{1}{2}\sqrt{x}; x = 36$

FIGURE 8 Producers' surplus.

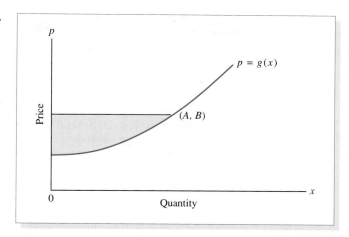

For a particular commodity, the quantity produced and the unit price are given by the coordinates of the point where the supply and demand curves intersect. For each pair of supply and demand curves, determine the point of intersection (A, B) and the consumers' and producers' surplus. (See Fig. 9.)

FIGURE 9

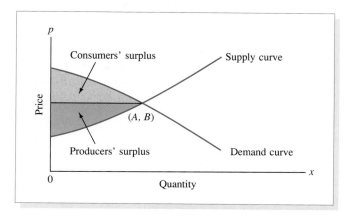

19. Demand curve: $p = 12 - (x/50)$; supply curve: $p = (x/20) + 5$.

20. Demand curve: $p = \sqrt{25 - .1x}$; supply curve: $p = \sqrt{.1x + 9} - 2$.

21. Suppose that money is deposited daily into a savings account at an annual rate of $1000. If the account pays 5% interest compounded continuously, estimate the balance in the account at the end of 3 years.

22. Suppose that money is deposited daily into a savings account at an annual rate of $2000. If the account pays 6% interest compounded continuously, approximately how much will be in the account at the end of 2 years?

23. Suppose that money is deposited steadily into a savings account at the rate of $16,000 per year. Determine the balance at the end of 4 years if the account pays 8% interest compounded continuously.

24. Suppose that money is deposited steadily into a savings account at the rate of $14,000 per year. Determine the balance at the end of 6 years if the account pays 7% interest compounded continuously.

25. A savings account pays 10% interest compounded continuously. If money is deposited steadily at the rate of $5000 per year, how much time is required until the balance reaches $140,000?

26. A savings account pays 7.5% interest compounded continuously. At what rate per year must money be deposited steadily into the account in order to accumulate a balance of $100,000 after 10 years?

27. Suppose money is to be deposited daily for 5 years into a savings account at an annual rate of $1000 and the account pays 6% interest compounded continuously. Let the interval from 0 to 5 be divided into daily subintervals, with each subinterval of duration $\Delta t = \frac{1}{365}$ years. Let $t_1, \ldots, t_n$ be points chosen from the subintervals.

    (a) Show that the present value of a daily deposit at time t_i is $1000 \, \Delta t e^{-.06 t_i}$.

    (b) Find the Riemann sum corresponding to the sum of the present values of all the deposits.

    (c) What is the function and interval corresponding to the Riemann sum in (b)?

    (d) Give the definite integral that approximates the Riemann sum in (b).

    (e) Evaluate the definite integral in (d). This number is the *present value of a continuous income stream.*

28. Use the result of Exercise 27 to calculate the present value of a continuous income stream of $5000 per year for 10 years at an interest rate of 5% compounded continuously.

Find the volume of the solid of revolution generated by revolving about the x-axis the region under each of the following curves.

29. $y = \sqrt{r^2 - x^2}$ from $x = -r$ to $x = r$ (generates a sphere of radius r)

30. $y = kx$ from $x = 0$ to $x = h$ (generates a cone)

31. $y = x^2$ from $x = 1$ to $x = 2$

32. $y = \dfrac{1}{x}$ from $x = 1$ to $x = 100$

33. $y = \sqrt{x}$ from $x = 0$ to $x = 4$ (The solid generated is called a *paraboloid.*)

34. $y = 2x - x^2$ from $x = 0$ to $x = 2$

35. $y = e^{-x}$ from $x = 0$ to $x = r$

36. $y = 2x + 1$ from $x = 0$ to $x = 1$ (The solid generated is called a *truncated cone.*)

For the Riemann sums in Exercises 37–40, determine n, b, and $f(x)$.

37. $[(8.25)^3 + (8.75)^3 + (9.25)^3 + (9.75)^3](.5); a = 8$

38. $\left[\dfrac{3}{1} + \dfrac{3}{1.5} + \dfrac{3}{2} + \dfrac{3}{2.5} + \dfrac{3}{3} + \dfrac{3}{3.5} \right](.5); a = 1$

39. $[(5 + e^5) + (6 + e^6) + (7 + e^7)](1); a = 4$

40. $[3(.3)^2 + 3(.9)^2 + 3(1.5)^2 + 3(2.1)^2 + 3(2.7)^2](.6); a = 0$

41. Suppose that the interval $0 \leq x \leq 3$ is divided into 100 subintervals of width $\Delta x = .03$. Let $x_1, x_2, \ldots, x_{100}$ be points in these subintervals. Suppose that in a particular application, one needs to estimate the sum

$$(3 - x_1)^2 \, \Delta x + (3 - x_2)^2 \, \Delta x + \cdots + (3 - x_{100})^2 \, \Delta x.$$

Show that this sum is close to 9.

42. Suppose that the interval $0 \leq x \leq 1$ is divided into 100 subintervals of width $\Delta x = .01$. Show that the following sum is close to $5/4$.

$$[2(.01) + (.01)^3]\Delta x + [2(.02) + (.02)^3]\Delta x + \cdots + [2(1.0) + (1.0)^3]\Delta x.$$

SOLUTIONS TO PRACTICE PROBLEMS 6.5

1. By definition, the average value of the function $v(t) = -32t$ for $t = 0$ to $t = 3$ is

$$\frac{1}{3 - 0} \int_0^3 -32t \, dt = \frac{1}{3}(-16t^2)\Big|_0^3 = \frac{1}{3}(-16 \cdot 3^2)$$

$$= -48 \text{ feet per second}$$

Note: There is another way to approach this problem:

$$[\text{average velocity}] = \frac{[\text{distance traveled}]}{[\text{time elapsed}]}.$$

As we discussed in Section 6.2, distance traveled equals the area under the velocity curve. Therefore,

$$[\text{average velocity}] = \frac{\int_0^3 -32t \, dt}{3}.$$

2. According to the formula developed in the text, the future value of the income stream after 10 years is equal to

$$\int_0^{10} 300e^{.1(10-t)} \, dt = -\frac{300}{.1}e^{.1(10-t)}\Big|_0^{10}$$

$$= -3000e^0 - (-3000e^1)$$

$$= 3000e - 3000$$

$$\approx \$5154.85.$$

CHAPTER 6 CHECKLIST

✓ Antiderivative

✓ Indefinite integral

✓ $\displaystyle\int x^r \, dx = \frac{1}{r + 1}x^{r+1} + C, \quad r \neq -1$

✓ $\displaystyle\int \frac{1}{x} \, dx = \ln|x| + C, \quad x \neq 0$

✓ $\displaystyle\int e^{kx}\,dx = \frac{1}{k}e^{kx} + C, \quad k \neq 0$

✓ $\displaystyle\int [f(x) + g(x)]\,dx = \int f(x)\,dx + \int g(x)\,dx$

✓ $\displaystyle\int kf(x)\,dx = k\int f(x)\,dx$

✓ Area under a graph

✓ $\displaystyle\Delta x = \frac{b - a}{n}$

✓ Partition of an interval

✓ Riemann sum approximation of area

✓ Area under rate of change function gives amount of increase of a quantity

✓ Definite integral: $\displaystyle\int_a^b f(x)\,dx$

✓ Fundamental theorem of calculus: $\displaystyle\int_a^b f(x)\,dx = F(b) - F(a),\ F'(x) = f(x)$

$$\int_a^b F'(x)\,dx = F(b) - F(a)$$

✓ Net change: $\displaystyle F(x)\Big|_a^b = F(b) - F(a)$

✓ Area between two curves: Assume graph of $f(x)$ lies above graph of $g(x)$,

$$\int_a^b [f(x) - g(x)]\,dx$$

✓ Average value of $f(x)$: $\displaystyle\frac{1}{b - a}\int_a^b f(x)\,dx$

✓ Consumers' surplus: $\displaystyle\int_0^A [f(x) - B]\,dx$

✓ Future value of a continuous income stream: $\displaystyle\int_0^N Pe^{r(N-t)}\,dt$

✓ Volume of a solid of revolution: $\displaystyle\int_a^b \pi[g(x)]^2\,dx$

CHAPTER 6   SUPPLEMENTARY EXERCISES

Calculate the following integrals.

1. $\displaystyle\int e^{-x/2}\,dx$

2. $\displaystyle\int \frac{5}{\sqrt{x - 7}}\,dx$

3. $\displaystyle\int (3x^4 - 4x^3)\,dx$

4. $\displaystyle\int (2x + 3)^7\,dx$

5. $\displaystyle\int \sqrt{4 - x}\,dx$

6. $\displaystyle\int \left(\frac{5}{x} - \frac{x}{5}\right)\,dx$

7. $\displaystyle\int_{1}^{4} \frac{1}{x^2}\, dx$ **8.** $\displaystyle\int_{3}^{6} e^{2-(x/3)}\, dx$ **9.** $\displaystyle\int_{0}^{5} (5 + 3x)^{-1}\, dx$

10. Find the area under the curve $y = 1 + \sqrt{x}$ from $x = 1$ to $x = 9$.

11. Find the area under the curve $y = (3x - 2)^{-3}$ from $x = 1$ to $x = 2$.

12. Find the area of the region bounded by the curves $y = 16 - x^2$ and $y = 10 - x$.

13. Find the area of the region bounded by the curves $y = x^3 - 3x + 1$ and $y = x + 1$.

14. Find the area of the region between the curves $y = 2x^2 + x$ and $y = x^2 + 2$ from $x = 0$ to $x = 2$.

15. Find the function $f(x)$ for which $f'(x) = (x - 5)^2$, $f(8) = 2$.

16. Find the function $f(x)$ for which $f'(x) = e^{-5x}$, $f(0) = 1$.

17. Describe all solutions of the following differential equations, where y represents a function of t.

  (a) $y' = 4t$ **(b)** $y' = 4y$ **(c)** $y' = e^{4t}$

18. Let k be a constant, and let $y = f(t)$ be a function such that $y' = kty$. Show that $y = Ce^{kt^2/2}$, for some constant C. [*Hint:* Use the product rule to evaluate $\dfrac{d}{dt}[f(t)e^{-kt^2/2}]$, and then apply Theorem II of Section 6.1.]

19. An airplane tire plant finds that its marginal cost of producing tires is $.04x + 150$ dollars at a production level of x tires per day. If fixed costs are \$500 per day, find the cost of producing x tires per day.

20. Suppose that the marginal revenue function for a company is $400 - 3x^2$. Find the additional revenue received from doubling production if currently 10 units are being produced.

21. A drug is injected into a patient at the rate of $f(t)$ cubic centimeters per minute at time t. What does the area under the graph of $y = f(t)$ from $t = 0$ to $t = 4$ represent?

22. A rock thrown straight up into the air has a velocity of $v(t) = -9.8t + 20$ meters per second after t seconds.

  (a) Determine the distance the rock travels during the first 2 seconds.

  (b) Represent the answer to part (a) as an area.

23. Suppose that money is deposited steadily into a savings account at the rate of \$4500 per year. Determine the balance at the end of 1 year if the account pays 9% interest compounded continuously.

24. The marginal revenue from producing x units of a certain commodity is $MR = -.02x + 5$ dollars. Find the demand equation for the commodity.

25. Use a Riemann sum with $n = 2$ and midpoints to estimate the area under the graph of $f(x) = \dfrac{1}{x + 2}$ on the interval $0 \le x \le 2$. Then use a definite integral to find the exact value of the area to five decimal places.

26. Use a Riemann sum with $n = 5$ and midpoints to estimate the area under the graph of $f(x) = e^{2x}$ on the interval $0 \le x \le 1$. Then use a definite integral to find the exact value of the area to five decimal places.

27. Find the consumers' surplus for the demand curve $p = \sqrt{25 - .04x}$ at the sales level $x = 400$.

28. Three thousand dollars is deposited in the bank at 6% interest compounded continuously. What will be the average value of the money in the account during the next 10 years?

29. Find the average value of $f(x) = 1/x^3$ from $x = \frac{1}{3}$ to $x = \frac{1}{2}$.

30. Suppose that the interval $0 \le x \le 1$ is divided into 100 subintervals with a width of $\Delta x = .01$. Show that the sum

$$[3e^{-.01}]\Delta x + [3e^{-.02}]\Delta x + [3e^{-.03}]\Delta x + \cdots + [3e^{-1}]\Delta x$$

is close to $3(1 - e^{-1})$.

31. Use an appropriate definite integral to estimate to two decimal places the value of the following sum.

$$\left(e^2 - \frac{3}{2}\right)(.01) + \left(e^{2.01} - \frac{3}{2.01}\right)(.01) + \left(e^{2.02} - \frac{3}{2.02}\right)(.01)$$

$$+ \cdots + \left(e^{2.99} - \frac{3}{2.99}\right)(.01)$$

32. Find the volume of the solid of revolution generated by revolving about the x-axis the region under the curve $y = 1 - x^2$ from $x = 0$ to $x = 1$.

33. A store has an inventory of Q units of a certain product at time $t = 0$. The store sells the product at the steady rate of Q/A units per week, and exhausts the inventory in A weeks.
    (a) Find a formula $f(t)$ for the amount of product in inventory at time t.
    (b) Find the average inventory level during the period $0 \le t \le A$.

34. A retail store sells a certain product at the rate of $g(t)$ units per week at time t, where $g(t) = rt$. At time $t = 0$, the store has Q units of the product in inventory.
    (a) Find a formula $f(t)$ for the amount of product in inventory at time t.
    (b) Determine the value of r in part (a) such that the inventory is exhausted in A weeks.
    (c) Using $f(t)$, with r as in part (b), find the average inventory level during the period $0 \le t \le A$.

35. Let x be any positive number, and define $g(x)$ to be the number determined by the definite integral

$$g(x) = \int_0^x \frac{1}{1 + t^2} \, dt.$$

    (a) Give a geometric interpretation of the number $g(3)$.
    (b) Find the derivative $g'(x)$.

36. For each number x satisfying $-1 \le x \le 1$, define $h(x)$ by

$$h(x) = \int_{-1}^x \sqrt{1 - t^2} \, dt.$$

    (a) Give a geometric interpretation of the values $h(0)$ and $h(1)$.
    (b) Find the derivative $h'(x)$.

37. Suppose that the interval $0 \le t \le 3$ is divided into 1000 subintervals of width Δt. Let $t_1, t_2, \ldots, t_{1000}$ denote the right endpoints of these subintervals. Suppose that in some problem one needs to estimate the sum:

$$5000e^{-.1t_1} \, \Delta t + 5000e^{-.1t_2} \, \Delta t + \cdots + 5000e^{-.1t_{1000}} \, \Delta t.$$

Show that this sum is close to 13,000. [*Note:* A sum such as this would arise if one wanted to compute the present value of a continuous stream of income of $5000 per year for 3 years, with interest compounded continuously at 10%.]

38. What number does $\left[e^0 + e^{1/n} + e^{2/n} + e^{3/n} + \cdots + e^{(n-1)/n}\right] \cdot \dfrac{1}{n}$ approach as n gets very large?

39. What number does the sum $\left[1^3 + \left(1 + \dfrac{1}{n}\right)^3 + \left(1 + \dfrac{2}{n}\right)^3 + \left(1 + \dfrac{3}{n}\right)^3 + \cdots + \left(1 + \dfrac{n-1}{n}\right)^3\right] \cdot \dfrac{1}{n}$ approach as n gets very large?

40. The rectangle in Fig. 1(a) has the same area as the region under the graph of $f(x)$ in Fig. 1(b). What is the average value of $f(x)$ on the interval $2 \le x \le 6$?

FIGURE 1 **Rectangle and region having the same area.**

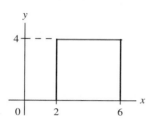 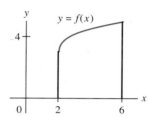

41. *True or false:* If $3 \le f(x) \le 4$ whenever $0 \le x \le 5$, then $3 \le \frac{1}{5} \int_0^5 f(x)\, dx \le 4$.

42. Suppose that water is flowing into a tank at a rate of $r(t)$ gallons per hour, where the rate depends on the time t according to the formula

$$r(t) = 20 - 4t, \qquad 0 \le t \le 5.$$

(a) Consider a brief period of time, say from t_1 to t_2. The length of this time period is $\Delta t = t_2 - t_1$. During this period the rate of flow does not change much and is approximately $20 - 4t_1$ (the rate at the beginning of the brief time interval). Approximately how much water flows into the tank during the time from t_1 to t_2?

(b) Explain why the total amount of water added to the tank during the time interval from $t = 0$ to $t = 5$ is given by $\int_0^5 r(t)\, dt$.

43. The annual world rate of water use t years after 1940, for $t \le 40$, was approximately $860e^{.04t}$ cubic kilometers per year. How much water was used between 1940 and 1980?

CHAPTER **7**

FUNCTIONS OF SEVERAL VARIABLES

Until now, most of our applications of calculus have involved functions of one variable. In real life, however, a quantity of interest often depends on more than one variable. For instance, the sales level of a product may depend not only on its price but also on the prices of competing products, the amount spent on advertising, and perhaps the time of year. The total cost of manufacturing the product depends on the cost of raw materials, labor, plant maintenance, and so on.

This chapter introduces the basic ideas of calculus for functions of more than one variable. Section 7.1 presents several examples that will be used throughout the chapter. Derivatives are treated in Section 7.2 and then used in Sections 7.3 and 7.4 to solve optimization problems more general than those in Chapter 2. Later sections are devoted to approximating the change in a function, least-squares problems, and a brief introduction to integration of functions of two variables.

441

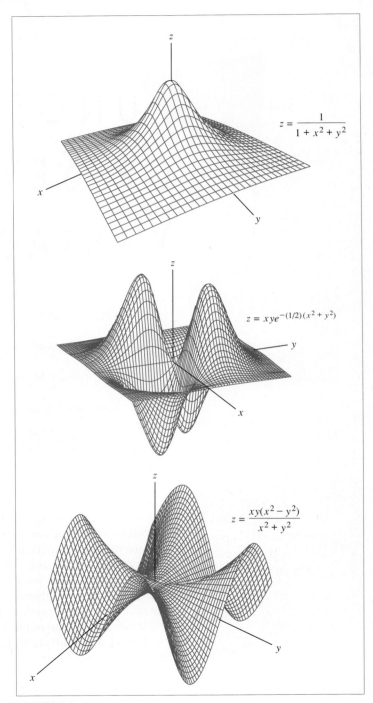

$$z = \frac{1}{1 + x^2 + y^2}$$

$$z = xye^{-(1/2)(x^2 + y^2)}$$

$$z = \frac{xy(x^2 - y^2)}{x^2 + y^2}$$

FIGURE 2

FIGURE 1   **Graph of $f(x, y)$.**

Surface
$z = f(x, y)$

$(x, y, f(x, y))$

(x, y)

7.1 EXAMPLES OF FUNCTIONS OF SEVERAL VARIABLES

A function $f(x, y)$ of the two variables x and y is a rule that assigns a number to each pair of values for the variables; for instance,

$$f(x, y) \ = \ e^x(x^2 + 2y).$$

An example of a function of three variables is

$$f(x, y, z) = 5xy^2z.$$

EXAMPLE I   A store sells butter at \$2.50 per pound and margarine at \$1.40 per pound. The revenue from the sale of x pounds of butter and y pounds of margarine is given by the function

$$f(x, y) = 2.50x + 1.40y.$$

Determine and interpret $f(200, 300)$.

Solution $f(200, \ 300) = 2.50(200) + 1.40(300) = 500 + 420 = 920$.   The revenue from the sale of 200 pounds of butter and 300 pounds of margarine is \$920. ●

A function $f(x, y)$ of two variables may be graphed in a manner analogous to that for functions of one variable. It is necessary to use a three-dimensional coordinate system, where each point is identified by three coordinates (x, y, z). For each choice of x, y, the graph of $f(x, y)$ includes the point $(x, y, f(x, y))$. This graph is a surface in three-dimensional space, with equation $z = f(x, y)$. See Fig. 1. Three graphs of specific functions are shown in Fig. 2.*

TECHNOLOGY PROJECT 7.1
Graphing Functions of Two Variables (Symbolic Program Needed)

Most symbolic math programs incorporate 3D graphics, which allow you to do fairly sophisticated imaging of graphs of functions of two variables. These 3D graphing programs allow you not only to display graphs, but to shade them in color, illuminate them from imaginary light sources placed at particular points, display them as true surfaces or as wire frame diagrams, and view them from a particular point in space. All these features are in addition to all the usual graph features that one finds in two-dimensional graphs (grids, axis labels, tick marks, axis scales, axis titles, and so forth).

In this technology project, we will concern ourselves with graphing functions of two variables using the default settings for most of the graphics parame-

---

* These graphs were drawn by Norton Starr at the University of Waterloo Computing Centre.

ters. The graphs should give you some feel for the nature of graphs of functions of two variables.

We will graph functions $f(x, y)$, where x ranges over the interval $a \le x \le b$ and y over the interval $c \le y \le d$. The *Mathematica* command to display this graph is:

```
Plot3D [f(x, y), {x, a, b}, {y, c, d}
```

For example, to display the graph of

$$f(x, y) = x^2 + y^2 - 8, \ -10 \le x \le 10, \ -10 \le y \le 10$$

we would use the command:

```
Plot3D [x^2+y^2-8, {x, -10, 10}, {y, -10, 10}]
```

The program displays the graph shown in Fig. 3.

FIGURE 3

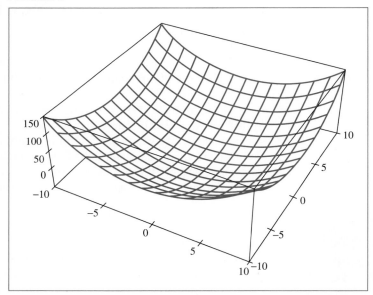

Use a symbolic math program to display the graphs of the following functions of two variables.

1. $x - y - 1, \ -5 \le x \le 5, \ -5 \le y \le 5$

2. $3x + 2x - 6, \ -10 \le x \le 10, \ -10 \le y \le 10$

3. $x - 4, \ -5 \le x \le 5, \ -5 \le y \le 5$

4. $y + 1, \ -5 \le x \le 5, \ -5 \le y \le 5$

5. $x^2 + y^2, \ -3 \le x \le 3, \ -3 \le y \le 3$

6. $x - y^2 - 4, 0 \le x \le 5, -2 \le y \le 2$

7. $xy, -2 \le x \le 2, -2 \le y \le 2$

8. $x^3y, -2 \le x \le 2, -2 \le y \le 2$

9. $\dfrac{1}{xy}, 0 \le x \le 10, 0 \le y \le 10$

10. $\dfrac{x}{y}, 0 \le x \le 10, 0 \le y \le 10$

Application to Architectural Design   When designing a building, one would like to know, at least approximately, how much heat the building loses per day. The heat loss affects many aspects of the design, such as the size of the heating plant, the size and location of duct work, and so on. A building loses heat through its sides, roof, and floor. How much heat is lost will generally differ for each face of the building and will depend on such factors as insulation, materials used in construction, exposure (north, south, east, or west), and climate. It is possible to estimate how much heat is lost per square foot of each face. Using these data, one can construct a heat-loss function as in the following example.

EXAMPLE 2   A rectangular industrial building of dimensions x, y, and z is shown in Fig. 4(a). In Fig. 4(b) we give the amount of heat lost per day by each side of the building, measured in suitable units of heat per square foot. Let $f(x, y, z)$ be the total daily heat loss for such a building.

(a) Find a formula for $f(x, y, z)$.

(b) Find the total daily heat loss if the building has length 100 feet, width 70 feet, and height 50 feet.

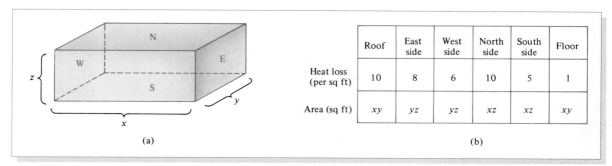

	Roof	East side	West side	North side	South side	Floor
Heat loss (per sq ft)	10	8	6	10	5	1
Area (sq ft)	xy	yz	yz	xz	xz	xy

(a)   (b)

FIGURE 4   **Heat loss from an industrial building.**

Solution   (a) The total heat loss is the sum of the amount of heat loss through each side of the building. The heat loss through the roof is

[heat loss per square foot of roof] · [area of roof in square feet] $= 10xy$.

Similarly, the heat loss through the east side is $8yz$. Continuing in this way, we see that the total daily heat loss is

$$f(x, y, z) = 10xy + 8yz + 6yz + 10xz + 5xz + 1 \cdot xy.$$

We collect terms to obtain

$$f(x, y, z) = 11xy + 14yz + 15xz.$$

(b) The amount of heat loss when $x = 100$, $y = 70$, and $z = 50$ is given by $f(100, 70, 50)$, which equals

$$f(100, 70, 50) = 11(100)(70) + 14(70)(50) + 15(100)(50)$$
$$= 77{,}000 + 49{,}000 + 75{,}000 = 201{,}000. \quad \bullet$$

In Section 7.3 we will determine the dimensions x, y, z that minimize the heat loss for a building of specified volume.

Production Functions in Economics   The costs of a manufacturing process can generally be classified as one of two types: cost of labor and cost of capital. The meaning of the cost of labor is clear. By the cost of capital, we mean the cost of buildings, tools, machines, and similar items used in the production process. A manufacturer usually has some control over the relative portions of labor and capital utilized in his production process. He can completely automate production so that labor is at a minimum, or he can utilize mostly labor and little capital. Suppose that x units of labor and y units of capital are used.* Let $f(x, y)$ denote the number of units of finished product that are manufactured. Economists have found that $f(x, y)$ is often a function of the form

$$f(x, y) = Cx^A y^{1-A},$$

where A and C are constants, $0 < A < 1$. Such a function is called a *Cobb-Douglas production function*.

EXAMPLE 3   (*Production in a Firm*)   Suppose that during a certain time period the number of units of goods produced when utilizing x units of labor and y units of capital is $f(x, y) = 60x^{3/4}y^{1/4}$.

(a) How many units of goods will be produced by using 81 units of labor and 16 units of capital?

(b) Show that whenever the amounts of labor and capital being used are doubled, so is the production. (Economists say that the production function has "constant returns to scale.")

Solution   (a) $f(81, 16) = 60(81)^{3/4} \cdot (16)^{1/4} = 60 \cdot 27 \cdot 2 = 3240$. There will be 3240 units of goods produced.

(b) Utilization of a units of labor and b units of capital results in the production of $f(a, b) = 60a^{3/4}b^{1/4}$ units of goods. Utilizing $2a$ and $2b$ units of labor and

---

*Economists normally use L and K, respectively, for labor and capital. However, for simplicity, we use x and y.

capital, respectively, results in $f(2a, 2b)$ units produced. Set $x = 2a$ and $y = 2b$. Then we see that

$$f(2a, 2b) = 60(2a)^{3/4}(2b)^{1/4}$$

$$= 60 \cdot 2^{3/4} \cdot a^{3/4} \cdot 2^{1/4} \cdot b^{1/4}$$

$$= 60 \cdot 2^{(3/4+1/4)} \cdot a^{3/4}b^{1/4}$$

$$= 2^1 \cdot 60a^{3/4}b^{1/4}$$

$$= 2f(a, b). \quad \bullet$$

Level Curves   It is possible graphically to depict a function $f(x, y)$ of two variables using a family of curves called level curves. Let c be any number. Then the graph of the equation $f(x, y) = c$ is a curve in the xy-plane called the *level curve of height c*. This curve describes all points of height c on the graph of the function $f(x, y)$. As c varies, we have a family of level curves indicating the sets of points on which $f(x, y)$ assumes various values c. In Fig. 5, we have drawn the graph and various level curves for the function $f(x, y) = x^2 + y^2$.

FIGURE 5   Level curves.

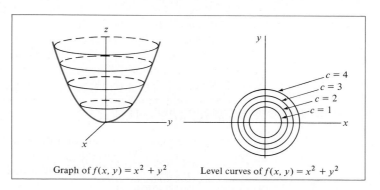

Graph of $f(x, y) = x^2 + y^2$ Level curves of $f(x, y) = x^2 + y^2$

Level curves often have interesting physical interpretations. For example, surveyors draw *topographic maps* that use level curves to represent points having equal altitude. Here $f(x, y) = $ the altitude at point (x, y). Figure 6(a) shows the graph of $f(x, y)$ for a typical hilly region. Figure 6(b) shows the level curves corresponding to various altitudes.

EXAMPLE 4   Determine the level curve at height 600 for the production function $f(x, y) = 60x^{3/4}y^{1/4}$ of Example 3.

Solution   The level curve is the graph of $f(x, y) = 600$, or

$$60x^{3/4}y^{1/4} = 600$$

$$y^{1/4} = \frac{10}{x^{3/4}}$$

$$y = \frac{10,000}{x^3}.$$

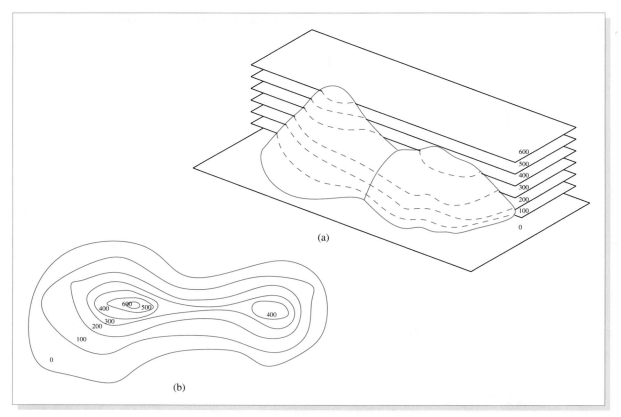

FIGURE 6 Topographic level curves show altitudes.

Of course, since x and y represent quantities of labor and capital, they must both be positive. We have sketched the graph of the level curve in Fig. 7. The points on the curve are precisely those combinations of capital and labor which yield 600 units of production.

TECHNOLOGY PROJECT 7.2
Graphing Level Curves (Symbolic Program Needed)

Use a symbolic math program to draw the level curves $f(x, y) = c$ for the following functions and the specified values of c. You will need to use the implicit plotting feature of your program.

1. $f(x, y) = 3x + 2y, \quad c = -1, 0, 1, 2$

2. $f(x, y) = xy, \quad c = 0, 1, 2, 3$

3. $f(x, y) = x^2 + 3y^2, \quad c = 4, 10, 25$

4. $f(x, y) = x - y^2, \quad c = 0, 1, 2, 5$

5. $f(x, y) = \dfrac{x}{y}, \quad c = -2, -1, 0, 1, 2$

6. $f(x, y) = x^3 + y^2, \quad c = 1, 3, 6, 10$

FIGURE 7 Isoquant of a production function.

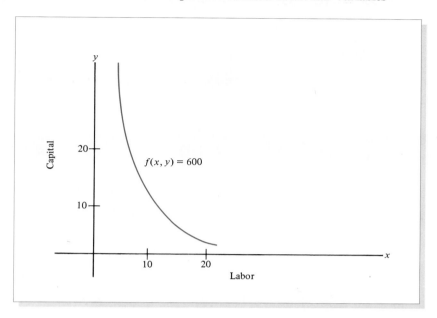

TECHNOLOGY PROJECT 7.3

Defining and Evaluating Functions of Two Variables Using Technology (Symbolic Program Needed)

You may define functions of two (or more variables) in a symbolic math program and use the resulting functions in computations, such as evaluation and differentiation. For example, to define the function $f(x, y) = 3x^3 + y^2$ in *Mathematica*, you would use the command:

```
f[x,y] := 3x^3+y^2
```

Then to evaluate this function for $(x, y) = (-1, 2)$, you would use the command:

```
f[-1,2]
```

Note that functions must be defined using brackets [ ] rather than parentheses, since parentheses in *Mathematica* (and in most symbolic math programs) are used strictly as a grouping symbol.

Determine the values of the following functions using a symbolic math program.

1. $f(x, y) = 5x^3 - 34y, \quad (x, y) = (14.1, 12.6)$

2. $f(x, y) = \dfrac{x + y}{x - y}, \quad (x, y) = (-15.3, 101.5)$

3. $f(x, y) = 379x^{1/3}y^{2/3}, \quad (x, y) = (89, 188)$

4. $f(x, y) = 3x^2 - 12xy + 4y^3 + 3y, \quad (x, y) = (-59, 103)$

1. Let $f(x, y, z) = x^2 + y/(x - z) - 4$. Compute $f(3, 5, 2)$.
2. Suppose that in a certain country the daily demand for coffee is given by $f(p_1, p_2) = 16p_1/p_2$ thousand pounds, where p_1 and p_2 are the respective prices of tea and coffee per pound. Compute and interpret $f(3, 4)$.

EXERCISES 7.1

1. Let $f(x, y) = x^2 + 8y$. Compute $f(1, 0), f(0, 1)$, and $f(3, 2)$.

2. Let $g(x, y) = 3xe^y$. Compute $g(2, 1), g(1, 0)$, and $g(0, 0)$.

3. Let $f(L, K) = 3\sqrt{LK}$. Compute $f(0, 1), f(3, 12)$, and $f(a, b)$.

4. Let $f(p, q) = pe^{q/p}$. Compute $f(1, 0), f(3, 12)$, and $f(a, b)$.

5. Let $f(x, y, z) = x/(y - z)$. Compute $f(2, 3, 4)$ and $f(7, 46, 44)$.

6. Let $f(x, y, z) = x^2 e^{\sqrt{y^2 + z^2}}$. Compute $f(1, 0, 1)$ and $f(5, 2, 3)$.

7. Let $f(x, y) = xy$. Show that $f(2 + h, 3) - f(2, 3) = 3h$.

8. Let $f(x, y) = xy$. Show that $f(2, 3 + K) - f(2, 3) = 2K$.

9. Let $f(x, y) = \dfrac{x^2 + 3xy + 3y^2}{x + y}$. Show that $f(2a, 2b) = 2f(a, b)$.

10. Let $f(x, y) = 75x^A y^{1-A}$, where $0 < A < 1$. Show that $f(2a, 2b) = 2f(a, b)$.

11. The present value of A dollars to be paid t years in the future (assuming a 5% continuous interest rate) is $P(A, t) = Ae^{-.05t}$. Find and interpret $P(100, 13.8)$.

12. Refer to Example 3. Suppose that labor costs \$100 per unit and capital costs \$200 per unit. Express as a function of two variables, $C(x, y)$, the cost of utilizing x units of labor and y units of capital.

13. The value of residential property for tax purposes is usually much lower than its actual market value. If v is the market value, then the *assessed value* for real estate taxes might be only 40% of v. Suppose the property tax, T, in a community is given by the function

$$T = f(r, v, x) = \frac{r}{100}(.40v - x),$$

where v is the estimated market value of a property (in dollars), x is a *homeowner's exemption* (a number of dollars depending on the type of property), and r is the tax rate (stated in dollars per hundred dollars) of net assessed value.

   (a) Determine the real estate tax on a property valued at \$200,000 with a homeowner's exemption of \$5000, assuming a tax rate of \$2.50 per hundred dollars of net assessed value.

   (b) Determine the tax due if the tax rate increases by 20% to \$3.00 per hundred dollars of net assessed value. Assume the same property value and homeowner's exemption. Does the tax due also increase by 20%?

14. Let $f(r, v, x)$ be the real estate tax function of Exercise 13.

(a) Determine the real estate tax on a property valued at $100,000 with a home-owner's exemption of $5000, assuming a tax rate of $2.20 per hundred dollars of net assessed value.

(b) Determine the real estate tax when the market value rises 20% to $120,000. Assume the same homeowner's exemption and a tax rate of $2.20 per hundred dollars of net assessed value. Does the tax due also increase by 20%?

Draw the level curves of heights 0, 1, and 2 for the functions in Exercises 15 and 16.

15. $f(x, y) = 2x + y$ 16. $f(x, y) = -x^2 + y$

17. Draw the level curve of the function $f(x, y) = x - y$ containing the point $(5, 3)$.

18. Draw the level curve of the function $f(x, y) = xy$ containing the point $(2, \frac{1}{2})$.

19. Find a function $f(x, y)$ that has the line $y = 3x - 4$ as a level curve.

20. Find a function $f(x, y)$ that has the curve $y = 3/x^2$ as a level curve.

21. Suppose that a topographic map is viewed as the graph of a certain function $f(x, y)$. What are the level curves?

22. A certain production process uses labor and capital. If the quantities of these commodities are x and y, respectively, then the total cost is $100x + 200y$ dollars. Draw the level curves of height 600, 800, and 1000 for this function. Explain the significance of these curves. (Economists frequently refer to these lines as *budget lines* or *isocost lines*.)

Match the graphs of the functions in Exercises 23–26 to the systems of level curves shown in Fig. 8(a)–(d).

23.   24.

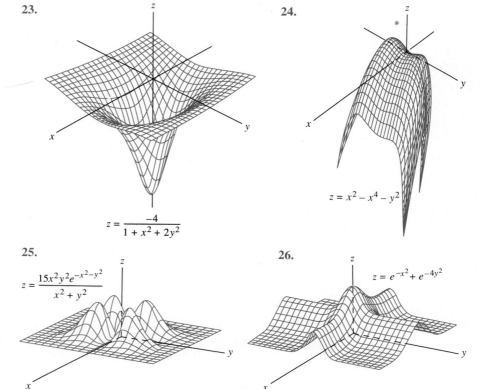

$$z = \frac{-4}{1 + x^2 + 2y^2}$$

$$z = x^2 - x^4 - y^2$$

25.   26.

$$z = \frac{15x^2y^2e^{-x^2-y^2}}{x^2 + y^2}$$

$$z = e^{-x^2} + e^{-4y^2}$$

FIGURE 8

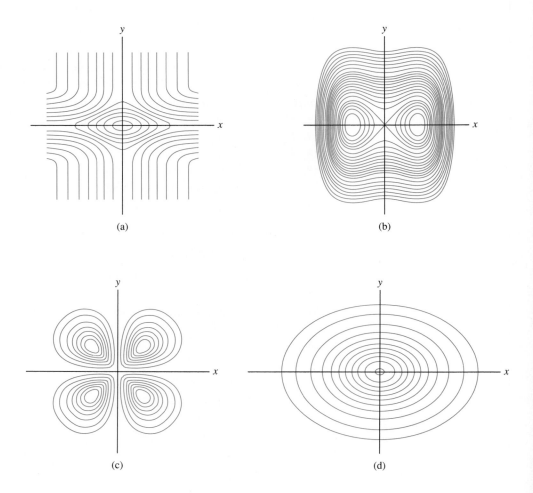

(a)

(b)

(c)

(d)

SOLUTIONS TO PRACTICE PROBLEMS 7.1

1. Substitute 3 for x, 5 for y, and 2 for z.

$$f(3, 5, 2) = 3^2 + \frac{5}{3 - 2} - 4 = 10.$$

2. To compute $f(3, 4)$, substitute 3 for p_1 and 4 for p_2 into $f(p_1, p_2) = 16 p_1/p_2$. Thus

$$f(3, 4) = 16 \cdot \tfrac{3}{4} = 12.$$

Therefore, if the price of tea is $3 per pound and the price of coffee is $4 per pound, then 12,000 pounds of coffee will be sold each day. (Notice that as the price of coffee increases, the demand decreases.)

7.2 PARTIAL DERIVATIVES

In Chapter 1 we introduced the notion of a derivative to measure the rate at which a function $f(x)$ is changing with respect to changes in the variable x. Let us now study the analog of the derivative for functions of two (or more) variables.

Let $f(x, y)$ be a function of the two variables x and y. Since we want to know how $f(x, y)$ changes with respect to both changes in the variable x and changes in the variable y, we shall define two derivatives of $f(x, y)$ (to be called "partial derivatives"), one with respect to each of the variables. The *partial derivative of $f(x, y)$ with respect to x,* written $\dfrac{\partial f}{\partial x}$, is the derivative of $f(x, y)$, where y is treated as a constant and $f(x, y)$ is considered as a function of x alone. The *partial derivative of $f(x, y)$ with respect to y,* written $\dfrac{\partial f}{\partial y}$, is the derivative of $f(x, y)$, where x is treated as a constant.

EXAMPLE I   Let $f(x, y) = 5x^3y^2$. Compute $\dfrac{\partial f}{\partial x}$ and $\dfrac{\partial f}{\partial y}$.

Solution   To compute $\dfrac{\partial f}{\partial x}$, we think of $f(x, y)$ written as

$$f(x, y) = [5y^2]x^3,$$

where the brackets emphasize that $5y^2$ is to be treated as a constant. Therefore, when differentiating with respect to $x, f(x, y)$ is just a constant times x^3. Recall that if k is any constant, then

$$\frac{d}{dx}(kx^3) = 3 \cdot k \cdot x^2.$$

Thus

$$\frac{\partial f}{\partial x} = 3 \cdot [5y^2] \cdot x^2 = 15x^2y^2.$$

After some practice, it is unnecessary to place the y^2 in front of the x^3 before differentiating.

Now, in order to compute $\dfrac{\partial f}{\partial y}$, we think of

$$f(x, y) = [5x^3]y^2.$$

When differentiating with respect to $y, f(x, y)$ is simply a constant (namely, $5x^3$) times y^2. Hence

$$\frac{\partial f}{\partial y} = 2 \cdot [5x^3] \cdot y = 10x^3y. \quad \bullet$$

EXAMPLE 2   Let $f(x, y) = 3x^2 + 2xy + 5y$. Compute $\dfrac{\partial f}{\partial x}$ and $\dfrac{\partial f}{\partial y}$.

Solution   To compute $\dfrac{\partial f}{\partial x}$, we think of

$$f(x, y) = 3x^2 + [2y]x + [5y].$$

Now we differentiate $f(x, y)$ as if it were a quadratic polynomial in x:

$$\frac{\partial f}{\partial x} = 6x + [2y] + 0 = 6x + 2y.$$

Note that $5y$ is treated as a constant when differentiating with respect to x, so the partial derivative of $5y$ with respect to x is zero.

To compute $\dfrac{\partial f}{\partial y}$, we think of

$$f(x, y) = [3x^2] + [2x]y + 5y.$$

Then

$$\frac{\partial f}{\partial y} = 0 + [2x] + 5 = 2x + 5.$$

Note that $3x^2$ is treated as a constant when differentiating with respect to y, so the partial derivative of $3x^2$ with respect to y is zero. •

EXAMPLE 3   Compute $\dfrac{\partial f}{\partial x}$ and $\dfrac{\partial f}{\partial y}$ for each of the following.

(a) $f(x, y) = (4x + 3y - 5)^8$

(b) $f(x, y) = e^{xy^2}$

(c) $f(x, y) = y/(x + 3y)$

Solution   (a) To compute $\dfrac{\partial f}{\partial x}$, we think of

$$f(x, y) = (4x + [3y - 5])^8.$$

By the general power rule,

$$\frac{\partial f}{\partial x} = 8 \cdot (4x + [3y - 5])^7 \cdot 4 = 32(4x + 3y - 5)^7.$$

Here we used the fact that the derivative of $4x + 3y - 5$ with respect to x is just 4.

To compute $\dfrac{\partial f}{\partial y}$, we think of

$$f(x, y) = ([4x] + 3y - 5)^8.$$

Then

$$\frac{\partial f}{\partial y} = 8 \cdot ([4x] + 3y - 5)^7 \cdot 3 = 24(4x + 3y - 5)^7.$$

(b) To compute $\dfrac{\partial f}{\partial x}$, we observe that

$$f(x, y) = e^{x[y^2]},$$

so that

$$\frac{\partial f}{\partial x} = [y^2]e^{x[y^2]} = y^2 e^{xy^2}.$$

To compute $\dfrac{\partial f}{\partial y}$, we think of

$$f(x, y) = e^{[x]y^2}.$$

Thus

$$\frac{\partial f}{\partial y} = e^{[x]y^2} \cdot 2[x]y = 2xye^{xy^2}.$$

(c) To compute $\dfrac{\partial f}{\partial x}$, we use the general power rule to differentiate $[y](x + [3y])^{-1}$ with respect to x:

$$\frac{\partial f}{\partial x} = (-1) \cdot [y](x + [3y])^{-2} \cdot 1 = -\frac{y}{(x + 3y)^2}.$$

To compute $\dfrac{\partial f}{\partial y}$, we use the quotient rule to differentiate

$$f(x, y) = \frac{y}{[x] + 3y}$$

with respect to y. We find that

$$\frac{\partial f}{\partial y} = \frac{([x] + 3y) \cdot 1 - y \cdot 3}{([x] + 3y)^2} = \frac{x}{(x + 3y)^2}.$$

The use of brackets to highlight constants is helpful initially in order to compute partial derivatives. From now on we shall merely form a mental picture of those terms to be treated as constants and dispense with brackets. ●

A partial derivative of a function of several variables is also a function of several variables and hence can be evaluated at specific values of the variables. We write

$$\frac{\partial f}{\partial x}(a, b)$$

for $\dfrac{\partial f}{\partial x}$ evaluated at $x = a, y = b$. Similarly,

$$\frac{\partial f}{\partial y}(a, b)$$

denotes the function $\dfrac{\partial f}{\partial y}$ evaluated at $x = a, y = b$.

EXAMPLE 4   Let $f(x, y) = 3x^2 + 2xy + 5y$.

(a) Calculate $\dfrac{\partial f}{\partial x}(1, 4)$.   (b) Evaluate $\dfrac{\partial f}{\partial y}$ at $(x, y) = (1, 4)$.

Solution   (a) $\dfrac{\partial f}{\partial x} = 6x + 2y, \dfrac{\partial f}{\partial x}(1, 4) = 6 \cdot 1 + 2 \cdot 4 = 14.$

(b) $\dfrac{\partial f}{\partial y} = 2x + 5, \dfrac{\partial f}{\partial y}(1, 4) = 2 \cdot 1 + 5 = 7.$ ●

Geometric Interpretation of Partial Derivatives   Consider the three-dimensional surface $z = f(x, y)$ in Fig. 1. If y is held constant at b and x is

FIGURE 1 $\dfrac{\partial f}{\partial x}$ **gives the slope of a curve formed by holding y constant.**

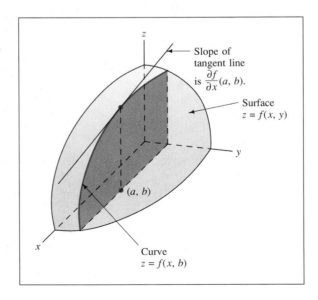

Slope of tangent line is $\dfrac{\partial f}{\partial x}(a, b)$.

Surface $z = f(x, y)$

(a, b)

Curve $z = f(x, b)$

allowed to vary, the equation

$$z = f(x, b)$$

└ constant

describes a curve on the surface. (The curve is formed by cutting the surface $z = f(x, y)$ with a vertical plane parallel to the xz-plane.) The value of $\dfrac{\partial f}{\partial x}(a, b)$ is the slope of the tangent line to the curve at the point where $x = a$ and $y = b$.

Likewise, if x is held constant at a and y is allowed to vary, the equation

$$z = f(a, y)$$

└ constant

describes the curve on the surface $z = f(x, y)$ shown in Fig. 2. The value of the partial derivative $\dfrac{\partial f}{\partial y}(a, b)$ is the slope of this curve at the point where $x = a$ and $y = b$.

FIGURE 2 $\dfrac{\partial f}{\partial y}$ **gives the slope of a curve formed by holding x constant.**

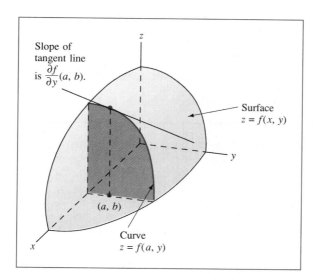

Partial Derivatives and Rates of Change Since $\dfrac{\partial f}{\partial x}$ is simply the ordinary derivative with y held constant, $\dfrac{\partial f}{\partial x}$ gives the rate of change of $f(x, y)$ with respect to x for y held constant. In other words, keeping y constant and increasing x by one (small) unit produces a change in $f(x, y)$ that is approximately given by $\dfrac{\partial f}{\partial x}$. An analogous interpretation holds for $\dfrac{\partial f}{\partial y}$.

EXAMPLE 5   Interpret the partial derivatives of $f(x, y) = 3x^2 + 2xy + 5y$ calculated in Example 4.

Solution   We showed in Example 4 that

$$\frac{\partial f}{\partial x}(1, 4) = 14, \qquad \frac{\partial f}{\partial y}(1, 4) = 7.$$

The fact that $\frac{\partial f}{\partial x}(1, 4) = 14$ means that if y is kept constant at 4 and x is allowed to vary near 1, then $f(x, y)$ changes at a rate 14 times the change in x. That is, if x increases by one small unit, then $f(x, y)$ increases by approximately 14 units. If x increases by h units (where h is small), then $f(x, y)$ increases by approximately $14 \cdot h$ units. That is,

$$f(1 + h, 4) - f(1, 4) \approx 14 \cdot h.$$

Similarly, the fact that $\frac{\partial f}{\partial y}(1, 4) = 7$ means that if we keep x constant at 1 and let y vary near 4, then $f(x, y)$ changes at a rate equal to seven times the change in y. So for a small value of k, we have

$$f(1, 4 + k) - f(1, 4) \approx 7 \cdot k. \quad \bullet$$

We can generalize the interpretations of $\frac{\partial f}{\partial x}$ and $\frac{\partial f}{\partial y}$ given in Example 5 to yield the following general fact.

Let $f(x, y)$ be a function of two variables. Then if h and k are small, we have

$$f(a + h, b) - f(a, b) \approx \frac{\partial f}{\partial x}(a, b) \cdot h$$

$$f(a, b + k) - f(a, b) \approx \frac{\partial f}{\partial y}(a, b) \cdot k.$$

Partial derivatives can be computed for functions of any number of variables. When taking the partial derivative with respect to one variable, we treat the other variables as constants.

EXAMPLE 6   Let $f(x, y, z) = x^2yz - 3z$.

(a) Compute $\frac{\partial f}{\partial x}, \frac{\partial f}{\partial y}$, and $\frac{\partial f}{\partial z}$.   (b) Calculate $\frac{\partial f}{\partial z}(2, 3, 1)$.

Solution   (a) $\frac{\partial f}{\partial x} = 2xyz, \quad \frac{\partial f}{\partial y} = x^2z, \quad \frac{\partial f}{\partial z} = x^2y - 3.$

(b) $\dfrac{\partial f}{\partial z}(2, 3, 1) = 2^2 \cdot 3 - 3 = 12 - 3 = 9.$ ●

EXAMPLE 7 Let $f(x, y, z)$ be the heat-loss function computed in Example 2 of Section 7.1. That is, $f(x, y, z) = 11xy + 14yz + 15xz$. Calculate and interpret $\dfrac{\partial f}{\partial x}(10, 7, 5)$.

Solution We have

$$\frac{\partial f}{\partial x} = 11y + 15z$$

$$\frac{\partial f}{\partial x}(10, 7, 5) = 11 \cdot 7 + 15 \cdot 5 = 77 + 75 = 152.$$

The quantity $\dfrac{\partial f}{\partial x}$ is commonly referred to as the *marginal heat loss with respect to change in x*. Specifically, if x is changed from 10 by h units (where h is small) and the values of y and z remain fixed at 7 and 5, then the amount of heat loss will change by approximately $152 \cdot h$ units. ●

EXAMPLE 8 (*Production*) Consider the production function $f(x, y) = 60x^{3/4}y^{1/4}$, which gives the number of units of goods produced when utilizing x units of labor and y units of capital.

(a) Find $\dfrac{\partial f}{\partial x}$ and $\dfrac{\partial f}{\partial y}$.

(b) Evaluate $\dfrac{\partial f}{\partial x}$ and $\dfrac{\partial f}{\partial y}$ at $x = 81$, $y = 16$.

(c) Interpret the numbers computed in part (b).

Solution (a) $\dfrac{\partial f}{\partial x} = 60 \cdot \dfrac{3}{4}x^{-1/4} \cdot y^{1/4} = 45x^{-1/4}y^{1/4} = 45\dfrac{y^{1/4}}{x^{1/4}}$

$$\frac{\partial f}{\partial y} = 60 \cdot \frac{1}{4}x^{3/4}y^{-3/4} = 15x^{3/4}y^{-3/4} = 15\frac{x^{3/4}}{y^{3/4}}.$$

(b) $\dfrac{\partial f}{\partial x}(81, 16) = 45 \cdot \dfrac{16^{1/4}}{81^{1/4}} = 45 \cdot \dfrac{2}{3} = 30$

$$\frac{\partial f}{\partial y}(81, 16) = 15 \cdot \frac{81^{3/4}}{16^{3/4}} = 15 \cdot \frac{27}{8} = \frac{405}{8} = 50\tfrac{5}{8}.$$

(c) The quantities $\dfrac{\partial f}{\partial x}$ and $\dfrac{\partial f}{\partial y}$ are referred to as the *marginal productivity of labor* and the *marginal productivity of capital*. If the amount of capital is held fixed at $y = 16$ and the amount of labor increases by 1 unit, then the quantity of goods produced will increase by approximately 30 units. Similarly, an increase in capital of 1 unit (with labor fixed at 81) results in an increase in production of approximately $50\frac{5}{8}$ units of goods. ●

Just as we formed second derivatives in the case of one variable, we can form second partial derivatives of a function $f(x, y)$ of two variables. Since $\dfrac{\partial f}{\partial x}$ is a function of x and y, we can differentiate it with respect to x or y. The partial derivative of $\dfrac{\partial f}{\partial x}$ with respect to x is denoted by $\dfrac{\partial^2 f}{\partial x^2}$. The partial derivative of $\dfrac{\partial f}{\partial x}$ with respect to y is denoted by $\dfrac{\partial^2 f}{\partial y\, \partial x}$. Similarly, the partial derivative of the function $\dfrac{\partial f}{\partial y}$ with respect to x is denoted by $\dfrac{\partial^2 f}{\partial x\, \partial y}$, and the partial derivative of $\dfrac{\partial f}{\partial y}$ with respect to y is denoted by $\dfrac{\partial^2 f}{\partial y^2}$. Almost all functions $f(x, y)$ encountered in applications (and all functions $f(x, y)$ in this text) have the property that

$$\frac{\partial^2 f}{\partial y\, \partial x} = \frac{\partial^2 f}{\partial x\, \partial y}.$$

EXAMPLE 9   Let $f(x, y) = x^2 + 3xy + 2y^2$. Calculate $\dfrac{\partial^2 f}{\partial x^2}, \dfrac{\partial^2 f}{\partial y^2}, \dfrac{\partial^2 f}{\partial x\, \partial y}$, and $\dfrac{\partial^2 f}{\partial y\, \partial x}$.

Solution   First we compute $\dfrac{\partial f}{\partial x}$ and $\dfrac{\partial f}{\partial y}$.

$$\frac{\partial f}{\partial x} = 2x + 3y, \qquad \frac{\partial f}{\partial y} = 3x + 4y.$$

To compute $\dfrac{\partial^2 f}{\partial x^2}$, we differentiate $\dfrac{\partial f}{\partial x}$ with respect to x:

$$\frac{\partial^2 f}{\partial x^2} = 2.$$

Similarly, to compute $\dfrac{\partial^2 f}{\partial y^2}$, we differentiate $\dfrac{\partial f}{\partial y}$ with respect to y:

$$\frac{\partial^2 f}{\partial y^2} = 4.$$

To compute $\dfrac{\partial^2 f}{\partial x\, \partial y}$, we differentiate $\dfrac{\partial f}{\partial y}$ with respect to x:

$$\frac{\partial^2 f}{\partial x\, \partial y} = 3.$$

Finally, to compute $\dfrac{\partial^2 f}{\partial y\,\partial x}$, we differentiate $\dfrac{\partial f}{\partial x}$ with respect to y:

$$\frac{\partial^2 f}{\partial y\,\partial x} = 3. \quad \bullet$$

TECHNOLOGY PROJECT 7.4
Calculating Partial Derivatives Using Technology (Symbolic Program Needed)

A symbolic math program can be used to calculate partial derivatives. For example, suppose that `f` is a function of two variables `x` and `y`. The partial derivative with respect to `x` may be calculated in *Mathematica* using the command:

`D [f,x]`

The second partial derivatives $\dfrac{\partial^2 f}{\partial x^2}$ and $\dfrac{\partial^2 f}{\partial x\,\partial y}$ may be calculated, respectively, using the commands:

`D [f, {x,2}]`

`D [f, x,y]`

Use a symbolic math program to calculate the following derivatives of the function

$$f(x, y) = x^3 y^2 e^{-x^2 y}$$

1. $\dfrac{\partial f}{\partial x}$ 2. $\dfrac{\partial f}{\partial y}$ 3. $\dfrac{\partial^2 f}{\partial x^2}$ 4. $\dfrac{\partial^2 f}{\partial y^2}$

5. $\dfrac{\partial^2 f}{\partial x\,\partial y}$ 6. $\dfrac{\partial^2 f}{\partial y\,\partial x}$ 7. $\dfrac{\partial^3 f}{\partial^2 x\,\partial y}$ 8. $\dfrac{\partial^4 f}{\partial x^2\,\partial y^2}$

PRACTICE PROBLEMS 7.2

1. The number of TV sets sold per week by an appliance store is given by a function of two variables, $f(x, y)$, where x is the price per TV set and y is the amount of money spent weekly on advertising. Suppose that the current price is $400 per set and that currently $2000 per week is being spent for advertising.

   (a) Would you expect $\dfrac{\partial f}{\partial x}(400, 2000)$ to be positive or negative?

   (b) Would you expect $\dfrac{\partial f}{\partial y}(400, 2000)$ to be positive or negative?

2. The monthly mortgage payment for a house is a function of two variables, $f(A, r)$, where A is the amount of the mortgage and the interest rate is $r\%$. For a 30-year mortgage, $f(92{,}000, 9) = 740.25$ and $\dfrac{\partial f}{\partial r}(92{,}000, 9) = 66.20$. What is the significance of the number 66.20?

EXERCISES 7.2

Find $\dfrac{\partial f}{\partial x}$ and $\dfrac{\partial f}{\partial y}$ for each of the following functions.

1. $f(x, y) = 5xy$

2. $f(x, y) = 3x^2 + 2y + 1$

3. $f(x, y) = 2x^2 e^y$

4. $f(x, y) = x + e^{xy}$

5. $f(x, y) = \dfrac{y^2}{x}$

6. $f(x, y) = \dfrac{x}{1 + e^y}$

7. $f(x, y) = (2x - y + 5)^2$

8. $f(x, y) = (9x^2 y + 3x)^{12}$

9. $f(x, y) = x^2 e^{3x} \ln y$

10. $f(x, y) = (x - \ln y)e^{xy}$

11. $f(x, y) = \dfrac{x - y}{x + y}$

12. $f(x, y) = \dfrac{2xy}{e^x}$

13. Let $f(L, K) = 3\sqrt{LK}$. Compute $\dfrac{\partial f}{\partial L}$.

14. Let $f(p, q) = e^{q/p}$. Compute $\dfrac{\partial f}{\partial q}$ and $\dfrac{\partial f}{\partial p}$.

15. Let $f(x, y, z) = (1 + x^2 y)/z$. Compute $\dfrac{\partial f}{\partial x}, \dfrac{\partial f}{\partial y}$, and $\dfrac{\partial f}{\partial z}$.

16. Let $f(x, y, z) = x^2 y + 3yz - z^2$. Compute $\dfrac{\partial f}{\partial x}, \dfrac{\partial f}{\partial y}$, and $\dfrac{\partial f}{\partial z}$.

17. Let $f(x, y, z) = xze^{yz}$. Find $\dfrac{\partial f}{\partial x}, \dfrac{\partial f}{\partial y}$, and $\dfrac{\partial f}{\partial z}$.

18. Let $f(x, y, z) = ze^{z/xy}$. Find $\dfrac{\partial f}{\partial x}, \dfrac{\partial f}{\partial y}$, and $\dfrac{\partial f}{\partial z}$.

19. Let $f(x, y) = x^2 + 2xy + y^2 + 3x + 5y$. Compute $\dfrac{\partial f}{\partial x}(2, -3)$ and $\dfrac{\partial f}{\partial y}(2, -3)$.

20. Let $f(x, y) = xye^{2x-y}$. Evaluate $\dfrac{\partial f}{\partial x}$ and $\dfrac{\partial f}{\partial y}$ at $(x, y) = (1, 2)$.

21. Let $f(x, y, z) = xy^2 z + 5$. Evaluate $\dfrac{\partial f}{\partial y}$ at $(x, y, z) = (2, -1, 3)$.

22. Let $f(x, y, z) = \dfrac{x}{y - z}$. Compute $\dfrac{\partial f}{\partial y}(2, -1, 3)$.

23. Let $f(x, y) = x^3 y + 2xy^2$. Find $\dfrac{\partial^2 f}{\partial x^2}, \dfrac{\partial^2 f}{\partial y^2}, \dfrac{\partial^2 f}{\partial x\, \partial y}$, and $\dfrac{\partial^2 f}{\partial y\, \partial x}$.

24. Let $f(x, y) = xe^y + x^4 y + y^3$. Find $\dfrac{\partial^2 f}{\partial x^2}, \dfrac{\partial^2 f}{\partial y^2}, \dfrac{\partial^2 f}{\partial x\, \partial y}$, and $\dfrac{\partial^2 f}{\partial y\, \partial x}$.

25. A farmer can produce $f(x, y) = 200\sqrt{6x^2 + y^2}$ units of produce by utilizing x

units of labor and y units of capital. (The capital is used to rent or purchase land, materials, and equipment.)

   (a) Calculate the marginal productivities of labor and capital when $x = 10$ and $y = 5$.

   (b) Use the result of part (a) to determine the approximate effect on production of utilizing 5 units of capital but cutting back to $9\frac{1}{2}$ units of labor.

26. The productivity of a country is given by $f(x, y) = 300x^{2/3}y^{1/3}$, where x and y are the amounts of labor and capital.

   (a) Compute the marginal productivites of labor and capital when $x = 125$ and $y = 64$.

   (b) What would be the approximate effect of utilizing 125 units of labor but cutting back to 62 units of capital?

27. In a certain suburban community commuters have the choice of getting into the city by bus or train. The demand for these modes of transportation varies with their cost. Let $f(p_1, p_2)$ be the number of people who will take the bus when p_1 is the price of the bus ride and p_2 is the price of the train ride. For example, if $f(4.50, 6) = 7000$, then 7000 commuters will take the bus when the price of a bus ticket is \$4.50 and the price of a train ticket is \$6.00. Explain why $\dfrac{\partial f}{\partial p_1} < 0$ and $\dfrac{\partial f}{\partial p_2} > 0$.

28. Refer to Exercise 27. Let $g(p_1, p_2)$ be the number of people who will take the train when p_1 is the price of the bus ride and p_2 is the price of the train ride. Would you expect $\dfrac{\partial g}{\partial p_1}$ to be positive or negative? How about $\dfrac{\partial g}{\partial p_2}$?

29. Let p_1 be the average price of VCRs, p_2 the average price of video tape, $f(p_1, p_2)$ the demand for VCRs, and $g(p_1, p_2)$ the demand for video tape. Explain why $\dfrac{\partial f}{\partial p_2} < 0$ and $\dfrac{\partial g}{\partial p_1} < 0$.

30. The demand for a certain gas-guzzling car is given by $f(p_1, p_2)$, where p_1 is the price of the car and p_2 is the price of gasoline. Explain why $\dfrac{\partial f}{\partial p_1} < 0$ and $\dfrac{\partial f}{\partial p_2} < 0$.

31. The volume (V) of a certain amount of a gas is determined by the temperature (T) and the pressure (P) by the formula $V = .08(T/P)$. Calculate and interpret $\dfrac{\partial V}{\partial P}$ and $\dfrac{\partial V}{\partial T}$ when $P = 20, T = 300$.

32. Using data collected from 1929–1941, Richard Stone* determined that the yearly quantity Q of beer consumed in the United Kingdom was approximately given by the formula $Q = f(m, p, r, s)$, where

$$f(m, p, r, s) = (1.058)m^{.136}p^{-.727}r^{.914}s^{.816}$$

and m is the aggregate real income (personal income after direct taxes, adjusted for retail price changes), p is the average retail price of the commodity (in this case,

---

   *Richard Stone, "The Analysis of Market Demand," *Journal of the Royal Statistical Society,* CVIII (1945), 286–391.

beer), r is the average retail price level of all other consumer goods and services, and s is a measure of the strength of the beer. Determine which partial derivatives are positive and which are negative and give interpretations. (For example, since $\frac{\partial f}{\partial r} > 0$, people buy more beer when the prices of other goods increase and the other factors remain constant.)

33. Richard Stone (see Exercise 32) determined that the yearly consumption of food in the United States was given by

$$f(m, p, r) = (2.186)m^{.595}p^{-.543}r^{.922}.$$

Determine which partial derivatives are positive and which are negative and give interpretations of these facts.

34. For the production function $f(x, y) = 60x^{3/4}y^{1/4}$ considered in Example 8, think of $f(x, y)$ as the revenue when utilizing x units of labor and y units of capital. Under actual operating conditions, say $x = a$ and $y = b$, $\frac{\partial f}{\partial x}(a, b)$ is referred to as the *wage per unit of labor* and $\frac{\partial f}{\partial y}(a, b)$ is referred to as the *wage per unit of capital.* Show that

$$f(a, b) = a \cdot \left[\frac{\partial f}{\partial x}(a, b)\right] + b \cdot \left[\frac{\partial f}{\partial y}(a, b)\right].$$

(This equation shows how the revenue is distributed between labor and capital.)

35. Compute $\frac{\partial^2 f}{\partial x^2}$ where $f(x, y) = 60x^{3/4}y^{1/4}$, a production function (where x is units of labor). Explain why $\frac{\partial^2 f}{\partial x^2}$ is always negative.

36. Compute $\frac{\partial^2 f}{\partial y^2}$ where $f(x, y) = 60x^{3/4}y^{1/4}$, a production function (where y is units of capital). Explain why $\frac{\partial^2 f}{\partial y^2}$ is always negative.

37. Let $f(x, y) = 3x^2 + 2xy + 5y$, as in Example 5. Show that

$$f(1 + h, 4) - f(1, 4) = 14h + 3h^2.$$

Thus the error in approximating $f(1 + h, 4) - f(1, 4)$ by $14h$ is $3h^2$. (If $h = .01$, for instance, the error is only $.0003$.)

38. Physicians, particularly pediatricians, sometimes need to know the body surface area of a patient. For instance, the surface area is used to adjust the results of certain tests of kidney performance. Tables are available that give the approximate body surface area A in square meters of a person who weighs W kilograms and is H centimeters tall. The following empirical formula* is also used:

$$A = .007W^{.425}H^{.725}.$$

Evaluate $\frac{\partial A}{\partial W}$ and $\frac{\partial A}{\partial H}$ when $W = 54$, $H = 165$, and give a physical interpretation

---

* See J. Routh, *Mathematical Preparation for Laboratory Technicians* (Philadelphia, P.A.: W. B. Saunders Co., 1971), p. 92.

of your answers. You may use the approximations $(54)^{.425} \approx 5.4$, $(54)^{-.575} \approx .10$, $(165)^{.725} \approx 40.5$, $(165)^{-.275} \approx .25$.

SOLUTIONS TO PRACTICE PROBLEMS 7.2

1. (a) Negative. $\dfrac{\partial f}{\partial x}$ (400, 2000) is approximately the change in sales due to a $1 increase in x (price). Since raising prices lowers sales, we would expect $\dfrac{\partial f}{\partial x}$ (400, 2000) to be negative.

   (b) Positive. $\dfrac{\partial f}{\partial y}$ (400, 2000) is approximately the change in sales due to a $1 increase in advertising. Since spending more money on advertising brings in more customers, we would expect sales to increase; that is, $\dfrac{\partial f}{\partial y}$ (400, 2000) is most likely positive.

2. If the interest rate is raised from 9% to 10%, then the monthly payment will increase by about $66.20. [An increase to $9\frac{1}{2}\%$ causes an increase in the monthly payment of $\frac{1}{2} \cdot (66.20)$ or $33.10, and so on.]

7.3 MAXIMA AND MINIMA OF FUNCTIONS OF SEVERAL VARIABLES

Previously, we studied how to determine the maxima and minima of functions of a single variable. Let us extend that discussion to functions of several variables.

If $f(x, y)$ is a function of two variables, then we say that $f(x, y)$ has a *relative maximum* when $x = a$, $y = b$ if $f(x, y)$ is at most equal to $f(a, b)$ whenever x is near a and y is near b. Geometrically, the graph of $f(x, y)$ has a peak at the point (a, b). [See Fig. 1(a).] Similarly, we say that $f(x, y)$ has a *relative minimum* when $x = a$, $y = b$ if $f(x, y)$ is at least equal to $f(a, b)$ whenever x is near a and y is near b. Geometrically, the graph of $f(x, y)$ has a pit with bottom at the point (a, b). [See Fig. 1(b).]

Suppose the function $f(x, y)$ has a relative minimum at $(x, y) = (a, b)$, as in Fig. 2. When y is held constant at b, $f(x, y)$ is a function of x with a relative minimum at $x = a$. Therefore, the tangent line to the curve $z = f(x, b)$ is horizontal at $x = a$ and hence has slope 0. That is,

$$\frac{\partial f}{\partial x} (a, b) = 0.$$

Likewise, when x is held constant at a, $f(x, y)$ is a function of y with a relative minimum at $y = b$. Therefore, its derivative with respect to y is zero at $y = b$. That is,

$$\frac{\partial f}{\partial y} (a, b) = 0.$$

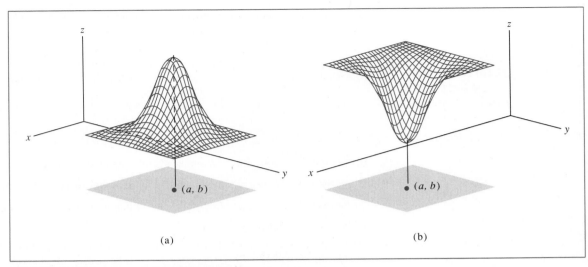

(a)

(b)

FIGURE I   Maximum and minimum points.

FIGURE 2   Horizontal tangent lines at a relative minimum.

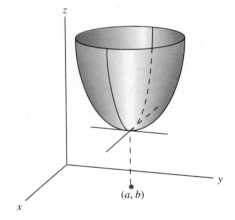

Similar considerations apply when $f(x, y)$ has a relative maximum at $(x, y) = (a, b)$.

First-Derivative Test for Functions of Two Variables   If $f(x, y)$ has either a relative maximum or minimum at $(x, y) = (a, b)$, then

$$\frac{\partial f}{\partial x}(a, b) = 0$$

and

$$\frac{\partial f}{\partial y}(a, b) = 0.$$

A relative maximum or minimum may or may not be an absolute maximum or minimum. However, to simplify matters in this text, the examples and exercises have been chosen so that if an absolute extremum of $f(x, y)$ exists, it will occur at a point where $f(x, y)$ has a relative extremum.

EXAMPLE 1 The function $f(x, y) = 3x^2 - 4xy + 3y^2 + 8x - 17y + 30$ has the graph pictured in Fig. 2. Find the point (a, b) at which $f(x, y)$ attains its minimum.

Solution We look for those values of x and y at which both partial derivatives are zero. The partial derivatives are

$$\frac{\partial f}{\partial x} = 6x - 4y + 8$$

$$\frac{\partial f}{\partial y} = -4x + 6y - 17.$$

Setting $\dfrac{\partial f}{\partial x} = 0$ and $\dfrac{\partial f}{\partial y} = 0$, we obtain

$$6x - 4y + 8 = 0 \quad \text{or} \quad y = \frac{6x + 8}{4}$$

$$-4x + 6y - 17 = 0 \quad \text{or} \quad y = \frac{4x + 17}{6}.$$

By equating these two expressions for y, we have

$$\frac{6x + 8}{4} = \frac{4x + 17}{6}.$$

Cross-multiplying, we see that

$$36x + 48 = 16x + 68$$

$$20x = 20$$

$$x = 1.$$

When we substitute this value for x into our first equation for y in terms of x, we obtain

$$y = \frac{6x + 8}{4} = \frac{6 \cdot 1 + 8}{4} = \frac{7}{2}.$$

If $f(x, y)$ has a minimum, it must occur where $\dfrac{\partial f}{\partial x} = 0$ and $\dfrac{\partial f}{\partial y} = 0$. We have determined that the partial derivatives are zero only when $x = 1$, $y = \frac{7}{2}$. From Fig. 2 we know that $f(x, y)$ has a minimum, so it must be at $(x, y) = (1, \frac{7}{2})$. ●

EXAMPLE 2 (*Price Discrimination*) A monopolist markets his product in two countries and can charge different amounts in each country. Let x be the number of units to be sold in the first country and y the number of units to be sold in the second country. Due to the laws of demand, the monopolist must set the price at $97 - (x/10)$ dollars in the first country and $83 - (y/20)$ dollars in the second country in order to sell all the units. The cost of producing these units is $20,000 + 3(x + y)$. Find the values of x and y that maximize the profit.

Solution   Let $f(x, y)$ be the profit derived from selling x units in the first country and y in the second. Then

$f(x, y)$

= [revenue from first country] + [revenue from second country] − [cost]

$$= \left(97 - \frac{x}{10}\right)x + \left(83 - \frac{y}{20}\right)y - [20,000 + 3(x + y)]$$

$$= 97x - \frac{x^2}{10} + 83y - \frac{y^2}{20} - 20,000 - 3x - 3y$$

$$= 94x - \frac{x^2}{10} + 80y - \frac{y^2}{20} - 20,000.$$

To find where $f(x, y)$ has its maximum value, we look for those values of x and y at which both partial derivatives are zero.

$$\frac{\partial f}{\partial x} = 94 - \frac{x}{5}$$

$$\frac{\partial f}{\partial y} = 80 - \frac{y}{10}.$$

We set $\dfrac{\partial f}{\partial x} = 0$ and $\dfrac{\partial f}{\partial y} = 0$ to obtain

$$94 - \frac{x}{5} = 0 \quad \text{or} \quad x = 470$$

$$80 - \frac{y}{10} = 0 \quad \text{or} \quad y = 800.$$

Therefore, the firm should adjust its prices to levels where it will sell 470 units in the first country and 800 units in the second country. ●

EXAMPLE 3   Suppose that we want to design a rectangular building having volume 147,840 cubic feet. Assuming that the daily loss of heat is given by

$$11xy + 14yz + 15xz,$$

where x, y, and z are, respectively, the length, width, and height of the building, find the dimensions of the building for which the daily heat loss is minimal.

Solution   We must minimize the function

$$11xy + 14yz + 15xz, \tag{1}$$

where x, y, z satisfy the constraint equation

$$xyz = 147,840.$$

For simplicity, let us denote 147,840 by V. Then $xyz = V$, so that $z = V/xy$. We substitute this expression for z into the objective function (1) to obtain a heat-loss function $g(x, y)$ of two variables—namely,

$$g(x, y) = 11xy + 14y\frac{V}{xy} + 15x\frac{V}{xy}$$

$$= 11xy + \frac{14V}{x} + \frac{15V}{y}.$$

To minimize this function, we first compute the partial derivatives with respect to x and y; then we equate them to zero.

$$\frac{\partial g}{\partial x} = 11y - \frac{14V}{x^2} = 0$$

$$\frac{\partial g}{\partial y} = 11x - \frac{15V}{y^2} = 0.$$

These two equations yield

$$y = \frac{14V}{11x^2} \tag{2}$$

$$11xy^2 = 15V. \tag{3}$$

If we substitute the value of y from (2) into (3), we see that

$$11x\left(\frac{14V}{11x^2}\right)^2 = 15V$$

$$\frac{14^2V^2}{11x^3} = 15V$$

$$x^3 = \frac{14^2 \cdot V^2}{11 \cdot 15 \cdot V} = \frac{14^2 \cdot V}{11 \cdot 15}$$

$$= \frac{14^2 \cdot 147,840}{11 \cdot 15}$$

$$= 175,616.$$

Therefore, we see (using a calculator) that

$$x = 56.$$

From equation (2) we find that

$$y = \frac{14 \cdot V}{11x^2} = \frac{14 \cdot 147,840}{11 \cdot 56^2} = 60.$$

Finally,

$$z = \frac{V}{xy} = \frac{147{,}840}{56 \cdot 60} = 44.$$

Thus the building should be 56 feet long, 60 feet wide, and 44 feet high in order to minimize the heat loss.* •

When considering a function of two variables, we find points (x, y) at which $f(x, y)$ has a potential relative maximum or minimum by setting $\dfrac{\partial f}{\partial x}$ and $\dfrac{\partial f}{\partial y}$ equal to zero and solving for x and y. However, if we are given no additional information about $f(x, y)$, it may be difficult to determine whether we have found a maximum or a minimum (or neither). In the case of functions of one variable, we studied concavity and deduced the second-derivative test. There is an analog of the second derivative test for functions of two variables, but it is much more complicated than the one-variable test. We state it without proof.

Second-Derivative Test for Functions of Two Variables   Suppose $f(x, y)$ is a function and (a, b) is a point at which

$$\frac{\partial f}{\partial x}(a, b) = 0 \quad \text{and} \quad \frac{\partial f}{\partial y}(a, b) = 0,$$

and let

$$D(x, y) = \frac{\partial^2 f}{\partial x^2} \cdot \frac{\partial^2 f}{\partial y^2} - \left(\frac{\partial^2 f}{\partial x \, \partial y}\right)^2.$$

1. If

$$D(a, b) > 0 \quad \text{and} \quad \frac{\partial^2 f}{\partial x^2}(a, b) > 0,$$

then $f(x, y)$ has a relative minimum at (a, b).
2. If

$$D(a, b) > 0 \quad \text{and} \quad \frac{\partial^2 f}{\partial x^2}(a, b) < 0,$$

then $f(x, y)$ has a relative maximum at (a, b).
3. If

$$D(a, b) < 0,$$

then $f(x, y)$ has neither a relative maximum nor a relative minimum at (a, b).
4. If $D(a, b) = 0$, then no conclusion can be drawn from this test.

---

*For further discussion of this heat-loss problem, as well as other examples of optimization in architectural design, see L. March, "Elementary Models of Built Forms," Chapter 3 in *Urban Space and Structures,* L. Martin and L. March, eds. (Cambridge: Cambridge University Press, 1972).

The saddle-shaped graph in Fig. 3 illustrates a function $f(x, y)$ for which $D(a, b) < 0$. Both partial derivatives are zero at $(x, y) = (a, b)$ and yet the function has neither a relative maximum nor a relative minimum there. (Observe that the function has a relative maximum with respect to x when y is held constant and a relative minimum with respect to y when x is held constant.)

FIGURE 3

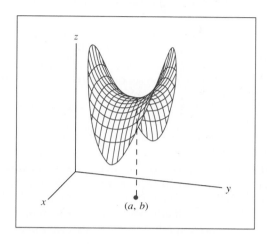

EXAMPLE 4　Let $f(x, y) = y^3 - x^2 + 6x - 12y + 5$. Find all possible relative maximum and minimum points of $f(x, y)$. Use the second-derivative test to determine the nature of each such point.

Solution　Since

$$\frac{\partial f}{\partial x} = -2x + 6, \qquad \frac{\partial f}{\partial y} = 3y^2 - 12,$$

we find that $f(x, y)$ has a potential relative extreme point when

$$-2x + 6 = 0$$
$$3y^2 - 12 = 0.$$

The first equation implies that $x = 3$. From the second equation we have

$$3y^2 = 12$$
$$y^2 = 4$$
$$y = \pm 2.$$

Thus $\dfrac{\partial f}{\partial x}$ and $\dfrac{\partial f}{\partial y}$ are both zero when $(x, y) = (3, 2)$ and when $(x, y) = (3, -2)$.

To apply the second-derivative test, we compute

$$\frac{\partial^2 f}{\partial x^2} = -2, \qquad \frac{\partial^2 f}{\partial y^2} = 6y, \qquad \frac{\partial^2 f}{\partial x \, \partial y} = 0,$$

and

$$D(x, y) = \frac{\partial^2 f}{\partial x^2} \cdot \frac{\partial^2 f}{\partial y^2} - \left(\frac{\partial^2 f}{\partial x \, \partial y}\right)^2 = (-2)(6y) - 0 = -12y. \qquad (4)$$

Since $D(3, 2) = -24$ is negative, case 3 of the second-derivative test says that $f(x, y)$ has neither a relative maximum nor a relative minimum at $(3, 2)$. Next, note that

$$D(3, -2) = 24 > 0, \qquad \frac{\partial^2 f}{\partial x^2}(3, -2) = -2 < 0.$$

Thus, by case 2 of the second-derivative test, the function $f(x, y)$ has a relative maximum at $(3, -2)$. ●

TECHNOLOGY PROJECT 7.5
Solving Optimization Problems in Several Variables (Symbolic Program Needed)

In solving optimization problems in several variables, it is often difficult to solve the system of equations obtained by setting the partial derivatives equal to 0. Moreover, it is often difficult to determine whether a point is a relative maximum or relative minimum. We can solve systems of equations in several variables using symbolic math programs. For example, to solve the system

$$\begin{cases} x^2 + y^2 = 10 \\ 3x - 7y = 9 \end{cases}$$

using *Mathematica,* we can use the command:

```
Solve [{x^2+y^2, 3x-7y} == {10,9}, {x,y}]
```

Note the use of the symbol ==, which is necessary to indicate that there are equations to solve. The symbols $\{x, y\}$ tell which variables are being solved for.

Once you determine solutions that are potential extreme points, you can determine whether they are relative maxima, relative minima, or neither by graphing the function in a region near the point.

Use a symbolic math program and the approach indicated above to determine the relative maxima and relative minima of the following functions:

1. $5x^2 + 4y^2 - 13x + 10y + 10$

2. $(3x - 1)^4 + 10y^2 + 6y - 4$

3. $x^4 + 10xy + y^6 - 50$

4. $x^2y + 10x^3y^2 + x + y - 3$

In this section we have restricted ourselves to functions of two variables, but the case of three or more variables is handled in a similar fashion. For instance, here is the first-derivative test for a function of three variables.

If $f(x, y, z)$ has a relative maximum or minimum at $(x, y, z) = (a, b, c)$, then

$$\frac{\partial f}{\partial x}(a, b, c) = 0$$

$$\frac{\partial f}{\partial y}(a, b, c) = 0$$

$$\frac{\partial f}{\partial z}(a, b, c) = 0.$$

PRACTICE PROBLEMS 7.3

1. Find all points (x, y) where $f(x, y) = x^3 - 3xy + \frac{1}{2}y^2 + 8$ has a possible relative maximum or minimum.
2. Apply the second-derivative test to the function $g(x, y)$ of Example 3 to confirm that a relative minimum actually occurs when $x = 56$ and $y = 60$.

EXERCISES 7.3

Find all points (x, y) where $f(x, y)$ has a possible relative maximum or minimum.

1. $f(x, y) = x^2 - 3y^2 + 4x + 6y + 8$
2. $f(x, y) = \frac{1}{2}x^2 + y^2 - 3x + 2y - 5$
3. $f(x, y) = x^2 - 5xy + 6y^2 + 3x - 2y + 4$
4. $f(x, y) = -3x^2 + 7xy - 4y^2 + x + y$
5. $f(x, y) = x^3 + y^2 - 3x + 6y$
6. $f(x, y) = x^2 - y^3 + 5x + 12y + 1$
7. $f(x, y) = \frac{1}{3}x^3 - 2y^3 - 5x + 6y - 5$
8. $f(x, y) = x^4 - 8xy + 2y^2 - 3$
9. The function $f(x, y) = 2x + 3y + 9 - x^2 - xy - y^2$ has a maximum at some point (x, y). Find the values of x and y where this maximum occurs.
10. The function $f(x, y) = \frac{1}{2}x^2 + 2xy + 3y^2 - x + 2y$ has a minimum at some point (x, y). Find the values of x and y where this minimum occurs.

In Exercises 11–16, both first partial derivatives of the function $f(x, y)$ are zero at the given points. Use the second-derivative test to determine the nature of $f(x, y)$ at each of these points. If the second-derivative test is inconclusive, so state.

11. $f(x, y) = 3x^2 - 6xy + y^3 - 9y; (3, 3), (-1, -1)$
12. $f(x, y) = 6xy^2 - 2x^3 - 3y^4; (0, 0), (1, 1), (1, -1)$

13. $f(x, y) = 2x^2 - x^4 - y^2; (-1, 0), (0, 0), (1, 0)$

14. $f(x, y) = x^4 - 4xy + y^4; (0, 0), (1, 1), (-1, -1)$

15. $f(x, y) = ye^x - 3x - y + 5; (0, 3)$

16. $f(x, y) = \dfrac{1}{x} + \dfrac{1}{y} + xy; (1, 1)$

Find all points (x, y) where $f(x, y)$ has a possible relative maximum or minimum. Then use the second-derivative test to determine, if possible, the nature of $f(x, y)$ at each of these points. If the second-derivative test is inconclusive, so state.

17. $f(x, y) = x^2 - 2xy + 4y^2$

18. $f(x, y) = 2x^2 + 3xy + 5y^2$

19. $f(x, y) = -2x^2 + 2xy - y^2 + 4x - 6y + 5$

20. $f(x, y) = -x^2 - 8xy - y^2$

21. $f(x, y) = x^2 + 2xy + 5y^2 + 2x + 10y - 3$

22. $f(x, y) = x^2 - 2xy + 3y^2 + 4x - 16y + 22$

23. $f(x, y) = x^3 - y^2 - 3x + 4y$

24. $f(x, y) = x^3 - 2xy + 4y$

25. $f(x, y) = 2x^2 + y^3 - x - 12y + 7$

26. $f(x, y) = x^2 + 4xy + 2y^4$

27. Find the possible values of x, y, z at which

$$f(x, y, z) = 2x^2 + 3y^2 + z^2 - 2x - y - z$$

assumes its minimum value.

28. Find the possible values of x, y, z at which

$$f(x, y, z) = 5 + 8x - 4y + x^2 + y^2 + z^2$$

assumes its minimum value.

29. U.S. postal rules require that the length plus the girth of a package cannot exceed 84 inches in order to be mailed. Find the dimensions of the rectangular package of greatest volume that can be mailed. [*Note:* From Fig. 4 we see that $84 =$ (length) + (girth) $= l + (2x + 2y)$.]

FIGURE 4

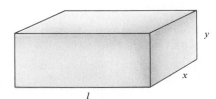

30. Find the dimensions of the rectangular box of least surface area that has a volume of 1000 cubic inches.

31. A company manufactures and sells two products, call them I and II, that sell for $10 and $9 per unit, respectively. The cost of producing x units of product I and y units of product II is

$$400 + 2x + 3y + .01(3x^2 + xy + 3y^2).$$

Find the values of x and y that maximize the company's profit. [*Note:* Profit = (revenue) − (cost).]

32. A monopolist manufactures and sells two competing products, call them I and II, that cost $30 and $20 per unit, respectively, to produce. The revenue from marketing x units of product I and y units of product II is $98x + 112y - .04xy - .1x^2 - .2y^2$. Find the values of x and y that maximize the monopolist's profits.

33. A company manufactures and sells two products, call them I and II, that sell for p_I and p_{II} per unit, respectively. Let $C(x, y)$ be the cost of producing x units of product I and y units of product II. Show that if the company's profit is maximized when $x = a, y = b,$ then

$$\frac{\partial C}{\partial x}(a, b) = p_I \quad \text{and} \quad \frac{\partial C}{\partial y}(a, b) = p_{II}.$$

34. A monopolist manufactures and sells two competing products, call them I and II, that cost p_I and p_{II} per unit, respectively, to produce. Let $R(x, y)$ be the revenue from marketing x units of product I and y units of product II. Show that if the monopolist's profit is maximized when $x = a, y = b,$ then

$$\frac{\partial R}{\partial x}(a, b) = p_I \quad \text{and} \quad \frac{\partial R}{\partial y}(a, b) = p_{II}.$$

SOLUTIONS TO PRACTICE PROBLEMS 7.3

1. Compute the first partial derivatives of $f(x, y)$ and solve the system of equations that results from setting the partials equal to zero.

$$\frac{\partial f}{\partial x} = 3x^2 - 3y = 0$$

$$\frac{\partial f}{\partial y} = -3x + y = 0.$$

Solve each equation for y in terms of x.

$$\begin{cases} y = x^2 \\ y = 3x. \end{cases}$$

Equate expressions for y and solve for x.

$$x^2 = 3x$$

$$x^2 - 3x = 0$$

$$x(x - 3) = 0$$

$$x = 0 \quad \text{or} \quad x = 3.$$

When $x = 0, y = 0^2 = 0$. When $x = 3, y = 3^2 = 9$. Therefore, the possible relative maximum or minimum points are $(0, 0)$ and $(3, 9)$.

2. We have

$$g(x, y) = 11xy + \frac{14V}{x} + \frac{15V}{y},$$

$$\frac{\partial g}{\partial x} = 11y - \frac{14V}{x^2}, \quad \text{and} \quad \frac{\partial g}{\partial y} = 11x - \frac{15V}{y^2}.$$

Now,

$$\frac{\partial^2 g}{\partial x^2} = \frac{28V}{x^3}, \quad \frac{\partial^2 g}{\partial y^2} = \frac{30V}{y^3}, \quad \text{and} \quad \frac{\partial^2 g}{\partial x\, \partial y} = 11.$$

Therefore,

$$D(x, y) = \frac{28V}{x^3} \cdot \frac{30V}{y^3} - (11)^2$$

$$D(56, 60) = \frac{28(147{,}840)}{(56)^3} \cdot \frac{30(147{,}840)}{(60)^3} - 121$$

$$= 484 - 121 = 363 > 0,$$

and

$$\frac{\partial^2 g}{\partial x^2}(56, 60) = \frac{28(147{,}840)}{(56)^3} > 0.$$

It follows that $g(x, y)$ has a relative minimum at $x = 56$, $y = 60$.

7.4 LAGRANGE MULTIPLIERS AND CONSTRAINED OPTIMIZATION

We have seen a number of optimization problems in which we were required to minimize (or maximize) an objective function where the variables were subject to a constraint equation. For instance, in Example 4 of Section 2.5, we minimized the cost of a rectangular enclosure by minimizing the objective function $21x + 14y$, where x and y were subject to the constraint equation $600 - xy = 0$. In the preceding section (Example 3) we minimized the daily heat loss from a building by minimizing the objective function $11xy + 14yz + 15xz$, subject to the constraint equation $147{,}840 - xyz = 0$.

Figure 1 gives a graphical illustration of what happens when an objective function is maximized subject to a constraint. The graph of the objective function is the cone-shaped surface $z = 36 - x^2 - y^2$, and the colored curve on that surface consists of those points whose x- and y-coordinates satisfy the constraint equation $x + 7y - 25 = 0$. The constrained maximum is at the highest point on this curve. Of course, the surface itself has a higher "unconstrained maximum" at $(x, y, z) = (0, 0, 36)$, but these values of x and y do not satisfy the constraint equation.

FIGURE I   A constrained optimization problem.

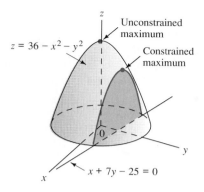

In this section we introduce a powerful technique for solving problems of this type. Let us begin with the following general problem, which involves two variables.

Problem   Let $f(x, y)$ and $g(x, y)$ be functions of two variables. Find values of x and y that maximize (or minimize) the objective function $f(x, y)$ and that also satisfy the constraint equation $g(x, y) = 0$.

Of course, if we can solve the equation $g(x, y) = 0$ for one variable in terms of the other and substitute the resulting expression into $f(x, y)$, we arrive at a function of a single variable that can be maximized (or minimized) by using the methods of Chapter 2. However, this technique can be unsatisfactory for two reasons. First, it may be difficult to solve the equation $g(x, y) = 0$ for x or for y. For example, if $g(x, y) = x^4 + 5x^3y + 7x^2y^3 + y^5 - 17 = 0$, then it is difficult to write y as a function of x or x as a function of y. Second, even if $g(x, y) = 0$ can be solved for one variable in terms of the other, substitution of the result into $f(x, y)$ may yield a complicated function.

One clever idea for handling the preceding problem was discovered by the eighteenth-century mathematician Lagrange, and the technique that he pioneered today bears his name—the method of *Lagrange multipliers*. The basic idea of this method is to replace $f(x, y)$ by an auxiliary function of three variables $F(x, y, \lambda)$, defined as

$$F(x, y, \lambda) = f(x, y) + \lambda g(x, y).$$

The new variable λ (lambda) is called a *Lagrange multiplier* and always multiplies the constraint function $g(x, y)$. The following theorem is stated without proof.

Theorem   Suppose that, subject to the constraint $g(x, y) = 0$, the function $f(x, y)$ has a relative maximum or minimum at $(x, y) = (a, b)$. Then there is a value of λ, say $\lambda = c$, such that the partial derivatives of $F(x, y, \lambda)$ all equal zero at $(x, y, \lambda) = (a, b, c)$.

The theorem implies that if we locate all points (x, y, λ) where the partial derivatives of $F(x, y, \lambda)$ are all zero, then among the corresponding points (x, y) we will find all possible places where $f(x, y)$ may have a constrained relative maximum or minimum. Thus the first step in the method of Lagrange multipliers is to set the partial derivatives of $F(x, y, \lambda)$ equal to zero and solve for x, y, and λ:

$$\frac{\partial F}{\partial x} = 0 \tag{L-1}$$

$$\frac{\partial F}{\partial y} = 0 \tag{L-2}$$

$$\frac{\partial F}{\partial \lambda} = 0. \tag{L-3}$$

From the definition of $F(x, y, \lambda)$, we see that $\dfrac{\partial F}{\partial \lambda} = g(x, y)$. Thus the third equation (L-3) is just the original constraint equation $g(x, y) = 0$. So when we find a point (x, y, λ) that satisfies (L-1), (L-2), and (L-3), the coordinates x and y will automatically satisfy the constraint equation.

The first example applies this method to the problem described in Fig. 1.

EXAMPLE 1   Maximize $36 - x^2 - y^2$ subject to the constraint $x + 7y - 25 = 0$.

Solution   Here $f(x, y) = 36 - x^2 - y^2$, $g(x, y) = x + 7y - 25$, and

$$F(x, y, \lambda) = 36 - x^2 - y^2 + \lambda(x + 7y - 25).$$

Equations (L-1) to (L-3) read

$$\frac{\partial F}{\partial x} = -2x + \lambda = 0 \tag{1}$$

$$\frac{\partial F}{\partial y} = -2y + 7\lambda = 0 \tag{2}$$

$$\frac{\partial F}{\partial \lambda} = x + 7y - 25 = 0. \tag{3}$$

We solve the first two equations for λ:

$$\lambda = 2x$$

$$\lambda = \frac{2}{7}y. \tag{4}$$

If we equate these two expressions for λ, we obtain

$$2x = \frac{2}{7}y$$

$$x = \frac{1}{7}y. \tag{5}$$

Substituting this expression for x into equation (3), we have

$$\frac{1}{7}y + 7y - 25 = 0$$

$$\frac{50}{7}y = 25$$

$$y = \frac{7}{2}.$$

With this value for y, equations (4) and (5) produce the values of x and λ:

$$x = \frac{1}{7}y = \frac{1}{7}\left(\frac{7}{2}\right) = \frac{1}{2}$$

$$\lambda = \frac{2}{7}y = 1.$$

Therefore, the partial derivatives of $F(x, y, \lambda)$ are zero when $x = \frac{1}{2}$, $y = \frac{7}{2}$, and $\lambda = 1$. So the minimum value of $36 - x^2 - y^2$ subject to the constraint $x + 7y - 25 = 0$ is

$$36 - \left(\frac{1}{2}\right)^2 - \left(\frac{7}{2}\right)^2 = \frac{47}{2}. \quad \bullet$$

The preceding technique for solving three equations in the three variables x, y, and λ can usually be applied to solve Lagrange multiplier problems. Here is the basic procedure.

1. Solve (L-1) and (L-2) for λ in terms of x and y; then equate the resulting expressions for λ.
2. Solve the resulting equation for one of the variables.
3. Substitute the expression so derived into the equation (L-3) and solve the resulting equation of one variable.
4. Use the one known variable and the equations of steps 1 and 2 to determine the other two variables.

In most applications we know that an absolute (constrained) maximum or minimum exists. In the event that the method of Lagrange multipliers produces exactly one possible relative extreme value, we will assume that it is indeed the sought-after absolute extreme value. For instance, the statement of Example 1 is meant to imply that there is an absolute maximum value. Since we determined that there was just one possible relative extreme value, we concluded that it was the absolute maximum value.

EXAMPLE 2   Using Lagrange multipliers, minimize $42x + 28y$, subject to the constraint $600 - xy = 0$, where x and y are restricted to positive values. (This problem arose in Example 4 of Section 2.5, where $42x + 28y$ was the cost of building a 600-square-foot enclosure having dimensions x and y.)

Solution   We have $f(x, y) = 42x + 28y$, $g(x, y) = 600 - xy$, and

$$F(x, y, \lambda) = 42x + 28y + \lambda(600 - xy).$$

The equations (L-1) to (L-3), in this case, are

$$\frac{\partial F}{\partial x} = 42 - \lambda y = 0$$

$$\frac{\partial F}{\partial y} = 28 - \lambda x = 0$$

$$\frac{\partial F}{\partial \lambda} = 600 - xy = 0.$$

From the first two equations we see that

$$\lambda = \frac{42}{y} = \frac{28}{x}.$$ **(step 1)**

Therefore,

$$42x = 28y$$

and

$$x = \frac{2}{3}y.$$ **(step 2)**

Substituting this expression for x into the third equation, we derive

$$600 - \left(\frac{2}{3}y\right)y = 0$$

$$y^2 = \frac{3}{2} \cdot 600 = 900$$

$$y = \pm 30$$ **(step 3)**

We discard the case $y = -30$ because we are interested only in positive values of x and y. Using $y = 30$, we find that

$$\left. \begin{array}{l} x = \dfrac{2}{3}(30) = 20 \\[2mm] \lambda = \dfrac{14}{20} = \dfrac{7}{10} \end{array} \right\}$$ **(step 4).**

So the minimum value of $42x + 28y$ with x and y subject to the constraint occurs when $x = 20$, $y = 30$, and $\lambda = \frac{7}{10}$. That minimum value is

$$42 \cdot (20) + 28 \cdot (30) = 1680. \quad \bullet$$

EXAMPLE 3   (*Production*)   Suppose that x units of labor and y units of capital can produce $f(x, y) = 60x^{3/4}y^{1/4}$ units of a certain product. Also suppose that each unit of labor costs $100, whereas each unit of capital costs $200. Assume that $30,000 is available to spend on production. How many units of labor and how many of capital should be utilized in order to maximize production?

Solution   The cost of x units of labor and y units of capital equals $100x + 200y$. Therefore, since we want to use all the available money ($30,000), we must satisfy the constraint equation

$$100x + 200y = 30,000$$

or

$$g(x, y) = 30,000 - 100x - 200y = 0.$$

Our objective function is $f(x, y) = 60x^{3/4}y^{1/4}$. In this case, we have

$$F(x, y, \lambda) = 60x^{3/4}y^{1/4} + \lambda(30,000 - 100x - 200y).$$

The equations (L-1) to (L-3) read

$$\frac{\partial F}{\partial x} = 45x^{-1/4}y^{1/4} - 100\lambda = 0 \qquad \text{(L-1)}$$

$$\frac{\partial F}{\partial y} = 15x^{3/4}y^{-3/4} - 200\lambda = 0 \qquad \text{(L-2)}$$

$$\frac{\partial F}{\partial \lambda} = 30,000 - 100x - 200y = 0. \qquad \text{(L-3)}$$

By solving the first two equations for λ, we see that

$$\lambda = \frac{45}{100}x^{-1/4}y^{1/4} = \frac{9}{20}x^{-1/4}y^{1/4}$$

$$\lambda = \frac{15}{200}x^{3/4}y^{-3/4} = \frac{3}{40}x^{3/4}y^{-3/4}.$$

Therefore, we must have

$$\frac{9}{20}x^{-1/4}y^{1/4} = \frac{3}{40}x^{3/4}y^{-3/4}.$$

To solve for y in terms of x, let us multiply both sides of this equation by $x^{1/4}y^{3/4}$:

$$\frac{9}{20}y = \frac{3}{40}x$$

or

$$y = \frac{1}{6}x.$$

Inserting this result in (L-3), we find that

$$100x + 200(\tfrac{1}{6}x) = 30,000$$

$$\frac{400x}{3} = 30,000$$

$$x = 225.$$

Hence

$$y = \frac{225}{6} = 37.5.$$

So maximum production is achieved by using 225 units of labor and 37.5 units of capital. ●

In Example 3 it turns out that, at the optimum value of x and y,

$$\lambda = \frac{9}{20}x^{-1/4}y^{1/4} = \frac{9}{20}(225)^{-1/4}(37.5)^{1/4} \approx .2875$$

$$\frac{\partial f}{\partial x} = 45x^{-1/4}y^{1/4} = 45(225)^{-1/4}(37.5)^{1/4} \tag{6}$$

$$\frac{\partial f}{\partial y} = 15x^{3/4}y^{-3/4} = 15(225)^{3/4}(37.5)^{-3/4}. \tag{7}$$

It can be shown that the Lagrange multiplier λ can be interpreted as the marginal productivity of money. That is, if one additional dollar is available, then approximately .2875 additional units of the product can be produced.

Recall that the partial derivatives $\dfrac{\partial f}{\partial x}$ and $\dfrac{\partial f}{\partial y}$ are called the marginal productivity of labor and capital, respectively. From (6) and (7) we have

$$\frac{[\text{marginal productivity of labor}]}{[\text{marginal productivity of capital}]} = \frac{45(225)^{-1/4}(37.5)^{1/4}}{15(225)^{3/4}(37.5)^{-3/4}}$$

$$= \frac{45}{15}(225)^{-1}(37.5)^{1}$$

$$= \frac{3(37.5)}{225} = \frac{37.5}{75} = \frac{1}{2}.$$

On the other hand,

$$\frac{[\text{cost per unit of labor}]}{[\text{cost per unit of capital}]} = \frac{100}{200} = \frac{1}{2}.$$

This result illustrates the following law of economics. *If labor and capital are at their optimal levels, then the ratio of their marginal productivities equals the ratio of their unit costs.*

The method of Lagrange multipliers generalizes to functions of any number of variables. For instance, we can maximize $f(x, y, z)$, subject to the constraint equation $g(x, y, z) = 0$, by considering the Lagrange function

$$F(x, y, z, \lambda) = f(x, y, z) + \lambda g(x, y, z).$$

The analogs of equations (L-1) to (L-3) are

$$\frac{\partial F}{\partial x} = 0$$

$$\frac{\partial F}{\partial y} = 0$$

$$\frac{\partial F}{\partial z} = 0$$

$$\frac{\partial F}{\partial \lambda} = 0.$$

Let us now show how we can solve the heat-loss problem of Section 7.3 by using this method.

EXAMPLE 4   Use Lagrange multipliers to find the values of x, y, z that minimize the objective function

$$f(x, y, z) = 11xy + 14yz + 15xz,$$

subject to the constraint

$$xyz = 147{,}840.$$

Solution   The Lagrange function is

$$F(x, y, z, \lambda) = 11xy + 14yz + 15xz + \lambda(147{,}840 - xyz).$$

The conditions for a relative minimum are

$$\frac{\partial F}{\partial x} = 11y + 15z - \lambda yz = 0$$

$$\frac{\partial F}{\partial y} = 11x + 14z - \lambda xz = 0$$

$$\frac{\partial F}{\partial z} = 14y + 15x - \lambda xy = 0$$

$$\frac{\partial F}{\partial \lambda} = 147{,}840 - xyz = 0. \tag{8}$$

From the first three equations we have

$$\left. \begin{array}{l} \lambda = \dfrac{11y + 15z}{yz} = \dfrac{11}{z} + \dfrac{15}{y} \\[2mm] \lambda = \dfrac{11x + 14z}{xz} = \dfrac{11}{z} + \dfrac{14}{x} \\[2mm] \lambda = \dfrac{14y + 15x}{xy} = \dfrac{14}{x} + \dfrac{15}{y} \end{array} \right\}. \tag{9}$$

Let us equate the first two expressions for λ:

$$\frac{11}{z} + \frac{15}{y} = \frac{11}{z} + \frac{14}{x}$$

$$\frac{15}{y} = \frac{14}{x}$$

$$x = \frac{14}{15}y.$$

Next, we equate the second and third expressions for λ in (9):

$$\frac{11}{z} + \frac{14}{x} = \frac{14}{x} + \frac{15}{y}$$

$$\frac{11}{z} = \frac{15}{y}$$

$$z = \frac{11}{15}y.$$

We now substitute the expressions for x and z into the constraint equation (8) and obtain

$$\frac{14}{15}y \cdot y \cdot \frac{11}{15}y = 147{,}840$$

$$y^3 = \frac{(147{,}840)(15)^2}{(14)(11)} = 216{,}000$$

$$y = 60.$$

From this, we find that

$$x = \frac{14}{15}(60) = 56 \quad \text{and} \quad z = \frac{11}{15}(60) = 44.$$

We conclude that the heat loss is minimized when $x = 56$, $y = 60$, and $z = 44$. ●

In the solution of Example 4, we found that at the optimal values of x, y, and z,

$$\frac{14}{x} = \frac{15}{y} = \frac{11}{z}.$$

Referring to Example 2 of Section 7.1, we see that 14 is the combined heat loss through the east and west sides of the building, 15 is the heat loss through the north and south sides of the building, and 11 is the heat loss through the floor and roof. Thus we have that under optimal conditions

$$\frac{[\text{heat loss through east and west sides}]}{[\text{distance between east and west sides}]}$$

$$= \frac{[\text{heat loss through north and south sides}]}{[\text{distance between north and south sides}]}$$

$$= \frac{[\text{heat loss through floor and roof}]}{[\text{distance between floor and roof}]}.$$

This is a principle of optimal design: minimal heat loss occurs when the distance between each pair of opposite sides is some fixed constant times the heat loss from the pair of sides.

The value of λ in Example 4 corresponding to the optimal values of x, y, and z is

$$\lambda = \frac{11}{z} + \frac{15}{y} = \frac{11}{44} + \frac{15}{60} = \frac{1}{2}.$$

One can show that the Lagrange multiplier λ is the marginal heat loss with respect to volume. That is, if a building of volume slightly more than 147,840 cubic feet is optimally designed, then $\frac{1}{2}$ unit of additional heat will be lost for each additional cubic foot of volume.

PRACTICE PROBLEMS 7.4

1. Let $F(x, y, \lambda) = 2x + 3y + \lambda(90 - 6x^{1/3}y^{2/3})$. Find $\dfrac{\partial F}{\partial x}$.

2. Refer to Exercise 29 of Section 7.3. What is the function $F(x, y, \lambda)$ when the exercise is solved using the method of Lagrange multipliers?

EXERCISES 7.4

Solve the following exercises by the method of Lagrange multipliers.

1. Minimize the function $x^2 + 3y^2 + 10$, subject to the constraint $8 - x - y = 0$.
2. Maximize the function $x^2 - y^2$, subject to the constraint $2x + y - 3 = 0$.
3. Maximize $x^2 + xy - 3y^2$, subject to the constraint $2 - x - 2y = 0$.
4. Minimize $\frac{1}{2}x^2 - 3xy + y^2 + \frac{1}{2}$, subject to the constraint $3x - y - 1 = 0$.
5. Find the values of x, y that maximize the function

$$-2x^2 - 2xy - \tfrac{3}{2}y^2 + x + 2y,$$

subject to the constraint $x + y - \frac{5}{2} = 0$.

6. Find the values of x, y that minimize the function

$$x^2 + xy + y^2 - 2x - 5y,$$

subject to the constraint $1 - x + y = 0$.

7. Find the two positive numbers whose product is 25 and whose sum is as small as possible.

8. Four hundred eighty dollars are available to fence in a rectangular garden. The fencing for the north and south sides of the garden costs $10 per foot and the fencing for the east and west sides costs $15 per foot. Find the dimensions of the largest possible garden.

9. Three hundred square inches of material are available to construct an open rectangular box with a square base. Find the dimensions of the box that maximize the volume.

10. The amount of space required by a particular firm is $f(x, y) = 1000\sqrt{6x^2 + y^2}$, where x and y are, respectively, the number of units of labor and capital utilized. Suppose that labor costs $480 per unit and capital costs $40 per unit and that the firm has $5000 to spend. Determine the amounts of labor and capital that should be utilized in order to minimize the amount of space required.

11. Find the dimensions of the rectangle of maximum area that can be inscribed in the unit circle. [See Fig. 2(a).]

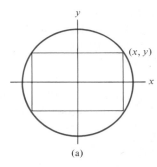

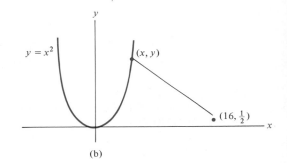

(a)   (b)

FIGURE 2

12. Find the point on the parabola $y = x^2$ that has minimal distance from the point $(16, \frac{1}{2})$. [See Fig. 2(b).] [*Suggestion:* If d denotes the distance from (x, y) to $(16, \frac{1}{2})$, then $d^2 = (x - 16)^2 + (y - \frac{1}{2})^2$. If d^2 is minimized, then d will be minimized.]

13. Suppose that a firm makes two products A and B that use the same raw materials. Given a fixed amount of raw materials and a fixed amount of manpower, the firm must decide how much of its resources should be allocated to the production of A and how much to B. If x units of A and y units of B are produced, suppose that x and y must satisfy

$$9x^2 + 4y^2 = 18,000.$$

The graph of this equation (for $x \geq 0$, $y \geq 0$) is called a *production possibilities curve* (Fig. 3). A point (x, y) on this curve represents a *production schedule* for the firm, committing it to produce x units of A and y units of B. The reason for the relationship between x and y involves the limitations on personnel and raw materials available to the firm. Suppose that each unit of A yields a $3 profit, whereas each

FIGURE 3 A production possibilities curve.

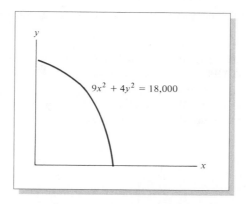

$9x^2 + 4y^2 = 18{,}000$

unit of B yields a \$4 profit. Then the profit of the firm is

$$P(x, y) = 3x + 4y.$$

Find the production schedule that maximizes the profit function $P(x, y)$.

14. A firm makes x units of product A and y units of product B and has a production possibilities curve given by the equation $4x^2 + 25y^2 = 50{,}000$ for $x \geq 0$, $y \geq 0$. (See Exercise 13.) Suppose profits are \$2 per unit for product A and \$10 per unit for product B. Find the production schedule that maximizes the total profit.

15. The production function for a firm is $f(x, y) = 64x^{3/4}y^{1/4}$, where x and y are the number of units of labor and capital utilized. Suppose that labor costs \$96 per unit and capital costs \$162 per unit and that the firm decides to produce 3456 units of goods.

(a) Determine the amounts of labor and capital that should be utilized in order to minimize the cost. That is, find the values of x, y that minimize $96x + 162y$, subject to the constraint $3456 - 64x^{3/4}y^{1/4} = 0$.

(b) Find the value of λ at the optimal level of production.

(c) Show that, at the optimal level of production, we have

$$\frac{[\text{marginal productivity of labor}]}{[\text{marginal productivity of capital}]} = \frac{[\text{unit price of labor}]}{[\text{unit price of capital}]}.$$

16. Consider the monopolist of Example 2, Section 7.3, who sells his goods in two countries. Suppose that he must set the same price in each country. That is, $97 - (x/10) = 83 - (y/20)$. Find the values of x and y that maximize profits under this new restriction.

17. Find the values of x, y, and z that maximize the function xyz subject to the constraint $36 - x - 6y - 3z = 0$.

18. Find the values of x, y, and z that maximize the function $xy + 3xz + 3yz$ subject to the constraint $9 - xyz = 0$.

19. Find the values of x, y, z that maximize the function

$$3x + 5y + z - x^2 - y^2 - z^2,$$

subject to the constraint $6 - x - y - z = 0$.

20. Find the values of x, y, z that minimize the function

$$x^2 + y^2 + z^2 - 3x - 5y - z,$$

subject to the constraint $20 - 2x - y - z = 0$.

21. The material for a rectangular box costs $2 per square foot for the top and $1 per square foot for the sides and bottom. Using Lagrange multipliers, find the dimensions for which the volume of the box is 12 cubic feet and the cost of the materials is minimized. (Referring to Fig. 4(a), the cost will be $3xy + 2xz + 2yz$.)

FIGURE 4

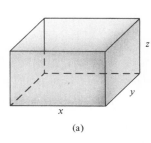

(a)

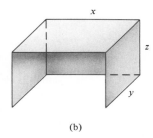

(b)

22. Use Lagrange multipliers to find the three positive numbers whose sum is 15 and whose product is as large as possible.

23. Find the dimensions of an open rectangular glass tank of volume 32 cubic feet for which the amount of material needed to construct the tank is minimized. [See Fig. 4(a).]

24. A shelter for use at the beach has a back, two sides, and a top made of canvas. [See Fig. 4(b).] Find the dimensions that maximize the volume and require 96 square feet of canvas.

25. Let $f(x, y)$ be any production function where x represents labor (costing $a per unit) and y represents capital (costing $b per unit). Assuming that $c is available, show that, at the values of x, y that maximize production,

$$\frac{\dfrac{\partial f}{\partial x}}{\dfrac{\partial f}{\partial y}} = \frac{a}{b}.$$

Note: Let $F(x, y, \lambda) = f(x, y) + \lambda(c - ax - by)$. The result follows from (L-1) and (L-2).

26. By applying the result in Exercise 25 to the production function $f(x, y) = kx^\alpha y^\beta$, show that, for the values of x, y that maximize production, we have

$$\frac{y}{x} = \frac{a\beta}{b\alpha}.$$

(This tells us that the ratio of capital to labor does not depend on the amount of money available nor on the level of production but only on the numbers a, b, α, and β.)

**SOLUTIONS TO
PRACTICE
PROBLEMS 7.4**

1. The function can be written as

$$F(x, y, \lambda) = 2x + 3y + \lambda \cdot 90 - \lambda \cdot 6x^{1/3}y^{2/3}.$$

   When differentiating with respect to x, both y and λ should be treated as constants (so $\lambda \cdot 90$ and $\lambda \cdot 6$ are also regarded as constants).

$$\frac{\partial F}{\partial x} = 2 - \lambda \cdot 6 \cdot \frac{1}{3}x^{-2/3} \cdot y^{2/3}$$

$$= 2 - 2\lambda x^{-2/3}y^{2/3}.$$

   [*Note:* It is not necessary to write out the multiplication by λ as we did. Most people just do this mentally and then differentiate.]

2. The quantity to be maximized is the volume xyl. The constraint is that length plus girth is 84. This translates to $84 = l + 2x + 2y$ or $84 - l - 2x - 2y = 0$. Therefore,

$$F(x, y, l, \lambda) = xyl + \lambda(84 - l - 2x - 2y).$$

7.5 THE METHOD OF LEAST SQUARES

Modern people compile graphs of literally thousands of different quantities: the purchasing value of the dollar as a function of time, the pressure of a fixed volume of air as a function of temperature, the average income of people as a function of their years of formal education, or the incidence of strokes as a function of blood pressure. The observed points on such graphs tend to be irregularly distributed due to the complicated nature of the phenomena underlying them as well as to errors made in observation (for example, a given procedure for measuring average income may not count certain groups). In spite of the imperfect nature of the data, we are often faced with the problem of making assessments and predictions based on them. Roughly speaking, this problem amounts to filtering the sources of errors in the data and isolating the basic underlying trend. Frequently, on the basis of a suspicion or a working hypothesis, we may suspect that the underlying trend is linear—that is, the data should lie on a straight line. But which straight line? This is the problem that the *method of least squares* attempts to answer. To be more specific, let us consider the following problem:

Problem   Given observed data points (x_1, y_1), (x_2, y_2), . . . , (x_N, y_N) on a graph, find the straight line that "best" fits these points.

In order to completely understand the statement of the problem being considered, we must define what it means for a line to "best" fit a set of points.

If (x_i, y_i) is one of our observed points, then we will measure how far it is from a given line $y = Ax + B$ by the vertical distance from the point to the line. Since the point on the line with x-coordinate x_i is $(x_i, Ax_i + B)$, this vertical distance is the distance between the y-coordinates $Ax_i + B$ and y_i. (See Fig. 1.) If $E_i = (Ax_i + B) - y_i$, then either E_i or $-E_i$ is the vertical distance from (x_i, y_i) to the line. To avoid this ambiguity we work with the square of this vertical distance, namely,

$$E_i^2 = (Ax_i + B - y_i)^2.$$

The total error in approximating the data points $(x_1, y_1), \ldots, (x_N, y_N)$ by the line $y = Ax + B$ is usually measured by the sum E of the squares of the vertical distances from the points to the line,

$$E = E_1^2 + E_2^2 + \cdots + E_N^2.$$

E is called the *least-squares error* of the observed points with respect to the line. If all the observed points lie on the line $y = Ax + B$, then all E_i are zero and the error E is zero. If a given observed point is far away from the line, the corresponding E_i^2 is large and hence makes a large contribution to the error E.

FIGURE I   Fitting a line to data points.

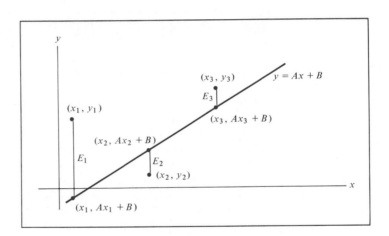

In general, we cannot expect to find a line $y = Ax + B$ that fits the observed points so well that the error E is zero. Actually, this situation will occur only if the observed points lie on a straight line. However, we can rephrase our original problem as follows:

Problem   Given observed data points $(x_1, y_1), (x_2, y_2), \ldots, (x_N, y_N)$, find a straight line $y = Ax + B$ for which the error E is as small as possible.

It turns out that this problem is a minimization problem in the two variables A and B and so can be solved by using the methods of Section 7.3. Let us consider an example.

EXAMPLE 1 Find the straight line that minimizes the least-squares error for the points $(1, 4)$, $(2, 5)$, $(3, 8)$.

Solution Let the straight line be $y = Ax + B$. When $x = 1, 2, 3$, the y-coordinate of the corresponding point of the line is $A + B, 2A + B, 3A + B$, respectively. Therefore, the squares of the vertical distances from the points $(1, 4), (2, 5), (3, 8)$ are, respectively,

$$E_1^2 = (A + B - 4)^2$$
$$E_2^2 = (2A + B - 5)^2$$
$$E_3^2 = (3A + B - 8)^2.$$

(See Fig. 2.) Thus the least-squares error is

$$E_1^2 + E_2^2 + E_3^2 = (A + B - 4)^2 + (2A + B - 5)^2 + (3A + B - 8)^2.$$

FIGURE 2

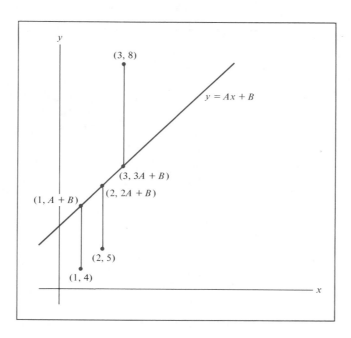

This error obviously depends on the choice of A and B. Let $f(A, B)$ denote this least-squares error. We want to find values of A and B that minimize $f(A, B)$. To do so, we take partial derivatives with respect to A and B and set the partial derivatives equal to zero:

$$\frac{\partial f}{\partial A} = 2(A + B - 4) + 2(2A + B - 5) \cdot 2 + 2(3A + B - 8) \cdot 3$$

$$= 28A + 12B - 76 = 0,$$

$$\frac{\partial f}{\partial B} = 2(A + B - 4) + 2(2A + B - 5) + 2(3A + B - 8)$$

$$= 12A + 6B - 34 = 0.$$

In order to find A and B, we must solve the system of simultaneous linear equations

$$28A + 12B = 76$$

$$12A + 6B = 34.$$

Multiplying the second equation by 2 and subtracting from the first equation, we have $4A = 8$ or $A = 2$. Therefore, $B = \frac{5}{3}$, and the straight line that minimizes the least-squares error is $y = 2x + \frac{5}{3}$. ●

We may follow a similar procedure for finding the line $y = Ax + B$ when more observed points are given.

EXAMPLE 2   The following table* gives the crude male death rate for lung cancer in 1950 and the per capita consumption of cigarettes in 1930 in various countries.

Country	Cigarette Consumption (Per Capita)	Lung Cancer Deaths (Per Million Males)
Norway	250	95
Sweden	300	120
Denmark	350	165
Australia	470	170

(a) Use the method of least squares to obtain the straight line that best fits these data.

(b) In 1930 the per capita cigarette consumption in Finland was 1100. Use the straight line found in part (a) to estimate the male lung cancer death rate in Finland in 1950.

---

*These data were obtained from *Smoking and Health,* Report of the Advisory Committee to the Surgeon General of the Public Health Service, U.S. Department of Health, Education, and Welfare, Washington, D.C., Public Health Service Publication No. 1103, p. 176.

FIGURE 3 Lung cancer data for least-squares analysis.

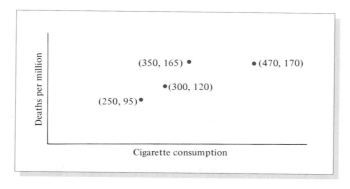

Solution (a) The points are plotted in Fig. 3. Let the line that best fits these points have equation $y = Ax + B$. The least-squares error of the given data points is

$$E = (250A + B - 95)^2 + (300A + B - 120)^2$$
$$+ (350A + B - 165)^2 + (470A + B - 170)^2.$$

We determine the values of A and B for which E is minimized by taking the partial derivatives with respect to A and B and setting them equal to zero. Computing the partial derivatives, we get

$$\frac{\partial E}{\partial A} = 2(250A + B - 95) \cdot 250 + 2(300A + B - 120) \cdot 300$$

$$+ 2(350A + B - 165) \cdot 350 + 2(470A + B - 170) \cdot 470$$

$$\frac{\partial E}{\partial B} = 2(250A + B - 95) + 2(300A + B - 120)$$

$$+ 2(350A + B - 165) + 2(470A + B - 170).$$

If we set $\frac{\partial E}{\partial A} = 0$ and $\frac{\partial E}{\partial B} = 0$, the following system of equations results:

$$495{,}900A + 1370B = 197{,}400 \tag{1}$$
$$1370A + 4B = 550. \tag{2}$$

There are several ways to solve this system of equations. We will solve equation (2) for B and then insert this expression for B into equation (1) so as to obtain an equation in A alone. Solving (2) for B gives

$$B = \frac{550 - 1370A}{4} = 137.5 - 342.5A. \tag{3}$$

Now substitute (3) into (1) to obtain

$$495,900A + 1370(137.5 - 342.5A) = 197,400$$

$$[495,900 - (1370)(342.5)]A = 197,400 - (1370)(137.5)$$

$$26,675A = 9025$$

$$A = \frac{9025}{26,675} \approx .338.$$

Then substitute this value for A into (3) to find

$$B = 137.5 - 342.5(.338) = 21.735.$$

Therefore, the straight line that best fits these points is

$$y = .338x + 21.735.$$

(b) We use the straight line to estimate the lung cancer death rate in Finland by setting $x = 1100$. Then we get

$$y = .338(1100) + 21.735 = 393.535 \approx 394.$$

Therefore, we estimate the lung cancer death rate in Finland to be 394.

(*Note:* The actual rate was 350.) ●

TECHNOLOGY PROJECT 7.6
Calculated Least-Squares Lines Using Technology

Most scientific calculators have built-in facilities for calculating the coefficents A and B of a least-squares line $y = Ax + B$ corresponding to a list of data points you key in.

1. Consult your calculator manual and learn to operate the least-squares facility on your calculator.
2. Use your calculator to determine the least-squares line corresponding to the lung cancer data in Example 2.
3. Use your calculator to determine the least-squares line corresponding to the ecology data from Exercise 15 below.

PRACTICE PROBLEMS 7.5

1. Let $E = (A + B + 2)^2 + (3A + B)^2 + (6A + B - 8)^2$. What is $\frac{\partial E}{\partial A}$?

2. Find the formula (of the type in Problem 1) that gives the least-squares error for the points $(1, 10)$, $(5, 8)$, $(7, 0)$.

EXERCISES 7.5

Find the straight line that best fits the following data points, where "best" is meant in the sense of least squares.

1. $(1, 0), (2, -1), (3, -4)$ **2.** $(2, 2), (3, 0), (7, -1)$

3. $(1, 6), (3, 0), (6, 10)$ **4.** $(1, 1), (3, 2), (5, 4)$

5. $(0, 1), (1,1), (2, 2), (3, 2)$ **6.** $(0, 2), (1, 2), (2, 2), (3, 3)$

7. $(1, 0), (2, 1), (4, 2), (5, 3)$ **8.** $(2, 3), (3, 2), (5, 1), (6, 0)$

9. $(1, 0), (2, 1), (3, 5), (15, 11), (-1, 4)$

10. $(0, 7), (-1, 5), (1, 9), (3, 14), (-2, 0)$

Refer to Example 1. Figure 4 shows the coordinates of the points on the least-squares line corresponding to the three given points. What is the least-squares error for this line?

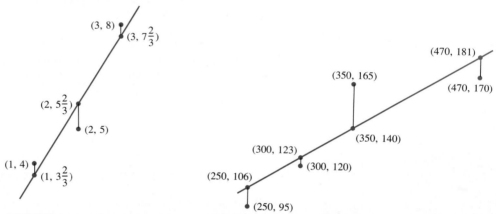

FIGURE 4 Least-squares line from Example 1.

FIGURE 5 Least-squares line from Example 2.

12. Refer to Example 2. Figure 5 shows the coordinates of the points on the least-squares line corresponding to the four given points. What is the least-squares error for this line?

13. The trend of sales of a certain new car dealer is as shown in the table below.

Week Number	Number of Cars Sold
1	38
2	40
3	41
4	39
5	45

Make a prediction of the number of cars sold during the seventh week.

14. A company analyzes production and profit figures and determines that in recent years productivity and profits have been related as shown in the table below.

Productivity (Thousands of Units)	Profits
78	$52,000
83	55,000
100	68,000
110	77,000
129	97,000

For the current year, their plant has been increased to allow production of 150,000 units. Estimate the profits.

15. An ecologist wished to know whether certain species of aquatic insects have their ecological range limited by temperature. He collected the following data, relating the average daily temperature at different portions of a creek with the elevation of that portion of the creek (above sea level).*

Elevation (Kilometers)	Average Temperature (Degree Celsius)
2.7	11.2
2.8	10
3.0	8.5
3.5	7.5

(a) Find the straight line that provides the best least-squares fit to these data.

(b) Use the linear function to estimate the average daily temperature for this creek at altitude 3.2 kilometers.

16. A soap manufacturer hopes to determine the demand curve for her soap by setting different prices in each of three cities of the same size and observing the volume of sales in each city. The results are given in the following table.

Price (Cents)	Volume of Sales (Thousands of Cases)
30	120
35	100
40	90

(a) Let x be the price and y the volume. Find the straight line that best fits these data.

---

* The authors express their thanks to Dr. J. David Allen, Department of Zoology, University of Maryland, for suggesting this exercise.

(b) Use the demand curve of part (a) to estimate the volume of sales if the price is 50 cents.

17. The accompanying table shows the 1988 price of a gallon of fuel and the consumption of motor fuel for several countries.

    (a) Find the straight line that provides the best least-squares fit to these data.

    (b) In 1988, the price of gas in Holland was \$3.00 per gallon. Use the straight line of part (a) to estimate the amount of motor fuel used per 1000 people in Holland.

Country	Price per Gallon in U.S. Dollars	Tons of Motor Fuel per 1000 Persons
United States	\$1.00	1400
England	\$2.20	620
Sweden	\$2.80	700
France	\$3.10	580
Italy	\$3.85	420

SOLUTIONS TO PRACTICE PROBLEMS 7.5

1. $\dfrac{\partial E}{\partial A} = 2(A + B + 2) \cdot 1 + 2(3A + B) \cdot 3 + 2(6A + B - 8) \cdot 6$

   $= (2A + 2B + 4) + (18A + 6B) + (72A + 12B - 96)$

   $= 92A + 20B - 92.$

   (Notice that we used the general power rule when differentiating and so had to always multiply by the derivative of the quantity inside the parentheses. Also, you might be tempted to first square the terms in the expression for E and then differentiate. We recommend that you resist that temptation.)

2. $E = (A + B - 10)^2 + (5A + B - 8)^2 + (7A + B)^2$. In general, E is a sum of squares, one for each point being fitted. The point (a, b) gives rise to the term $(aA + B - b)^2$.

7.6 DOUBLE INTEGRALS

Up to this point, our discussion of calculus of several variables has been confined to the study of differentiation. Let us now take up the topic of integration of functions of several variables. As has been the case throughout most of this chapter, we restrict our discussion to functions $f(x, y)$ of two variables.

Let us begin with some motivation. Before we define the concept of integral for functions of several variables, let us review the essential features of the integral in one variable.

Consider the definite integral $\int_a^b f(x)\, dx$. To write down this integral takes two pieces of information. The first is the function $f(x)$. The second is the interval over which the integration is to be performed. In this case, the interval is the portion of the x-axis from $x = a$ to $x = b$. The value of the definite integral is a number. In case the function $f(x)$ is nonnegative throughout the interval from $x = a$ to $x = b$, this number equals the area under the graph of $f(x)$ from $x = a$ to $x = b$. (See Fig. 1.) If $f(x)$ is negative for some values of x in the interval, then the integral still equals the area under the graph, but areas below the x-axis are counted as negative.

FIGURE 1

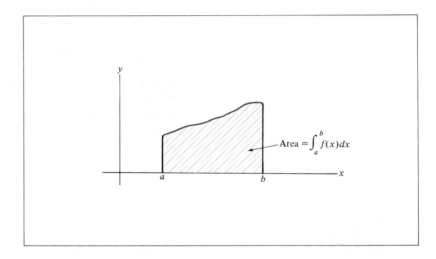

Let us generalize the above ingredients to a function $f(x, y)$ of two variables. First, we must provide a two-dimensional analog of the interval from $x = a$ to $x = b$. This is easy. We take a two-dimensional region R of the plane, such as the region shown in Fig. 2. As our generalization of $f(x)$, we take a function $f(x, y)$ of two variables. Our generalization of the definite integral is

FIGURE 2   A region in the xy-plane.

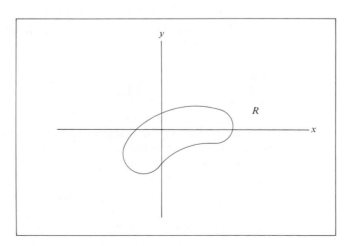

denoted

$$\iint\limits_{R} f(x, y) \, dx \, dy$$

and is called the *double integral of* $f(x, y)$ *over the region R*. The value of the double integral is a number defined as follows. For the sake of simplicity, let us begin by assuming that $f(x, y) \geq 0$ for all points (x, y) in the region R. (This is the analog of the assumption that $f(x) \geq 0$ for all x in the interval from $x = a$ to $x = b$.) This means that the graph of f lies above the region R in three-dimensional space. (See Fig. 3.) The portion of the graph over R determines a solid figure. (See Fig. 4.) This figure is called the *solid bounded by* $f(x, y)$ *over the region R*. We define the double integral $\iint\limits_{R} f(x, y) \, dx \, dy$ to be the volume of this solid. In case the graph of $f(x, y)$ lies partially above the region R and partially below, we define the double integral to be the volume of the solid above the region minus the volume of the solid below the region. That is, we count volumes below the xy-plane as negative.

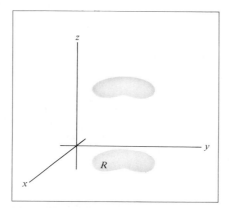

FIGURE 3 Graph of $f(x, y)$ above the region R.

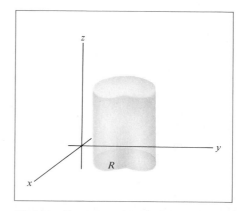

FIGURE 4 Solid bounded by $f(x, y)$ over R.

Now that we have defined the notion of a double integral, we must learn how to calculate its value. To do so, let us introduce the notion of an iterated integral. Let $f(x, y)$ be a function of two variables, let $g(x)$ and $h(x)$ be two functions of x alone, and let a and b be numbers. Then an *iterated integral* is an expression of the form

$$\int_{a}^{b} \left(\int_{g(x)}^{h(x)} f(x, y) \, dy \right) dx.$$

To explain the meaning of this collection of symbols, we proceed from the inside out. We evaluate the integral

$$\int_{g(x)}^{h(x)} f(x, y)\, dy$$

by considering $f(x, y)$ as a function of y alone. We treat x as a constant in this integration. So we evaluate the integral by first finding an antiderivative $F(x, y)$ with respect to y. The integral above is then evaluated as

$$F(x, h(x)) - F(x, g(x)).$$

That is, we evaluate the antiderivative between the limits $y = g(x)$ and $y = h(x)$. This gives us a function of x alone. To complete the evaluation of the iterated integral, we integrate this function from $x = a$ to $x = b$. The next two examples illustrate the procedure for evaluating iterated integrals.

EXAMPLE 1   Evaluate the iterated integral $\displaystyle\int_{1}^{2} \left(\int_{3}^{4} (y - x)\, dy \right) dx$.

Solution   Here $g(x)$ and $h(x)$ are constant functions: $g(x) = 3$ and $h(x) = 4$. We evaluate the inner integral first. The variable in this integral is y, so we treat x as a constant.

$$\int_{3}^{4} (y - x)\, dy = \left(\frac{1}{2}y^2 - xy \right) \Big|_{3}^{4}$$

$$= \left(\frac{1}{2} \cdot 16 - x \cdot 4 \right) - \left(\frac{1}{2} \cdot 9 - x \cdot 3 \right)$$

$$= 8 - 4x - \frac{9}{2} + 3x$$

$$= \frac{7}{2} - x.$$

Now we carry out the integration with respect to x:

$$\int_{1}^{2} \left(\frac{7}{2} - x \right) dx = \frac{7}{2}x - \frac{1}{2}x^2 \Big|_{1}^{2}$$

$$= \left(\frac{7}{2} \cdot 2 - \frac{1}{2} \cdot 4 \right) - \left(\frac{7}{2} - \frac{1}{2} \cdot 1 \right)$$

$$= (7 - 2) - (3) = 2.$$

So the value of the iterated integral is 2. ●

EXAMPLE 2   Evaluate the iterated integral

$$\int_{0}^{1} \left(\int_{\sqrt{x}}^{x+1} 2xy\, dy \right) dx.$$

Solution We evaluate the inner integral first.

$$\int_{\sqrt{x}}^{x+1} 2xy \, dy = xy^2 \Big|_{\sqrt{x}}^{x+1}$$

$$= x(x+1)^2 - x(\sqrt{x})^2$$

$$= x(x^2 + 2x + 1) - x \cdot x$$

$$= x^3 + 2x^2 + x - x^2$$

$$= x^3 + x^2 + x.$$

Now we evaluate the outer integral.

$$\int_0^1 (x^3 + x^2 + x) \, dx = \frac{1}{4}x^4 + \frac{1}{3}x^3 + \frac{1}{2}x^2 \Big|_0^1$$

$$= \frac{1}{4} + \frac{1}{3} + \frac{1}{2} = \frac{13}{12}.$$

So the value of the iterated integral is $\frac{13}{12}$. ●

Let us now return to the discussion of the double integral $\iint_R f(x, y) \, dx \, dy$. When the region R has a special form, the double integral may be expressed as an iterated integral. Namely, let us suppose that R is bounded by the graphs of $y = g(x)$, $y = h(x)$ and by the vertical lines $x = a$ and $x = b$. (See Fig. 5.) In this case, we have the following fundamental result, which we cite without proof.

Let R be the region in the xy-plane bounded by the graphs of $y = g(x)$, $y = h(x)$, and the vertical lines $x = a, x = b$. Then

$$\iint_R f(x, y) \, dx \, dy = \int_a^b \left(\int_{g(x)}^{h(x)} f(x, y) \, dy \right) dx.$$

FIGURE 5

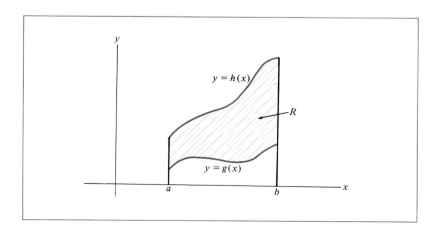

Since the value of the double integral gives the volume of the solid bounded by the graph $f(x, y)$ over the region R, the result above may be used to calculate volumes, as the next two examples show.

EXAMPLE 3   Calculate the volume of the solid bounded above by the function $y - x$ and lying over the rectangular region R: $1 \leq x \leq 2$, $3 \leq y \leq 4$. (See Fig. 6.)

Solution   The desired volume is given by the double integral $\iint\limits_R (y - x) \, dx \, dy$. By the result just cited, this double integral is equal to the iterated integral

$$\int_1^2 \left(\int_3^4 (y - x) \, dy \right) dx.$$

The value of this iterated integral was shown in Example 1 to be 2, so the volume of the solid shown in Fig. 6 is 2. ●

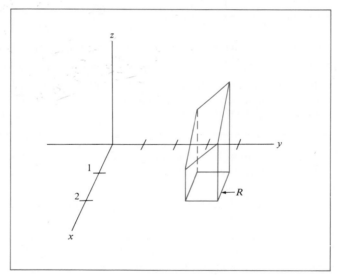

FIGURE 6   **FIGURE 7**

EXAMPLE 4   Calculate $\iint\limits_R 2xy \, dx \, dy$, where R is the region shown in Fig. 7.

Solution   The region R is bounded below by $y = \sqrt{x}$, above by $y = x + 1$, on the left by $x = 0$, and on the right by $x = 1$. Therefore,

$$\iint\limits_R 2xy \, dx \, dy = \int_0^1 \left(\int_{\sqrt{x}}^{x+1} 2xy \, dy \right) dx$$

$$= \frac{13}{12} \quad \text{(by Example 2).} ●$$

In our discussion, we have confined ourselves to iterated integrals in which the inner integral was with respect to y. In a completely analogous manner, we may treat iterated integrals in which the inner integral is with respect to x. Such iterated integrals may be used to evaluate double integrals over regions R bounded by curves of the form $x = g(y)$, $x = h(y)$, and horizontal lines $y = a$, $y = b$. The computations are analogous to those given above.

PRACTICE PROBLEMS 7.6

1. Calculate the iterated integral

$$\int_0^2 \left(\int_0^{x/2} e^{2y-x} \, dy \right) dx.$$

2. Calculate

$$\iint_R e^{2y-x} \, dx \, dy$$

where R is the region in Fig. 8.

FIGURE 8

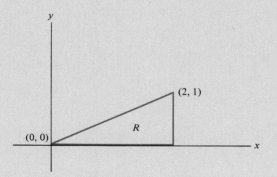

EXERCISES 7.6

Calculate the following iterated integrals.

1. $\displaystyle\int_0^1 \left(\int_0^1 e^{x+y} \, dy \right) dx$

2. $\displaystyle\int_{-1}^1 \left(\int_0^2 (3x^3 + y^2) \, dy \right) dx.$

3. $\displaystyle\int_{-2}^0 \left(\int_{-1}^1 xe^{xy} \, dy \right) dx$

4. $\displaystyle\int_1^3 \left(\int_2^5 (2y - 3x) \, dy \right) dx$

5. $\displaystyle\int_1^4 \left(\int_x^{x^2} xy \, dy \right) dx$

6. $\displaystyle\int_0^2 \left(\int_x^{5x} y \, dy \right) dx$

7. $\displaystyle\int_{-1}^1 \left(\int_x^{2x} (x + y) \, dy \right) dx$

8. $\displaystyle\int_0^1 \left(\int_0^x (x + e^y) \, dy \right) dx$

Let R be the rectangle consisting of all points (x, y) such that $0 \le x \le 2, 2 \le y \le 3$. Calculate the following double integrals. Interpret each as a volume.

9. $\iint\limits_R xy^2 \, dx \, dy$

10. $\iint\limits_R (xy + y^2) \, dx \, dy$

11. $\iint\limits_R e^{-x-y} \, dx \, dy$

12. $\iint\limits_R e^{y-x} \, dx \, dy$

Calculate the volumes over the following regions R bounded above by the graph of $f(x, y) = x^2 + y^2$.

13. R is the rectangle bounded by the lines $x = 1, x = 3, y = 0, y = 1$.

14. R is the region bounded by the lines $x = 0$, $x = 1$ and the curves $y = 0$ and $y = \sqrt[3]{x}$.

SOLUTIONS TO PRACTICE PROBLEMS 7.6

1. $\displaystyle\int_0^2 \left(\int_0^{x/2} e^{2y-x} \, dy \right) dx = \int_0^2 \left(\frac{1}{2} e^{2y-x} \Big|_0^{x/2} \right) dx$

$$= \int_0^2 \left(\frac{1}{2} e^{2(x/2)-x} - \frac{1}{2} e^{2\cdot 0 - x} \right) dx$$

$$= \int_0^2 \left(\frac{1}{2} - \frac{1}{2} e^{-x} \right) dx$$

$$= \frac{1}{2} x + \frac{1}{2} e^{-x} \Big|_0^2$$

$$= \frac{1}{2} \cdot 2 + \frac{1}{2} e^{-2} - \frac{1}{2} \cdot 0 - \frac{1}{2} e^{-0}$$

$$= 1 + \frac{1}{2} e^{-2} - 0 - \frac{1}{2}$$

$$= \frac{1}{2} + \frac{1}{2} e^{-2}.$$

2. The line passing through the points $(0, 0)$ and $(2, 1)$ has equation $y = x/2$. Hence the region R is bounded below by $y = 0$, above by $y = x/2$, on the left by $x = 0$, and on the right by $x = 2$. Therefore,

$$\iint\limits_R e^{2y-x} \, dx \, dy = \int_0^2 \left(\int_0^{x/2} e^{2y-x} \, dy \right) dx = \frac{1}{2} + \frac{1}{2} e^{-2}$$

by Problem 1.

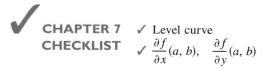 **CHAPTER 7 CHECKLIST**
✓ Level curve
✓ $\dfrac{\partial f}{\partial x}(a, b)$, $\dfrac{\partial f}{\partial y}(a, b)$

✓ Partial derivative
✓ Second partial derivatives
✓ First-derivative test

✓ Relative maxima and minima in several variables
✓ Second-derivative test in two variables
✓ Method of least squares
✓ Iterated integral

✓ Method of Lagrange multipliers:
   objective function
   constraint equation
✓ Double integral

CHAPTER 7   SUPPLEMENTARY EXERCISES

1. Let $f(x, y) = x\sqrt{y}/(1 + x)$. Compute $f(2, 9), f(5, 1)$, and $f(0, 0)$.

2. Let $f(x, y, z) = x^2 e^{y/z}$. Compute $f(-1, 0, 1), f(1, 3, 3)$, and $f(5, -2, 2)$.

3. If A dollars are deposited in a bank (assuming a 6% continuous interest rate), the amount in the account after t years is $f(A, t) = Ae^{.06t}$. Find and interpret $f(10, 11.5)$.

4. Let $f(x, y, \lambda) = xy + \lambda(5 - x - y)$. Find $f(1, 2, 3)$.

5. Let $f(x, y) = 3x^2 + xy + 5y^2$. Find $\dfrac{\partial f}{\partial x}$ and $\dfrac{\partial f}{\partial y}$.

6. Let $f(x, y) = 3x - \frac{1}{2}y^4 + 1$. Find $\dfrac{\partial f}{\partial x}$ and $\dfrac{\partial f}{\partial y}$.

7. Let $f(x, y) = e^{x/y}$. Find $\dfrac{\partial f}{\partial x}$ and $\dfrac{\partial f}{\partial y}$.

8. Let $f(x, y) = x/(x - 2y)$. Find $\dfrac{\partial f}{\partial x}$ and $\dfrac{\partial f}{\partial y}$.

9. Let $f(x, y, z) = x^3 - yz^2$. Find $\dfrac{\partial f}{\partial x}, \dfrac{\partial f}{\partial y}$, and $\dfrac{\partial f}{\partial z}$.

10. Let $f(x, y, \lambda) = xy + \lambda(5 - x - y)$. Find $\dfrac{\partial f}{\partial x}, \dfrac{\partial f}{\partial y}$, and $\dfrac{\partial f}{\partial \lambda}$.

11. Let $f(x, y) = x^3 y + 8$. Compute $\dfrac{\partial f}{\partial x}(1, 2)$ and $\dfrac{\partial f}{\partial y}(1, 2)$.

12. Let $f(x, y, z) = (x + y)z$. Evaluate $\dfrac{\partial f}{\partial y}$ at $(x, y, z) = (2, 3, 4)$.

13. Let $f(x, y) = x^5 - 2x^3 y + \dfrac{1}{2}y^4$. Find $\dfrac{\partial^2 f}{\partial x^2}, \dfrac{\partial^2 f}{\partial y^2}, \dfrac{\partial^2 f}{\partial x \partial y}$, and $\dfrac{\partial^2 f}{\partial y \partial x}$.

14. Let $f(x, y) = 2x^3 + x^2 y - y^2$. Compute $\dfrac{\partial^2 f}{\partial x^2}, \dfrac{\partial^2 f}{\partial y^2}$, and $\dfrac{\partial^2 f}{\partial x \partial y}$ at $(x, y) = (1, 2)$.

15. A dealer in a certain brand of electronic calculator finds that (within certain limits) the number of calculators she can sell per week given by $f(p, t) = -p + 6t - .02pt$,

where p is the price of the calculator and t is the number of dollars spent on advertising. Compute $\dfrac{\partial f}{\partial p}(25, 10{,}000)$ and $\dfrac{\partial f}{\partial t}(25, 10{,}000)$ and interpret these numbers.

16. Suppose that the crime rate in a certain city can be approximated by a function $f(x, y, z)$, where x is the unemployment rate, y is the amount of social services available, and z is the size of the police force. Explain why $\dfrac{\partial f}{\partial x} > 0, \dfrac{\partial f}{\partial y} < 0$, and $\dfrac{\partial f}{\partial z} < 0$.

In Exercises 17–20, find all points (x, y) where $f(x, y)$ has a possible relative maximum or minimum.

17. $f(x, y) = -x^2 + 2y^2 + 6x - 8y + 5$.

18. $f(x, y) = x^2 + 3xy - y^2 - x - 8y + 4$.

19. $f(x, y) = x^3 + 3x^2 + 3y^2 - 6y + 7$.

20. $f(x, y) = \frac{1}{2}x^2 + 4xy + y^3 + 8y^2 + 3x + 2$.

In Exercises 21–23, find all points (x, y) where $f(x, y)$ has a possible relative maximum or minimum. Then use the second-derivative test to determine, if possible, the nature of $f(x, y)$ at each of these points. If the second-derivative test is inconclusive, so state.

21. $f(x, y) = x^2 + 3xy + 4y^2 - 13x - 30y + 12$.

22. $f(x, y) = 7x^2 - 5xy + y^2 + x - y + 6$.

23. $f(x, y) = x^3 + y^2 - 3x - 8y + 12$.

24. Find the values of x, y, z at which

$$f(x, y, z) = x^2 + 4y^2 + 5z^2 - 6x + 8y + 3$$

assumes its minimum value.

25. Using the method of Lagrange multipliers, maximize the function $3x^2 + 2xy - y^2$, subject to the constraint $5 - 2x - y = 0$.

26. Using the method of Lagrange multipliers, find the values of x, y that minimize the function $-x^2 - 3xy - \frac{1}{2}y^2 + y + 10$, subject to the constraint $10 - x - y = 0$.

27. Using the method of Lagrange multipliers, find the values of x, y, z that minimize the function $3x^2 + 2y^2 + z^2 + 4x + y + 3z$, subject to the constraint $4 - x - y - z = 0$.

28. Using the method of Lagrange multipliers, find the dimensions of the rectangular box of volume 1000 cubic inches for which the sum of the dimensions is minimized.

29. A person wants to plant a rectangular garden along one side of a house and put a fence on the other three sides of the garden. (See Fig. 1.) Using the method of Lagrange multipliers, find the dimensions of the garden of greatest area that can be enclosed by using 40 feet of fencing.

FIGURE I   A garden.

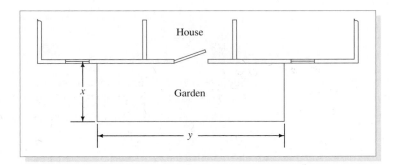

House

Garden

x

y

30. The solution to Exercise 29 is $x = 10, y = 20, \lambda = 10$. Suppose that one additional foot of fencing becomes available. Show that now the optimal dimensions for the garden are $10 + h$ and $20 + k$, where $2h + k = 1$.

In Exercise 31–33, find the straight line that best fits the following data points, where "best" is meant in the sense of least squares.

31. $(1, 1), (2, 3), (3, 6)$ 32. $(1, 1), (3, 4), (5, 7)$

33. $(0, 1), (1, -1), (2, -3), (3, -5)$

In Exercises 34 and 35, calculate the iterated integral.

34. $\displaystyle\int_0^4 \left(\int_0^1 (x\sqrt{y} + y)\, dx \right) dy$ 35. $\displaystyle\int_1^4 \left(\int_0^5 (2xy^4 + 3)\, dx \right) dy$

In Exercises 36 and 37, let R be the rectangle consisting of all points (x, y) such that $0 \le x \le 4, 1 \le y \le 3$, and calculate the double integral.

36. $\displaystyle\iint_R (2x + 3y)\, dx\, dy$ 37. $\displaystyle\iint_R 5\, dx\, dy$

38. The present value of y dollars after x years at 15% continuous interest is $f(x, y) = ye^{-.15x}$. Sketch some sample level curves. (Economists call this collection of level curves a "discount system.")

CHAPTER 8

THE TRIGONOMETRIC FUNCTIONS

In this chapter we expand the collection of functions to which we can apply calculus by introducing the trigonometric functions. As we shall see, these functions are *periodic*. That is, after a certain point their graphs repeat themselves. This repetitive phenomenon is not displayed by any of the functions that we have considered until now. Yet many natural phenomena are repetitive or cyclical—for example, the motion of the planets in our solar system, earthquake vibrations, and the natural rhythm of the heart. Thus the functions introduced in this chapter add considerably to our capacity to describe physical processes.

8.1 RADIAN MEASURE OF ANGLES

The ancient Babylonians introduced angle measurement in terms of degrees, minutes, and seconds, and these units are still generally used today for navigation and practical measurements. In calculus, however, it is more convenient to measure angles in terms of *radians,* for in this case the differentiation formulas for the trigonometric functions are easier to remember and use. Also, the radian is becoming more widely used today in scientific work because it is the unit of angle measurement in the international metric system (Système Internationale des Unités).

In order to define a radian, we consider a circle of radius 1 and measure angles in terms of distances around the circumference. The central angle determined by an arc of length 1 along the circumference is said to have a measure of *one radian.* (See Fig. 1.) Since the circumference of the circle of radius 1 has length 2π, there are 2π radians in one full revolution of the circle. Equivalently,

$$360° = 2\pi \text{ radians.} \tag{1}$$

FIGURE 1

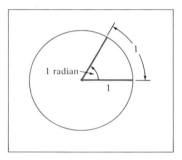

The following important relations should be memorized (see Fig. 2):

$$90° = \frac{\pi}{2} \text{ radians} \qquad \text{(one quarter-revolution)}$$

$$180° = \pi \text{ radians} \qquad \text{(one half-revolution)}$$

$$270° = \frac{3\pi}{2} \text{ radians} \qquad \text{(three quarter-revolutions)}$$

$$360° = 2\pi \text{ radians} \qquad \text{(one full revolution).}$$

FIGURE 2

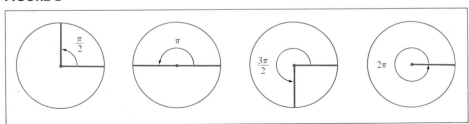

From formula (1) we see that

$$1° = \frac{2\pi}{360} \text{ radians} = \frac{\pi}{180} \text{ radians}.$$

If d is any number, then

$$d° = d \times \frac{\pi}{180} \text{ radians}. \tag{2}$$

That is, in order to convert degrees to radians, multiply the number of degrees by $\pi/180$.

EXAMPLE 1 Convert 45°, 60°, and 135° to radians.

Solution

$$45° = \overset{1}{\cancel{45}} \times \frac{\pi}{\underset{4}{\cancel{180}}} \text{ radians} = \frac{\pi}{4} \text{ radians}$$

$$60° = \overset{1}{\cancel{60}} \times \frac{\pi}{\underset{3}{\cancel{180}}} \text{ radians} = \frac{\pi}{3} \text{ radians}$$

$$135° = \overset{3}{\cancel{135}} \times \frac{\pi}{\underset{4}{\cancel{180}}} \text{ radians} = \frac{3\pi}{4} \text{ radians}.$$

These three angles are shown in Fig. 3. ●

FIGURE 3

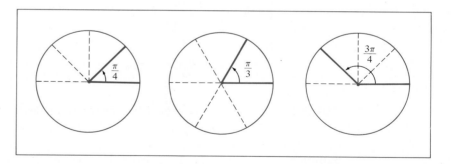

We usually omit the word "radian" when measuring angles because all our angle measurements will be in radians unless degrees are specifically indicated.

For our purposes, it is important to be able to speak of negative as well as positive angles, so let us define what we mean by a negative angle. We shall usually consider angles that are in *standard position* on a coordinate system, with the vertex of the angle at $(0, 0)$ and one side, called the "initial side," along

FIGURE 4

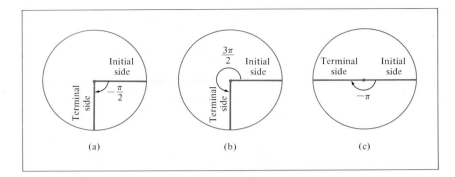

(a)   (b)   (c)

the positive x-axis. We measure such an angle from the initial side to the "terminal side," where a *counterclockwise angle is positive* and a *clockwise angle is negative.* Some examples are given in Fig. 4.

Notice in Fig. 4(a) and (b) how essentially the same picture can describe more than one angle.

By considering angles formed from more than one revolution (in the positive or negative direction), we can construct angles whose measure is of arbitrary size (i.e., not necessarily between -2π and 2π). Some examples are illustrated in Fig. 5.

FIGURE 5

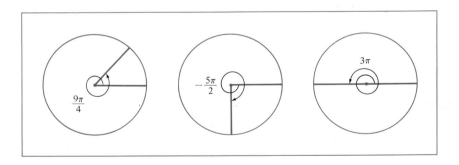

EXAMPLE 2   (a) What is the radian measure of the angle in Fig. 6?

(b) Construct an angle of $5\pi/2$ radians.

Solution   (a) The angle described in Fig. 6 consists of one full revolution (2π radians) plus three quarter-revolutions [$3 \times (\pi/2)$ radians]. That is,

$$t = 2\pi + 3 \times \frac{\pi}{2} = 4 \times \frac{\pi}{2} + 3 \times \frac{\pi}{2} = \frac{7\pi}{2}.$$

(b) Think of $5\pi/2$ radians as $5 \times (\pi/2)$ radians—that is, five quarter-revolutions of the circle. This is one full revolution plus one quarter-revolution. An angle of $5\pi/2$ radians is shown in Fig. 7. ●

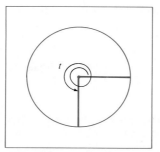

FIGURE 6

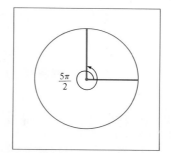

FIGURE 7

1. A right triangle has one angle of $\pi/3$ radians. What are the other angles?
2. How many radians are there in an angle of $-780°$? Draw the angle.

EXERCISES 8.1

Convert the following to radian measure.

1. $30°, 120°, 315°$ **2.** $18°, 72°, 150°$

3. $450°, -210°, -90°$ **4.** $990°, -270°, -540°$

Give the radian measure of each angle described below.

5.

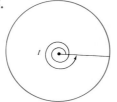

6.

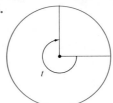

7.

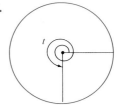

8.

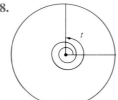

9.

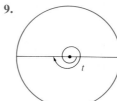

10.

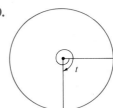

11.

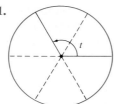

12.

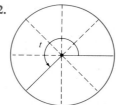

Construct angles with the following radian measure.

13. $3\pi/2, 3\pi/4, 5\pi$ 14. $\pi/3, 5\pi/2, 6\pi$

15. $-\pi/3, -3\pi/4, -7\pi/2$ 16. $-\pi/4, -3\pi/2, -3\pi$

17. $\pi/6, -2\pi/3, -\pi$ 18. $2\pi/3, -\pi/6, 7\pi/2$

**SOLUTIONS TO
PRACTICE
PROBLEMS 8.1**

1. The sum of the angles of a triangle is $180°$ or π radians. Since a right angle is $\pi/2$ radians and one angle is $\pi/3$ radians, the remaining angle is $\pi - (\pi/2 + \pi/3) = \pi/6$ radians.

2. $-780° = -780 \times (\pi/180)$ radians $= -13\pi/3$ radians. Since $-13\pi/3 = -4\pi - \pi/3$, we draw the angle by first making two revolutions in the negative direction and then a rotation of $\pi/3$ in the negative direction. See Fig. 8.

FIGURE 8

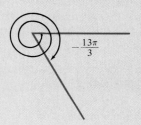

$$-\frac{13\pi}{3}$$

◢ 8.2 THE SINE AND THE COSINE

Given a number t, we consider an angle of t radians placed in standard position, as in Fig. 1, and we let P be a point on the terminal side of this angle. Denote the coordinates of P by (x, y) and let r be the length of the segment OP; that is, $r = \sqrt{x^2 + y^2}$. The *sine* and *cosine* of t, denoted by $\sin t$ and $\cos t$, respectively, are defined by the ratios

$$\sin t = \frac{y}{r}$$

$$\cos t = \frac{x}{r}. \tag{1}$$

It does not matter which point on the ray through P we use to define $\sin t$ and $\cos t$. If $P' = (x', y')$ is another point on the same ray and if r' is the length of OP' (Fig. 2), then, by properties of similar triangles, we have

$$\frac{y'}{r'} = \frac{y}{r} = \sin t$$

$$\frac{x'}{r'} = \frac{x}{r} = \cos t.$$

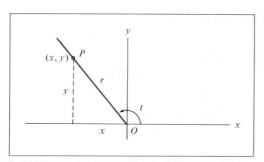

FIGURE 1 Diagram for definitions of sine and cosine.

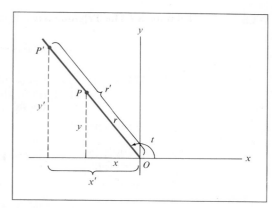

FIGURE 2

Three examples that illustrate the definition of sin t and cos t are shown in Fig. 3.

When $0 < t < \pi/2$, the values of sin t and cos t may be expressed as ratios of the lengths of the sides of a right triangle. Indeed, if we are given a right triangle as in Fig. 4, then we have

$$\sin t = \frac{\text{opposite}}{\text{hypotenuse}}, \qquad \cos t = \frac{\text{adjacent}}{\text{hypotenuse}}. \qquad (2)$$

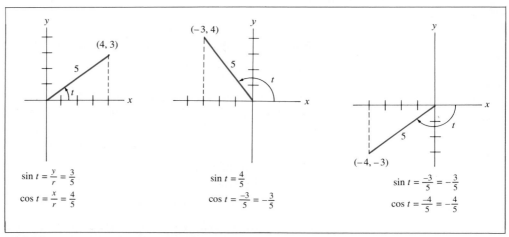

FIGURE 3 Calculation of sin t and cos t.

FIGURE 4

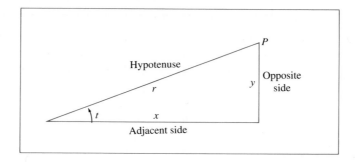

A typical application of (2) appears in Example 1.

EXAMPLE 1   The hypotenuse of a right triangle is four units and one angle is .7 radian. Determine the length of the side opposite this angle.

Solution   See Fig. 5. Since $y/4 = \sin .7$, we have,

$$y = 4 \sin .7$$

$$y = 4(.64422) = 2.57688. \; \bullet$$

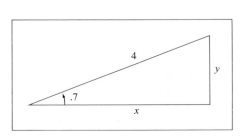

FIGURE 5   **FIGURE 6**

Another way to describe the sine and cosine functions is to choose the point P in Fig. 1 so that $r = 1$. That is, choose P on the unit circle. (See Fig. 6.) In this case

$$\sin t = \frac{y}{1} = y$$

$$\cos t = \frac{x}{1} = x.$$

So the y-coordinate of P is $\sin t$ and the x-coordinate of P is $\cos t$. Thus we have the following result.

Alternative Definition of Sine and Cosine Functions   We can think of cos t and sin t as the x- and y-coordinates of the point P on the unit circle that is determined by an angle of t radians (Fig. 7).

FIGURE 7

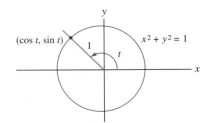

EXAMPLE 2   Find a value of t such that $0 < t < \pi/2$ and $\cos t = \cos(-\pi/3)$.

Solution   On the unit circle we locate the point P that is determined by an angle of $-\pi/3$ radians. The x-coordinate of P is $\cos(-\pi/3)$. There is another point Q on the unit circle with the same x-coordinate. (See Fig. 8.) Let t be the radian measure of the angle determined by Q. Then

$$\cos t = \cos\left(-\frac{\pi}{3}\right)$$

because Q and P have the same x-coordinate. Also, $0 < t < \pi/2$. From the symmetry of the diagram it is clear that $t = \pi/3$. ●

FIGURE 8

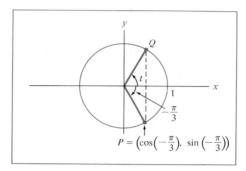

$$P = \left(\cos\left(-\tfrac{\pi}{3}\right), \ \sin\left(-\tfrac{\pi}{3}\right)\right)$$

Properties of the Sine and Cosine Functions   Each number t determines a point $(\cos t, \sin t)$ on the unit circle $x^2 + y^2 = 1$ as in Fig. 7. Therefore, $(\cos t)^2 + (\sin t)^2 = 1$. It is convenient (and traditional) to write $\sin^2 t$ instead of $(\sin t)^2$ and $\cos^2 t$ instead of $(\cos t)^2$. Thus we can write the last formula as follows:

$$\cos^2 t + \sin^2 t = 1. \tag{3}$$

The numbers t and $t \pm 2\pi$ determine the same point on the unit circle (because 2π represents a full revolution of the circle). But $t + 2\pi$ and $t - 2\pi$ correspond to the points $(\cos(t + 2\pi), \sin(t + 2\pi))$ and $(\cos(t - 2\pi), \sin(t - 2\pi))$, respectively. Hence

$$\cos(t \pm 2\pi) = \cos t, \qquad \sin(t \pm 2\pi) = \sin t. \tag{4}$$

Figure 9(a) illustrates another property of the sine and cosine—namely,

$$\cos(-t) = \cos t, \qquad \sin(-t) = -\sin t. \tag{5}$$

FIGURE 9 Diagrams for two identities.

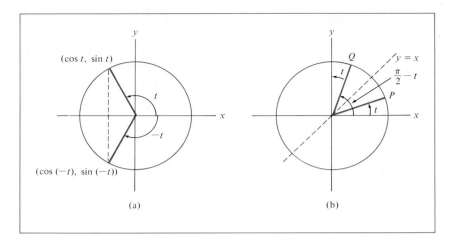

(a)   (b)

Figure 9(b) shows that the points P and Q corresponding to t and to $\pi/2 - t$ are reflections of each other through the line $y = x$. Consequently, the coordinates of Q are obtained by interchanging the coordinates of P. This means that

$$\cos\left(\frac{\pi}{2} - t\right) = \sin t, \qquad \sin\left(\frac{\pi}{2} - t\right) = \cos t. \qquad (6)$$

The equations in (3) to (6) are called *identities* because they hold for all values of t. Another identity that holds for all numbers s and t is

$$\sin(s + t) = \sin s \cos t + \cos s \sin t. \qquad (7)$$

A proof of (7) may be found in any introductory book on trigonometry. There are a number of other identities concerning trigonometric functions, but we shall not need them here.

The Graph of sin t   Let us analyze what happens to $\sin t$ as t increases from 0 to π. When $t = 0$, the point $P = (\cos t, \sin t)$ is at $(1, 0)$, as in Fig. 10(a). As t increases, P moves counterclockwise around the unit circle. The y-coordinate of P—that is, $\sin t$—increases until $t = \pi/2$, where $P = (0, 1)$. [See Fig. 10(c).] As t increases from $\pi/2$ to π, the y-coordinate of P—that is, $\sin t$—decreases from 1 to 0. [See Fig. 10(d) and (e).]

Part of the graph of $\sin t$ is sketched in Fig. 11. Notice that, for t between 0 and π, the values of $\sin t$ increase from 0 to 1 and then decrease back to 0, just as we predicted from Fig. 10. For t between π and 2π, the values of $\sin t$ are negative. Can you explain why? The graph of $y = \sin t$ for t between 2π and 4π is exactly the same as the graph for t between 0 and 2π. This result follows from formula (5). We say that the sine function is *periodic with period 2π* because the

FIGURE 10   Movement along the unit circle.

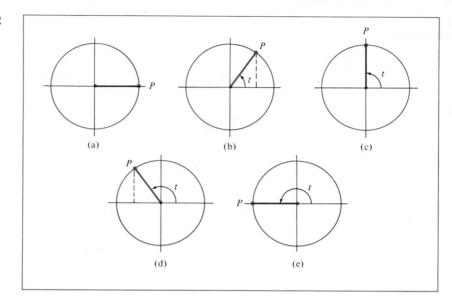

(a)   (b)   (c)

(d)   (e)

FIGURE 11

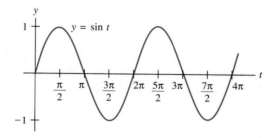

FIGURE 12   Graph of the sine function.

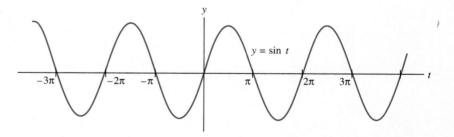

graph repeats itself every 2π units. We can use this fact to make a quick sketch of part of the graph for negative values of t (Fig. 12).

The Graph of cos t   By analyzing what happens to the first coordinate of the point $(\cos t, \sin t)$ as t varies, we obtain the graph of $\cos t$. Note from Fig. 13 that the graph of the cosine function is also periodic with period 2π.

FIGURE 13   Graph of the cosine function.

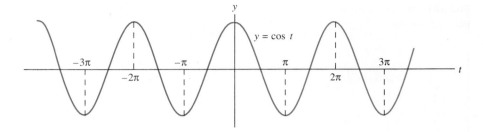

A Remark About Notation   The sine and cosine functions assign to each number *t* the values sin *t* and cos *t*, respectively. There is nothing special, however, about the letter *t*. Although we chose to use the letters *t*, *x*, *y*, and *r* in the *definition* of the sine and cosine, other letters could have been used as well. Now that the sine and cosine of every number are defined, we are free to use *any* letter to represent the independent variable.

TECHNOLOGY PROJECT 8.1
Amplitude and Period of Sine and Cosine Functions

In this technology project, we will explore some properties of the functions $f(x) = a \sin (bx)$ and $g(x) = a \cos (bx)$ for constants *a* and *b*.

1. On a single coordinate system, graph the functions sin *x*, sin 2*x*, sin 3*x* in the range $-2\pi \leq x \leq 2\pi$. How many repetitions of each graph are there?

2. On a single coordinate system, graph the functions sin *x*, sin (*x*/2), sin (*x*/3) in range $-6\pi \leq x \leq 6\pi$. How many repetitions of each graph are there?

3. A number *c* is called a **period** of the function *f* provided that the graph of *f* repeats every *c* units along the *x*-axis; that is,

$$f(x + c) = f(x)$$

for all *x*. Based on your observations in 1 and 2 what are the periods of the functions sin *x*, sin 2*x*, sin 3*x* and sin *x*, sin (*x*/2), sin (*x*/3)?

4. Based on your observations, can you guess the period of the function sin *bx*?

5. Repeat 1–4, substituting the cosine function for the sine.

6. On a single coordinate system, graph the functions, sin *x*, 2 sin *x*, $\frac{1}{2}$ sin *x* in the range $-2\pi \leq x \leq 2\pi$. How are the graphs related to one another?

7. Repeat 6 replacing the sine with the cosine.

8. On the basis of 6 and 7 can you make a guess as to the geometric significance of the constant *a* in the graphs of the functions $f(x) = a \sin (bx)$ and $g(x) = a \cos (bx)$? The number *a* is called the **amplitude.**

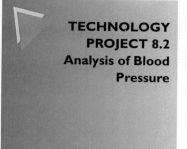

TECHNOLOGY PROJECT 8.2

Analysis of Blood Pressure

The pressure of the blood being pumped by the heart can be described by a model involving the cosine function. Suppose that the blood pressure at time t minutes is given by the function:

$$f(t) = 100 + 30 \cos 150\pi t$$

1. Draw the graph of $f(t)$. Choose a suitable interval of time over which to graph the function so that you can easily read the graph.
2. Describe in words the behavior of the blood pressure.
3. How many times per minute is the heart beating? (One heartbeat corresponds to one cycle of blood pressure.)
4. What are the maximum and minimum blood pressure readings?

PRACTICE PROBLEMS 8.2

1. Find $\cos t$, where t is the radian measure of the angle shown in Fig. 14.
2. Assume that $\cos(1.17) = .390$. Use properties of cosine and sine to determine $\sin(1.17)$, $\cos(1.17 + 4\pi)$, $\cos(-1.17)$, and $\sin(-1.17)$.

FIGURE 14

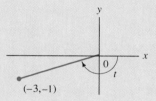

EXERCISES 8.2

In Exercises 1–12, give the values of sin t and cos t, where t is the radian measure of the angle shown.

1.

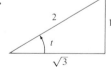

2.

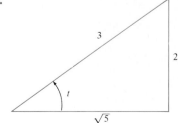

3.

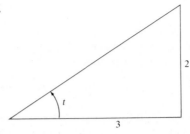

4.

5.

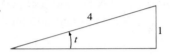

6.

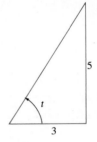

7.

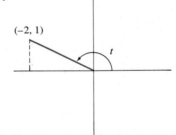

8.

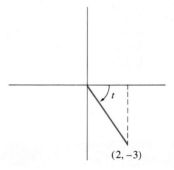

9.

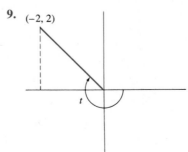

10.

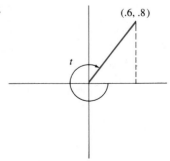

11.

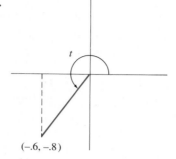

(-.6, -.8)

12.

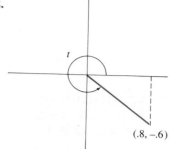

(.8, -.6)

Exercises 13–20 refer to various right triangles whose sides and angles are labeled as in Fig. 15. Round off all lengths of sides to one decimal place.

FIGURE 15

13. Estimate t if $a = 12$, $b = 5$, and $c = 13$.

14. If $t = 1.1$ and $c = 10.0$, find b.   15. If $t = 1.1$ and $b = 3.2$, find c.

16. If $t = .4$ and $c = 5.0$, find a.   17. If $t = .4$ and $a = 10.0$, find c.

18. If $t = .9$ and $c = 20.0$, find a and b.

19. If $t = .5$ and $a = 2.4$, find b and c.

20. If $t = 1.1$ and $b = 3.5$, find a and c.

Find t such that $0 \le t \le \pi$ and t satisfies the stated condition.

21. $\cos t = \cos(-\pi/6)$ 22. $\cos t = \cos(3\pi/2)$

23. $\cos t = \cos(5\pi/4)$ 24. $\cos t = \cos(-4\pi/6)$

25. $\cos t = \cos(-5\pi/8)$ 26. $\cos t = \cos(-3\pi/4)$

Find t such that $-\pi/2 \le t \le \pi/2$ and t satisfies the stated condition.

27. $\sin t = \sin(3\pi/4)$ 28. $\sin t = \sin(7\pi/6)$

29. $\sin t = \sin(-4\pi/3)$ 30. $\sin t = -\sin(3\pi/8)$

31. $\sin t = -\sin(\pi/6)$ 32. $\sin t = -\sin(-\pi/3)$

33. $\sin t = \cos t$ 34. $\sin t = -\cos t$

35. By referring to Fig. 10, describe what happens to $\cos t$ as t increases from 0 to π.

36. Use the unit circle to describe what happens to $\sin t$ as t increases from π to 2π.

37. Determine the value of $\sin t$ when $t = 5\pi, -2\pi, 17\pi/2, -13\pi/2$.

38. Determine the value of $\cos t$ when $t = 5\pi, -2\pi, 17\pi/2, -13\pi/2$.

39. Assume that $\cos(.19) = .98$. Use properties of cosine and sine to determine $\sin(.19), \cos(.19 - 4\pi), \cos(-.19)$, and $\sin(-.19)$.

40. Assume that $\sin(.42) = .41$. Use properties of cosine and sine to determine $\sin(-.42), \sin(6\pi - .42)$, and $\cos(.42)$.

SOLUTIONS TO PRACTICE PROBLEMS 8.2

1. Here $P = (x, y) = (-3, -1)$. The length of the line segment OP is
$$r = \sqrt{x^2 + y^2} = \sqrt{(-3)^2 + (-1)^2} = \sqrt{10}$$
Then
$$\cos t = \frac{x}{r} = \frac{-3}{\sqrt{10}} \approx -.94868.$$

2. Given $\cos(1.17) = .390$, use the relation $\cos^2 t + \sin^2 t = 1$ with $t = 1.17$ to solve for $\sin(1.17)$:
$$\cos^2(1.17) + \sin^2(1.17) = 1$$
$$\sin^2(1.17) = 1 - \cos^2(1.17)$$
$$= 1 - (.390)^2 = .8479$$
So
$$\sin(1.17) = \sqrt{.8479} \approx .921$$
Also, from properties (4) and (5),
$$\cos(1.17 + 4\pi) = \cos(1.17) = .390$$
$$\cos(-1.17) = \cos(1.17) = .390$$
$$\sin(-1.17) = -\sin(1.17) = -.921.$$

8.3 DIFFERENTIATION OF sin t AND cos t

In this section we study the two differentiation rules

$$\frac{d}{dt} \sin t = \cos t \tag{1}$$

$$\frac{d}{dt} \cos t = -\sin t. \tag{2}$$

It is not difficult to see why these rules might be true. Formula (1) says that the slope of the curve $y = \sin t$ at a particular value of t is given by the corresponding value of $\cos t$. To check it, we draw a careful graph of $y = \sin t$ and estimate the slope at various points, as indicated in Fig. 1. Let us plot the slope as a

FIGURE 1   Slope estimates along the graph of y = sin t.

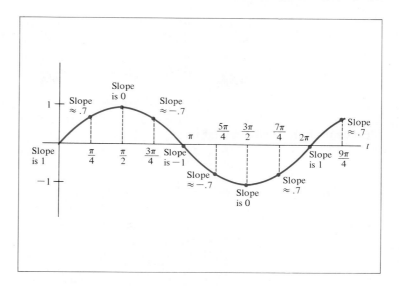

FIGURE 2   Graph of the slope function for y = sin t.

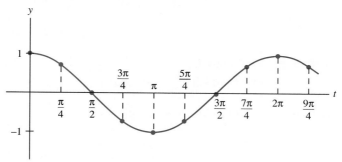

function of t (Fig. 2). As can be seen, the "slope function" (i.e., the derivative) of sin t has a graph similar to the curve $y = \cos t$. Thus formula (1) seems to be reasonable. A similar analysis of the graph of $y = \cos t$ would show why (2) might be true. Proofs of these differentiation rules are outlined in an appendix at the end of this section.

Combining (1), (2), and the chain rule, we obtain the following general rules.

$$\frac{d}{dt}\sin g(t) = [\cos g(t)]g'(t) \qquad (3)$$

$$\frac{d}{dt}\cos g(t) = [-\sin g(t)]g'(t). \qquad (4)$$

EXAMPLE 1   Differentiate.

(a) $\sin 3t$ (b) $(t^2 + 3\sin t)^5$

Solution   (a) $\dfrac{d}{dt} (\sin 3t) = (\cos 3t)\dfrac{d}{dt} (3t) = (\cos 3t) \cdot 3 = 3 \cos 3t.$

(b) $\dfrac{d}{dt} (t^2 + 3 \sin t)^5 = 5(t^2 + 3 \sin t)^4 \cdot \dfrac{d}{dt} (t^2 + 3 \sin t)$

$$= 5(t^2 + 3 \sin t)^4 (2t + 3 \cos t). \;\bullet$$

EXAMPLE 2   Differentiate.

(a) $\cos(t^2 + 1)$ (b) $\cos^2 t$

Solution   (a) $\dfrac{d}{dt} \cos(t^2 + 1) = -\sin(t^2 + 1)\dfrac{d}{dt}(t^2 + 1) = -\sin(t^2 + 1) \cdot (2t)$

$$= -2t \sin(t^2 + 1).$$

(b) Recall that the notation $\cos^2 t$ means $(\cos t)^2$.

$$\dfrac{d}{dt} \cos^2 t = \dfrac{d}{dt} (\cos t)^2 = 2(\cos t)\dfrac{d}{dt} \cos t$$

$$= -2 \cos t \sin t. \;\bullet$$

EXAMPLE 3   Differentiate.

(a) $t^2 \cos 3t$ (b) $(\sin 2t)/t$

Solution   (a) From the product rule we have

$$\dfrac{d}{dt} (t^2 \cos 3t) = t^2 \dfrac{d}{dt} \cos 3t + (\cos 3t)\dfrac{d}{dt} t^2$$

$$= t^2(-3 \sin 3t) + (\cos 3t)(2t)$$

$$= -3t^2 \sin 3t + 2t \cos 3t.$$

(b) From the quotient rule we have

$$\dfrac{d}{dt} \left(\dfrac{\sin 2t}{t} \right) = \dfrac{t\dfrac{d}{dt} \sin 2t - (\sin 2t) \cdot 1}{t^2}$$

$$= \dfrac{2t \cos 2t - \sin 2t}{t^2}. \;\bullet$$

EXAMPLE 4   A V-shaped trough is to be constructed with sides that are 200 centimeters long and 30 centimeters wide (Fig. 3). Find the angle t between the sides that maximizes the capacity of the trough.

Solution   The volume of the trough is its length times its cross-sectional area. Since the length is constant, it suffices to maximize the cross-sectional area. Let us rotate

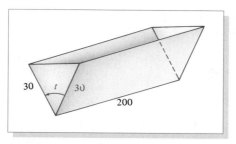

FIGURE 3

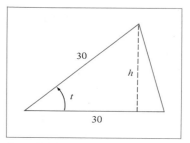

FIGURE 4

the diagram of a cross section so that one side is horizontal (Fig. 4). Note that $h/30 = \sin t$, so $h = 30 \sin t$. Thus the area A of the cross section is

$$A = \frac{1}{2} \cdot \text{base} \cdot \text{height}$$

$$A = \frac{1}{2}(30)(h) = 15(30 \sin t) = 450 \sin t.$$

To find where A is a maximum, we set the derivative equal to zero and solve for t.

$$\frac{dA}{dt} = 0$$

$$450 \cos t = 0$$

Physical considerations force us to consider only values of t between 0 and π. From the graph of $y = \cos t$ we see that $t = \pi/2$ is the only value of t between 0 and π that makes $\cos t = 0$. So in order to maximize the volume of the trough, the two sides should be perpendicular to one another. ●

EXAMPLE 5　Calculate the following indefinite integrals.

　　(a) $\displaystyle\int \sin t \, dt$　　　　　　　　(b) $\displaystyle\int \sin 3t \, dt$

Solution　(a) Since $\dfrac{d}{dt}(-\cos t) = \sin t$, we have

$$\int \sin t \, dt = -\cos t + C$$

where C is an arbitrary constant.

　　(b) From part (a) we guess that an antiderivative of $\sin 3t$ should resemble the function $-\cos 3t$. However, if we differentiate, we find that

$$\frac{d}{dt}(-\cos 3t) = (\sin 3t) \cdot \frac{d}{dt}(3t)$$

$$= 3 \sin 3t,$$

which is three times too much. So we multiply this last equation by $\frac{1}{3}$ to derive that

$$\frac{d}{dt}\left(-\frac{1}{3}\cos 3t\right) = \sin 3t,$$

so

$$\int \sin 3t \, dt = -\frac{1}{3}\cos 3t + C. \quad \bullet$$

EXAMPLE 6   Find the area under the curve $y = \sin 3t$ from $t = 0$ to $t = \pi/3$.

Solution   The area is shaded in Fig. 5.

$$[\text{shaded area}] = \int_0^{\pi/3} \sin 3t \, dt$$

$$= -\frac{1}{3}\cos 3t \Big|_0^{\pi/3}$$

$$= -\frac{1}{3}\cos\left(3 \cdot \frac{\pi}{3}\right) - \left(-\frac{1}{3}\cos 0\right)$$

$$= -\frac{1}{3}\cos \pi + \frac{1}{3}\cos 0$$

$$= \frac{1}{3} + \frac{1}{3} = \frac{2}{3}. \quad \bullet$$

FIGURE 5

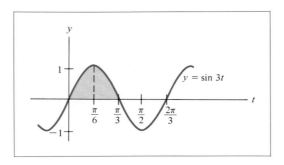

As we mentioned earlier, the trigonometric functions are required to model situations which are repetitive (or periodic). The next example illustrates such a situation.

EXAMPLE 7   In many mathematical models used to study the interaction between predators and prey, both the number of predators and the number of prey are described by periodic functions. Suppose that in one such model, the number of predators (in a particular geographical region) at time t is given by the equation $N(t) = 5000 + 2000 \cos(2\pi t/36)$, where t is measured in months from June 1, 1990.

(a) At what rate is the number of predators changing on August 1, 1990?

(b) What is the average number of predators during the time interval from June 1, 1990, to June 1, 1993?

Solution   (a) The date August 1, 1990, corresponds to $t = 2$. The rate of change of $N(t)$ is given by the derivative $N'(t)$:

$$N'(t) = \frac{d}{dt}\left[5000 + 2000 \cos\left(\frac{2\pi t}{36}\right) \right]$$

$$= 2000\left[-\sin\left(\frac{2\pi t}{36}\right) \cdot \left(\frac{2\pi}{36}\right) \right]$$

$$= -\frac{1000\pi}{9} \sin\left(\frac{2\pi t}{36}\right),$$

$$N'(2) = -\frac{1000\pi}{9} \sin\left(\frac{\pi}{9}\right)$$

$$\approx -119.$$

Thus, on August 1, 1990, the number of predators is decreasing at the rate of 119 per month.

(b) The time interval from June 1, 1990, to June 1, 1993, corresponds to $t = 0$ to $t = 36$. The average value of $N(t)$ over this interval is

$$\frac{1}{36 - 0}\int_0^{36} N(t)\, dt = \frac{1}{36}\int_0^{36}\left[5000 + 2000 \cos\left(\frac{2\pi t}{36}\right) \right] dt$$

$$= \frac{1}{36}\left[5000t + \frac{2000}{2\pi/36} \sin\left(\frac{2\pi t}{36}\right) \right]\Bigg|_0^{36}$$

$$= \frac{1}{36}\left[5000 \cdot 36 + \frac{2000}{2\pi/36} \sin(2\pi) \right]$$

$$- \frac{1}{36}\left[5000 \cdot 0 + \frac{2000}{2\pi/36} \sin(0) \right]$$

$$= 5000.$$

We have sketched the graph of $N(t)$ in Fig. 6. Note how $N(t)$ oscillates around 5000, the average value. ●

FIGURE 6   **Periodic fluctuation of a predator population.**

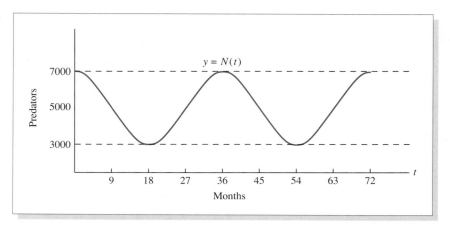

<table>
<tr><td rowspan="2" style="vertical-align:top">

TECHNOLOGY PROJECT 8.3

Further Analysis of Predator-Prey Model

</td><td>

Refer to Example 7.

1. Graph the number of predators versus time. Choose an appropriate range for t so that all important geometric facts about the graph are displayed.
2. What is the period of the graph?
3. During a single period, beginning June 1, 1990, when is the number of predators increasing?
4. During a single period, beginning June 1, 1990, when is the number of predators decreasing?
5. What is the maximum number of predators? When is the maximum achieved in the first period beginning June 1, 1990?
6. What is the minimum number of predators? When is the minimum achieved in the first period beginning June 1, 1990?

</td></tr>
</table>

TECHNOLOGY PROJECT 8.4

Numbers of Hours of Daylight

The number of hours of daylight D depends on the latitude and the day t of the year and is given by the function

$$D(t) = 12 + A \sin \left[\frac{2\pi}{365}(t - 80) \right]$$

where A depends only on the latitude (and not on t). For latitude 30° the value of A is approximately 2.3.

1. Graph the function $D(t)$ for latitude 30°.
2. When is the number of hours of daylight greatest? What is the length of that day?
3. When is the number of hours of daylight least? What is the length of that day?

4. When is the length of the day 11 hours?
5. When is the length of the day 13 hours?
6. When is the number of hours of daylight increasing at the rate of 2 minutes per day?
7. When is the number of hours of daylight decreasing at the rate of 2 minutes per day?
8. When is the number of hours of daylight equal to the number of hours of darkness? (These times are called **equinoxes.**)

 APPENDIX HEURISTIC JUSTIFICATION OF THE DIFFERENTIATION RULES FOR sin *t* AND cos *t*

First, let us examine the derivatives of cos *t* and sin *t* at *t* = 0. The function cos *t* has a maximum at *t* = 0; consequently, its derivative there must be zero. [See Fig. 7(a).] If we approximate the tangent line at *t* = 0 by a secant line, as in Fig. 7(b), then the slope of the secant line must approach 0 as $h \rightarrow 0$. Since the slope of the secant line is $(\cos h - 1)/h$, we conclude that

$$\lim_{h \to 0} \frac{\cos h - 1}{h} = 0. \tag{5}$$

It appears from the graph of $y = \sin t$ that the tangent line at *t* = 0 has slope 1 [Fig. 8(a)]. If it does, then the slope of the approximating secant line

FIGURE 7

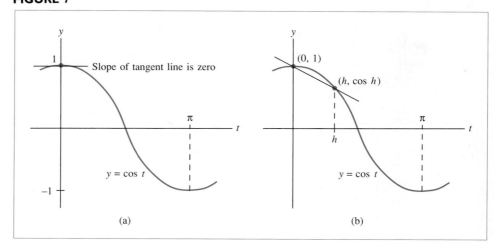

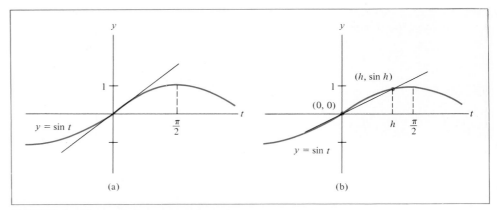

FIGURE 8   Slope of sin t at t = 0.

[Fig. 8(b)] must approach 1. Since the slope of the secant line in Fig. 8(b) is $(\sin h)/h$, this would imply that

$$\lim_{h \to 0} \frac{\sin h}{h} = 1. \tag{6}$$

We can check (6) for small values of h with a calculator.

h	.1	.01	.001
$\sin h$	.099833417	.009999833	.0009999998
$\dfrac{\sin h}{h}$	.99833417	.9999833	.9999998

The numerical evidence does not prove (6), but it should be sufficiently convincing for our purposes.

To obtain the differentiation formula for sin t, we approximate the slope of a tangent line by the slope of a secant line. (See Fig. 9.) The slope of a secant line is

$$\frac{\sin(t + h) - \sin t}{h}.$$

From formula (7) of Section 8.2 we note that $\sin(t + h) = \sin t \cos h + \cos t \sin h$. Thus

$$[\text{slope of secant line}] = \frac{(\sin t \cos h + \cos t \sin h) - \sin t}{h}$$

$$= \frac{\sin t(\cos h - 1) + \cos t \sin h}{h}$$

$$= (\sin t)\frac{\cos h - 1}{h} + (\cos t)\frac{\sin h}{h}.$$

FIGURE 9 Secant line approximation for y = sin t.

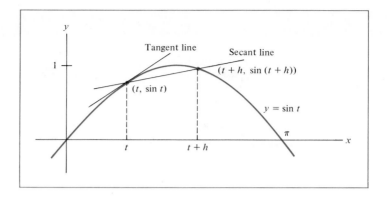

From (5) and (6) it follows that

$$\frac{d}{dt} \sin t = \lim_{h \to 0} \left[(\sin t) \frac{\cos h - 1}{h} + (\cos t) \frac{\sin h}{h} \right]$$

$$= (\sin t) \lim_{h \to 0} \frac{\cos h - 1}{h} + (\cos t) \lim_{h \to 0} \frac{\sin h}{h}$$

$$= (\sin t) \cdot 0 + (\cos t) \cdot 1$$

$$= \cos t.$$

A similar argument may be given to verify the formula for the derivative of cos t. Here is a shorter proof that uses the chain rule and the two identities

$$\cos t = \sin\left(\frac{\pi}{2} - t\right), \qquad \sin t = \cos\left(\frac{\pi}{2} - t\right).$$

[See formula (6) of Section 8.2.] We have

$$\frac{d}{dt} \cos t = \frac{d}{dt} \sin\left(\frac{\pi}{2} - t\right)$$

$$= \cos\left(\frac{\pi}{2} - t\right) \cdot \frac{d}{dt}\left(\frac{\pi}{2} - t\right)$$

$$= \cos\left(\frac{\pi}{2} - t\right) \cdot (-1)$$

$$= -\sin t.$$

PRACTICE PROBLEMS 8.3

1. Differentiate $y = 2 \sin[t^2 + (\pi/6)]$.
2. Differentiate $y = e^t \sin 2t$.

EXERCISES 8.3

Differentiate (with respect to t or x).

1. $\sin 4t$
2. $-3 \cos t$
3. $4 \sin t$
4. $\cos(-3t)$
5. $2 \cos 3t$
6. $2 \sin \pi t$
7. $t + \cos \pi t$
8. $t^2 - 2 \sin 4t$
9. $\sin(\pi - t)$
10. $2 \cos(t + \pi)$
11. $\cos^3 t$
12. $\sin t^3$
13. $\sin \sqrt{x - 1}$
14. $\cos(e^x)$
15. $\sqrt{\sin(x - 1)}$
16. $e^{\cos x}$
17. $(1 + \cos t)^8$
18. $\sqrt[3]{\sin \pi t}$
19. $\cos^2 x^3$
20. $\sin^3 x + 4 \sin^2 x$
21. $e^x \sin x$
22. $x\sqrt{\cos x}$
23. $\sin 2x \cos 3x$
24. $\sin^3 x \cos x$
25. $\dfrac{\sin t}{\cos t}$
26. $\dfrac{e^t}{\cos 2t}$
27. $\ln(\cos t)$
28. $\ln(\sin 2t)$
29. $\sin(\ln t)$
30. $(\cos t) \ln t$

31. Find the slope of the line tangent to the graph of $y = \cos 3x$ at $x = 13\pi/6$.
32. Find the slope of the line tangent to the graph of $y = \sin 2x$ at $x = 5\pi/4$.
33. Find the equation of the line tangent to the graph of $y = 3 \sin x + \cos 2x$ at $x = \pi/2$.
34. Find the equation of the line tangent to the graph of $y = 3 \sin 2x - \cos 2x$ at $x = 3\pi/4$.

Find the following indefinite integrals.

35. $\displaystyle\int \cos 2x \, dx$
36. $\displaystyle\int \sin \frac{x}{3} \, dx$
37. $\displaystyle\int \sin(4x + 1) \, dx$
38. $\displaystyle\int \cos(5 - x) \, dx$

39. Suppose that a person's blood pressure P at time t (in seconds) is given by $P = 100 + 20 \cos 6t$.

   (a) Find the maximum value of P (called the systolic pressure) and the minimum value of P (called the diastolic pressure) and give one or two values of t where these maximum and minimum values of P occur.

   (b) If time is measured in seconds, approximately how many heartbeats per minute are predicted by the equation for P?

40. The *basal metabolism* (BM) of an organism over a certain time period may be described as the total amount of heat in kilocalories (kcal) the organism produces during that period, assuming that the organism is at rest and not subject to stress. The *basal metabolic rate* (BMR) is the rate in kcal per hour at which the organism produces heat. The BMR of an animal such as a desert rat fluctuates in response to changes in temperature and other environmental factors. The BMR generally follows a *diurnal* cycle—rising at night during low temperatures and decreasing during the warmer daytime temperatures. Find the BM for 1 day if $\text{BMR}(t) = .4 + .2 \sin(\pi t/12)$ kcal per hour ($t = 0$ corresponds to 3 A.M.). See Fig. 10.

FIGURE 10   Diurnal cycle of the basal metabolic rate.

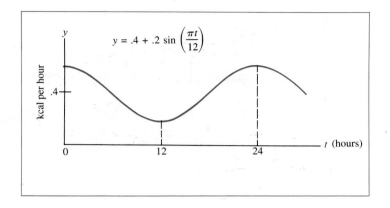

41. In any given locality, tap water temperature varies during the year. In Dallas, Texas, the tap water temperature (in degrees Fahrenheit) *t* days after the beginning of a year is given approximately by the formula*

$$T = 59 + 14 \cos\left[\frac{(t - 208)\pi}{183}\right], \qquad 0 \le t \le 365.$$

Find the maximum and minimum tap water temperature during the year and the dates at which they occur.

42. In any given locality, the length of daylight varies during the year. In Des Moines, Iowa, the number of minutes of daylight on a day *t* days after the beginning of a year is given approximately by the formula**

$$D = 720 + 200 \sin\left[\frac{(t - 79.5)\pi}{183}\right], \qquad 0 \le t \le 365.$$

Find the maximum and minimum lengths of daylight and the dates at which they occur.

43. As h approaches 0, what value is approached by $\dfrac{\sin\left(\dfrac{\pi}{2} + h\right) - 1}{h}$?

$$\left[\text{Hint: } \sin\frac{\pi}{2} = 1.\right]$$

44. As *h* approaches 0, what value is approached by $\dfrac{\cos(\pi + h) + 1}{h}$?

[*Hint:* cos π = −1.]

45. The average weekly temperature in Washington, DC *t* weeks after the beginning of the year is:

$$f(t) = 54 + 23 \sin\left(\frac{2\pi(t - 12)}{52}\right)$$

The graph of this function is sketched in Fig. 11.

*See D. Rapp, *Solar Energy* (Englewood Cliffs, N.J.: Prentice-Hall, Inc., 1981), p. 171.

** See D. R. Duncan et al., "Climate Curves," *School Science and Mathematics,* vol. LXXVI (January 1976), pp. 41–49.

FIGURE 11

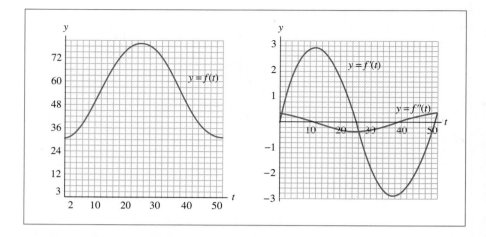

(a) What is the average weekly temperature at week 18?

(b) At week 20, how fast is the temperature changing?

(c) When is the average weekly temperature 39 degrees?

(d) When is the average weekly temperature falling at the rate of 1 degree per week?

(e) When is average weekly temperature greatest? Least?

(f) When is average weekly temperature increasing fastest? Decreasing fastest?

46. The number of hours of daylight per day in Washington, DC t weeks after the beginning of the year is

$$f(t) = 12.18 + 2.725 \sin\left(\frac{2\pi(t - 12)}{52}\right)$$

The graph of this function is sketched in Fig. 12.

FIGURE 12

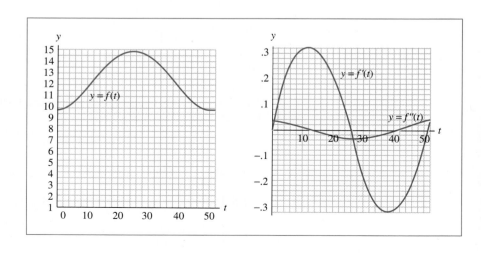

(a) How many hours of daylight are there after 40 weeks?

(b) After 32 weeks, how fast is the number of hours of daylight decreasing?

(c) When is there 14 hours of daylight per day?

(d) When is the number of hours of daylight increasing at the rate of 15 minutes per week?

(e) When are the days longest? Shortest?

(f) When is the number of hours of daylight increasing fastest? Decreasing fastest?

SOLUTIONS TO PRACTICE PROBLEMS 8.3

1. By the chain rule,

$$y' = 2\cos\left(t^2 + \frac{\pi}{6}\right) \cdot \frac{d}{dt}\left(t^2 + \frac{\pi}{6}\right)$$

$$= 2\cos\left(t^2 + \frac{\pi}{6}\right) \cdot 2t$$

$$= 4t\cos\left(t^2 + \frac{\pi}{6}\right).$$

2. By the product rule,

$$y' = e^t \frac{d}{dt}(\sin 2t) + (\sin 2t)\frac{d}{dt}e^t$$

$$= 2e^t \cos 2t + e^t \sin 2t.$$

8.4 THE TANGENT AND OTHER TRIGONOMETRIC FUNCTIONS

Certain functions involving the sine and cosine functions occur so frequently in applications that they have been given special names. The *tangent* (tan), *cotangent* (cot), *secant* (sec), and *cosecant* (csc) are such functions and are defined as follows:

$$\tan t = \frac{\sin t}{\cos t}, \qquad \cot t = \frac{\cos t}{\sin t},$$

$$\sec t = \frac{1}{\cos t}, \qquad \csc t = \frac{1}{\sin t}.$$

They are defined only for t such that the denominators in the preceding quotients are not zero. These four functions, together with the sine and cosine, are called the *trigonometric functions*. Our main interest in this section is with the tangent function. Some properties of the cotangent, secant, and cosecant are developed in the exercises.

Many identities involving the trigonometric functions can be deduced from the identities given in Section 8.2. We shall mention just one:

$$\tan^2 t + 1 = \sec^2 t. \tag{1}$$

[Here $\tan^2 t$ means $(\tan t)^2$ and $\sec^2 t$ means $(\sec t)^2$.] This identity follows from the identity $\sin^2 t + \cos^2 t = 1$ when we divide everything by $\cos^2 t$.

An important interpretation of the tangent function can be given in terms of the diagram used to define the sine and cosine. For a given t, let us construct an angle of t radians, as in Fig. 1. Since $\sin t = y/r$ and $\cos t = x/r$, we have

$$\frac{\sin t}{\cos t} = \frac{y/r}{x/r} = \frac{y}{x},$$

FIGURE I

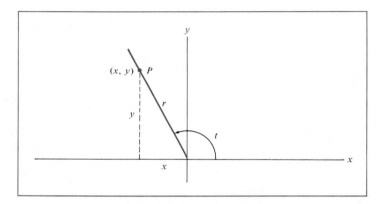

where this formula holds provided that $x \neq 0$. Thus

$$\tan t = \frac{y}{x}. \tag{2}$$

Three examples that illustrate this property of the tangent appear in Fig. 2.

When $0 < t < \pi/2$, the value of $\tan t$ is a ratio of the lengths of the sides of a right triangle. In other words, suppose that we are given a triangle as in Fig. 3. Then we have

$$\tan t = \frac{\text{opposite}}{\text{adjacent}}. \tag{3}$$

EXAMPLE I   The angle of elevation from an observer to the top of a building is 29° (Fig. 4). If the observer is 100 meters from the base of the building, how high is the building?

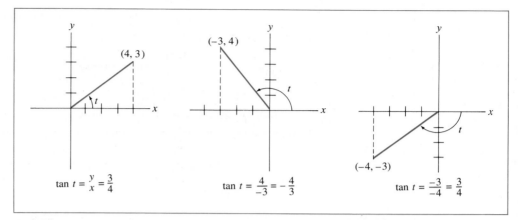

FIGURE 2

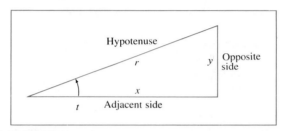

FIGURE 3

FIGURE 4   Surveying the height of a building.

Solution   Let h denote the height of the building. Then (3) implies that

$$\frac{h}{100} = \tan 29°$$

$$h = 100 \tan 29°.$$

We convert $29°$ into radians. We find that $29° = (\pi/180) \cdot 29$ radians $\approx .5$ radian, and $\tan .5 = .54630$. Hence

$$h \approx 100(.54630) = 54.63 \text{ meters.} \quad \bullet$$

The Derivative of tan t   Since $\tan t$ is defined in terms of $\sin t$ and $\cos t$, we can compute the derivative of $\tan t$ from our rules of differentiation. That is, by applying the quotient rule for differentiation, we have

$$\frac{d}{dt}(\tan t) = \frac{d}{dt}\left(\frac{\sin t}{\cos t}\right) = \frac{(\cos t)(\cos t) - (\sin t)(-\sin t)}{(\cos t)^2}$$

$$= \frac{\cos^2 t + \sin^2 t}{\cos^2 t} = \frac{1}{\cos^2 t}.$$

Now

$$\frac{1}{\cos^2 t} = \frac{1}{(\cos t)^2} = \left(\frac{1}{\cos t}\right)^2 = (\sec t)^2 = \sec^2 t.$$

So the derivative of tan t can be expressed in two equivalent ways:

$$\frac{d}{dt}(\tan t) = \frac{1}{\cos^2 t} = \sec^2 t. \qquad (4)$$

Combining (4) with the chain rule, we have

$$\frac{d}{dt}(\tan g(t)) = [\sec^2 g(t)]g'(t). \qquad (5)$$

EXAMPLE 2   Differentiate.

(a) $\tan(t^3 + 1)$ (b) $\tan^3 t$

Solution   (a) From (5) we find that

$$\frac{d}{dt}[\tan(t^3 + 1)] = \sec^2(t^3 + 1) \cdot \frac{d}{dt}(t^3 + 1)$$

$$= 3t^2 \sec^2(t^3 + 1).$$

(b) We write $\tan^3 t$ as $(\tan t)^3$ and use the chain rule (in this case, the general power rule):

$$\frac{d}{dt}(\tan t)^3 = (3 \tan^2 t) \cdot \frac{d}{dt}\tan t$$

$$= 3 \tan^2 t \sec^2 t. \quad \bullet$$

FIGURE 5   Graph of the tangent function.

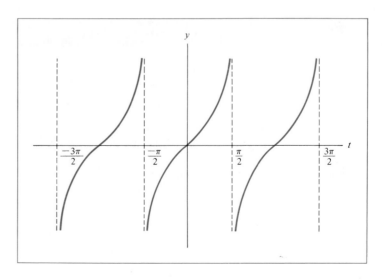

The Graph of tan t   Recall that tan t is defined for all t except where cos $t = 0$. (We cannot have zero in the denominator of sin t/cos t.) The graph of tan t is sketched in Fig. 5, using Table 3. Note that tan t is periodic with period π.

TECHNOLOGY PROJECT 8.5
Graphing of Trigonometric Functions

Graph the following functions. Determine if the graphs are periodic and if so determine the period.

1. $f(x) = \tan 3x$

2. $f(x) = \sec 2x$

3. $f(x) = \tan \left(\dfrac{x}{2} + \dfrac{\pi}{8} \right)$

4. $f(x) = \sec \dfrac{x}{4}$

5. $f(x) = 3 \sin 2x + \cos x$

6. $f(x) = 4 \sin x + 5 \sin \dfrac{x}{3}$

7. $f(x) = x \sin x$

8. $f(x) = \dfrac{\sin x}{x}$

9. $f(x) = e^{-x} \sin x$

10. $f(x) = x + \sin x$

PRACTICE PROBLEMS 8.4

1. Show that the slope of a straight line is equal to the tangent of the angle that the line makes with the x-axis.
2. Calculate $\int_0^{\pi/4} \sec^2 t \, dt$.

EXERCISES 8.4

1. Suppose that $0 < t < \pi/2$ and use Fig. 3 to describe sec t as a ratio of the lengths of the sides of a right triangle.
2. Describe cot t for $0 < t < \pi/2$ as a ratio of the lengths of the sides of a right triangle.

In Exercises 3–10, give the value of tan t and sec t, where t is the radian measure of the angle shown.

3.

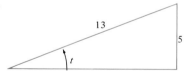

4.

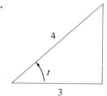

5.

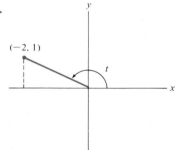

6.

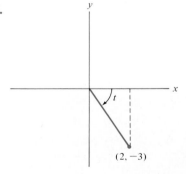

7.

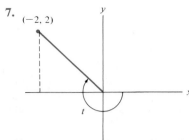

8.

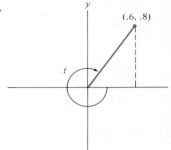

9.

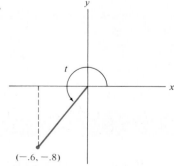

10.

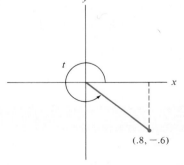

11. Find the width of a river at points A and B if the angle BAC is $90°$, the angle ACB is $40°$, and the distance from A to C is 75 feet. See Fig. 6.

FIGURE 6

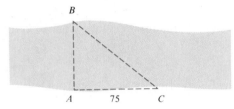

12. The angle of elevation from an observer to the top of a church is .3 radian, while the angle of elevation from the observer to the top of the church spire is .4 radian. If the observer is 70 meters from the church, how tall is the spire on top of the church?

Differentiate (with respect to t or x).

13. $f(t) = \sec t$ **14.** $f(t) = \csc t$ **15.** $f(t) = \cot t$

16. $f(t) = \cot 3t$ **17.** $f(t) = \tan 4t$ **18.** $f(t) = \tan \pi t$

19. $f(x) = 3 \tan(\pi - x)$ **20.** $f(x) = 5 \tan(2x + 1)$

21. $f(x) = 4 \tan(x^2 + x + 3)$ **22.** $f(x) = 3 \tan(1 - x^2)$

23. $y = \tan \sqrt{x}$ **24.** $y = 2 \tan \sqrt{x^2 - 4}$

25. $y = x \tan x$ **26.** $y = e^{3x} \tan 2x$ **27.** $y = \tan^2 x$

28. $y = \sqrt{\tan x}$ **29.** $y = (1 + \tan 2t)^3$ **30.** $y = \tan^4 3t$

31. $y = \ln(\tan t + \sec t)$ **32.** $y = \ln(\tan t)$

SOLUTIONS TO PRACTICE PROBLEMS 8.4

1. A line of positive slope m is shown in Fig. 7(a). Here, $\tan \theta = m/1 = m$. Suppose that $y = mx + b$ where the slope m is negative. The line $y = mx$ has the same slope and makes the same angle with the x-axis. [See Fig. 7(b).] We see that $\tan \theta = -m/-1 = m$.

2. $\displaystyle\int_0^{\pi/4} \sec^2 t \, dt = \tan t \Big|_0^{\pi/4} = \tan \frac{\pi}{4} - \tan 0 = 1 - 0 = 1.$

FIGURE 7

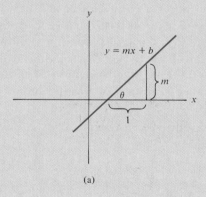

(a)

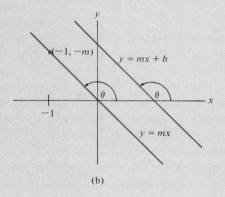

(b)

 **CHAPTER 8 CHECKLIST**

✓ Radian

✓ 2π radians $= 360°$

✓ $\sin t$, $\cos t$, $\tan t$

✓ Triangle interpretation of $\sin t$, $\cos t$, $\tan t$ for $0 < t < \pi/2$

✓ $\sin^2 t + \cos^2 t = 1$

✓ $\sin t$ and $\cos t$ are periodic with period 2π

✓ $\dfrac{d}{dt} \sin t = \cos t$

✓ $\dfrac{d}{dt} \cos t = -\sin t$

✓ $\dfrac{d}{dt} \tan t = \sec^2 t$

CHAPTER 8   SUPPLEMENTARY EXERCISES

Determine the radian measure of the angles shown in Exercises 1–3.

1.    **2.** 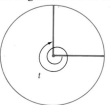   **3.**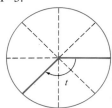

Construct angles with the following radian measure.

4. $-\pi$ **5.** $\dfrac{5\pi}{4}$ **6.** $-\dfrac{9\pi}{2}$

In Exercises 7–10, the point with the given coordinates determines an angle of t radians, where $0 \le t \le 2\pi$. Find $\sin t$, $\cos t$, $\tan t$.

7. $(3, 4)$ **8.** $(-.6, .8)$ **9.** $(-.6, -.8)$ **10.** $(3, -4)$

11. If $\sin t = \frac{1}{5}$, what are the possible values for $\cos t$?

12. If $\cos t = -\frac{2}{3}$, what are the possible values for $\sin t$?

13. Find four values of t between -2π and 2π at which $\sin t = \cos t$.

14. Find four values of t between -2π and 2π at which $\sin t = -\cos t$.

15. When $-\pi/2 < t < 0$, is $\tan t$ positive or negative?

16. When $\pi/2 < t < \pi$, is $\sin t$ positive or negative?

17. A gabled roof is to be built on a house that is 30 feet wide, so that the roof rises at a pitch of 23°. Determine the length of the rafters needed to support the roof.

18. A tree casts a 60-foot shadow when the angle of elevation of the sun (measured from the horizontal) is 53°. How tall is the tree?

Differentiate (with respect to t or x).

19. $f(t) = 3 \sin t$

20. $f(t) = \sin 3t$

21. $f(t) = \sin \sqrt{t}$

22. $f(t) = \cos t^3$

23. $g(x) = x^3 \sin x$

24. $g(x) = \sin(-2x) \cos 5x$

25. $f(x) = \dfrac{\cos 2x}{\sin 3x}$

26. $f(x) = \dfrac{\cos x - 1}{x^3}$

27. $f(x) = \cos^3 4x$

28. $f(x) = \tan^3 2x$

29. $y = \tan(x^4 + x^2)$

30. $y = \tan e^{-2x}$

31. $y = \sin(\tan x)$

32. $y = \tan(\sin x)$

33. $y = \sin x \tan x$

34. $y = \ln x \cos x$

35. $y = \ln(\sin x)$

36. $y = \ln(\cos x)$

37. $y = e^{3x} \sin^4 x$

38. $y = \sin^4 e^{3x}$

39. $f(t) = \dfrac{\sin t}{\tan 3t}$

40. $f(t) = \dfrac{\tan 2t}{\cos t}$

41. $f(t) = e^{\tan t}$

42. $f(t) = e^t \tan t$

43. If $f(t) = \sin^2 t$, find $f''(t)$.

44. Show that $y = 3 \sin 2t + \cos 2t$ satisfies the differential equation $y'' = -4y$.

45. If $f(s, t) = \sin s \cos 2t$, find $\dfrac{\partial f}{\partial s}$ and $\dfrac{\partial f}{\partial t}$.

46. If $z = \sin wt$, find $\dfrac{\partial z}{\partial w}$ and $\dfrac{\partial z}{\partial t}$.

47. If $f(s, t) = t \sin st$, find $\dfrac{\partial f}{\partial s}$ and $\dfrac{\partial f}{\partial t}$.

48. The identity

$$\sin (s + t) = \sin s \cos t + \cos s \sin t$$

was given in Section 8.2. Compute the partial derivative of each side with respect to t and obtain an identity involving $\cos(s + t)$.

49. Find the equation of the line tangent to the graph of $y = \tan t$ at $t = \pi/4$.

50. Sketch the graph of $f(t) = \sin t + \cos t$ for $-2\pi \le t \le 2\pi$, using the following steps:

  (a) Find all t (between -2π and 2π) such that $f'(t) = 0$. Plot the corresponding points on the graph of $y = f(t)$.

  (b) Check the concavity of $f(t)$ at the points in part (a). Make sketches of the graph near these points.

  (c) Determine any inflection points and plot them. Then complete the sketch of the graph.

51. Sketch the graph of $y = t + \sin t$ for $0 \le t \le 2\pi$.

52. Find the area under the curve $y = 2 + \sin 3t$ from $t = 0$ to $t = \pi/2$.

53. Find the area of the region between the curve $y = \sin t$ and the t-axis from $t = 0$ to $t = 2\pi$.

54. Find the area of the region between the curve $y = \cos t$ and the t-axis from $t = 0$ to $t = 3\pi/2$.

55. Find the area of the region bounded by the curves $y = x$ and $y = \sin x$ from $x = 0$ to $x = \pi$.

A spirogram is a device that records on a graph the volume of air in a person's lungs as a function of time. If a person undergoes spontaneous hyperventilation, the spirogram trace will closely approximate a sine curve. A typical trace is given by

$$V(t) = 3 + .05 \sin\left(160\pi t - \frac{\pi}{2}\right),$$ where t is measured in minutes and $V(t)$ is the lung volume in liters. (See Fig. 1.) Exercises 56–58 refer to this function.

56. (a) Compute $V(0)$, $V(\tfrac{1}{320})$, $V(\tfrac{1}{160})$, and $V(\tfrac{1}{80})$.

  (b) What is the maximum lung volume?

57. (a) Find a formula for the rate of flow of air into the lungs at time t.

FIGURE 1   **Spirogram trace.**

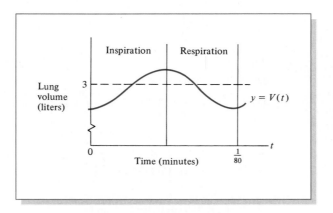

(b) Find the maximum rate of flow of air during inspiration (i.e., breathing in). This is called the *peak inspiratory flow*.

(c) Inspiration occurs during the time from $t = 0$ to $t = 1/160$. Find the average rate of flow of air during inspiration. This quantity is called the *mean inspiratory flow*.

58. The *minute volume* is defined to be the total amount of air inspired (breathed in) during one minute. According to a standard text on respiratory physiology, when a person undergoes spontaneous hyperventilation, the peak inspiratory flow equals π times the minute volume, and the mean inspiratory flow equals twice the minute volume.* Verify these assertions using the data from Exercise 57.

_____

* J. F. Nunn, *Applied Respiratory Physiology,* 2nd ed. (London: Butterworths, 1977), p. 122.

CHAPTER 9

TECHNIQUES OF INTEGRATION

In this chapter we develop techniques for calculating integrals, both indefinite and definite. The need for these techniques has been justified in the preceding chapters. In addition to adding to our fund of applications, we will see even more clearly how the need to calculate integrals arises in physical problems.

As noted, integration is the reverse process of differentiation. However, integration is much harder to carry out. If a function is an expression involving elementary functions (such as x^r, $\sin x$, e^x, . . .), then so is its derivative. Moreover, we were able to develop methods of calculation that enable us to differentiate, with comparative ease, almost any function that we can write down. Although many integration problems have these characteristics, certain ones do not. There are some elementary functions (e.g., e^{x^2}) for which an antiderivative cannot be expressed in terms of elementary functions. Even where an elementary antiderivative exists, the techniques for finding the antiderivative are often complicated. For this reason, we must be prepared with a broad range of tools in order to cope with the problem of calculating integrals. Among the ideas to be discussed are:

1. Techniques for evaluating indefinite integrals. We will concentrate on two methods—integration by substitution and integration by parts.
2. Evaluation of definite integrals.
3. Approximation of definite integrals. We will develop two new techniques for obtaining numerical approximations to $\int_a^b f(x)\ dx$. These techniques are especially useful in those cases in which we cannot find an antiderivative for $f(x)$.

Let us review the most elementary facts about integration. The indefinite integral

$$\int f(x)\ dx$$

is, by definition, a function whose derivative is $f(x)$. If $F(x)$ is one such function, then the most general function whose derivative is $f(x)$ is simply $F(x) + C$, where C is any constant. We write

$$\int f(x)\ dx = F(x) + C$$

to mean that all antiderivatives of $f(x)$ are precisely the functions $F(x) + C$, where C is any constant.

Each time we differentiate a function, we also derive an integration formula. For example, the fact that

$$\frac{d}{dx}\ (3x^2) = 6x$$

can be turned into the integration formula

$$\int 6x\ dx = 3x^2 + C.$$

Some of the formulas that follow immediately from differentiation formulas are reviewed in the table below.

This table illustrates the need for techniques of integration. For although $\sin x$, $\cos x$, and $\sec^2 x$ occur as derivatives of simple trigonometric functions, the functions $\tan x$ and $\cot x$ are not on our list. In fact, if we experiment with various elementary combinations of the trigonometric functions, it is easy to convince ourselves that antiderivatives of $\tan x$ and $\cot x$ are not easy to compute. In this chapter we develop techniques for calculating such antiderivatives (among others).

Differentiation Formula	Corresponding Integration Formula
$\dfrac{d}{dx}(x^r) = rx^{r-1}$	$\displaystyle\int rx^{r-1}\,dx = x^r + C \text{ or}$
	$\displaystyle\int x^r\,dx = \dfrac{x^{r+1}}{r+1} + C, r \neq -1$
$\dfrac{d}{dx}(e^x) = e^x$	$\displaystyle\int e^x\,dx = e^x + C$
$\dfrac{d}{dx}(\ln|x|) = \dfrac{1}{x}$	$\displaystyle\int \dfrac{1}{x}\,dx = \ln|x| + C$
$\dfrac{d}{dx}(\sin x) = \cos x$	$\displaystyle\int \cos x\,dx = \sin x + C$
$\dfrac{d}{dx}(\cos x) = -\sin x$	$\displaystyle\int \sin x\,dx = -\cos x + C$
$\dfrac{d}{dx}(\tan x) = \sec^2 x$	$\displaystyle\int \sec^2 x\,dx = \tan x + C$

 9.1 INTEGRATION BY SUBSTITUTION

As noted earlier, every differentiation formula can be turned into a corresponding integration formula. This point is true even for the chain rule. The resulting formula is called *integration by substitution* and is often used to transform a complicated integral into a simpler one.

Let $f(x)$ and $g(x)$ be two given functions and let $F(x)$ be an antiderivative for $f(x)$. The chain rule asserts that

$$\frac{d}{dx}[F(g(x))] = F'(g(x))g'(x)$$

$$= f(g(x))g'(x) \qquad [\text{since } F'(x) = f(x)].$$

Turning this formula into an integration formula, we have

$$\int f(g(x))g'(x)\,dx = F(g(x)) + C, \tag{1}$$

where C is any constant.

EXAMPLE I   Determine $\int (x^2 + 1)^3 \cdot 2x \, dx$.

Solution   If we set $f(x) = x^3$, $g(x) = x^2 + 1$, then $f(g(x)) = (x^2 + 1)^3$ and $g'(x) = 2x$. Therefore, we can apply formula (1). An antiderivative $F(x)$ of $f(x)$ is given by

$$F(x) = \frac{1}{4}x^4,$$

so that, by formula (1), we have

$$\int (x^2 + 1)^3 \cdot 2x \, dx = F(g(x)) + C$$

$$= \frac{1}{4}(x^2 + 1)^4 + C. \quad \bullet$$

Formula (1) can be elevated from the status of a sometimes-useful formula to a technique of integration by the introduction of a simple mnemonic device. Suppose that we are faced with integrating a function of the form $f(g(x))g'(x)$. Of course, we know the answer from formula (1). However, let us proceed somewhat differently. Replace the expression $g(x)$ by a new variable u, and replace $g'(x) \, dx$ by du. Such a substitution has the advantage that it reduces the generally complex expression $f(g(x))$ to the simpler form $f(u)$. In terms of u, the integration problem may be written

$$\int f(g(x))g'(x) \, dx = \int f(u) \, du.$$

However, the integral on the right is easy to evaluate, since

$$\int f(u) \, du = F(u) + C.$$

Since $u = g(x)$, we then obtain

$$\int f(g(x))g'(x) \, dx = F(u) + C = F(g(x)) + C,$$

which is the correct answer by (1). Remember, however, that replacing $g'(x) \, dx$ by du only has status as a correct mathematical statement because doing so leads to the correct answers. We do not, in this book, seek to explain in any deeper way what this replacement means.

Let us rework Example 1 using this method.

Second Solution of Example 1   Set $u = x^2 + 1$. Then $du = \dfrac{d}{dx}(x^2 + 1) \, dx = 2x \, dx$, and

$$\int (x^2 + 1)^3 \cdot 2x \, dx = \int u^3 \, du$$

$$= \frac{1}{4}u^4 + C$$

$$= \frac{1}{4}(x^2 + 1)^4 + C \qquad \text{(since } u = x^2 + 1\text{)}.$$

EXAMPLE 2   Evaluate $\displaystyle\int 2xe^{x^2}\, dx$.

Solution   Let $u = x^2$, so that $du = \dfrac{d}{dx}(x^2)\, dx = 2x\, dx$. Therefore,

$$\int 2xe^{x^2}\, dx = \int e^{x^2} \cdot 2x\, dx$$

$$= \int e^u\, du$$

$$= e^u + C$$

$$= e^{x^2} + C. \quad \bullet$$

From Examples 1 and 2 we can deduce the following method for integration of functions of the form $f'(g(x))g'(x)$.

Integration by Substitution

1. Define a new variable $u = g(x)$, where $g(x)$ is chosen in such a way that, when written in terms of u, the integrand is simpler than when written in terms of x.
2. Transform the integral with respect to x into an integral with respect to u by replacing $g(x)$ everywhere by u and $g'(x)\, dx$ by du.
3. Integrate the resulting function of u.
4. Rewrite the answer in terms of x by replacing u by $g(x)$.

Let us try a few more examples.

EXAMPLE 3   Evaluate $\displaystyle\int 3x^2\sqrt{x^3 + 1}\, dx$.

Solution   The first problem facing us is to find an appropriate substitution that will simplify the integral. An immediate possibility is offered by setting $u = x^3 + 1$.

Then $\sqrt{x^3 + 1}$ will become $\sqrt{u}$, a significant simplification. If $u = x^3 + 1$, then $du = \dfrac{d}{dx}(x^3 + 1)\,dx = 3x^2\,dx$, so that

$$\int 3x^2\sqrt{x^3 + 1}\,dx = \int \sqrt{u}\,du$$

$$= \frac{2}{3}u^{3/2} + C$$

$$= \frac{2}{3}(x^3 + 1)^{3/2} + C. \quad \bullet$$

EXAMPLE 4   Find $\displaystyle\int \frac{(\ln x)^2}{x}\,dx$.

Solution   Let $u = \ln x$. Then $du = (1/x)\,dx$ and

$$\int \frac{(\ln x)^2}{x}\,dx = \int (\ln x)^2 \cdot \frac{1}{x}\,dx$$

$$= \int u^2\,du$$

$$= \frac{u^3}{3} + C$$

$$= \frac{(\ln x)^3}{3} + C \qquad \text{(since } u = \ln x\text{).} \quad \bullet$$

Knowing the correct substitution to make is a skill that develops through practice. Basically, we look for an occurrence of function composition, $f(g(x))$, where $f(x)$ is a function that we know how to integrate and where $g'(x)$ also appears in the integrand. Sometimes $g'(x)$ does not appear exactly, but can be obtained by multiplying by a constant. Such a shortcoming is easily remedied, as is illustrated in Examples 5 and 6.

EXAMPLE 5   Find $\displaystyle\int x^2 e^{x^3}\,dx$.

Solution   Let $u = x^3$; then $du = 3x^2\,dx$. The integrand involves $x^2\,dx$, not $3x^2\,dx$. To introduce the missing factor "3", we write

$$\int x^2 e^{x^3}\,dx = \int \frac{1}{3}\cdot 3x^2 e^{x^3}\,dx = \frac{1}{3}\int e^{x^3}3x^2\,dx.$$

(Recall from Section 6.1 that constant multiples may be moved through the integral sign.) Substituting, we obtain

$$\int x^2 e^{x^3} \, dx = \frac{1}{3} \int e^u \, du = \frac{1}{3} e^u + C$$

$$= \frac{1}{3} e^{x^3} + C \quad \text{(since } u = x^3 \text{)}.$$

Another way to handle the missing factor "3" is to write

$$u = x^3, \quad du = 3x^2 \, dx, \quad \text{and} \quad \frac{1}{3} du = x^2 \, dx.$$

Then substitution yields

$$\int x^2 e^{x^3} \, dx = \int e^{x^3} \cdot x^2 \, dx = \int e^u \cdot \frac{1}{3} \, du = \frac{1}{3} \int e^u \, du$$

$$= \frac{1}{3} e^u + C = \frac{1}{3} e^{x^3} + C. \; \bullet$$

EXAMPLE 6   Find $\displaystyle \int \frac{2 - x}{\sqrt{2x^2 - 8x + 1}} \, dx$.

Solution   Let $u = 2x^2 - 8x + 1$; then $du = (4x - 8) \, dx$. Observe that $4x - 8 = -4(2 - x)$. So we multiply the integrand by -4 and compensate by placing a factor of $-\frac{1}{4}$ in front of the integral.

$$\int \frac{1}{\sqrt{2x^2 - 8x + 1}} \cdot (2 - x) \, dx = -\frac{1}{4} \int \frac{1}{\sqrt{2x^2 - 8x + 1}} \cdot (-4)(2 - x) \, dx$$

$$= -\frac{1}{4} \int \frac{1}{\sqrt{u}} \, du = -\frac{1}{4} \int u^{-1/2} \, du$$

$$= -\frac{1}{4} \cdot 2u^{1/2} + C = -\frac{1}{2} u^{1/2} + C$$

$$= -\frac{1}{2} (2x^2 - 8x + 1)^{1/2} + C. \; \bullet$$

EXAMPLE 7   Find $\displaystyle \int \frac{2x}{x^2 + 1} \, dx$.

Solution   We note that the derivative of $x^2 + 1$ is $2x$. Thus we make the substitution $u = x^2 + 1, du = 2x \, dx$, in order to derive

$$\int \frac{2x}{x^2 + 1} \, dx = \int \frac{1}{u} \, du = \ln|u| + C = \ln|x^2 + 1| + C. \; \bullet$$

EXAMPLE 8   Evaluate $\int \tan x \, dx$.

Solution   Since $\tan x = \dfrac{\sin x}{\cos x}$, we have

$$\int \tan x \, dx = \int \frac{\sin x}{\cos x} \, dx.$$

Let $u = \cos x$, so that $du = -\sin x \, dx$. Then

$$\int \frac{\sin x}{\cos x} \, dx = -\int \frac{-\sin x}{\cos x} \, dx$$

$$= -\int \frac{1}{u} \, du$$

$$= -\ln|u| + C$$

$$= -\ln|\cos x| + C. \;\bullet$$

**TECHNOLOGY
PROJECT 9.1
More
Antidifferentiation
Using Symbolic
Math Program**

Use your symbolic math program to calculate the following antiderivatives.

1. $\displaystyle\int (3x^2 + 4)(x^3 + 4x)^{-1/2} e^{-\sqrt{x^3 + 4x}} \, dx$

2. $\displaystyle\int \cot x \cdot \ln(\sin x) \, dx$

3. $\displaystyle\int \sin^2(\cos x) \sin x \, dx$

4. $\displaystyle\int (3x - 1)^3 (4x + 1)^7 (3x + 1)^9 \, dx$

**PRACTICE
PROBLEMS 9.1**

1. (*Review*) Differentiate the following functions.
   (a) $e^{(2x^3 + 3x)}$ (b) $\ln x^5$ (c) $\ln \sqrt{x}$ (d) $\ln 5|x|$
   (e) $x \ln x$ (f) $\ln(x^4 + x^2 + 1)$ (g) $\sin x^3$ (h) $\tan x$

2. Use the substitution $u = \dfrac{3}{x}$ to determine $\displaystyle\int \frac{e^{3/x}}{x^2} \, dx$.

EXERCISES 9.1

Determine the integrals in Exercises 1–30 by making appropriate substitutions.

1. $\displaystyle\int 2x(x^2 + 4)^5\,dx$

2. $\displaystyle\int 3x^2(x^3 + 1)^2\,dx$

3. $\displaystyle\int (x^2 - 5x)^3(2x - 5)\,dx$

4. $\displaystyle\int 2x\sqrt{x^2 + 3}\,dx$

5. $\displaystyle\int 3x^2 e^{(x^3-1)}\,dx$

6. $\displaystyle\int 2xe^{-x^2}\,dx$

7. $\displaystyle\int x\sqrt{4 - x^2}\,dx$

8. $\displaystyle\int \frac{(\ln x)^3}{x}\,dx$

9. $\displaystyle\int \frac{1}{\sqrt{2x + 1}}\,dx$

10. $\displaystyle\int (x^3 - 6x)^7(x^2 - 2)\,dx$

11. $\displaystyle\int xe^{3x^2}\,dx$

12. $\displaystyle\int \frac{e^{\sqrt{x}}}{\sqrt{x}}\,dx$

13. $\displaystyle\int \frac{\ln(2x)}{x}\,dx$

14. $\displaystyle\int \frac{\sqrt{\ln x}}{x}\,dx$

15. $\displaystyle\int \frac{x^4}{x^5 + 1}\,dx$

16. $\displaystyle\int \frac{x}{\sqrt{x^2 + 1}}\,dx$

17. $\displaystyle\int \frac{x^3}{(x^4 + 4)^4}\,dx$

18. $\displaystyle\int x^{-2}\left(\frac{1}{x} + 2\right)^5\,dx$

19. $\displaystyle\int \frac{\ln\sqrt{x}}{x}\,dx$

20. $\displaystyle\int \frac{x^2}{3 - x^3}\,dx$

21. $\displaystyle\int \frac{x^2 - 2x}{x^3 - 3x^2 + 1}\,dx$

22. $\displaystyle\int \frac{\ln(3x)}{3x}\,dx$

23. $\displaystyle\int \frac{8x}{e^{x^2}}\,dx$

24. $\displaystyle\int \frac{x + 4}{(1 - 8x - x^2)^3}\,dx$

25. $\displaystyle\int \frac{1}{x \ln x^2}\,dx$

26. $\displaystyle\int \frac{1}{x \ln x}\,dx$

27. $\displaystyle\int (3 - x)(x^2 - 6x)^4\,dx$

28. $\displaystyle\int \frac{e^x}{1 + 3e^x}\,dx$

29. $\displaystyle\int \frac{1}{x(\ln x)^3}\,dx$

30. $\displaystyle\int \frac{e^{2x} - e^{-2x}}{e^{2x} + e^{-2x}}\,dx$

31. Figure 1 shows graphs of several functions $f(x)$ whose slope at each x is $x/\sqrt{x^2 + 9}$. Find the expression for the function $f(x)$ whose graph passes through $(4, 8)$.

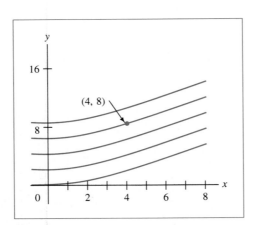

FIGURE I

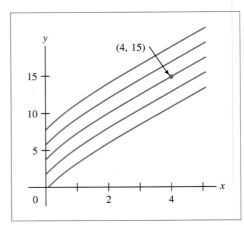

FIGURE 2

32. Figure 2 shows graphs of several functions $f(x)$ whose slope at each x is $(2\sqrt{x} + 1)/\sqrt{x}$. Find the expression for the function $f(x)$ whose graph passes through $(4, 15)$.

Determine the following integrals using the indicated substitution:

33. $\displaystyle\int (x + 5)^{-1/2} e^{\sqrt{x+5}} \, dx; \quad u = \sqrt{x + 5}$

34. $\displaystyle\int \frac{x^4}{x^5 - 7} \ln(x^5 - 7) \, dx; \quad u = \ln(x^5 - 7)$

35. $\displaystyle\int x \sec^2 x^2 \, dx; \quad u = x^2$

36. $\displaystyle\int (1 + \ln x) \sin(x \ln x) \, dx; \quad u = x \ln x$

Determine the following integrals by making an appropriate substitution.

37. $\displaystyle\int \sin x \cos x \, dx$

38. $\displaystyle\int 2x \cos x^2 \, dx$

39. $\displaystyle\int \frac{\cos \sqrt{x}}{\sqrt{x}} \, dx$

40. $\displaystyle\int \frac{\cos x}{(2 + \sin x)^3} \, dx$

41. $\displaystyle\int \cos^3 x \sin x \, dx$

42. $\displaystyle\int (\sin 2x) e^{\cos 2x} \, dx$

43. $\displaystyle\int (\cos 5x)\sqrt{1 - \sin 5x}\ dx$

44. $\displaystyle\int \frac{\sin 2x}{\sqrt{4 - \cos 2x}}\ dx$

45. $\displaystyle\int (2x - 1)\sin(x^2 - x + 1)\ dx$

46. $\displaystyle\int x\cos(x^2)e^{\sin(x^2)}\ dx$

47. Determine $\int 2x(x^2 + 5)\ dx$ by making a substitution. Then determine the integral by multiplying out the integrand and antidifferentiating. Account for the difference in the two results.

SOLUTIONS TO PRACTICE PROBLEMS 9.1

1. (a) $\dfrac{d}{dx} e^{(2x^3+3x)} = e^{(2x^3+3x)} \cdot (6x^2 + 3)$ (chain rule)

(b) $\dfrac{d}{dx} \ln x^5 = \dfrac{d}{dx} 5 \ln x = 5 \cdot \dfrac{1}{x}$ (Logarithm Property LIV)

(c) $\dfrac{d}{dx} \ln \sqrt{x} = \dfrac{d}{dx} \dfrac{1}{2} \ln x = \dfrac{1}{2} \cdot \dfrac{1}{x} = \dfrac{1}{2x}$ (Logarithm Property LIV)

(d) $\dfrac{d}{dx} \ln 5|x| = \dfrac{d}{dx} [\ln 5 + \ln|x|] = 0 + \dfrac{1}{x} = \dfrac{1}{x}$ (Logarithm Property LI)

(e) $\dfrac{d}{dx} x \ln x = x \cdot \dfrac{1}{x} + (\ln x) \cdot 1 = 1 + \ln x$ (product rule)

(f) $\dfrac{d}{dx} \ln(x^4 + x^2 + 1) = \dfrac{4x^3 + 2x}{x^4 + x^2 + 1}$ (chain rule)

(g) $\dfrac{d}{dx} \sin x^3 = (\cos x^3) \cdot (3x^2)$ (chain rule)

(h) $\dfrac{d}{dx} \tan x = \sec^2 x$ (formula for the derivative of $\tan x$)

2. Let $u = 3/x$, $du = (-3/x^2)\ dx$. Then

$$\int \frac{e^{3/x}}{x^2}\ dx = -\frac{1}{3}\int e^{3/x} \cdot \left(-\frac{3}{x^2}\right) dx$$

$$= -\frac{1}{3}\int e^u\ du$$

$$= -\frac{1}{3}e^u + C$$

$$= -\frac{1}{3}e^{3/x} + C.$$

9.2 INTEGRATION BY PARTS

In the preceding section we developed the method of integration by substitution by turning the chain rule into an integration formula. Let us do the same for the product rule. Let $f(x)$ and $g(x)$ be any two functions and let $G(x)$ be an anti-derivative of $g(x)$. The product rule asserts that

$$\frac{d}{dx}[f(x)G(x)] = f(x)G'(x) + f'(x)G(x)$$

$$= f(x)g(x) + f'(x)G(x) \qquad [\text{since } G'(x) = g(x)].$$

Therefore,

$$f(x)G(x) = \int f(x)g(x)\, dx + \int f'(x)G(x)\, dx.$$

This last formula can be rewritten in the following more useful form.

$$\int f(x)g(x)\, dx = f(x)G(x) - \int f'(x)G(x)\, dx. \qquad (1)$$

Equation (1) is the principle of *integration by parts* and is one of the most important techniques of integration.

EXAMPLE 1   Evaluate $\int xe^x\, dx$.

Solution   Set $f(x) = x$, $g(x) = e^x$. Then $f'(x) = 1$, $G(x) = e^x$, and equation (1) yields

$$\int xe^x\, dx = xe^x - \int 1 \cdot e^x\, dx$$

$$= xe^x - e^x + C. \; \bullet$$

The following principles underlie Example 1 and also illustrate general features of situations to which integration by parts may be applied:

1. The integrand is the product of two functions $f(x) = x$ and $g(x) = e^x$.
2. It is easy to compute $f'(x)$ and $G(x)$. That is, we can differentiate $f(x)$ and integrate $g(x)$.
3. The integral $\int f'(x)G(x)\, dx$ can be calculated.

Let us consider another example to see how these three principles work.

EXAMPLE 2   Evaluate $\int x(x + 5)^8\, dx$.

Solution   Our calculations can be set up as follows:

$$f(x) = x, \qquad g(x) = (x + 5)^8,$$

$$f'(x) = 1, \qquad G(x) = \frac{1}{9}(x + 5)^9.$$

Then

$$\int x(x + 5)^8 \, dx = x \cdot \frac{1}{9}(x + 5)^9 - \int 1 \cdot \frac{1}{9}(x + 5)^9 \, dx$$

$$= \frac{1}{9}x(x + 5)^9 - \frac{1}{9}\int (x + 5)^9 \, dx$$

$$= \frac{1}{9}x(x + 5)^9 - \frac{1}{9} \cdot \frac{1}{10}(x + 5)^{10} + C$$

$$= \frac{1}{9}x(x + 5)^9 - \frac{1}{90}(x + 5)^{10} + C. \quad \bullet$$

We were led to try integration by parts because our integrand is the product of two functions. We were led to choose $f(x) = x$ [and not $(x + 5)^8$] because $f'(x) = 1$, so that the factor x in the integrand is made to disappear, thereby simplifying the integral.

EXAMPLE 3   Evaluate $\int x \sin x \, dx$.

Solution   Let us set

$$f(x) = x, \qquad g(x) = \sin x,$$

$$f'(x) = 1, \qquad G(x) = -\cos x.$$

Then

$$\int x \sin x \, dx = -x \cos x - \int 1 \cdot (-\cos x) \, dx$$

$$= -x \cos x + \int \cos x \, dx$$

$$= -x \cos x + \sin x + C. \quad \bullet$$

EXAMPLE 4   Evaluate $\int x^2 \ln x \, dx$.

Solution   Set

$$f(x) = \ln x, \qquad g(x) = x^2,$$

$$f'(x) = \frac{1}{x}, \qquad G(x) = \frac{x^3}{3}.$$

Then

$$\int x^2 \ln x \, dx = \frac{x^3}{3} \ln x - \int \frac{1}{x} \cdot \frac{x^3}{3} \, dx$$

$$= \frac{x^3}{3} \ln x - \frac{1}{3} \int x^2 \, dx$$

$$= \frac{x^3}{3} \ln x - \frac{1}{9} x^3 + C. \quad \bullet$$

The next example shows how integration by parts can be used to compute a reasonably complicated integral.

EXAMPLE 5   Find $\displaystyle\int \frac{xe^x}{(x + 1)^2} \, dx$.

Solution   Let $f(x) = xe^x$, $g(x) = \dfrac{1}{(x + 1)^2}$. Then

$$f'(x) = xe^x + e^x \cdot 1 = (x + 1)e^x, \qquad G(x) = \frac{-1}{x + 1}.$$

As a result, we have

$$\int \frac{xe^x}{(x + 1)^2} \, dx = xe^x \cdot \frac{-1}{x + 1} - \int (x + 1)e^x \cdot \frac{-1}{x + 1} \, dx$$

$$= -\frac{xe^x}{x + 1} + \int e^x \, dx$$

$$= -\frac{xe^x}{x + 1} + e^x + C = \frac{e^x}{x + 1} + C. \bullet$$

Sometimes we must use integration by parts more than once.

EXAMPLE 6   Find $\displaystyle\int x^2 \sin x \, dx$.

Solution   Let $f(x) = x^2$, $g(x) = \sin x$. Then $f'(x) = 2x$ and $G(x) = -\cos x$. Applying our formula for integration by parts, we have

$$\int x^2 \sin x \, dx = -x^2 \cos x - \int 2x \cdot (-\cos x)$$

$$= -x^2 \cos x + 2 \int x \cos x \, dx. \qquad (2)$$

The integral $\int x \cos x \, dx$ can itself be handled by integration by parts. Let $f(x) = x$, $g(x) = \cos x$. Then $f'(x) = 1$ and $G(x) = \sin x$, so that

$$\int x \cos x \, dx = x \sin x - \int 1 \cdot \sin x \, dx$$

$$= x \sin x + \cos x. \qquad (3)$$

Combining (2) and (3), we see that

$$\int x^2 \sin x \, dx = -x^2 \cos x + 2(x \sin x + \cos x) + C$$

$$= -x^2 \cos x + 2x \sin x + 2 \cos x + C. \; \bullet$$

EXAMPLE 7 Evaluate $\int \ln x \, dx$.

Solution Since $\ln x = 1 \cdot \ln x$, we may view $\ln x$ as a product $f(x)g(x)$, where $f(x) = \ln x$, $g(x) = 1$. Then

$$f'(x) = \frac{1}{x}, \qquad G(x) = x.$$

Finally,

$$\int \ln x \, dx = x \ln x - \int \frac{1}{x} \cdot x \, dx$$

$$= x \ln x - \int 1 \, dx$$

$$= x \ln x - x + C. \; \bullet$$

**TECHNOLOGY
PROJECT 9.2
Still More
Antidifferentiation**

Use your symbolic math program to calculate the following antiderivatives. Some of them can be calculated using the methods of this section, although with more computation than required by the examples or exercises. Others can be calculated using methods not covered in this text, but which are "known" to your symbolic math program.

1. $\int x^5 e^{-4x} \, dx$

2. $\int (3x - 1)^5 \ln x \, dx$

3. $\int \sin 5x \cos x \, dx$

4. $\int x \tan x \, dx$

5. $\int \sin^3 x \, dx$

6. $\int e^x \sin x \, dx$

7. $\int \sin^4 x \cos^6 x \, dx$

8. $\int \frac{1}{(x - 1)(x - 2)(2x + 1)} dx$

9. $\int (2x - 1)^4 \ln x \, dx$

10. $\int \frac{1}{3x^2 - 2x - 1} dx$

PRACTICE
PROBLEMS 9.2

Evaluate the following integrals.

1. $\displaystyle \int \frac{x}{e^{3x}}\, dx$

2. $\displaystyle \int \ln \sqrt{x}\, dx$

EXERCISES 9.2

Evaluate the following integrals.

1. $\displaystyle \int xe^{5x}\, dx$

2. $\displaystyle \int xe^{-x/2}\, dx$

3. $\displaystyle \int x(2x + 1)^4\, dx$

4. $\displaystyle \int (x + 1)e^x\, dx$

5. $\displaystyle \int xe^{-x}\, dx$

6. $\displaystyle \int x(x + 5)^{-3}\, dx$

7. $\displaystyle \int \frac{x}{\sqrt{x + 1}}\, dx$

8. $\displaystyle \int \frac{x}{\sqrt{3 + 2x}}\, dx$

9. $\displaystyle \int x(1 - 3x)^4\, dx$

10. $\displaystyle \int x(1 + x)^{10}\, dx$

11. $\displaystyle \int \frac{3x}{e^x}\, dx$

12. $\displaystyle \int \frac{x + 1}{e^x}\, dx$

13. $\displaystyle \int x\sqrt{x + 1}\, dx$

14. $\displaystyle \int x\sqrt{2 - x}\, dx$

15. $\displaystyle \int \sqrt{x}\, \ln \sqrt{x}\, dx$

16. $\displaystyle \int x^5 \ln x\, dx$

17. $\displaystyle \int x^2 \cos x\, dx$

18. $\displaystyle \int x \sin 8x\, dx$

19. $\displaystyle \int x \ln 5x\, dx$

20. $\displaystyle \int x^{-3} \ln x\, dx$

21. $\displaystyle \int \ln x^4\, dx$

22. $\displaystyle \int x^2 \sin 3x\, dx$

23. $\displaystyle \int x^2 e^{-x}\, dx$

24. $\displaystyle \int \frac{x^2}{\sqrt{x + 4}}\, dx$

Evaluate the following integrals using techniques studied thus far.

25. $\displaystyle \int x(x + 5)^4\, dx$

26. $\displaystyle \int 4x \cos(x^2 + 1)\, dx$

27. $\displaystyle \int x(x^2 + 5)^4\, dx$

28. $\displaystyle \int 4x \cos(x + 1)\, dx$

29. $\displaystyle \int (3x + 1)e^{x/3}\, dx$

30. $\displaystyle \int \frac{(\ln x)^5}{x}\, dx$

31. $\displaystyle \int x \sec^2(x^2 + 1)\, dx$

32. $\displaystyle \int x^{3/2} \ln 2x\, dx$

33. $\displaystyle \int (xe^{2x} + x^2)\, dx$

34. $\displaystyle \int (x^{3/2} + \ln 2x)\, dx$

35. $\displaystyle\int (xe^{x^2} - 2x)\, dx$

36. $\displaystyle\int (x^2 - x \sin 2x)\, dx$

37. Figure 1 shows graphs of several functions $f(x)$ whose slope at each x is $x/\sqrt{x+9}$. Find the expression for the function $f(x)$ whose graph passes through $(0, 2)$.

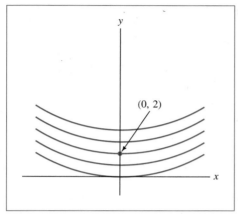

FIGURE 1

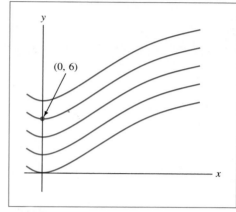

FIGURE 2

38. Figure 2 shows graphs of several functions $f(x)$ whose slope at each x is $\dfrac{x}{e^{x/3}}$. Find the expression for the function $f(x)$ whose graph passes through $(0, 6)$.

SOLUTIONS TO PRACTICE PROBLEMS 9.2

1. $\dfrac{x}{e^{3x}}$ is the same as xe^{-3x}, a product of two familiar functions. Set $f(x) = x$, $g(x) = e^{-3x}$. Then

$$f'(x) = 1, \qquad G(x) = -\frac{1}{3}e^{-3x},$$

so that

$$\int \frac{x}{e^{3x}}\, dx = x \cdot \left(-\frac{1}{3}e^{-3x}\right) - \int 1 \cdot \left(-\frac{1}{3}e^{-3x}\right) dx$$

$$= -\frac{1}{3}xe^{-3x} + \frac{1}{3}\int e^{-3x}\, dx$$

$$= -\frac{1}{3}xe^{-3x} + \frac{1}{3}\left[-\frac{1}{3}e^{-3x}\right] + C$$

$$= -\frac{1}{3}xe^{-3x} - \frac{1}{9}e^{-3x} + C.$$

2. This problem is similar to Example 7, which asks for $\int \ln x \, dx$ and can be approached in the same way by letting $f(x) = \ln \sqrt{x}$ and $g(x) = 1$. Another approach is to use a property of logarithms to simplify the integrand.

$$\int \ln \sqrt{x} \, dx = \int \ln (x)^{1/2} \, dx = \int \frac{1}{2} \ln x \, dx = \frac{1}{2} \int \ln x \, dx.$$

Since we know $\int \ln x \, dx$ from Example 7,

$$\int \ln \sqrt{x} \, dx = \frac{1}{2} \int \ln x \, dx = \frac{1}{2}(x \ln x - x) + C.$$

 # 9.3 EVALUATION OF DEFINITE INTEGRALS

Earlier we discussed techniques for determining antiderivatives (indefinite integrals). One of the most important applications of such techniques concerns the computation of definite integrals. For if $F(x)$ is an antiderivative of $f(x)$, then

$$\int_a^b f(x) \, dx = F(b) - F(a).$$

Thus the techniques of the previous sections can be used to evaluate definite integrals. Here we will simplify the method of evaluating definite integrals in those cases where the antiderivative is found by integration by substitution or parts.

EXAMPLE I   Evaluate $\displaystyle\int_0^1 2x(x^2 + 1)^5 \, dx$.

Solution—First Method   Let $u = x^2 + 1$, $du = 2x \, dx$. Then

$$\int 2x(x^2 + 1)^5 \, dx = \int u^5 \, du$$

$$= \frac{u^6}{6} + C$$

$$= \frac{(x^2 + 1)^6}{6} + C.$$

Consequently,

$$\int_0^1 2x(x^2 + 1)^5 \, dx = \frac{(x^2 + 1)^6}{6}\bigg|_0^1 = \frac{2^6}{6} - \frac{1^6}{6} = \frac{21}{2}. \ \bullet$$

Solution—Second Method Again we make the substitution $u = x^2 + 1$, $du = 2x\,dx$; however, we also apply the substitution to the limits of integration. When $x = 0$ (the lower limit of integration), we have $u = 0^2 + 1 = 1$; and when $x = 1$ (the upper limit of integration), we have $u = 1^2 + 1 = 2$. Therefore,

$$\int_0^1 2x(x^2 + 1)^5\,dx = \int_1^2 u^5\,du$$

$$= \frac{u^6}{6}\Big|_1^2$$

$$= \frac{2^6}{6} - \frac{1^6}{6} = \frac{21}{2}.$$

Notice that, in utilizing the second method, we did not need to reexpress the function $u^6/6$ in terms of x. •

The foregoing computation is an example of a general computational method, which can be expressed as follows:

Change of Limits Rule Suppose that the integral $\int f(g(x))g'(x)\,dx$ is subjected to the substitution $u = g(x)$, so that $\int f(g(x))g'(x)\,dx$ becomes $\int f(u)\,du$. Then

$$\int_a^b f(g(x))g'(x)\,dx = \int_{g(a)}^{g(b)} f(u)\,du.$$

Justification of Change of Limits Rule If $F(x)$ is an antiderivative of $f(x)$, then

$$\frac{d}{dx}[F(g(x))] = F'(g(x))g'(x) = f(g(x))g'(x).$$

Therefore,

$$\int_a^b f(g(x))g'(x)\,dx = F(g(x))\Big|_a^b = F(g(b)) - F(g(a)) = \int_{g(a)}^{g(b)} f(u)\,du.$$

EXAMPLE 2 Evaluate $\int_3^5 x\sqrt{x^2 - 9}\,dx$.

Solution   Let $u = x^2 - 9$, then $du = 2x \, dx$. When $x = 3$, we have $u = 3^2 - 9 = 0$. When $x = 5$, we have $u = 5^2 - 9 = 16$. Thus

$$\int_3^5 x\sqrt{x^2 - 9} \, dx = \frac{1}{2} \int_3^5 2x \sqrt{x^2 - 9} \, dx$$

$$= \frac{1}{2} \int_0^{16} \sqrt{u} \, du$$

$$= \frac{1}{2} \cdot \frac{2}{3} u^{3/2} \Big|_0^{16}$$

$$= \frac{1}{3} \cdot [16^{3/2} - 0] = \frac{1}{3} \cdot 16^{3/2}$$

$$= \frac{1}{3} \cdot 64 = \frac{64}{3}. \quad \bullet$$

EXAMPLE 3   Determine the area of the ellipse $x^2/a^2 + y^2/b^2 = 1$. (See Fig. 1.)

FIGURE 1

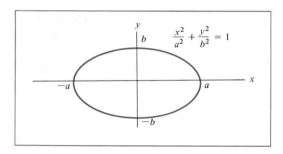

Solution   Owing to the symmetry of the ellipse, the area is equal to twice the area of the upper half of the ellipse. Solving for y,

$$\frac{y^2}{b^2} = 1 - \frac{x^2}{a^2}$$

$$\frac{y}{b} = \pm\sqrt{1 - \left(\frac{x}{a}\right)^2}$$

$$y = \pm b\sqrt{1 - \left(\frac{x}{a}\right)^2}.$$

Since the area of the upper half-ellipse is the area under the curve

$$y = b\sqrt{1 - \left(\frac{x}{a}\right)^2},$$

the area of the ellipse is given by the integral

$$2\int_{-a}^a b\sqrt{1 - \left(\frac{x}{a}\right)^2} \, dx.$$

Let $u = x/a$; then $du = 1/a \, dx$. When $x = -a$, we have $u = -a/a = -1$. So when $x = a$, we have $u = a/a = 1$.

$$2 \int_{-a}^{a} b \sqrt{1 - \left(\frac{x}{a}\right)^2} \, dx = 2b \cdot a \int_{-a}^{a} \frac{1}{a} \sqrt{1 - \left(\frac{x}{a}\right)^2} \, dx$$

$$= 2ba \int_{-1}^{1} \sqrt{1 - u^2} \, du. \ \bullet$$

Although we cannot evaluate this integral by our existing techniques, we obtain its value immediately by recognizing that since the area under the curve $y = \sqrt{1 - x^2}$ is the area of the top half of the unit circle, this integral has value $\pi/2$. And so the area of the ellipse is $2ba \cdot (\pi/2) = \pi ab$.

Integration by Parts in Definite Integrals

EXAMPLE 4   Evaluate $\displaystyle\int_{0}^{\pi/2} x \cos x \, dx$.

Solution—First Method   We use integration by parts to find an antiderivative of $x \cos x$. Let $f(x) = x$, $g(x) = \cos x, f'(x) = 1, G(x) = \sin x$. Then

$$\int x \cos x \, dx = x \sin x - \int 1 \cdot \sin x \, dx = x \sin x + \cos x + C.$$

Hence

$$\int_{0}^{\pi/2} x \cos x \, dx = (x \sin x + \cos x)\Big|_{0}^{\pi/2}$$

$$= \left(\frac{\pi}{2} \sin \frac{\pi}{2} + \cos \frac{\pi}{2}\right) - (0 + \cos 0)$$

$$= \frac{\pi}{2} - 1. \ \bullet$$

Solution—Second Method   Note that the antiderivative of $x \cos x$ consists of two terms. Instead of evaluating the entire antiderivative of $x \cos x$ at $\pi/2$ and then subtracting its value at 0, we may evaluate each term of the antiderivative separately. The computations may be written in the form

$$\int_{0}^{\pi/2} x \cos x \, dx = x \sin x \Big|_{0}^{\pi/2} - \int_{0}^{\pi/2} 1 \cdot \sin x \, dx$$

$$= x \sin x \Big|_{0}^{\pi/2} - (-\cos x)\Big|_{0}^{\pi/2}$$

$$= \left(\frac{\pi}{2} \sin \frac{\pi}{2} - 0\right) - \left(-\cos \frac{\pi}{2} + \cos 0\right) = \frac{\pi}{2} - 1.$$

The second method involved less writing because we never had to consider the *in*definite integral $\int x \cos x \, dx$. ●

The second method in Example 4 illustrates the following rule for integrating a definite integral by parts.

Integration by Parts   If $G(x)$ is an antiderivative of $g(x)$, then

$$\int_a^b f(x)g(x) \, dx = f(x)G(x)\Big|_a^b - \int_a^b f'(x)G(x) \, dx.$$

Justification of Integration-by-Parts Rule   Using the integration-by-parts formula, we have

$$\int_a^b f(x)g(x) \, dx = \left[f(x)G(x) - \int f'(x)G(x) \, dx\right]\Big|_a^b$$

$$= f(x)G(x)\Big|_a^b - \left[\int f'(x)G(x) \, dx\right]\Big|_a^b$$

$$= f(x)G(x)\Big|_a^b - \int_a^b f'(x)G(x) \, dx.$$

EXAMPLE 5   Evaluate $\displaystyle\int_0^5 \frac{x}{\sqrt{x+4}} \, dx$.

Solution   Let $f(x) = x$, $g(x) = (x+4)^{-1/2}$, $f'(x) = 1$, and $G(x) = 2(x+4)^{1/2}$. Then

$$\int_0^5 \frac{x}{\sqrt{x+4}} \, dx = 2x(x+4)^{1/2}\Big|_0^5 - \int_0^5 1 \cdot 2(x+4)^{1/2} \, dx$$

$$= 2x(x+4)^{1/2}\Big|_0^5 - \frac{4}{3}(x+4)^{3/2}\Big|_0^5$$

$$= [10(9)^{1/2} - 0] - \left[\frac{4}{3}(9)^{3/2} - \frac{4}{3}(4)^{3/2}\right]$$

$$= [30] - \left[36 - \frac{32}{3}\right] = 4\frac{2}{3}. \quad ●$$

PRACTICE PROBLEMS 9.3   Evaluate the following definite integrals.

1. $\displaystyle\int_0^1 (2x + 3)e^{x^2+3x+6} \, dx$

2. $\displaystyle\int_e^{e^{\pi/2}} \frac{\sin(\ln x)}{x} \, dx$

EXERCISES 9.3

Evaluate the following definite integrals.

1. $\displaystyle\int_{3/2}^{2} 2(2x - 3)^{17}\, dx$

2. $\displaystyle\int_{\sqrt{3}}^{2} [(x^2 - 3)^{17} - (x^2 - 3)]x\, dx$

3. $\displaystyle\int_{1}^{3} (4 - 2x)\sin(4x - x^2)\, dx$

4. $\displaystyle\int_{5}^{13} x\sqrt{x^2 - 25}\, dx$

5. $\displaystyle\int_{1}^{e} \frac{\ln x}{x}\, dx$

6. $\displaystyle\int_{0}^{\pi} e^{(\sin x)^2} \cdot \cos x \sin x\, dx$

7. $\displaystyle\int_{3}^{5} x\sqrt{x^2 - 9}\, dx$

8. $\displaystyle\int_{1}^{e^{\pi/2}} \frac{\sin(\ln x)}{x}\, dx$

9. $\displaystyle\int_{\ln 1}^{\ln 2} [(e^x)^2 + e^x]e^x\, dx$

10. $\displaystyle\int_{\pi/4}^{\pi/2} \ln(\sin x) \cos x\, dx$

11. $\displaystyle\int_{-1}^{2} (x^3 - 2x - 1)^2 \cdot (3x^2 - 2)\, dx$

12. $\displaystyle\int_{-1}^{1} x \sin x^4\, dx$ (Let $u = x^2$.)

13. $\displaystyle\int_{1}^{3} x^2 e^{x^3}\, dx$

14. $\displaystyle\int_{0}^{1} (2x - 1)(x^2 - x)^{10}\, dx$

15. $\displaystyle\int_{-\pi}^{2\pi} \sin(8x - \pi)\, dx$

16. $\displaystyle\int_{-2}^{2} 2x \sin(x^2)\, dx$

17. $\displaystyle\int_{0}^{2} xe^{x/2}\, dx$

18. $\displaystyle\int_{0}^{4} 8x(x + 4)^{-3}\, dx$

19. $\displaystyle\int_{0}^{1} x \sin \pi x\, dx$

20. $\displaystyle\int_{1}^{4} \ln x\, dx$

Use substitutions and the fact that a circle of radius r has area πr^2 to evaluate the following integrals.

21. $\displaystyle\int_{-\pi/2}^{\pi/2} \sqrt{1 - \sin^2 x}\, \cos x\, dx$

22. $\displaystyle\int_{0}^{\sqrt{2}} \sqrt{4 - x^4} \cdot 2x\, dx$

23. $\displaystyle\int_{-6}^{0} \sqrt{-x^2 - 6x}\, dx$ (Complete the square: $-x^2 - 6x = 9 - (x + 3)^2$.)

In Exercises 24 and 25, find the area of the shaded regions.

24.

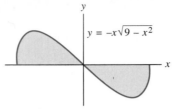

$y = -x\sqrt{9 - x^2}$

25.

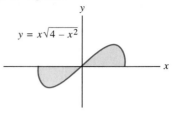

$y = x\sqrt{4 - x^2}$

SOLUTIONS TO PRACTICE PROBLEMS 9.3

1. Let $u = x^2 + 3x + 6$ so that $du = 2x + 3$. When $x = 0$, $u = 6$; when $x = 1$, $u = 10$. Thus

$$\int_0^1 (2x + 3)e^{x^2+3x+6} \, dx = \int_6^{10} e^u \, du$$

$$= e^u \Big|_6^{10}$$

$$= e^{10} - e^6.$$

2. Let $u = \ln x$, $du = (1/x) \, dx$. When $x = e$, $u = \ln e = 1$; when $x = e^{\pi/2}$, $u = \ln e^{\pi/2} = \pi/2$. Thus

$$\int_e^{e^{\pi/2}} \frac{\sin(\ln x)}{x} \, dx = \int_1^{\pi/2} \sin u \, du$$

$$= -\cos u \Big|_1^{\pi/2}$$

$$= -\cos \frac{\pi}{2} + \cos 1$$

$$\approx .54030.$$

9.4 APPROXIMATION OF DEFINITE INTEGRALS

The definite integrals that arise in practical problems cannot always be evaluated by computing the net change in an antiderivative, as in the preceding section. In some cases we may not know enough techniques for finding antiderivatives, whereas in other cases there may actually be no way to express antiderivatives in terms of elementary functions. In this section we discuss three methods for approximating the numerical value of a definite integral,

$$\int_a^b f(x) \, dx.$$

Given a positive integer n, divide the interval $a \le x \le b$ into n equal subintervals, each of length $\Delta x = (b - a)/n$. Denote the endpoints of the subintervals by $a_0, a_1, \ldots, a_n$, and denote the midpoints of the subintervals by $x_1, x_2, \ldots, x_n$, as in Fig. 1. Recall from Chapter 6 that the definite integral is the limit of Riemann sums. When the midpoints of the subintervals in Fig. 1 are used to construct a Riemann sum, the resulting approximation to $\int_a^b f(x)\,dx$ is called the *midpoint rule*.

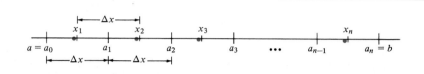

FIGURE 1

> *Midpoint Rule*
>
> $$\int_a^b f(x)\,dx \approx f(x_1)\Delta x + f(x_2)\Delta x + \cdots + f(x_n)\Delta x$$
>
> $$= [f(x_1) + f(x_2) + \cdots + f(x_n)]\Delta x.$$
>
> (1)

EXAMPLE 1 Use the midpoint rule with $n = 4$ to approximate $\int_0^2 \dfrac{1}{1 + e^x}\,dx$.

Solution We have $\Delta x = (b - a)/n = (2 - 0)/4 = .5$. The endpoints of the four subintervals begin at $a = 0$ and are spaced .5 unit apart. The first midpoint is at

```
        .25        .75       1.25       1.75
   +-----●-----+-----●-----+-----●-----+-----●-----+
   0          .5         1.0        1.5        2.0
```

$a + \Delta x/2 = .25$. The midpoints are also spaced .5 unit apart. According to the midpoint rule, the integral is approximately equal to

$$\left[\frac{1}{1 + e^{.25}} + \frac{1}{1 + e^{.75}} + \frac{1}{1 + e^{1.25}} + \frac{1}{1 + e^{1.75}}\right](.5)$$

$$= .5646961 \qquad \text{(to seven decimal places).} \quad \bullet$$

A second method of approximation, the trapezoidal rule, uses the values of $f(x)$ at the endpoints of the subintervals of the interval $a \le x \le b$.

> *Trapezoidal Rule*
>
> $$\int_a^b f(x)\,dx \approx [f(a_0) + 2f(a_1) + \cdots + 2f(a_{n-1}) + f(a_n)]\frac{\Delta x}{2}.$$
>
> (2)

EXAMPLE 2 Use the trapezoidal rule with $n = 4$ to approximate $\int_0^2 \frac{1}{1 + e^x}\, dx$.

Solution As in Example 1, $\Delta x = .5$ and the endpoints of the subintervals are $a_0 = 0$, $a_1 = .5$, $a_2 = 1$, $a_3 = 1.5$, and $a_4 = 2$. The trapezoidal rule gives

$$\left[\frac{1}{1 + e^0} + 2 \cdot \frac{1}{1 + e^{.5}} + 2 \cdot \frac{1}{1 + e^1} + 2 \cdot \frac{1}{1 + e^{1.5}} + \frac{1}{1 + e^2}\right] \frac{.5}{2}$$

$$= .5692545 \qquad \text{(to seven decimal places).} \quad \bullet$$

When the function $f(x)$ is given explicitly, either the midpoint rule or the trapezoidal rule may be used to approximate the definite integral. However, occasionally the values of $f(x)$ may be known only at the endpoints of the subintervals. This may happen, for instance, when the values of $f(x)$ are obtained from experimental data. In this case, the midpoint rule cannot be used.

EXAMPLE 3 (*Measuring Cardiac Output*[*]) Five milligrams of dye is injected into a vein leading to the heart. The concentration of the dye in the aorta, an artery leading from the heart, is determined every 2 seconds for 22 seconds. (See Table 1.) Let $c(t)$ be the concentration in the aorta after t seconds. Use the trapezoidal rule to estimate $\int_0^{22} c(t)\, dt$.

TABLE I	Concentration of Dye in the Aorta										
Seconds After Injection	0	2	4	6	8	10	12	14	16	18	20	22
Concentration (mg/liter)	0	0	.6	1.4	2.7	3.7	4.1	3.8	2.9	1.5	.9	.5

Solution Let $n = 11$. Then $a = 0$, $b = 22$, and $\Delta t = (22 - 0)/11 = 2$. The endpoints of the subintervals are $a_0 = 0$, $a_1 = 2$, $a_2 = 4$, . . . , $a_{10} = 20$, and $a_{11} = 22$. By the trapezoidal rule,

$$\int_0^{22} c(t)\, dt \approx [c(0) + 2c(2) + 2c(4) + 2c(6) + \cdots + 2c(20) + c(22)]\left(\frac{2}{2}\right)$$

$$= [0 + 2(0) + 2(.6) + 2(1.4) + \cdots + 2(.9) + .5](1)$$

$$\approx 43.7. \quad \bullet$$

---

* Data from B. Horelick and S. Koont, Project UMAP, *Measuring Cardiac Output* (Newton, Mass.: Educational Development Center, Inc., 1978).

Note Cardiac output is the rate (usually measured in liters per minute) at which the heart pumps blood, and it may be computed by the formula

$$R = \frac{60D}{\displaystyle\int_0^{22} c(t)\, dt},$$

where D is the quantity of dye injected. For the data above, $R = 60(5)/43.7 = 6.9$ liters per minute.

Let us return to the approximations to $\displaystyle\int_0^2 \frac{1}{1 + e^x}\, dx$ found in Examples 1 and 2. These numbers are shown in Fig. 2 along with the exact value of the definite integral to seven decimal places and the error of the two approximations. (The scale is greatly enlarged.) It can be shown that, in general, the error from the midpoint rule is about one-half the error from the trapezoidal rule, and the estimates from these two rules are usually on opposite sides of the actual value of the definite integral. These observations suggest that we might improve our estimate of the value of a definite integral by using a "weighted average" of these two estimates. Let M and T denote the estimates from the midpoint and trapezoidal rules, respectively, and define

$$S = \frac{2}{3}M + \frac{1}{3}T = \frac{2M + T}{3}. \tag{3}$$

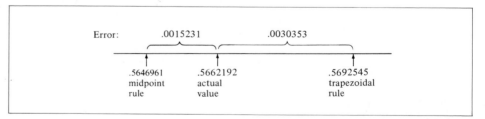

FIGURE 2

The use of S as an estimate of the value of a definite integral is called *Simpson's rule*. If we use Simpson's rule to estimate the definite integral in Example 1, we find that

$$S = \frac{2(.5646961) + .5692545}{3} = .5662156.$$

The error here is only .0000036. The error from the trapezoidal rule, in this example, is over 800 times as large!

As the number n of subintervals increases, Simpson's rule becomes far more accurate than both the midpoint rule and the trapezoidal rule. For a given definite integral, the error in the midpoint and trapezoidal rules is proportional to $1/n^2$, so doubling n will divide the error by 4. However, the error in Simpson's rule is proportional to $1/n^4$, so doubling n will divide the error by 16, and multiplying n by a factor of 10 will divide the error by 10,000.

It is possible to combine the formulas for the midpoint and trapezoidal rules into a single formula for Simpson's rule, by using the fact that $S = (4M + 2T)/6$.

Simpson's Rule

$$\int_a^b f(x)\, dx \approx [\, f(a_0) + 4f(x_1) + 2f(a_1) + 4f(x_2) + 2f(a_2)$$

$$+ \cdots + 2f(a_{n-1}) + 4f(x_n) + f(a_n)]\frac{\Delta x}{6}. \qquad (4)$$

Distribution of IQs   Psychologists use various standardized tests to measure intelligence. The method most commonly used to describe the results of such tests is an intelligence quotient (or IQ). An IQ is a positive number that, in theory, indicates how a person's mental age compares with the person's chronological age. The median IQ is arbitrarily set at 100, so that half the population has an IQ less than 100 and half greater. IQs are distributed according to a bell-shaped curve called the *normal curve*, pictured in Fig. 3. The proportion of all people having IQs between A and B is given by the area under the curve from A to B, that is, by the integral

$$\frac{1}{16\sqrt{2\pi}} \int_A^B e^{-(1/2)[(x-100)/16]^2}\, dx.$$

EXAMPLE 4   Estimate the proportion of all people having IQs between 120 and 126.

FIGURE 3   Proportion of IQs between 120 and 126.

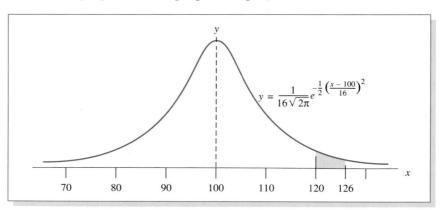

Solution   We have seen that this proportion is given by

$$\frac{1}{16\sqrt{2\pi}} \int_{120}^{126} f(x)\, dx, \quad \text{where } f(x) = e^{-(1/2)[(x-100)/16]^2}.$$

Let us approximate the definite integral by Simpson's rule with $n = 3$. Then $\Delta x = (126 - 120)/3 = 2$. The endpoints of the subintervals are 120, 122, 124, and 126; the midpoints of these subintervals are 121, 123, and 125. Simpson's rule gives

$$[f(120) + 4f(121) + 2f(122) + 4f(123) + 2f(124) + 4f(125) + f(126)]\frac{\Delta x}{6}$$

$$\approx [.4578 + 1.6904 + .7771 + 1.4235 + .6493 + 1.1801 + .2671]\left(\frac{2}{6}\right)$$

$$\approx 2.1484.$$

Multiplying this estimate by $1/(16\sqrt{2\pi})$, the constant in front of the integral, we get .0536. Thus approximately 5.36% of the population have IQs between 120 and 126. ●

Geometric Interpretation of the Approximation Rules   Let $f(x)$ be a continuous nonnegative function on $a \le x \le b$. The approximation rules discussed above may be interpreted as methods for estimating the area under the graph of $f(x)$. The midpoint rule arises from replacing this area by a collection of n rectangles, one lying over each subinterval, the first of height $f(x_1)$, the second of height $f(x_2)$, and so on. (See Fig. 4.)

   If we approximate the area under the graph of $f(x)$ by trapezoids, as in Fig. 5, then the total area of these trapezoids turns out to be the number given by the trapezoidal rule (hence the name of the rule).

FIGURE 4   **Approximation by rectangles.**

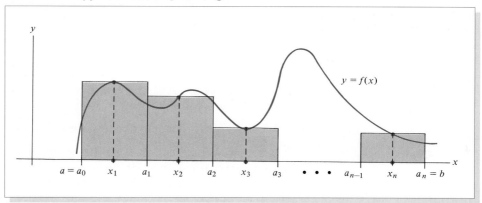

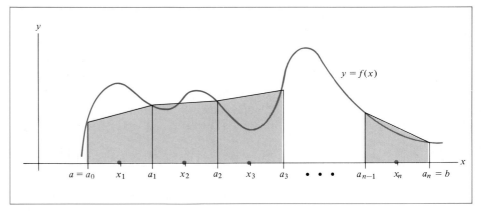

FIGURE 5   Approximation by trapezoids.

Simpson's rule corresponds to approximating the graph of $f(x)$ on each subinterval by a parabola instead of a straight line as in the midpoint and trapezoidal rules. On each subinterval, the parabola is chosen so that it intersects the graph of $f(x)$ at the midpoint and both endpoints of the subinterval. (See Fig. 6.) It can be shown that the sum of the areas under these parabolas is the number given by Simpson's rule. Even more powerful approximation rules may be obtained by approximating the graph of $f(x)$ on each subinterval by cubic curves, or graphs of higher-order polynomials.

FIGURE 6   Approximation by parabolas.

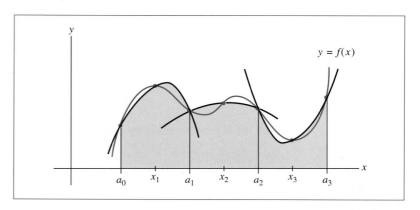

Error Analysis   A simple measure of the error of an approximation to a definite integral is the quantity

$$|\,[\text{approximate value}] - [\text{actual value}]\,|.$$

The following theorem gives an idea of how small this error must be for the various approximation rules. In a concrete example, the actual error of an approximation may be even substantially less than the "error bound" given in the theorem.

> **Error of Approximation Theorem** Let n be the number of subintervals used in an approximation of the definite integral
>
> $$\int_a^b f(x)\, dx.$$
>
> 1. The error for the midpoint rule is at most $\dfrac{A(b-a)^3}{24n^2}$, where A is a number such that $|\,f''(x)\,| \le A$ for all x satisfying $a \le x \le b$.
>
> 2. The error for the trapezoidal rule is at most $\dfrac{A(b-a)^{3+1}}{12n^2}$, where A is a number such that $|\,f''(x)\,| \le A$ for all x satisfying $a \le x \le b$.
>
> 3. The error for Simpson's rule is at most $\dfrac{A(b-a)^5}{2880n^4}$, where A is a number such that $|\,f^{(''')}(x)\,| \le A$ for all x satisfying $a \le x \le b$.

EXAMPLE 5 Obtain a bound on the error of using the trapezoidal rule with $n = 20$ to approximate $\int_0^1 e^{x^2}\, dx$.

Solution Here $a = 0$, $b = 1$, and $f(x) = e^{x^2}$. Differentiating twice, we find that

$$f''(x) = (4x^2 + 2)e^{x^2}.$$

How large could $|\,f''(x)\,|$ be if x satisfies $0 \le x \le 1$? Since x is between 0 and 1, so is x^2. Hence $4x^2 + 2 \le 6$ and $e^{x^2} \le e^1 = e$. Therefore $|(4x^2 + 2)e^{x^2}| \le 6e$, so we may take $A = 6e$ in the theorem above. The error of approximation using the trapezoidal rule is at most

$$\frac{6e(1-0)^3}{12(20)^2} = \frac{e}{800} \approx \frac{2.71828}{800} = .003398. \ \bullet$$

TECHNOLOGY PROJECT 9.3

Approximating Definite Integrals Using the Trapezoid and Simpson's Rules

Approximate the following integrals using (a) the trapezoidal rule for $n = 100$, 500, 1000, 10,000, (b) Simpson's rule for $n = 100$, 500, 1000, 10,000. In each case, estimate the number of decimal places of accuracy provided by each approximation. Based on the data collected, what can you say about the relative accuracy of the two rules?

1. $\displaystyle\int_0^2 x\, dx$

2. $\displaystyle\int_1^3 \frac{1}{x}\, dx$

3. $\displaystyle\int_0^{\pi/4} \sin x\, dx$

4. $\displaystyle\int_{-1}^1 e^{-x^2}\, dx$

TECHNOLOGY PROJECT 9.4
Approximating Definite Integrals with a Specified Degree of Accuracy

You may obtain an upper bound for the number A in the *Error of Approximation Theorem* as follows: Calculate the second or fourth derivative. Then graph the resulting function. Determine A by examining the graph. Use this approach to determine the number n of subintervals required by (a) the trapezoidal rule and (b) Simpson's rule to approximate the following integrals to within an error 10^{-5}:

1. $\displaystyle\int_0^1 (5x - 3)^7 e^{-x}\, dx$

2. $\displaystyle\int_0^{\pi/4} \sqrt{\sin x}\, dx$

3. $\displaystyle\int_1^{100} (\ln x)^5\, dx$

4. $\displaystyle\int_0^1 \frac{x}{x + 1}\, dx$

PRACTICE PROBLEMS 9.4

Consider $\displaystyle\int_1^{3.4} (5x - 9)^2\, dx$.

1. Divide the interval $1 \le x \le 3.4$ into three subintervals. List Δx and the endpoints and midpoints of the subintervals.
2. Approximate the integral by the midpoint rule with $n = 3$.
3. Approximate the integral by the trapezoidal rule with $n = 3$.
4. Approximate the integral by Simpson's rule with $n = 3$.
5. Find the exact value of the integral by integration.

EXERCISES 9.4

In Exercises 1 and 2, divide the given interval into n subintervals and list the value of Δx and the endpoints $a_0, a_1, \ldots, a_n$ of the subintervals.

1. $3 \le x \le 5$; $n = 5$ 2. $-1 \le x \le 2$; $n = 5$

In Exercises 3–6, divide the interval into n subintervals and list the value of Δx and the midpoints $x_1, \ldots, x_n$ of the subintervals.

3. $0 \le x \le 2$; $n = 4$ 4. $0 \le x \le 3$; $n = 6$

5. $1 \le x \le 4$; $n = 5$ 6. $3 \le x \le 5$; $n = 10$

Approximate the following integrals by the midpoint rule, and then find the exact value by integration. Express your answers to five decimal places.

7. $\displaystyle\int_0^4 (x^2 + 5)\, dx$; $n = 2, 4$ 8. $\displaystyle\int_1^5 (x - 1)^2\, dx$; $n = 2, 4$

9. $\displaystyle\int_0^1 e^{-x}\, dx$; $n = 5$ 10. $\displaystyle\int_1^2 \frac{1}{x + 1}\, dx$; $n = 5$

Approximate the following integrals by the trapezoidal rule, and then find the exact value by integration. Express your answers to five decimal places.

11. $\int_0^1 (x - \frac{1}{2})^2 \, dx; n = 4$

12. $\int_4^9 \frac{1}{x - 3} \, dx; n = 5$

13. $\int_1^5 \frac{1}{x^2} \, dx; n = 3$

14. $\int_{-1}^1 e^{2x} \, dx; n = 2, 4$

Approximate the following integrals by the midpoint rule, the trapezoidal rule, and Simpson's rule. Then find the exact value by integration. Express your answers to five decimal places.

15. $\int_1^4 (2x - 3)^3 \, dx; n = 3$

16. $\int_{10}^{20} \frac{\ln x}{x} \, dx; n = 5$

17. $\int_0^2 2xe^{x^2} \, dx; n = 4$

18. $\int_0^3 x\sqrt{4 - x} \, dx; n = 5$

19. $\int_2^5 xe^x \, dx; n = 5$

20. $\int_1^5 (4x^3 - 3x^2) \, dx; n = 2$

The following integrals cannot be evaluated in terms of elementary antiderivatives. Find an approximate value by Simpson's rule. Express your answers to five decimal places.

21. $\int_0^2 \sqrt{1 + x^3} \, dx; n = 4$

22. $\int_0^1 \frac{1}{x^3 + 1} \, dx; n = 2$

23. $\int_0^2 \sqrt{\sin x} \, dx; n = 5$

24. $\int_{-1}^1 \sqrt{1 + x^4} \, dx; n = 4$

25. In a survey of a piece of oceanfront property, measurements of the distance to the water were made every 50 feet along a 200-foot side (Fig. 7). Use the trapezoidal rule to estimate the area of the property.

FIGURE 7   Survey of an oceanfront property.

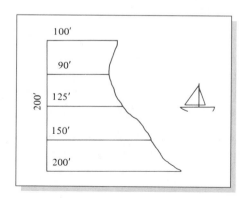

26. In order to determine the amount of water flowing down a certain 100-yard-wide river, engineers need to know the area of a vertical cross section of the river.

Measurements of the depth of the river were made every 20 yards from one bank to the other. The readings in fathoms were 0, 1, 2, 3, 1, 0. (One fathom equals 2 yards.) Use the trapezoidal rule to estimate the area of the cross section.

27. Upon takeoff, the velocity readings of a rocket noted every second for 10 seconds were 0, 30, 75, 115, 155, 200, 250, 300, 360, 420, and 490 feet per second. Use the trapezoidal rule to estimate the distance traveled by the rocket during the first 10 seconds. [*Hint*: If $s(t)$ is the distance traveled by time t and $v(t)$ is the velocity at time t, then $s(10) = \int_0^{10} v(t)\, dt$.]

28. In a drive along a country road, the speedometer readings are recorded each minute during a 5-minute interval.

Time (minutes)	0	1	2	3	4	5
Velocity (mph)	33	32	28	30	32	35

Use the trapezoidal rule to estimate the distance traveled during the 5 minutes. [*Hint*: If time is measured in minutes, then velocity should be expressed in distance per minute. For example, 35 mph is $\frac{35}{60}$ miles per minute. Also, see the hint for Exercise 27.]

29. Obtain a bound on the error of using the midpoint rule with $n = 100$ to approximate $\int_0^2 (x^4 + 3x^2)\, dx$.

30. Obtain a bound on the error of using Simpson's rule with $n = 5$ to approximate $\int_1^2 3 \ln x\, dx$.

31. (a) Show that the area of the trapezoid in Fig. 8(a) is $\frac{1}{2}(h + k)l$. [*Hint*: Divide the trapezoid into a rectangle and a triangle.]

    (b) Show that the area of the first trapezoid on the left in Fig. 8(b) is $\frac{1}{2}[f(a_0) + f(a_1)]\Delta x$.

    (c) Derive the trapezoidal rule for the case $n = 4$.

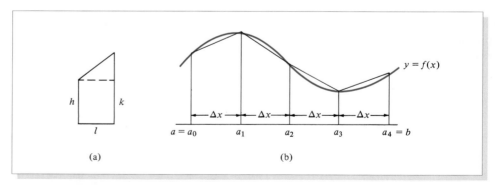

FIGURE 8   Derivation of the trapezoidal rule.

32. Approximate the value of $\int_a^b f(x)\, dx$, where $f(x) \geq 0$, by dividing the interval $a \leq x \leq b$ into four subintervals and constructing five rectangles as shown in Fig. 9. Note that the width of the three inside rectangles is Δx, while the width of the two

outside rectangles is $\Delta x/2$. Compute the sum of the areas of these five rectangles and compare this sum with the trapezoidal rule for $n = 4$.

FIGURE 9   Another view of the trapezoidal rule.

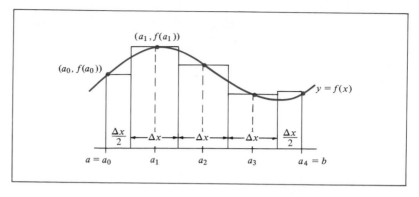

33. **(a)** Suppose that the graph of $f(x)$ is above the x-axis and concave down on the interval $a_0 \leq x \leq a_1$. Let x_1 be the midpoint of that interval, let $\Delta x = a_1 - a_0$, and construct the line tangent to the graph of $f(x)$ at $(x_1, f(x_1))$, as in Fig. 10(a). Show that the area of the shaded trapezoid in Fig. 10(a) is the same as the area of the shaded rectangle in Fig. 10(c), namely, $f(x_1) \, \Delta x$. [*Hint:* Look at Fig. 10(b).] This shows that the area of the rectangle in Fig. 10(c) exceeds the area under the graph of $f(x)$ on the interval $a_0 \leq x \leq a_1$.

   (b) Suppose that the graph of $f(x)$ is above the x-axis and concave down for all x in the interval $a \leq x \leq b$. Explain why $T \leq \int_a^b f(x) \, dx \leq M$, where T and M are the approximations given by the trapezoidal and midpoint rules, respectively.

FIGURE 10

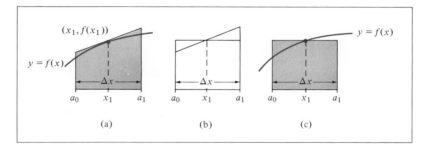

34. (*Riemann Sum Derivation of Formula for Cardiac Output*, see example 3.) Subdivide the interval $0 \leq t \leq 22$ into n subintervals of length $\Delta t = 22/n$ seconds. Let t_i be a point in the ith subinterval.

   (a) Show that $(R/60) \, \Delta t \approx$ [number of liters of blood flowing past the monitoring point during the ith time interval].

   (b) Show that $c(t_i)(R/60) \, \Delta t \approx$ [quantity of dye flowing past the monitoring point during the ith time interval].

   (c) Assume that basically all of the dye will have flowed past the monitoring point during the 22 seconds. Explain why $D \approx (R/60)[c(t_1) + c(t_2) + \cdots + c(t_n)] \, \Delta t$, where the approximation improves as n gets large.

   (d) Conclude that $D = \int_0^{22} (R/60)c(t) \, dt$, and solve for R.

**SOLUTIONS TO
PRACTICE
PROBLEMS 9.4**

1. $\Delta x = (3.4 - 1)/3 = 2.4/3 = .8$. Each subinterval will have length .8. A good way to proceed is first to draw two hatchmarks that subdivide the interval into three equal subintervals [Fig. 11(a)]. Then label the hatchmarks by successively adding .8 to the left endpoint [Fig. 11(b)]. The first midpoint can be obtained by adding one-half of .8 to the left endpoint. Then add .8 to get the next midpoint, and so on [Fig. 11(c)].

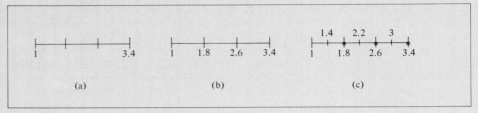

(a)   (b)   (c)

FIGURE 11

2. The midpoint rule uses only the midpoints of the subintervals:

$$\int_1^{3.4} (5x - 9)^2 \, dx \approx \{(5[1.4] - 9)^2 + (5[2.2] - 9)^2 + (5[3] - 9)^2\}(.8)$$

$$= \{(-2)^2 + 2^2 + 6^2\}(.8) = 35.2.$$

3. The trapezoidal rule uses only the endpoints of the subintervals:

$$\int_1^{3.4} (5x - 9)^2 \, dx$$

$$\approx \{(5[1] - 9)^2 + 2(5[1.8] - 9)^2 + 2(5[2.6] - 9)^2$$

$$+ (5[3.4] - 9)^2\}\left(\frac{.8}{2}\right)$$

$$= \{(-4)^2 + 2(0)^2 + 2(4)^2 + 8^2\}(.4) = 44.8.$$

4. Using the formula $S = \dfrac{2M + T}{3}$, we obtain

$$\int_1^{3.4} (5x - 9)^2 \, dx \approx \frac{2(35.2) + 44.8}{3} = \frac{115.2}{3} = 38.4.$$

This approximation may also be obtained directly with formula (4), but the arithmetic requires about the same effort as calculating the midpoint and trapezoidal approximations separately and then combining them, as we did here.

5. $$\int_1^{3.4} (5x - 9)^2 \, dx = \frac{1}{15}(5x - 9)^3 \Big|_1^{3.4} = \frac{1}{15}[8^3 - (-4)^3] = 38.4.$$

(Notice that here Simpson's rule gives the exact answer. This was so since the function to be integrated was a quadratic polynomial. Actually, Simpson's rule gives the exact value of the definite integral of any polynomial of degree 3 or less. The reason for this can be discovered from the "error of approximation theorem.")

9.5 SOME APPLICATIONS OF THE INTEGRAL

Recall that the integral

$$\int_a^b f(t)\, dt$$

can be approximated by a Riemann sum as follows: We divide the t-axis from a to b into n subintervals by adding intermediate points $t_0 = a,\ t_1,\ \ldots,\ t_{n-1}$, $t_n = b$.

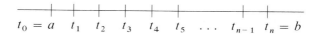

$t_0 = a \quad t_1 \quad t_2 \quad t_3 \quad t_4 \quad t_5 \quad \ldots \quad t_{n-1} \quad t_n = b$

We assume that the points are equally spaced, so that each subinterval has length $\Delta t = (b - a)/n$. For large n, the integral is very closely approximated by the Riemann sum

$$\int_a^b f(t)\, dt \approx f(t_1)\, \Delta t + f(t_2)\, \Delta t + \cdots + f(t_n)\, \Delta t. \tag{1}$$

The approximation in (1) works both ways. If we encounter a Riemann sum like the one in (1), we can approximate it by the corresponding integral. The approximation becomes better as the number of subintervals increases—that is, as n gets large. Thus, as n gets large, the sum approaches the value of the integral. This will be our approach in the examples below.

Our first two examples involve the concept of the present value of money. Suppose that we make an investment that promises to repay A dollars at time t (measuring the present as time 0). How much should we be willing to pay for such an investment? Clearly, we would not want to pay as much as A dollars. For if we had A dollars now, we could invest it at the current rate of interest and, at time t, we would get back our original A dollars plus the accrued interest. Instead, we should only be willing to pay an amount P that, if invested for t years, would yield A dollars. We call P the *present value of A dollars in t years*. We shall assume continuous compounding of interest. If the current (annual) rate of interest is r, then P dollars invested for t years will yield Pe^{rt} dollars (see Section 5.2). That is,

$$Pe^{rt} = A.$$

Thus the formula for the present value of A dollars in t years at interest rate r is

$$P = Ae^{-rt}.$$

EXAMPLE 1   (*Present Value of an Income Stream*) Consider a small printing company that does most of its work on one printing press. The firm's profits are directly influenced by the amount of material that the press can produce (assuming that other factors, such as wages, are held constant). We may say that the press is producing a continuous stream of income for the company. Of course, the efficiency of the press may decline as it gets older. At time t, let $K(t)$ be the annual rate of income from the press. [This means that the press is producing $K(t) \cdot \frac{1}{365}$ dollars per day at time t.] Find a model for the present value of the income generated by the printing press over the next T years, assuming an interest rate r (with interest compounded continuously).

Solution Let us divide the T-year period into n small subintervals of time, each of duration Δt years. (If each subinterval were 1 day, for example, then Δt would equal $\frac{1}{365}$.)

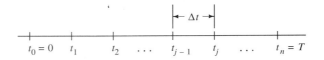

We now consider the income produced by the printing press during a small time interval from t_{j-i} to t_j. Since Δt is small, the rate $K(t)$ of income production changes by only a negligible amount in that interval and can be considered approximately equal to $K(t_j)$. Since $K(t_j)$ gives an annual rate of income, the actual income produced during the period of Δt years is $K(t_j)\,\Delta t$. This income will be produced at approximately time t_j (i.e., t_j years from $t = 0$), so its present value is

$$[K(t_j)\,\Delta t]e^{-rt_j}$$

The present value of the total income produced over the T-year period is the sum of the present values of the amounts produced during each time subinterval; that is

$$K(t_1)e^{-rt_1}\,\Delta t \; + \; K(t_2)e^{-rt_2}\,\Delta t \; + \; \cdots \; + \; K(t_n)e^{-rt_n}\,\Delta t. \tag{2}$$

As the number of subintervals gets large, the length Δt of each subinterval becomes small, and the sum in (2) approaches the integral

$$\int_0^T K(t)e^{-rt}\,dt. \tag{3}$$

We call this quantity the *present value of the stream of income* produced by the printing press over the period from $t = 0$ to $t = T$ years. (The interest rate r used to compute present value is often called the company's *internal rate of return*.) ●

The concept of the present value of a continuous stream of income is an important tool in management decision processes involving the selection or replacement of equipment. It is also useful when analyzing various investment opportunities. Even when $K(t)$ is a simple function, the evaluation of the integral in (3) usually requires special techniques, such as integration by parts, as we see in the next example.

EXAMPLE 2 A company estimates that the rate of revenue produced by a machine at time t will be $5000 - 100t$ dollars per year. Find the present value of this continuous stream of income over the next 4 years at a 16% interest rate.

Solution We use (3) with $K(t) = 5000 - 100t$, $T = 4$, and $r = .16$. The present value of this income stream is

$$\int_0^4 (5000 - 100t)e^{-.16t}\, dt.$$

Using integration by parts, with $f(t) = 5000 - 100t$ and $g(t) = e^{-.16t}$, we find that the preceding integral equals

$$(5000 - 100t)\frac{1}{-.16}e^{-.16t}\Big|_0^4 - \int_0^4 (-100)\frac{1}{-.16}e^{-.16t}\, dt$$

$$\approx 16{,}090 - \frac{100}{.16}\cdot\frac{1}{-.16}e^{-.16t}\Big|_0^4$$

$$\approx 16{,}090 - 1847 = \$14{,}243. \quad \bullet$$

A slight modification of the discussion in Example 1 gives the following general result.

Present Value of a Continuous Stream of Income

$$[\text{present value}] = \int_{T_1}^{T_2} K(t)e^{-rt}\, dt,$$

where

1. $K(t)$ dollars per year is the annual rate of income at time t:
2. r is the annual interest rate of invested money:
3. T_1 to T_2 (years) is the time period of income stream:

EXAMPLE 3 (*A Demographic Model*) It has been determined* that in 1940 the population density t miles from the center of New York City was approximately $120e^{-.2t}$ thousand people per square mile. Estimate the number of people who lived within 2 miles of the center of New York in 1940.

Solution Let us choose a fixed line emanating from the center of the city along which to measure distance. Subdivide this line from $t = 0$ to $t = 2$ into a large number

---

*C. Clark, "Urban Population Densities," *Journal of the Royal Statistical Society,* Series A, 114 (1951), 490–496. See also M. J. White, "On Cumulative Urban Growth and Urban Density Functions," *Journal of Urban Economics,* 4 (1977), 104–112.

FIGURE I **A ring around the city center.**

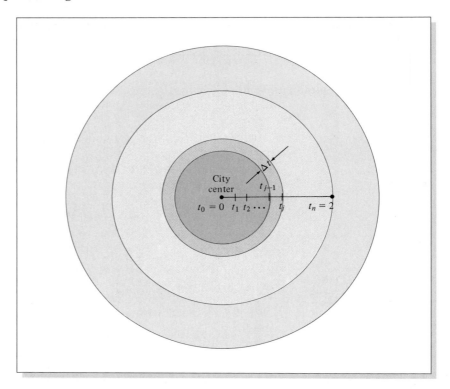

of subintervals, each of length Δt. Each subinterval determines a ring. (See Fig. 1.) Let us estimate the population of each ring and then add these populations together to obtain the total population. Suppose that j is an index ranging from 1 to n. If the outer circle of the jth ring is at a distance t_j from the center, the inner circle of that ring is at a distance $t_{j-1} = t_j - \Delta t$ from the center. The area of the jth ring is

$$\pi t_j^2 - \pi t_{j-1}^2 = \pi t_j^2 - \pi(t_j - \Delta t)^2$$
$$= \pi t_j^2 - \pi[t_j^2 - 2t_j \Delta t + (\Delta t)^2]$$
$$= 2\pi t_j \Delta t - \pi(\Delta t)^2.$$

Assume that Δt is very small. Then $\pi(\Delta t)^2$ is much smaller than $2\pi t_j \Delta t$. Hence the area of this ring is very close to $2\pi t_j \Delta t$.

Within the jth ring, the density of people is about $120e^{-.2t_j}$ thousand people per square mile. So the number of people in this ring is approximately

[population density] $\cdot$ [area of ring] $\approx 120e^{-.2t_j} \cdot 2\pi t_j \Delta t$

$$= 240\pi t_j e^{-.2t_j} \Delta t.$$

Adding up the populations of all the rings, we obtain a total of

$$240\pi t_1 e^{-.2t_1} \Delta t + 240\pi t_2 e^{-.2t_2} \Delta t + \cdots + 240\pi t_n e^{-.2t_n} \Delta t,$$

which is a Riemann sum for the function $f(t) = 240\pi t e^{-.2t}$ over the interval from $t = 0$ to $t = 2$. This approximation to the population improves as the number n increases. Thus the number of people (in thousands) who lived within 2 miles of the center of the city was

$$\int_0^2 240\pi t e^{-.2t} \, dt = 240\pi \int_0^2 t e^{-.2t} \, dt.$$

The last integral can be computed using integration by parts to obtain

$$240\pi \int_0^2 t e^{-.2t} \, dt = 240\pi \left. \frac{t e^{-.2t}}{-.2} \right|_0^2 - 240\pi \int_0^2 \frac{e^{-.2t}}{-.2} \, dt$$

$$= -2400\pi e^{-.4} + 1200\pi \left. \left(\frac{e^{-.2t}}{-.2}\right) \right|_0^2$$

$$= -2400\pi e^{-.4} + (-6000\pi e^{-.4} + 6000\pi)$$

$$\approx 1160.$$

Thus, in 1940, approximately 1,160,000 people lived within 2 miles of the center of the city. ●

An argument analogous to that in Example 3 leads to the following result.

Total Population in a Ring Around the City Center

$$[\text{population}] = \int_a^b 2\pi t D(t) \, dt,$$

where

1. $D(t)$ is density of population (in persons per square mile) at distance t miles from city center,
2. Ring includes all persons who live between a and b miles from the city center.

EXAMPLE 4 (*A Second Demographic Model*)   Let $P(t)$ denote the number of people living at time t in a certain metropolitan area. Suppose that the dynamics of population change are governed by two rules:

1. At time t, new people are arriving in the area (by birth and by moving van) at a rate $r(t)$ per year.
2. During any time interval, say from T_1 to T_2, a certain fraction of the original $P(T_1)$ people present at time T_1 will still be present at time T_2 (the others having died or moved away). Suppose that we write the number still present as $h(T_2 - T_1)P(T_1)$, where $h(T_2 - T_1)$ is a fraction between 0 and 1 that depends only on the length $T_2 - T_1$ of the time interval.

Construct a model for the size of the population at an arbitrary time T.

Solution   We divide the time interval from 0 to T into n equal subintervals, each of length Δt. Of the original population of $P(0)$ people, only $h(T)P(0)$ will remain at time T, by rule 2. Let us analyze what happens during the time interval from t_{j-1} to t_j. If Δt is small, the rate of new arrivals in the time interval from t_{j-1} to t_j will be nearly constant and equal to $r(t_j)$. The number of arrivals in the time period of length Δt will be $r(t_j)\,\Delta t$. How many of these people will remain from time t_j until the time T? By rule 2, this number is

$$h(T - t_j)r(t_j)\Delta t.$$

Therefore, the total population at time T is approximately

$$h(T)P(0) + h(T - t_1)r(t_1)\,\Delta t + h(T - t_2)r(t_2)\,\Delta t + \cdots$$

$$+ \, h(T - t_n)r(t_n)\,\Delta t \approx h(T)P(0) + \int_0^T h(T - t)r(t)\,dt.$$

So, letting n get large, we see that

$$P(T) = h(T)P(0) + \int_0^T h(T - t)r(t)\,dt. \tag{4}$$

•

EXAMPLE 5   Suppose that in Example 4 we have $r(t) = 50{,}000 + 2000t$ and $h(t) = e^{-t/40}$. Determine the population at the end of 10 years if the initial population is 10^6.

Solution   Using (4), we have

$$P(10) = h(10)P(0) + \int_0^{10} h(10 - t)r(t)\,dt$$

$$= e^{-1/4}\cdot 10^6 + \int_0^{10} e^{-(10-t)/40}(50{,}000 + 2000t)\,dt$$

$$\approx 778{,}800 + \int_0^{10} e^{(t-10)/40}(50{,}000 + 2000t)\,dt.$$

We evaluate the integral by using integration by parts, with $f(t) = 50{,}000 + 2000t$ and $g(t) = e^{(t-10)/40}$. After some calculations we obtain

$$P(10) \approx 778{,}800 + 534{,}560 = 1{,}313{,}360. \quad \bullet$$

EXERCISES 9.5

1. Find the present value of a continuous stream of income over 5 years when the rate of income is constant at 40 thousand dollars per year and the interest rate is 8%.

2. A continuous stream of income is being produced at the constant rate of $60,000 per year. Find the present value of the income generated during the time from $t = 2$ to $t = 5$ years, with a 12% interest rate.

3. Find the present value of a continuous stream of income over 3 years if the rate of income, $K(t)$, is $80e^{-.09t}$ thousand dollars per year at time t and the interest rate is 11%.

4. Find the present value of a continuous stream of income over 4 years if the rate of income is $25e^{-.02t}$ thousand dollars per year at time t and the interest rate is 8%.

5. A "growth company" is one whose net earnings tend to increase each year. Suppose that the net earnings of a company at time t are being generated at the rate of $30 + 5t$ million dollars per year. Find the present value of these earnings over the next 2 years, using a 10% interest rate.

6. Find the present value of a stream of earnings generated over the next 2 years at the rate of $50 + 7t$ thousand dollars per year at time t, assuming the interest rate is 10%.

7. In 1900 the population density of Philadelphia t miles from the city center was $120e^{-.65t}$ thousand people per square mile. Calculate the number of people who lived within 5 miles of the city center.

8. Use the population density function from Exercise 7 to calculate the number of people who lived between 3 and 5 miles of the city center.

9. The population density of Philadelphia in 1940 was given by the function $60e^{-.4t}$. Calculate the number of people who lived within 5 miles of the city center. Sketch the graphs of the population densities for 1900 and 1940 on a common graph. What trend do the graphs exhibit?

10. A volcano erupts and spreads lava in all directions. The density of the deposits at a distance t kilometers from the center is $D(t)$ thousand tons per square kilometer, where $D(t) = 11(t^2 + 10)^{-2}$. Find the tonnage of lava deposited between the distances of 1 and 10 kilometers from the center.

11. Suppose that in Example 4 we have $r(t) = 10,000 - 100t$ and $h(t) = e^{-t/50}$. Determine the population at the end of 10 years if the initial population is 250,000.

12. Suppose that in Example 4 the rate $r(t)$ of new arrivals is a constant 10,000 per year (no matter how large the population becomes) and also suppose that $h(t) = e^{-t/40}$.

    (a) Suppose that the initial population is 300,000 and use (4) to determine a formula for $P(T)$, where T is arbitrary.

    (b) What happens to the size of the population in part (a) as T gets larger and larger?

13. Suppose the population density function for a city is $40e^{-.5t}$ thousand people per square mile. Let $P(t)$ be the total population that lives within t miles of the city center, and let Δt be a small positive number.

    (a) Consider the ring about the city whose inner circle is at t miles and outer circle is at $t + \Delta t$ miles. The text shows that the area of this ring is approximately $2\pi t \, \Delta t$ square miles. Approximately how many people live within this ring? (Your answer will involve t and Δt.)

(b) What does $\dfrac{P(t + \Delta t) - P(t)}{\Delta t}$ approach as Δt tends to zero?

(c) What does the quantity $P(5 + \Delta t) - P(5)$ represent?

(d) Use (a) and (c) to find a formula for $\dfrac{P(t + \Delta t) - P(t)}{\Delta t}$ and from that obtain an approximate formula for the derivative $P'(t)$. This formula gives the rate of change of total population with respect to the distance t from the city center.

(e) Given two positive numbers a and b, find a formula, involving a definite integral, for the number of people who live in the city between a miles and b miles of the city center. [*Hint:* Use (d) and the fundamental theorem of calculus to compute $P(b) - P(a)$.]

9.6 IMPROPER INTEGRALS

In applications of calculus, especially to statistics, it is often necessary to consider the area of a region that extends infinitely far to the right or left along the x- axis. We have drawn several such regions in Fig. 1. The areas of such "infinite" regions may be computed using *improper integrals*.

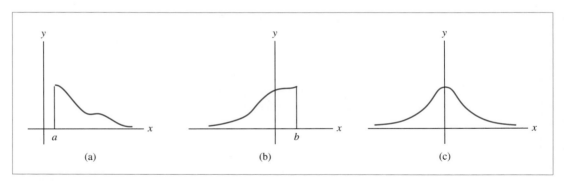

FIGURE I

In order to motivate the idea of an improper integral, let us attempt to calculate the area under the curve $y = 3/x^2$ to the right of $x = 1$ (Fig. 2).

First, we shall compute the area under the graph of this function from $x = 1$ to $x = b$, where b is some number greater than 1. [See Fig. 3(a).] Then we shall examine how the area increases as we let b get larger, as in Fig. 3(b) and 3(c). The area from 1 to b is given by

$$\int_1^b \frac{3}{x^2}\, dx = -\frac{3}{x}\Big|_1^b = \left(-\frac{3}{b}\right) - \left(-\frac{3}{1}\right) = 3 - \frac{3}{b}.$$

When b is large, $3/b$ is small and the integral nearly equals 3. That is, the area under the curve from 1 to b nearly equals 3. (See Table 1.) In fact, the area gets

FIGURE 2

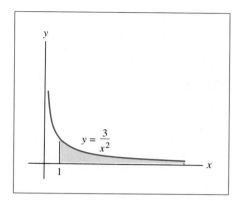

FIGURE 3

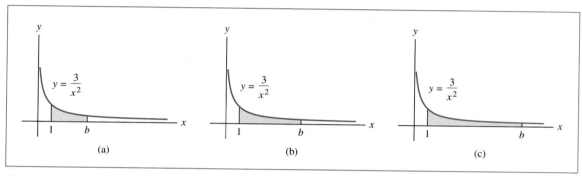

(a)   (b)   (c)

TABLE I

b	Area $= \displaystyle\int_1^b \frac{3}{x^2}\,dx = 3 - \frac{3}{b}$
10	2.7000
100	2.9700
1,000	2.9970
10,000	2.9997

arbitrarily close to 3 as b gets larger. Thus it is reasonable to say that the region under the curve $y = 3/x^2$ for $x \geq 1$ has area 3.

Recall from Chapter 1 that we write $b \to \infty$ as shorthand for "b gets arbitrarily large, without bound." Then, to express the fact that the value of $\displaystyle\int_1^b \frac{3}{x^2}\,dx$ approaches 3 as $b \to \infty$, we write

$$\int_1^\infty \frac{3}{x^2}\,dx = \lim_{b \to \infty} \int_1^b \frac{3}{x^2}\,dx = 3.$$

We call $\int_1^\infty \dfrac{3}{x^2}\,dx$ an *improper* integral because the upper limit of the integral is ∞ (infinity) rather than a finite number.

Definition   Let a be fixed and suppose that $f(x)$ is a nonnegative function for $x \ge a$. If $\lim\limits_{b \to \infty} \int_a^b f(x)\,dx = L$, we define

$$\int_a^\infty f(x)\,dx = \lim_{b \to \infty} \int_a^b f(x)\,dx = L.$$

We say that the improper integral $\int_a^\infty f(x)\,dx$ is *convergent* and that the region under the curve $y = f(x)$ for $x \ge a$ has area L. (See Fig. 4.)

It is possible to consider improper integrals in which $f(x)$ is both positive and negative. However, we shall consider only nonnegative functions, since this is the case occurring in most applications.

FIGURE 4   Area defined by an improper integral.

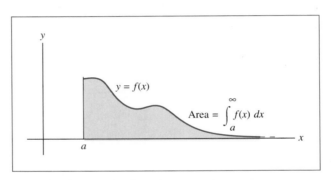

EXAMPLE 1   Find the area under the curve $y = e^{-x}$ for $x \ge 0$ (Fig. 5).

FIGURE 5

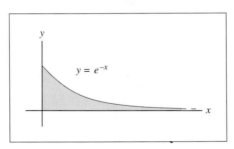

Solution   We must calculate the improper integral

$$\int_0^\infty e^{-x}\,dx.$$

We take $b > 0$ and compute

$$\int_0^b e^{-x}\,dx = -e^{-x}\Big|_0^b = (-e^{-b}) - (-e^0) = 1 - e^{-b} = 1 - \frac{1}{e^b}.$$

We now consider the limit as $b \to \infty$ and note that $1/e^b$ approaches zero. Thus

$$\int_0^\infty e^{-x} \, dx = \lim_{b \to \infty} \int_0^b e^{-x} \, dx = \lim_{b \to \infty} \left(1 - \frac{1}{e^b}\right) = 1.$$

Therefore, the region in Fig. 5 has area 1. ●

EXAMPLE 2   Evaluate the improper integral $\displaystyle\int_7^\infty \frac{1}{(x-5)^2} \, dx$.

Solution $\displaystyle\int_7^b \frac{1}{(x-5)^2} \, dx = -\frac{1}{x-5} \Big|_7^b = -\frac{1}{b-5} - \left(-\frac{1}{7-5}\right) = \frac{1}{2} - \frac{1}{b-5}.$

As $b \to \infty$, the fraction $1/(b-5)$ approaches zero, so

$$\int_7^\infty \frac{1}{(x-5)^2} \, dx = \lim_{b \to \infty} \int_7^b \frac{1}{(x-5)^2} \, dx = \lim_{b \to \infty} \left(\frac{1}{2} - \frac{1}{b-5}\right) = \frac{1}{2}. \ ●$$

Not every improper integral is convergent. If the value $\int_a^b f(x) \, dx$ does not have a limit as $b \to \infty$, we cannot assign any numerical value to $\int_a^\infty f(x) \, dx$, and we say that the improper integral $\int_a^\infty f(x) \, dx$ is *divergent*.

EXAMPLE 3   Show that $\displaystyle\int_1^\infty \frac{1}{\sqrt{x}} \, dx$ is divergent.

Solution   For $b > 1$ we have

$$\int_1^b \frac{1}{\sqrt{x}} \, dx = 2\sqrt{x} \Big|_1^b = 2\sqrt{b} - 2. \tag{1}$$

As $b \to \infty$, the quantity $2\sqrt{b} - 2$ increases without bound. That is, $2\sqrt{b} - 2$ can be made larger than any specific number. Therefore, $\displaystyle\int_1^b \frac{1}{\sqrt{x}} \, dx$ has no limit as $b \to \infty$, so $\displaystyle\int_1^\infty \frac{1}{\sqrt{x}} \, dx$ is divergent. ●

In some cases it is necessary to consider improper integrals of the form

$$\int_{-\infty}^b f(x) \, dx.$$

Let b be fixed and examine the value of $\int_a^b f(x) \, dx$ as $a \to -\infty$, that is, as a moves arbitrarily far to the left on the number line. If $\lim_{a \to -\infty} \int_a^b f(x) \, dx = L$, we say that the improper integral $\int_{-\infty}^b f(x) \, dx$ is *convergent* and we write

$$\int_{-\infty}^b f(x) \, dx = L.$$

Otherwise, the improper integral is divergent. An integral of the form $\int_{-\infty}^{b} f(x)\, dx$ may be used to compute the area of a region such as that shown in Fig. 1(b).

EXAMPLE 4   Determine if $\int_{-\infty}^{0} e^{5x}\, dx$ is convergent. If convergent, find its value.

Solution

$$\int_{-\infty}^{0} e^{5x}\, dx = \lim_{a \to -\infty} \int_{a}^{0} e^{5x}\, dx$$

$$= \lim_{a \to -\infty} \left. \frac{1}{5} e^{5x} \right|_{a}^{0}$$

$$= \lim_{a \to -\infty} \left(\frac{1}{5} - \frac{1}{5} e^{5a} \right).$$

As $a \to -\infty$, e^{5a} approaches 0 so that $\frac{1}{5} - \frac{1}{5} e^{5a}$ approaches $\frac{1}{5}$. Thus the improper integral converges and has value $\frac{1}{5}$. ●

Areas of regions that extend infinitely far to the left *and* right, such as the region in Fig. 1(c), are calculated using improper integrals of the form

$$\int_{-\infty}^{\infty} f(x)\, dx.$$

We define such an integral to have the value

$$\int_{-\infty}^{0} f(x)\, dx + \int_{0}^{\infty} f(x)\, dx,$$

provided that both of the latter improper integrals are convergent.

An important area that arises in probability theory is the area under the so-called normal curve, whose equation is

$$y = \frac{1}{\sqrt{2\pi}} e^{-x^2/2}.$$

FIGURE 6   **The standard normal curve.**

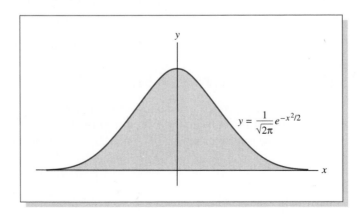

$$y = \frac{1}{\sqrt{2\pi}} e^{-x^2/2}$$

(See Fig. 6.) It is of fundamental importance for probability theory that this area is 1. In terms of an improper integral, this fact may be written

$$\int_{-\infty}^{\infty} \frac{1}{\sqrt{2\pi}} e^{-x^2/2} \, dx = 1.$$

The proof of this result is beyond the scope of this book.

TECHNOLOGY PROJECT 9.5
Calculating Improper Integrals Using a Symbolic Math Program

Most symbolic math programs allow you to use ∞, where appropriate, within commands. In *Mathematica*, ∞ is denoted by `Infinity`. So, for example, to calculate the definite integral

$$\int_{2}^{\infty} \frac{\ln x}{x^2} \, dx$$

you could use the command:

```
Integrate[Log[x]/x^2,{x,2,Infinity}]
```

Use a symbolic math program to calculate the values of the following improper integrals:

1. $\displaystyle\int_{1}^{\infty} 41e^{-.03x} \, dx$

2. $\displaystyle\int_{1}^{\infty} (3x^{-2} + 4x^{-3})^4 \, dx$

3. $\displaystyle\int_{0}^{\infty} x^3 e^{-5x} \, dx$

4. $\displaystyle\int_{1}^{\infty} \frac{\ln x}{x^3} \, dx$

PRACTICE PROBLEMS 9.6

1. Does $1 - 2(1 - 3b)^{-4}$ approach a limit as $b \to \infty$?

2. Evaluate $\displaystyle\int_{1}^{\infty} \frac{x^2}{x^3 + 8} \, dx$.

3. Evaluate $\displaystyle\int_{-\infty}^{-2} \frac{1}{x^4} \, dx$.

EXERCISES 9.6

In Exercises 1–14, determine if the given expression approaches a limit as $b \to \infty$, and find that number when it exists.

1. $\dfrac{5}{b}$

2. b^2

3. $-3e^{2b}$

4. $\dfrac{1}{b} + \dfrac{1}{3}$

5. $\dfrac{1}{4} - \dfrac{1}{b^2}$

6. $\dfrac{1}{2}\sqrt{b}$

7. $2 - (b + 1)^{-1/2}$

8. $\dfrac{2}{b} - \dfrac{3}{b^{3/2}}$

9. $5 - \dfrac{1}{b - 1}$ **10.** $5(b^2 + 3)^{-1}$ **11.** $6 - 3b^{-2}$ **12.** $e^{-b/2} + 5$

13. $2(1 - e^{-3b})$ **14.** $4(1 - b^{-3/4})$

15. Find the area under the graph of $y = 1/x^2$ for $x \geq 2$.

16. Find the area under the graph of $y = (x + 1)^{-2}$ for $x \geq 0$.

17. Find the area under the graph of $y = e^{-x/2}$ for $x \geq 0$.

18. Find the area under the graph of $y = 4e^{-4x}$ for $x \geq 0$.

19. Find the area under the graph of $y = (x + 1)^{-3/2}$ for $x \geq 3$.

20. Find the area under the graph of $y = (2x + 6)^{-4/3}$. See Fig. 7.

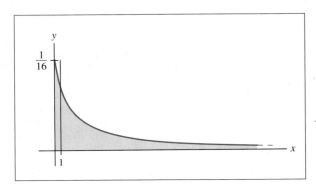

FIGURE 7

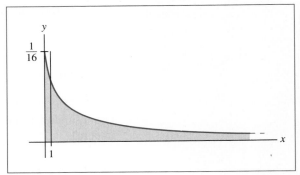

FIGURE 8

21. Show that the region under the graph of $y = (14x + 18)^{-4/5}$ for $x \geq 1$ cannot be assigned any finite number as its area. See Fig. 8.

22. Show that the region under the graph of $y = (x - 1)^{-1/3}$ for $x \geq 2$ cannot be assigned any finite number as its area.

Evaluate the following improper integrals whenever they are convergent.

23. $\displaystyle\int_1^\infty \frac{1}{x^3}\, dx$ **24.** $\displaystyle\int_1^\infty \frac{2}{x^{3/2}}\, dx$ **25.** $\displaystyle\int_0^\infty e^{2x}\, dx$

26. $\displaystyle\int_0^\infty (x^2 + 1)\, dx$ **27.** $\displaystyle\int_0^\infty \frac{1}{(2x + 3)^2}\, dx$ **28.** $\displaystyle\int_0^\infty e^{-3x}\, dx$

29. $\displaystyle\int_2^\infty \frac{1}{(x - 1)^{5/2}}\, dx$ **30.** $\displaystyle\int_2^\infty e^{2-x}\, dx$ **31.** $\displaystyle\int_0^\infty .01e^{-.01x}\, dx$

32. $\displaystyle\int_0^\infty \frac{4}{(2x + 1)^3}\, dx$ **33.** $\displaystyle\int_0^\infty 6e^{1-3x}\, dx$ **34.** $\displaystyle\int_1^\infty e^{-.2x}\, dx$

35. $\displaystyle\int_3^\infty \frac{x^2}{\sqrt{x^3 - 1}}\, dx$ **36.** $\displaystyle\int_2^\infty \frac{1}{x \ln x}\, dx$ **37.** $\displaystyle\int_0^\infty xe^{-x^2}\, dx$

38. $\displaystyle\int_0^\infty \frac{x}{x^2 + 1}\, dx$ **39.** $\displaystyle\int_0^\infty 2x(x^2 + 1)^{-3/2}\, dx$ **40.** $\displaystyle\int_1^\infty (5x + 1)^{-4}\, dx$

41. $\displaystyle\int_{-\infty}^0 e^{4x}\, dx$ **42.** $\displaystyle\int_{-\infty}^0 \frac{8}{(x - 5)^2}\, dx$ **43.** $\displaystyle\int_{-\infty}^0 \frac{6}{(1 - 3x)^2}\, dx$

44. $\displaystyle\int_{-\infty}^0 \frac{1}{\sqrt{4 - x}}\, dx$ **45.** $\displaystyle\int_0^\infty \frac{e^{-x}}{(e^{-x} + 2)^2}\, dx$ **46.** $\displaystyle\int_{-\infty}^\infty \frac{e^{-x}}{(e^{-x} + 2)^2}\, dx$

47. If $k > 0$, show that $\displaystyle\int_0^\infty ke^{-kx}\, dx = 1$.

48. If $k > 0$, show that $\displaystyle\int_1^\infty \frac{k}{x^{k+1}}\, dx = 1$.

49. The *capital value* of an asset such as a machine is sometimes defined as the present value of all future net earnings of the asset. (See Section 9.5.) The actual lifetime of the asset may not be known, and since some assets may last indefinitely, the capital value of the asset may be written in the form $\int_0^\infty K(t)e^{-rt}\, dt$, where $K(t)$ is the annual rate of earnings produced by the asset at time t, and where r is the annual rate of interest, compounded continuously. Find the capital value of an asset that generates income at the rate of \$5000 per year, assuming an interest rate of 10%.

50. Find the capital value of an asset that at time t is producing income at the rate of $6000e^{.04t}$ dollars per year, assuming an interest rate of 16%.

**SOLUTIONS TO
PRACTICE
PROBLEMS 9.6**

1. The expression $1 - 2(1 - 3b)^{-4}$ may also be written in the form

$$1 - \frac{2}{(1 - 3b)^4}.$$

When b is large, $(1 - 3b)^4$ is very large, so $2/(1 - 3b)^4$ is very small. Thus $1 - 2(1 - 3b)^{-4}$ approaches 1 as $b \to \infty$.

2. The first step is to find an antiderivative of $x^2/(x^3 + 8)$. Using the substitution $u = x^3 + 8, du = 3x^2\, dx$, we obtain

$$\int \frac{x^2}{x^3 + 8}\, dx = \frac{1}{3}\int \frac{1}{u}\, du = \frac{1}{3}\ln|u| + C = \frac{1}{3}\ln|x^3 + 8| + C.$$

Now,

$$\int_1^b \frac{x^2}{x^3 + 8}\, dx = \frac{1}{3}\ln|x^3 + 8|\,\Big|_1^b = \frac{1}{3}\ln(b^3 + 8) - \frac{1}{3}\ln 9.$$

Finally, we examine what happens as $b \to \infty$. Certainly, $b^3 + 8$ gets arbitrarily large, so $\ln(b^3 + 8)$ must also get arbitrarily large. Hence

$$\int_1^b \frac{x^2}{x^3 + 8}\, dx$$

has no limit as $b \to \infty$, so the improper integral

$$\int_1^\infty \frac{x^2}{x^3 + 8}\, dx$$

is divergent.

3.
$$\int_a^{-2} \frac{1}{x^4}\, dx = \int_a^{-2} x^{-4}\, dx = \left. \frac{x^{-3}}{-3} \right|_a^{-2} = \left. \frac{1}{-3x^3} \right|_a^{-2}$$

$$= \frac{1}{-3(-2)^3} - \left(\frac{1}{-3 \cdot a^3} \right)$$

$$= \frac{1}{24} + \frac{1}{3a^3}$$

$$\int_{-\infty}^{2} \frac{1}{x^4}\, dx = \lim_{a \to -\infty} \int_a^{-2} \frac{1}{x^4}\, dx = \lim_{a \to -\infty} \left(\frac{1}{24} + \frac{1}{3a^3} \right) = \frac{1}{24}.$$

✓

CHAPTER 9 CHECKLIST

✓ Integration by substitution:

$$\int f(g(x))g'(x)\, dx = \int f(u)\, du, \qquad u = g(x)$$

$$\int_a^b f(g(x))g'(x)\, dx = \int_{g(a)}^{g(b)} f(u)\, du$$

✓ Integration by parts:

$$\int f(x)g(x)\, dx = f(x)G(x) - \int f'(x)G(x)\, dx, \qquad G'(x) = g(x)$$

$$\int_a^b f(x)g(x)\, dx = \left. f(x)G(x) \right|_a^b - \int_a^b f'(x)G(x)\, dx$$

✓ Midpoint rule (using midpoints of the subintervals):

$$\int_a^b f(x)\, dx \approx [f(x_1) + f(x_2) + \cdots + f(x_n)]\, \Delta x$$

✓ Trapezoidal rule (using endpoints of the subintervals):

$$\int_a^b f(x)\, dx \approx [f(a_0) + 2f(a_1) + \cdots + 2f(a_{n-1}) + f(a_n)]\frac{\Delta x}{2}$$

✓ Simpson's rule: $\displaystyle \int_a^b f(x)\, dx \approx \frac{2M + T}{3}$

✓ Improper integrals: $\displaystyle \int_a^{\infty} f(x)\, dx = \lim_{b \to \infty} \int_a^b f(x)\, dx$

CHAPTER 9   SUPPLEMENTARY EXERCISES

Determine the following indefinite integrals.

1. $\displaystyle \int x \sin 3x^2\, dx$

2. $\displaystyle \int \sqrt{2x + 1}\, dx$

3. $\displaystyle \int x(1 - 3x^2)^5\, dx$

4. $\int \dfrac{(\ln x)^5}{x} \, dx$

5. $\int \dfrac{(\ln x)^2}{x} \, dx$

6. $\int \dfrac{1}{\sqrt{4x + 3}} \, dx$

7. $\int x\sqrt{4 - x^2} \, dx$

8. $\int x \sin 3x \, dx$

9. $\int x^2 e^{-x^3} \, dx$

10. $\int \dfrac{x \ln (x^2 + 1)}{x^2 + 1} \, dx$

11. $\int x^2 \cos 3x \, dx$

12. $\int \dfrac{\ln(\ln x)}{x \ln x} \, dx$

13. $\int \ln x^2 \, dx$

14. $\int x\sqrt{x + 1} \, dx$

15. $\int \dfrac{x}{\sqrt{3x - 1}} \, dx$

16. $\int x^2 \ln x^2 \, dx$

17. $\int \dfrac{x}{(1 - x)^5} \, dx$

18. $\int x(\ln x)^2 \, dx$

In Exercises 19–36, decide whether integration by parts or a substitution should be used to compute the indefinite integral. If substitution, indicate the substitution to be made. If by parts, indicate the functions $f(x)$ and $g(x)$ to be used in formula (1) of Section 9.2.

19. $\int xe^{2x} \, dx$

20. $(x - 3)e^{-x} \, dx$

21. $\int (x + 1)^{-1/2} e^{\sqrt{x+1}} \, dx$

22. $\int x^2 \sin (x^3 - 1) \, dx$

23. $\int \dfrac{x - 2x^3}{x^4 - x^2 + 4} \, dx$

24. $\int \ln \sqrt{5 - x} \, dx$

25. $\int e^{-x}(3x - 1)^2 \, dx$

26. $\int xe^{3 - x^2} \, dx$

27. $\int (500 - 4x)e^{-x/2} \, dx$

28. $x^{5/2} \ln x \, dx$

29. $\int \sqrt{x + 2} \ln(x + 2) \, dx$

30. $\int (x + 1)^2 e^{3x} \, dx$

31. $\int (x + 3)e^{x^2 + 6x} \, dx$

32. $\int \sin^2 x \cos x \, dx$

33. $\int x \cos(x^2 - 9) \, dx$

34. $\int (3 - x) \sin 3x \, dx$

35. $\int \dfrac{2 - x^2}{x^3 - 6x} \, dx$

36. $\int \dfrac{1}{x(\ln x)^{3/2}} \, dx$

Evaluate the following definite integrals.

37. $\displaystyle\int_0^1 \dfrac{2x}{(x^2 + 1)^3} \, dx$

38. $\displaystyle\int_0^{\pi/2} x \sin 8x \, dx$

39. $\displaystyle\int_0^2 xe^{-(1/2)x^2} \, dx$

40. $\displaystyle\int_{1/2}^1 \dfrac{\ln(2x + 3)}{2x + 3} \, dx$

41. $\displaystyle\int_1^2 xe^{-2x} \, dx$

42. $\displaystyle\int_1^2 x^{-3/2} \ln x \, dx$

Approximate the following definite integrals by the midpoint rule, the trapezoidal rule, and Simpson's rule.

43. $\int_1^9 \dfrac{1}{\sqrt{x}}\, dx; \; n = 4$

44. $\int_0^{10} e^{\sqrt{x}}\, dx; \; n = 5$

45. $\int_1^4 \dfrac{e^x}{x + 1}\, dx; \; n = 5$

46. $\int_{-1}^{1} \dfrac{1}{1 + x^2}\, dx; \; n = 5$

Evaluate the following improper integrals whenever they are convergent.

47. $\int_0^{\infty} e^{6-3x}\, dx$

48. $\int_1^{\infty} x^{-2/3}\, dx$

49. $\int_1^{\infty} \dfrac{x + 2}{x^2 + 4x - 2}\, dx$

50. $\int_0^{\infty} x^2 e^{-x^3}\, dx$

51. $\int_{-1}^{\infty} (x + 3)^{-5/4}\, dx$

52. $\int_{-\infty}^{0} \dfrac{8}{(5 - 2x)^3}\, dx$

53. It can be shown that $\lim\limits_{b \to \infty} b e^{-b} = 0$. Use this fact to compute $\int_1^{\infty} x e^{-x}\, dx$.

54. Let k be a positive number. It can be shown that $\lim\limits_{b \to \infty} b e^{-kb} = 0$. Use this fact to compute $\int_0^{\infty} x e^{-kx}\, dx$.

55. Find the present value of a continuous stream of income over the next 4 years, where the rate of income is $50e^{-.08t}$ thousand dollars per year at time t, and the interest rate is 12%.

56. Suppose that t miles from the center of a certain city, the property tax revenue is approximately $R(t)$ thousand dollars per square mile, where $R(t) = 50e^{-t/20}$. Use this model to predict the total property tax revenue that will be generated by property within 10 miles of the city center.

57. Suppose that a machine requires daily maintenance, and let $M(t)$ be the *annual* rate of maintenance expense at time t. Suppose that the interval $0 \le t \le 2$ is divided into n subintervals, with endpoints $t_0 = 0, t_1, \ldots, t_n = 2$.

(a) Give a Riemann sum that approximates the total maintenance expense over the next 2 years. Then write the integral that the Riemann sum approximates for large values of n.

(b) Give a Riemann sum that approximates the present value of the total maintenance expense over the next 2 years, using a 10% annual interest rate, compounded continuously. Then write the integral that the Riemann sum approximates for large values of n.

58. The *capitalized cost* of an asset is the total of the original cost and the present value of all future "renewals" or replacements. This concept is useful, for example, when selecting equipment that is manufactured by several different companies. Suppose that a corporation computes the present value of future expenditures using an annual interest rate r, with continuous compounding of interest. Assume that the original cost of an asset is $80,000 and the annual renewal expense will be $50,000, spread more or less evenly throughout each year, for a large but indefinite number of years. Find a formula, involving an integral, that gives the capitalized cost of the asset.

DIFFERENTIAL EQUATIONS

A *differential equation* is an equation in which derivatives of an unknown function $y = f(t)$ occur. Examples of such equations are

$$y' = 6t + 3$$
$$y' = 6y$$
$$y' + 3y + t = 0.$$

As we shall see, many physical processes can be described by differential equations. In this chapter we explore some topics in differential equations and use our resulting knowledge to study problems from many different fields, including business, genetics, and ecology.

10.1 SOLUTIONS OF DIFFERENTIAL EQUATIONS

A differential equation is an equation involving an unknown function y and one or more of the derivatives y', y'', y''', and so on. Suppose that y is a function of the variable t. A *solution* of a differential equation is any function $f(t)$ such that the differential equation becomes a true statement when y is replaced by $f(t)$, y' by $f'(t)$, y'' by $f''(t)$, and so forth.

EXAMPLE 1   Show that the function $f(t) = 5e^{-2t}$ is a solution of the differential equation

$$y' + 2y = 0. \tag{1}$$

Solution   The differential equation (1) says that $y' + 2y$ equals zero for all values of t. We must show that this result holds if y is replaced by $5e^{-2t}$ and y' is replaced by $(5e^{-2t})' = -10e^{-2t}$. But

$$\overbrace{(5e^{-2t})'}^{y'} + 2\overbrace{(5e^{-2t})}^{y} = -10e^{-2t} + 10e^{-2t} = 0.$$

Therefore, $y = 5e^{-2t}$ is a solution of the differential equation (1) $\bullet$

EXAMPLE 2   Show that the function $f(t) = \frac{1}{9}t + \sin 3t$ is a solution of the differential equation

$$y'' + 9y = t. \tag{2}$$

Solution   If $f(t) = \frac{1}{9}t + \sin 3t$, then

$$f'(t) = \frac{1}{9} + 3\cos 3t$$

$$f''(t) = -9\sin 3t.$$

Substituting $f(t)$ for y and $f''(t)$ for y'' in the left side of (2), we obtain

$$\overbrace{-9\sin 3t}^{y''} + 9\overbrace{(\tfrac{1}{9}t + \sin 3t)}^{y} = -9\sin 3t + t + 9\sin 3t = t.$$

Therefore, $y'' + 9y = t$ if $y = \frac{1}{9}t + \sin 3t$, and hence $y = \frac{1}{9}t + \sin 3t$ is a solution to $y'' + 9y = t$. $\bullet$

The process of determining all the functions that are solutions of a differential equation is called *solving the differential equation*. The process of antidifferentiation amounts to solving a simple type of differential equation. For example, a solution of the differential equation

$$y' = 3t^2 - 4 \tag{3}$$

is a function y whose derivative is $3t^2 - 4$. Thus, solving (3) consists of finding all antiderivatives of $3t^2 - 4$. Clearly, y must be of the form $y = t^3 - 4t + C$, where C is a constant. The solutions of (3) corresponding to several values of C are sketched in Fig. 1.

We encountered differential equations such as

$$y' = 2y \tag{4}$$

in our discussion of exponential functions. Unlike (3), this equation does not

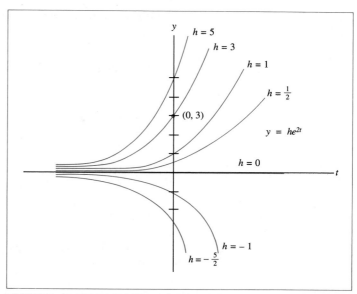

FIGURE 2   Typical solutions of $y' = 2y$.

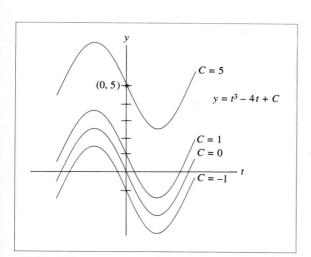

FIGURE I   Typical solutions of $y' = 3t^2 - 4$.

give a specific formula for y' but instead describes a property of y', namely that y' is proportional to y (with 2 as the constant of proportionality). At the moment, the only way we can "solve" (4) is simply to know in advance what the solutions are. Recall from Chapter 4 that the solutions of (4) have the form $y = he^{2t}$ for any constant h. Some typical solutions of (4) are sketched in Fig. 2. In the next section we shall discuss a method for solving a class of differential equations that includes both (3) and (4) as special cases.

Figures 1 and 2 illustrate two important differences between differential equations and algebraic equations (e.g., $ax^2 + bx + c = 0$). First, a solution of a differential equation is a *function* rather than a number. Second, a differential equation usually has infinitely many solutions.

Sometimes we want to find a particular solution that satisfies certain additional conditions called *initial conditions*. Initial conditions specify the values of a solution and a certain number of its derivatives at some specific value of t, often $t = 0$. If the solution to a differential equation is $y = f(t)$, we will often write $y(0)$ for $f(0)$, $y'(0)$ for $f'(0)$, and so on. The problem of determining a solution of a differential equation that satisfies given initial conditions is called an *initial value problem*.

EXAMPLE 3   (a)   Solve the initial value problem $y' = 3t^2 - 4$, $y(0) = 5$.

(b)   Solve the initial value problem $y' = 2y$, $y(0) = 3$.

Solution   (a)   We have already noted that the general solution of $y' = 3t^2 - 4$ is $f(t) = t^3 - 4t + C$. We want the particular solution that satisfies $f(0) = 5$. Geometrically, we are looking for the curve in Fig. 1 that passes through the point $(0, 5)$. Using the general formula for $f(t)$, we have

$$5 = f(0) = (0)^3 - 4(0) + C = C$$

$$C = 5.$$

Thus $f(t) = t^3 - 4t + 5$ is the desired solution.

(b) The general solution of $y' = 2y$ is $y = he^{2t}$. The condition $y(0) = 3$ means that y must be 3 when $t = 0$; that is, the point $(0, 3)$ must be on the graph of the solution to $y' = 2y$. (See Fig. 2.) We have

$$3 = y(0) = he^{2(0)} = h \cdot 1 = h$$

$$h = 3.$$

Thus $y = 3e^{2t}$ is the desired solution. •

A constant function that satisfies a differential equation is called a *constant solution* of the differential equation. Constant solutions occur in many of the applied problems considered later in the chapter.

EXAMPLE 4   Find a constant solution of $y' = 3y - 12$.

Solution   Let $f(t) = c$ for all t. Then $f'(t)$ is zero for all t. If $f(t)$ satisfies the differential equation

$$f'(t) = 3 \cdot f(t) - 12,$$

then

$$0 = 3 \cdot c - 12,$$

and so $c = 4$. This is the only possible value for a constant solution. Substitution shows that the function $f(t) = 4$ is indeed a solution of the differential equation.

•

We conclude this section with an example of how a differential equation may be used to describe a physical process. (The equation can be solved by the method of the next section.) The example should be studied carefully, for it contains the key to understanding many similar problems that will appear in exercises.

EXAMPLE 5   (*Newton's Law of Cooling*) Suppose that a red-hot steel rod is plunged into a bath of cool water. Let $f(t)$ be the temperature of the rod at time t, and suppose that the water is maintained at a constant temperature of 10°C. According to Newton's law of cooling, the rate of change of $f(t)$ is proportional to the difference between the two temperatures 10° and $f(t)$. Find a differential equation that describes this physical law.

Solution   The two key ideas are "rate of change" and "proportional." The rate of change of $f(t)$ is the derivative $f'(t)$. Since this is proportional to the difference $10 - f(t)$, there exists a constant k such that

$$f'(t) = k[10 - f(t)]. \tag{5}$$

The term "proportional" does not tell us whether k is positive or negative (or zero). We must decide this, if possible, from the context of the problem. In the present situation, the steel rod is hotter than the water, so $10 - f(t)$ is negative. Also, $f(t)$ will decrease as time passes, so $f'(t)$ should be negative. Thus, to make $f'(t)$ negative in (5), k must be a positive number. From (5) we see that $y = f(t)$ satisfies a differential equation of the form

$$y' = k(10 - y), \qquad k \text{ a positive constant.} \quad \bullet$$

Throughout this section we have assumed that our solution functions y are functions of the variable t, which in applications usually stands for time. In most of this chapter we will continue to use the variable t. However, if we use another variable occasionally, such as x, we will make the variable explicit by writing $\dfrac{dy}{dx}$ instead of y'.

PRACTICE PROBLEMS 10.1

1. Show that any function of the form $y = Ae^{t^3/3}$, where A is a constant, is a solution of the differential equation

$$y' - t^2y = 0.$$

2. If the function $f(t)$ is a solution of the initial value problem

$$y' = (t + 2)y, \quad y(0) = 3,$$

find $f(0)$ and $f'(0)$.

EXERCISES 10.1

1. Show that the function $f(t) = \frac{3}{2}e^{t^2} - \frac{1}{2}$ is a solution of the differential equation $y' - 2ty = t$.

2. Show that the function $f(t) = t^2 - \frac{1}{2}$ is a solution of the differential equation $(y')^2 - 4y = 2$.

3. Show that the function $f(t) = (e^{-t} + 1)^{-1}$ satisfies $y' + y^2 = y, y(0) = \frac{1}{2}$.

4. Show that the function $f(t) = 5e^{2t}$ satisfies $y'' - 3y' + 2y = 0$, $y(0) = 5$, $y'(0) = 10$.

5. Solve $y' = \frac{1}{2}y, y(0) = 1$.

6. Solve $y' = e^{t/2}, y(0) = 1$.

7. Is the constant function $f(t) = 3$ a solution of the differential equation $y' = 6 - 2y$?

8. Is the constant function $f(t) = -4$ a solution of the differential equation $y' = t^2(y + 4)$?

9. Find a constant solution of $y' = t^2y - 5t^2$.

10. Find two constant solutions of $y' = 4y(y - 7)$.

11. If the function $f(t)$ is a solution of the initial value problem

$$y' = 2y - 3, \qquad y(0) = 4,$$

find $f(0)$ and $f'(0)$.

12. If the function $f(t)$ is a solution to the initial value problem $y' = e^t + y$, $y(0) = 0$, find $f(0)$ and $f'(0)$.

13. Let $f(t)$ denote the amount of capital invested by a certain business firm at time t. The rate of change of invested capital, $f'(t)$, is sometimes called the *rate of net investment*. Suppose that the management of the firm decides that the optimum level of investment should be C dollars and that, at any time, the rate of net investment should be proportional to the difference between C and the total capital invested. Construct a differential equation that describes this situation.

14. A cool object is placed in a room that is maintained at a constant temperature of $20°C$. The rate at which the temperature of the object rises is proportional to the difference between the room temperature and the temperature of the object. Let $y = f(t)$ be the temperature of the object at time t, and give a differential equation that describes the rate of change of $f(t)$.

SOLUTIONS TO PRACTICE PROBLEMS 10.1

1. If $y = Ae^{t^3/3}$, then

$$\overbrace{(Ae^{t^3/3})'}^{y'} - t^2 \overbrace{(Ae^{t^3/3})}^{y} = At^2 e^{t^3/3} - t^2 Ae^{t^3/3} = 0.$$

Therefore, $y' - t^2 y = 0$ if $y = Ae^{t^3/3}$.

2. The initial condition $y(0) = 3$ says that $f(0) = 3$. Since $f(t)$ is a solution to $y' = (t + 2)y$,

$$f'(t) = (t + 2)f(t)$$

and hence

$$f'(0) = (0 + 2)f(0)$$

$$= 2 \cdot 3$$

$$= 6.$$

10.2 SEPARATION OF VARIABLES

Here we describe a technique for solving an important class of differential equations—those of the form

$$y' = p(t)q(y),$$

where $p(t)$ is a function of t only and $q(y)$ is a function of y only. Two equations of this type are non zero

$$y' = \frac{3t^2}{y^2} \qquad\qquad \left[p(t) = 3t^2, \, q(y) = \frac{1}{y^2} \right], \qquad (1)$$

$$y' = e^{-y}(2t + 1) \qquad [p(t) = 2t + 1, \, q(y) = e^{-y}]. \qquad (2)$$

The main feature of such equations is that we may *separate the variables;* that is, we may rewrite the equations so that y occurs only on one side of the equation and t on the other. For example, if we multiply both sides of equation (1) by y^2, the equation becomes

$$y^2 y' = 3t^2;$$

if we multiply both sides of equation (2) by e^y, the equation becomes

$$e^y y' = 2t + 1.$$

It should be pointed out that the differential equation

$$y' = 3t^2 - 4$$

is of the preceding type. Here $p(t) = 3t^2 - 4$ and $q(y) = 1$. The variables are already separated, however. Similarly, the differential equation

$$y' = 5y$$

is of the preceding type, with $p(t) = 5, q(y) = y$. We can separate the variables by writing the equation as

$$\frac{1}{y} y' = 5.$$

In the next example we present a procedure for solving differential equations in which the variables are separated.

EXAMPLE I   Find all solutions of the differential equation $y^2 y' = 3t^2$.

Solution   (a) Write y' as $\dfrac{dy}{dt}$.

$$y^2 \frac{dy}{dt} = 3t^2.$$

(b) Integrate both sides with respect to t.

$$\int y^2 \frac{dy}{dt} \, dt = \int 3t^2 \, dt.$$

(c) Rewrite the left-hand side, "canceling the dt."

$$\int y^2 \, dy = \int 3t^2 \, dt.$$

(d) Calculate the antiderivatives.

$$\frac{1}{3}y^3 + C_1 = t^3 + C_2.$$

(e) Solve for y in terms of t.

$$y^3 = 3(t^3 + C_2 - C_1)$$
$$y = \sqrt[3]{3t^3 + C}, \quad C \text{ a constant.}$$

We can check that this method works by showing that $\dot{y} = \sqrt[3]{3t^3 + C}$ is a solution to $y^2y' = 3t^2$. Since $y = (3t^3 + C)^{1/3}$, we have

$$y' = \frac{1}{3}(3t^3 + C)^{-2/3} \cdot 3 \cdot 3t^2 = 3t^2(3t^3 + C)^{-2/3}$$

$$y^2y' = [(3t^3 + C)^{1/3}]^2 \cdot 3t^2(3t^3 + C)^{-2/3}$$
$$= 3t^2.$$

Discussion of Step (c)   Suppose that $y = f(t)$ is a solution of the differential equation $y^2y' = 3t^2$. Then

$$[f(t)]^2 f'(t) = 3t^2.$$

Integrating, we have

$$\int [f(t)]^2 f'(t) \, dt = \int 3t^2 \, dt.$$

Let us make the substitution $y = f(t)$, $dy = f'(t) \, dt$ in the left-hand side in order to get

$$\int y^2 \, dy = \int 3t^2 \, dt.$$

This is just the result of step (c). The process of "canceling the dt" and integrating with respect to y is actually just equivalent to making the substitution $y = f(t)$, $dy = f'(t) \, dt$. ●

The technique used in Example 1 can be used for any differential equation with separated variables. Suppose that we are given such an equation:

$$h(y)y' = p(t),$$

where $h(y)$ is a function of y only and $p(t)$ is a function of t only. Our method of solution can be summarized as follows:

a. Write y' as $\dfrac{dy}{dt}$.

$$h(y)\frac{dy}{dt} = p(t).$$

b. Integrate both sides with respect to t.

$$\int h(y) \frac{dy}{dt} \, dt = \int p(t) \, dt.$$

c. Rewrite the left-hand side by "canceling the dt."

$$\int h(y) \, dy = \int p(t) \, dt.$$

d. Calculate the antiderivatives $H(y)$ for $h(y)$ and $P(t)$ for $p(t)$.

$$H(y) = P(t) + C.$$

e. Solve for y in terms of t.

$$y = \ldots .$$

Note   In step (d) there is no need to write two constants of integration (as we did in Example 1), since they will be combined into one in step (e).

EXAMPLE 2   Solve $e^y y' = 2t + 1$, $y(0) = 1$.

Solution   (a) $e^y \dfrac{dy}{dt} = 2t + 1$

(b) $\displaystyle \int e^y \frac{dy}{dt} \, dt = \int (2t + 1) \, dt$

(c) $\displaystyle \int e^y \, dy = \int (2t + 1) \, dt$

(d) $e^y = t^2 + t + C$

(e) $y = \ln(t^2 + t + C)$ [Take logarithms of both sides of the equation in step (d).]

If $y = \ln(t^2 + t + C)$ is to satisfy the initial condition $y(0) = 1$, then $1 = y(0) = \ln(0^2 + 0 + C) = \ln C$, so that $C = e$ and $y = \ln(t^2 + t + e)$. ●

EXAMPLE 3   Solve $y' = t^3 y^2 + y^2$.

Solution   As the equation is given, the right-hand side is not in the form $p(t)q(y)$. However, we may rewrite the equation in the form $y' = (t^3 + 1)y^2$. Now we may separate the variables, dividing both sides by y^2, to get

$$\frac{1}{y^2} y' = t^3 + 1. \tag{3}$$

Then we apply our method of solution:

(a) $\dfrac{1}{y^2} \dfrac{dy}{dt} = t^3 + 1$

(b) $\displaystyle \int \frac{1}{y^2} \frac{dy}{dt} \, dt = \int (t^3 + 1) \, dt$

(c) $\displaystyle \int \frac{1}{y^2} \, dy = \int (t^3 + 1) \, dt$

(d) $\displaystyle -\frac{1}{y} = \frac{1}{4} t^4 + t + C, \quad C \text{ a constant}$

(e) $\displaystyle y = -\frac{1}{\frac{1}{4} t^4 + t + C}$

The method that we have applied yields all the solutions of equation (3). However, we have ignored an important point. We wish to solve $y' = y^2(t^3 + 1)$ and not equation (3). Do the two equations have precisely the same solutions? We obtained equation (3) from the given equation by dividing by y^2. This is a permissible operation, provided that y is not equal to zero for all t. (Of course, if y is zero for some t, the resulting differential equation is understood to hold only for some limited range of t.) Thus in dividing by y^2, we must assume that y is not the zero function. However, note that $y = 0$ is a solution of the original equation because

$$0 = (0)' = t^3 \cdot 0^2 + 0^2.$$

So, when we divided by y^2, we "lost" the solution $y = 0$. Finally, we see that the solutions of the differential equation $y' = t^3 y^2 + y^2$ are

$$y = -\frac{1}{\frac{1}{4} t^4 + t + C}, \quad C \text{ a constant}$$

and

$$y = 0. \; \bullet$$

Warning   If the equation in Example 3 had been

$$y' = t^3 y^2 + 1,$$

we would not have been able to use the method of separation of variables because the expression $t^3 y^2 + 1$ cannot be written in the form $p(t)q(y)$.

EXAMPLE 4   Solve $y' = 5 - 2y$.

Solution   The algebra involved in solving an equation of the form $y' = a + by$ is always simplified by factoring out the coefficient of y. Thus we rewrite $y' = 5 - 2y$ as

$$y' = -2(y - \tfrac{5}{2}). \tag{4}$$

Clearly, the constant function $y = \frac{5}{2}$ is a solution of the differential equation because it makes both sides of (4) zero for all t. (The left-hand side is zero because the derivative of a constant function is zero.) Now, supposing that $y \neq \frac{5}{2}$, we may divide by $y - \frac{5}{2}$, obtaining

$$\frac{1}{y - \frac{5}{2}} y' = -2$$

$$\int \frac{1}{y - \frac{5}{2}} \frac{dy}{dt} \, dt = \int -2 \, dt$$

$$\int \frac{1}{y - \frac{5}{2}} \, dy = \int -2 \, dt$$

$$\ln\left| y - \tfrac{5}{2} \right| = -2t + C, \quad C \text{ a constant}$$

$$\left| y - \tfrac{5}{2} \right| = e^{-2t+C} = e^C \cdot e^{-2t}. \tag{5}$$

In Section 10.4 we shall see that if $y \neq \frac{5}{2}$, then $y - \frac{5}{2}$ will be either positive for all t or negative for all t. Consequently, either

$$y - \tfrac{5}{2} = e^C \cdot e^{-2t} \quad \text{or} \quad y - \tfrac{5}{2} = -e^C \cdot e^{-2t},$$

that is,

$$y = \tfrac{5}{2} + e^C \cdot e^{-2t} \quad \text{or} \quad y = \tfrac{5}{2} - e^C \cdot e^{-2t}. \tag{6}$$

These two types of solutions and the constant solution $y = \frac{5}{2}$ may all be written in the form

$$y = \tfrac{5}{2} + Ae^{-2t}, \quad A \text{ any constant.}$$

The two types of solutions shown in (6) correspond to positive and negative values of A, respectively. The constant solution is given by setting $A = 0$. ●

Warning

1. In Example 4 the "extra" solution $y = \frac{5}{2}$ corresponded to the value $A = 0$ of our final solution. Such is not always the case. In Example 3 the solution $y = 0$ does not correspond to any value of the constant C.

2. A common mistake in solving $y' = 5 - 2y$ is to *add* $2y$ to both sides to obtain $y' + 2y = 5$. In a sense, the variables are then separated but not in the proper way. For if we integrate both sides, we get

$$\int \left(\frac{dy}{dt} + 2y \right) dt = \int 5 \, dt,$$

and there is no substitution that will simplify the left-hand side, because $\dfrac{dy}{dt}$ does not appear as a factor.

EXAMPLE 5   Solve $y' = te^t/y$, $y(0) = -5$.

Solution   Separating the variables, we have

$$yy' = te^t$$

$$\int y\frac{dy}{dt}\,dt = \int te^t\,dt$$

$$\int y\,dy = \int te^t\,dt.$$

The integral $\int te^t\,dt$ may be found by integration by parts:

$$\int te^t\,dt = te^t - \int 1\cdot e^t\,dt = te^t - e^t + C.$$

Therefore,

$$\frac{1}{2}y^2 = te^t - e^t + C$$

$$y^2 = 2te^t - 2e^t + C_1$$

$$y = \pm\sqrt{2te^t - 2e^t + C_1}.$$

Note that the $\pm$ appears because there are two square roots of $2te^t - 2e^t + C_1$, differing from one another by a minus sign. Thus the solutions are of two sorts—namely,

$$y = +\sqrt{2te^t - 2e^t + C_1}$$

$$y = -\sqrt{2te^t - 2e^t + C_1}.$$

We must choose C_1 so that $y(0) = -5$. Since the values of y for the first solution are always positive, the given initial condition must correspond to the second solution, and we must have

$$-5 = y(0) = -\sqrt{2\cdot 0\cdot e^0 - 2e^0 + C_1} = -\sqrt{-2 + C_1}$$

$$-2 + C_1 = 25$$

$$C_1 = 27.$$

Hence the desired solution is

$$y = -\sqrt{2te^t - 2e^t + 27}. \quad \bullet$$

When working the exercises at the end of the section, it is a good practice first to find the constant solution(s), if any. A constant function $y = c$ is a solution of $y' = p(t)q(y)$ if and only if $q(c) = 0$. (For $y = c$ implies that $y' = (c)' = 0$, and $p(t)q(y)$ will be zero for all t if and only if $q(y) = 0$—that is, $q(c) = 0$.) After listing the constant solutions, one may assume that $q(y) \neq 0$ and go on to divide both sides of the equation $y' = p(t)q(y)$ by $q(y)$ to separate the variables.

**TECHNOLOGY
PROJECT 10.1
Solving Differential
Equations Using a
Symbolic Math
Program**

In this section, we learned to solve differential equations using a single method, separation of variables. This technique applies to a limited class of differential equations. In a course in differential equations, you learn many other techniques, which considerably expand the types of differential equations you can solve. The more sophisticated symbolic math programs have built-in commands that implement most of the methods you would learn in a course in differential equations. In this technology project, we will explore how to use such commands to solve a variety of differential equations, some of the separation-of-variables type and some of different types.

For the sake of simplicity, assume that the variable in our differential equations is x and that the variable function is $y = y(x)$. In *Mathematica*, the solution is denoted $y[x]$ and derivatives in differential equations are written using apostrophes. Thus, for example, the differential equation

$$y' + 3xy = 0$$

is written as:

`y[x]'+3x y[x] == 0`

Note the space between `3x` and `y[x]`. This is necessary for the program to correctly interpret the product. Also note the use of `==` to indicate an equation.

Initial conditions are listed, within braces, along with the differential equations, with equations separated by commas. For example, to consider the above differential equation along with the initial condition

$$y(1) = 3$$

you would enter:

`{y[x]'+3x y[x] == 0, y[1] == 3}`

Mathematica uses the command `DSolve` to solve differential equations. For example, to solve the first differential equation above (without the boundary value condition), you would enter the command:

`DSolve [y[x]' +3x y[x] == 0, y[x],x]`

Note that we must list the variable x and the unknown function $y[x]$ as part of the command. The program responds with the result:

$$y[x] = \frac{c[1]}{E^{3x^2/2}}$$

Here $c[1]$ is a constant that may assume any real value and E denotes the exponential function.

Use a symbolic math program to solve the following differential equations:

1. $y' + xy = 0$

2. $xy' = 3y, \, y(0) = 4$

3. $y' + y^2 = 1$

4. $y' = x^3 e^{-5y}, \, y(0) = 10$

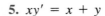

5. $xy' = x + y$

6. $x^2y' = 2y^2 - xy$

7. $y' - 3y = 1$

8. $y' + e^x y = 0$

9. $y' + y = e^{-x}$

10. $y' + 2xy = 4e^x$

PRACTICE PROBLEMS 10.2

1. Solve the initial value problem $y' = 5y$, $y(0) = 2$, by separation of variables.

2. Solve $y' = \sqrt{ty}$, $y(1) = 4$.

EXERCISES 10.2

Solve the following differential equations.

1. $\dfrac{dy}{dt} = \dfrac{5 - t}{y^2}$

2. $\dfrac{dy}{dt} = \dfrac{e^t}{e^y}$

3. $\dfrac{dy}{dt} = te^{2y}$

4. $\dfrac{dy}{dt} = -\dfrac{1}{t^2y^2}$

5. $\dfrac{dy}{dt} = t^{1/2}y^2$

6. $\dfrac{dy}{dt} = \dfrac{t^2y^2}{t^3 + 8}$

7. $y' = \left(\dfrac{t}{y}\right)^2 e^{t^3}$

8. $y' = e^{4y}t^3 - e^{4y}$

9. $y' = \sqrt{\dfrac{y}{t}}$

10. $y' = \left(\dfrac{e^t}{y}\right)^2$

11. $y' = 4(y - 3)$

12. $y' = -\dfrac{1}{2}(y - 4)$

13. $y' = 2 - y$

14. $y' = \dfrac{1}{ty + y}$

15. $y' = \dfrac{\ln t}{ty}$

16. $y' = 4t - ty$

17. $y' = (y - 3)^2 \ln t$

18. $yy' = t\cos(t^2 + 1)$

Solve the following differential equations with given initial conditions.

19. $y' = 2te^{-2y} - e^{-2y}$, $y(0) = 3$

20. $y' = y^2 - e^{3t}y^2$, $y(0) = 1$

21. $y^2y' = t\sin t$, $y(0) = 2$

22. $y' = t^2e^{-3y}$, $y(0) = 2$

23. $y' = 5 - 3y$, $y(0) = 1$

24. $y' = \dfrac{1}{2}y - 3$, $y(0) = 4$

25. $\dfrac{dy}{dt} = \dfrac{t+1}{ty}$, $t > 0$, $y(1) = -3$

26. $\dfrac{dy}{dt} = \left(\dfrac{1+t}{1+y}\right)^2$, $y(0) = 2$

27. $y' = 5ty - 2t$, $y(0) = 1$

28. $y' = \dfrac{t^2}{y}$, $y(0) = -5$

29. $\dfrac{dy}{dx} = \dfrac{\ln x}{\sqrt{xy}}$, $y(1) = 4$

30. $\dfrac{dN}{dt} = 2tN^2$, $N(0) = 5$

31. A model that describes the relationship between the price and the weekly sales of a product might have a form such as

$$\frac{dy}{dp} = -\frac{1}{2}\left(\frac{y}{p+3}\right),$$

where y is the volume of sales and p is the price per unit. That is, at any time the rate of decrease of sales with respect to price is directly proportional to the sales level and inversely proportional to the sales price plus a constant. Solve this differential equation. (Figure 1 shows several typical solutions.)

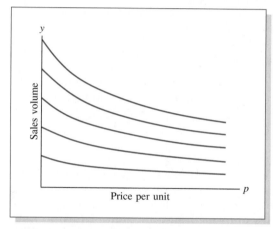

FIGURE 1   Demand curves.

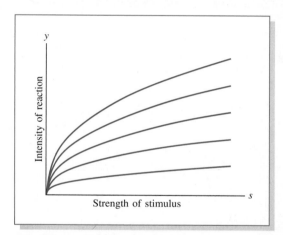

FIGURE 2   Reaction to stimuli.

32. One problem in psychology is to determine the relation between some physical stimulus and the corresponding sensation or reaction produced in a subject. Suppose that, measured in appropriate units, the strength of a stimulus is s and the intensity of the corresponding sensation is some function of s, say $f(s)$. Some experimental data suggest that the rate of change of intensity of the sensation with respect to the stimulus is directly proportional to the intensity of the sensation and inversely proportional to the strength of the stimulus; that is, $f(s)$ satisfies the differential equation

$$\frac{dy}{ds} = k\frac{y}{s}$$

for some positive constant k. Solve this differential equation. (Figure 2 shows several solutions corresponding to $k = .4$.)

33. Let t represent the total number of hours that a truck driver spends during a year driving on a certain highway connecting two cities, and let $p(t)$ represent the probability that the driver will have at least one accident during these t hours. Then $0 \leq p(t) \leq 1$, and $1 - p(t)$ represents the probability of not having an accident. Under ordinary conditions the rate of increase in the probability of an accident (as a function of t) is proportional to the probability of not having an accident. Construct and solve a differential equation for this situation.

34. In certain learning situations there is a maximum amount, M, of information that can be learned, and at any time the rate of learning is proportional to the amount yet to be learned. Let $y = f(t)$ be the amount of information learned up to time t. Construct and solve a differential equation that is satisfied by $f(t)$.

35. Mothballs tend to evaporate at a rate proportional to their surface area. If V is the volume of a mothball, then its surface area is roughly a constant times $V^{2/3}$. So the mothball's volume decreases at a rate proportional to $V^{2/3}$. Suppose that initially a mothball has a volume of 27 cubic centimeters and 4 weeks later has volume 15.625 cubic centimeters. Construct and solve a differential equation satisfied by the volume at time t. Then determine if and when the mothball will vanish ($V = 0$).

36. Some homeowner's insurance policies include automatic inflation coverage based on the U.S. Commerce Department's construction cost index (CCI). Each year the property insurance coverage is increased by an amount based on the change in the CCI. Let $f(t)$ be the CCI at time t years since January 1, 1990, and let $f(0) = 100$. Suppose the construction cost index is rising at a rate proportional to the CCI and the index was 115 on January 1, 1992. Construct and solve a differential equation satisfied by $f(t)$. Then determine when the CCI will reach 200.

37. The Gompertz growth equation is

$$\frac{dy}{dt} = -ay \ln \frac{y}{b},$$

where a and b are positive constants. This equation is used in biology to describe the growth of certain populations. Find the general form of solutions to this differential equation. (Figure 3 shows several solutions corresponding to $a = .04$ and $b = 90$.)

FIGURE 3   Gompertz growth curves.

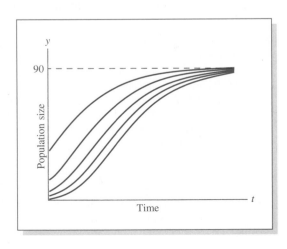

38. When a certain liquid substance A is heated in a flask, it decomposes into a substance B at a rate (measured in units of A per hour) that at any time t is proportional to the square of the amount of substance A present. Let $y = f(t)$ be the amount of substance A present at time t. Construct and solve a differential equation that is satisfied by $f(t)$.

SOLUTIONS TO PRACTICE PROBLEMS 10.2

1. The constant function $y = 0$ is a solution of $y' = 5y$. If $y \neq 0$, we may divide by y and obtain

$$\frac{1}{y}y' = 5$$

$$\int \frac{1}{y}\frac{dy}{dt}\,dt = \int 5\,dt$$

$$\int \frac{1}{y}\,dy = \int 5\,dt$$

$$\ln |y| = 5t + C$$

$$|y| = e^{5t+C} = e^C \cdot e^{5t}$$

$$y = \pm e^C \cdot e^{5t}.$$

These two types of solutions and the constant solution may all be written in the form

$$y = Ae^{5t},$$

where A is an arbitrary constant (positive, negative, or zero). The initial condition $y(0) = 2$ implies that

$$2 = y(0) = Ae^{5(0)} = A.$$

Hence the solution of the initial value problem is $y = 2e^{5t}$.

2. We rewrite $y' = \sqrt{ty}$ as $y' = \sqrt{t} \cdot \sqrt{y}$. The contant function $y = 0$ is one solution. To find the others, we suppose that $y \neq 0$ and divide by $\sqrt{y}$ to obtain

$$\frac{1}{\sqrt{y}}y' = \sqrt{t}$$

$$\int y^{-1/2}\frac{dy}{dt}\,dt = \int t^{1/2}\,dt$$

$$\int y^{-1/2}\,dy = \int t^{1/2}\,dt$$

$$2y^{1/2} = \frac{2}{3}t^{3/2} + C$$

$$y^{1/2} = \frac{1}{3}t^{3/2} + C_1 \tag{7}$$

$$y = \left(\frac{1}{3}t^{3/2} + C_1\right)^2. \tag{8}$$

We must choose C_1 so that $y(1) = 4$. The quickest method is to use (7) instead of (8). We have $y = 4$ when $t = 1$, so

$$4^{1/2} = \frac{1}{3}(1)^{3/2} + C_1$$

$$2 = \frac{1}{3} + C_1$$

$$C_1 = \frac{5}{3}.$$

Hence the desired solution is

$$y = \left(\frac{1}{3}t^{3/2} + \frac{5}{3}\right)^2.$$

10.3 NUMERICAL SOLUTION OF DIFFERENTIAL EQUATIONS

Many differential equations that arise in real-life applications cannot be solved by *any* known method. However, approximate solutions may be obtained by several different numerical techniques. In this section we describe what is known as *Euler's method* for approximating solutions to initial value problems of the form

$$y' = g(t, y), \qquad y(a) = y_0 \tag{1}$$

for values of t in some interval $a \le t \le b$. Here $g(t, y)$ is some reasonably well-behaved function of two variables. Equations of the form $y' = p(t)q(y)$ studied in the preceding section are a special case of (1).

In the discussion below we assume that $f(t)$ is a solution of (1) for $a \le t \le b$. The basic idea on which Euler's method rests is the following: *If the graph of $y = f(t)$ passes through some given point (t, y), the slope of the graph (i.e., the value of y') at that point is just $g(t, y)$, because $y' = g(t, y)$.* Euler's method uses this observation to approximate the graph of $f(t)$ by a polygonal path, such as in Fig. 1.

The t-axis from a to b is subdivided by the equally spaced points t_0, $t_1, \ldots, t_n$. Each of the n subintervals has length $h = (b - a)/n$. The initial condition $y(a) = y_0$ in (1) implies that the graph of the solution $f(t)$ passes through the point (t_0, y_0). As noted above, the slope of this graph at (t_0, y_0) must be $g(t_0, y_0)$. Thus on the first subinterval, Euler's method approximates the graph of $f(t)$ by the straight line

$$y = y_0 + g(t_0, y_0) \cdot (t - t_0)$$

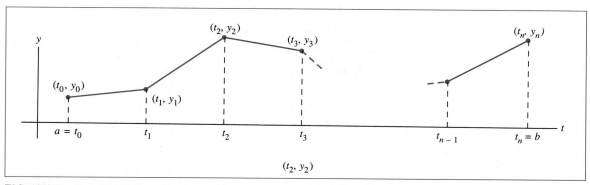

FIGURE I   A polygonal path.

which passes through (t_0, y_0) and has slope $g(t_0, y_0)$. When $t = t_1$, the y-coordinate on this line is

$$y_1 = y_0 + g(t_0, y_0) \cdot (t_1 - t_0)$$
$$= y_0 + g(t_0, y_0) \cdot h.$$

Since the graph of $f(t)$ is close to the point (t_1, y_1) on the line, the slope of the graph of $f(t)$ when $t = t_1$ will be close to $g(t_1, y_1)$. So we draw the straight line

$$y = y_1 + g(t_1, y_1) \cdot (t - t_1) \qquad (2)$$

through (t_1, y_1) with slope $g(t_1, y_1)$ (Fig. 2), and we use this line to approximate $f(t)$ on the second subinterval. From (2) we determine an estimate y_2 for the value of $f(t)$ at $t = t_2$:

$$y_2 = y_1 + g(t_1, y_1) \cdot h.$$

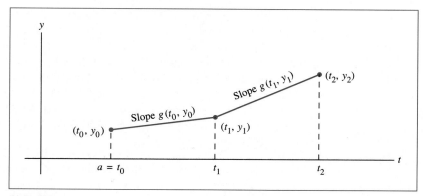

FIGURE 2

The slope of the graph of $f(t)$ at t_2 is now estimated by $g(t_2, y_2)$, and so on. Let us summarize this procedure:

> *Euler's Method*   The endpoints $(t_0, y_0), \ldots, (t_n, y_n)$ of the line segments approximating the solution of (1) on the interval $a \le t \le b$ are given by the following formulas, where $h = (b - a)/n$:
>
> $$t_0 = a \text{ (given)}, \qquad y_0 \text{ (given)},$$
> $$t_1 = t_0 + h, \qquad y_1 = y_0 + g(t_0, y_0) \cdot h,$$
> $$t_2 = t_1 + h, \qquad y_2 = y_1 + g(t_1, y_1) \cdot h,$$
> $$\vdots \qquad\qquad\qquad \vdots$$
> $$t_n = t_{n-1} + h. \qquad y_n = y_{n-1} + g(t_{n-1}, y_{n-1}) \cdot h.$$

EXAMPLE I   Use Euler's method with $n = 4$ to approximate the solution $f(t)$ to $y' = 2t - 3y$, $y(0) = 4$, for t in the interval $0 \le t \le 2$. In particular, estimate $f(2)$.

FIGURE 3

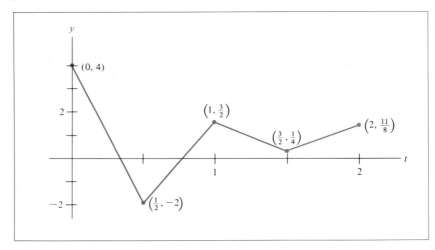

Solution   Here $g(t, y) = 2t - 3y$, $a = 0$, $b = 2$, $y_0 = 4$ and $h = (2 - 0)/4 = \frac{1}{2}$. Starting with $(t_0, y_0) = (0, 4)$, we find that $g(0, 4) = -12$. Thus

$$t_1 = \frac{1}{2}, \qquad y_1 = 4 + (-12) \cdot \frac{1}{2} = -2.$$

Next, $g(\frac{1}{2}, -2) = 7$, so

$$t_2 = 1, \qquad y_2 = -2 + 7 \cdot \frac{1}{2} = \frac{3}{2}.$$

Next, $g(1, \frac{3}{2}) = -\frac{5}{2}$, so

$$t_3 = \frac{3}{2}, \qquad y_3 = \frac{3}{2} + \left(-\frac{5}{2}\right) \cdot \frac{1}{2} = \frac{1}{4}.$$

Finally, $g(\frac{3}{2}, \frac{1}{4}) = \frac{9}{4}$, so

$$t_4 = 2, \qquad y_4 = \frac{1}{4} + \frac{9}{4} \cdot \frac{1}{2} = \frac{11}{8}.$$

Thus the approximation to the solution $f(t)$ is given by the polygonal path shown in Fig. 3. The last point $(2, \frac{11}{8})$ is close to the graph of $f(t)$ at $t = 2$, so $f(2) \approx \frac{11}{8}$. ●

Actually, this polygonal path is somewhat misleading. The accuracy can be improved dramatically by increasing the value of n. Figure 4 shows the Euler approximations for $n = 8$ and $n = 20$. The graph of the exact solution is shown for comparison.

FIGURE 4   Approximating an exact solution by polygonal paths.

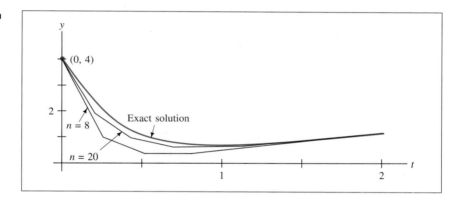

For many purposes, satisfactory graphs can be obtained by running Euler's method on a computer with large values of n. There is a limit to the accuracy obtainable, however, because each computer calculation involves a slight "round-off" error. When n is very large, the cumulative round-off error can become significant. Several more sophisticated methods for approximating the solutions of differential equations are discussed in Chapter 8 of *Elementary Differential Equations,* 4th ed., by William Boyce and Richard DiPrima (New York: John Wiley & Sons, 1986).

TECHNOLOGY PROJECT 10.2
Euler's Method for Approximating Solutions to Differential Equations

Analyze further the differential equation of Example 1.

1. Use Euler's method with $n = 10$ to determine an approximate solution to the differential equation.
2. Use graphical analysis to determine the maximum difference between the approximate solution and the actual solution in each of the subintervals.
3. Use the data in 2 to determine the accuracy of the approximate solution.

**PRACTICE
PROBLEMS 10.3**   Let $f(c)$ be the solution of $y' = \sqrt{ty}$, $y(1) = 4$.

1. Use Euler's method with $n = 2$ on the interval $1 \leq t \leq 2$ to estimate $f(2)$.

2. Draw the polygonal path corresponding to the application of Euler's method in Problem 1.

EXERCISES 10.3

1. Suppose that $f(t)$ is a solution of the differential equation $y' = ty - 5$, and the graph of $f(t)$ passes through the point $(2, 4)$. What is the slope of the graph at this point?

2. Suppose that $f(t)$ is a solution of $y' = t^2 - y^2$ and the graph of $f(t)$ passes through the point $(2, 3)$. Find the slope of the graph when $t = 2$.

3. Suppose that $f(t)$ satisfies the initial value problem $y' = y^2 + ty - 7$, $y(0) = 3$. Is $f(t)$ increasing or decreasing at $t = 0$?

4. Suppose that $f(t)$ satisfies the initial value problem $y' = y^2 + ty - 7$, $y(0) = 2$. Is the graph of $f(t)$ increasing or decreasing when $t = 0$?

5. Use Euler's method with $n = 2$ on the interval $0 \leq t \leq 1$ to approximate the solution $f(t)$ to $y' = t^2 y$, $y(0) = -2$. In particular, estimate $f(1)$.

6. Use Euler's method with $n = 2$ on the interval $2 \leq t \leq 3$ to approximate the solution $f(t)$ to $y' = t - 2y$, $y(2) = 3$. Estimate $f(3)$.

7. Use Euler's method with $n = 4$ to approximate the solution $f(t)$ to $y' = 2t - y + 1$, $y(0) = 5$ for $0 \leq t \leq 2$. Estimate $f(2)$.

8. Let $f(t)$ be the solution of $y' = y(2t - 1)$, $y(0) = 8$. Use Euler's method with $n = 4$ to estimate $f(1)$.

9. Let $f(t)$ be the solution of $y' = -(t + 1)y^2$, $y(0) = 1$. Use Euler's method with $n = 5$ to estimate $f(1)$. Then solve the differential equation, find an explicit formula for $f(t)$, and compute $f(1)$. How accurate is the estimated value of $f(1)$?

10. Let $f(t)$ be the solution of $y' = 10 - y$, $y(0) = 1$. Use Euler's method with $n = 5$ to estimate $f(1)$. Then solve the differential equation and find the exact value of $f(1)$.

11. Suppose that the Consumer Products Safety Commission issues new regulations that affect the toy manufacturing industry. Every toy manufacturer will have to make certain changes in its manufacturing process. Let $f(t)$ be the fraction of manufacturers that have complied with the regulations within t months. Note that $0 \leq f(t) \leq 1$. Suppose that the rate at which new companies comply with the regulations is proportional to the fraction of companies who have not yet complied.

    (a) Construct a differential equation satisfied by $f(t)$.

    (b) Use Euler's method with $n = 3$ to estimate the fraction of companies that comply with the regulations within the first 3 months. (The answer will involve an unknown constant of proportionality, k.)

(c) Solve the differential equation in part (a) and compute $f(3)$.

(d) Compare the answers in parts (b) and (c) when $k = .1$.

12. The Los Angeles Zoo plans to transport a California sea lion to the San Diego Zoo. The animal will be wrapped in a wet blanket during the trip. At any time t the blanket will lose water (owing to evaporation) at a rate proportional to the amount $f(t)$ of water in the blanket. Initially, the blanket will contain 2 gallons of seawater.

(a) Set up the differential equation satisfied by $f(t)$.

(b) Use Euler's method with $n = 2$ to estimate the amount of moisture in the blanket after 1 hour. (The answer will involve an unknown constant of proportionality, k.)

(c) Solve the differential equation in part (a) and compute $f(1)$.

(d) Compare the answers in parts (b) and (c) when $k = -.3$.

Exercises 13–18 review concepts that will be important in the next section. In each exercise, sketch the graph of a function with the stated properties.

13. Domain: $0 \leq t \leq 3$; $(0, 1)$ is on the graph; the slope is always positive, and the slope becomes less positive (as t increases).

14. Domain: $0 \leq t \leq 4$; $(0, 2)$ is on the graph; the slope is always positive, and the slope becomes more positive (as t increases).

15. Domain: $0 \leq t \leq 5$; $(0, 3)$ is on the graph; the slope is always negative, and the slope becomes less negative.

16. Domain: $0 \leq t \leq 6$; $(0, 4)$ is on the graph; the slope is always negative, and the slope becomes more negative.

17. Domain: $0 \leq t \leq 7$; $(0, 2)$ is on the graph; the slope is always positive, the slope becomes more positive as t increases from 0 to 3, and the slope becomes less positive as t increases from 3 to 7.

18. Domain: $0 \leq t \leq 8$; $(0, 6)$ is on the graph; the slope is always negative, the slope becomes more negative as t increases from 0 to 3, and the slope becomes less negative as t increases from 3 to 8.

SOLUTIONS TO PRACTICE PROBLEMS 10.3

1. Here $g(t, y) = \sqrt{ty}$, $a = 1$, $b = 2$, $y_0 = 4$, and $h = (2 - 1)/2 = \frac{1}{2}$. We have

$$t_0 = 1, \qquad y_0 = 4, \qquad\qquad g(1, 4) = \sqrt{1 \cdot 4} = 2,$$
$$t_1 = \tfrac{3}{2}, \qquad y_1 = 4 + 2(\tfrac{1}{2}) = 5, \qquad g(\tfrac{3}{2}, 5) = \sqrt{\tfrac{3}{2} \cdot 5} \approx 2.7386,$$
$$t_2 = 2, \qquad y_2 = 5 + (2.7386)(\tfrac{1}{2}) = 6.3693.$$

Hence $f(2) \approx y_2 = 6.3693$. [In Practice Problems 10.2 we found the solution of $y' = \sqrt{ty}$, $y(1) = 4$, to be $f(t) = (\frac{1}{3}t^{3/2} + \frac{5}{3})^2$. We find that $f(2) = 6.8094$ (to four decimal places). The error $6.8094 - 6.3693 = .4401$ in the approximation above is about 6.5%.]

2. To find the polygonal path, plot the points (t_0, y_0), (t_1, y_1), and (t_2, y_2) and join them by line segments. See Fig. 5.

FIGURE 5

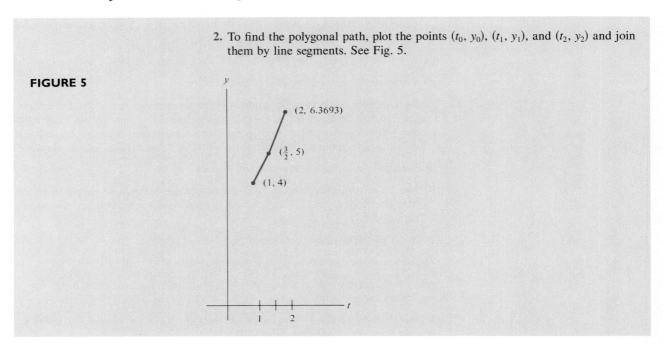

10.4 QUALITATIVE THEORY OF DIFFERENTIAL EQUATIONS

In this section we present a technique for sketching solutions to differential equations of the form $y' = g(y)$ *without having to solve the differential equation.* This technique is valuable for three reasons. First, there are many differential equations for which explicit solutions cannot be written down. Second, even when an explicit solution is available, we still face the problem of determining its behavior. For example, does the solution increase or decrease? If it increases, does it approach an asymptote or does it grow arbitrarily large? Third, and probably most significantly, in many applications the explicit formula for a solution is unnecessary; only a general knowledge of the behavior of the solution is needed. That is, a qualitative understanding of the solution is sufficient.

The theory introduced in this section is part of what is called the *qualitative theory of differential equations.* We shall limit our attention to differential equations of the form $y' = g(y)$. Such differential equations are called *autonomous.* The term "autonomous" here means "independent of time" and refers to the fact that the right-hand side of $y' = g(y)$ depends only on y and not on t. All the applications studied in the next section involve autonomous differential equations.

Throughout this section we consider the values of each solution $y = f(t)$ only for $t \geq 0$. To introduce the qualitative theory, let us examine the graphs of the various typical solutions of the differential equation $y' = \frac{1}{2}(1 - y)(4 - y)$. The solution curves in Fig. 1 illustrate the following properties.

FIGURE 1 Solutions of $y' = \frac{1}{2}(1 - y)(4 - y)$.

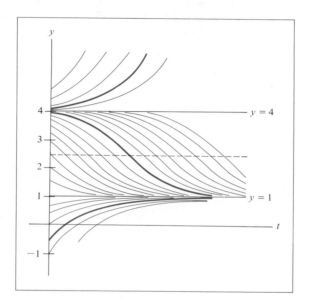

Property I   Corresponding to each zero of $g(y)$ there is a constant solution of the differential equation. Specifically, if $g(c) = 0$, the constant function $y = c$ is a solution. (The constant solutions in Fig. 1 are $y = 1$ and $y = 4$.)

Property II   The constant solutions divide the ty-plane into horizontal strips. Each nonconstant solution lies completely in one strip.

Property III   Each nonconstant solution is either strictly increasing or decreasing.

Property IV   Each nonconstant solution either is asymptotic to a constant solution or else increases or decreases without bound.

It can be shown that Properties I–IV are valid for the solutions of any autonomous differential equation $y' = g(y)$ provided that $g(y)$ is a "sufficiently well-behaved" function. We shall assume these properties in this chapter.

Using Properties I–IV, we can sketch the general shape of any solution curve by looking at the graph of the function $g(y)$ and the behavior of that graph near $y(0)$. The procedure for doing this is illustrated in the following example.

EXAMPLE 1   Sketch the solution to $y' = e^{-y} - 1$ that satisfies $y(0) = -2$.

Solution   Here $g(y) = e^{-y} - 1$. On a yz-coordinate system we draw the graph of the function $z = g(y) = e^{-y} - 1$ [Fig. 2(a)]. The function $g(y) = e^{-y} - 1$ has a zero when $y = 0$. Therefore, the differential equation $y' = e^{-y} - 1$ has the constant solution $y = 0$. We indicate this constant solution on a ty-coordinate system in Fig. 2(b). To begin the sketch of the solution satisfying $y(0) = -2$, we locate this initial value of y on the (horizontal) y-axis in Fig. 2(a) and on the (vertical) y-axis in Fig. 2(b).

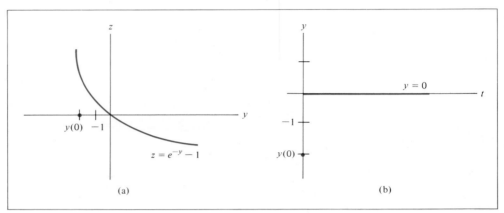

FIGURE 2

To determine whether the solution increases or decreases when it leaves the initial point $y(0)$ on the ty graph, we look at the yz graph and note that $z = g(y)$ is positive at $y = -2$ [Fig. 3(a)]. Consequently, since $y' = g(y)$, the derivative of the solution is positive, which implies that the solution is increasing. We indicate this by an arrow at the initial point in Fig. 3(b). Moreover, the solution

FIGURE 3

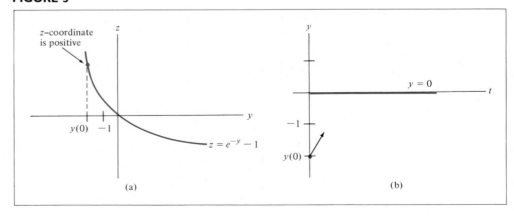

y will increase asymptotically to the constant solution $y = 0$, by Properties III and IV of autonomous differential equations.

Next, we place an arrow in Fig. 4(a) to remind us that y will move from $y = -2$ toward $y = 0$. As y moves to the right toward $y = 0$ in Fig. 4(a), the z-coordinate of points on the graph of $g(y)$ becomes less positive; that is, $g(y)$ becomes less positive. Consequently, since $y' = g(y)$, the slope of the solution curve becomes less positive. Thus the solution curve is concave down, as we have shown in Fig. 4(b). ●

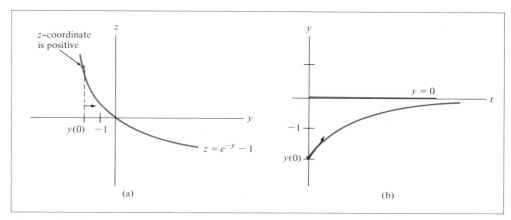

(a)   (b)

FIGURE 4

An important point to remember when sketching solutions is that z-coordinates on the yz graph are values of $g(y)$, and since $y' = g(y)$, a z-coordinate gives the *slope* of the solution curve at the corresponding point on the ty graph.

EXAMPLE 2   Sketch the graphs of the solutions to $y' = y + 2$ satisfying
(a) $y(0) = 1$,
(b) $y(0) = -3$.

Solution   Here $g(y) = y + 2$. The graph of $z = g(y)$ is a straight line of slope 1 and z-intercept 2. [See Fig. 5(a).] This line crosses the y-axis only where $y = -2$. Thus the differential equation $y' = y + 2$ has one constant solution, $y = -2$. [See Fig. 5(b).]

(a) We locate the initial value $y(0) = 1$ on the y-axes of both graphs in Fig. 5. The corresponding z-coordinate on the yz graph is positive; therefore, the solution on the ty graph has positive slope and is increasing as it leaves the initial point. We indicate this by an arrow in Fig. 5(b). Now, Property IV of autonomous differential equations implies that y will increase without bound from its initial value. As we let y increase from 1 in Fig. 6(a), we see that the z-coordinates [i.e., values of $g(y)$] increase. Consequently, y' is

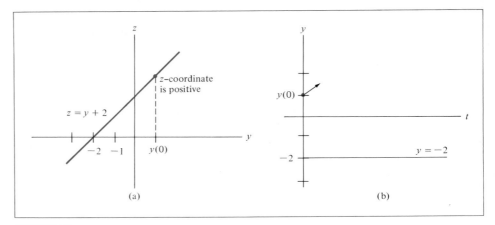

FIGURE 5

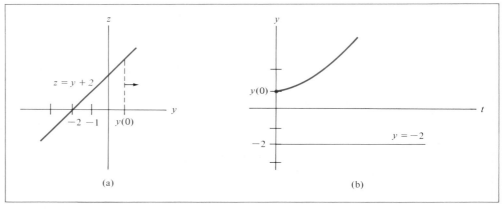

FIGURE 6

increasing, so the graph of the solution must be concave up. We have sketched the solution in Fig. 6(b).

(b) Next we graph the solution for which $y(0) = -3$. From the graph of $z = y + 2$, we see that z is negative when $y = -3$. This implies that the solution is decreasing as it leaves the initial point. (See Fig. 7.) It follows that the values of y will continue to decrease without bound and become more and more negative. This means that on the yz graph y must move to the *left* [Fig. 8(a)]. We now examine what happens to $g(y)$ as y moves to the left. (This is the opposite of the ordinary way to read a graph.) The z-coordinate becomes more negative; hence the slopes on the solution curve will become more negative. Thus the solution curve must be concave down, as in Fig. 8(b). ●

FIGURE 7

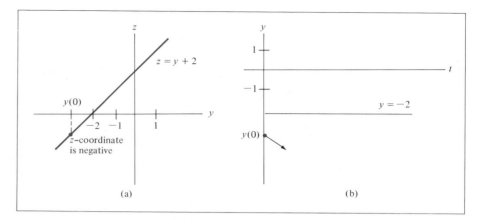

(a)　　　　　　　　　　　　　　　　　(b)

FIGURE 8

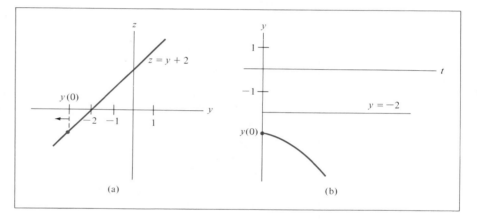

(a)　　　　　　　　　　　　　　　　　(b)

From the preceding examples we can state a few rules for sketching a solution to $y' = g(y)$ with $y(0)$ given:

1. Sketch the graph of $z = g(y)$ on a yz-coordinate system. Find and label the zeros of $g(y)$.
2. For each zero c of $g(y)$ draw the constant solution $y = c$ on the ty-coordinate system.
3. Plot $y(0)$ on the y-axes of the two coordinate systems.
4. Determine whether the value of $g(y)$ is positive or negative when $y = y(0)$. This tells us whether the solution is increasing or decreasing. On the ty graph, indicate the direction of the solution through $y(0)$.
5. On the yz graph, indicate which direction y should move. (*Note:* If y is moving *down* on the ty graph, y moves to the *left* on the yz graph.) As y moves in the proper direction on the yz graph, determine whether $g(y)$ becomes more positive, less positive, more negative, or less negative. This tells us the concavity of the solution.

6. Beginning at $y(0)$ on the ty graph, sketch the solution, being guided by the principle that the solution will grow (positively or negatively) without bound unless it encounters a constant solution. In this case, it will approach the constant solution asymptotically.

EXAMPLE 3   Sketch the solutions to $y' = y^2 - 4y$ satisfying $y(0) = 4.5$ and $y(0) = 3$.

Solution   Refer to Fig. 9. Since $g(y) = y^2 - 4y = y(y - 4)$, the zeros of $g(y)$ are 0 and 4; hence the constant solutions are $y = 0$ and $y = 4$. The solution satisfying $y(0) = 4.5$ is increasing, because the z-coordinate is positive when $y = 4.5$ on the yz graph. This solution continues to increase without bound. The solution satisfying $y(0) = 3$ is decreasing because the z-coordinate is negative when $y = 3$ on the yz graph. This solution will decrease and approach asymptotically the constant solution $y = 0$.

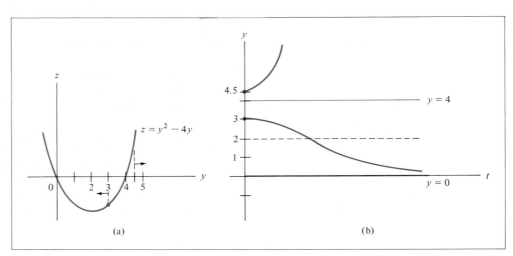

(a)   (b)

FIGURE 9

An additional piece of information about the solution satisfying $y(0) = 3$ may be obtained from the graph of $z = g(y)$. We know that y decreases from 3 and approaches 0. From the graph of $z = g(y)$ in Fig. 9 it appears that at first the z-coordinates become more negative until y reaches 2 and then become less negative as y moves on toward 0. Since these z-coordinates are slopes on the solution curve, we conclude that as the solution moves downward from its initial point on the ty-coordinate system, its slope becomes more negative until the y-coordinate is 2 and then the slope becomes less negative as the y-coordinate approaches 0. Hence the solution is concave down until $y = 2$ and then is concave up. Thus there is an inflection point at $y = 2$, where the concavity changes. ●

We saw in Example 3 that the inflection point at $y = 2$ was produced by the fact that $g(y)$ had a minimum at $y = 2$. A generalization (see below) of the argument in Example 3 shows that inflection points of solution curves occur at each value of y where $g(y)$ has a nonzero relative maximum or minimum point. Thus we may formulate an additional rule for sketching a solution of $y' = g(y)$.

7. On the ty-coordinate system draw dashed horizontal lines at all values of y at which $g(y)$ has a *nonzero* relative maximum or minimum point. A solution curve will have an inflection point whenever it crosses such a dashed line.

It is useful to note that when $g(y)$ is a quadratic function, as in Example 3, its maximum or minimum point occurs at a value of y halfway between the zeros of $g(y)$. This is because the graph of a quadratic function is a parabola, which is symmetric about a vertical line through its vertex.

EXAMPLE 4   Sketch a solution to $y' = e^{-y}$ with $y(0) > 0$.

Solution   Refer to Fig. 10. Since $g(y) = e^{-y}$ is always positive, there are no constant solutions to the differential equation and every solution will increase without bound. When drawing solutions that asymptotically approach a horizontal line, we nave no choice as to whether to draw it concave up or concave down. This decision will be obvious from its increasing or decreasing nature and from knowledge of inflection points. However, for solutions that grow without bound, we must look at $g(y)$ in order to determine concavity. In this example, as t increases, the values of y increase. As y increases, $g(y)$ becomes less positive. Since $g(y) = y'$, we deduce that the slope of the solution curve becomes less positive; therefore, the solution curve is concave down. ●

FIGURE 10

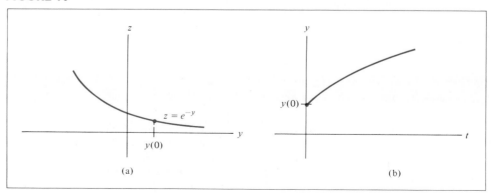

(a)

(b)

TECHNOLOGY PROJECT 10.3
Quantitative Alternatives to Qualitative Methods

Rework Examples 2 and 3 using a symbolic math program as follows:

1. Determine the actual solution of the differential equation satisfying each initial condition.

2. Graph the solution and confirm the general shape of the curve derived in the example.

3. What advantage do qualitative methods have over the method just developed in 1 and 2?

PRACTICE PROBLEMS 10.4

Consider the differential equation $y' = g(y)$, where $g(y)$ is the function whose graph is drawn in Fig. 11.

FIGURE 11

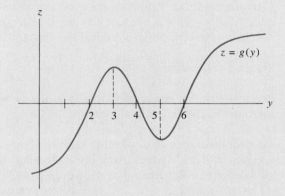

1. How many constant solutions are there to the differential equation $y' = g(y)$?
2. For what initial values $y(0)$ will the corresponding solution of the differential equation be an increasing function?
3. Is it true that if the initial value $y(0)$ is near 4, the corresponding solution will be asymptotic to the constant solution $y = 4$?
4. For what initial values $y(0)$ will the corresponding solution of the differential equation have an inflection point?

EXERCISES 10.4

One or more initial conditions are given for each differential equation below. Use the qualitative theory of autonomous differential equations to sketch the graphs of the corresponding solutions. Include a yz graph if one is not already provided. Always indicate the constant solutions on the ty graph whether they are mentioned or not.

1. $y' = 2y - 6$, $y(0) = 1$, $y(0) = 4$. (The graph of $z = g(y)$ is drawn in Fig. 12.)

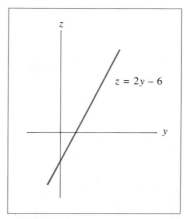

FIGURE 12

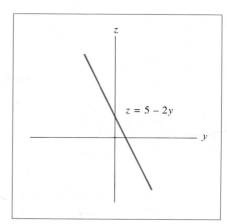

FIGURE 13

2. $y' = 5 - 2y$, $y(0) = 1$, $y(0) = 4$. (See Fig. 13.)

3. $y' = 4 - y^2$, $y(0) = -3$, $y(0) = -1$, $y(0) = 3$. (See Fig. 14.)

4. $y' = y^2 - 5$, $y(0) = -4$, $y(0) = 2$, $y(0) = 3$. (See Fig. 15.)

5. $y' = y^2 - 6y + 5$ or $y' = (y - 1)(y - 5)$, $y(0) = -1$, $y(0) = 2$, $y(0) = 4$, $y(0) = 6$. (See Fig. 16.)

6. $y' = -\frac{1}{3}(y + 2)(y - 4)$, $y(0) = -3$, $y(0) = -1$, $y(0) = 6$. (See Fig. 17.)

7. $y' = y^3 - 9y$ or $y' = y(y^2 - 9)$, $y(0) = -4$, $y(0) = -1$, $y(0) = 2$, $y(0) = 4$. (See Fig. 18.)

8. $y' = 9y - y^3$, $y(0) = -4$, $y(0) = -1$, $y(0) = 2$, $y(0) = 4$. (See Fig. 19.)

FIGURE 14

FIGURE 15

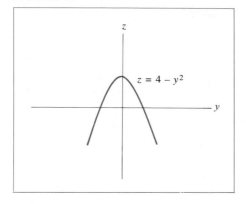

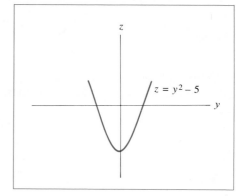

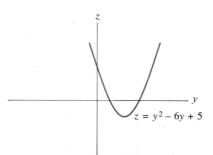

FIGURE 16

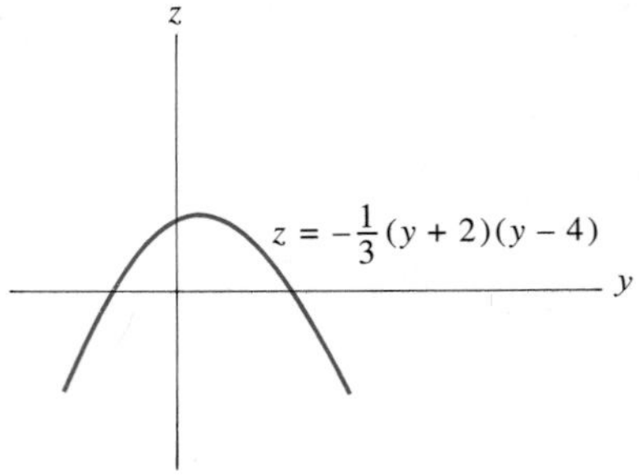

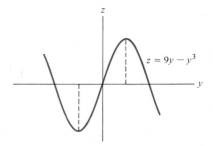

FIGURE 17

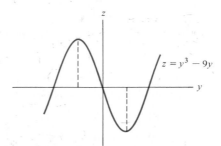

FIGURE 18

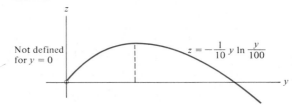

FIGURE 19

9. Use the graph in Fig. 20 to sketch the solutions to the Gompertz growth equation

$$\frac{dy}{dt} = -\frac{1}{10}y \ln \frac{y}{100}$$

satisfying $y(0) = 10$ and $y(0) = 150$.

10. The graph of $z = -\frac{1}{2}y \ln(y/30)$ has the same general shape as the graph in Fig. 20 with relative maximum point at $y \approx 11.0364$ and y-intercept at $y = 30$. Sketch the solutions to the Gompertz growth equation

$$\frac{dy}{dt} = -\frac{1}{2}y \ln \frac{y}{30}$$

satisfying $y(0) = 1$, $y(0) = 20$, and $y(0) = 40$.

FIGURE 20

11. $y' = g(y), y(0) = -.5, y(0) = .5$, where $g(y)$ is the function whose graph is given in Fig. 21.

12. $y' = g(y), y(0) = 0, y(0) = 4$, where the graph of $g(y)$ is given in Fig. 22.

13. $y' = g(y), y(0) = 0, y(0) = 1.2, y(0) = 5, y(0) = 7$, where the graph of $g(y)$ is given in Fig. 23.

14. $y' = g(y), y(0) = 1, y(0) = 3, y(0) = 11$, where the graph of $g(y)$ is given in Fig. 24.

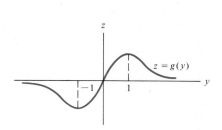

FIGURE 21

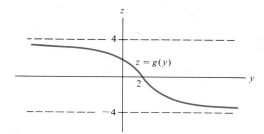

FIGURE 22

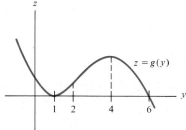

FIGURE 23

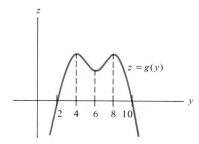

FIGURE 24

15. $y' = 3 - \frac{1}{2}y, y(0) = 2, y(0) = 8$.

16. $y' = 3y, y(0) = -2, y(0) = 2$.

17. $y' = 5y - y^2, y(0) = 1, y(0) = 7$.

18. $y' = -y^2 + 10y - 21, y(0) = 1, y(0) = 4$.

19. $y' = y^2 - 3y - 4, y(0) = 0, y(0) = 3$.

20. $y' = \frac{1}{2}y^2 - 3y, y(0) = 3, y(0) = 6, y(0) = 9$.

21. $y' = y^2 + 2, y(0) = -1, y(0) = 1$.

22. $y' = y - \frac{1}{4}y^2, y(0) = -1, y(0) = 1$.

23. $y' = \sin y, y(0) = -\pi/6, y(0) = \pi/6, y(0) = 7\pi/4$.

24. $y' = 1 + \sin y, y(0) = 0, y(0) = \pi$.

25. $y' = 1/y, y(0) = -1, y(0) = 1$.

26. $y' = y^3, y(0) = -1, y(0) = 1$.

27. $y' = ky^2$, where k is a negative constant, $y(0) = -2$, $y(0) = 2$.

28. $y' = ky(M - y)$, where $k > 0, M > 10$, and $y(0) = 1$.

29. $y' = ky - A$, where k and A are positive constants. Sketch solutions where $0 < y(0) < A/k$ and $y(0) > A/k$.

30. $y' = k(y - A)$, where $k < 0$ and $A > 0$. Sketch solutions where $y(0) < A$ and $y(0) > A$.

31. Suppose that once a sunflower plant has started growing, the rate of growth at any time is proportional to the product of its height and the difference between its height at maturity and its current height. Give a differential equation that is satisfied by $f(t)$, the height at time t, and sketch the solution.

32. A parachutist has a terminal velocity of -176 feet per second. That is, no matter how long a person falls, his or her speed will not exceed 176 feet per second, but it will get arbitrarily close to that value. The velocity in feet per second, $v(t)$, after t seconds satisfies the differential equation $v'(t) = 32 - k \cdot v(t)$. What is the value of k?

SOLUTIONS TO PRACTICE PROBLEMS 10.4

1. Three. The function $g(y)$ has zeros when y is 2, 4, and 6. Therefore, $y' = g(y)$ has the constant functions $y = 2$, $y = 4$, and $y = 6$ as solutions.

2. For $2 < y(0) < 4$ and $y(0) > 6$. Since nonconstant solutions are either strictly increasing or strictly decreasing, a solution is an increasing function provided that it is increasing at time $t = 0$. This is the case when the first derivative is positive at $t = 0$. When $t = 0$, $y' = g(y(0))$. Therefore, the solution corresponding to $y(0)$ is increasing whenever $g(y(0))$ is positive.

3. Yes. If $y(0)$ is slightly to the right of 4, then $g(y(0))$ is negative, so the corresponding solution will be a decreasing function with values moving to the left closer and closer to 4. If $y(0)$ is slightly to the left of 4, then $g(y(0))$ is positive, so the corresponding solution will be an increasing function with values moving to the right closer and closer to 4. (The constant solution $y = 4$ is referred to as a *stable* constant solution. The solution with initial value 4 stays at 4, and solutions with initial values near 4 move toward 4. The constant solution $y = 2$ is *unstable*. Solutions with initial values near 2 move away from 2.)

4. For $2 < y(0) < 3$ and $5 < y(0) < 6$. Inflection points of solutions correspond to relative maximum and relative minimum points of the function $g(y)$. If $2 < y(0) < 3$, the corresponding solution will be an increasing function. The values of y will move to the right (toward 4) and therefore will cross 3, a place at which $g(y)$ has a relative maximum point. Similarly, if $5 < y(0) < 6$, the corresponding solution will be decreasing. The values of y on the yz graph will move to the left and cross 5.

10.5 APPLICATIONS OF DIFFERENTIAL EQUATIONS

Equations describing conditions in a physical process are often referred to as *mathematical models.* In this section we study real-life situations that may be modeled by an autonomous differential equation $y' = g(y)$. Here y will represent some quantity that is changing with time, and the equation $y' = g(y)$ will be obtained from a description of the rate of change of y.

We have already encountered two situations where the rate of change of y is *proportional* to some quantity:

1. $y' = ky$: "the rate of change of y is proportional to y" (exponential growth or decay).
2. $y' = k(M - y)$: "the rate of change of y is proportional to the difference between M and y" (Newton's law of cooling, for example).

The following example is of the same general type. It concerns the rate at which a technological innovation may spread through an industry, a subject of concern to both sociologists and economists.

EXAMPLE I The by-product coke oven was first introduced into the iron and steel industry in 1894. It took about 30 years before all the major steel producers had adopted this innovation. Let $f(t)$ be the percentage of the producers that had installed the new coke ovens by time t. Then a reasonable model* for the way $f(t)$ increased is given by the assumption that the rate of change of $f(t)$ at time t was proportional to the product of $f(t)$ and the percentage of firms that had not yet installed the new coke ovens at time t. Write a differential equation that is satisfied by $f(t)$.

Solution Since $f(t)$ is the *percentage* of firms that have the new coke oven, $100 - f(t)$ is the percentage of firms that still have not installed any new coke ovens. We are told that the rate of change of $f(t)$ is proportional to the product of $f(t)$ and $100 - f(t)$. Hence there is a constant of proportionality k such that

$$f'(t) = kf(t)[100 - f(t)].$$

Replacing $f(t)$ by y and $f'(t)$ by y', we obtain the desired differential equation,

$$y' = ky(100 - y).$$

Note that both y and $100 - y$ are nonnegative quantities. Clearly, y' must be positive, because $y = f(t)$ is an increasing function. Hence the constant k must be positive. ●

---

*See E. Mansfield, "Technical Change and the Rate of Imitation," *Econometrica,* **29** (1961), 741–766.

The differential equation obtained in Example 1 is a special case of the *logistic differential equation,*

$$y' = ky(a - y), \tag{1}$$

where k and a are positive constants. This equation is used as a simple mathematical model of a wide variety of physical phenomena. In Section 5.4, we described applications of the logistic equation to restricted population growth and to the spread of an epidemic. Let us use the qualitative theory of differential equations to gain more insight into this important equation.

The first step in sketching solutions of (1) is to draw the yz graph. Rewriting the equation $z = ky(a - y)$ in the form

$$z = -ky^2 + kay,$$

we see that the equation is quadratic in y and hence its graph will be a parabola. The parabola is concave down because the coefficient of y^2 is negative (since k is a positive constant). The zeros of the quadratic expression $ky(a - y)$ occur where $y = 0$ and $y = a$. Since a represents some positive constant, we select an arbitrary point on the positive y-axis and label it "a." With this information we can sketch a representative graph, as in Fig. 1. Note that the vertex of the parabola occurs at $y = a/2$, halfway between the y-intercepts. (The reader should review how we obtained this graph, given only that k and a are positive constants. Similar situations will arise in the exercises.)

We begin the ty graph as in Fig. 2, showing the constant solutions and placing a dashed line at $y = a/2$, where certain solution curves will have an inflection point. On either side of the constant solutions we choose initial values for y—say, y_1, y_2, y_3, y_4. Then we use the yz graph to sketch the corresponding solution curves, as in Fig. 3.

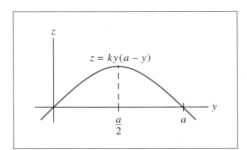

FIGURE 1   **yz graph for a logistic differential equation.**

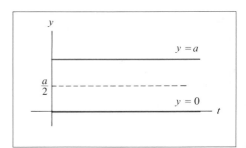

FIGURE 2

The solution in Fig. 3(b) beginning at y_2 has the general shape usually referred to as a *logistic curve.* This is the type of solution that would model the situation described in Example 1. The solution in Fig. 3(b) beginning at y_1 usually has no physical significance. The other solutions shown in Fig. 3(b) can occur in practice, particularly in the study of population growth.

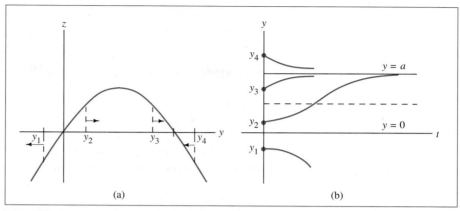

FIGURE 3

In ecology, the growth of a population is often described by a logistic equation written in the form

$$\frac{dN}{dt} = rN\left(\frac{K - N}{K}\right) \tag{2}$$

or, equivalently,

$$\frac{dN}{dt} = \frac{r}{K}N(K - N),$$

where N is used instead of y to denote the size of the population at time t. Typical solutions of this equation are sketched in Fig. 4. The constant K is called the *carrying capacity* of the environment. When the initial population is close to zero, the population curve has the typical S-shaped appearance, and N approaches the carrying capacity asymptotically. When the initial population is

FIGURE 4   A logistic model for population change.

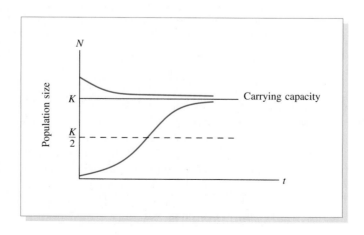

greater than K, the population decreases in size, again asymptotically approaching the carrying capacity.

The quantity $(K - N)/K$ in (2) is a fraction between 0 and 1. It reflects the limiting effect of the environment upon the population and is close to 1 when N is close to 0. If this fraction were replaced by the constant 1, then (2) would become

$$\frac{dN}{dt} = rN.$$

This is the equation for ordinary exponential growth, where r is the growth rate. For this reason the parameter r in (2) is called the *intrinsic rate of growth* of the population. It expresses how the population would grow if the environment were to permit unrestricted exponential growth.

We now turn to applications that involve a different sort of autonomous differential equation. The main idea is illustrated in the following example.

EXAMPLE 2   A savings account earns 6% interest per year, compounded continuously, and continuous withdrawals are made from the account at the rate of $900 per year. Set up a differential equation that is satisfied by the amount $f(t)$ of money in the account at time t. Sketch typical solutions of the differential equation.

Solution   At first, let us ignore the withdrawals from the account. In Section 5.2 we discussed continuous compounding of interest and showed that if no deposits or withdrawals are made, then $f(t)$ satisfies the equation

$$y' = .06y.$$

That is, the savings account grows at a rate proportional to the size of the account. Since this growth comes from the interest, we conclude that *interest is being added to the account at a rate proportional to the amount in the account.*

Now suppose that continuous withdrawals are made from this same account at the rate of $900 per year. Then there are two influences on the way the amount of money in the account changes—the rate at which interest is added and the rate at which money is withdrawn. The rate of change of $f(t)$ is the *net effect* of these two influences. That is, $f(t)$ now satisfies the equation

$$y' = .06y - 900,$$

$$\begin{bmatrix} \text{rate of change} \\ \text{of } y \end{bmatrix} = \begin{bmatrix} \text{rate at which} \\ \text{interest is added} \end{bmatrix} - \begin{bmatrix} \text{rate at which} \\ \text{money is withdrawn} \end{bmatrix}.$$

The qualitative sketches for this differential equation are given in Fig. 5. The constant solution is found by solving $.06y - 900 = 0$, which gives $y = 900/.06 = 15{,}000$. If the initial amount $y(0)$ in the account is $15,000, the balance in the account will always be $15,000. If the initial amount is greater than $15,000, the savings account will accumulate money without bound. If the initial amount is less than $15,000, the account balance will decrease. Presumably, the bank will stop withdrawals when the account balance reaches zero. ●

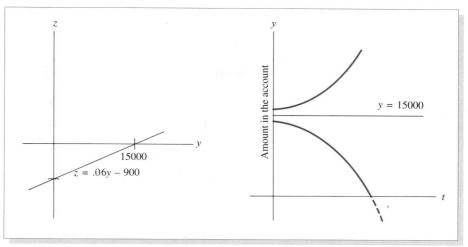

FIGURE 5   A differential equation model for a savings account:
$y' = .06y - 900.$

We may think of the savings account in Example 2 as a compartment or container into which money (interest) is being steadily added and also from which money is being steadily withdrawn. (See Fig. 6.)

FIGURE 6   A one-compartment model in economics.

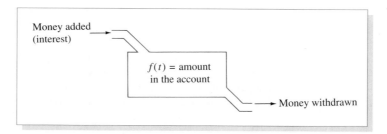

A similar situation arises frequently in physiology in what are called "one-compartment problems."* Typical examples of a compartment are a person's lungs, the digestive system, and the cardiovascular system. A common problem is to study the rate at which some substance in the compartment is changing when two or more processes are increasing or decreasing the substance in the compartment. In many important cases, each of these processes changes the substance either at a constant rate or at a rate proportional to the amount in the compartment.

An earlier example of such a one-compartment problem, discussed in Section 5.4, concerned the continuous infusion of glucose into a patient's bloodstream. A similar situation is discussed in the next example.

---

*See William Simon, *Mathematical Techniques for Physiology and Medicine* (New York: Academic Press, 1972), Chap. V.

EXAMPLE 3   (*A One-Compartment Mixing Process*) Consider a flask that contains 3 liters of salt water. Suppose that water containing 25 grams per liter of salt is pumped into the flask at the rate of 2 liters per hour, and the mixture, being steadily stirred, is pumped out of the flask at the same rate. Find a differential equation satisfied by the amount of salt $f(t)$ in the flask at time t.

Solution   Let $f(t)$ be the amount of salt measured in grams. Since the volume of the mixture in the flask is being held constant at 3 liters, the concentration of salt in the flask at time t is

$$[\text{concentration}] = \frac{[\text{amount of salt}]}{[\text{volume of mixture}]} = \frac{f(t) \text{ grams}}{3 \text{ liters}} = \frac{1}{3} f(t) \frac{\text{grams}}{\text{liter}}.$$

Next we compute the rates at which salt is entering and leaving the flask at time t:

$$[\text{rate of salt entering}] = [\text{concentration}] \times [\text{flow rate}]$$

$$= \left[25 \frac{\text{grams}}{\text{liter}} \right] \times \left[2 \frac{\text{liters}}{\text{hour}} \right]$$

$$= 50 \frac{\text{grams}}{\text{hour}}.$$

$$[\text{rate of salt leaving}] = [\text{concentration}] \times [\text{flow rate}]$$

$$= \left[\frac{1}{3} f(t) \frac{\text{grams}}{\text{liter}} \right] \times \left[2 \frac{\text{liters}}{\text{hour}} \right]$$

$$= \frac{2}{3} f(t) \frac{\text{grams}}{\text{hour}}.$$

The *net* rate of change of salt (in grams per hour) at time t is $f'(t) = 50 - \frac{2}{3} f(t)$. Hence the desired differential equation is

$$y' = 50 - \tfrac{2}{3} y. \quad \bullet$$

Differential Equations in Population Genetics   In population genetics, hereditary phenomena are studied on a populational level rather than on an individual level. Consider a particular hereditary feature of an animal, such as the length of the hair. Suppose that basically there are two types of hair for a certain animal—long hair and short hair. Also suppose that long hair is the dominant type. Let A denote the gene responsible for long hair and a the gene responsible for short hair. Each animal has a pair of these genes—either AA ("dominant" individuals), or aa ("recessive" individuals), or Aa ("hybrid" individuals). If there are N animals in the population, then there are $2N$ genes in the population controlling hair length. Each Aa individual has one a gene, and each aa individual has two a genes. The total number of a genes in the population divided by $2N$ gives the fraction of a genes. This fraction is called the *gene frequency of a* in the population. Similarly, the fraction of A genes is called the gene frequency of A. Note that

$$\begin{bmatrix} \text{gene} \\ \text{frequency} \\ \text{of } a \end{bmatrix} + \begin{bmatrix} \text{gene} \\ \text{frequency} \\ \text{of } A \end{bmatrix} = \frac{[\text{number of } a \text{ genes}]}{2N} + \frac{[\text{number of } A \text{ genes}]}{2N}$$

$$= \frac{2N}{2N} = 1. \tag{3}$$

We shall denote the gene frequency of a by q. From (3) it follows that the gene frequency of A is $1 - q$.

   An important problem in population genetics involves the way the gene frequency q changes as the animals in the population reproduce. If each unit on the time axis represents one "generation," we can consider q as a function of time t (Fig. 7). In general, many hundreds or thousands of generations are studied, so the time for one generation is small compared with the overall time period. For many purposes, q is considered to be a differentiable function of t. In what follows we assume that the population mates at random and that the distribution of a and A genes is the same for males and females. In this case one can show from elementary probability theory that the gene frequency is essentially constant from one generation to the next when no "disturbing factors" are present, such as mutation or external influences on the population. We shall discuss differential equations that describe the effect of such disturbing factors on $q(t)$.*

FIGURE 7   Gene frequency in a population.

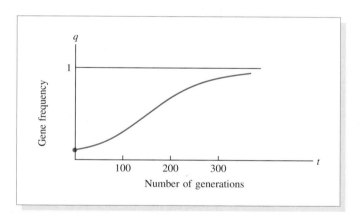

Gene frequency

q

1

100   200   300 t

Number of generations

   Suppose that in every generation a fraction ν of the a genes mutate and become A genes. Then the rate of change of the gene frequency q due to this mutation is

$$\frac{dq}{dt} = -\nu q. \tag{4}$$

*See C. C. Li, *Population Genetics* (Chicago: University of Chicago Press, 1955), pp. 240–263, 283–286.

To understand this equation, think of q as a measure of the number of a genes, and think of the a genes as a population that is losing members at a constant percentage rate of $100\nu\%$ per generation (i.e., per unit time). This is an exponential decay process, so q satisfies the exponential decay equation (4). Now suppose, instead, that in every generation a fraction μ of the A genes mutate into a genes. Since the gene frequency of A is $1 - q$, the decrease in the gene frequency of A due to mutation will be $\mu(1 - q)$ per generation. But this must match the increase in the gene frequency of a, since from (3) the sum of the two gene frequencies is constant. Thus the rate of change of the gene frequency of a per generation, due to the mutation from A to a, is given by

$$\frac{dq}{dt} = \mu(1 - q).$$

When mutation occurs, it frequently happens that in each generation a fraction μ of A mutate to a and at the same time a fraction ν of a mutate to A. The *net effect* of these two influences on the gene frequency q is described by the equation

$$\frac{dq}{dt} = \mu(1 - q) - \nu q. \tag{5}$$

(The situation here is analogous to the one-compartment problems discussed earlier.)

Let us make a qualitative analysis of equation (5). To be specific, we take $\mu = .00003$ and $\nu = .00001$. Then

$$\frac{dq}{dt} = .00003(1 - q) - .00001q$$

$$= .00003 - .00004q$$

or

$$\frac{dq}{dt} = -.00004(q - .75). \tag{6}$$

Figure 8(a) shows the graph of $z = -.00004(q - .75)$ with the z-axis scale greatly enlarged. Typical solution curves are sketched in Fig. 8(b). We see that the gene frequency $q = .75$ is an equilibrium value. If the initial value of q is smaller than .75, the value of q will rise under the effect of the mutations; after many generations it will be approximately .75. If the initial value of q is between .75 and 1.00, then q will eventually decrease to .75. The equilibrium value is completely determined by the magnitudes of the two opposing rates of mutation μ and v. From (6) we see that the rate of change of gene frequency is proportional to the difference between q and the equilibrium value .75.

In the study of how a population adapts to an environment over a long period, geneticists assume that some hereditary types have an advantage over others in survival and reproduction. Suppose that the adaptive ability of the hybrid (Aa) individuals is slightly greater than that of both the dominant (AA) and the recessive (aa) individuals. In this case it turns out that the rate of change

FIGURE 8

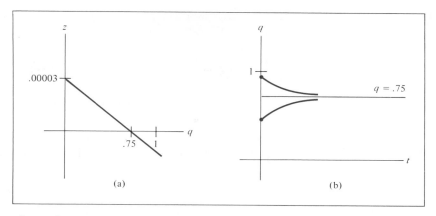

(a)   (b)

of gene frequency due to this "selection pressure" is

$$\frac{dq}{dt} = q(1 - q)(c - eq), \tag{7}$$

where c and e are positive constants with $c < e$. On the other hand, if the adaptive ability of the hybrid individuals is slightly less than that of both the dominant and the recessive individuals, it can be shown that

$$\frac{dq}{dt} = kq(1 - q)(2q - 1), \tag{8}$$

where k is a constant between 0 and 1, called the *coefficient of selection against hybrids*.

It is possible to consider the joint effects of mutation and natural selection. Suppose that mutations from A to a occur at a rate μ per generation and from a to A at a rate ν per generation. Suppose also that selection is against recessive individuals (i.e., recessives do not adapt as well as the rest of the population). Then the net rate of change in gene frequency turns out to be

$$\frac{dq}{dt} = \mu(1 - q) - \nu q - kq^2(1 - q).$$

Here $\mu(1 - q)$ represents the gain in a genes from mutations $A \rightarrow a$, the term νq is the loss in a genes from mutations $a \rightarrow A$, and the term $kq^2(1 - q)$ represents the loss in a genes due to natural selection pressures.

PRACTICE PROBLEMS 10.5

1. Refer to Example 3, involving the flow of salt water through a flask. Will $f(t)$ be an increasing or a decreasing function?

2. (*Rate of Litter Accumulation*) In a certain tropical forest, "litter" (mainly dead vegetation such as leaves and vines) forms on the ground at the rate of 10 grams per square centimeter per year. At the same time, however, the litter is decomposing at the rate of 80% per year. Let $f(t)$ be the amount of litter (in grams per square centimeter) present at time t. Find a differential equation satisfied by $f(t)$.

EXERCISES 10.5

1. (*Social Diffusion*) For information being spread by mass media, rather than through individual contact, the rate of spread of the information at any time is proportional to the percentage of the population not having the information at that time. Give the differential equation that is satisfied by $y = f(t)$, the percentage of the population having the information at time t. Assume that $f(0) = 1$. Sketch the solution.

2. (*Gravity*) At one point in his study of a falling body starting from rest, Galileo conjectured that its velocity at any time is proportional to the distance it has dropped. Using this hypothesis, set up the differential equation whose solution is $y = f(t)$, the distance fallen by time t. By making use of the initial value, show why Galileo's original conjecture is invalid.

3. (*Autocatalytic Reaction*) In an autocatalytic reaction, one substance is converted into a second substance in such a way that the second substance catalyzes its own formation. This is the process by which trypsinogen is converted into the enzyme trypsin. The reaction starts only in the presence of some trypsin, and each molecule of trypsinogen yields one molecule of trypsin. The rate of formation of trypsin is proportional to the product of the amounts of the two substances present. Set up the differential equation that is satisfied by $y = f(t)$, the amount (number of molecules) of trypsin present at time t. Sketch the solution. For what value of y is the reaction proceeding the fastest? [*Note:* Letting M be the total amount of the two substances, the amount of trypsinogen present at time t is $M - f(t)$.]

4. (*Drying*) A porous material dries outdoors at a rate that is proportional to the moisture content. Set up the differential equation whose solution is $y = f(t)$, the amount of water at time t in a towel on a clothesline. Sketch the solution.

5. (*Movement of Solutes Through a Cell Membrane*) Let c be the concentration of a solute outside a cell that we assume to be constant throughout the process—that is, unaffected by the small influx of the solute across the membrane due to a difference in concentration. The rate of change of the concentration of the solute inside the cell at any time t is proportional to the difference between the outside concentration and the inside concentration. Set up the differential equation whose solution is $y = f(t)$, the concentration of the solute inside the cell at time t. Sketch a solution.

6. An experimenter reports that a certain strain of bacteria grows at a rate proportional to the square of the size of the population. Set up a differential equation which describes the growth of the population. Sketch a solution.

7. (*Chemical Reaction*) Suppose that substance A is converted into substance B at a rate that, at any time t, is proportional to the square of the amount of A. This situation occurs, for instance, when it is necessary for two molecules of A to collide in order to create one molecule of B. Set up the differential equation that is satisfied by $y = f(t)$, the amount of substance A at time t. Sketch a solution.

8. (*War Fever*) L. F. Richardson proposed the following model to describe the spread of war fever.* If $y = f(t)$ is the percent of the population advocating war at time t,

---

*See L. F. Richardson, "War Moods I," *Psychometrica,* 1948, p. 13.

then the rate of change of $f(t)$ at any time is proportional to the product of the percentage of the population advocating war and the percentage not advocating war. Set up a differential equation that is satisfied by $y = f(t)$ and sketch a solution.

9. (*Capital Investment Model*) In economic theory, the following model is used to describe a possible capital investment policy. Let $f(t)$ represent the total invested capital of a company at time t. Additional capital is invested whenever $f(t)$ is below a certain equilibrium value E, and capital is withdrawn whenever $f(t)$ exceeds E. The rate of investment is proportional to the difference between $f(t)$ and E. Construct a differential equation whose solution is $f(t)$ and sketch two or three typical solution curves.

10. (*Evans Price Adjustment Model*) Consider a certain commodity that is produced by many companies and purchased by many other firms. Over a relatively short period there tends to be an equilibrium price p_0 per unit of the commodity that balances the supply and the demand. Suppose that, for some reason, the price is different from the equilibrium price. The Evans price adjustment model says that the rate of change of price with respect to time is proportional to the difference between the actual market price p and the equilibrium price. Write a differential equation that expresses this relation.

11. (*Continuous Annuity*) A *continuous annuity* is a steady stream of money that is paid to some person. Such an annuity may be established, for example, by making an initial deposit in a savings account and then making steady withdrawals to pay the continuous annuity. Suppose that an initial deposit of $5400 is made into a savings account that earns $5\frac{1}{2}\%$ interest compounded continuously, and immediately continuous withdrawals are begun at the rate of $300 per year. Set up the differential equation that is satisfied by the amount $f(t)$ of money in the account at time t. Sketch the solution.

12. An initial deposit of $10,000 is made into a savings account that earns 6% interest compounded continuously. Six months later continuous withdrawals are begun at the rate of $600 per year. Set up the differential equation that is satisfied by the amount $f(t)$ of money in the account at time t for $t \geq \frac{1}{2}$. Sketch the solution.

13. A company wishes to set aside funds for future expansion and so arranges to make continuous *deposits* into a savings account at the rate of $10,000 per year. The savings account earns 5% interest compounded continuously.

   (a) Set up the differential equation that is satisfied by the amount $f(t)$ of money in the account at time t.

   (b) Solve the differential equation in part (a), assuming that $f(0) = 0$, and determine how much money will be in the account at the end of 5 years.

14. A company arranges to make continuous deposits into a savings account at the rate of P dollars per year. The savings account earns 5% interest compounded continuously. Find the approximate value of P that will make the savings account balance amount to $50,000 in 4 years.

15. The air in a crowded room full of people contains .25% carbon dioxide (CO_2). An air conditioner is turned on that blows fresh air into the room at the rate of 500 cubic feet per minute. The fresh air mixes the stale air and the mixture leaves the room at the rate of 500 cubic feet per minute. The fresh air contains .01% CO_2, and the room has a volume of 2500 cubic feet.

(a) Find a differential equation satisfied by the amount $f(t)$ of CO_2 in the room at time t.

(b) The model developed in part (a) ignores the CO_2 produced by the respiration of the people in the room. Suppose that the people generate .08 cubic foot of CO_2 per minute. Modify the differential equation in (a) to take into account this additional source of CO_2.

16. A certain drug is administered intravenously to a patient at the continuous rate of 5 milligrams per hour. The patient's body removes the drug from the bloodstream at a rate proportional to the amount of the drug in the blood. Write a differential equation that is satisfied by the amount $f(t)$ of the drug in the blood at time t. Sketch a typical solution.

17. A single dose of iodine is injected intravenously into a patient. Suppose that the iodine mixes thoroughly in the blood before any is lost as a result of metabolic processes (and ignore the time required for this mixing process). Iodine will leave the blood and enter the thyroid gland at a rate proportional to the amount of iodine in the blood. Also, iodine will leave the blood and pass into the urine at a (different) rate proportional to the amount of iodine in the blood. Suppose that the iodine enters the thyroid at the rate of 4% per hour, and the iodine enters the urine at the rate of 10% per hour. Let $f(t)$ denote the amount of iodine in the blood at time t. Write a differential equation satisfied by $f(t)$.

18. Show that the mathematical model in Problem 2 predicts that the amount of litter in the forest will eventually stabilize. What is the "equilibrium level" of litter in that problem? [*Note:* Today most forests are close to their equilibrium levels. This was not so during the Carboniferous Period when the great coal deposits were formed.]

19. In the study of the effect of natural selection on a population, one encounters the differential equation

$$\frac{dq}{dt} = -.0001q^2(1 - q),$$

where q is the frequency of a gene a and the selection pressure is against the recessive genotype aa. Sketch a solution of this equation when $q(0)$ is close to but slightly less than 1.

20. Typical values of c and d in equation (7) are $c = .15$, $d = .50$, and a typical value of k in equation (8) is $k = .05$. Sketch representative solutions for the equations

(a) $\dfrac{dq}{dt} = q(1 - q)(.15 - .50q)$ (selection favoring hybrids),

(b) $\dfrac{dq}{dt} = .05q(1 - q)(2q - 1)$ (selection against hybrids).

Consider various initial conditions with $q(0)$ between 0 and 1. Discuss possible genetic interpretations of these curves; that is, describe the effect of "selection" on the gene frequency q in terms of the various initial conditions.

SOLUTIONS TO PRACTICE PROBLEMS 10.5

1. The nature of the function $f(t)$ depends on the initial amount of salt water in the flask. Figure 9 contains solutions for three different initial amounts, $y(0)$. If the initial amount is less than 75 grams, the amount of salt in the flask will increase asymptotically to 75. If the initial concentration is greater than 75 grams, the amount of salt in the flask will decrease asymptotically to 75. Of course, if the initial concentration is exactly 75 grams, the amount of salt in the flask will remain constant.

FIGURE 9

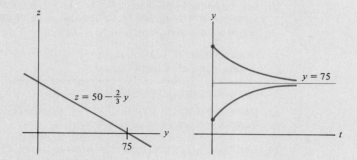

2. This problem resembles a one-compartment problem, where the forest floor is the compartment. We have

$$\begin{bmatrix} \text{rate of change of} \\ \text{litter} \end{bmatrix} = \begin{bmatrix} \text{rate of litter} \\ \text{formation} \end{bmatrix} - \begin{bmatrix} \text{rate of litter} \\ \text{decomposition} \end{bmatrix}$$

If $f(t)$ is the amount of litter (in grams per square centimeter) at time t, then the 80% decomposition rate means that at time t the litter is decaying at the rate of $.80 f(t)$ grams per square centimeter per year. Thus the net rate of change of litter is $f'(t) = 10 - .80 f(t)$. The desired differential equation is

$$y' = 10 - .80y.$$

 **CHAPTER 10 CHECKLIST**

✓ Solution of a differential equation
✓ Separation of variables
✓ Euler's method
✓ Autonomous differential equation $y' = g(y)$
✓ Qualitative theory of differential equations
✓ Logistic differential equation $y' = ky(a - y)$

CHAPTER 10   SUPPLEMENTARY EXERCISES

Solve the differential equations in Exercises 1–8.

1. $y^2 y' = 4t^3 - 3t^2 + 2$

2. $\dfrac{y'}{t + 1} = y + 1$

3. $y' = \dfrac{y}{t} - 3y, \ t > 0$

4. $(y')^2 = t$

5. $y = 7y' + ty', \ y(0) = 3$

6. $y' = te^{t+y}, \ y(0) = 0$

7. $yy' + t = 6t^2, \ y(0) = 7$

8. $y' = 5 - 8y, \ y(0) = 1$

9. Let $f(t)$ be the solution to $y' = 2e^{2t-y}, \ y(0) = 0$. Use Euler's method with $n = 4$ on $0 \le t \le 2$ to estimate $f(2)$. Then show that Euler's method gives the exact value of $f(2)$ by solving the differential equation.

10. Let $f(t)$ be the solution to $y' = (t + 1)/y, \ y(0) = 1$. Use Euler's method with $n = 3$ on $0 \le t \le 1$ to estimate $f(1)$. Then show that Euler's method gives the exact value of $f(1)$ by solving the differential equation.

11. Suppose that $f(t)$ is a solution of $y' = (2 - y)e^{-y}$. Is $f(t)$ increasing or decreasing at some value of t where $f(t) = 3$?

12. Solve the initial value problem $y' = e^{y^2}(\cos y)(1 - e^{y-1}), \ y(0) = 1$.

13. Use Euler's method with $n = 6$ on the interval $0 \le t \le 3$ to approximate the solution $f(t)$ to $y' = .1y(20 - y), \ y(0) = 2$.

14. Use Euler's method with $n = 5$ on the interval $0 \le t \le 1$ to approximate the solution $f(t)$ to $y' = \frac{1}{2}y(y - 10), \ y(0) = 9$.

Sketch the solutions of the differential equations in Exercises 15–24. In each case, also indicate the constant solutions.

15. $y' = 2 \cos y, \ y(0) = 0$

16. $y' = 5 + 4y - y^2, \ y(0) = 1$

17. $y' = y^2 + y, \ y(0) = -\frac{1}{3}$

18. $y' = y^2 - 2y + 1, \ y(0) = -1$

19. $y' = \ln y, \ y(0) = 2$

20. $y' = 1 + \cos y, \ y(0) = -\frac{3}{4}$

21. $y' = \dfrac{1}{y^2 + 1}, \ y(0) = -1$

22. $y' = \dfrac{3}{y + 3}, \ y(0) = 2$

23. $y' = .4y^2(1 - y), \ y(0) = -1, y(0) = .1, \ y(0) = 2$

24. $y' = y^3 - 6y^2 + 9y, \ y(0) = -\frac{1}{4}, y(0) = \frac{1}{4}, \ y(0) = 4$

25. The birth rate in a certain city is 3.5% per year and the death rate is 2% per year. Also, there is a net movement of population out of the city at a steady rate of 3000 people per year. Let $N = f(t)$ be the city's populatoin at time t.

    (a) Write a differential equation satisfied by N.

    (b) Use a qualitative analysis of the equation to determine if there is a size at which the population would remain constant. Is it likely that a city would have such a constant population?

26. Suppose that in a chemical reaction, each graph of substance A combines with 3 grams of substance B to form 4 grams of substance C. Suppose that the reaction begins with 10 grams of A, 15 grams of B, and 0 grams of C present. Let $y = f(t)$ be the amount of C present at time t. Suppose that the rate at whch substance C is formed is proportional to the product of the unreacted amounts of A and B present.

That is, suppose that $f(t)$ satisfies the differential equation

$$y' = k(10 - \frac{1}{4}y)(15 - \frac{3}{4}y), \qquad y(0) = 0,$$

where k is a constant.

(a) What do the quantities $10 - \frac{1}{4}f(t)$ and $15 - \frac{3}{4}f(t)$ represent?

(b) Should the constant k be positive or negative?

(c) Make a qualitative sketch of the solution of the differential equation above.

27. A bank account has \$20,000 earning 5% interest compounded continuously. A pensioner uses the account to pay himself an annuity, drawing continuously at a \$2000 annual rate. How long will it take for the balance in the account to drop to zero?

28. A continuous annuity of \$12,000 per year is to be funded by steady withdrawals from a savings account that earns 6% interest compounded continuously.

(a) What is the smallest initial amount in the account that will fund such an annuity forever?

(b) What initial amount will fund such an annuity for exactly 20 years (at which time the savings account balance will be zero)?

PROBABILITY AND CALCULUS

In this chapter we shall survey a few applications of calculus to the theory of probability. Since we do not intend this chapter to be a self-contained course in probability, we shall select only a few salient ideas to present a taste of probability theory and provide a starting point for further study.

11.1 DISCRETE RANDOM VARIABLES

We will motivate the concepts of mean, variance, standard deviation, and random variable by analyzing examination grades.

Suppose that the grades on an exam taken by 10 people are 50, 60, 60, 70, 70, 90, 100, 100, 100, 100. This information is displayed in a frequency table in Fig. 1.

One of the first things we do when looking over the results of an exam is to compute the *mean* or *average* of the grades. We do this by totaling the grades and dividing by the number of people.

$$[\text{mean}] = \frac{50 \cdot 1 + 60 \cdot 2 + 70 \cdot 2 + 90 \cdot 1 + 100 \cdot 4}{10} = \frac{800}{10} = 80.$$

Grade	50	60	70	90	100
Frequency	1	2	2	1	4

FIGURE 1

[Grade] − [Mean]	−30	−20	−10	10	20
Frequency	1	2	2	1	4

FIGURE 2

In order to get an idea of how spread out the grades are, we can compute the difference between each grade and the average grade. We have tabulated these differences in Fig. 2. For example, if a person received a 50, then [grade] − [mean] is $50 - 80 = -30$. As a measure of the spread of the grades, statisticians compute the average of the squares of these differences and call it the *variance* of the grade distribution. We have

$$[\text{variance}] = \frac{(-30)^2 \cdot 1 + (-20)^2 \cdot 2 + (-10)^2 \cdot 2 + (10)^2 \cdot 1 + (20)^2 \cdot 4}{10}$$

$$= \frac{900 + 800 + 200 + 100 + 1600}{10} = \frac{3600}{10} = 360.$$

The square root of the variance is called the *standard deviation* of the grade distribution. In this case, we have

$$[\text{standard deviation}] = \sqrt{360} \approx 18.97.$$

There is another way of looking at the grade distribution and its mean and variance. This new point of view is useful because it can be generalized to other situations. We begin by converting the frequency table to a relative frequency table. (See Fig. 3.) Below each grade we list the fraction of the class receiving that grade. The grade of 50 occurred $\frac{1}{10}$ of the time, the grade of 60 occurred $\frac{2}{10}$ of the time, and so on. Note that the relative frequencies add up to 1, because they represent the various fractions of the class grouped by test scores.

It is sometimes helpful to display the data in the relative frequency table by constructing a *relative frequency histogram* as in Fig. 4. Over each grade we place a rectangle whose height equals the relative frequency of that grade.

An alternative way to compute the mean grade is

$$[\text{mean}] = \frac{50 \cdot 1 + 60 \cdot 2 + 70 \cdot 2 + 90 \cdot 1 + 100 \cdot 4}{10}$$

$$= 50 \cdot \frac{1}{10} + 60 \cdot \frac{2}{10} + 70 \cdot \frac{2}{10} + 90 \cdot \frac{1}{10} + 100 \cdot \frac{4}{10}$$

$$= 5 + 12 + 14 + 9 + 40 = 80.$$

FIGURE 3

Grade	50	60	70	90	100
Relative frequency	$\frac{1}{10}$	$\frac{2}{10}$	$\frac{2}{10}$	$\frac{1}{10}$	$\frac{4}{10}$

FIGURE 4   A relative frequency histogram.

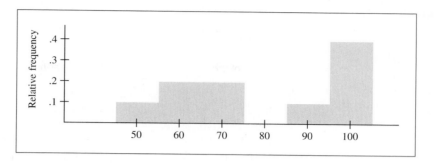

FIGURE 4   A relative fre-quency histogram.

Looking at the second line of this computation, we see that the mean is a sum of the various grades times their relative frequencies. We say that the mean is the *weighted sum* of the grades. (Grades are weighted by their relative frequencies.)

In a similar manner we see that the variance is also a weighted sum.

$$[\text{variance}] = [(50 - 80)^2 \cdot 1 + (60 - 80)^2 \cdot 2 + (70 - 80)^2 \cdot 2$$

$$+ (90 - 80)^2 \cdot 1 + (100 - 80)^2 \cdot 4]\frac{1}{10}$$

$$= (50 - 80)^2 \cdot \frac{1}{10} + (60 - 80)^2 \cdot \frac{2}{10} + (70 - 80)^2 \cdot \frac{2}{10}$$

$$+ (90 - 80)^2 \cdot \frac{1}{10} + (100 - 80)^2 \cdot \frac{4}{10}$$

$$= 90 + 80 + 20 + 10 + 160 = 360.$$

A relative frequency table such as is shown in Fig. 3 is also called a *probability table.* The reason for this terminology is as follows. Suppose that we perform an *experiment* that consists of picking an exam paper at random from among the ten papers. If the experiment is repeated many times, we expect the grade of 50 to occur about one-tenth of the time, the grade of 60 about two-tenths of the time, and so on. We say that the *probability* of the grade of 50 being chosen is $\frac{1}{10}$, the probability of the grade of 60 being chosen is $\frac{2}{10}$, and so on. In other words, the probability associated with a given grade measures the likelihood that an exam having that grade is chosen.

In this section we consider various experiments described by probability tables similar to the one in Fig. 3. The results of these experiments will be numbers (such as the exam scores above) called the *outcomes* of the experiment. We will also be given the probability of each outcome, indicating the relative frequency with which the given outcome is expected to occur if the experiment is repeated very often. If the outcomes of an experiment are $a_1, a_2, \ldots, a_n$, with respective probabilities $p_1, p_2, \ldots, p_n$, then we describe the experiment by a probability table as in Fig. 5. Since the probabilities indicate relative frequencies, we see that

$$0 \leq p_i \leq 1$$

FIGURE 5

Outcome	a_1	a_2	a_3	$\cdots$	a_n
Probability	p_1	p_2	p_3	$\cdots$	p_n

and

$$p_1 + p_2 + \cdots + p_n = 1.$$

The last equation indicates that the outcomes $a_1, \ldots, a_n$ comprise all possible results of the experiment. We will usually list the outcomes of our experiments in ascending order, so that $a_1 < a_2 < \cdots < a_n$.

We may display the data of a probability table in a histogram that has a rectangle of height p_i over the outcome a_i. (See Fig. 6.)

FIGURE 6

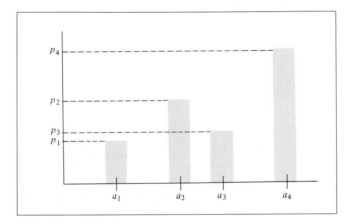

Let us define the *expected value* (or *mean*) of the probability table of Fig. 5 to be the weighted sum of the outcomes $a_1, \ldots, a_n$, each outcome weighted by the probability of its occurrence. That is,

$$[\text{expected value}] = a_1 p_1 + a_2 p_2 + \cdots + a_n p_n.$$

Similarly, let us define the variance of the probability table to be the weighted sum of the squares of the differences between each outcome and the expected value. That is, if m denotes the expected value, then

$$[\text{variance}] = (a_1 - m)^2 p_1 + (a_2 - m)^2 p_2 + \cdots + (a_n - m)^2 p_n.$$

To keep from writing the word "outcome" so many times, we shall abbreviate by X the outcome of our experiment. That is, X is a variable which takes on the values $a_1, a_2, \ldots, a_n$ with respective probabilities $p_1, p_2, \ldots, p_n$. We will assume that our experiment is performed many times, being repeated in an unbiased (or random) way. Then X is a variable whose value depends on chance, and for this reason we say that X is a *random variable*. Instead of speaking of the expected value (mean) and the variance of a probability table, let us speak of the *expected value* and the *variance of the random variable* X which is associated with the probability table. We shall denote the expected value of X by $E(X)$ and the variance of X by $\text{Var}(X)$. The *standard deviation* of X is defined to be $\sqrt{\text{Var}(X)}$.

EXAMPLE I   One bet in roulette is to wager \$1 on "red." The two possible outcomes are: "lose \$1" and "win \$1." These outcomes and their probabilities are given in Fig. 7. (*Note:* A roulette wheel in Las Vegas has 18 red numbers, 18 black numbers, and two green numbers.) Compute the expected value and the variance of the amount won.

Solution   Let X be the random variable "amount won." Then

$$E(X) = -1 \cdot \frac{20}{38} + 1 \cdot \frac{18}{38} = -\frac{2}{38} \approx -.0526,$$

$$\text{Var}(X) = \left[-1 - \left(-\frac{2}{38}\right)\right]^2 \cdot \frac{20}{38} + \left[1 - \left(-\frac{2}{38}\right)\right]^2 \cdot \frac{18}{38}$$

$$= \left(-\frac{36}{38}\right)^2 \cdot \frac{20}{38} + \left(\frac{40}{38}\right)^2 \cdot \frac{18}{38} \approx .997.$$

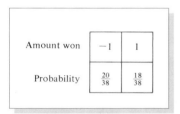

FIGURE 7   Las Vegas roulette.

Amount won	−1	1
Probability	$\frac{20}{38}$	$\frac{18}{38}$

The expected value of the amount won is approximately $-5\frac{1}{4}$ cents. In other words, sometimes we will win \$1 and sometimes we will lose \$1, but in the long run we can expect to lose an average of about $5\frac{1}{4}$ cents for each time we bet. ●

EXAMPLE 2   An experiment consists of selecting a number at random from the set of integers $\{1, 2, 3\}$. The probabilities are given by the table in Fig. 8. Let X designate the outcome. Find the expected value and the variance of X.

Solution

FIGURE 8

Number	1	2	3
Probability	$\frac{1}{3}$	$\frac{1}{3}$	$\frac{1}{3}$

$$E(X) = 1 \cdot \frac{1}{3} + 2 \cdot \frac{1}{3} + 3 \cdot \frac{1}{3} = 2,$$

$$\text{Var}(X) = (1-2)^2 \cdot \frac{1}{3} + (2-2)^2 \cdot \frac{1}{3} + (3-2)^2 \cdot \frac{1}{3}$$

$$= (-1)^2 \cdot \frac{1}{3} + 0 + (1)^2 \cdot \frac{1}{3} = \frac{2}{3}. \; ●$$

EXAMPLE 3   A cement company plans to bid on a contract for constructing the foundations of new homes in a housing development. The company is considering two bids: a high bid that will produce $75,000 profit (if the bid is accepted), and a low bid that will produce $40,000 profit. From past experience the company estimates that the high bid has a 30% chance of acceptance and the low bid a 50% chance. Which bid should the company make?

Solution   The standard method of decision is to choose the bid that has the higher expected value. Let X be the amount the company makes if it submits the high bid, and let Y be the amount it makes if it submits the low bid. Then the company must analyze the following probability tables.

TABLE I	Bids on a Cement Contract					
	High Bid				**Low Bid**	
	Accepted	Rejected			Accepted	Rejected
Value of X	75,000	0	Value of Y		40,000	0
Probability	.30	.70	Probability		.50	.50

The expected values are

$$E(X) = (75,000)(.30) + 0(.70) = 22,500,$$

$$E(Y) = (40,000)(.50) + 0(.50) = 20,000.$$

If the cement company has many opportunities to bid on similar contracts, then a "high" bid each time will be accepted sufficiently often to produce an average profit of $22,500 per bid. A consistently "low" bid will produce an average profit of $20,000 per bid. Thus the company should submit the high bid. ●

When a probability table contains a large number of possible outcomes of an experiment, the associated histogram for the random variable X becomes a valuable aid for "visualizing" the data in the table. Look at Fig. 9, for example. Since the rectangles that make up the histogram all have the same width, their areas are in the same ratios as their heights. By an appropriate change of scale on the y-axis, we may assume that the *area* (instead of the height) of each rectangle gives the associated probability of X. Such a histogram is sometimes referred to as a *probability density histogram.*

A histogram that displays probabilities as areas is useful when one wishes to visualize the probability that X has a value between two specified numbers. For example, in Fig. 9 suppose that the probabilities associated with $X = 5$, $X = 6, \ldots, X = 10$ are $p_5, p_6, \ldots, p_{10}$, respectively. Then the probability

FIGURE 9 Probabilities displayed as areas.

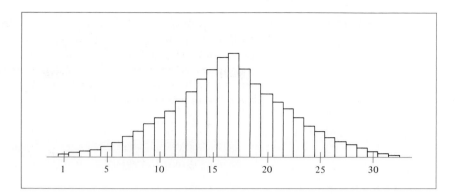

FIGURE 10 Probability that $5 \le X \le 10$.

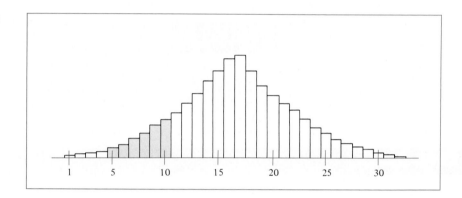

that X lies between 5 and 10 inclusive is $p_5 + p_6 + \cdots + p_{10}$. In terms of areas, this probability is just the total area of those rectangles over the values 5, 6, . . . , 10. (See Fig. 10.) We will consider analogous situations in the next section.

PRACTICE PROBLEMS 11.1

1. Compute the expected value and the variance of the random variable X with Table 2 as its probability table.

TABLE 2				
Value of X	-1	0	1	2
Probability	$\frac{1}{8}$	$\frac{1}{8}$	$\frac{3}{8}$	$\frac{3}{8}$

2. The production department at a radio factory sends CB radios to the inspection department in lots of 100. There an inspector examines three radios at random from each lot. If at least one of the three radios is defective and needs adjustment, the entire lot is sent back to the production department. Records of the inspection department show that the number X of defective radios in a sample of three radios has Table 3 as its probability table.

TABLE 3	Quality Control Data			
Defectives	0	1	2	3
Probability	.7265	.2477	.0251	.0007

(a) What percentage of the lots does the inspection department reject?

(b) Find the mean number of defective radios in the samples of three radios.

(c) Based on the evidence in part (b), estimate the average number of defective radios in each lot of 100 radios.

EXERCISES 11.1

1. Table 4 is the probability table for a random variable X. Find $E(X)$, $Var(X)$, and the standard deviation of X.

TABLE 4		
Outcome	0	1
Probability	$\frac{1}{5}$	$\frac{4}{5}$

TABLE 5			
Outcome	1	2	3
Probability	$\frac{4}{9}$	$\frac{4}{9}$	$\frac{1}{9}$

2. Find $E(X)$, $Var(X)$, and the standard deviation of X, where X is the random variable whose probability table is given in Table 5.

3. Compute the variances of the three random variables whose probability tables are given in Table 6. Relate the sizes of the variances to the "spread" of the values of the random variable.

TABLE 6								
Outcome	4	6	Outcome	3	7	Outcome	1	9
Probability	.5	.5	Probability	.5	.5	Probability	.5	.5
	(a)			(b)			(c)	

4. Compute the variances of the two random variables whose probability tables are given in Table 7. Relate the sizes of the variances to the "spread" of the values of the random variables.

TABLE 7

Outcome	2	4	6	8	Outcome	2	4	6	8
Probability	.1	.4	.4	.1	Probability	.3	.2	.2	.3
	(a)					(b)			

5. The number of accidents per week at a busy intersection was recorded for a year. There were 11 weeks with no accidents, 26 weeks with one accident, 13 weeks with two accidents, and 2 weeks with three accidents. A week is to be selected at random and the number of accidents noted. Let X be the outcome. Then X is a random variable taking on the values 0, 1, 2, and 3.

   (a) Write out a probability table for X.

   (b) Compute $E(X)$.

   (c) Interpret $E(X)$.

6. The number of phone calls coming into a telephone switchboard during each minute was recorded during an entire hour. During 30 of the 1-minute intervals there were no calls, during 20 intervals there was one call, and during 10 intervals there were two calls. A 1-minute interval is to be selected at random and the number of calls noted. Let X be the outcome. Then X is a random variable taking on the values 0, 1, and 2.

   (a) Write out a probability table for X.

   (b) Compute $E(X)$.

   (c) Interpret $E(X)$.

7. Consider a circle with radius 1.

   (a) What percentage of the points lie within $\frac{1}{2}$ unit of the center?

   (b) Let c be a constant with $0 < c < 1$. What percentage of the points lie within c units of the center?

8. Consider a circle with circumference 1. An arrow (i.e., a spinner) is attached at the center so that, when flicked, it spins freely. Upon stopping, it points to a particular point on the circumference of the circle. Determine the likelihood that the point is:

   (a) On the top half of the circumference.

   (b) On the top quarter of the circumference.

   (c) On the top one-hundredth of the circumference.

   (d) Exactly at the top of the circumference.

9. A citrus grower anticipates a profit of $100,000 this year if the nightly temperatures remain mild. Unfortunately, the weather forecast indicates a 25% chance that the temperatures will drop below freezing during the next week. Such freezing weather will destroy 40% of the crop and reduce the profit to $60,000. However, the grower can protect the citrus fruit against the possible freezing (using smudge pots, electric fans, and so on) at a cost of $5000. Should the grower spend the $5000 and thereby reduce the profit to $95,000? [*Hint:* Compute $E(X)$, where X is the profit the grower will get if he does nothing to protect the fruit.]

10. Suppose that the weather forecast in Exercise 9 indicates a 10% chance that cold weather will reduce the citrus grower's profit from $100,000 to $85,000, and a 10% chance that cold weather will reduce the profit to $75,000. Should the grower spend $5000 to protect the citrus fruit against the possible bad weather?

SOLUTIONS TO PRACTICE PROBLEMS 11.1

1.
$$E(X) = (-1) \cdot \tfrac{1}{8} + 0 \cdot \tfrac{1}{8} + 1 \cdot \tfrac{3}{8} + 2 \cdot \tfrac{3}{8} = 1,$$

$$\text{Var}(X) = (-1 - 1)^2 \cdot \tfrac{1}{8} + (0 - 1)^2 \cdot \tfrac{1}{8} + (1 - 1)^2 \cdot \tfrac{3}{8} + (2 - 1)^2 \cdot \tfrac{3}{8}$$

$$= 4 \cdot \tfrac{1}{8} + 1 \cdot \tfrac{1}{8} + 0 + 1 \cdot \tfrac{3}{8} = 1.$$

2. (a) There are three cases where a lot will be rejected: $X = 1, 2,$ or 3. Adding the corresponding probabilities, we find that the probability of rejecting a lot is .2477 + .0251 + .0007 = .2735, or 27.35%. (An alternative method of solution uses the fact that the sum of the probabilities for *all* possible cases must be 1. From the table we see that the probability of accepting a lot is .7265, so the probability of rejecting a lot is 1 − .7265 = .2735.)

   (b) $E(X) = 0(.7265) + 1(.2477) + 2(.0251) + 3(.0007) = .3000.$

   (c) In part (b) we found that an average of .3 radio in every sample of three radios is defective. Thus about 10% of the radios in the sample are defective. Since the samples are chosen at random, we may assume that about 10% of *all* the radios are defective. Thus we estimate that on the average 10 out of each lot of 100 radios will be defective.

11.2 CONTINUOUS RANDOM VARIABLES

Consider a cell population that is growing vigorously. Suppose that when a cell is T days old it divides and forms two new "daughter" cells. If the population is sufficiently large, it will contain cells of many different ages between 0 and T. It turns out that the proportion of cells of various ages remains constant. That is, if a and b are any two numbers between 0 and T, with $a < b$, the proportion of cells whose ages lie between a and b is essentially constant from one moment to the next, even though individual cells are aging and new cells are being formed all the time. In fact, biologists have found that under the ideal circumstances described, the proportion of cells whose ages are between a and b is given by the area under the graph of the function $f(x) = 2ke^{-kx}$ from $x = a$ to $x = b$, where $k = (\ln 2)/T$.* (See Fig. 1.)

Now consider an experiment where we select a cell at random from the population and observe its age, X. Then the probability that X lies between a and b is given by the area under the graph of $f(x) = 2ke^{-kx}$ from a to b, as in Fig. 1.

---

* See J. R. Cook and T. W. James, "Age Distribution of Cells in Logarithmically Growing Cell Populations," in *Synchrony in Cell Division and Growth,* Erik Zeuthen, ed. (New York: John Wiley & Sons, 1964), pp. 485–495.

FIGURE I   Age distribution in a cell population.

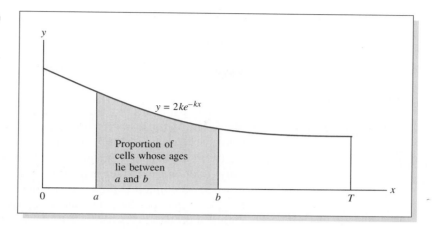

Let us denote this probability by $\Pr(a \leq X \leq b)$. Using the fact that the area under the graph of $f(x)$ is given by a definite integral, we have

$$\Pr(a \leq X \leq b) = \int_a^b f(x) \, dx = \int_a^b 2ke^{-kx} \, dx. \tag{1}$$

Since X can assume any one of the (infinitely many) numbers in the continuous interval from 0 to T, we say that X is a *continuous random variable*. The function $f(x)$ that determines the probability in (1) for each a and b is called the *(probability) density function* of X (or of the experiment whose outcome is X).

Suppose that we consider an experiment whose outcome may be any value between A and B. The outcome of the experiment, denoted X, is called a *continuous random variable*. For the cell population described above, $A = 0$ and $B = T$. Another typical experiment might consist of choosing a number X at random between $A = 5$ and $B = 6$. Or one could observe the duration X of a random telephone call passing through a given telephone switchboard. If we have no way of knowing how long a call might last, then X might be any nonnegative number. In this case it is convenient to say that X lies between 0 and ∞ and to take $A = 0$ and $B = \infty$. On the other hand, if the possible values of X for some experiment include rather large negative numbers, one sometimes takes $A = -\infty$.

Given an experiment whose outcome is a continuous random variable X, the probability $\Pr(a \leq X \leq b)$ is a measure of the likelihood that an outcome of the experiment will lie between a and b. If the experiment is repeated many times, the proportion of times X has a value between a and b should be close to $\Pr(a \leq X \leq b)$. In experiments of practical interest involving a continuous random variable X it is usually possible to find a function $f(x)$ such that

$$\Pr(a \leq X \leq b) = \int_a^b f(x) \, dx \tag{2}$$

for all a and b in the range of possible values of X. Such a function $f(x)$ is called a *probability density function* and satisfies the following properties:

I. $f(x) \geq 0, \qquad A \leq x \leq B.$

II. $\displaystyle\int_A^B f(x)\,dx = 1.$

Indeed, Property I means that for x between A and B, the graph of $f(x)$ must lie on or above the x-axis. Property II simply says that there is probability 1 that X has a value between A and B. (Of course, if $B = \infty$ and/or $A = -\infty$, the integral in Property II is an improper integral.) Properties I and II characterize probability density functions, in the sense that any function $f(x)$ satisfying I and II is the probability density function for some continuous random variable X. Moreover, $\Pr(a \leq X \leq b)$ can then be calculated using equation (2).

Unlike a probability table for a discrete random variable, a density function $f(x)$ does *not* give the probability that X has a certain value. Instead, $f(x)$ can be used to find the probability that X is *near* a specific value in the following sense. If x_0 is a number between A and B and if Δx is the width of a small interval centered at x_0, the probability that X is between $x_0 - \frac{1}{2}\Delta x$ and $x_0 + \frac{1}{2}\Delta x$ is approximately $f(x_0)\,\Delta x$—that is, the area of the rectangle shown in Fig. 2.

EXAMPLE I   Consider the cell population described earlier. Let $f(x) = 2ke^{-kx}$, where $k = (\ln 2)/T$. Show that $f(x)$ is indeed a probability density function on $0 \leq x \leq T$.

Solution   Clearly, $f(x) \geq 0$, since $\ln 2$ is positive and the exponential function is never negative. Thus Property I is satisfied. For Property II we check that

$$\int_0^T f(x)\,dx = \int_0^T 2ke^{-kx}\,dx = -2e^{-kx}\Big|_0^T = -2e^{-kT} + 2e^0$$

$$= -2e^{-[(\ln 2)/T]T} + 2 = -2e^{-\ln 2} + 2$$

$$= -2e^{\ln(1/2)} + 2 = -2\left(\frac{1}{2}\right) + 2 = 1. \quad\bullet$$

FIGURE 2   Area of rectangle gives approximate probability that X is near x_0.

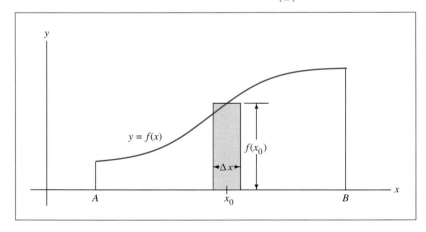

EXAMPLE 2   Let $f(x) = kx^2$.

(a) Find the value of k that makes $f(x)$ a probability density function on $0 \le x \le 4$.

(b) Let X be a continuous random variable whose density function is $f(x)$. Compute $\Pr(1 \le X \le 2)$.

Solution   (a) We must have $k \ge 0$ so that Property I is satisfied. For Property II, we calculate

$$\int_0^4 f(x)\,dx = \int_0^4 kx^2\,dx = \frac{1}{3}kx^3 \Big|_0^4 = \frac{1}{3}k(4)^3 - 0$$

$$= \frac{64}{3}k.$$

To satisfy Property II we must have $\dfrac{64}{3}k = 1$, or $k = \dfrac{3}{64}$. Thus $f(x) = \dfrac{3}{64}x^2$.

(b) $\Pr(1 \le X \le 2) = \displaystyle\int_1^2 f(x)\,dx$

$$= \int_1^2 \frac{3}{64}x^2\,dx$$

$$= \frac{1}{64}x^3 \Big|_1^2$$

$$= \frac{8}{64} - \frac{1}{64}$$

$$= \frac{7}{64}.$$

The area corresponding to this probability is shown in Fig. 3. ●

FIGURE 3

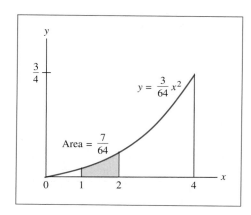

The density function in the next example is a special case of what statisticians sometimes call a *beta* probability density.

EXAMPLE 3   The parent corporation for a franchised chain of fast-food restaurants claims that the proportion of their new restaurants that make a profit during their first year of operation has the probability density

$$f(x) = 12x(1 - x)^2, \qquad 0 \le x \le 1.$$

(a) What is the probability that less than 40% of the restaurants opened this year will make a profit during their first year of operation?

(b) What is the probability that more than 50% of the restaurants will make a profit during their first year of operation?

Solution   Let X be the proportion of new restaurants opened this year that make a profit during their first year of operation. Then the possible values of X range between 0 and 1.

(a) The probability that X is less than .4 equals the probability that X is between 0 and .4. We note that $f(x) = 12x(1 - 2x + x^2) = 12x - 24x^2 + 12x^3$ and therefore,

$$\Pr(0 \le X \le .4) = \int_0^{.4} f(x)\, dx = \int_0^{.4} (12x - 24x^2 + 12x^3)\, dx$$

$$= (6x^2 - 8x^3 + 3x^4)\Big|_0^{.4} = .5248.$$

(b) The probability that X is greater than .5 equals the probability that X is between .5 and 1. Thus

$$\Pr(.5 \le X \le 1) = \int_{.5}^1 (12x - 24x^2 + 12x^3)\, dx$$

$$= (6x^2 - 8x^3 + 3x^4)\Big|_{.5}^1 = .3125. \quad \bullet$$

Each probability density function is closely related to another important function called a cumulative distribution function. To describe this relationship, let us consider an experiment whose outcome is a continuous random variable X, with values between A and B, and let $f(x)$ be the associated density function. For each number x between A and B, let $F(x)$ be the probability that X is less than or equal to the number x. Sometimes we write $F(x) = \Pr(X \le x)$; however, since X is never less than A, we may also write

$$F(x) = \Pr(A \le X \le x). \qquad (3)$$

Graphically, $F(x)$ is the area under the graph of the density function $f(x)$ from A to x. (See Fig. 4.) The function $F(x)$ is called the *cumulative distribution function* of the random variable X (or of the experiment whose outcome is X).

FIGURE 4 The cumulative distribution function F(x).

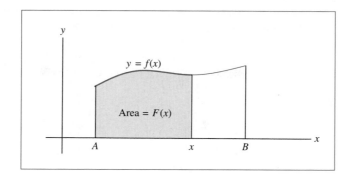

Note that $F(x)$ also has the properties

$$F(A) = \Pr(A \le X \le A) = 0, \tag{4}$$

$$F(B) = \Pr(A \le X \le B) = 1. \tag{5}$$

Since $F(x)$ is an "area function" that gives the area under the graph of $f(x)$ from A to x, we know from Section 6.3 that $F(x)$ is an antiderivative of $f(x)$. That is,

$$F'(x) = f(x), \qquad A \le x \le B. \tag{6}$$

It follows that one may use $F(x)$ to compute probabilities, since

$$\Pr(a \le X \le b) = \int_a^b f(x)\,dx = F(b) - F(a), \tag{7}$$

for any a and b between A and B.

The relation (6) between $F(x)$ and $f(x)$ makes it possible to find one of these functions when the other is known, as we see in the following two examples.

EXAMPLE 4 Let X be the age of a cell selected at random from the cell population described earlier. The density function for X is $f(x) = 2ke^{-kx}$, where $k = (\ln 2)/T$. See Fig. 5. Find the cumulative distribution function $F(x)$ for X.

Solution Since $F(x)$ is an antiderivative of $f(x) = 2ke^{-kx}$, we have $F(x) = -2e^{-kx} + C$ for some constant C. Now $F(x)$ is defined for $0 \le x \le T$. Thus (4) implies that $F(0) = 0$. Setting $F(0) = -2e^0 + C = 0$, we find that $C = 2$, so

$$F(x) = -2e^{-kx} + 2.$$

See Fig. 6. ●

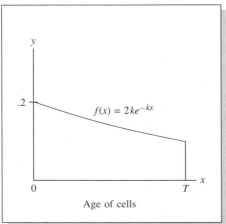

FIGURE 5   Density function.

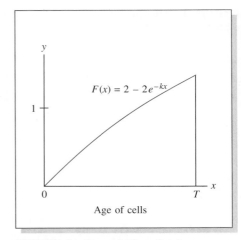

FIGURE 6   Cumulative distribution function.

EXAMPLE 5   Let X be the random variable associated with the experiment which consists of selecting a point at random from a circle of radius 1 and observing its distance from the center. Find the probability density function $f(x)$ and cumulative distribution function $F(x)$ of X.

Solution   The distance of a point from the center of the unit circle is a number between 0 and 1. Suppose that $0 \leq x \leq 1$. Let us first compute the cumulative distribution function $F(x) = \Pr(0 \leq X \leq x)$. That is, let us find the probability that a point selected at random lies within x units of the center of the circle—in other words, lies inside the circle of radius x. See the shaded region in Fig. 7(b). Since the area of this shaded region is πx^2 and the area of the entire unit circle is $\pi \cdot 1^2 = \pi$, the proportion of points inside the shaded region is $\pi x^2/\pi = x^2$. Thus the probability is x^2 that a point selected at random will be in this shaded region. Hence

$$F(x) = x^2.$$

FIGURE 7

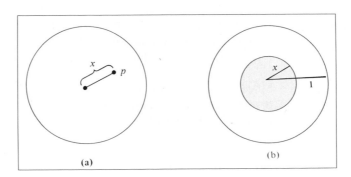

Differentiating, we find that the probability density function for X is

$$f(x) = F'(x) = 2x. \bullet$$

Our final example involves a continuous random variable X whose possible values lie between $A = 1$ and $B = \infty$; that is, X is any number greater than or equal to 1.

EXAMPLE 6   Let $f(x) = 3x^{-4}$, $x \geq 1$.

(a) Show that $f(x)$ is the probability density function of some random variable X.

(b) Find the cumulative distribution function $F(x)$ of X.

(c) Compute $\Pr(X \leq 4)$, $\Pr(4 \leq X \leq 5)$, and $\Pr(4 \leq X)$.

Solution   (a) It is clear that $f(x) \geq 0$ for $x \geq 1$. Thus Property I holds. In order to check Property II, we must compute

$$\int_1^\infty 3x^{-4}\, dx.$$

But

$$\int_1^b 3x^{-4}\, dx = -x^{-3}\Big|_1^b = -b^{-3} + 1 \to 1$$

as $b \to \infty$. Thus

$$\int_1^\infty 3x^{-4}\, dx = 1,$$

and Property II holds.

(b) Since $F(x)$ is an antiderivative of $f(x) = 3x^{-4}$, we have

$$F(x) = \int 3x^{-4}\, dx = -x^{-3} + C.$$

Since X has values greater than or equal to 1, we must have $F(1) = 0$. Setting $F(1) = -1 + C = 0$, we find that $C = 1$, so

$$F(x) = 1 - x^{-3}.$$

(c) $\Pr(X \leq 4) = F(4) = 1 - 4^{-3} = 1 - \frac{1}{64} = \frac{63}{64}.$

Since we know $F(x)$, we may use it to compute $\Pr(4 \leq X \leq 5)$, as follows:

$$\Pr(4 \leq X \leq 5) = F(5) - F(4) = (1 - 5^{-3}) - (1 - 4^{-3})$$

$$= \frac{1}{4^3} - \frac{1}{5^3} \approx .0076.$$

We may compute $\Pr(4 \le X)$ directly by evaluating the improper integral

$$\int_4^\infty 3x^{-4} \, dx.$$

However, there is a simpler method. We know that

$$\int_1^4 3x^{-4} \, dx + \int_4^\infty 3x^{-4} \, dx = \int_1^\infty 3x^{-4} \, dx = 1. \tag{8}$$

In terms of probabilities, (8) may be written as

$$\Pr(X \le 4) + \Pr(4 \le X) = 1.$$

Hence

$$\Pr(4 \le X) = 1 - \Pr(X \le 4) = 1 - \frac{63}{64} = \frac{1}{64}. \quad \bullet$$

TECHNOLOGY PROJECT 11.1

The Beta Probability Density Functions

A *beta probability density function* has the form $f(x) = kx^c(1 - x)^d$ for $0 \le x \le 1$, where the parameters c and d are positive, frequently integers. The graph of such a function has one peak (relative maximum) between 0 and 1. Beta density functions are widely used in statistical models, partly because the parameters c and d can be adjusted to make the peak occur anywhere between 0 and 1. In Exercises 1–3, X denotes a continuous random variable with the beta density function, $f(x)$.

1. In each part, find the value of k that makes $f(x)$ a probability density function. Use a program that computes definite integrals. Suggestion: Start with $k = 1$, compute the appropriate definite integral, and then adjust k appropriately. Recompute the definite integral with the new k to check your work.

   (a) $c = 1, d = 1$ (b) $c = 2, d = 2$ (c) $c = 3, d = 3$

   (d) $c = 1, d = 2$ (e) $c = 1, d = 3$ (f) $c = 1, d = 4$

2. Sketch the graphs of $f(x)$, $0 \le x \le 1$, for parts (a)–(c) in Exercise 1.

   (a) How are the graphs similar? How are they different?

   (b) How do you think $E(X)$ and $\mathrm{Var}(X)$ will change from one graph to the next?

   (c) Verify your conjectures by computing $E(X)$ and $\mathrm{Var}(X)$ for the three density functions from Exercise 1(a)–(c).

3. (a) Sketch the graphs of $f(x)$, $0 \le x \le 1$, for parts (d)–(f) in Exercise 1.

   (b) In each case, estimate where the peak in the graph occurs. Describe the method you used to do this.

   (c) Where do you think the peak occurs for the density function $f(x) = kx(1 - x)^9$? Verify your conjecture by calculus methods.

(d) Where will the peak occur for the density function $f(x) = kx(1 - x)^d$, if $d > 0$? Justify your answer.

(e) Compute $E(X)$ for parts (d)–(f) in Exercise 1 and for each case compare $E(X)$ with the location of the peak in the graph of $f(x)$.

Exercise 4 examines the fact that, for any n, $x^n e^{-x}$ approaches 0 as x gets large. This property is used in the calculation of $E(X)$ and $Var(X)$, for an exponential random variable X.

4. (a) Graph $y = xe^{-x}$ with window $0 \leq x \leq 5, 0 \leq y \leq 1$.

(b) Graph $y = x^2 e^{-x}$ with window $0 \leq x \leq 5, 0 \leq y \leq 5$.

(c) Graph $y = x^5 e^{-x}$ with window $0 \leq x \leq 15, 0 \leq y \leq 50$.

(d) Graph $y = x^{10} e^{-x}$ with window $0 \leq x \leq 30, 0 \leq y \leq 500{,}000$.

[*Note:* Since $x^n e^{-x}$ is the same as $\dfrac{x^n}{e^x}$, we conclude that e^x is significantly larger than x^n for large values of x.]

TECHNOLOGY PROJECT 11.2
The Chi-Square Probability Density Functions

The chi-square probability distribution is used extensively in statistics work. It depends on a parameter d, which indicates the number of degrees of freedom (defined in a statistics course). The density function for the chi-squared distribution is given by the function:

$$f_d(x) = \frac{1}{2^{d/2}\Gamma(d/2)} e^{-x/2} x^{d/2 - 1}$$

Here $x \geq 0$ and Γ denotes the gamma function (which can be calculated by symbolic math programs). In *Mathematica*, this function can be defined using the formula:

```
f[x, d] := 1/2^[d/2]Gamma(d/2)Exp[−x/2]x^(d/2−1)
```

1. Determine the probability distribution function corresponding to 2 degrees of freedom.

2. Graph the density function for a chi-square random variable with 2 degrees of freedom.

3. Determine the probability that a chi-square random variable with 2 degrees of freedom is between 0 and 4; is greater than or equal to 8.

4. Determine the probability distribution function corresponding to 6 degrees of freedom.

5. Graph the density function for a chi-square random variable with 6 degrees of freedom.

6. Determine the probability that a chi-square random variable with 6 degrees of freedom is less than or equal to 5; greater than or equal to 10.

1. Suppose that in a certain farming region, and in a certain year, the number of bushels of wheat produced per acre is a random variable X with a density function

$$f(x) = \frac{x - 30}{50}, \qquad 30 \le x \le 40.$$

(a) What is the probability that an acre selected at random produced less than 35 bushels of wheat?

(b) If the farming region had 20,000 acres of wheat, how many acres produced less than 35 bushels of wheat?

2. The density function for a continuous random variable X on the interval $1 \le x \le 2$ is $f(x) = 8/(3x^3)$. Find the corresponding cumulative distribution function for X.

EXERCISES 11.2

Verify that each of the following functions is a probability density function.

1. $f(x) = \frac{1}{18}x, 0 \le x \le 6$

2. $f(x) = 2(x - 1), 1 \le x \le 2$

3. $f(x) = \frac{1}{4}, 1 \le x \le 5$

4. $f(x) = \frac{8}{9}x, 0 \le x \le \frac{3}{2}$

5. $f(x) = 5x^4, 0 \le x \le 1$

6. $f(x) = \frac{3}{2}x - \frac{3}{4}x^2, 0 \le x \le 2$

In Exercises 7–12, find the value of k that makes the given function a probability density function on the specified interval.

7. $f(x) = kx, 1 \le x \le 3$

8. $f(x) = kx^2, 0 \le x \le 2$

9. $f(x) = k, 5 \le x \le 20$

10. $f(x) = k/\sqrt{x}, 1 \le x \le 4$

11. $f(x) = kx^2(1 - x), 0 \le x \le 1$

12. $f(x) = k(3x - x^2), 0 \le x \le 3$

13. The density function of a continuous random variable X is $f(x) = \frac{1}{8}x, 0 \le x \le 4$. Sketch the graph of $f(x)$ and shade in the areas corresponding to (a) $\Pr(X \le 1)$; (b) $\Pr(2 \le X \le 2.5)$; (c) $\Pr(3.5 \le X)$.

14. The density function of a continuous random variable X is $f(x) = 3x^2, 0 \le x \le 1$. Sketch the graph of $f(x)$ and shade in the areas corresponding to (a) $\Pr(X \le .3)$; (b) $\Pr(.5 \le X \le .7)$; (c) $\Pr(.8 \le X)$.

15. Find $\Pr(1 \le X \le 2)$ when X is a random variable whose density function is given in Exercise 1.

16. Find $\Pr(1.5 \le X \le 1.7)$ when X is a random variable whose density function is given in Exercise 2.

17. Find $\Pr(X \le 3)$ when X is a random variable whose density function is given in Exercise 3.

18. Find $\Pr(1 \le X)$ when X is a random variable whose density function is given in Exercise 4.

19. Suppose that the lifetime X (in hours) of a certain type of flashlight battery is a random variable on the interval $30 \leq x \leq 50$ with density function $f(x) = \frac{1}{20}$, $30 \leq x \leq 50$. Find the probability that a battery selected at random will last at least 35 hours.

20. Suppose that at a certain supermarket the amount of time one must wait at the express lane is a random variable with density function $f(x) = 11/[10(x + 1)^2]$, $0 \leq x \leq 10$. See Fig. 8. Find the probability of having to wait less than 4 minutes at the express lane.

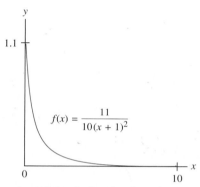

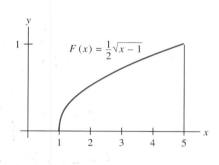

FIGURE 8   A density function.

FIGURE 9   A cumulative distribution function.

21. The cumulative distribution function for a random variable X on the interval $1 \leq x \leq 5$ is $F(x) = \frac{1}{2}\sqrt{x} - 1$. See Fig. 9. Find the corresponding density function.

22. The cumulative distribution function for a random variable X on the interval $1 \leq x \leq 2$ is $F(x) = \frac{4}{3} - 4/(3x^2)$. Find the corresponding density function.

23. Compute the cumulative distribution function corresponding to the density function $f(x) = \frac{1}{5}, 2 \leq x \leq 7$.

24. Compute the cumulative distribution function corresponding to the density function $f(x) = \frac{1}{2}(3 - x), 1 \leq x \leq 3$.

25. The time (in minutes) required to complete a certain subassembly is a random variable X with density function $f(x) = \frac{1}{21}x^2, 1 \leq x \leq 4$.

   (a) Use $f(x)$ to compute $\Pr(2 \leq X \leq 3)$.
   (b) Find the corresponding cumulative distribution function $F(x)$.
   (c) Use $F(x)$ to compute $\Pr(2 \leq X \leq 3)$.

26. The density function for a continuous random variable X on the interval $1 \leq x \leq 4$ is $f(x) = \frac{4}{9}x - \frac{1}{9}x^2$.

   (a) Use $f(x)$ to compute $\Pr(3 \leq X \leq 4)$.
   (b) Find the corresponding cumulative distribution function $F(x)$.
   (c) Use $F(x)$ to compute $\Pr(3 \leq X \leq 4)$.

An experiment consists of selecting a point at random from the square in Fig. 10(a). Let X be the maximum of the coordinates of the point.

27. Show that the cumulative distribution function of X is $F(x) = x^2/4, 0 \leq x \leq 2$.

28. Find the corresponding density function of X.

FIGURE 10

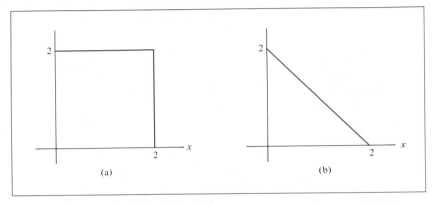

(a)   (b)

An experiment consists of selecting a point at random from the triangle in Fig. 10(b). Let X be the sum of the coordinates of the point.

29. Show that the cumulative distribution function of X is $F(x) = x^2/4, 0 \leq x \leq 2$.

30. Find the corresponding density function of X.

Suppose that in a certain cell population, cells divide every 10 days, and the age of a cell selected at random is a random variable X with the density function $f(x) = 2ke^{-kx}$, $0 \leq x \leq 10$, $k = (\ln 2)/10$.

31. Find the probability that a cell is at most 5 days old.

32. Upon examination of a slide, 10% of the cells are found to be undergoing mitosis (a change in the cell leading to division). Compute the length of time required for mitosis; that is, find the number M such that

$$\int_{10-M}^{10} 2ke^{-kx}\, dx = .10.$$

33. A random variable X has a density function $f(x) = \frac{1}{3}, 0 \leq x \leq 3$. Find b such that $\Pr(0 \leq X \leq b) = .6$.

34. A random variable X has a density function $f(x) = \frac{2}{3}x$ on $1 \leq x \leq 2$. Find a such that $\Pr(a \leq X) = \frac{1}{3}$.

35. A random variable X has a cumulative distribution function $F(x) = \frac{1}{4}x^2$ on $0 \leq x \leq 2$. Find b such that $\Pr(X \leq b) = .09$.

36. A random variable X has a cumulative distribution function $F(x) = (x - 1)^2$ on $1 \leq x \leq 2$. Find b such that $\Pr(X \leq b) = \frac{1}{4}$.

37. Let X be a continuous random variable with values between $A = 1$ and $B = \infty$, and with the density function $f(x) = 4x^{-5}$.

   (a) Verify that $f(x)$ is a probability density function for $x \geq 1$.

   (b) Find the corresponding cumulative distribution function $F(x)$.

   (c) Use $F(x)$ to compute $\Pr(1 \leq X \leq 2)$ and $\Pr(2 \leq X)$.

38. Let X be a continuous random variable with the density function $f(x) = 2(x + 1)^{-3}$, $x \geq 0$.

    (a) Verify that $f(x)$ is a probability density function for $x \geq 0$.
    (b) Find the cumulative distribution function for X.
    (c) Compute $\Pr(1 \leq X \leq 2)$ and $\Pr(3 \leq X)$.

SOLUTIONS TO PRACTICE PROBLEMS 11.2

1. (a) $\Pr(X \leq 35) = \int_{30}^{35} \dfrac{x - 30}{50}\, dx = \dfrac{(x - 30)^2}{100}\Bigg|_{30}^{35}$

    $= \dfrac{5^2}{100} - 0 = .25.$

    (b) Using part (a), we see that 25% of the 20,000 acres, or 5000 acres, produced less than 35 bushels of wheat.

2. The cumulative distribution function $F(x)$ is an antiderivative of $f(x) = 8/(3x^3) = \frac{8}{3}x^{-3}$. Thus $F(x) = -\frac{4}{3}x^{-2} + C$ for some constant C. Since X varies over the interval $1 \leq x \leq 2$, we must have $F(1) = 0$; that is, $-\frac{4}{3}(1)^{-2} + C = 0$. Thus $C = \frac{4}{3}$, and

$$F(x) = \frac{4}{3} - \frac{4}{3}x^{-2}.$$

 ## 11.3 EXPECTED VALUE AND VARIANCE

When studying the cell population described in Section 11.2, one might reasonably ask for the average age of the cells. In general, if one is given an experiment described by a random variable X and a probability density function $f(x)$, it is often important to know the "average" outcome of the experiment, and the degree to which the experimental outcomes are spread out around the average. To provide this information in Section 11.1 we introduced the concepts of expected value and variance of a discrete random variable. Let us now examine the analogous definitions for a continuous random variable.

Definition   Let X be a continuous random variable whose possible values lie between A and B, and let $f(x)$ be the probability density function for X. Then the *expected value* (or *mean*) of X is the number $E(X)$ defined by

$$E(X) = \int_A^B xf(x)\, dx. \qquad (1)$$

The *variance* of X is the number $\text{Var}(X)$ defined by

$$\text{Var}(X) = \int_A^B [x - E(X)]^2 f(x)\, dx. \qquad (2)$$

The expected value of X has the same interpretation as in the discrete case—namely, if the experiment whose outcome is X is performed many times, then the average of all the outcomes will approximately equal $E(X)$. As in the case of a discrete random variable, the variance of X is a quantitative measure of the likely spread of the values of X about the mean $E(X)$ when the experiment is performed many times.

To explain why the definition (1) of $E(X)$ is analogous to the definition in Section 11.1, let us approximate the integral in (1) by a Riemann sum of the form

$$x_1 f(x_1) \, \Delta x + x_2 f(x_2) \, \Delta x + \cdots + x_n f(x_n) \, \Delta x. \tag{3}$$

Here $x_1, \ldots, x_n$ are the midpoints of subintervals of the interval from A to B, each subinterval of width $\Delta x = (B - A)/n$. (See Fig. 1.) Now recall from Section 11.2 that, for $i = 1, \ldots, n$, the quantity $f(x_i) \Delta x$ is approximately the probability that X is close to x_i—that is, the probability that X lies in the subinterval centered at x_i. If we write $\Pr(X \approx x_i)$ for this probability, then (3) is nearly the same as

$$x_1 \cdot \Pr(X \approx x_1) + x_2 \cdot \Pr(X \approx x_2) + \cdots + x_n \cdot \Pr(X \approx x_n). \tag{4}$$

FIGURE I

As the number of subintervals increases, the sum becomes closer and closer to the integral in (1) defining $E(X)$. Furthermore, each approximating sum in (4) resembles the sum in the definition of the expected value of a discrete random variable, where one computes the weighted sum over all possible outcomes, with each outcome weighted by the probability of its occurrence.

A similar analysis would show that the definition (2) of variance is analogous to the definition for the discrete case.

EXAMPLE I   Let us consider the experiment of selecting a number at random from among the numbers between 0 and B. Let X denote the associated random variable. Determine the cumulative distribution function of X, the density function of X, and the mean and variance of X.

Solution

$$F(x) = \frac{[\text{length of the interval from 0 to } x]}{[\text{length of the interval from 0 to } B]}$$

$$= \frac{x}{B}.$$

Since $f(x) = F'(x)$, we see that $f(x) = 1/B$. Thus we have

$$E(X) = \int_0^B x \cdot \frac{1}{B}\,dx = \frac{1}{B}\int_0^B x\,dx = \frac{1}{B} \cdot \frac{B^2}{2} = \frac{B}{2},$$

$$\begin{aligned}
\text{Var}(X) &= \int_0^B \left(x - \frac{B}{2}\right)^2 \cdot \frac{1}{B}\,dx \\
&= \frac{1}{B}\int_0^B \left(x - \frac{B}{2}\right)^2 dx \\
&= \frac{1}{B} \cdot \frac{1}{3}\left(x - \frac{B}{2}\right)^3 \Big|_0^B \\
&= \frac{1}{3B}\left[\left(\frac{B}{2}\right)^3 - \left(-\frac{B}{2}\right)^3\right] = \frac{B^2}{12}.
\end{aligned}$$

The graph of the density function $f(x)$ is shown in Fig. 2. Since the density function has a flat graph, the random variable X is called the *uniform random variable* on the interval from 0 to B. ●

FIGURE 2   **A uniform probability density function.**

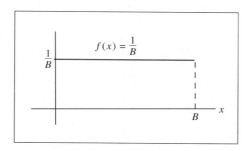

EXAMPLE 2   Let X be the age of a cell chosen at random from the population described in Section 11.2, where the density function for X was given as

$$f(x) = 2ke^{-kx}, \qquad 0 \le x \le T,$$

and $k = (\ln 2)/T$. Find the average age, $E(X)$, of the cell population.

Solution   By definition
$$E(X) = \int_0^T x \cdot 2ke^{-kx}\,dx.$$

To calculate this integral we need integration by parts, with $f(x) = 2x$, $g(x) = ke^{-kx}$, $f'(x) = 2$, and $G(x) = -e^{-kx}$. We have

$$\begin{aligned}
\int_0^T 2xke^{-kx}\,dx &= -2xe^{-kx}\Big|_0^T - \int_0^T -2e^{-kx}\,dx \\
&= -2Te^{-kT} - \left(\frac{2}{k}e^{-kx}\right)\Big|_0^T \\
&= -2Te^{-kT} - \frac{2}{k}e^{-kT} + \frac{2}{k}.
\end{aligned}$$

This formula for $E(X)$ may be simplified by noting that $e^{-kT} = e^{-\ln 2} = \frac{1}{2}$. Thus

$$E(X) = -2T\left(\frac{1}{2}\right) - \frac{2}{k}\left(\frac{1}{2}\right) + \frac{2}{k} = \frac{1}{k} - T$$

$$= \frac{T}{\ln 2} - T = \left(\frac{1}{\ln 2} - 1\right)T$$

$$\approx .4427T. \quad \bullet$$

EXAMPLE 3   Consider the experiment of selecting a point at random in a circle of radius 1, and let X be the distance from this point to the center. Compute the expected value and variance of the random variable X.

Solution   We showed in Example 5 of Section 11.2 that the density function for X is given by $f(x) = 2x, 0 \le x \le 1$. Therefore, we see that

$$E(X) = \int_0^1 x \cdot 2x \, dx = \int_0^1 2x^2 \, dx$$

$$= \frac{2x^3}{3}\bigg|_0^1 = \frac{2}{3}$$

and

$$Var(X) = \int_0^1 \left(x - \frac{2}{3}\right)^2 \cdot 2x \, dx \qquad \left[\text{since } E(X) = \frac{2}{3}\right]$$

$$= \int_0^1 \left(x^2 - \frac{4}{3}x + \frac{4}{9}\right) \cdot 2x \, dx$$

$$= \int_0^1 \left(2x^3 - \frac{8}{3}x^2 + \frac{8}{9}x\right) dx$$

$$= \left(\frac{1}{2}x^4 - \frac{8}{9}x^3 + \frac{4}{9}x^2\right)\bigg|_0^1$$

$$= \frac{1}{2} - \frac{8}{9} + \frac{4}{9} = \frac{1}{18}.$$

From our first calculation, we see that if a large number of points are chosen randomly from a circle of radius 1, their average distance to the center should be about $\frac{2}{3}$. $\bullet$

The following alternative formula for the variance of a random variable is usually easier to use than the actual definition of $Var(X)$.

Let X be a continuous random variable whose values lie between A and B, and let $f(x)$ be the density function for X. Then

$$Var(X) = \int_A^B x^2 f(x) \, dx - E(X)^2. \qquad (5)$$

To prove (5), we let $m = E(X) = \int_A^B xf(x)\,dx$. Then

$$\text{Var}(X) = \int_A^B (x - m)^2 f(x)\,dx = \int_A^B (x^2 - 2xm + m^2)f(x)\,dx$$

$$= \int_A^B x^2 f(x)\,dx - 2m \int_A^B xf(x)\,dx + m^2 \int_A^B f(x)\,dx$$

$$= \int_A^B x^2 f(x)\,dx - 2m \cdot m + m^2 \cdot 1 \qquad \text{(by Property II)}$$

$$= \int_A^B x^2 f(x)\,dx - m^2.$$

EXAMPLE 4 A college library has found that, in any given month during a school year, the proportion of students who make some use of the library is a random variable X with the cumulative distribution function

$$F(x) = 4x^3 - 3x^4, \qquad 0 \le x \le 1.$$

(a) Compute $E(X)$ and give an interpretation of this quantity.
(b) Compute $\text{Var}(X)$.

Solution (a) To compute $E(X)$ we first find the probability density function $f(x)$. From Section 11.2 we know that

$$f(x) = F'(x) = 12x^2 - 12x^3.$$

Hence

$$E(X) = \int_0^1 xf(x)\,dx = \int_0^1 (12x^3 - 12x^4)\,dx$$

$$= \left(3x^4 - \frac{12}{5}x^5 \right)\Big|_0^1 = 3 - \frac{12}{5} = \frac{3}{5}.$$

The meaning of $E(X)$ in this example is that over a period of many months (during school years) the average proportion of students each month who make some use of the library should be close to $\frac{3}{5}$.

(b) We first compute

$$\int_0^1 x^2 f(x)\,dx = \int_0^1 (12x^4 - 12x^5)\,dx = \left(\frac{12}{5}x^5 - 2x^6 \right)\Big|_0^1$$

$$= \frac{12}{5} - 2 = \frac{2}{5}.$$

Then, from the alternative formula (5) for the variance, we find that

$$\text{Var}(X) = \frac{2}{5} - E(X)^2 = \frac{2}{5} - \left(\frac{3}{5} \right)^2 = \frac{1}{25}. \quad \bullet$$

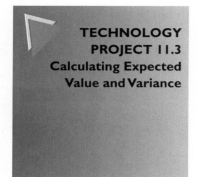

TECHNOLOGY PROJECT 11.3
Calculating Expected Value and Variance

Refer to Technology Project 11.2.

1. Calculate the expected value of a chi-square random variable with 2 degrees of freedom.

2. Calculate the variance of a chi-square random variable with 2 degrees of freedom.

3. Calculate the expected value of a chi-square random variable with 6 degrees of freedom.

4. Calculate the variance of a chi-square random variable with 6 degrees of freedom.

PRACTICE PROBLEMS 11.3

1. Find the expected value and variance of the random variable X whose density function is $f(x) = 1/(2\sqrt{x})$, $1 \leq x \leq 4$.

2. An insurance company finds that the proportion X of their salespeople who sell more than \$25,000 worth of insurance in a given week is a random variable with the beta probability density function

$$f(x) = 60x^3(1 - x)^2, \qquad 0 \leq x \leq 1.$$

   (a) Compute $E(X)$ and give an interpretation of this quantity.
   (b) Compute $\text{Var}(X)$.

EXERCISES 11.3

Find the expected value and variance for each random variable whose probability density function is given below. When computing the variance, use formula (5).

1. $f(x) = \frac{1}{18}x$, $0 \leq x \leq 6$
2. $f(x) = 2(x - 1)$, $1 \leq x \leq 2$
3. $f(x) = \frac{1}{4}$, $1 \leq x \leq 5$
4. $f(x) = \frac{8}{9}x$, $0 \leq x \leq \frac{3}{2}$
5. $f(x) = 5x^4$, $0 \leq x \leq 1$
6. $f(x) = \frac{3}{2}x - \frac{3}{4}x^2$, $0 \leq x \leq 2$
7. $f(x) = 12x(1 - x)^2$, $0 \leq x \leq 1$
8. $f(x) = \dfrac{3\sqrt{x}}{16}$, $0 \leq x \leq 4$

9. A newspaper publisher estimates that the proportion X of space devoted to advertising on a given day is a random variable with the beta probability density $f(x) = 30x^2(1 - x)^2$, $0 \leq x \leq 1$.
   (a) Find the cumulative distribution function for X.
   (b) Find the probability that less than 25% of the newspaper's space on a given day contains advertising.
   (c) Find $E(X)$ and give an interpretation of this quantity.
   (d) Compute $\text{Var}(X)$.

10. Let X be the proportion of new restaurants in a given year that make a profit during their first year of operation, and suppose that the density function for X is $f(x) = 20x^3(1 - x)$, $0 \leq x \leq 1$.
    (a) Find $E(X)$ and give an interpretation of this quantity.
    (b) Compute $\text{Var}(X)$.

11. The useful life (in hundreds of hours) of a certain machine component is a random variable X with the cumulative distribution function $F(x) = \frac{1}{9}x^2, 0 \le x \le 3$.

    (a) Find $E(X)$ and give an interpretation of this quantity.

    (b) Compute $\text{Var}(X)$.

12. The time (in minutes) required to complete an assembly on a production line is a random variable X with the cumulative distribution function $F(x) = \frac{1}{125}x^3$, $0 \le x \le 5$.

    (a) Find $E(X)$ and give an interpretation of this quantity.

    (b) Compute $\text{Var}(X)$.

13. Suppose that the amount of time (in minutes) a person spends reading the editorial page of the newspaper is a random variable with the density function $f(x) = \frac{1}{72}x$, $0 \le x \le 12$. Find the average time spent reading the editorial page.

14. Suppose that at a certain bus stop the time between buses is a random variable X with the density function $f(x) = 6x(10 - x)/1000, 0 \le x \le 10$. Find the average time between buses.

15. When preparing a bid on a large construction project, a contractor analyzes how long each phase of the construction will take. Suppose the contractor estimates that the time required for the electrical work will be X hundred worker-hours, where X is a random variable with the density function $f(x) = x(6 - x)/18, 3 \le x \le 6$. See Fig. 3.

    (a) Find the cumulative distribution function $F(x)$.

    (b) What is the likelihood that the electrical work will take less than 500 worker-hours?

    (c) Find the mean time to complete the electrical work.

    (d) Find $\text{Var}(X)$.

16. The amount of milk (in thousands of gallons) a dairy sells each week is a random variable X with the density function $f(x) = 4(x - 1)^3, 1 \le x \le 2$. See Fig. 4.

    (a) What is the likelihood that the dairy will sell more than 1500 gallons?

    (b) What is the average amount of milk the dairy sells each week?

FIGURE 3 Density function for construction bid.

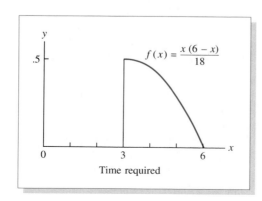

FIGURE 4 Density function for sales of milk.

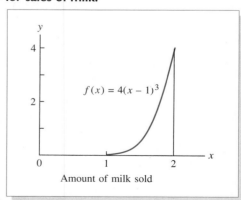

17. Let X be a continuous random variable with values between $A = 1$ and $B = \infty$, and with the density function $f(x) = 4x^{-5}$. Compute $E(X)$ and $\text{Var}(X)$.

18. Let X be a continuous random variable with density function $f(x) = 3x^{-4}$, $x \geq 1$. Compute $E(X)$ and $\text{Var}(X)$.

If X is a random variable with a density function $f(x)$ on $A \leq x \leq B$, the *median* of X is that number M such that $\int_A^M f(x)\,dx = \frac{1}{2}$. In other words, $\text{Pr}(X \leq M) = \frac{1}{2}$.

19. Find the median of the random variable whose density function is $f(x) = \frac{1}{18}x$, $0 \leq x \leq 6$.

20. Find the median of the random variable whose density function is $f(x) = 2(x - 1)$, $1 \leq x \leq 2$.

21. A machine component described in Exercise 11 has a 50% chance of lasting at least how long?

22. In Exercise 12, find the length of time T such that half of the assemblies are completed in T minutes or less.

23. In Exercise 20 of Section 11.2, find the length of time T such that about half of the time one waits only T minutes or less in the express lane at the supermarket.

24. Find the number M such that half of the time the dairy in Exercise 16 sells M thousand gallons of milk or less.

25. Show that $E(X) = B - \int_A^B F(x)\,dx$, where $F(x)$ is the cumulative distribution function for X on $A \leq x \leq B$.

26. Use the formula in Exercise 25 to compute $E(X)$ for the random variable X in Exercise 12.

SOLUTIONS TO PRACTICE PROBLEMS 11.3

1. $E(X) = \int_1^4 x \cdot \frac{1}{2\sqrt{x}}\,dx = \int_1^4 \frac{1}{2}x^{1/2}\,dx = \frac{1}{3}x^{3/2}\Big|_1^4$

$= \frac{1}{3}(4)^{3/2} - \frac{1}{3} = \frac{8}{3} - \frac{1}{3} = \frac{7}{3}.$

To find $\text{Var}(X)$, we first compute

$\int_1^4 x^2 \cdot \frac{1}{2\sqrt{x}}\,dx = \int_1^4 \frac{1}{2}x^{3/2}\,dx = \frac{1}{5}x^{5/2}\Big|_1^4$

$= \frac{1}{5}(4)^{5/2} - \frac{1}{5} = \frac{32}{5} - \frac{1}{5} = \frac{31}{5}.$

Then, from formula (5),

$$\text{Var}(X) = \frac{31}{5} - \left(\frac{7}{3}\right)^2 = \frac{34}{45}.$$

2. (a) First note that $f(x) = 60x^3(1 - x)^2 = 60x^3(1 - 2x + x^2) = 60x^3 - 120x^4 + 60x^5$. Then

$$E(X) = \int_0^1 xf(x)\, dx = \int_0^1 (60x^4 - 120x^5 + 60x^6)\, dx$$

$$= \left(12x^5 - 20x^6 + \frac{60}{7}x^7\right)\Bigg|_0^1 = 12 - 20 + \frac{60}{7} = \frac{4}{7}.$$

Thus, in an "average" week, about four-sevenths of the salespeople sell more than \$25,000 worth of insurance. More precisely, over a period of many weeks, we expect an average of four-sevenths of the salespeople each week to sell more than \$25,000 worth of insurance.

(b) $$\int_0^1 x^2 f(x)\, dx = \int_0^1 (60x^5 - 120x^6 + 60x^7)\, dx$$

$$= \left(10x^6 - \frac{120}{7}x^7 + \frac{60}{8}x^8\right)\Bigg|_0^1$$

$$= 10 - \frac{120}{7} + \frac{60}{8} = \frac{5}{14}.$$

Hence

$$\mathrm{Var}(X) = \frac{5}{14} - \left(\frac{4}{7}\right)^2 = \frac{3}{98}.$$

11.4 EXPONENTIAL AND NORMAL RANDOM VARIABLES

This section is devoted to the two most important types of probability density functions—the exponential and normal density functions. These functions are associated with random variables that arise in a wide variety of applications. We will describe some typical examples.

Exponential Density Functions   Let k be a positive constant. Then the function

$$f(x) = ke^{-kx}, \qquad x \geq 0,$$

is called an *exponential density function*. (See Fig. 1.) This function is indeed a probability density function. First, $f(x)$ is clearly greater than or equal to 0. Second,

$$\int_0^b ke^{-kx}\, dx = -e^{-kx}\Bigg|_0^b = 1 - e^{-kb} \to 1 \quad \text{as } b \to \infty,$$

FIGURE 1   Exponential density function.

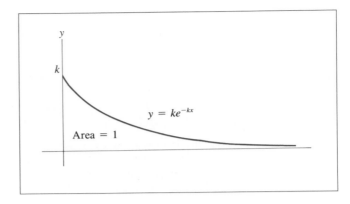

so that

$$\int_0^\infty ke^{-kx}\,dx = 1.$$

A random variable X with an exponential density function is called an *exponential random variable,* and the values of X are said to be *exponentially distributed.* Exponential random variables are used in reliability calculations to represent the lifetime (or "time to failure") of electronic components such as computer chips. They are used to describe the length of time between two successive random events, such as the interarrival times between successive telephone calls at a switchboard. Also, exponential random variables can arise in the study of service times (e.g., the length of time a person spends in a doctor's office or at a gas station).

Let us compute the expected value of an exponential random variable X, namely,

$$E(X) = \int_0^\infty xf(x)\,dx = \int_0^\infty xke^{-kx}\,dx.$$

We may approximate this improper integral by a definite integral and use integration by parts to find that

$$\int_0^b xke^{-kx}\,dx = -xe^{-kx}\Big|_0^b - \int_0^b -e^{-kx}\,dx$$

$$= (-be^{-kb} + 0) - \frac{1}{k}e^{-kx}\Big|_0^b$$

$$= -be^{-kb} - \frac{1}{k}e^{-kb} + \frac{1}{k}. \tag{1}$$

As $b \to \infty$, this quantity approaches $1/k$, because the numbers $-be^{-kb}$ and $-(1/k)e^{-kb}$ both approach 0. (See Section 12.6, Exercise 42.) Thus

$$E(X) = \int_0^\infty xke^{-kx}\,dx = \frac{1}{k}.$$

Let us now compute the variance of X. From the alternate formula for $\text{Var}(X)$ given in Section 11.3, we have

$$\text{Var}(X) = \int_0^\infty x^2 f(x) \, dx - E(X)^2$$

$$= \int_0^\infty x^2 k e^{-kx} \, dx - \frac{1}{k^2}. \tag{2}$$

Using integration by parts, we obtain

$$\int_0^b x^2 k e^{-kx} \, dx = x^2(-e^{-kx})\Big|_0^b - \int_0^b 2x(-e^{-kx}) \, dx$$

$$= (-b^2 e^{-kb} + 0) + 2 \int_0^b x e^{-kx} \, dx$$

$$= -b^2 e^{-kb} + \frac{2}{k} \int_0^b x k e^{-kx} \, dx. \tag{3}$$

Now let $b \to \infty$. We know from our calculation (1) of $E(X)$ that the integral in the second term of (3) approaches $1/k$; also, it can be shown that $-b^2 e^{-kb}$ approaches 0 (see Section 12.6, Exercise 44). Therefore,

$$\int_0^\infty x^2 k e^{-kx} \, dx = \frac{2}{k} \cdot \frac{1}{k} = \frac{2}{k^2}.$$

And by equation (2), we have

$$\text{Var}(X) = \frac{2}{k^2} - \frac{1}{k^2} = \frac{1}{k^2}.$$

Let us summarize our results:

Let X be a random variable with an exponential density function $f(x) = k e^{-kx} \, (x \geq 0)$. Then

$$E(X) = \frac{1}{k} \quad \text{and} \quad \text{Var}(X) = \frac{1}{k^2}.$$

EXAMPLE 1   Suppose that the number of days of continuous use provided by a certain brand of light bulb is an exponential random variable X with expected value 100 days.

(a) Find the density function of X.

(b) Find the probability that a randomly chosen bulb will last between 80 and 90 days.

(c) Find the probability that a randomly chosen bulb will last for more than 40 days.

Solution   (a) Since X is an exponential random variable, its density function must be of the form $f(x) = ke^{-kx}$ for some $k > 0$. Since the expected value of such a density function is $1/k$ and is equal to 100 in this case, we see that

$$\frac{1}{k} = 100,$$

$$k = \frac{1}{100} = .01.$$

Thus

$$f(x) = .01e^{-.01x}.$$

(b)
$$\Pr(80 \leq X \leq 90) = \int_{80}^{90} .01e^{-.01x}\, dx$$

$$= -e^{-.01x}\bigg|_{80}^{90}$$

$$= -e^{-.9} + e^{-.8} \approx .04276.$$

(c)
$$\Pr(X \geq 40) = \int_{40}^{\infty} .01e^{-.01x}\, dx = 1 - \int_{0}^{40} .01e^{-.01x}\, dx$$

[since $\int_0^\infty f(x)\, dx = 1$], so that

$$\Pr(X \geq 40) = 1 + (e^{-.01x})\bigg|_{0}^{40}$$

$$= 1 + (e^{-.4} - 1)$$

$$= e^{-.4} \approx .67032. \quad \bullet$$

EXAMPLE 2   During a certain part of the day, the interarrival time between successive phone calls at a central telephone exchange is an exponential random variable X with expected value $\frac{1}{3}$ second.

(a) Find the density function of X.

(b) Find the probability that between $\frac{1}{3}$ and $\frac{2}{3}$ second elapse between consecutive phone calls.

(c) Find the probability that the time between successive phone calls is more than 2 seconds.

Solution   (a) Since X is an exponential random variable, its density function is $f(x) = ke^{-kx}$ for some $k > 0$. Since the expected value of X is $1/k = \frac{1}{3}$, we have $k = 3$ and $f(x) = 3e^{-3x}$.

(b) $\Pr\left(\dfrac{1}{3} \leq X \leq \dfrac{2}{3}\right) = \displaystyle\int_{1/3}^{2/3} 3e^{-3x}\, dx = -e^{-3x}\bigg|_{1/3}^{2/3} = -e^{-2} + e^{-1}$

$$\approx .23254.$$

(c) $\Pr(X \geq 2) = \displaystyle\int_2^\infty 3e^{-3x}\, dx = 1 - \int_0^2 3e^{-3x}\, dx$

$$= 1 + (e^{-3x})\Big|_0^2$$

$$= e^{-6} \approx .00248.$$

In other words, about .25% of the time, the waiting time between consecutive calls is at least 2 seconds. ●

Normal Density Functions Let μ, σ be given numbers, with $\sigma > 0$. Then the function

$$f(x) = \frac{1}{\sigma\sqrt{2\pi}}\, e^{-(1/2)[(x-\mu)/\sigma]^2} \tag{4}$$

is called a *normal density function*. A random variable X whose density function has this form is called a *normal random variable,* and the values of X are said to be *normally distributed*. Many random variables in applications are approximately normal. For example, errors which occur in physical measurements and various manufacturing processes, as well as many human physical and mental characteristics, are all conveniently modeled by normal random variables.

The graph of the density function in (4) is called a *normal curve* (Fig. 2). A normal curve is symmetric about the line $x = \mu$ and has inflection points at $\mu - \sigma$ and $\mu + \sigma$. Figure 3 shows three normal curves corresponding to different values of σ. The parameters μ and σ determine the shape of the curve. The value of μ determines the point where the curve reaches its maximum height, and the value of σ determines how sharp a peak the curve has.

It can be shown that the constant $1/(\sigma\sqrt{2\pi})$ in the definition (4) of a normal density function $f(x)$ is needed in order to make the area under the normal curve equal to 1 (i.e., to make $f(x)$ a probability density function). The theoretical values of a normal random variable X include all positive and negative numbers, but the normal curve approaches the horizontal axis so rapidly beyond the inflection points that the probabilities associated with intervals on the x-axis far to the left or right of $x = \mu$ are negligible.

FIGURE 2   A normal density function.

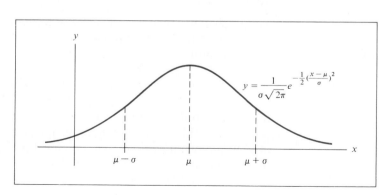

FIGURE 3   **Several normal curves.**

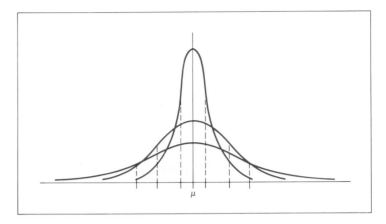

Using techniques outside the scope of this book, one can verify the following basic facts about a normal random variable.

Let X be a random variable with the normal density function

$$f(x) = \frac{1}{\sigma \sqrt{2\pi}} \; e^{-(1/2)[(x-\mu)/\sigma]^2}.$$

Then the expected value (mean), variance, and standard deviation of X are given by

$$\mathrm{E}(X) = \mu, \qquad \mathrm{Var}(X) = \sigma^2, \qquad \text{and} \qquad \sqrt{\mathrm{Var}(X)} = \sigma.$$

A normal random variable with expected value $\mu = 0$ and standard deviation $\sigma = 1$ is called a *standard normal random variable* and is often denoted by the letter Z. Using these values for μ and σ in (4), and writing z in place of the variable x, we see that the density function for Z is

$$f(z) = \frac{1}{\sqrt{2\pi}} \; e^{-(1/2)z^2}.$$

The graph of this function is called the *standard normal curve*. (See Fig. 4.)

Probabilities involving a standard normal random variable Z may be written in the form

$$\Pr(a \le Z \le b) = \int_a^b \frac{1}{\sqrt{2\pi}} \; e^{-(1/2)z^2} \, dz.$$

Such an integral cannot be evaluated directly because the density function for Z cannot be antidifferentiated in terms of elementary functions. However, tables of such probabilities have been compiled, using numerical approximations to the definite integrals. For $z \ge 0$, let $A(z) = \Pr(0 \le Z \le z)$ and $A(-z) = \Pr(-z \le Z \le 0)$. That is, let $A(z)$ and $A(-z)$ be the areas of the regions shown

FIGURE 4 The standard normal curve.

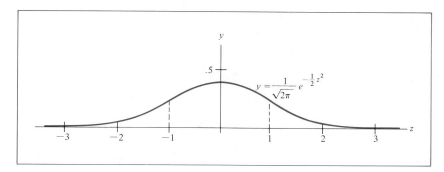

FIGURE 5 Areas under the standard normal curve.

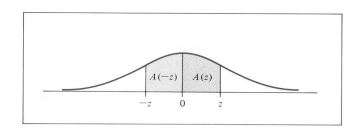

in Fig. 5. From the symmetry of the standard normal curve, it is clear that $A(-z) = A(z)$. The values of $A(z)$ for $z \geq 0$ are listed in Table 1 of the Appendix.

EXAMPLE 3 Let Z be a standard normal random variable. Use Table 1 to compute the following probabilities:

(a) $\Pr(0 \leq Z \leq 1.84)$ (b) $\Pr(-1.65 \leq Z \leq 0)$

(c) $\Pr(.7 \leq Z)$ (d) $\Pr(.5 \leq Z \leq 2)$

(e) $\Pr(-.75 \leq Z \leq 1.46)$

Solution (a) $\Pr(0 \leq Z \leq 1.84) = A(1.84)$. In Table 1 we move down the column under "z" until we reach 1.8; then we move to the right in the same row to the column with the heading ".04." There we find that $A(1.84) = .4671$.

(b) $\Pr(-1.65 \leq Z \leq 0) = A(-1.65) = A(1.65) = .4505$ (from Table 1).

(c) Since the area under the normal curve is 1, the symmetry of the curve implies that the area to the right of the y-axis is .5. Now $\Pr(.7 \leq Z)$ is the area under the curve to the right of .7, and so we can find this area by subtracting from .5 the area between 0 and .7. [See Fig. 6(a).] Thus

$$\Pr(.7 \leq Z) = .5 - \Pr(0 \leq Z \leq .7)$$

$$= .5 - A(.7) = .5 - .2580 \quad \text{(from Table 1)}$$

$$= .2420.$$

(d) The area under the standard normal curve from .5 to 2 equals the area from

FIGURE 6

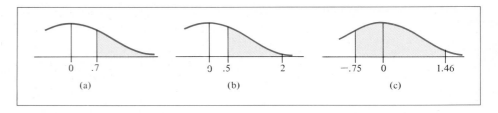

(a)   (b)   (c)

0 to 2 minus the area from 0 to .5. [See Fig. 6(b).] Thus we have

$$\Pr(.5 \leq Z \leq 2) = A(2) - A(.5)$$

$$= .4772 - .1915 = .2857.$$

(e) The area under the standard normal curve from $-.75$ to 1.46 equals the area from $-.75$ to 0 plus the area from 0 to 1.46. [See Fig. 6(c).] Thus

$$\Pr(-.75 \leq Z \leq 1.46) = A(-.75) + A(1.46)$$

$$= A(.75) + A(1.46)$$

$$= .2734 + .4279 = .7013. \quad \bullet$$

When X is an *arbitrary* normal random variable, with mean μ and standard deviation σ, one may compute a probability such as $\Pr(a \leq X \leq b)$ by making the change of variable $z = (x - \mu)/\sigma$. This converts the integral for $\Pr(a \leq X \leq b)$ into an integral involving the standard normal density function. The following example illustrates this procedure.

EXAMPLE 4   A metal flange on a truck must be between 92.1 and 94 millimeters long in order to fit properly. Suppose that the lengths of the flanges supplied to the truck manufacturer are normally distributed with mean $\mu = 93$ millimeters and standard deviation $\sigma = .4$ millimeter.

(a) What percentage of the flanges have an acceptable length?

(b) What percentage of the flanges are too long?

Solution   Let X be the length of a metal flange selected at random from the supply of flanges.

(a) We have

$$\Pr(92.1 \leq X \leq 94) = \int_{92.1}^{94} \frac{1}{(.4)\sqrt{2\pi}}\, e^{-(1/2)[(x-93)/.4]^2}\, dx.$$

Using the substitution $z = (x - 93)/.4$, $dz = (1/.4)\, dx$, we note that if $x = 92.1$, then $z = (92.1 - 93)/.4 = -.9/.4 = -2.25$, and if $x = 94$, then $z = (94 - 93)/.4 = 1/.4 = 2.5$. Hence

$$\Pr(92.1 \leq X \leq 94) = \int_{-2.25}^{2.5} \frac{1}{\sqrt{2\pi}}\, e^{-(1/2)z^2}\, dz.$$

The value of this integral is the area under the standard normal curve from -2.25 to 2.5, which equals the area from -2.25 to 0 plus the area from 0 to 2.5. Thus

$$\Pr(92.1 \le X \le 94) = A(-2.25) + A(2.5)$$
$$= A(2.25) + A(2.5)$$
$$= .4878 + .4938 = .9816.$$

From this probability we conclude that about 98% of the flanges will have an acceptable length.

(b)
$$\Pr(94 \le X) = \int_{94}^{\infty} \frac{1}{(.4)\sqrt{2\pi}}\, e^{-(1/2)[(x-93)/.4]^2}\, dx.$$

This integral is approximated by an integral from $x = 94$ to $x = b$, where b is large. If we substitute $z = (x - 93)/.4$, we find that

$$\int_{94}^{b} \frac{1}{(.4)\sqrt{2\pi}}\, e^{-(1/2)[(x-93)/.4]^2}\, dx = \int_{2.5}^{(b-93)/.4} \frac{1}{\sqrt{2\pi}}\, e^{-(1/2)z^2}\, dz. \qquad (5)$$

Now when $b \to \infty$, the quantity $(b - 93)/.4$ also becomes arbitrarily large. Since the left integral in (5) approaches $\Pr(94 \le X)$, we conclude that

$$\Pr(94 \le X) = \int_{2.5}^{\infty} \frac{1}{\sqrt{2\pi}}\, e^{-(1/2)z^2}\, dz.$$

To calculate this integral we use the method of Example 3(c). The area under the standard normal curve to the right of 2.5 equals the area to the right of 0 minus the area from 0 to 2.5. That is,

$$\Pr(94 \le X) = .5 - A(2.5)$$
$$= .5 - .4938 = .0062.$$

Approximately .6% of the flanges exceed the maximum acceptable length.

●

TECHNOLOGY PROJECT 11.4
Calculating Normal Probabilities

Use Simpson's rule with $n = 100$ to estimate the probabilities (a)–(e) of Example 3. Compare your results with those obtained in the solution above.

PRACTICE PROBLEMS 11.4

1. The emergency flasher on an automobile is warranted for the first 12,000 miles that the car is driven. During that period a defective flasher will be replaced free. Suppose that the time before failure of the emergency flasher (measured in thousands of miles) is an exponential random variable X with mean 50 (thousand miles). What percentage of the flashers will have to be replaced during the warranty period?

2. Suppose that the lead time between ordering furniture from a certain company and receiving delivery is a normal random variable with $\mu = 18$ weeks and $\sigma = 5$ weeks. Find the likelihood that a customer will have to wait more than 16 weeks.

EXERCISES 11.4

Find (by inspection) the expected values and variances of the exponential random variables with the density functions given in Exercises 1–4.

1. $3e^{-3x}$ **2.** $\frac{1}{4}e^{-x/4}$ **3.** $.2e^{-.2x}$ **4.** $1.5e^{-1.5x}$

Suppose that in a large factory there is an average of two accidents per day and the time between accidents has an exponential density function with expected value of $\frac{1}{2}$ day.

5. Find the probability that the time between two accidents will be more than $\frac{1}{2}$ day and less than 1 day.

6. Find the probability that the time between accidents will be less than 8 hours (i.e., $\frac{1}{3}$ day).

Suppose that the amount of time required to serve a customer at a bank has an exponential density function with mean 3 minutes.

7. Find the probability that a customer is served in less than 2 minutes.

8. Find the probability that a customer will require more than 5 minutes.

During a certain part of the day, the time between arrivals of automobiles at the tollgate on a turnpike is an exponential random variable with expected value 20 seconds.

9. Find the probability that the time between successive arrivals is more than 60 seconds.

10. Find the probability that the time between successive arrivals is greater than 10 seconds and less than 30 seconds.

Upon studying the vacancies occurring in the U.S. Supreme Court, it has been determined that the time elapsed between successive resignations is an exponential random variable with expected value 2 years.*

11. A new president takes office at the same time a justice retires. Find the probability that the next vacancy on the court will take place during the president's 4-year term.

12. Find the probability that the composition of the U.S. Supreme Court will remain unchanged for a period of 5 years or more.

13. Suppose that the average life span of an electronic component is 72 months and that the life spans are exponentially distributed.

   (a) Find the probability that a component lasts for more than 24 months.

   (b) The *reliability function* $r(t)$ gives the probability that a component will last for more than t months. Compute $r(t)$ in this case.

---

* See W. A. Wallis, "The Poisson Distribution and the Supreme Court," *Journal of the American Statistical Association*, **31** (1936), 376–380.

14. Consider a group of patients that have been treated for an acute disease such as cancer, and let X be the number of years a person lives after receiving the treatment (the "survival time"). Under suitable conditions, the density function for X will be $f(x) = ke^{-kx}$ for some constant k.

   (a) The *survival function* $S(x)$ is the probability that a person chosen at random from the group of patients survives until at least time x. Explain why $S(x) = 1 - F(x)$, where $F(x)$ is the cumulative distribution function for X, and compute $S(x)$.

   (b) Suppose that the probability is .90 that a patient will survive at least 5 years [i.e., $S(5) = .90$]. Find the constant k in the exponential density function $f(x)$.

Find the expected values and the standard deviations (by inspection) of the normal random variables with the density functions given in Exercises 15–18.

15. $\dfrac{1}{\sqrt{2\pi}} e^{-(1/2)(x-4)^2}$

16. $\dfrac{1}{\sqrt{2\pi}} e^{-(1/2)(x+5)^2}$

17. $\dfrac{1}{3\sqrt{2\pi}} e^{-(1/18)x^2}$

18. $\dfrac{1}{5\sqrt{2\pi}} e^{-(1/2)[(x-3)/5]^2}$

19. Show that the function $f(x) = e^{-x^2/2}$ has a relative maximum at $x = 0$.

20. Show that the function $f(x) = e^{-(1/2)[(x-\mu)/\sigma]^2}$ has a relative maximum at $x = \mu$.

21. Show that the function $f(x) = e^{-x^2/2}$ has inflection points at $x = \pm 1$.

22. Show that the function $f(x) = e^{-(1/2)[(x-\mu)/\sigma]^2}$ has inflection points at $x = \mu \pm \sigma$.

23. Let Z be a standard normal random variable. Calculate:

   (a) $\Pr(-1.3 \le Z \le 0)$ (b) $\Pr(.25 \le Z)$

   (c) $\Pr(-1 \le Z \le 2.5)$ (d) $\Pr(Z \le 2)$

24. Calculate the area under the standard normal curve for values of z

   (a) between .5 and 1.5,   (b) between $-.75$ and .75,

   (c) to the left of $-.3$,   (d) to the right of -1.

25. The gestation period (length of pregnancy) of pregnant females of a certain species is approximately normally distributed with a mean of 6 months and standard deviation of $\frac{1}{2}$ month.

   (a) Find the percentage of births that occur after a gestation period of between 6 and 7 months.

   (b) Find the percentage of births that occur after a gestation period of between 5 and 6 months.

26. Suppose that the life span of a certain automobile tire is normally distributed with $\mu = 25{,}000$ miles and $\sigma = 2000$ miles.

   (a) Find the probability that a tire will last between 28,000 and 30,000 miles.

   (b) Find the probability that a tire will last more than 29,000 miles.

27. Suppose that the amount of milk in a gallon container is a normal random variable with $\mu = 128.2$ ounces and $\sigma = .2$ ounce. Find the probability that a random bottle of milk contains less than 128 ounces.

28. The amount of weight required to break a certain brand of twine has a normal density function with $\mu = 43$ kilograms and $\sigma = 1.5$ kilograms. Find the probability that the breaking weight of a piece of the twine is less than 40 kilograms.

29. A student with an eight o'clock class at the University of Maryland commutes to school by car. She has discovered that along each of two possible routes her traveling time to school (including the time to get to class) is approximately a normal random variable. If she uses the Capital Beltway for most of her trip, $\mu = 25$ minutes and $\sigma = 5$ minutes. If she drives a longer route over local city streets, then $\mu = 28$ minutes and $\sigma = 3$ minutes. Which route should the student take if she leaves home at 7:30 A.M.? (Assume that the best route is one that minimizes the probability of being late to class.)

30. Which route should the student in Exercise 29 take if she leaves home at 7:26 A.M.?

31. A certain type of bolt must fit through a 20-millimeter test hole, or else it is discarded. If the diameters of the bolts are normally distributed with $\mu = 18.2$ millimeters and $\sigma = .8$ millimeter, what percentage of the bolts will be discarded?

32. The Math SAT scores of a recent freshman class at a university were normally distributed with $\mu = 535$ and $\sigma = 100$.

    (a) What percent of the scores were between 500 and 600?

    (b) Find the minimum score needed to be in the top 10% of the class.

33. Let X be the time to failure (in years) of a transistor, and suppose the transistor has been operating properly for a years. Then it can be shown that the probability that the transistor will fail within the next b years is

$$\frac{\text{Pr}\,(a \leq X \leq a + b)}{\text{Pr}\,(a \leq X)}. \tag{6}$$

Compute this probability for the case when X is an exponential random variable with density function $f(x) = ke^{-kx}$, and show that this probability equals $\text{Pr}(0 \leq X \leq b)$. This means that the probability given by (6) does not depend on how long the transistor has already been operating. Exponential random variables are therefore said to be *memoryless*.

34. Recall that the *median* of an exponential density function is that number M such that $\text{Pr}(X \leq M) = \frac{1}{2}$. Show that $M = (\ln 2)/k$. (We see that the median is less than the mean.)

SOLUTIONS TO PRACTICE PROBLEMS 11.4

1. The density function for X is $f(x) = ke^{-kx}$, where $1/k = 50$ (thousand miles), and $k = 1/50 = .02$. Then

$$\text{Pr}(X \leq 12) = \int_0^{12} .02e^{-.02x}\, dx = -e^{-.02x}\Big|_0^{12}$$

$$= 1 - e^{-.24} \approx .21337.$$

About 21% of the flashers will have to be replaced during the warranty period.

2. Let X be the time between ordering and receiving the furniture. Since $\mu = 18$ and $\sigma = 5$, we have

$$\text{Pr}(16 \leq X) = \int_{16}^{\infty} \frac{1}{5\sqrt{2\pi}}\, e^{-(1/2)[(x-18)/5]^2}\, dx.$$

If we substitute $z = (x - 18)/5$, then $dz = \frac{1}{5} dx$, and $z = -.4$ when $x = 16$.

$$\Pr(16 \le X) = \int_{-.4}^{\infty} \frac{1}{\sqrt{2\pi}} e^{-(1/2)z^2} \, dz.$$

[A similar substitution was made in Example 4(b).] The integral above gives the area under the standard normal curve to the right of $-.4$. Since the area between $-.4$ and 0 is $A(-.4) = A(.4)$, and the area to the right of 0 is $.5$, we have

$$\Pr(16 \le X) = A(.4) + .5 = .1554 + .5 = .6554.$$

 **CHAPTER 11 CHECKLIST**

- ✓ Relative frequency histogram
- ✓ Probability table
- ✓ Discrete random variable
- ✓ Expected value (mean) of a discrete random variable
- ✓ Variance of a discrete random variable
- ✓ Standard deviation
- ✓ Probability density histogram
- ✓ Continuous random variable
- ✓ Probability density function
- ✓ Cumulative distribution function

- ✓ Expected value (mean) of a continuous random variable
- ✓ Variance of a continuous random variable
- ✓ Alternate formula for variance
- ✓ Exponential density function
- ✓ Expected value and variance of an exponential random variable
- ✓ Normal density function
- ✓ Expected value and variance of a normal random variable
- ✓ Standard normal density function

CHAPTER 11 SUPPLEMENTARY EXERCISES

1. Let X be a continuous random variable on $0 \le x \le 2$, with the density function $f(x) = \frac{3}{8}x^2$.

   (a) Calculate $\Pr(X \le 1)$ and $\Pr(1 \le X \le 1.5)$.

   (b) Find $E(X)$ and $\text{Var}(X)$.

2. Let X be a continuous random variable on $3 \le x \le 4$, with the density function $f(x) = 2(x - 3)$.

   (a) Calculate $\Pr(3.2 \le X)$ and $\Pr(3 \le X)$.

   (b) Find $E(X)$ and $\text{Var}(X)$.

3. Verify that for any number A, $f(x) = e^{A-x}$, $x \ge A$, is a density function. Compute the associated cumulative distribution function for X.

4. Verify that for any positive constants k and A, the function $f(x) = kA^k/x^{k+1}$, $x \ge A$, is a density function. The associated cumulative distribution function $F(x)$ is called a *Pareto distribution*. Compute $F(x)$.

5. For any positive integer n, the function $f_n(x) = c_n x^{(n-2)/2} e^{-x/2}$, $x \geq 0$, where c_n is an appropriate constant, is called the *chi-square density function* with n degrees of freedom. Find c_2 and c_4 such that $f_2(x)$ and $f_4(x)$ are probability density functions.

6. Verify that for any positive number k, $f(x) = 1/(2k^3) x^2 e^{-x/k}$, $x \geq 0$, is a density function.

7. A medical laboratory tests many blood samples for a certain disease that occurs in about 5% of the samples. The lab collects samples from 10 persons and mixes together some blood from each sample. If a test on the mixture is positive, an additional 10 tests must be run, one on each individual sample. But if the test on the mixture is negative, no other tests are needed. It can be shown that the test of the mixture will be negative with probability $(.95)^{10} = .599$, because each of the 10 samples has a 95% chance of being free of the disease. If X is the total number of tests required, then X has the following probability table.

   (a) Find $E(X)$.

   (b) If the laboratory uses the procedure described on 200 blood samples (that is, 20 batches of 10 samples), about how many tests can it expect to run?

TABLE 1   Blood Test Probabilities for Batches of 10 Samples

	Test of Mixture	
	Negative	Positive
Total Tests	1	11
Probability	.599	.401

---

8. Suppose that the laboratory in Exercise 7 uses batches of 5 instead of 10 samples. The probability of a negative test on the mixture of 5 samples is $(.95)^5 = .774$. Thus the following table gives the probabilities for the number X of tests required.

TABLE 2   Blood Test Probabilities for Batches of 5 Samples

	Test of Mixture	
	Negative	Positive
Total Tests	1	6
Probability	.774	.226

(a) Find E(X).

(b) If the laboratory uses this procedure on 200 blood samples (that is, 40 batches of 5 samples), about how many tests can it expect to run?

9. A certain gas station sells X thousand gallons of gas each week. Suppose that the cumulative distribution function for X is $F(x) = 1 - \frac{1}{4}(2 - x)^2$, $0 \le x \le 2$.

   (a) If the tank contains 1.6 thousand gallons at the beginning of the week, find the probability that the gas station will have enough gas for its customers throughout the week.

   (b) How much gas must be in the tank at the beginning of the week in order to have a probability of .99 that there will be enough gasoline for the week?

   (c) Compute the density function for X.

10. A service contract on computer costs $100 per year. The contract covers all necessary maintenance and repairs on the typewriter. Suppose that the actual cost to the manufacturer for providing this service is a random variable X (measured in hundreds of dollars) whose probability density function is $f(x) = (x - 5)^4/625$, $0 \le x \le 5$. Compute E(X) and determine how much money the manufacturer expects to make on each service contract on the average.

11. A random variable X has a uniform density function $f(x) = \frac{1}{5}$ on $20 \le X \le 25$.

    (a) Find E(X) and Var(X).

    (b) Find b such that $\Pr(X \le b) = .3$.

12. A random variable X has a cumulative distribution function $F(x) = (x^2 - 9)/16$ on $3 \le x \le 5$.

    (a) Find the density function for X.

    (b) Find a such that $\Pr(a \le X) = \frac{1}{4}$.

13. The annual incomes of the households in a certain community range between 5 and 25 thousand dollars. Let X represent the annual income (in thousands of dollars) of a household chosen at random in this community, and suppose that the probability density function for X is $f(x) = kx$, $5 \le x \le 25$.

    (a) Find the value of k that makes $f(x)$ a density function.

    (b) Find the fraction of households whose incomes exceeds $20,000.

    (c) Find the mean annual income of the households in the community.

14. For each positive integer k, the function $f(x) = k(k + 1)x^{k-1}(1 - x)$ is a probability density function on $0 \le x \le 1$.

    (a) Find the associated cumulative distribution function.

    (b) Show that E(X) = $k/(k + 2)$.

15. A point is selected at random from the rectangle of Fig. 1(a); call its coordinates (θ, y). Find the probability that $y \le \sin \theta$.

FIGURE I

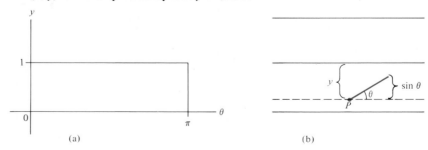

(a)   (b)

16. (*Buffon Needle Problem*) A needle of length one unit is dropped on a floor which is ruled with parallel lines, one unit apart. [See Fig. 1(b).] Let P be the lowest point of the needle, y the distance of P from the ruled line above it, and θ the angle the needle makes with a line parallel to the ruled lines. Show that the needle touches a ruled line if and only if $y \leq \sin \theta$. Conclude that the probability of the needle touching a ruled line is the probability found in Exercise 15.

17. The lifetime of a certain TV picture tube is an exponential random variable with an expected value of 5 years. Suppose that the tube manufacturer sells the tube for $100 but will give a complete refund if the tube burns out within 3 years. Then the revenue the manufacturer receives on each tube is a discrete random variable Y with values 100 and 0. Determine the expected revenue per tube.

18. The condenser motor in an air conditioner costs $300 to replace, but a home air conditioning service will guarantee to replace it free when it burns out if you will pay an annual insurance premium of $25. Suppose that the life span of the motor is an exponential random variable with an expected life of 10 years. Should you take out the insurance for the first year? [*Hint:* Consider the random variable Y such that $Y = 300$ if the motor burns out during the year and $Y = 0$ otherwise. Compare $E(Y)$ with the cost of 1 year's insurance.]

19. An exponential random variable X has been used to model the relief times (in minutes) of arthritic patients who have taken an analgesic for their pain. Suppose that the density function for X is $f(x) = ke^{-kx}$ and that a certain analgesic provides relief within 4 minutes for 75% of a large group of patients. Then one may estimate that $\Pr(X \leq 4) = .75$. Use this estimate to find an approximate value for k. [*Hint:* First show that $\Pr(X \leq 4) = 1 - e^{-4k}$.]

20. A piece of new equipment has a useful life of X thousand hours, where X is a random variable with the density function $f(x) = .01xe^{-x/10}$, $x \geq 0$. A manufacturer expects the machine to generate $5000 of additional income for every thousand hours of use, but the machine costs $60,000. Should the manufacturer purchase the new equipment? [*Hint:* Compute the expected value of the additional earnings generated by the machine.]

21. Extensive records are kept of the life spans (in months) of a certain product, and a relative frequency histogram is constructed from the data, using areas to represent relative frequencies (as in Fig. 4 in Section 11.1). It turns out that the upper boundary of the relative frequency histogram is approximated closely by the graph of the function

$$f(x) = \frac{1}{8\sqrt{2\pi}} e^{-(1/2)[(x-50)/8]^2}.$$

Determine the probability that the life span of such a product is between 30 and 50 months.

22. A certain machine part has a nominal length of 80 millimeters, with a tolerance of $\pm.05$ millimeter. Suppose that the actual length of the parts supplied is a normal random variable with mean 79.99 millimeters and standard deviation .02 millimeter. How many parts in a lot of 1000 should you expect to lie outside the tolerance limits?

23. The men hired by a certain city police department must be at least 69 inches tall. Suppose that the heights of adult men in the city are normally distributed with $\mu = 70$ inches and $\sigma = 2$ inches. What percentage of the men are tall enough to be eligible for recruitment by the police department?

24. Suppose that the police force in Exercise 23 maintains the same height requirements for women as men and that the heights of women in the city are normally distributed, with $\mu = 65$ inches and $\sigma = 1.6$ inches. What percentage of the women are eligible for recruitment?

25. Let Z be a standard normal random variable. Find the number a such that $\Pr(a \leq Z) = .40$.

26. Scores on a school's entrance exam are normally distributed, with $\mu = 500$ and $\sigma = 100$. If the school wishes to admit only the students in the top 40%, what should be the cutoff grade?

27. It is useful in some applications to know that about 68% of the area under the standard normal curve lies between -1 and 1.

    (a) Verify this statement.

    (b) Let X be a normal random variable with expected value μ and variance σ^2. Compute $\Pr(\mu - \sigma \leq X \leq \mu + \sigma)$.

28. (a) Show that about 95% of the area under the standard normal curve lies between -2 and 2.

    (b) Let X be a normal random variable with expected value μ and variance σ^2. Compute $\Pr(\mu - 2\sigma \leq X \leq \mu + 2\sigma)$.

29. The Chebychev inequality says that for any random variable X with expected value μ and standard deviation σ,

$$\Pr(\mu - n\sigma \leq X \leq \mu + n\sigma) \geq 1 - \frac{1}{n^2}.$$

    (a) Take $n = 2$. Apply the Chebychev inequality to an exponential random variable.

    (b) By integrating, find the exact value of the probability in part (a).

30. Do the same as in Exercise 29 with a normal random variable.

TAYLOR POLYNOMIALS AND INFINITE SERIES

In earlier chapters we introduced the functions e^x, $\ln x$, $\sin x$, $\cos x$, and $\tan x$. Whenever we needed the value of one of these functions for a particular value of x, such as $e^{.023}$, $\ln 5.8$, or $\sin .25$, we had to use a scientific calculator. Now we shall take up the problem of numerically computing the values of such functions for particular choices of the variable x. The computational methods developed have many applications—for example, to differential equations and probability theory.

12.1 TAYLOR POLYNOMIALS

A polynomial of degree n is a function of the form

$$p(x) = a_0 + a_1x + \cdots + a_nx^n$$

where $a_0, a_1, \ldots, a_n$ are given numbers and $a_n \neq 0$. There are many instances in mathematics and its applications where calculations are much simpler for polynomials than for other functions. In this section we show how to approximate a given function $f(x)$ by a polynomial $p(x)$ for all values of x near some specified number, say $x = a$. To simplify matters, we begin by considering values of x near $x = 0$.

In Fig. 1 we have drawn the graph of the function $f(x) = e^x$ together with the tangent line through $(0, f(0)) = (0, 1)$. The slope of the tangent line is $f'(0) = 1$. So the equation of the tangent line is

$$y - f(0) = f'(0)(x - 0)$$
$$y = f(0) + f'(0)x$$
$$y = 1 + x.$$

FIGURE 1   A linear approximation of e^x at $x = 0$.

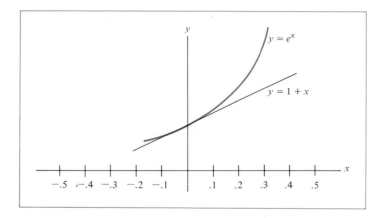

FIGURE 1   A linear approximation of e^x at $x = 0$.

From our discussion of the derivative, we know that the tangent line at $x = 0$ closely approximates the graph of $y = e^x$ for values of x near 0. Thus, if we let $p_1(x) = 1 + x$, then the values of $p_1(x)$ are close to the corresponding values of $f(x) = e^x$ for x near 0.

In general, a given function $f(x)$ may be approximated for values of x near 0 by the polynomial

$$p_1(x) = f(0) + f'(0)x,$$

which is called the *first Taylor polynomial of $f(x)$ at $x = 0$*. The graph of $p_1(x)$ is just the tangent line to $y = f(x)$ at $x = 0$.

The first Taylor polynomial "resembles" $f(x)$ near $x = 0$ in the sense that

$p_1(0) = f(0)$ (both graphs go through the same point at $x = 0$),

$p_1'(0) = f'(0)$ (both graphs have the same slope at $x = 0$).

That is, $p_1(x)$ coincides with $f(x)$ in both its value at $x = 0$ and the value of its first derivative at $x = 0$. This suggests that to approximate $f(x)$ even more closely at $x = 0$, we look for a polynomial that coincides with $f(x)$ in its value at $x = 0$ and in the values of its first *and second* derivatives at $x = 0$. Yet a further approximation can be obtained by going out to the third derivative, and so on.

EXAMPLE 1   Given a function $f(x)$, suppose that $f(0) = 1$, $f'(0) = -2$, $f''(0) = 7$, and $f'''(0) = -5$. Find a polynomial of degree 3,

$$p(x) = a_0 + a_1 x + a_2 x^2 + a_3 x^3,$$

such that $p(x)$ coincides with $f(x)$ up to the third derivative at $x = 0$; that is,

$p(0) = f(0) = 1$ (same value at $x = 0$),

$p'(0) = f'(0) = -2$ (same first derivative at $x = 0$),

$p''(0) = f''(0) = 7$ (same second derivative at $x = 0$),

$p'''(0) = f'''(0) = -5$ (same third derivative at $x = 0$).

Solution   To find the coefficients $a_0, \ldots, a_3$ of $p(x)$, we first compute the values of $p(x)$ and its derivatives at $x = 0$:

$$p(x) = a_0 + a_1 x + a_2 x^2 + a_3 x^3, \qquad p(0) = a_0,$$
$$p'(x) = 0 + a_1 + 2a_2 x + 3a_3 x^2, \qquad p'(0) = a_1,$$
$$p''(x) = 0 + 0 + 2a_2 + 2 \cdot 3a_3 x, \qquad p''(0) = 2a_2,$$
$$p'''(x) = 0 + 0 + 0 + 2 \cdot 3a_3, \qquad p'''(0) = 2 \cdot 3a_3.$$

Since we want $p(x)$ and its derivatives to coincide with the given values of $f(x)$ and its derivatives, we must have

$$a_0 = 1, \qquad a_1 = -2, \qquad 2a_2 = 7, \quad \text{and} \quad 2 \cdot 3a_3 = -5.$$

So

$$a_0 = 1, \qquad a_1 = -2, \qquad a_2 = \frac{7}{2}, \quad \text{and} \quad a_3 = \frac{-5}{2 \cdot 3}.$$

Rewriting the coefficients slightly, we have

$$p(x) = 1 + \frac{(-2)}{1}x + \frac{7}{1 \cdot 2}x^2 + \frac{-5}{1 \cdot 2 \cdot 3}x^3.$$

The form in which we have written $p(x)$ clearly exhibits the values $1, -2, 7, -5$ of $f(x)$ and its derivatives at $x = 0$. In fact, we could also write this formula for $p(x)$ in the form

$$p(x) = f(0) + \frac{f'(0)}{1}x + \frac{f''(0)}{1 \cdot 2}x^2 + \frac{f'''(0)}{1 \cdot 2 \cdot 3}x^3. \quad \bullet$$

Given a function $f(x)$, we may use the formula in Example 1 to find a polynomial that coincides with $f(x)$ up to the third derivative at $x = 0$. To describe the general formula for higher-order polynomials, we let $f^{(n)}(x)$ denote the nth derivative of $f(x)$, and we let $n!$ (read "n factorial") denote the product of all the integers from 1 to n, so that $n! = 1 \cdot 2 \cdot \ldots \cdot (n-1) \cdot n$. (Thus $1! = 1$, $2! = 1 \cdot 2$, $3! = 1 \cdot 2 \cdot 3$, and so forth.)

Given a function $f(x)$, the nth *Taylor polynomial of $f(x)$ at $x = 0$* is the polynomial $p_n(x)$ defined by

$$p_n(x) = f(0) + \frac{f'(0)}{1!}x + \frac{f''(0)}{2!}x^2 + \cdots + \frac{f^{(n)}(0)}{n!}x^n.$$

This polynomial coincides with $f(x)$ up to the nth derivative at $x = 0$ in the sense that

$$p_n(0) = f(0), \quad p_n'(0) = f'(0), \ldots, p_n^{(n)}(0) = f^{(n)}(0).$$

The next example shows how Taylor polynomials are used to calculate the values of e^x for x near 0. The choice of which polynomial to use depends on how accurate one wants the values of e^x.

EXAMPLE 2   Determine the first three Taylor polynomials of $f(x) = e^x$ at $x = 0$ and sketch their graphs.

Solution   Since all derivatives of e^x are e^x, we see that

$$f(0) = f'(0) = f''(0) = f'''(0) = e^0 = 1.$$

Thus the desired Taylor polynomials are

$$p_1(x) = 1 + \frac{1}{1} \cdot x = 1 + x,$$

$$p_2(x) = 1 + \frac{1}{1!}x + \frac{1}{2!}x^2 = 1 + x + \frac{1}{2}x^2,$$

$$p_3(x) = 1 + \frac{1}{1!}x + \frac{1}{2!}x^2 + \frac{1}{3!}x^3 = 1 + x + \frac{1}{2}x^2 + \frac{1}{6}x^3.$$

The relative accuracy of these approximations to e^x may be seen using a calculator and from the graphs in Fig. 2. Notice that for x between $-.3$ and $.3$, the

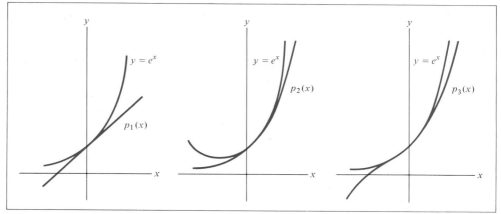

FIGURE 2   Taylor polynomials of e^x at $x = 0$.

difference between $p_1(x)$ and e^x is less than .05, the difference between $p_2(x)$ and e^x is less than .005, and the difference between $p_3(x)$ and e^x is less than .0004. ●

EXAMPLE 3   Determine the nth Taylor polynomial of $f(x) = \dfrac{1}{1 - x}$ at $x = 0$.

Solution

$$f(x) = (1 - x)^{-1}, \qquad\qquad\qquad f(0) = 1$$
$$f'(x) = 1(1 - x)^{-2}, \qquad\qquad\qquad f'(0) = 1$$
$$f''(x) = 1 \cdot 2(1 - x)^{-3} = 2!(1 - x)^{-3}, \qquad f''(0) = 2!$$
$$f'''(x) = 1 \cdot 2 \cdot 3(1 - x)^{-4} = 3!(1 - x)^{-4}, \qquad f'''(0) = 3!$$
$$f^{(4)}(x) = 1 \cdot 2 \cdot 3 \cdot 4(1 - x)^{-5} = 4!(1 - x)^{-5}, \qquad f^{(4)}(0) = 4!$$

From the pattern of calculations, it is clear that $f^{(k)}(0) = k!$ for each k. Therefore,

$$p_n(x) = 1 + \frac{1}{1!}x + \frac{2!}{2!}x^2 + \frac{3!}{3!}x^3 + \cdots + \frac{n!}{n!}x^n$$

$$= 1 + x + x^2 + x^3 + \cdots + x^n. \;●$$

We have already mentioned the possibility of using a polynomial to approximate the values of a function near $x = 0$. Here is another way to use a polynomial approximation.

EXAMPLE 4   It can be shown that the second Taylor polynomial of $\sin x^2$ at $x = 0$ is $p_2(x) = x^2$. Use this polynomial to approximate the area under the graph of $y = \sin x^2$ from $x = 0$ to $x = 1$. (See Fig. 3.)

FIGURE 3 The second Taylor polynomial of sin x^2 at $x = 0$.

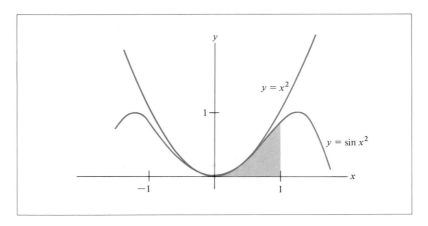

Solution   Since the graph of $p_2(x)$ is very close to the graph of sin x^2 for x near 0, the areas under the two graphs should be almost the same. The area under the graph of $p_2(x)$ is

$$\int_0^1 p_2(x)\, dx = \int_0^1 x^2\, dx = \tfrac{1}{3}x^3 \Big|_0^1 = \tfrac{1}{3}. \ \ \bullet$$

In Example 4, the exact area under the graph of sin x^2 is given by

$$\int_0^1 \sin x^2\, dx.$$

However, this integral cannot be computed by the usual method because there is no way to construct an antiderivative of sin x^2 consisting of elementary functions. Using an approximation technique from Chapter 9, one can find that the value of the integral is .3103 to four decimal places. Thus the error in using $p_2(x)$ as an approximation for sin x^2 is about .023. (The error can be reduced further by using a Taylor polynomial of higher degree. In this particular example, the effort involved is far less than that required by the approximation methods of Chapter 9.)

Suppose now that we wish to approximate a given function $f(x)$ by a polynomial for values of x near some number a. Since the behavior of $f(x)$ near $x = a$ is determined by the values of $f(x)$ and its derivatives at $x = a$, we should try to approximate $f(x)$ by a polynomial $p(x)$ for which the values of $p(x)$ and its derivatives at $x = a$ are the same as those of $f(x)$. This is easily done if we use a polynomial that has the form

$$p(x) = a_0 + a_1(x - a) + a_2(x - a)^2 + \cdots + a_n(x - a)^n.$$

We call this a *polynomial in x − a*. In this form it is easy to compute $p(a)$, $p'(a)$, and so on, because setting $x = a$ in $p(x)$ or one of its derivatives makes most of the terms equal zero. The following result is easily verified.

> Given a function $f(x)$, the nth *Taylor polynomial of* $f(x)$ *at* $x = a$ is the polynomial $p_n(x)$ defined by
>
> $$p_n(x) = f(a) + \frac{f'(a)}{1!}(x - a) + \frac{f''(a)}{2!}(x - a)^2 + \cdots + \frac{f^{(n)}(a)}{n!}(x - a)^n.$$
>
> This polynomial coincides with $f(x)$ up to the nth derivative at $x = a$ in the sense that
>
> $$p_n(a) = f(a), \quad p'_n(a) = f'(a), \ldots, \quad p_n^{(n)}(a) = f^{(n)}(a).$$

When $a = 0$, of course, these Taylor polynomials are just the same as those introduced earlier.

EXAMPLE 5 Calculate the second Taylor polynomial of $f(x) = \sqrt{x}$ at $x = 1$ and use this polynomial to estimate $\sqrt{1.02}$.

Solution Here $a = 1$. Since we want the second Taylor polynomial, we must calculate the values of $f(x)$ and of its first two derivatives at $x = 1$.

$$f(x) = x^{1/2}, \qquad f'(x) = \tfrac{1}{2}x^{-1/2}, \qquad f''(x) = -\tfrac{1}{4}x^{-3/2},$$
$$f(1) = 1, \qquad f'(1) = \tfrac{1}{2}, \qquad f''(1) = -\tfrac{1}{4}.$$

Therefore, the desired Taylor polynomial is

$$p_2(x) = 1 + \frac{\tfrac{1}{2}}{1!}(x - 1) + \frac{-\tfrac{1}{4}}{2!}(x - 1)^2$$
$$= 1 + \tfrac{1}{2}(x - 1) - \tfrac{1}{8}(x - 1)^2.$$

Since 1.02 is close to 1, $p_2(1.02)$ gives a good approximation to $f(1.02)$, that is, to $\sqrt{1.02}$.

$$p_2(1.02) = 1 + \tfrac{1}{2}(1.02 - 1) - \tfrac{1}{8}(1.02 - 1)^2$$
$$= 1 + \tfrac{1}{2}(.02) - \tfrac{1}{8}(.02)^2$$
$$= 1 + .01 - .00005$$
$$= 1.00995. \ \bullet$$

The solution to Example 5 is incomplete in a practical sense, for it offers no information about how close 1.00995 is to the true value of $\sqrt{1.02}$. In general, when we obtain an approximation to some quantity, we also want an indication of the quality of the approximation.

In order to measure the accuracy of an approximation to a function $f(x)$ by its Taylor polynomial at $x = 1$, we define

$$R_n(x) = f(x) - p_n(x).$$

This difference between $f(x)$ and $p_n(x)$ is called the *nth remainder of* $f(x)$ at $x = a$. The following formula is derived in advanced texts:

The Remainder Formula   Suppose that the function $f(x)$ can be differentiated $n + 1$ times on an interval containing the number a. Then for each x in this interval, there exists a number c between a and x such that

$$R_n(x) = \frac{f^{(n+1)}(c)}{(n + 1)!}(x - a)^{n+1}. \tag{1}$$

Usually the precise value of c is unknown. However, if we can find a number M such that $\left| f^{(n+1)}(c) \right| \le M$ for all c between a and x, then we don't need to know which c appears in (1), because we have

$$\left| f(x) - p_n(x) \right| = \left| R_n(x) \right| \le \frac{M}{(n + 1)!}\left| x - a \right|^{n+1}.$$

EXAMPLE 6   Determine the accuracy of the estimate in Example 5.

Solution   The second remainder for a function $f(x)$ at $x = 1$ is

$$R_2(x) = \frac{f^{(3)}(c)}{3!}(x - 1)^3,$$

where c is between 1 and x (and where c depends on x). Here $f(x) = \sqrt{x}$, and therefore $f^{(3)}(c) = \frac{3}{8}c^{-5/2}$. We are interested in $x = 1.02$, and so $1 \le c \le 1.02$. We observe that since $c^{5/2} \ge 1^{5/2} = 1$, we have $c^{-5/2} \le 1$. Thus

$$\left| f^{(3)}(c) \right| \le \tfrac{3}{8} \cdot 1 = \tfrac{3}{8},$$

and

$$\left| R_2(1.02) \right| \le \frac{3/8}{3!}(1.02 - 1)^3$$

$$\le \tfrac{3}{8} \cdot \tfrac{1}{6}(.02)^3$$

$$= .0000005.$$

Thus the error in using $p_2(1.02)$ as an approximation of $f(1.02)$ is at most .0000005. ●

TECHNOLOGY PROJECT 12.1
Approximation by Taylor Polynomials (Symbolic Math Program Required)

You may use the differentiation capabilities of symbolic math programs to determine Taylor polynomials. Moreover, you may use a graphical approach to determine how closely the Taylor polynomials approximate the original function. These topics will be explored in this project.

1. Use a symbolic math program to determine the first four Taylor polynomials about $x = 0$ of the rational function

$$f(x) = \frac{1}{1 + x + x^2}.$$

2. Graph each Taylor polynomial and $f(x)$ on the same coordinate system. By examining the graphs, determine how closely each Taylor polynomial approximates $f(x)$ for x in the interval $-1 \le x \le 1$.

3. Use a graphical approach to determining upper bounds on the first five derivatives of $f(x)$ for $-1 \le x \le 1$. Use the bounds to estimate the error in the Taylor polynomial approximations using the result in the text.

4. Compare the results of 3 and the results from 2. Which method yields better results? Why?

TECHNOLOGY PROJECT 12.2
Bernoulli Numbers (Symbolic Math Program Required)

Use a symbolic math program to determine the first ten Taylor polynomials for

$$f(x) = \frac{x}{e^x - 1}.$$

The coefficients of the Taylor polynomials are an interesting set of fractions called **Bernoulli numbers.** These numbers have many important properties which are derived in more advanced mathematics courses.

PRACTICE PROBLEMS 12.1

1. (a) Determine the third Taylor polynomial of $f(x) = \cos x$ at $x = 0$.

(b) Use the result of part (a) to estimate $\cos .12$.

2. Determine all Taylor polynomials of $f(x) = 3x^2 - 17$ at $x = 3$.

EXERCISES 12.1

In Exercises 1–8, determine the third Taylor polynomial of the given function at $x = 0$.

1. $f(x) = \sin x$ 2. $f(x) = e^{-x/2}$ 3. $f(x) = 5e^{2x}$

4. $f(x) = \cos(\pi - 5x)$ 5. $f(x) = \sqrt{4x + 1}$ 6. $f(x) = \dfrac{1}{x + 2}$

7. $f(x) = xe^{3x}$ 8. $f(x) = \sqrt{1 - x}$

9. Determine the fourth Taylor polynomial of $f(x) = e^x$ at $x = 0$ and use it to estimate $e^{.01}$.

10. Determine the fourth Taylor polynomial of $f(x) = \ln(1 - x)$ at $x = 0$ and use it to estimate $\ln(.9)$.

11. Sketch the graphs of $f(x) = \dfrac{1}{1 - x}$ and its first three Taylor polynomials at $x = 0$.

12. Sketch the graphs of $f(x) = \sin x$ and its first three Taylor polynomials at $x = 0$.

13. Determine the nth Taylor polynomial for $f(x) = e^x$ at $x = 0$.

14. Determine all Taylor polynomials for $f(x) = x^2 + 2x + 1$ at $x = 0$.

15. Use a second Taylor polynomial at $x = 0$ to estimate the area under the curve $y = \ln(1 + x^2)$ from $x = 0$ to $x = \frac{1}{2}$.

16. Use a second Taylor polynomial at $x = 0$ to estimate the area under the curve $y = \sqrt{\cos x}$ from $x = -1$ to $x = 1$. (The exact answer to three decimal places is 1.828.)

17. Determine the third Taylor polynomial of $\dfrac{1}{5 - x}$ at $x = 4$.

18. Determine the fourth Taylor polynomial of $\ln x$ at $x = 1$.

19. Determine the third and fourth Taylor polynomials of $\cos x$ at $x = \pi$.

20. Determine the third and fourth Taylor polynomials of $x^3 + 3x - 1$ at $x = -1$.

21. Use the second Taylor polynomial of $f(x) = \sqrt{x}$ at $x = 9$ to estimate $\sqrt{9.3}$.

22. Use the second Taylor polynomial of $f(x) = \ln x$ at $x = 1$ to estimate $\ln .8$.

23. Determine all Taylor polynomials of $f(x) = x^4 + x + 1$ at $x = 2$.

24. Determine the nth Taylor polynomial of $f(x) = 1/x$ at $x = 1$.

25. Compute the second Taylor polynomial at $z = 0$ for the standard normal density function $f(z) = \dfrac{1}{\sqrt{2\pi}} e^{-z^2/2}$.

26. Use the result of Exercise 25 to estimate $\Pr(0 \le Z \le .2)$, where Z is the standard normal random variable. Compare your answer with that found by using Table 1 of the Appendix.

27. If $f(x) = 3 + 4x - \dfrac{5}{2!}x^2 + \dfrac{7}{3!}x^3$, what are $f''(0)$ and $f'''(0)$?

28. If $f(x) = 2 - 6(x - 1) + \dfrac{3}{2!}(x - 1)^2 - \dfrac{5}{3!}(x - 1)^3 + \dfrac{1}{4!}(x - 1)^4$, what are $f''(1)$ and $f'''(1)$?

29. The third remainder for $f(x)$ at $x = 0$ is

$$R_3(x) = \frac{f^{(4)}(c)}{4!}x^4$$

where c is a number between 0 and x. Let $f(x) = \cos x$, as in Practice Problem 1.

(a) Find a number M such that $|f^{(4)}(c)| \le M$ for all values of c.

(b) In Practice Problem 1, the error in using $p_3(.12)$ as an approximation to $f(.12) = \cos .12$ is given by $R_3(.12)$. Show that this error does not exceed 8.64×10^{-6}.

30. Let p_4 be the fourth Taylor polynomial of $f(x) = e^x$ at $x = 0$. Show that the error in using $p_4(.1)$ as an approximation for $e^{.1}$ is at most 2.5×10^{-7}. [*Hint:* Observe that if $x = .1$ and if c is a number between 0 and .1, then $|f^{(5)}(c)| \le f^{(5)}(.1) = e^{.1} \le e^1 \le 3$.]

31. Let $p_2(x)$ be the second Taylor polynomial of $f(x) = \sqrt{x}$ at $x = 9$, as in Exercise 21.

(a) Give the second remainder for $f(x)$ at $x = 9$.

(b) Show that $|f^{(3)}(c)| \le \frac{1}{648}$ if $c \ge 9$.

(c) Show that the error in using $p_2(9.3)$ as an approximation for $\sqrt{9.3}$ is at most $\frac{1}{144} \times 10^{-3} < 7 \times 10^{-6}$.

32. Let $p_2(x)$ be the second Taylor polynomial of $f(x) = \ln x$ at $x = 1$, as in Exercise 22.

(a) Show that $|f^{(3)}(c)| < 4$ if $c \ge .8$.

(b) Show that the error in using $p_2(.8)$ as an approximation for $\ln .8$ is at most $\frac{16}{3} \times 10^{-3} < .0054$.

SOLUTIONS TO PRACTICE PROBLEMS 12.1

1. (a) $f(x) = \cos x$, $\quad f'(x) = -\sin x$, $\quad f''(x) = -\cos x$, $\quad f'''(x) = \sin x$,
$f(0) = 1$, $\quad f'(0) = 0$, $\quad f''(0) = -1$, $\quad f'''(0) = 0$.

Therefore,

$$p_3(x) = 1 + \frac{0}{1!}x + \frac{-1}{2!}x^2 + \frac{0}{3!}x^3$$
$$= 1 - \tfrac{1}{2}x^2.$$

[Notice that here the third Taylor polynomial is actually a polynomial of degree 2. The important thing about $p_3(x)$ is not its degree but rather the fact that it agrees with $f(x)$ at $x = 0$ up to its third derivative.]

(b) By part (a), $\cos x \approx 1 - \tfrac{1}{2}x^2$ when x is near 0. Therefore,

$$\cos .12 \approx 1 - \tfrac{1}{2}(.12)^2 = .99280.$$

[*Note:* To five decimal places, $\cos .12 = .99281$.]

2. $f(x) = 3x^2 - 17$, $\quad f'(x) = 6x$, $\quad f''(x) = 6$, $\quad f^{(3)}(x) = 0$,
$f(3) = 10$, $\quad f'(3) = 18$, $\quad f''(3) = 6$, $\quad f^{(3)}(3) = 0$.

The derivatives $f^{(n)}(x)$ for $n \ge 3$ are all the zero constant function. In particular, $f^{(n)}(3) = 0$ for $n \ge 3$. Therefore,

$$p_1(x) = 10 + 18(x - 3),$$
$$p_2(x) = 10 + 18(x - 3) + \frac{6}{2!}(x - 3)^2,$$
$$p_3(x) = 10 + 18(x - 3) + \frac{6}{2!}(x - 3)^2 + \frac{0}{3!}(x - 3)^3.$$

For $n \geq 3$, we have

$$p_n(x) = p_2(x) = 10 + 18(x - 3) + 3(x - 3)^2.$$

[This is the appropriate form of the Taylor polynomial at $x = 3$. However, it is instructive to multiply out the terms in $p_2(x)$ and collect the like powers of x:

$$p_2(x) = 10 + 18x - 18 \cdot 3 + 3(x^2 - 6x + 9)$$

$$= 10 + 18x - 54 + 3x^2 - 18x + 27 = 3x^2 - 17.$$

That is, $p_2(x)$ is $f(x)$, but written in a different form. This is not too surprising, since $f(x)$ itself is a polynomial that agrees with $f(x)$ and all of its derivatives at $x = 3$.]

12.2 THE NEWTON-RAPHSON ALGORITHM

Many applications of mathematics involve the solution of equations. Often one has a function $f(x)$ and must find a value of x, say $x = r$, such that $f(r) = 0$. Such a value of x is called a *zero* of the function or, equivalently, a *root* of the equation $f(x) = 0$. Graphically, a zero of $f(x)$ is a value of x where the graph of $y = f(x)$ crosses the x-axis. (See Fig. 1.) When $f(x)$ is a polynomial, it is sometimes possible to factor $f(x)$ and quickly discover the zeros of $f(x)$. Unfortunately, in most realistic applications there is no simple way to locate zeros. However, there are several methods for finding an approximate value of a zero to any desired degree of accuracy. We shall describe one such method—the Newton-Raphson algorithm.

FIGURE 1

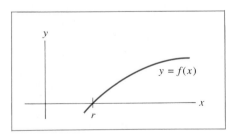

Suppose that we know that a zero of $f(x)$ is approximately x_0. The idea of the Newton-Raphson algorithm is to obtain an even better approximation of the zero by replacing $f(x)$ by its first Taylor polynomial at x_0—that is, by

$$p(x) = f(x_0) + \frac{f'(x_0)}{1}(x - x_0).$$

Since $p(x)$ closely resembles $f(x)$ near $x = x_0$, the zero of $f(x)$ should be close to the zero of $p(x)$. But solving the equation $p(x) = 0$ for x gives

$$f(x_0) + f'(x_0)(x - x_0) = 0$$

$$xf'(x_0) = f'(x_0)x_0 - f(x_0)$$

$$x = x_0 - \frac{f(x_0)}{f'(x_0)}.$$

That is, if x_0 is an approximation to the zero r, then the number

$$x_1 = x_0 - \frac{f(x_0)}{f'(x_0)} \tag{1}$$

generally provides an improved approximation.

We may visualize the situation geometrically as in Fig. 2. The first Taylor polynomial $p(x)$ at x_0 has as its graph the tangent line to $y = f(x)$ at the point $(x_0, f(x_0))$. The value of x for which $p(x) = 0$—that is, $x = x_1$—corresponds to the point where the tangent line crosses the x-axis.

FIGURE 2   Obtaining x_1 from x_0.

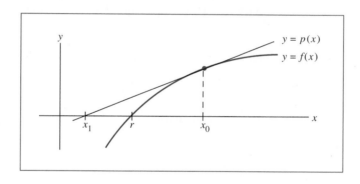

Now let us use x_1 in place of x_0 as an approximation to the zero r. We obtain a new approximation x_2 from x_1 in the same way we obtained x_1 from x_0—namely,

$$x_2 = x_1 - \frac{f(x_1)}{f'(x_1)}.$$

We may repeat this process over and over. At each stage a new approximation x_{new} is obtained from the old approximation x_{old} by the formula

$$x_{new} = x_{old} - \frac{f(x_{old})}{f'(x_{old})}.$$

In this way we obtain a sequence of approximations $x_0, x_1, x_2, \ldots$, which usually approach as close to r as desired. (See Fig. 3.)

FIGURE 3   A sequence of approximations to r.

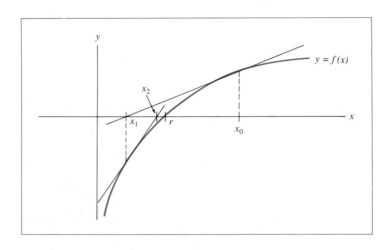

EXAMPLE 1   The polynomial $f(x) = x^3 - x - 2$ has a zero between 1 and 2. Let $x_0 = 1$ and find the next three approximations of the zero of $f(x)$ using the Newton-Raphson algorithm.

Solution   Since $f'(x) = 3x^2 - 1$, formula (1) becomes

$$x_1 = x_0 - \frac{x_0^3 - x_0 - 2}{3x_0^2 - 1}.$$

With $x_0 = 1$, we have

$$x_1 = 1 - \frac{1^3 - 1 - 2}{3(1)^2 - 1} = 1 - \frac{-2}{2} = 2,$$

$$x_2 = 2 - \frac{2^3 - 2 - 2}{3(2)^2 - 1} = 2 - \tfrac{4}{11} = \tfrac{18}{11},$$

$$x_3 = \tfrac{18}{11} - \frac{(\tfrac{18}{11})^3 - \tfrac{18}{11} - 2}{3(\tfrac{18}{11})^2 - 1} \approx 1.530.$$

The actual value of r to three decimal places is 1.521. ●

EXAMPLE 2   Use four repetitions of the Newton-Raphson algorithm to approximate $\sqrt{2}$.

Solution $\sqrt{2}$ is a zero of the function $f(x) = x^2 - 2$. Since $\sqrt{2}$ clearly lies between 1 and 2, let us take our initial approximation as $x_0 = 1$. ($x_0 = 2$ would do just as well.) Since $f'(x) = 2x$, we have

$$x_1 = x_0 - \frac{x_0^2 - 2}{2x_0}$$

$$= 1 - \frac{1^2 - 2}{2(1)} = 1 - \left(-\frac{1}{2}\right) = 1.5,$$

$$x_2 = 1.5 - \frac{(1.5)^2 - 2}{2(1.5)} \approx 1.4167,$$

$$x_3 = 1.4167 - \frac{(1.4167)^2 - 2}{2(1.4167)} \approx 1.41422,$$

$$x_4 = 1.41422 - \frac{(1.41422)^2 - 2}{2(1.41422)} \approx 1.41421.$$

This approximation to $\sqrt{2}$ is correct to five decimal places. ●

EXAMPLE 3   Approximate the zeros of the polynomial $x^3 + x + 3$.

Solution   By applying our curve-sketching techniques, we can make a rough sketch of the graph of $y = x^3 + x + 3$, as in Fig. 4. The graph crosses the x-axis between $x = -2$ and $x = -1$. So the polynomial has one zero lying between -2 and -1. Let us therefore set $x_0 = -1$. Since $f'(x) = 3x^2 + 1$, we have

$$x_1 = x_0 - \frac{x_0^3 + x_0 + 3}{3x_0^2 + 1}$$

$$= -1 - \frac{(-1)^3 + (-1) + 3}{3(-1)^2 + 1} = -1.25,$$

$$x_2 = -1.25 - \frac{(-1.25)^3 + (-1.25) + 3}{3(-1.25)^2 + 1} \approx -1.21429,$$

$$x_3 \approx -1.21341,$$

$$x_4 \approx -1.21342.$$

FIGURE 4

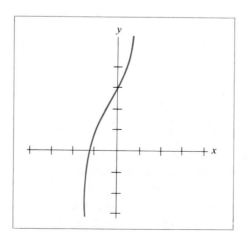

Therefore, the zero of the given polynomial is approximately -1.21342. ●

EXAMPLE 4 Approximate the positive solution of $e^x - 4 = x$.

Solution The rough sketches of the two graphs in Fig. 5 indicate that the solution lies near 2. Let $f(x) = e^x - 4 - x$. Then the solution of the original equation will be a zero of $f(x)$. Let us apply the Newton-Raphson algorithm to $f(x)$ with $x_0 = 2$. Since $f'(x) = e^x - 1$,

$$x_1 = x_0 - \frac{e^{x_0} - 4 - x_0}{e^{x_0} - 1}$$

$$= 2 - \frac{e^2 - 4 - 2}{e^2 - 1} \approx 2 - \frac{1.38906}{6.38906} \approx 1.78,$$

$$x_2 = 1.78 - \frac{e^{1.78} - 4 - (1.78)}{e^{1.78} - 1} \approx 1.78 - \frac{.14986}{4.92986} \approx 1.75,$$

$$x_3 = 1.75 - \frac{e^{1.75} - 4 - (1.75)}{e^{1.75} - 1} \approx 1.75 - \frac{.0046}{4.7546} \approx 1.749.$$

Therefore, an approximate solution is $x = 1.749$. ●

FIGURE 5

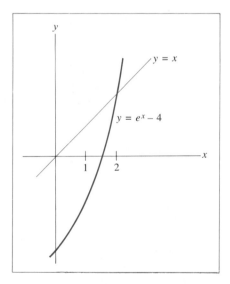

EXAMPLE 5 (*Internal Rate of Return*)   Suppose that an investment of $100 yields the following returns:

$2 at the end of the first month

$15 at the end of the second month

$45 at the end of the third month

$50 at the end of the fourth (and last) month

The total of these returns is $112. This represents the initial investment of $100, plus earnings of $12 during the four months. The *internal rate of return* on this investment is the interest rate (per month) for which the sum of the present values of the returns equals the initial investment, $100. Determine the internal rate of return.

Solution Let i be the *monthly* rate of interest. The present value of an amount A to be received in k months is $A(1 + i)^{-k}$. Therefore, we must solve

$$\begin{bmatrix} \text{amount of initial} \\ \text{investment} \end{bmatrix} = \begin{bmatrix} \text{sum of present} \\ \text{values of returns} \end{bmatrix}$$

$$100 = 2(1 + i)^{-1} + 15(1 + i)^{-2} + 45(1 + i)^{-3} + 50(1 + i)^{-4}.$$

Multiplying both sides of the equation by $(1 + i)^4$ and taking all terms to the left, we obtain

$$100(1 + i)^4 - 2(1 + i)^3 - 15(1 + i)^2 - 45(1 + i) - 50 = 0.$$

Let $x = 1 + i$, and solve the resulting equation by the Newton-Raphson algorithm with $x_0 = 1.1$.

$$100x^4 - 2x^3 - 15x^2 - 45x - 50 = 0$$

$$f(x) = 100x^4 - 2x^3 - 15x^2 - 45x - 50$$

$$f'(x) = 400x^3 - 6x^2 - 30x - 45$$

$$x_1 = x_0 - \frac{100x_0^4 - 2x_0^3 - 15x_0^2 - 45x_0 - 50}{400x_0^3 - 6x_0^2 - 30x_0 - 45}$$

$$= 1.1 - \frac{100(1.1)^4 - 2(1.1)^3 - 15(1.1)^2 - 45(1.1) - 50}{400(1.1)^3 - 6(1.1)^2 - 30(1.1) - 45}$$

$$= 1.1 - \frac{26.098}{447.14} \approx 1.042$$

$$x_2 \approx 1.035$$

$$x_3 \approx 1.035.$$

Therefore, an approximate solution is $x = 1.035$. Hence $i = .035$ and the investment had an internal rate of return of 3.5% per month. ●

In general, if an investment of P dollars produces the returns

R_1 at the end of the first period,

R_2 at the end of the second period,

$\vdots$

R_N at the end of the Nth (and last) period,

then the internal rate of return, i, is obtained by solving* the equation

$$P(1 + i)^N - R_1(1 + i)^{N-1} - R_2(1 + i)^{N-2} - \cdots - R_N = 0$$

for its positive root.

When a loan of P is paid back with N equal periodic payments of R dollars at interest rate i per period, the equation to be solved becomes

$$P(1 + i)^N - R(1 + i)^{N-1} - R(1 + i)^{N-2} - \cdots - R = 0.$$

This equation can be simplified to

$$Pi + R[(1 + i)^{-N} - 1] = 0$$

(See Exercise 41 in Section 12.3.)

EXAMPLE 6   (*Amortization of a Loan*)   A mortgage of $60,050 is repaid in 360 monthly payments of $905. Determine the monthly rate of interest.

Solution   Here $P = 60{,}050$, $R = 905$, and $N = 360$. Therefore, we must solve the equation

$$60{,}050i + 905[(1 + i)^{-360} - 1] = 0.$$

Let $f(i) = 60{,}050i + 905[(1 + i)^{-360} - 1]$.   Then $f'(i) = 60{,}050 - 325{,}800 \cdot (1 + i)^{-361}$. Apply the Newton-Raphson algorithm to $f(i)$ with $i_0 = .02$.

$$i_1 = i_0 - \frac{60{,}050i_0 + 905[(1 + i_0)^{-360} - 1]}{60{,}050 - 325{,}800(1 + i_0)^{-361}}$$

$$\approx .015$$

$$i_2 \approx .015.$$

Therefore, the monthly interest rate is approximately 1.5%. ●

Comments

1. The values of successive approximations in the Newton-Raphson algorithm depend on the extent of round-off used during the calculation. It is best to use a computer or hand calculator, since they carry out numbers to eight or more places of accuracy.

2. The Newton-Raphson algorithm is an excellent computational tool. However, in some cases it will not work. For instance, if $f'(x_n) = 0$ for some approximation x_n, then there is no way to compute the next approximation. Other instances in which the algorithm fails are presented in Exercises 25 and 26.

---

* We are assuming that all of the returns are nonnegative and add up to at least P. An analysis of the general case can be found in H. Paley, P. Colwell, and R. Cannaday, *Internal Rates of Return,* UMAP Module 640 (Lexington, Mass.: COMAP, Inc., 1984).

3. It can be shown that if $f(x)$, $f'(x)$, and $f''(x)$ are continuous near r [a zero of $f(x)$] and $f'(r) \neq 0$, then the Newton-Raphson algorithm will definitely work provided that the initial approximation x_0 is not too far away.

TECHNOLOGY PROJECT 12.3
Programming the Newton-Raphson Algorithm (for Students with Some Programming Experience)
(Symbolic Math Program Required)

Most symbolic math programs and graphing calculators allow you to create programs to automatically execute a sequence of commands. You may write a program to carry out the Newton-Raphson algorithm. The main steps in the program are as follows:

a. Ask the user to specify the function $f(x)$, the initial guess x_0, and the desired accuracy of the answer.

b. Calculate $f'(x)$.

c. Determine if $f(x_0)$ is sufficiently close to 0. If so, display it as the answer. Otherwise compute a new value of x_0 as $x_0 - f'(x_0)/f(x_0)$.

d. Repeat step c until an answer is displayed or until a certain number of steps have been executed. (This is to prevent the program from running indefinitely.)

Explore the programming capabilities of your symbolic math program or graphing calculator and write a program as outlined above. Use your program to determine all zeros of the following functions to 5 decimal places accuracy.

1. $x^3 + x^2 - 4x + 1$ 2. $-x^3 + 3x^2 + 36x + 25$

3. $x^2 - \sin 2x$ 4. $e^{-.1x} - x^3$

PRACTICE PROBLEMS 12.2

1. Use three repetitions of the Newton-Raphson algorithm to estimate $\sqrt[3]{7}$.
2. Use three repetitions of the Newton-Raphson algorithm to estimate the zeros of $f(x) = 2x^3 + 3x^2 + 6x - 3$.

EXERCISES 12.2

In Exercises 1–8, use three repetitions of the Newton-Raphson algorithm to approximate the following:

1. $\sqrt{5}$ 2. $\sqrt{7}$ 3. $\sqrt[3]{6}$ 4. $\sqrt[3]{11}$

5. The zero of $x^2 - x - 5$ between 2 and 3.

6. The zero of $x^2 + 3x - 11$ between -5 and -6.

7. The zero of $\sin x + x^2 - 1$ near $x_0 = 0$.

8. The zero of $e^x + 10x - 3$ near $x_0 = 0$.

9. Sketch the graph of $y = x^3 + 2x + 2$ and use the Newton-Raphson algorithm (three repetitions) to approximate all x-intercepts.

10. Sketch the graph of $y = x^3 + x - 1$ and use the Newton-Raphson algorithm (three repetitions) to approximate all x-intercepts.

11. Use the Newton-Raphson algorithm to find an approximate solution to $e^{-x} = x^2$.

12. Use the Newton-Raphson algorithm to find an approximate solution to $e^{5-x} = 10 - x$.

13. Suppose that an investment of $500 yields returns of $100, $200, and $300 at the end of the first, second, and third months, respectively. Determine the internal rate of return on this investment.

14. An investor buys a bond for $1000. She receives $10 at the end of each month for 2 months and then sells the bond at the end of second month for $1040. Determine the internal rate of return on this investment.

15. A $663 television set is purchased with a down payment of $100 and a loan of $563 to be repaid in five monthly installments of $116. Determine the monthly rate of interest on the loan.

16. A mortgage of $100,050 is repaid in 240 monthly payments of $900. Determine the monthly rate of interest.

17. A function $f(x)$ has the graph given in Fig. 6. Let x_1 and x_2 be the estimates of a root of $f(x)$ obtained by applying the Newton-Raphson algorithm using an initial approximation of $x_0 = 5$. Draw the appropriate tangent lines and estimate the numerical values of x_1 and x_2.

FIGURE 6

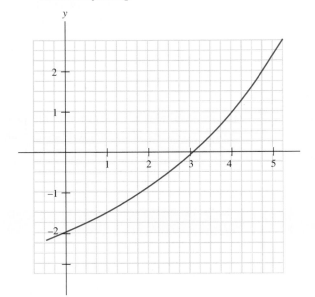

18. Redo Exercise 17 with $x_0 = 1$.

19. Suppose the line $y = 4x + 5$ is tangent to the graph of the function $f(x)$ at $x = 3$. If the Newton-Raphson algorithm is used to find a root of $f(x) = 0$ with the initial guess $x_0 = 3$, what is x_1?

20. Suppose the graph of the function $f(x)$ has slope -2 at the point $(1, 2)$. If the Newton-Raphson algorithm is used to find a root of $f(x) = 0$ with the initial guess $x_0 = 1$, what is x_1?

21. Figure 7 contains the graph of the function $f(x) = x^2 - 2$. The function has zeros at $x = \sqrt{2}$ and $x = -\sqrt{2}$. When the Newton-Raphson algorithm is applied to find a zero, what values of x_0 lead to the zero $\sqrt{2}$?

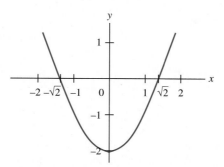

FIGURE 7 Graph of $f(x) = x^2 - 2$.

FIGURE 8 Graph of $f(x) = x^4 - 2x^2$.

22. Figure 8 contains the graph of the function $f(x) = x^4 - 2x^2$. The function has zeros at $x = -\sqrt{2}$, 0, and $\sqrt{2}$. When the Newton-Raphson algorithm is applied to find a zero, what values of x_0 lead to the zero at 0?

23. What special occurrence takes place when the Newton-Raphson algorithm is applied to a linear function, $f(x) = mx + b$ with $m \neq 0$?

24. What happens when the first approximation, x_0, is actually a zero of $f(x)$?

Exercises 25 and 26 present two examples in which successive repetitions of the Newton-Raphson algorithm do not approach a root.

25. Apply the Newton-Raphson algorithm to the function $f(x) = x^{1/3}$ whose graph is drawn in Fig. 9(a). Use $x_0 = 1$.

26. Apply the Newton-Raphson algorithm to the function whose graph is drawn in Fig. 9(b). Use $x_0 = 1$.

FIGURE 9

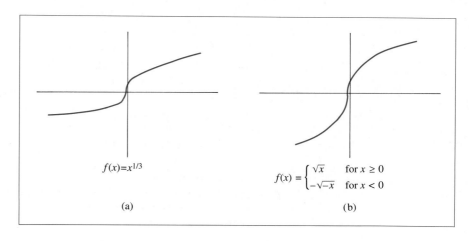

$f(x) = x^{1/3}$

(a)

$f(x) = \begin{cases} \sqrt{x} & \text{for } x \geq 0 \\ -\sqrt{-x} & \text{for } x < 0 \end{cases}$

(b)

**SOLUTIONS TO
PRACTICE
PROBLEMS 12.2**

1. We wish to approximate a zero of $f(x) = x^3 - 7$. Since $f(1) = -6 < 0$ and $f(2) = 1 > 0$, the graph of $f(x)$ crosses the x-axis somewhere between $x = 1$ and $x = 2$. Take $x_0 = 2$ as the initial approximation to the zero. Since $f'(x) = 3x^2$, we have

$$x_1 = x_0 - \frac{x_0^3 - 7}{3x_0^2}$$

$$= 2 - \frac{2^3 - 7}{3(2)^2} = \tfrac{23}{12} \approx 1.9167,$$

$$x_2 = 1.9167 - \frac{(1.9167)^3 - 7}{3(1.9167)^2} \approx 1.91294,$$

$$x_3 = 1.91294 - \frac{(1.91294)^3 - 7}{3(1.91294)^2} \approx 1.91293.$$

2. As a preliminary step, we use the methods of Chapter 2 to sketch the graph of $f(x)$ (Fig. 10). We see that $f(x)$ has a zero which occurs for a positive value of x. Since $f(0) = -3$ and $f(1) = 8$, the graph crosses the x-axis between 0 and 1. Let us choose $x_0 = 0$ as our initial approximation to the zero of $f(x)$. Since $f'(x) = 6x^2 + 6x + 6$, we then have

$$x_1 = x_0 - \frac{2x_0^3 + 3x_0^2 + 6x_0 - 3}{6x_0^2 + 6x_0 + 6}$$

$$= 0 - \tfrac{-3}{6} = \tfrac{1}{2},$$

$$x_2 = \tfrac{1}{2} - \frac{1}{\tfrac{21}{2}} = \tfrac{1}{2} - \tfrac{2}{21}$$

$$= \tfrac{17}{42} \approx .40476.$$

Continuing, we find that $x_3 \approx .39916$.

FIGURE 10

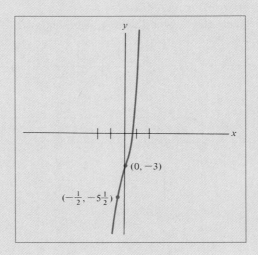

$(0, -3)$

$\left(-\tfrac{1}{2}, -5\tfrac{1}{2}\right)$

12.3 INFINITE SERIES

An *infinite series* is an infinite addition of numbers

$$a_1 + a_2 + a_3 + a_4 + \cdots .$$

Here are some examples:

$$1 + \tfrac{1}{2} + \tfrac{1}{4} + \tfrac{1}{8} + \tfrac{1}{16} + \cdots , \tag{1}$$

$$1 + 1 + 1 + 1 + \cdots , \tag{2}$$

$$1 - 1 + 1 - 1 + \cdots . \tag{3}$$

To certain infinite series it is possible to associate a "sum." To illustrate how this is done, let us consider the infinite series (1). If we add up the first two, three, four, five, and six terms of the infinite series (1), we obtain

$$1 + \tfrac{1}{2} = 1\tfrac{1}{2},$$

$$1 + \tfrac{1}{2} + \tfrac{1}{4} = 1\tfrac{3}{4},$$

$$1 + \tfrac{1}{2} + \tfrac{1}{4} + \tfrac{1}{8} = 1\tfrac{7}{8},$$

$$1 + \tfrac{1}{2} + \tfrac{1}{4} + \tfrac{1}{8} + \tfrac{1}{16} = 1\tfrac{15}{16},$$

$$1 + \tfrac{1}{2} + \tfrac{1}{4} + \tfrac{1}{8} + \tfrac{1}{16} + \tfrac{1}{32} = 1\tfrac{31}{32}.$$

Each total lies halfway between the preceding total and the number 2. It appears from the calculations above that by increasing the number of terms, we bring the total arbitrarily close to 2. Indeed, this is supported by further calculation. For example,

$$\underbrace{1 + \frac{1}{2} + \frac{1}{4} + \cdots + \frac{1}{2^9}}_{10 \text{ terms}} = 2 - \frac{1}{2^9} \approx 1.998047,$$

$$\underbrace{1 + \frac{1}{2} + \frac{1}{4} + \cdots + \frac{1}{2^{19}}}_{20 \text{ terms}} = 2 - \frac{1}{2^{19}} \approx 1.999998,$$

$$\underbrace{1 + \frac{1}{2} + \frac{1}{4} + \cdots + \frac{1}{2^{n-1}}}_{n \text{ terms}} = 2 - \frac{1}{2^{n-1}}.$$

Therefore, it seems reasonable to assign the infinite series (1) the "sum" 2:

$$1 + \tfrac{1}{2} + \tfrac{1}{4} + \tfrac{1}{8} + \tfrac{1}{16} + \cdots = 2. \tag{4}$$

The sum of the first n terms of an infinite series is called its nth *partial sum* and is denoted S_n. In series (1) we were very fortunate that the partial sums approached a limiting value, 2. This is not always the case. For example, consider the infinite series (2). If we form the first few partial sums, we get

$$S_2 = 1 + 1 \qquad\qquad = 2,$$
$$S_3 = 1 + 1 + 1 \qquad = 3,$$
$$S_4 = 1 + 1 + 1 + 1 = 4.$$

We see that these sums do not approach any limit. Rather, they become larger and larger, eventually exceeding any specified number.

The partial sums need not grow without bound in order that an infinite series not have a sum. For example, consider the infinite series (3). Here the sums of initial terms are

$$S_2 = 1 - 1 \qquad\qquad\qquad = 0,$$
$$S_3 = 1 - 1 + 1 \qquad\qquad = 1,$$
$$S_4 = 1 - 1 + 1 - 1 \qquad = 0,$$
$$S_5 = 1 - 1 + 1 - 1 + 1 = 1,$$

and so forth. The partial sums alternate between 0 and 1 and do not approach a limit. So the infinite series (3) has no sum.

An infinite series whose partial sums approach a limit is called *convergent*. The limit is then called the *sum* of the infinite series. An infinite series whose partial sums do not approach a limit is called *divergent*. From our discussion above, we know that the infinite series (1) is convergent, whereas (2) and (3) are divergent.

It is often an extremely difficult task to determine whether or not a given infinite series is convergent. And intuition is not always an accurate guide. For example, one might at first suspect that the infinite series

$$1 + \tfrac{1}{2} + \tfrac{1}{3} + \tfrac{1}{4} + \tfrac{1}{5} + \cdots$$

(the so-called *harmonic series*) is convergent. However, it is not. The sums of its initial terms increase without bound, although they do so very slowly. For example, it takes about 12,000 terms before the sum exceeds 10 and about 2.7×10^{43} terms before the sum exceeds 100. Nevertheless, the sum eventually exceeds any prescribed number. (See Exercise 42 and Section 12.4.)

There is an important type of infinite series whose convergence or divergence is easily determined. Let a and r be given nonzero numbers. A series of the form

$$a + ar + ar^2 + ar^3 + ar^4 + \cdots$$

is called a *geometric series with ratio r*. (The "ratio" of consecutive terms is r.)

The Geometric Series   The infinite series

$$a + ar + ar^2 + ar^3 + ar^4 + \cdots$$

converges if and only if $|\, r \,| < 1$. When $|\, r \,| < 1$, the sum of the series is

$$\frac{a}{1 - r}. \tag{5}$$

For example, if $a = 1$ and $r = \frac{1}{2}$, we obtain the infinite series (1). In this case $\dfrac{a}{1 - r} = \dfrac{1}{1 - \frac{1}{2}} = \dfrac{1}{\frac{1}{2}} = 2$, in agreement with our previous observation. Also, series (2) and (3) are divergent geometric series, with $r = 1$ and $r = -1$, respectively. A proof of the foregoing result is outlined in Exercise 41.

EXAMPLE 1   Calculate the sums of the following geometric series:

(a) $1 + \dfrac{1}{5} + \dfrac{1}{5^2} + \dfrac{1}{5^3} + \dfrac{1}{5^4} + \cdots$

(b) $\dfrac{2}{3^2} + \dfrac{2}{3^4} + \dfrac{2}{3^6} + \dfrac{2}{3^8} + \dfrac{2}{3^{10}} + \cdots$

(c) $\dfrac{5}{2^2} - \dfrac{5^2}{2^5} + \dfrac{5^3}{2^8} - \dfrac{5^4}{2^{11}} + \dfrac{5^5}{2^{14}} - \cdots$

Solution   (a)   Here $a = 1$ and $r = \frac{1}{5}$. The sum of the series is

$$\frac{a}{1 - r} = \frac{1}{1 - \frac{1}{5}} = \frac{1}{\frac{4}{5}} = \frac{5}{4}.$$

(b)   We find r by dividing any term by the preceding term. So

$$r = \frac{\dfrac{2}{3^4}}{\dfrac{2}{3^2}} = \frac{2}{3^4} \cdot \frac{3^2}{2} = \frac{1}{3^2} = \frac{1}{9}.$$

Since the series is a geometric series, we obtain the same result using any other pair of successive terms. For instance,

$$\frac{\dfrac{2}{3^8}}{\dfrac{2}{3^6}} = \frac{2}{3^8} \cdot \frac{3^6}{2} = \frac{1}{3^2} = \frac{1}{9}.$$

The first term of the series is $a = \dfrac{2}{3^2} = \dfrac{2}{9}$, so the sum of the series is

$$\frac{a}{1 - r} = a \cdot \frac{1}{1 - r} = \frac{2}{9} \cdot \frac{1}{1 - \dfrac{1}{9}} = \frac{2}{9} \cdot \frac{9}{8} = \frac{1}{4}.$$

(c) We may find r as in part (b), or we may observe that the numerator of each fraction in the series (c) is increasing by a factor of 5, while the denominator is increasing by a factor of $2^3 = 8$. So the ratio of successive fractions is $\frac{5}{8}$. However, the terms in the series are alternately positive and negative, so the ratio of successive terms must be negative. Hence $r = -\frac{5}{8}$. Next, $a = \dfrac{5}{2^2} = \dfrac{5}{4}$, so the sum of series (c) is

$$a \cdot \frac{1}{1 - r} = \tfrac{5}{4} \cdot \frac{1}{1 - (-\tfrac{5}{8})} = \tfrac{5}{4} \cdot \frac{1}{\tfrac{13}{8}} = \tfrac{5}{4} \cdot \tfrac{8}{13} = \tfrac{10}{13}. \;\bullet$$

Sometimes a rational number is expressed as an infinite repeating decimal, such as $.12\overline{12}\ldots$. The value of such a "decimal expansion" is best explained as the sum of an infinite series.

EXAMPLE 2   What rational number has the decimal expansion $.121\overline{212}$.?

Solution   This number denotes the infinite series

$$.12 + .0012 + .000012 + \cdots = \frac{12}{100} + \frac{12}{100^2} + \frac{12}{100^3} + \cdots,$$

a geometric series with $a = \frac{12}{100}$ and $r = \frac{1}{100}$. The sum of the geometric series is

$$\frac{a}{1 - r} = a \cdot \frac{1}{1 - r} = \frac{12}{100} \cdot \frac{1}{1 - \frac{1}{100}} = \frac{12}{100} \cdot \frac{100}{99} = \frac{12}{99} = \frac{4}{33}.$$

Hence $.121\overline{212}$. $= \frac{4}{33}$. $\bullet$

EXAMPLE 3   (*The Multiplier Effect in Economics*)   Suppose that the federal government enacts an income tax cut of \$10 billion. Assume that each person will spend 93% of all resulting extra income and save the rest. Estimate the total effect of the tax cut on economic activity.

Solution   Express all amounts of money in billions of dollars. Of the increase in income created by the tax cut, $(.93)(10)$ billion dollars will be spent. These dollars become extra income to someone and hence 93% will be spent again and 7% saved, so additional spending of $(.93)(.93)(10)$ billion dollars is created. The recipients of those dollars will spend 93% of them, creating yet additional spending of

$$(.93)(.93)(.93)(10) = 10(.93)^3$$

billion dollars, and so on. The total amount of new spending created by the tax cut is thus given by the infinite series

$$10(.93) + 10(.93)^2 + 10(.93)^3 + \cdots.$$

This is a geometric series with initial term $10(.93)$ and ratio $.93$. Its sum is

$$\frac{a}{1 - r} = \frac{10(.93)}{1 - .93} = \frac{9.3}{.07} \approx 132.86.$$

Thus a \$10 billion tax cut creates new spending of about \$132.86 billion. ●

Example 3 illustrates the *multiplier effect*. The proportion of each extra dollar that a person will spend is called the *marginal propensity to consume*, denoted MPC. In Example 3, MPC $= .93$. As we observed, the total new spending generated by the tax cut is

$$[\text{total new spending}] = 10 \cdot \frac{.93}{1 - .93} = [\text{tax cut}] \cdot \frac{\text{MPC}}{1 - \text{MPC}}.$$

The tax cut is multiplied by the "multiplier" $\dfrac{\text{MPC}}{1 - \text{MPC}}$ to obtain its true effect.

EXAMPLE 4   (*Drug Therapy*)   Patients with certain heart problems are often treated with digitoxin, a derivative of the digitalis plant. The rate at which a person's body eliminates digitoxin is proportional to the amount of digitoxin present. In 1 day (24 hours) about 10% of any given amount of the drug will be eliminated. Suppose that a "maintenance dose" of .05 milligram (mg) is given daily to a patient. Estimate the total amount of digitoxin that should be present in the patient after several months of treatment.

Solution   For a moment, let us consider what happens to the initial dose of .05 mg, and disregard the subsequent doses. After 1 day, 10% of the .05 mg will have been eliminated and $(.90)(.05)$ mg will remain. By the end of the second day, this smaller amount will be reduced 10% to $(.90)(.90)(.05)$ mg, and so on, until after n days only $(.90)^n(.05)$ mg of the original dose will remain. (See Fig. 1.) To determine the cumulative effect of all the doses of digitoxin, we observe that at

FIGURE 1   Exponential decrease of the initial dose.

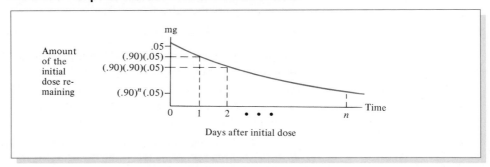

the time of the second dose (one day after the first dose), the patient's body will contain the second dose of .05 mg plus (.90)(.05) mg of the first dose. A day later, there will be the third dose of .05 mg, plus (.90)(.05) mg of the second dose, plus $(.90)^2(.05)$ of the first dose. At the time of any new dose, the patient's body will contain that dose plus the amounts that remain of earlier doses. Let us tabulate this.

<table>
<tr><td colspan="2">**Total Amount (mg) of Digitoxin Present**</td></tr>
<tr><td rowspan="6">**Days After Initial Dose**</td><td>0   .05</td></tr>
<tr><td>1   .05 + (.90)(.05)</td></tr>
<tr><td>2   .05 + (.90)(.05) $+ (.90)^2(.05)$</td></tr>
<tr><td>.</td></tr>
<tr><td>.</td></tr>
<tr><td>n .05 + (.90)(.05) $+ (.90)^2(.05) + \cdots + (.90)^n(.05)$</td></tr>
</table>

We can see that the amounts present at the time of each new dose correspond to the partial sums of the geometric series

$$.05 + (.90)(.05) + (.90)^2(.05) + (.90)^3(.05) + \cdots,$$

where $a = .05$ and $r = .90$. The sum of this series is

$$\frac{.05}{1 - .90} = \frac{.05}{.10} = .5.$$

Since the partial sums of the series approach the sum of .5, we may conclude that a daily maintenance dose of .05 mg will eventually raise the level of digitoxin in the patient to a "plateau" of .5 mg. Between doses the level will drop 10% down to $(.90)(.5) = .45$ mg. The use of a regular maintenance dose to sustain a certain level of a drug in a patient is an important technique in drug therapy.* ●

Sigma Notation   When studying series, it is often convenient to use the Greek capital letter sigma to indicate summation. For example, the sum

$$a_2 + a_3 + \cdots + a_{10}$$

is denoted by

$$\sum_{k=2}^{10} a_k$$

(read "the sum of a sub k from k equals 2 to 10"). The nth partial sum of a series, $a_1 + a_2 + \cdots + a_n$, is written as $\sum_{k=1}^{n} a_k$. In these examples the letter k is called the *index of summation.* Any letter not already in use may be used as the index of summation. For instance, both

$$\sum_{i=0}^{4} a_i \quad \text{and} \quad \sum_{j=0}^{4} a_j$$

---

* See "Principles of Drug Therapy" by John A. Oates and Grant R. Wilkinson, in *Principles of Internal Medicine,* T. R. Harrison, ed., 8th ed. (New York: McGraw-Hill Book Company, 1977), pp. 334–346.

indicate the sum $a_0 + a_1 + a_2 + a_3 + a_4$.

Finally, a formal infinite series

$$a_1 + a_2 + a_3 + \cdots$$

is written as

$$\sum_{k=1}^{\infty} a_k \quad \text{or} \quad \sum_{1}^{\infty} a_k.$$

We will also write $\sum_{k=1}^{\infty} a_k$ as the symbol for the numerical value of the series when it is convergent. Using this notation (and writing ar^0 in place of a) the main result about the geometric series may be written as follows:

$$\sum_{k=0}^{\infty} ar^k = \frac{a}{1 - r} \qquad \text{if } |r| < 1,$$

$$\sum_{k=0}^{\infty} ar^k \text{ is divergent} \qquad \text{if } |r| \geq 1.$$

EXAMPLE 5   Determine the sums of the following infinite series.

(a) $\displaystyle\sum_{k=0}^{\infty} \left(\frac{2}{3}\right)^k$ (b) $\displaystyle\sum_{j=0}^{\infty} 4^{-j}$ (c) $\displaystyle\sum_{i=3}^{\infty} \frac{2}{7^i}$

Solution   In each case, the first step is to write out the first few terms of the series.

(a) $\displaystyle\sum_{k=0}^{\infty} \left(\tfrac{2}{3}\right)^k = \quad 1 \quad + \quad \tfrac{2}{3} \quad + \quad \left(\tfrac{2}{3}\right)^2 \quad + \quad \left(\tfrac{2}{3}\right)^3 \quad + \cdots.$

$\qquad\qquad\qquad\quad [k = 0] \quad\ [k = 1] \quad\ [k = 2] \quad\ [k = 3]$

This is a geometric series with initial term $a = 1$ and ratio $r = \dfrac{2}{3}$; its sum is $\dfrac{1}{1 - \tfrac{2}{3}} = \dfrac{1}{\tfrac{1}{3}} = 3.$

(b) $\displaystyle\sum_{j=0}^{\infty} 4^{-j} = 4^0 + 4^{-1} + 4^{-2} + 4^{-3} + \cdots$

$\qquad\qquad = 1 + \tfrac{1}{4} + \left(\tfrac{1}{4}\right)^2 + \left(\tfrac{1}{4}\right)^3 + \cdots$

$\qquad\qquad = \dfrac{1}{1 - \tfrac{1}{4}} = \tfrac{4}{3}.$

(c) $\displaystyle\sum_{i=3}^{\infty} \frac{2}{7^i} = \frac{2}{7^3} + \frac{2}{7^4} + \frac{2}{7^5} + \frac{2}{7^6} + \cdots.$

This is a geometric series with $a = \dfrac{2}{7^3}$ and $r = \tfrac{1}{7}$; its sum is

$$a \cdot \frac{1}{1 - r} = \frac{2}{7^3} \cdot \frac{1}{1 - \tfrac{1}{7}} = \frac{2}{7^3} \cdot \frac{7}{6} = \tfrac{1}{147}. \quad \bullet$$

TECHNOLOGY PROJECT 12.4
Estimating Sums of Infinite Series

In this project, we will explore two series, one convergent and one divergent.

1. It can be shown that the series

$$1 - \frac{1}{3} + \frac{1}{5} - \frac{1}{7} + \cdots$$

converges and has the sum $\frac{\pi}{4}$, a fact proved by Gregory in 1607. Use a symbolic math program to calculate the partial sum of the first 100 terms; 1000 terms; 10,000 terms; 100,000 terms. Estimate how many terms it is necessary to use to obtain a given number of decimal places of $\frac{\pi}{4}$.

2. The series of problem 1. can be written as

$$\left(\frac{1}{1} - \frac{1}{3}\right) + \left(\frac{1}{5} - \frac{1}{7}\right) + \left(\frac{1}{9} - \frac{1}{11}\right) + \cdots$$

$$= \frac{2}{1 \cdot 3} + \frac{2}{5 \cdot 7} + \frac{2}{9 \cdot 11} + \cdots$$

Calculate the partial sum of the first ten terms of this series; 100 terms; 1000 terms. Compare each partial sum with the sum of the series. Estimate the number of terms required to obtain a given number of decimal places of $\frac{\pi}{4}$. (This example shows that by slightly altering an infinite series, we can make it converge to its sum much faster.)

3. Consider the **harmonic series:**

$$1 + \frac{1}{2} + \frac{1}{3} + \frac{1}{4} + \cdots$$

Calculate the sum of the first 1000 terms; 10,000 terms; 100,000 terms; 1,000,000 terms. What conclusion do these results suggest? (See Exercise 42.)

PRACTICE PROBLEMS 12.3

1. Determine the sum of the geometric series

$$8 - \frac{8}{3} + \frac{8}{9} - \frac{8}{27} + \frac{8}{81} - \cdots.$$

2. Find the value of $\sum_{k=0}^{\infty} (.7)^{-k+1}$.

EXERCISES 12.3

Determine the sums of the following geometric series, when they are convergent.

1. $1 + \frac{1}{6} + \frac{1}{6^2} + \frac{1}{6^3} + \frac{1}{6^4} + \cdots$

2. $1 + \dfrac{3}{4} + \left(\dfrac{3}{4}\right)^2 + \left(\dfrac{3}{4}\right)^3 + \left(\dfrac{3}{4}\right)^4 + \cdots$

3. $1 - \dfrac{1}{3^2} + \dfrac{1}{3^4} - \dfrac{1}{3^6} + \dfrac{1}{3^8} - \cdots$ **4.** $1 + \dfrac{1}{2^3} + \dfrac{1}{2^6} + \dfrac{1}{2^9} + \dfrac{1}{2^{12}} + \cdots$

5. $2 + \dfrac{2}{3} + \dfrac{2}{9} + \dfrac{2}{27} + \dfrac{2}{81} + \cdots$ **6.** $3 + \dfrac{6}{5} + \dfrac{12}{25} + \dfrac{24}{125} + \dfrac{48}{625} + \cdots$

7. $\dfrac{1}{5} + \dfrac{1}{5^4} + \dfrac{1}{5^7} + \dfrac{1}{5^{10}} + \dfrac{1}{5^{13}} + \cdots$ **8.** $\dfrac{1}{3^2} - \dfrac{1}{3^3} + \dfrac{1}{3^4} - \dfrac{1}{3^5} + \dfrac{1}{3^6} - \cdots$

9. $3 - \dfrac{3^2}{7} + \dfrac{3^3}{7^2} - \dfrac{3^4}{7^3} + \dfrac{3^5}{7^4} - \cdots$ **10.** $6 - 1.2 + .24 - .048 + .0096 - \cdots$

11. $\dfrac{2}{5^4} - \dfrac{2^4}{5^5} + \dfrac{2^7}{5^6} - \dfrac{2^{10}}{5^7} + \dfrac{2^{13}}{5^8} - \cdots$ **12.** $\dfrac{3^2}{2^5} + \dfrac{3^4}{2^8} + \dfrac{3^6}{2^{11}} + \dfrac{3^8}{2^{14}} + \dfrac{3^{10}}{2^{17}} + \cdots$

13. $5 + 4 + 3.2 + 2.56 + 2.048 + \cdots$ **14.** $\dfrac{5^3}{3} - \dfrac{5^5}{3^4} + \dfrac{5^7}{3^7} - \dfrac{5^9}{3^{10}} + \dfrac{5^{11}}{3^{13}} - \cdots$

Sum an appropriate infinite series to find the rational number whose decimal expansion is given.

15. $.272727\ldots$ **16.** $.173173\ldots$ **17.** $.222\ldots$

18. $.151515\ldots$ **19.** $4.011011\ldots$ **20.** $5.444\ldots$
                                            $(= 4 + .011011\ldots)$

21. Show that $.999\ldots = 1$.

22. Compute the value of $.12121212\ldots$ as a geometric series with $a = .1212$ and $r = .0001$. Compare your answer with the result of Example 2.

23. Compute the total new spending created by a \$10 billion federal income tax cut when the population's marginal propensity to consume is 95%. Compare your result with that of Example 3 and note how a small change in the MPC makes a dramatic change in the total spending generated by the tax cut.

24. Compute the effect of a \$20 billion federal income tax cut when the population's marginal propensity to consume is 98%. What is the "multiplier" in this case?

A perpetuity is a periodic sequence of payments that continue forever. The capital value of the perpetuity is the sum of the present values of all future payments.

25. Consider a perpetuity that promises to pay \$100 at the beginning of each month. Suppose that the interest rate is 12% compounded monthly. Then the present value of \$100 in k months is $100(1.01)^{-k}$.

   (a) Express the capital value of the perpetuity as an infinite series.

   (b) Find the sum of the infinite series.

26. Consider a perpetuity that promises to pay P dollars at the *end* of each month. (The first payment will be received in 1 month.) If the interest rate per month is r, the present value of P dollars in k months is $P(1 + r)^{-k}$. Find a simple formula for the capital value of the perpetuity.

27. A generous corporation not only gives its CEO a \$1,000,000 bonus, but gives her enough money to cover the taxes on the bonus, the taxes on the additional taxes, the taxes on the taxes on the additional taxes, and so on. If she is in the 39.6% tax bracket, how large is her bonus?

28. The *coefficient of restitution* of a ball, a number between 0 and 1, specifies how much energy is conserved when the ball hits a rigid surface. A coefficient of .9, for instance, means a bouncing ball will rise to 90% of its previous height after each bounce. The coefficients of restitution for a tennis ball, basketball, super ball, and softball are .7, .75, .9, and .3, respectively. Find the total distance traveled by a tennis ball dropped from a height of 6 feet.

29. A patient receives 6 milligrams of a certain drug daily. Each day the body eliminates 30% of the amount of the drug present in the system. Estimate the total amount of the drug that should be present after extended treatment, immediately after a dose is given.

30. A patient receives 2 milligrams of a certain drug each day. Each day the body eliminates 20% of the amount of drug present in the system. Estimate the total amount of the drug present after extended treatment, immediately *before* a dose is given.

31. A patient receives M milligrams of a certain drug each day. Each day the body eliminates 25% of the amount of drug present in the system. Determine the value of the maintenance dose M such that after many days approximately 20 milligrams of the drug are present immediately after a dose is given.

32. A patient receives M milligrams of a certain drug daily. Each day the body eliminates a fraction q of the amount of the drug present in the system. Estimate the total amount of the drug that should be present after extended treatment, immediately after a dose is given.

33. The infinite series $a_0 + a_1 + a_2 + a_3 + \cdots$ has partial sums given by

$$S_n = 3 - \frac{5}{n}.$$

   (a) Find $\displaystyle\sum_{k=0}^{10} a_k$.

   (b) Does the infinite series converge? If so, to what value does it converge?

34. The infinite series $a_0 + a_1 + a_2 + a_3 + \cdots$ has partial sums given by

$$S_n = n - \frac{1}{n}.$$

   (a) Find $\displaystyle\sum_{k=0}^{10} a_k$.

   (b) Does the infinite series converge? If so, to what value does it converge?

Determine the sums of the following infinite series.

35. $\displaystyle\sum_{k=0}^{\infty} \left(\frac{5}{6}\right)^k$

36. $\displaystyle\sum_{k=0}^{\infty} \frac{7}{10^k}$

37. $\displaystyle\sum_{j=1}^{\infty} 5^{-2j}$

38. $\displaystyle\sum_{j=0}^{\infty} \frac{(-1)^j}{3^j}$

39. $\displaystyle\sum_{k=0}^{\infty} (-1)^k \frac{3^{k+1}}{5^k}$

40. $\displaystyle\sum_{k=1}^{\infty} \left(\frac{1}{3}\right)^{2k}$

41. Let a and r be given nonzero numbers.

   (a) Show that $(1 - r)(a + ar + ar^2 + \cdots + ar^n) = a - ar^{n+1}$, and from this conclude that, for $r \neq 1$,

$$a + ar + ar^2 + \cdots + ar^n = \frac{a}{1-r} - \frac{ar^{n+1}}{1-r}.$$

(b) Use the result of part (a) to explain why the geometric series $\sum_0^\infty ar^k$ converges to $\frac{a}{1-r}$ when $|r| < 1$.

(c) Use the result of part (a) to explain why the geometric series diverges for $|r| > 1$.

(d) Explain why the geometric series diverges for $r = 1$ and $r = -1$.

42. Show that the infinite series

$$1 + \frac{1}{2} + \frac{1}{3} + \frac{1}{4} + \frac{1}{5} + \cdots$$

diverges. [*Hint:* $\frac{1}{3} + \frac{1}{4} > \frac{1}{2}$; $\frac{1}{5} + \frac{1}{6} + \frac{1}{7} + \frac{1}{8} > \frac{1}{2}$; $\frac{1}{9} + \cdots + \frac{1}{16} > \frac{1}{2}$; etc.]

SOLUTIONS TO PRACTICE PROBLEMS 12.3

1. Answer: 6. To obtain the sum of a geometric series, identify a and r, and (provided that $|r| < 1$) substitute these values into the formula $\frac{a}{1-r}$. The initial term a is just the first term of the series: $a = 8$. The ratio r is often obvious by inspection. However, if in doubt, divide any term by the *preceding* term. Here, the second term divided by the first term is $\frac{\frac{-8}{3}}{8} = -\frac{1}{3}$, so $r = -\frac{1}{3}$. Since $|r| = \frac{1}{3}$ the series is convergent and the sum is

$$\frac{a}{1-r} = \frac{8}{1-(-\frac{1}{3})} = \frac{8}{1+\frac{1}{3}} = \frac{8}{\frac{4}{3}} = 8 \cdot \frac{3}{4} = 6.$$

2. Write out the first few terms of the series and then proceed as in Problem 1.

$$\sum_{k=0}^\infty (.7)^{-k+1} = \underset{[k=0]}{(.7)^1} + \underset{[k=1]}{(.7)^0} + \underset{[k=2]}{(.7)^{-1}} + \underset{[k=3]}{(.7)^{-2}} + \underset{[k=4]}{(.7)^{-3}} + \cdots$$

$$= .7 + 1 + \frac{1}{.7} + \frac{1}{(.7)^2} + \frac{1}{(.7)^3} + \cdots.$$

Here $a = .7$ and $r = 1/.7 = \frac{10}{7}$. Since $|r| = \frac{10}{7} > 1$, the series is divergent and has no sum. (The formula $\frac{a}{1-r}$ yields $\frac{7}{3}$; however, this value is meaningless. The formula applies only to the case in which the series is convergent.)

12.4 TAYLOR SERIES

Consider the infinite series $1 + x + x^2 + x^3 + x^4 + \cdots$. This series is of a different type from those discussed in the preceding two sections. Its terms are not numbers but rather are powers of x. However, for some specific values of x, the series is convergent. In fact, for any value of x between -1 and 1, the series is a convergent geometric series with ratio x and sum $\dfrac{1}{1-x}$. We write

$$\frac{1}{1-x} = 1 + x + x^2 + x^3 + x^4 + \cdots, \qquad |x| < 1. \tag{1}$$

Looking at (1) from a different point of view, we see that the function $f(x) = \dfrac{1}{1-x}$ is represented as a series involving the powers of x. This representation is not valid throughout the entire domain of the function $\dfrac{1}{1-x}$ but just for values of x with $-1 < x < 1$.

In many important cases, a function $f(x)$ may be represented by a series of the form

$$f(x) = a_0 + a_1 x + a_2 x^2 + a_3 x^3 + \cdots, \tag{2}$$

where $a_0, a_1, a_2, \ldots$ are suitable constants, and where x ranges over values that make the series converge to $f(x)$. The series is called a *power series* (because it involves powers of x). It may be shown that when a function $f(x)$ has a representation by a power series as in (2), the coefficients $a_0, a_1, a_2, \ldots$ are uniquely determined by $f(x)$ and its derivatives at $x = 0$. In fact, $a_0 = f(0)$, and $a_k = f^{(k)}(0)/k!$ for $k = 1, 2, \ldots,$ so

$$f(x) = f(0) + \frac{f'(0)}{1!}x + \frac{f''(0)}{2!}x^2 + \cdots + \frac{f^{(k)}(0)}{k!}x^k + \cdots. \tag{3}$$

The series in (3) is often called the *Taylor series of $f(x)$ at $x = 0$* because the partial sums of the series are the Taylor polynomials of $f(x)$ at $x = 0$. The entire equation (3) is called the *Taylor series expansion of $f(x)$ at $x = 0$*.

EXAMPLE I Find the Taylor series expansion of $\dfrac{1}{1-x}$ at $x = 0$.

Solution We already know how to represent $\dfrac{1}{1-x}$ as a power series for $|x| < 1$. However, let us use the formula for the Taylor series to see if we get the same result.

$$
\begin{aligned}
f(x) &= \frac{1}{1-x} = (1-x)^{-1}, & f(0) &= 1 \\
f'(x) &= (1-x)^{-2}, & f'(0) &= 1 \\
f''(x) &= 2(1-x)^{-3}, & f''(0) &= 2 \\
f'''(x) &= 3\cdot 2(1-x)^{-4}, & f'''(0) &= 3\cdot 2 \\
f^{(4)}(x) &= 4\cdot 3\cdot 2(1-x)^{-5}, & f^{(4)}(0) &= 4\cdot 3\cdot 2 \\
&\quad\vdots & &\quad\vdots
\end{aligned}
$$

Therefore,

$$\frac{1}{1-x} = 1 + \frac{1}{1!}x + \frac{2}{2!}x^2 + \frac{3 \cdot 2}{3!}x^3 + \frac{4 \cdot 3 \cdot 2}{4!}x^4 + \cdots$$

$$= 1 + x + x^2 + x^3 + x^4 + \cdots.$$

We have verified that the Taylor series for $\frac{1}{1-x}$ is the familiar geometric power series. The Taylor series expansion is valid for $|x| < 1$. ●

EXAMPLE 2   Find the Taylor series at $x = 0$ for $f(x) = e^x$.

Solution

$$f(x) = e^x, \quad f'(x) = e^x, \quad f''(x) = e^x, \quad f'''(x) = e^x, \ldots$$

$$f(0) = 1, \quad f'(0) = 1, \quad f''(0) = 1, \quad f'''(0) = 1, \ldots.$$

Therefore,

$$e^x = 1 + x + \frac{1}{2!}x^2 + \frac{1}{3!}x^3 + \frac{1}{4!}x^4 + \cdots.$$

It can be shown that this Taylor series expansion of e^x is valid for all x. (*Note:* A Taylor polynomial of e^x gives only an approximation of e^x, but the infinite Taylor series actually *equals* e^x for all x, in the sense that for any given x the sum of the series is the same as the value of e^x.) ●

Operations of Taylor Series   It is often helpful to think of a Taylor series as a polynomial of infinite degree. Many operations on polynomials are also legitimate for Taylor series, provided that we restrict attention to values of x within an appropriate interval. For example, if we have a Taylor series expansion of $f(x)$, we may differentiate the series term by term to obtain the Taylor series expansion of $f'(x)$. An analogous result holds for antiderivatives. Other permissible operations that produce Taylor series include: multiplying a Taylor series expansion by a constant or a power of x; replacing x by a power of x or by a constant times a power of x; and adding or subtracting two Taylor series expansions. The use of such operations often makes it possible to find the Taylor series of a function without directly using the formal definition of a Taylor series. (The process of computing high order derivatives can become quite laborious when the product or quotient rule is involved.) Once a power series expansion of a function $f(x)$ is found, that series *must* be the Taylor series of the function, since the coefficients of the series are uniquely determined by $f(x)$ and its derivatives at $x = 0$.

EXAMPLE 3   Use the Taylor series at $x = 0$ for $\frac{1}{1-x}$ to find the Taylor series at $x = 0$ for the following functions.

(a) $\dfrac{1}{(1-x)^2}$ (b) $\dfrac{1}{(1-x)^3}$ (c) $\ln(1-x)$

Solution   We begin with the series expansion

$$\frac{1}{1-x} = 1 + x + x^2 + x^3 + x^4 + x^5 + \cdots, \quad |x| < 1.$$

(a) When we differentiate both sides of this equation, we obtain

$$\frac{1}{(1-x)^2} = 1 + 2x + 3x^2 + 4x^3 + 5x^4 + \cdots, \quad |x| < 1.$$

(b) Differentiating the series in part (a), we find that

$$\frac{2}{(1-x)^3} = 2 + 3 \cdot 2x + 4 \cdot 3x^2 + 5 \cdot 4x^3 + \cdots, \quad |x| < 1.$$

We may multiply a convergent series term by a constant. Multiplying by $\frac{1}{2}$, we have

$$\frac{1}{(1-x)^3} = 1 + 3x + 6x^2 + 10x^3 + \cdots + \frac{(n+2)(n+1)}{2}x^n + \cdots$$

for $|x| < 1$.

(c) For $|x| < 1$, we have

$$\int \frac{1}{1-x}\, dx = \int (1 + x + x^2 + x^3 + \cdots)\, dx$$

$$-\ln(1-x) + C = x + \tfrac{1}{2}x^2 + \tfrac{1}{3}x^3 + \tfrac{1}{4}x^4 + \cdots,$$

where C is the constant of integration. If we set $x = 0$ in both sides, we obtain

$$0 + C = 0,$$

so $C = 0$. Thus

$$\ln(1-x) = -x - \tfrac{1}{2}x^2 - \tfrac{1}{3}x^3 - \tfrac{1}{4}x^4 - \cdots, \quad |x| < 1. \ \bullet$$

EXAMPLE 4   Use the result of Example 3(c) to compute ln 1.1.

Solution   Take $x = -.1$ in the Taylor series expansion of $\ln(1 - x)$. Then

$$\ln(1 - (-.1)) = -(-.1) - \tfrac{1}{2}(-.1)^2 - \tfrac{1}{3}(-.1)^3 - \tfrac{1}{4}(-.1)^4 - \cdots$$

$$\ln 1.1 = .1 - \frac{.01}{2} + \frac{.001}{3} - \frac{.0001}{4} + \frac{.00001}{5} - \cdots.$$

This infinite series may be used to compute ln 1.1 to any degree of accuracy required. For instance, the fifth partial sum gives ln 1.1 ≈ .09531, which is correct to five decimal places. ●

EXAMPLE 5 Use the Taylor series at $x = 0$ for e^x to find the Taylor series at $x = 0$ for

(a) $x(e^x - 1)$ (b) e^{x^2}

Solution (a) If we subtract 1 from the Taylor series for e^x, we obtain a series that converges to $e^x - 1$:

$$e^x - 1 = \left(1 + x + \frac{1}{2!}x^2 + \frac{1}{3!}x^3 + \frac{1}{4!}x^4 + \cdots\right) - 1$$

$$= x + \frac{1}{2!}x^2 + \frac{1}{3!}x^3 + \frac{1}{4!}x^4 + \cdots .$$

Now we multiply this series by x, term by term:

$$x(e^x - 1) = x^2 + \frac{1}{2!}x^3 + \frac{1}{3!}x^4 + \frac{1}{4!}x^5 + \cdots .$$

(b) To obtain the Taylor series for e^{x^2}, we replace every occurrence of x by x^2 in the Taylor series for e^x,

$$e^{x^2} = 1 + (x^2) + \frac{1}{2!}(x^2)^2 + \frac{1}{3!}(x^2)^3 + \frac{1}{4!}(x^2)^4 + \cdots$$

$$= 1 + x^2 + \frac{1}{2!}x^4 + \frac{1}{3!}x^6 + \frac{1}{4!}x^8 + \cdots . \quad \bullet$$

EXAMPLE 6 Find the Taylor series at $x = 0$ for

(a) $\dfrac{1}{1 + x^3}$ (b) $\dfrac{x^2}{1 + x^3}$

Solution (a) In the Taylor series at $x = 0$ for $\dfrac{1}{1 - x}$, we replace x by $-x^3$, to obtain

$$\frac{1}{1 - (-x^3)} = 1 + (-x^3) + (-x^3)^2 + (-x^3)^3 + (-x^3)^4 + \cdots$$

$$\frac{1}{1 + x^3} = 1 - x^3 + x^6 - x^9 + x^{12} - \cdots .$$

(b) If we multiply the series in part (a) by x^2, we obtain

$$\frac{x^2}{1 + x^3} = x^2 - x^5 + x^8 - x^{11} + x^{14} - \cdots . \quad \bullet$$

Definite Integrals The standard normal curve of statistics has the equation $y = \dfrac{1}{\sqrt{2\pi}} e^{-x^2/2}$. Areas under the curve cannot be found by direct integration since there is no simple formula for an antiderivative of $e^{-x^2/2}$. However, Taylor series can be used to calculate these areas with a high degree of accuracy.

EXAMPLE 7 Find the area under the standard normal curve from $x = 0$ to $x = .8$; that is, calculate

$$\frac{1}{\sqrt{2\pi}} \int_0^{.8} e^{-x^2/2} \, dx.$$

Solution A Taylor series expansion for e^x was obtained in Example 2.

$$e^x = 1 + x + \frac{1}{2!}x^2 + \frac{1}{3!}x^3 + \frac{1}{4!}x^4 + \cdots.$$

Replace each occurrence of x by $-x^2/2$. Then

$$e^{-x^2/2} = 1 + \left(-\frac{x^2}{2}\right) + \frac{1}{2!}\left(-\frac{x^2}{2}\right)^2 + \frac{1}{3!}\left(-\frac{x^2}{2}\right)^3 + \frac{1}{4!}\left(-\frac{x^2}{2}\right)^4 + \cdots.$$

$$e^{-x^2/2} = 1 - \frac{1}{2\cdot 1!}x^2 + \frac{1}{2^2\cdot 2!}x^4 - \frac{1}{2^3\cdot 3!}x^6 + \frac{1}{2^4\cdot 4!}x^8 - \cdots.$$

Integrating, we obtain

$$\frac{1}{\sqrt{2\pi}} \int_0^{.8} e^{-x^2/2} \, dx =$$

$$\frac{1}{\sqrt{2\pi}}\left(x - \frac{1}{3\cdot 2\cdot 1!}x^3 + \frac{1}{5\cdot 2^2\cdot 2!}x^5 - \frac{1}{7\cdot 2^3\cdot 3!}x^7 + \frac{1}{9\cdot 2^4\cdot 4!}x^9 + \cdots\right)\Bigg|_0^{.8}$$

$$= \frac{1}{\sqrt{2\pi}}\left[.8 - \frac{1}{6}(.8)^3 + \frac{1}{40}(.8)^5 - \frac{1}{336}(.8)^7 + \frac{1}{3456}(.8)^9 - \cdots\right].$$

The infinite series on the right converges to the value of the definite integral. Summing up the five terms displayed gives the approximation .28815, which is accurate to four decimal places. This approximation can be made arbitrarily accurate by summing additional terms. ●

Convergence of Power Series When we differentiate, integrate, or perform algebraic operations on Taylor series, we are using the fact that Taylor series are *functions*. In fact, any power series in x is a function of x, whether or not its coefficients are obtained from the derivatives of some function. The domain of a "power series function" is the set of all x for which the series converges. The function value at a specific x in its domain is the number to which the series converges.

For instance, the geometric series $\sum_{k=0}^{\infty} x^k$ defines a function whose domain is the set of all x for which $|x| < 1$. The familiar Taylor series expansion

$$\frac{1}{1-x} = \sum_{k=0}^{\infty} x^k$$

simply states that the functions $\dfrac{1}{1-x}$ and $\sum_{k=0}^{\infty} x^k$ have the same value for each x such that $|x| < 1$.

Given any power series $\sum_{k=0}^{\infty} a_k x^k$, one of three possibilities must occur:

(i) There is a positive constant R such that the series converges for $|x| < R$ and diverges for $|x| > R$.

(ii) The series converges for all x.

(iii) The series converges only for $x = 0$.

In case (i) we call R the *radius of convergence* of the series. The series converges for all x in the interval $-R < x < R$ and may or may not converge at one or both of the endpoints of this interval. In case (ii), we say that the radius of convergence is ∞, and in case (iii), we say that the radius of convergence is 0.

When a power series with a positive radius of convergence is differentiated term by term, the new series will have the same radius of convergence. An analogous result holds for antiderivatives. Other operations, such as replacing x by a constant times a power of x, may affect the radius of convergence.

Suppose we begin with a function that has derivatives of all orders, and we write down its formal Taylor series at $x = 0$. Can we conclude that the Taylor series and the function have the same values for every x within the radius of convergence of the series? For all of the functions that we consider, the answer is yes. However, it is possible for the two values to differ. In this case we say that the function does not admit a power series expansion. In order to show that a function admits a power series expansion, it is necessary to show that the partial sums of the Taylor series converge to the function. The nth partial sum of the series is the nth Taylor polynomial p_n. Recall from Section 12.1 that we considered the nth remainder of $f(x)$,

$$R_n(x) = f(x) - p_n(x).$$

For a fixed x, the Taylor series converges to $f(x)$ if and only if $R_n(x) \to 0$ as $n \to \infty$. Exercises 57 and 58 illustrate how convergence can be verified by using the remainder formula from Section 12.1.

Taylor Series at $x = a$ In order to simplify the discussion in this section, we have restricted our attention to series that involve powers of x rather than powers of $x - a$. However, Taylor series, just like Taylor polynomials, can be formed as sums of powers of $x - a$. The Taylor series expansion of $f(x)$ at $x = a$ is

$$f(x) = f(a) + \frac{f'(a)}{1!}(x-a) + \frac{f''(a)}{2!}(x-a)^2 + \frac{f'''(a)}{3!}(x-a)^3 + \cdots$$
$$+ \frac{f^{(n)}(a)}{n!}(x-a)^n + \cdots .$$

1. Find the Taylor series expansion of $\sin x$ at $x = 0$.
2. Find the Taylor series expansion of $\cos x$ at $x = 0$.
3. Find the Taylor series expansion of $x^3 \cos 7x$ at $x = 0$.
4. If $f(x) = x^3 \cos 7x$, find $f^{(5)}(0)$. [*Hint:* How are the coefficients in the Taylor series of $f(x)$ related to $f(x)$ and its derivatives at $x = 0$?]

EXERCISES 12.4

In Exercises 1–4, find the Taylor series at $x = 0$ of the given function by computing three or four derivatives and using the definition of the Taylor series.

1. $\dfrac{1}{2x + 3}$ **2.** $\ln(1 - 3x)$ **3.** $\sqrt{1 + x}$ **4.** $(1 + x)^3$

In Exercises 5–20, find the Taylor series at $x = 0$ of the given function. Use suitable operations (differentiation, substitution, etc.) on the Taylor series at $x = 0$ of $\dfrac{1}{1 - x}$, e^x, or $\cos x$. These series are derived in Examples 1 and 2 and Practice Problem 2.

5. $\dfrac{1}{1 - 3x}$ **6.** $\dfrac{1}{1 + x}$ **7.** $\dfrac{1}{1 + x^2}$ **8.** $\dfrac{x}{1 + x^2}$

9. $\dfrac{1}{(1 + x)^2}$ **10.** $\dfrac{x}{(1 - x)^3}$ **11.** $5e^{x/3}$ **12.** $x^3 e^{x^2}$

13. $1 - e^{-x}$ **14.** $3(e^{-2x} - 2)$ **15.** $\ln(1 + x)$ **16.** $\ln(1 + x^2)$

17. $\cos 3x$ **18.** $\cos x^2$ **19.** $\sin 3x$ **20.** $x \sin x^2$

21. Find the Taylor series of xe^{x^2} at $x = 0$.

22. Show that $\ln\left(\dfrac{1 + x}{1 - x}\right) = 2x + \frac{2}{3}x^3 + \frac{2}{5}x^5 + \frac{2}{7}x^7 + \cdots$, $|x| < 1$. [*Hint:* Use Exercise 15 and Example 3.] This series converges much more quickly than the series for $\ln(1 - x)$ in Example 3, particularly for x close to zero. The series gives a formula for $\ln y$, where y is any number and $x = \dfrac{y - 1}{y + 1}$.

23. The *hyperbolic cosine* of x, denoted by $\cosh x$, is defined by

$$\cosh x = \tfrac{1}{2}(e^x + e^{-x}).$$

This function occurs often in physics and probability theory. The graph of $y = \cosh x$ is called a *catenary*.

(a) Use differentiation and the definition of a Taylor series to compute the first four nonzero terms in the Taylor series of $\cosh x$ at $x = 0$.

(b) Use the known Taylor series for e^x to obtain the Taylor series for $\cosh x$ at $x = 0$.

24. The *hyperbolic sine* of x is defined by

$$\sinh x = \tfrac{1}{2}(e^x - e^{-x}).$$

Repeat parts (a) and (b) of Exercise 23 for $\sinh x$.

25. Given the Taylor series expansion

$$\frac{1}{\sqrt{1+x}} = 1 - \frac{1}{2}x + \frac{1\cdot 3}{2\cdot 4}x^2 - \frac{1\cdot 3\cdot 5}{2\cdot 4\cdot 6}x^3 + \frac{1\cdot 3\cdot 5\cdot 7}{2\cdot 4\cdot 6\cdot 8}x^4 - \cdots,$$

find the first four terms in the Taylor series of $\dfrac{1}{\sqrt{1-x}}$ at $x = 0$.

26. Find the first four terms in the Taylor series of $\dfrac{1}{\sqrt{1-x^2}}$ at $x = 0$. (See Exercise 25.)

27. Use Exercise 25 and the fact that

$$\int \frac{1}{\sqrt{1+x^2}}\, dx = \ln(x + \sqrt{1+x^2}) + C$$

to find the Taylor series of $\ln(x + \sqrt{1+x^2})$ at $x = 0$.

28. Use the Taylor series expansion for $\dfrac{x}{(1-x)^2}$ to find the function whose Taylor series is $1 + 4x + 9x^2 + 16x^3 + 25x^4 + \cdots$.

29. Use the Taylor series for e^x to show that $\dfrac{d}{dx}e^x = e^x$.

30. Use the Taylor series for $\cos x$ (see Practice Problem 2) to show that $\cos(-x) = \cos x$.

31. The Taylor series at $x = 0$ for $f(x) = \ln\left(\dfrac{1+x}{1-x}\right)$ is given in Exercise 22. Find $f^{(5)}(0)$.

32. The Taylor series at $x = 0$ for $f(x) = \sec x$ is $1 + \frac{1}{2}x^2 + \frac{5}{24}x^4 + \frac{61}{720}x^6 + \cdots$. Find $f^{(4)}(0)$.

33. The Taylor series at $x = 0$ for $f(x) = \tan x$ is $x + \frac{1}{3}x^3 + \frac{2}{15}x^5 + \frac{17}{315}x^7 + \cdots$. Find $f^{(4)}(0)$.

34. The Taylor series at $x = 0$ for $\dfrac{1+x^2}{1-x}$ is $1 + x + 2x^2 + 2x^3 + 2x^4 + \cdots$. Find $f^{(4)}(0)$, where $f(x) = \dfrac{1+x^4}{1-x^2}$.

In Exercises 35–37, find the Taylor series expansion at $x = 0$ of the given antiderivative.

35. $\displaystyle\int e^{-x^2}\, dx$ 36. $\displaystyle\int xe^{x^3}\, dx$ 37. $\displaystyle\int \frac{1}{1+x^3}\, dx$

In Exercises 38–40, find an infinite series that converges to the value of the given definite integral.

38. $\displaystyle\int_0^1 \sin x^2\, dx$ 39. $\displaystyle\int_0^1 e^{-x^2}\, dx$ 40. $\displaystyle\int_0^1 xe^{x^3}\, dx$

41. (a) Use the Taylor series for e^x at $x = 0$ to show that $e^x > x^2/2$ for $x > 0$.
    (b) Deduce that $e^{-x} < 2/x^2$ for $x > 0$.
    (c) Show that xe^{-x} approaches 0 as $x \to \infty$.

42. Let k be a positive constant.

(a) Show that $e^{kx} > \dfrac{k^2 x^2}{2}$, for $x > 0$.

(b) Deduce that $e^{-kx} < \dfrac{2}{k^2 x^2}$, for $x > 0$.

(c) Show that xe^{-kx} approaches 0 as $x \to \infty$.

43. Show that $e^x > x^3/6$ for $x > 0$, and from this deduce that $x^2 e^{-x}$ approaches 0 as $x \to \infty$.

44. If k is a positive constant, show that $x^2 e^{-kx}$ approaches 0 as $x \to \infty$.

45. Let $R_n(x)$ be the nth remainder of $f(x) = \cos x$ at $x = 0$. See Section 12.1. Show that for any fixed value of x, $|R_n(x)| \le 1/(n+1)! \ |x|^{n+1}$, and hence conclude from Exercise 45 that $|R_n(x)| \to 0$ as $n \to \infty$. This shows that the Taylor series for $\cos x$ converges to $\cos x$ for every value of x.

46. Let $R_n(x)$ be the nth remainder of $f(x) = e^x$ at $x = 0$. Show that for any fixed positive value of x, $|R_n(x)| \le e^{|x|} \cdot 1/(n+1)! \ |x|^{n+1}$, and hence conclude that $|R_n(x)| \to 0$ as $n \to \infty$. This shows that the Taylor series for e^x converges to e^x for every value of x.

SOLUTIONS TO PRACTICE PROBLEMS 12.4

1. Use the definition of the Taylor series as an extended Taylor polynomial.

$$f(x) = \sin x, \qquad f'(x) = \cos x, \qquad f''(x) = -\sin x,$$
$$f(0) = 0, \qquad f'(0) = 1, \qquad f''(0) = 0.$$
$$f'''(x) = -\cos x, \qquad f^{(4)}(x) = \sin x, \qquad f^{(5)}(x) = \cos x$$
$$f'''(0) = -1, \qquad f^{(4)}(0) = 0, \qquad f^{(5)}(0) = 1$$

Therefore,

$$\sin x = 0 + 1 \cdot x + \frac{0}{2!}x^2 + \frac{-1}{3!}x^3 + \frac{0}{4!}x^4 + \frac{1}{5!}x^5 + \cdots$$

$$= x - \frac{1}{3!}x^3 + \frac{1}{5!}x^5 - \cdots .$$

2. Differentiate the Taylor series in Problem 1.

$$\frac{d}{dx}\sin x = \frac{d}{dx}\left(x - \frac{1}{3!}x^3 + \frac{1}{5!}x^5 - \cdots\right)$$

$$\cos x = 1 - \frac{1}{2!}x^2 + \frac{1}{4!}x^4 - \cdots .$$

$$\left[\textit{Note: } \text{We used the fact that } \frac{3}{3!} = \frac{3}{1 \cdot 2 \cdot 3} = \frac{1}{1 \cdot 2} = \frac{1}{2!} \text{ and } \frac{5}{5!} = \frac{1}{4!}.\right]$$

3. Replace x by $7x$ in the Taylor series for $\cos x$ and then multiply by x^3.

$$\cos x = 1 - \frac{1}{2!}x^2 + \frac{1}{4!}x^4 - \cdots$$

$$\cos 7x = 1 - \frac{1}{2!}(7x)^2 + \frac{1}{4!}(7x)^4 - \cdots$$

$$= 1 - \frac{7^2}{2!}x^2 + \frac{7^4}{4!}x^4 - \cdots$$

$$x^3 \cos 7x = x^3\left(1 - \frac{7^2}{2!}x^2 + \frac{7^4}{4!}x^4 - \cdots\right)$$

$$= x^3 - \frac{7^2}{2!}x^5 + \frac{7^4}{4!}x^7 - \cdots.$$

4. The coefficient of x^5 in the Taylor series of $f(x)$ is $\dfrac{f^{(5)}(0)}{5!}$. By Problem 3, this coefficient is $-\dfrac{7^2}{2!}$. Therefore,

$$\frac{f^{(5)}(0)}{5!} = -\frac{7^2}{2!}$$

$$f^{(5)}(0) = -\frac{7^2}{2!} \cdot 5! = -\frac{49}{2} \cdot 120 = -(49)(60) = -2940.$$

 ## 12.5 INFINITE SERIES AND PROBABILITY THEORY

One of the most important applications of infinite series is to the field of probability. A complete description of the connection between the two subjects is beyond the scope of this book. In this section we highlight a few of its most salient features.

In Chapter 11, we considered one class of experiments having an infinite number of possible outcomes. By the use of infinite series we can obtain results about another important class of such experiments—namely, those whose possible outcomes are the numbers 0, 1, 2, 3, ... Such experiments abound. Here are a few examples.

1. Flip a coin until a head appears and observe the number of consecutive tails which occurred. This number may be 0, 1, 2, 3, ...

2. Observe the number of telephone calls arriving at a switchboard during a given minute. This number may be 0, 1, 2, 3, ...

3. Observe the number of fire insurance claims filed (with a particular insurance company) during a given month. The number of claims is one of the numbers 0, 1, 2, 3, ...

Suppose that we consider an experiment whose possible outcomes are the numbers 0, 1, 2, 3, ... As in Chapter 11, we denote the outcome of the experiment by X and call X a *random variable*. To each possible value of X, there is an associated probability of occurrence, p_n. That is,

$$\Pr(X = 0) = p_0,$$
$$\Pr(X = 1) = p_1,$$
$$\vdots$$
$$\Pr(X = n) = p_n.$$
$$\vdots$$

Note that since $p_0, p_1, \ldots, p_n, \ldots$ are probabilities, each lies between 0 and 1. Furthermore, the sum of all these probabilities is 1. (One of the outcomes 0, 1, 2, 3, . . . always occurs.) That is,

$$p_0 + p_1 + \cdots + p_n + \cdots = 1.$$

In analogy with the case of experiments having a finite number of possible outcomes, we may define the *expected value* (or average value) of the random variable X (or of the experiment whose outcome is X) to be the number $E(X)$ given by the following formula:

$$E(X) = 0 \cdot p_0 + 1 \cdot p_1 + 2 \cdot p_2 + 3 \cdot p_3 + \cdots$$

(provided that the infinite series converges). That is, the expected value $E(X)$ is formed by adding the products of the possible outcomes by their respective probabilities of occurrence.

In a similar fashion, letting m denote $E(X)$, we define the *variance* of X by

$$\text{Var}(X) = (0 - m)^2 \cdot p_0 + (1 - m)^2 \cdot p_1 + (2 - m)^2 \cdot p_2$$
$$+ (3 - m)^2 \cdot p_3 + \cdots .$$

EXAMPLE I   Suppose that we toss a coin until a head appears and observe the number X of consecutive tails preceding it.

(a) Determine the probability p_n that exactly n consecutive tails occur.

(b) Determine the probability that an odd number of consecutive tails occur.

(c) Determine the average number of consecutive tails that occur.

(d) Write down the infinite series that gives the variance for the number of consecutive tails.

Solution   (a) The probability p_0 corresponds to tossing a head on the first throw. And since a head is just as likely as a tail, we have $p_0 = \frac{1}{2}$. To calculate p_1, notice that there are four possible results of tossing a coin twice:

heads heads;   tails tails;   heads tails;   tails heads.

Each of these results is equally likely, so each has probability $\frac{1}{4}$. The probability p_1 corresponds to the result "tails heads," so $p_1 = \frac{1}{4} = 1/2^2$. Similarly, to compute p_2 we note that there are eight possible results for tossing a coin three times, of which only one,

tails tails heads

relates to p_2. So $p_2 = \frac{1}{8} = 1/2^3$. In a similar fashion, we find that $p_n = 1/2^{n+1}$ for $n = 0, 1, 2, 3, \ldots$.

(b) The probability that an odd number of consecutive tails occurs is

$$
\begin{aligned}
p_1 + p_3 + p_5 + p_7 + \cdots &= \frac{1}{2^2} + \frac{1}{2^4} + \frac{1}{2^6} + \frac{1}{2^8} + \cdots \\
&= \frac{1}{2^2}\left(1 + \frac{1}{2^2} + \frac{1}{2^4} + \frac{1}{2^6} + \cdots\right) \\
&= \frac{1}{4}\left(1 + \frac{1}{4} + \frac{1}{4^2} + \frac{1}{4^3} + \cdots\right) \\
&= \frac{1}{4} \cdot \frac{1}{1 - \frac{1}{4}} = \frac{1}{4} \cdot \frac{4}{3} = \frac{1}{3}.
\end{aligned}
$$

(c) The average number of consecutive tails is

$$
\begin{aligned}
E(X) &= 0 \cdot p_0 + 1 \cdot p_1 + 2 \cdot p_2 + 3 \cdot p_3 + \cdots \\
&= 0 \cdot \frac{1}{2} + 1 \cdot \frac{1}{2^2} + 2 \cdot \frac{1}{2^3} + 3 \cdot \frac{1}{2^4} + \cdots \\
&= \frac{1}{2^2} + \frac{2}{2^3} + \frac{3}{2^4} + \cdots \\
&= \frac{1}{4}\left(1 + 2 \cdot \frac{1}{2} + 3 \cdot \frac{1}{2^2} + \cdots\right).
\end{aligned}
$$

Recall that we showed in the preceding section that for $|x| < 1$ we have

$$
1 + 2x + 3x^2 + \cdots = \frac{1}{(1 - x)^2}.
$$

Setting $x = \frac{1}{2}$ in the formula gives

$$
1 + 2 \cdot \frac{1}{2} + 3 \cdot \frac{1}{2^2} + \cdots = \frac{1}{(1 - \frac{1}{2})^2} = 4.
$$

Therefore,

$$
E(X) = \frac{1}{4} \cdot 4 = 1.
$$

So the average number of consecutive tails is 1, in agreement with our intuition.

(d) The variance for the number of consecutive tails is

$$\text{Var}(X) = (0-1)^2 \cdot \frac{1}{2} + (1-1)^2 \cdot \frac{1}{2^2} + (2-1)^2 \cdot \frac{1}{2^3}$$

$$+ (3-1)^2 \cdot \frac{1}{2^4} + (4-1)^2 \cdot \frac{1}{2^5} + \cdots$$

$$= \frac{1}{2} + 1^2 \cdot \frac{1}{2^3} + 2^2 \cdot \frac{1}{2^4} + 3^2 \cdot \frac{1}{2^5} + \cdots .$$

(It can be shown that this infinite series converges to 2. See Exercise 22.) ●

EXAMPLE 2   An electronics plant manufactures electronic calculators. After manufacture, each calculator is inspected for defects. The probability that exactly n nondefective calculators are observed before a defective one is observed is $p_n = \frac{1}{50}(.98)^n$. What is the average number of nondefective calculators between consecutive defective ones?

Solution   Our experiment consists of observing the number X of nondefective calculators in a row. We are given that $p_n = \frac{1}{50}(.98)^n$ and are asked for the expected value $E(X)$ of the experiment. But

$$E(X) = 0 \cdot \frac{1}{50}(.98)^0 + 1 \cdot \frac{1}{50}(.98)^1 + 2 \cdot \frac{1}{50}(.98)^2 + 3 \cdot \frac{1}{50}(.98)^3 + \cdots$$

$$= \frac{.98}{50}[1 + 2 \cdot (.98)^1 + 3 \cdot (.98)^2 + \cdots].$$

As we observed in the preceding example, the series in parentheses is the series

$$1 + 2x + 3x^2 + 4x^3 + \cdots = \frac{1}{(1-x)^2},$$

evaluated at $x = .98$. Thus

$$E(X) = \frac{.98}{50} \frac{1}{(1-.98)^2} = \frac{.98}{50} \frac{1}{(.02)^2} = 49.$$

In other words, on the average 49 nondefective calculators will occur between consecutive defective ones. ●

Poisson Experiments   In many experiments occurring in applications the probability p_n is given by a formula of the following type:

$$p_0 = e^{-\lambda},$$

$$p_1 = \frac{\lambda}{1}e^{-\lambda},$$

$$p_2 = \frac{\lambda^2}{1 \cdot 2}e^{-\lambda},$$

$$p_3 = \frac{\lambda^3}{1 \cdot 2 \cdot 3} e^{-\lambda},$$

$$\vdots$$

$$p_n = \frac{\lambda^n}{n!} e^{-\lambda}, \tag{1}$$

where λ is a constant depending on the particular experiment under consideration. An experiment for which equation (1) holds is called a *Poisson experiment*, and the outcome X is called a *Poisson random variable*. The probabilities for X given by equation (1) are said to be *Poisson distributed* (with parameter λ).

Poisson random variables constitute one of the most important classes of random variables studied in probability theory. This is because of the wide variety of physical phenomena which can be accurately modeled as Poisson experiments. For example, experiments 2 and 3 given at the beginning of this section are Poisson experiments. Here are some others:

4. Observe the annual number of deaths due to a particular disease.
5. Observe the monthly number of breakdowns of a particular type of machine in a given factory.
6. Observe the number of typographical errors per page in a given book.
7. Observe the number of persons arriving during 5-minute intervals at a supermarket checkout counter.
8. Observe the number of protozoa in various drop-sized samples of water drawn from a pond.

Note that for a Poisson experiment the requirement that

$$p_0 + p_1 + p_2 + p_3 + \cdots = 1$$

can be expressed by the equation

$$e^{-\lambda} + \frac{\lambda}{1!} e^{-\lambda} + \frac{\lambda^2}{2!} e^{-\lambda} + \frac{\lambda^3}{3!} e^{-\lambda} + \cdots = 1.$$

To verify this, we rewrite the left side in the form

$$e^{-\lambda} \left(1 + \frac{\lambda}{1!} + \frac{\lambda^2}{2!} + \frac{\lambda^3}{3!} + \cdots \right),$$

and we note that the quantity inside the parentheses is the Taylor series for e^{λ}. Thus

$$e^{-\lambda} \left(1 + \frac{\lambda}{1!} + \frac{\lambda^2}{2!} + \frac{\lambda^3}{3!} + \cdots \right) = e^{-\lambda} e^{\lambda} = e^0 = 1.$$

The following facts about Poisson experiments provide an interpretation for the parameter λ.

> Let X be a random variable whose probabilities are Poisson distributed with parameter λ, that is,
>
> $$p_0 = e^{-\lambda},$$
>
> $$p_n = \frac{\lambda^n}{n!} e^{-\lambda} \qquad (n = 1, 2, \ldots).$$
>
> Then the expected value and variance of X are given by
>
> $$\mathrm{E}(X) = \lambda, \qquad \mathrm{Var}(X) = \lambda.$$

We shall verify only the statement about $\mathrm{E}(X)$. The argument uses the Taylor series for e^λ. We have

$$\mathrm{E}(X) = 0 \cdot p_0 + 1 \cdot p_1 + 2 \cdot p_2 + 3 \cdot p_3 + 4 \cdot p_4 + \cdots$$

$$= 0 \cdot e^{-\lambda} + \not1 \cdot \frac{\lambda}{\not1} e^{-\lambda} + \not2 \cdot \frac{\lambda^2}{1 \cdot \not2} e^{-\lambda}$$

$$+ \not3 \cdot \frac{\lambda^3}{1 \cdot 2 \cdot \not3} e^{-\lambda} + \not4 \cdot \frac{\lambda^4}{1 \cdot 2 \cdot 3 \cdot \not4} e^{-\lambda} + \cdots$$

$$= \lambda e^{-\lambda} + \frac{\lambda^2}{1} e^{-\lambda} + \frac{\lambda^3}{1 \cdot 2} e^{-\lambda} + \frac{\lambda^4}{1 \cdot 2 \cdot 3} e^{-\lambda} + \cdots$$

$$= \lambda e^{-\lambda} \left(1 + \frac{\lambda}{1} + \frac{\lambda^2}{1 \cdot 2} + \frac{\lambda^3}{1 \cdot 2 \cdot 3} + \cdots \right)$$

$$= \lambda e^{-\lambda} \cdot e^\lambda$$

$$= \lambda.$$

The next two examples illustrate some applications of Poisson experiments.

EXAMPLE 3   Suppose that we observe the number X of calls received by a telephone switchboard during a 1-minute interval. Experience suggests that X is Poisson distributed with $\lambda = 3$.

(a) Determine the probability that zero, one, or two calls arrive during a particular minute.

(b) Determine the probability that three or more calls arrive during a particular minute.

(c) Determine the average number of calls received per minute.

Solution   (a) The probability that zero, one, or two calls arrive during a given minute is $p_0 + p_1 + p_2$. Moreover,

$$p_0 = e^{-\lambda} = e^{-3} \approx .04979,$$

$$p_1 = \frac{\lambda}{1} e^{-\lambda} = 3e^{-3} \approx .14937,$$

$$p_2 = \frac{\lambda^2}{1 \cdot 2} e^{-\lambda} = \frac{9}{2} e^{-3} \approx .22406.$$

Thus $p_0 + p_1 + p_2 \approx .42322$. That is, during approximately 42% of the minutes, either zero, one, or two calls are received.

(b) The probability of receiving three or more calls is the same as the probability of *not* receiving zero, one, or two calls and so is equal to

$$1 - (p_0 + p_1 + p_2) = 1 - .42322$$

$$= .57678.$$

(c) The average number of calls received per minute is equal to λ. That is, on the average the switchboard receives three calls per minute. ●

EXAMPLE 4 Drop-sized water samples are drawn from a New England pond. The number of protozoa in each sample is estimated and the average number is found to be 10. Suppose that a sample is chosen at random. What is the probability that it contains four or fewer protozoa?

Solution The probability that a sample contains exactly n protozoa is given by the Poisson distribution

$$p_n = \frac{\lambda^n}{n!} e^{-\lambda},$$

where $\lambda = 10$, the expected value. The probability that there are four or fewer protozoa equals

$$p_0 + p_1 + p_2 + p_3 + p_4.$$

To calculate this probability we use a calculator and find that it equals .02925. In other words, the probability that a random sample contains four or fewer protozoa is .02925. ●

PRACTICE PROBLEMS 12.5

1. Suppose that we toss a coin until a tail appears and observe the number of consecutive heads preceding it. What is the probability that this number is divisible by 3?

2. A Public Health Officer is tracking down the source of a bacterial infection in a certain city. She analyzes the reported incidence of the infection in each city block and finds an average of three cases per block. A certain block is found to have seven cases. What is the probability that a randomly chosen block has seven or more cases, assuming that the number of cases per block is Poisson distributed?

EXERCISES 12.5

Suppose that we toss a coin until a tail appears and observe the number of consecutive heads preceding the tail.

1. What is the probability that exactly three heads occur?
2. What is the probability that exactly three or five heads occur?
3. What is the probability that the number of consecutive heads is at least three?
4. What is the probability that the number of consecutive heads is even?
5. What is the probability that the number of consecutive heads is odd and at least three?
6. What is the probability that the number of consecutive heads is divisible by 5?

Suppose that in a certain town there are two competing taxi companies, Red Cab and Blue Cab. The taxis mix with downtown traffic in a random manner. The Red fleet is twice the size of the Blue fleet. Suppose that we stand on a downtown street corner and watch the taxis passing by, observing the number of consecutive Red taxis before a Blue taxi appears.

7. Determine the formula for p_n, the probability of exactly n consecutive Red taxis.
8. Determine the average number of consecutive Red taxis.
9. What is the probability that the number of consecutive Red taxis is divisible by 3?

The number of typographical errors per page of a certain newspaper is approximately Poisson distributed with $\lambda = 5$.

10. What is the probability that a given page is error-free?
11. What is the probability that a given page has exactly two errors?
12. What is the probability that a given page has either two or three errors?
13. What is the probability that a given page has at most four errors?
14. What is the average number of errors per page?
15. What is the probability that a given page has at least two errors?

The monthly number of fire insurance claims filed with the Firebug Insurance Company is Poisson distributed with $\lambda = 10$.

16. What is the probability that in a given month no claims are filed?
17. What is the probability that in a given month at least three claims are filed?
18. What is the average number of claims filed per month?
19. What is the probability that the number of claims is either one, two, or three?
20. A person shooting at a target has five successive hits and then a miss. If x is the probability of success on each shot, then the probability of having five successive hits followed by a miss is $p = x^5(1 - x)$. Take first and second derivatives to determine the value of x for which p has its maximum value.
21. In a production process a box of 100 fuses is examined and found to contain two defective fuses. Suppose that the probability of having two defective fuses in a box of 100 is $p = (\lambda^2/2)e^{-\lambda}$ for some λ. Take first and second derivatives to determine the value of λ for which p has its maximum value.

22. Use the result of Exercise 28 of Section 12.4 to show that the variance in Example 1 is 2.

23. When raisins are mixed into batter that is made into cookies, the number of raisins a particular cookie will have is Poisson distributed. If 100 cookies are being made, how many raisins should be used so that the probability of a cookie's having no raisins is .01?

Let X be a Poisson random variable with parameter λ. Exercises 24 and 25 involve the Taylor series of the hyperbolic cosine and sine functions. See Exercises 23 and 24 in Section 12.4.

24. Show that the probability that X is an even integer (including 0) is $e^{-\lambda} \cosh \lambda$.

25. Show that the probability that X is an odd integer is $e^{-\lambda} \sinh \lambda$.

SOLUTIONS TO PRACTICE PROBLEMS 12.5

1. If p_n is the probability that exactly n consecutive heads occur, then we know from Example 1 that $p_n = 1/2^{n+1}$. The probability that the number of consecutive heads is divisible by 3 equals

$$p_0 + p_3 + p_6 + p_9 + p_{12} + \cdots = \frac{1}{2} + \frac{1}{2^4} + \frac{1}{2^7} + \frac{1}{2^{10}} + \frac{1}{2^{13}} + \cdots$$

$$= \frac{1}{2}\left(1 + \frac{1}{2^3} + \frac{1}{2^6} + \frac{1}{2^9} + \frac{1}{2^{12}}\right) + \cdots$$

$$= \frac{1}{2}\left[\frac{1}{1 - (1/2^3)}\right]$$

$$= \frac{1}{2} \cdot \frac{8}{7}$$

$$= \frac{4}{7}.$$

2. The number of cases per block is Poisson distributed with $\lambda = 3$. So the probability of having seven or more cases in a given block is

$$P_7 + P_8 + P_9 + \cdots = 1 - (P_0 + P_1 + P_2 + P_3 + P_4 + P_5 + P_6).$$

However,

$$P_n = \frac{3^h}{1 \cdot 2 \cdots \cdots n} e^{-3},$$

so that

$$P_0 = .04979, \quad P_1 = .14937,$$
$$P_2 = .22406, \quad P_3 = .22406,$$
$$P_4 = .16804, \quad P_3 = .10082,$$

Therefore, the probability of seven or more cases in a given block equals $1 - (.04979 + .14937 + .22406 + .22406 + .22406 + .16804 + .10082 + .05041) = .03345$.

CHAPTER 12 CHECKLIST

✓ nth Taylor polynomial of $f(x)$ at $x = a$:

$$p_n(x) = f(a) + \frac{f'(a)}{1!}(x - a) + \frac{f''(a)}{2!}(x - a)^2 + \cdots + \frac{f^{(n)}(a)}{n!}(x - a)^n$$

✓ nth remainder of $f(x)$ at $x = a$

✓ Newton-Raphson algorithm:

$$x_{n+1} = x_n - \frac{f(x_n)}{f'(x_n)}$$

✓ Partial sum of an infinite series

✓ Sum of an infinite series

✓ Convergent and divergent series

✓ Geometric series:

$$\sum_{k=0}^{\infty} ar^k = \frac{a}{1 - r}, \qquad |r| < 1$$

✓ Taylor series expansion of $f(x)$ at $x = 0$:

$$f(x) = f(0) + \frac{f'(0)}{1!}x + \frac{f''(0)}{2!}x^2 + \cdots + \frac{f^{(n)}(0)}{n!}x^n + \cdots$$

✓ Poisson random variable X with parameter λ:

$$p_0 = e^{-\lambda}, \quad p_n = \frac{\lambda^n}{n!}e^{-\lambda}, \quad E(X) = \lambda, \quad \text{Var}(X) = \lambda$$

CHAPTER 12   SUPPLEMENTARY EXERCISES

1. Find the second Taylor polynomial of $x(x + 1)^{3/2}$ at $x = 0$.

2. Find the fourth Taylor polynomial of $(2x + 1)^{3/2}$ at $x = 0$.

3. Find the fifth Taylor polynomial of $x^3 - 7x^2 + 8$ at $x = 0$.

4. Find the nth Taylor polynomial of $\dfrac{2}{2 - x}$ at $x = 0$.

5. Find the third Taylor polynomial of x^2 at $x = 3$.

6. Find the third Taylor polynomial of e^x at $x = 2$.

7. Use a second Taylor polynomial at $t = 0$ to estimate the area under the graph of $y = -\ln(\cos 2t)$ between $t = 0$ and $t = \frac{1}{2}$.

8. Use a second Taylor polynomial at $x = 0$ to estimate the value of $\tan(.1)$.

9. (a) Find the second Taylor polynomial of $\sqrt{x}$ at $x = 9$.

   (b) Use part (a) to estimate $\sqrt{8.7}$ to six decimal places.

   (c) Use the Newton-Raphson algorithm with $n = 2$ and $x_0 = 3$ to approximate the solution of the equation $x^2 - 8.7 = 0$. Express your answer to six decimal places.

10. (a) Use the third Taylor polynomial of $\ln(1 - x)$ at $x = 0$ to approximate $\ln 1.3$ to four decimal places.

(b) Find an approximate solution of the equation $e^x = 1.3$ using the Newton-Raphson algorithm with $n = 2$ and $x_0 = 0$. Express your answer to four decimal places.

11. Use the Newton-Raphson algorithm with $n = 2$ to approximate the zero of $x^2 - 3x - 2$ near $x_0 = 4$.

12. Use the Newton-Raphson algorithm with $n = 2$ to approximate the solution of the equation $e^{2x} = 1 + e^{-x}$.

In Exercises 13–20, find the sum of the given infinite series if it is convergent.

13. $1 - \dfrac{3}{4} + \dfrac{9}{16} - \dfrac{27}{64} + \dfrac{81}{256} - \cdots$

14. $\dfrac{5^2}{6} + \dfrac{5^3}{6^2} + \dfrac{5^4}{6^3} + \dfrac{5^5}{6^4} + \dfrac{5^6}{6^5} + \cdots$

15. $\dfrac{1}{8} + \dfrac{1}{8^2} + \dfrac{1}{8^3} + \dfrac{1}{8^4} + \dfrac{1}{8^5} + \cdots$

16. $\dfrac{2^2}{7} - \dfrac{2^5}{7^2} + \dfrac{2^8}{7^3} - \dfrac{2^{11}}{7^4} + \dfrac{2^{14}}{7^5} - \cdots$

17. $\dfrac{1}{m+1} + \dfrac{m}{(m+1)^2} + \dfrac{m^2}{(m+1)^3} + \dfrac{m^3}{(m+1)^4} + \cdots$, where m is a positive number.

18. $\dfrac{1}{m} - \dfrac{1}{m^2} + \dfrac{1}{m^3} - \dfrac{1}{m^4} + \dfrac{1}{m^5} - \cdots$, where m is a positive number.

19. $1 + 2 + \dfrac{2^2}{2!} + \dfrac{2^3}{3!} + \dfrac{2^4}{4!} + \cdots$

20. $1 + \dfrac{1}{3} + \dfrac{1}{2!}\left(\dfrac{1}{3}\right)^2 + \dfrac{1}{3!}\left(\dfrac{1}{3}\right)^3 + \dfrac{1}{4!}\left(\dfrac{1}{3}\right)^4 + \cdots$

21. Use properties of convergent series to find $\displaystyle\sum_{k=0}^{\infty} \dfrac{1 + 2^k}{3^k}$.

22. Find $\displaystyle\sum_{k=0}^{\infty} \dfrac{3^k + 5^k}{7^k}$.

In Exercises 23–26, find the Taylor series at $x = 0$ of the given function. Use suitable operations on the Taylor series at $x = 0$ of $\dfrac{1}{1-x}$ and e^x.

23. $\dfrac{1}{1+x^3}$ 24. $\ln(1 + x^3)$ 25. $\dfrac{1}{(1-3x)^2}$ 26. $\dfrac{e^x - 1}{x}$

27. **(a)** Find the Taylor series of $\cos 2x$ at $x = 0$, either by direct calculation or by using the known series for $\cos x$.

(b) Use the trigonometric identity
$$\sin^2 x = \tfrac{1}{2}(1 - \cos 2x)$$
to find the Taylor series of $\sin^2 x$ at $x = 0$.

28. **(a)** Find the Taylor series of $\cos 3x$ at $x = 0$.

(b) Use the trigonometric identity
$$\cos^3 x = \tfrac{1}{4}(\cos 3x + 3\cos x)$$
to find the fourth Taylor polynomial of $\cos^3 x$ at $x = 0$.

29. Use the decomposition $\dfrac{1+x}{1-x} = \dfrac{1}{1-x} + \dfrac{x}{1-x}$ to find the Taylor series of $\dfrac{1+x}{1-x}$ at $x = 0$.

30. Find an infinite series that converges to

$$\int_0^{1/2} \frac{e^x - 1}{x}\,dx.$$

[*Hint*: Use Exercise 26.]

31. It can be shown that the sixth Taylor polynomial of $f(x) = \sin x^2$ at $x = 0$ is $x^2 - \frac{1}{6}x^6$. Use this fact in parts (a), (b), and (c).

    (a) What is the fifth Taylor polynomial of $f(x)$ at $x = 0$?

    (b) What is $f'''(0)$?

    (c) Estimate the area under the graph of $y = \sin x^2$ between $x = 0$ and $x = 1$. Use four decimal places, and compare your answer with the values given in Example 4 of Section 12.1.

32. Let $f(x) = \ln |\sec x + \tan x|$. It can be shown that $f'(0) = 1$, $f''(0) = 0$, $f'''(0) = 1$, and $f^{(4)}(0) = 0$. What is the fourth Taylor polynomial of $f(x)$ at $x = 0$?

33. Let $f(x) = 1 + x^2 + x^4 + x^6 + \cdots$.

    (a) Find the Taylor series expansion of $f'(x)$ at $x = 0$.

    (b) Find the simple formula for $f'(x)$ not involving a series. [*Hint:* First find a simple formula for $f(x)$.]

34. Let $f(x) = x - 2x^3 + 4x^5 - 8x^7 + 16x^9 - \cdots$.

    (a) Find the Taylor series expansion of $\int f(x)\,dx$ at $x = 0$.

    (b) Find a simple formula for $\int f(x)\,dx$ not involving a series. [*Hint:* First find a simple formula for $f(x)$.]

35. (*Fractional Reserve Banking*) Suppose that the Federal Reserve (the Fed) buys $\$100$ million of government debt obligations from private owners. This creates $\$100$ million of new money and sets off a chain reaction because of the "fractional reserve" banking system. When the $\$100$ million is deposited into private bank accounts, the banks keep only 15% in reserve and may loan out the remaining 85%, creating more new money—$(.85)(100)$ million dollars. The companies who borrow this money turn around and spend it, and the recipients deposit the money in their bank accounts. Assuming that all of the $(.85)(100)$ million is redeposited, the banks may again loan out 85% of this amount, creating $(.85)^2(100)$ million additional dollars. This process may be repeated indefinitely. Compute the total amount of new money that can be created theoretically by this process, beyond the original $\$100$ million. (In practice, only about an additional $\$300$ million is created, usually within a few weeks of the action of the Fed.)

36. Suppose that the Federal Reserve creates $\$100$ million of new money, as in Exercise 43, and the banks lend 85% of all new money they receive. However, suppose that out of each loan, only 80% is redeposited into the banking system. Thus, whereas the first set of loans total $(.85)(100)$ million dollars, the second set is only 85% of $(.80)(.85)(100)$, or $(.80)(.85)^2(100)$ million, and the next set is 85% of $(.80)^2(.85)^2(100)$, or $(.80)^2(.85)^3(100)$ million dollars, and so on. Compute the total theoretical amount that may be loaned in this situation.

Suppose that when you die, the proceeds of a life insurance policy will be deposited into a trust fund that will earn 8% interest, compounded continuously. According to the terms of your will, the trust fund must pay to your descendants and their heirs c_1 dollars (total) at the end of the first year, c_2 dollars at the end of the second year, c_3 dollars at the end of the third year, and so on, forever. The amount that must be in the trust fund initially to make the kth payment is $c_k e^{-.08k}$, the present value of the amount to be paid in k years. So the life insurance policy should pay a total of $\sum_{k=1}^{\infty} c_k e^{-.08k}$ dollars into the trust fund in order to provide for all the payments.

37. How large must the insurance policy be if $c_k = 10{,}000$ for all k? (Find the sum of the series.)

38. How large must the insurance policy be if $c_k = 10{,}000(.9)^k$ for all k?

39. Suppose that $c_k = 10{,}000(1.08)^k$ for all k. Find the sum of the series above if the series converges.

A pair of dice is rolled until a "seven" or "eleven" appears, and the number of rolls preceding the final roll is observed. The probability of rolling "seven or eleven" is $\frac{2}{9}$.

40. Determine the formula for p_n, the probability of exactly n consecutive rolls preceding the final roll.

41. What is the probability that an odd number of consecutive rolls precede the final roll?

42. Determine the average number of consecutive rolls preceding the final roll.

A small volume of blood is to be selected, examined under a microscope, and the number of white blood cells is counted. Suppose that for healthy people the number of white blood cells in such a specimen is Poisson distributed with $\lambda = 4$.

43. What is the probability that a specimen from a healthy person has exactly four white blood cells?

44. What is the probability that a specimen from a healthy person has eight or more white blood cells?

45. What is the average number of white blood cells per specimen from a healthy person?

APPENDIX TABLE 1 Areas Under the Standard Normal Curve

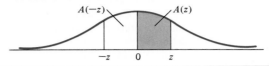

z	.00	.01	.02	.03	.04	.05	.06	.07	.08	.09
0.0	.0000	.0040	.0080	.0120	.0160	.0199	.0239	.0279	.0319	.0359
0.1	.0398	.0438	.0478	.0517	.0557	.0596	.0636	.0675	.0714	.0754
0.2	.0793	.0832	.0871	.0910	.0948	.0987	.1026	.1064	.1103	.1141
0.3	.1179	.1217	.1255	.1293	.1331	.1368	.1406	.1443	.1480	.1517
0.4	.1554	.1591	.1628	.1664	.1700	.1736	.1772	.1808	.1844	.1879
0.5	.1915	.1950	.1985	.2019	.2054	.2088	.2123	.2157	.2190	.2224
0.6	.2258	.2291	.2324	.2357	.2389	.2422	.2454	.2486	.2518	.2549
0.7	.2580	.2612	.2642	.2673	.2704	.2734	.2764	.2794	.2823	.2852
0.8	.2881	.2910	.2939	.2967	.2996	.3023	.3051	.3078	.3106	.3133
0.9	.3159	.3186	.3212	.3238	.3264	.3289	.3315	.3340	.3365	.3389
1.0	.3413	.3438	.3461	.3485	.3508	.3531	.3554	.3577	.3599	.3621
1.1	.3643	.3665	.3686	.3708	.3729	.3749	.3770	.3790	.3810	.3820
1.2	.3849	.3869	.3888	.3907	.3925	.3944	.3962	.3980	.3997	.4015
1.3	.4032	.4049	.4066	.4082	.4099	.4115	.4131	.4147	.4162	.4177
1.4	.4192	.4207	.4222	.4236	.4251	.4265	.4279	.4292	.4306	.4319
1.5	.4332	.4345	.4357	.4370	.4382	.4394	.4406	.4418	.4429	.4441
1.6	.4452	.4463	.4474	.4484	.4495	.4505	.4515	.4525	.4535	.4545
1.7	.4554	.4564	.4573	.4582	.4591	.4599	.4608	.4616	.4625	.4633
1.8	.4641	.4649	.4656	.4664	.4671	.4678	.4686	.4693	.4699	.4706
1.9	.4713	.4719	.4726	.4732	.4738	.4744	.4750	.4756	.4761	.4767
2.0	.4772	.4778	.4783	.4788	.4793	.4798	.4803	.4808	.4812	.4817
2.1	.4821	.4826	.4830	.4834	.4838	.4842	.4846	.4850	.4854	.4857
2.2	.4861	.4864	.4868	.4871	.4875	.4878	.4881	.4884	.4887	.4890
2.3	.4893	.4896	.4898	.4901	.4904	.4906	.4909	.4911	.4913	.4916
2.4	.4918	.4920	.4922	.4925	.4927	.4929	.4931	.4932	.4934	.4936
2.5	.4938	.4940	.4941	.4943	.4945	.4946	.4948	.4949	.4951	.4952
2.6	.4953	.4955	.4956	.4957	.4959	.4960	.4961	.4962	.4963	.4964
2.7	.4965	.4966	.4967	.4968	.4969	.4970	.4971	.4972	.4973	.4974
2.8	.4974	.4975	.4976	.4977	.4977	.4978	.4979	.4979	.4980	.4981
2.9	.4981	.4982	.4982	.4983	.4984	.4984	.4985	.4985	.4986	.4986
3.0	.4987	.4987	.4987	.4988	.4988	.4989	.4989	.4989	.4990	.4990
3.1	.4990	.4991	.4991	.4991	.4992	.4992	.4992	.4992	.4993	.4993
3.2	.4993	.4993	.4994	.4994	.4994	.4994	.4994	.4995	.4995	.4995
3.3	.4995	.4995	.4995	.4996	.4996	.4996	.4996	.4996	.4996	.4997
3.4	.4997	.4997	.4997	.4997	.4997	.4997	.4997	.4997	.4997	.4998
3.5	.4998	.4998	.4998	.4998	.4998	.4998	.4998	.4998	.4998	.4998

ANSWERS TO EXERCISES

CHAPTER 0
Exercises 0.1, page 18

1.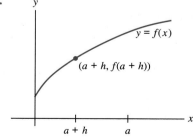
$-1\ 0 \qquad 4$

3.
$-2\ \ 0\ \sqrt{2}$

5.
$0 \qquad 3$

7. $[2, 3)$

9. $[-1, 0)$ **11.** $(-\infty, 3)$ **13.** 0, 10, 0, 70

15. $0, 0, -\frac{9}{8}, a^3 + a^2 - a - 1$ **17.** $\frac{1}{3}, 3, \frac{a + 1}{a + 2}$ **19.** $a^2 - 1, a^2 + 2a$

21. (a) 1990 sales **(b)** 60 **23.** $x \neq 1, 2$ **25.** $x < 3$

27. Function **29.** Not a function **31.** Not a function **33.** 1

35. 3 **37.** Positive **39.** Positive **41.** $-1, 5, 9$ **43.** .03

45. .04 **47.** No **49.** Yes **51.** $(a + 1)^3$ **53.** 1, 3, 4

55. π, 3, 12

57. $f(x) = \begin{cases} .06x & \text{for } 50 \leq x \leq 300 \\ .02x + 12 & \text{for } 300 < x \leq 600 \\ .015x + 15 & \text{for } 600 < x \end{cases}$

59.

y

$y = f(x)$

$(a + h, f(a + h))$

$a + h \qquad a$

x

Exercises 0.2, page 29

1. 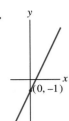   **3.** 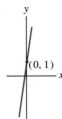   **5.**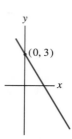

7. $\left(-\frac{1}{3}, 0\right), (0, 3)$ **9.** y-intercept $(0, 5)$ **11.** $(12, 0), (0, 3)$ **13.** (a) $K = \frac{1}{250}, V = \frac{1}{50}$

(b) $\left(-\frac{1}{K}, 0\right), \left(0, \frac{1}{V}\right)$ **15.** (a) $58   (b) $f(x) = .20x + 18$ **17.** $300x + 1500$, $x =$ number of days

19. The cost for another 5% is $25 million. The cost for the final 5% is 21 times as much.   **21.** $a = 3, b = -4$,

$c = 0$ **23.** $a = -2, b = 3, c = 1$ **25.** $a = -1, b = 0, c = 1$

27. 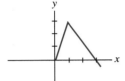   **29.** 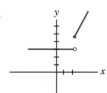   **31.**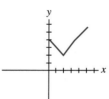

33. 1   **35.** 10^{-2} **37.** 2.5

Exercises 0.3, page 35

1. $x^2 + 9x + 1$ **3.** $9x^3 + 9x$ **5.** $\dfrac{t}{9} + \dfrac{1}{9t}$ **7.** $\dfrac{3x + 1}{x^2 - x - 6}$ **9.** $\dfrac{4x}{x^2 - 12x + 32}$

11. $\dfrac{2x^2 + 5x + 50}{x^2 - 100}$ **13.** $\dfrac{2x^2 - 2x + 10}{x^2 + 3x - 10}$ **15.** $\dfrac{-x^2 + 5x}{x^2 + 3x - 10}$ **17.** $\dfrac{x^2 + 5x}{-x^2 + 7x - 10}$

19. $\dfrac{-x^2 + 3x + 4}{x^2 + 5x - 6}$ **21.** $\dfrac{-x^2 - 3x}{x^2 + 15x + 50}$ **23.** $\dfrac{5u - 1}{5u + 1}, u \neq 0$ **25.** $\left(\dfrac{x}{1 - x}\right)^6$

27. $\left(\dfrac{x}{1 - x}\right)^3 - 5\left(\dfrac{x}{1 - x}\right)^2 + 1$ **29.** $\dfrac{t^3 - 5t^2 + 1}{-t^3 + 5t^2}$ **31.** $2xh + h^2$ **33.** $4 - 2t - h$

35. (a) $C(A(t)) = 3000 + 1600t - 40t^2$ (b) $6040   **37.** $h(x) = x + \frac{1}{8}$; $h(x)$ converts from British sizes to U.S.

sizes.

Exercises 0.4, page 45

1. $2, \frac{3}{2}$ **3.** $\frac{3}{2}$ **5.** No zeros   **7.** $1, -\frac{1}{5}$ **9.** $5, 4$ **11.** $2 + \sqrt{6}/3, 2 - \sqrt{6}/3$

13. $(x + 5)(x + 3)$ **15.** $(x - 4)(x + 4)$ **17.** $3(x + 2)^2$ **19.** $-2(x - 3)(x + 5)$ **21.** $x(3 - x)$

23. $-2x(x - \sqrt{3})(x + \sqrt{3})$ **25.** $(-1, 1), (5, 19)$ **27.** $(-1, 9), (4, 4)$ **29.** $(0, 0), (2, -2)$
31. $(0, 5), (2 - \sqrt{3}, 25 - 23\sqrt{3}/2), (2 + \sqrt{3}, 25 + 23\sqrt{3}/2)$ **33.** $-7, 3$ **35.** $-2, 3$ **37.** -7
39. 16,667 and 78,571 subscribers

Exercises 0.5, page 53

1. 27 **3.** 1 **5.** .0001 **7.** -16 **9.** 4 **11.** .01 **13.** $\frac{1}{6}$ **15.** 100 **17.** 16

19. 125 **21.** 1 **23.** 4 **25.** $\frac{1}{2}$ **27.** 1000 **29.** 10 **31.** 6 **33.** 16 **35.** 18

37. $\frac{4}{9}$ **39.** 7 **41.** $x^6 y^6$ **43.** $x^3 y^3$ **45.** $\dfrac{1}{\sqrt{x}}$ **47.** $\dfrac{x^{12}}{y^6}$ **49.** $x^{12} y^{20}$ **51.** $x^2 y^6$

53. $16x^4$ **55.** x^2 **57.** $\dfrac{1}{x^7}$ **59.** x **61.** $\dfrac{27x^6}{8y^3}$ **63.** $2\sqrt{x}$ **65.** $\dfrac{1}{8x^6}$ **67.** $\dfrac{1}{32x^2}$

69. $9x^3$ **71.** $x - 1$ **73.** $1 + 6\sqrt{x}$ **77.** 16 **79.** $\frac{1}{4}$ **81.** 8 **83.** $\frac{1}{32}$ **85.** \$709.26

87. \$127,857.61 **89.** \$164.70 **91.** \$1592.75 **93.** \$3268.00 **95.** $\frac{125}{64} r^4 + \frac{125}{4} r^3 +$
$\frac{375}{2} r^2 + 500r + 500$

Exercises 0.6, page 63

1.

3.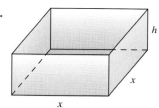

5.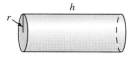

7. $P = 8x; 3x^2 = 25$ **9.** $A = \pi r^2; 2\pi r = 15$ **11.** $V = x^2 h; x^2 + 4xh = 65$ **13.** $\pi r^2 h = 100;$
$C = 11\pi r^2 + 14\pi rh$ **15.** $2x + 3h = 5000; A = xh$ **17.** $C = 36x + 20h$ **19.** 75 cm^2

21. (a) 38 **(b)** \$40 **23. (a)** 200 **(b)** 275 **(c)** 25 **25. (a)** $P(x) = 12x - 800$ **(b)** \$640 **(c)** \$3150
27. 270 cents **29.** A 100-in.^{3} cylinder of radius 3 in. costs \$1.62 to construct. **31.** \$1.08 **33.** Revenue:
\$1800; cost: \$1200 **35.** 40 **37.** $C(1000) = 4000$ **39.** Find the y-coordinate of the point on the graph
whose x-coordinate is 400. **41.** The greatest profit, \$52,500, occurs when 2500 units of goods are produced.
43. Find the point on the graph whose second coordinate is 30,000. **45.** Find $h(3)$. Find the y-coordinate of the
point on the graph whose t-coordinate is 3. **47.** Find the maximum value of $h(t)$. Find the second coordinate of the
highest point of the graph. **49.** Solve $h(t) = 100$. Find the t-coordinate of the point whose y-coordinate is 100.

Chapter 0: Supplementary Exercises, page 68

1. $2, 27\frac{1}{3}, -2, -2\frac{1}{8}, \dfrac{5\sqrt{2}}{2}$ **3.** $a^2 - 4a + 2$ **5.** $x \neq 0, -3$ **7.** All x **9.** Yes

11. $5x(x - 1)(x + 4)$ **13.** $(-1)(x - 6)(x + 3)$ **15.** $-\frac{2}{3}, 1$ **17.** $\left(\dfrac{5 + \sqrt{45}}{10}, \dfrac{\sqrt{45}}{5}\right)$,

$\left(\dfrac{5 - \sqrt{45}}{10}, -\dfrac{\sqrt{45}}{5}\right)$ **19.** $x^2 + x - 1$ **21.** $x^{5/2} - 2x^{3/2}$ **23.** $x^{3/2} - 2x^{1/2}$

25. $\dfrac{x^2 - x + 1}{x^2 - 1}$ **27.** $-\dfrac{3x^2 + 1}{3x^2 + 4x + 1}$ **29.** $\dfrac{-3x^2 + 9x - 10}{3x^2 - 5x - 8}$ **31.** $\dfrac{1}{x^4} - \dfrac{2}{x^2} + 4$ **33.** $(\sqrt{x} - 1)^2$

35. $\dfrac{1}{(\sqrt{x} - 1)^2} - \dfrac{2}{\sqrt{x} - 1} + 4$ **37.** $27, 32, 4$ **39.** $301 + 10t + .04t^2$ **41.** $x^2 + 2x + 1$ **43.** x

CHAPTER I
Exercises I.I, page 81

1. -5 **3.** 0 **5.** $\frac{2}{7}$ **7.** $-\frac{2}{3}$ **9.** $y = 3x - 1$ **11.** $y = x + 1$ **13.** $y = 35 - 7x$

15. $y = 4$ **17.** $y = \dfrac{x}{2}$ **19.** $y = -2x$ **21.** $y = 6 - 2x$ **23.** $y = \frac{1}{2}x - \frac{1}{2}$ **25.**   C

(b) B   **(c)** D   **(d)** A   **27.** 2   **29.** $-.75$ **31.** $(2, 5); (3, 7); (0, 1)$ **33.** $(0, -\frac{5}{4}); (1, -\frac{3}{2});$

$(-2, -\frac{3}{4})$ **35.** l_1

37.

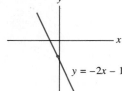

$y = -2x - 1$

39.

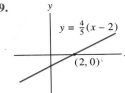

$y = \frac{4}{5}(x - 2)$

$(2, 0)$

41. $y = 5$ **43.** $4x + 5y = 7$ **45.** $y = -2x$ **47.** $y - 1 = 2(x - 1)$ **49.** $y - \frac{1}{4} = -(x + \frac{1}{2})$

51. If the monopolist wants to sell one more unit of goods, then the price per unit must be lowered by 2¢. No one is

willing to pay \$7 or more for a unit of goods.   **53. (a)** $-$**(c)**

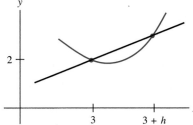

(d) $\dfrac{f(3 + h) - f(3)}{h}$

Exercises 1.2, page 88

1.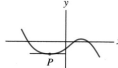

3.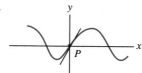

5.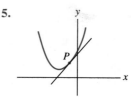

7. 1   9. -3 11. $\frac{2}{3}$ 13. Small positive slope   15. Zero slope   17. Zero slope

19. -4, $y - 4 = -4(x + 2)$ 21. $\frac{8}{3}$, $y - \frac{16}{9} = \frac{8}{3}(x - \frac{4}{3})$ 23. $y - 2.25 = 3(x - 1.5)$ 25. $(\frac{5}{6}, \frac{25}{36})$

27. $(-\frac{1}{4}, \frac{1}{16})$ 29. 12   31. $\frac{3}{4}$ 33. $y + 1 = 3(x + 1)$ 35. (a) 3, 9   (b) increase

Exercises 1.3, page 99

1. 2   3. $8x^7$ 5. $\frac{5}{2}x^{3/2}$ 7. $\frac{1}{3}x^{-2/3}$ 9. $-2x^{-3}$ 11. $-\frac{1}{4}x^{-5/4}$ 13. 0

15. $-3x^{-4}$ 17. -192 19. $-\frac{1}{9}$ 21. -1 23. $\frac{9}{2}$ 25. 108   27. $\frac{1}{6}$ 29. 25, -10

31. $\frac{1}{32}$, $-\frac{5}{64}$ 33. 16, $\frac{8}{3}$ 35. 48, $y - 64 = 48(x - 4)$ 37. $8x^7$ 39. $\frac{3}{4}x^{-1/4}$ 41. 0

43. $\frac{1}{5}x^{-4/5}$ 45. 4, $\frac{1}{3}$ 47. $a = 4$, $b = 1$ 49. 1, 1.5, 2   51. 6   53. $2x + 5$

57. $y - 5 = \frac{1}{2}(x - 4)$

Exercises 1.4, page 111

1. No limit   3. 1   5. No limit   7. -5 9. 5   11. No limit   13. 288   15. 0

17. 3   19. -4 21. -8 23. $\frac{6}{7}$ 25. No limit   27. $-\frac{2}{11}$ 29. 6   31. 3

33. $-\frac{2}{121}$ 35. $\frac{-\sqrt{3}}{6}$ 37. 0   39. 3   41. $f(x) = \sqrt{x}$; $a = 9$ 43. $f(x) = \frac{1}{x}$; $a = 10$

45. $f(x) = 3x^2 + 4$; $a = 1$ 47. 0   49. 0   51. 2

Exercises 1.5, page 119

1. No   3. Yes   5. No   7. No   9. Yes   11. No   13. Continuous, differentiable

15. Continuous, not differentiable   17. Continuous, not differentiable   19. Not continuous, not differentiable

21. $f(5) = 3$ 23. Not possible   25. $f(0) = 12$

Exercises 1.6, page 126

1. $3x^2 + 2x$ 3. $2x + 3$ 5. $5x^4 - \dfrac{1}{x^2}$ 7. $4x^3 + 3x^2 + 1$ 9. $6x$ 11. $3x^2 + 14x$

13. $-\dfrac{8}{x^3}$ 15. $3 + \dfrac{1}{x^2}$ 17. $x^2 - x$ 19. $\dfrac{1}{x^6}$ 21. $\dfrac{-1}{2\sqrt{x}}$ 23. $30(3x + 1)^9$

25. $\dfrac{45x^2 + 5}{2\sqrt{3x^3 + x}}$ 27. $6(4x - 1)(2x^2 - x + 4)^5$ 29. $\frac{1}{3} - 3x^{-2}$ 31. $10(1 - 5x)^{-2}$

33. $4x^3(1 - x^4)^{-2}$ **35.** $-2(x^2 + x)^{-3/2}(2x + 1)$ **37.** $\frac{3}{2}\left(\frac{\sqrt{x}}{2} + 1\right)^{1/2}\left(\frac{1}{4}x^{-1/2}\right)$ or $\frac{3}{8\sqrt{x}}\left(\frac{\sqrt{x}}{2} + 1\right)^{1/2}$

39. 4   **41.** 15   **43.** $f'(4) = 48, y = 48x - 191$ **45.** $f'(x) = 2(3x^2 + x - 2)(6x + 1) =$

$36x^3 + 18x^2 - 22x - 4$ **47.** 4.8; 1.8   **49.** 14; 11   **51.** 10; $\frac{15}{4}$ **53.** $(5, \frac{161}{3}), (3, 49)$

55. $f(4) = 5, f'(4) = \frac{1}{2}$

Exercises 1.7, page 132

1. $10t(t^2 + 1)^4$ **3.** $(2t - 1)^{-1/2}$ **5.** $\frac{2}{3}(T^3 + 5T)^{-1/3}(3T^2 + 5)$ **7.** $6P - \frac{1}{2}$ **9.** $2a^2t + b^2$

11. $f'(x) = x - 7, f''(x) = 1$ **13.** $y' = \frac{1}{2}x^{-1/2}, y'' = -\frac{1}{4}x^{-3/2}$ **15.** $f'(r) = 2\pi(hr + 1), f''(r) = 2\pi h$

17. $g'(x) = -5, g''(x) = 0$ **19.** $f'(P) = 15(3P + 1)^4, f''(P) = 180(3P + 1)^3$ **21.** 20   **23.** 54

25. 34   **27.** $8k(2P - 1)^{-3}$ **29.** $f'(3) = -\frac{1}{2}, f''(3) = -\frac{1}{8}$ **31.** 20   **33. (a)** $f'''(x) = 60x^2 - 24x$

(b) $f'''(x) = \dfrac{15}{2\sqrt{x}}$

Exercises 1.8, page 144

1. (a) 8; 4; 2   **(b)** 0   **3. (a)** 14   **(b)** 13   **5.** 63 units/h   **7. (a)** 1 g/week   **(b)** 25 weeks

9. (a) $-.015$ units/day, .0032 units/day   **(b)** increasing   **11.** True   **13.** True   **15. (a)** 5010

(b) 5005   **(c)** 4990   **(d)** 4980   **(e)** 4997.5   **17.** The manufacture of 2000 radios costs $50,000 and at that

level, the cost of manufacturing 1 additional radio is about $10.   **19.** The temperature of the coffee after 3 minutes

is 170° and is falling at the rate of 5°/minute.   **21.** An advertising expenditure of $100,000 results in the sale of

3,000,000 toys and, at that level, an additional dollar spent on advertising increases sales by about 30 toys.

23. (a) $.80 per unit   **(b)** 450 units   **25. (a)** $16.10   **(b)** $16 per unit   **27.** A–b, B–e, C–f, D–d, E–a,

F–c, G–g   **29.** 48 mi/hr, 40 mi/hr   **31. (a)** 160 ft/s   **(b)** 96 ft/s   **(c)** -32 ft/s^2 **(d)** after 10 s

(e) -160 ft/s   **33. (a)** 10 km/h   **(b)** 42 km   **(c)** after 2h   **35. (a)** 4400   **(b)** 4700   **(c)** 4100

(d) 4900   **37. (a)** An increase in the interest rate from 6% to 7% would produce about $168.95 additional

interest.   **(b)** $1824.64   **(c)** $1740.17   **39.** 2.002   **41. (a)** 60 ft   **(b)** 20 ft/sec   **(c)** 10 ft/sec^2

(d) 5.5 sec   **(e)** 7 sec   **(f)** 30 ft/sec, 4.5 sec

Chapter 1: Supplementary Exercises, page 151

1.
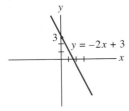

3.
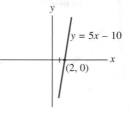

5.
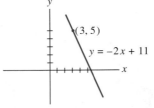

7.
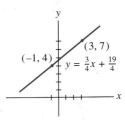

9.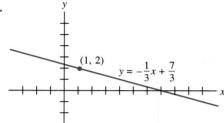

11.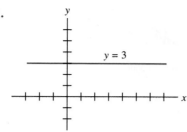

13.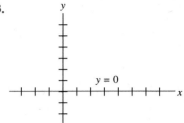

15. $7x^6 + 3x^2$ **17.** $\dfrac{3}{\sqrt{x}}$ **19.** $-\dfrac{3}{x^2}$

21. $48x(3x^2 - 1)^7$ **23.** $-\dfrac{5}{(5x - 1)^2}$ **25.** $\dfrac{x}{\sqrt{x^2 + 1}}$ **27.** $-\dfrac{1}{4x^{5/4}}$ **29.** 0

31. $10[x^5 - (x - 1)^5]^9 [5x^4 - 5(x - 1)^4]$ **33.** $\dfrac{3}{2}t^{-1/2} + \dfrac{3}{2}t^{-3/2}$ **35.** $\dfrac{2(9t^2 - 1)}{(t - 3t^3)^2}$ **37.** $\dfrac{9}{4}x^{1/2} - 4x^{-1/3}$

39. 28 **41.** 14, 3 **43.** $\frac{15}{2}$ **45.** 33 **47.** $4x^3 - 4x$ **49.** $-\frac{3}{2}(1 - 3P)^{-1/2}$ **51.** 29

53. $300(5x + 1)^2$ **55.** -2 **57.** $3x^{-1/2}$ **59.** Slope -4; tangent $y = -4x + 6$

61. 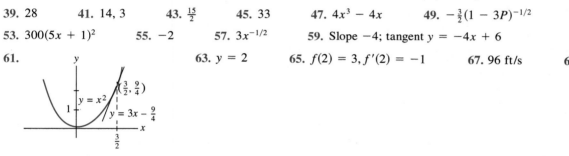 **63.** $y = 2$ **65.** $f(2) = 3, f'(2) = -1$ **67.** 96 ft/s **69.** 11 ft

71. $\frac{5}{3}$ ft/s **73.** Positive **(b)** limousine **(c)** $\frac{1}{5}$ **75.** $\frac{3}{4}$ in. **77.** 4 **79.** Does not exist

81. $-\frac{1}{50}$ **83.** The slope of a secant line at $(3, 9)$

CHAPTER 2
Exercises 2.1, page 167

1. (a), (e), (f) **3.** (b), (c), (d) **5.** Decreasing for $x < -2$, relative minimum point at $x= -2$, minimum value $= -2$, increasing for $x > -2$, concave up, y-intercept $(0, 0)$, x-intercepts $(0, 0)$ and $(-3.6, 0)$. **7.** Decreasing for $x < 0$, relative minimum point at $x = 0$, increasing for $0 < x < 2$, relative maximum point at $x = 2$, decreasing for $x > 2$, concave up for $x < 1$, concave down for $x > 1$, inflection point at $(1, 3)$, y-intercept $(0, 2)$, x-intercept $(3.4, 0)$.

9. Decreasing for $x < 2$, relative minimum at $x = 2$, minimum value $= 3$, increasing for $x > 2$, concave up for all x, no inflection points, defined for $x > 0$, the line $y = x$ is an asymptote, the y-axis is an asymptote.

11. Decreasing for $1 \le x < 3$, relative minimum point at $x = 3$, increasing for $x > 3$, maximum value $= 6$ (at $x = 1$), minimum value $= 1$ (at $x = 3$), inflection point at $x = 4$, concave up for $1 \le x < 4$, concave down for $x > 4$, the line $y = 4$ is an asymptote. **13.** Slope increases for all x. **15.** Slope decreases for $x < 3$, increases for $x > 3$. Minimum slope occurs at $x = 3$.   **17.** (a) C, F   (b) A, B, F   (c) C

19.
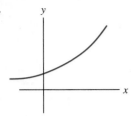

21.
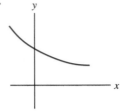

23.
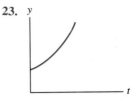

25.
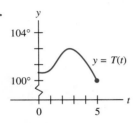

27.
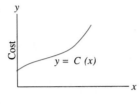

29. Oxygen content decreases until time a, at which time it reaches a minimum. After a, oxygen content steadily increases. The rate of increase increases until b, and then decreases. Time b is the time when oxygen content is increasing fastest.   **31.** 1960   **33.** The parachutist's speed levels off to 15 ft/s.

35.
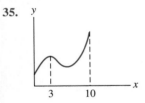

37.
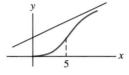

39. (a) Yes   (b) Yes   **41.** Relatively low

Exercises 2.2, page 180

1. (b), (c), (f)   **3.** (d), (e), (f)   **5.** (d)

7.
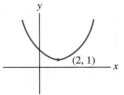

9.
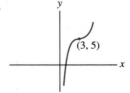

11.
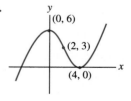

13.
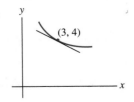

15.
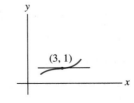

17.
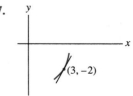

19.

	f	f'	f''
A	Pos.	Pos.	Neg.
B	0	Neg.	0
C	Neg.	0	Pos.

21. $t = 1$ **23.** The slope is positive because $f'(6) = 2$, a positive number.   **25.** The slope is 0 because $f'(3) = 0$. Also, $f'(x)$ is positive for x slightly less than 3, and $f'(x)$ is negative for x slightly greater than 3. Hence $f(x)$ changes from increasing to decreasing at $x = 3$.   **27.** $f'(x)$ is increasing at $x = 0$, so the graph of $f(x)$ is concave up.   **29.** At $x = 1$, $f'(x)$ changes from increasing to decreasing, so the slope of the graph of $f(x)$ changes from increasing to decreasing.   **31.** $y - 3 = 2(x - 6)$ **33.** 3.25   **35. (a)** $\frac{1}{6}$ in.   **(b)** ii, because the water level is falling.   **37.** The second curve   **39.** The second curve   **41. (a)** 3   **(b)** $t = 4.5, 6$ **(c)** $t = 1$ **(d)** $t = 5$ **(e)** 1   **(f)** $t = 2.5, 3.5$ **(g)** $t = 3$ **(h)** $t = 7$ **43. (a)** decreasing   **(b)** (2, 9)   **(c)** (10, 1)   **(d)** concave down   **(e)** $x = 6$ **(f)** (15, 14)   **45. (a)** 75,000 per year   **(b)** 1945   **(c)** 1950, 1975   **(d)** 1961   **47. (a)** 900 trillion kilowatt-hours   **(b)** 35 trillion kilowatt-hours/yr   **(c)** 1980   **(d)** 1935   **(e)** 1970, 1600 trillion kilowatt-hours

Exercises 2.3, page 195

1.
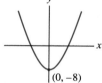
$(0, -8)$

3.
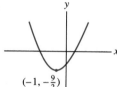
$(-1, -\frac{9}{2})$

5.
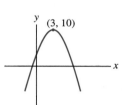
$(3, 10)$

7.
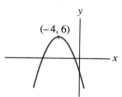
$(-4, 6)$

9.
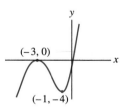
$(-3, 0)$ $(-1, -4)$

11.
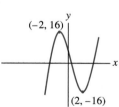
$(-2, 16)$ $(2, -16)$

13.
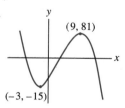
$(9, 81)$ $(-3, -15)$

15.
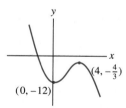
$(4, -\frac{4}{3})$ $(0, -12)$

17.
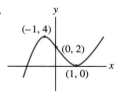
$(-1, 4)$ $(0, 2)$ $(1, 0)$

19.
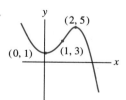
$(2, 5)$ $(0, 1)$ $(1, 3)$

21.
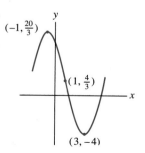
$(-1, \frac{20}{3})$ $(1, \frac{4}{3})$ $(3, -4)$

23.
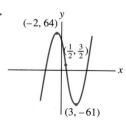
$(-2, 64)$ $(\frac{1}{2}, \frac{3}{2})$ $(3, -61)$

25. No, $f''(x) = 2a \neq 0$ **27.** $(4, 3)$ min   **29.** $(1, 5)$ max   **31.** $(-.1, -3.05)$ min   **33.** $f(x) = g'(x)$

35. (a) f has a relative minimum   **(b)** f has an inflection point   **37. (a)** 70%   **(b)** 1984   **(c)** increasing at 5% per year   **(d)** 1990   **(e)** 1987, 40%

Exercises 2.4, page 206

1. $\left(\dfrac{3 \pm \sqrt{5}}{2}, 0\right)$ **3.** $(-2, 0)$, $\left(-\dfrac{1}{2}, 0\right)$ **5.** $\left(\dfrac{1}{2}, 0\right)$ **7.** The derivative $x^2 - 4x + 5$ has no zeros.

9.
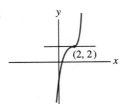

11.
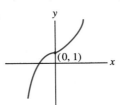

13.
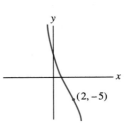

15.
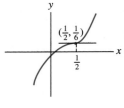

17.
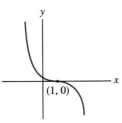

19.
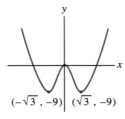

21.
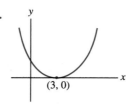

23.
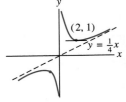

25.
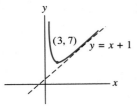

27.
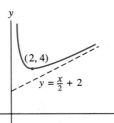

29.
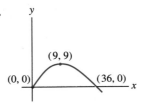

31. $g(x) = f'(x)$

Exercises 2.5, page 214

1. 20   **3.** $t = 4, f(4) = 8$ **5. (a)** Objective equation: $A = xy$, constraint equation: $8x + 4y = 320$

(b) $A = -2x^2 + 80x$ **(c)** $x = 20$ ft, $y = 40$ ft   **7. (a)**

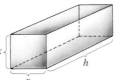

(b) $h + 4x$ **(c)** Obj.:

$V = x^2 h$; con.: $h + 4x = 84$ **(d)** $V = -4x^3 + 84x^2$ **(e)** $x = 14$ in., $h = 28$ in.   **9.** Let x be the length of the fence and y the other dimension. Obj.: $C = 15x + 20y$; con.: $xy = 75$; $x = 10$ ft, $y = 7.5$ ft

11. Let x be the length of each edge of the base and h the height. Obj.: $A = 2x^2 + 4xh$; con.: $x^2h = 8000$; 20 cm by 20 cm by 20 cm **13.** Let x be the length of the fence parallel to the river and y the length of each section perpendicular to the river. Obj.: $A = xy$; con.: $6x + 15y = 1500$; $x = 125$ ft, $y = 50$ ft **15.** Obj.: $P = xy$; con.: $x + y = 100$; $x = 50$, $y = 50$ **17.** Obj.: $A = \dfrac{\pi x^2}{2} + 2xh$; con.: $(2 + \pi)x + 2h = 14$; $x = \dfrac{14}{4 + \pi}$ ft **21.** $(3/2, \sqrt{3/2})$

Exercises 2.6, page 226

1. Let x be the number of prints and p the price per print. Obj.: $R = px$; con.: $p = 650 - 5x$; 65 prints **3.** Let x be the number of tables and p the profit per table. Obj.: $P = px$; con.: $p = 16 - (x/2)$; 16 tables **5.** Let x be the number of cases per order and r the number of orders per year. (a) \$4100 (b) Obj.: $C = 80r + 5x$; con.: $rx = 10{,}000$; 400 cases **7.** Let r be the number of production runs and x the number of microscopes manufactured per run. Obj.: $C = 2500r + 25x$; con.: $rx = 1600$; 4 runs **11.** Obj.: $A = (100 + x)w$; con.: $2x + 2w = 300$; $x = 25$ ft, $w = 125$ ft **13.** Obj.: $F = 2x + 3w$; con.: $xw = 54$; $x = 9$ m, $w = 6$ m **15.** Let x be the number of people and c the cost. Obj.: $R = xc$; con.: $c = 1040 - 20x$; 25 people **17.** Let x be the length of each edge of the base and h the height. Obj.: $C = 6x^2 + 10xh$; con.: $x^2h = 150$; 5 ft by 5 ft by 6 ft **19.** Let x be the length of each edge of the end and h the length. Obj.: $V = x^2h$; con.: $2x + h = 120$; 40 cm by 40 cm by 40 cm **21.** Obj.: $V = w^2x$; con.: $2x + w = 16$; $\frac{8}{3}$ in. **23.** $t = 20$ **25.** $2\sqrt{3}$ by 6 **27.** $10 \times 10 \times 4$

Exercises 2.7, page 240

1. \$1 **3.** 32 **5.** 5 **7.** $x = 20$ units, $p = \$133.33$ **9.** 2 million tons, \$156 per ton **11.** (a) \$2.00 (b) \$2.30 **13.** (a) $x = 15 \cdot 10^5$, $p = \$45$. (b) No. Profit is maximized when price is increased to \$50. **15.** 5% **17.** (a) \$79 thousand (b) \$2 thousand (c) 5 (d) 34 (e) 28

Chapter 2: Supplementary Exercises, page 244

1. b **3.** **5.** **7.** d, e **9.** c, d **11.** e

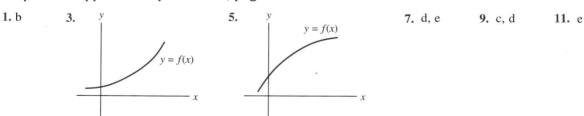

13. Graph goes through $(1, 2)$, increasing at $x = 1$. **15.** Increasing and concave up at $x = 3$. **17.** $(10, 2)$ is a relative minimum point. **19.** Graph goes through $(5, -1)$, decreasing at $x = 5$. **21.** $(-2, 0)$ is a relative maximum point.

23.

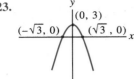
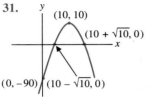

25.

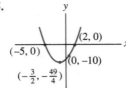
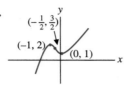

27.

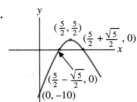
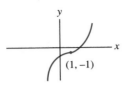

29.

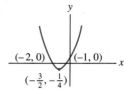
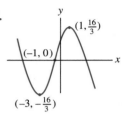

31.

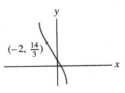

33.

35.

37.

39.

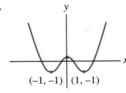

41.

43.

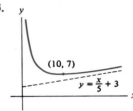

45. $f'(x) = 3x(x^2 + 2)^{1/2}, f'(0) = 0$ **47.** $f''(x) = -2x(1 + x^2)^{-2}$, $f''(x)$ is positive for $x < 0$ and negative for $x > 0$.   **49.** A–c, B–e, C–f, D–b, E–a, F–d   **51. (a)** The number of people living between $10 + h$ and 10 mi from the center of the city.   **(b)** If so, $f(x)$ would be decreasing at $x = 10$.   **53.** The endpoint maximum value of 2 occurs at $x = 0$.   **55.** Let x be the width and h the height. Obj.: $A = 4x + 2xh + 8h$; con.: $4xh = 200$; 4 by 10 by 5   **57.** $\frac{30}{4}$ in.   **59.** Let r be the number of production runs and x the number of books manufactured per run. Obj.: $C = 1000r + (.25)x$; con.: $rx = 400,000$; $x = 40,000$ **61.** $x = 3500$

CHAPTER 3
Exercises 3.1, page 257

1. $(x + 1) \cdot (3x^2 + 5) + (x^3 + 5x + 2) \cdot 1$, or $4x^3 + 3x^2 + 10x + 7$ **3.** $(3x^2 - x + 2) \cdot 4x + (2x^2 - 1) \cdot (6x - 1)$, or $24x^3 - 6x^2 + 2x + 1$ **5.** $(2x - 7) \cdot 5(x - 1)^4(1) + (x - 1)^5 \cdot 2$, or $(x - 1)^4(12x - 37)$
7. $(x^2 + 3) \cdot 10(x^2 - 3)^9(2x) + (x^2 - 3)^{10} \cdot 2x$, or $2x(x^2 - 3)^9(11x^2 + 27)$
9. $\frac{1}{3}(4 - x)^3 \cdot 3(4 + x)^2(1) + (4 + x)^3 \cdot (4 - x)^2(-1)$, or $-2x(4 - x)^2(4 + x)^2$
11. $\dfrac{(4 + x) \cdot (-1) - (4 - x) \cdot 1}{(4 + x)^2}$, or $\dfrac{-8}{(4 + x)^2}$ **13.** $\dfrac{(x^2 + 1) \cdot 2x - (x^2 - 1) \cdot 2x}{(x^2 + 1)^2}$, or $\dfrac{4x}{(x^2 + 1)^2}$
15. $(-1)(5x^2 + 2x + 5)^{-2}(10x + 2)$ **17.** $\dfrac{(x + 1) \cdot (2x + 2) - (x^2 + 2x) \cdot 1}{(x + 1)^2}$, or $\dfrac{x^2 + 2x + 2}{(x + 1)^2}$

19. $\dfrac{(3 - x^2) \cdot (6x + 5) - (3x^2 + 5x + 1) \cdot (-2x)}{(3 - x^2)^2}$, or $\dfrac{5(x + 1)(x + 3)}{(3 - x^2)^2}$ **21.** $\dfrac{(x^2 + 1)^2 \cdot 1 - x \cdot 2(x^2 + 1)(2x)}{(x^2 + 1)^4}$,

or $\dfrac{1 - 3x^2}{(x^2 + 1)^3}$ **23.** $\dfrac{(x + 2)^2 \cdot 4(x - 1)^3(1) - (x - 1)^4 \cdot 2(x + 2)(1)}{(x + 2)^4}$, or $\dfrac{2(x - 1)^3(x + 5)}{(x + 2)^3}$

25. $\dfrac{x(4x^3 - 8x) - (x^4 - 4x^2 + 3) \cdot 1}{x^2}$, or $\dfrac{3x^4 - 4x^2 - 3}{x^2}$ **27.** $2x^{1/2} \cdot 3(3x^2 - 1)^2(6x) + (3x^2 - 1)^3 \cdot x^{-1/2}$, or

$x^{-1/2}(3x^2 - 1)^2(39x^2 - 1)$

29. $(x + 3) \cdot \frac{1}{2}(2x - 3)^{-1/2}(2) + (2x - 3)^{1/2} \cdot 1$, or $3x(2x - 3)^{-1/2}$ **31.** $y - 16 = 88(x - 3)$ **33.** $0, \pm 2, \pm \frac{5}{4}$

35. $2, 7$ **37.** $(\frac{1}{2}, \frac{3}{2}), (-\frac{1}{2}, \frac{9}{2})$ **39.** $2 \times 3 \times 1$ **41.** $150; AC(150) = 35 = C'(150)$

43. AR is maximized where $0 = \dfrac{d}{dx}(AR) = \dfrac{x \cdot R'(x) - R(x) \cdot 1}{x^2}$. This happens when the production level x satisfies

$xR'(x) - R(x) = 0$, and hence $R'(x) = R(x)/x = AR$. **45.** $38 \text{ in.}^2/\text{s}$ **49.** $\dfrac{1 - 2x \cdot f(x)}{(1 + x^2)^2}$

51. $\frac{1}{8}$ **55.** $f(x)g(x)h'(x) + f(x)g'(x)h(x) + f'(x)g(x)h(x)$

Exercises 3.2, page 265

1. $\dfrac{x^3}{x^3 + 1}$ **3.** $(x^2 + 4)^5 + 3(x^2 + 4)$ **5.** $f(x) = x^5, g(x) = x^3 + 8x - 2$ **7.** $f(x) = \sqrt{x}$,

$g(x) = 4 - x^2$ **9.** $f(x) = \dfrac{1}{x}, g(x) = x^3 - 5x^2 + 1$ **11.** $30x(x^2 + 5)^{14}$ **13.** $6x^2 \cdot 3(x - 1)^2(1) +$

$(x - 1)^3 \cdot 12x$, or $6x(x - 1)^2(5x - 2)$ **15.** $2(x^3 - 1) \cdot 4(3x^2 + 1)^3(6x) + (3x^2 + 1)^4 \cdot 2(3x^2)$, or

$6x(3x^2 + 1)^3(11x^3 + x - 8)$

17. $\dfrac{d}{dx} 4^3(1 - x)^{-3} = 192(1 - x)^{-4}$ **19.** $3\left(\dfrac{4x - 1}{3x + 1}\right)^2 \cdot \dfrac{(3x + 1) \cdot 4 - (4x - 1) \cdot 3}{(3x + 1)^2}$, or $\dfrac{21(4x - 1)^2}{(3x + 1)^4}$

21. $3\left(\dfrac{4 - x}{x^2}\right)^2 \cdot \dfrac{x^2 \cdot (-1) - (4 - x) \cdot 2x}{x^4}$, or $\dfrac{3(4 - x)^2(x - 8)}{x^7}$ **23.**

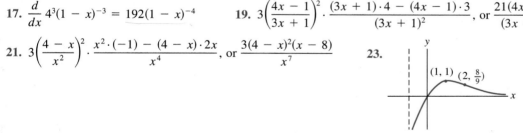

25. $5(6x - 1)^4 \cdot 6$ **27.** $-(1 - x^2)^{-2} \cdot (-2x)$, or $2x(1 - x^2)^{-2}$ **29.** $[4(x^2 - 4)^3 - 2(x^2 - 4)] \cdot (2x)$, or

$8x(x^2 - 4)^3 - 4x(x^2 - 4)$ **31.** $2((x^2 + 5)^3 + 1)3(x^2 + 5)^2 \cdot (2x)$, or $12x[(x^2 + 5)^3 + 1](x^2 + 5)^2$

33. $6(4x + 1)^{1/2}$ **35.** $[-8(x - 3x^2)^{-3} + \frac{1}{2}(x - 3x^2)] \cdot (1 - 6x)$ **37.** $(x^2 + x + 1)^4(6x^2 + 6x + 1)(2x + 1)$

39. $\dfrac{2}{(3 + \sqrt{x})^2} \cdot \dfrac{1}{2\sqrt{x}}$, or $\dfrac{1}{\sqrt{x}(3 + \sqrt{x})^2}$ **41.** $y = 62x - 300$ **43.** $1; 2; 3$ **45. (a)** $\dfrac{dV}{dt} = \dfrac{dV}{dx} \cdot \dfrac{dx}{dt}$

(b) 2 **47. (a)** $\dfrac{dy}{dt}, \dfrac{dP}{dy}, \dfrac{dP}{dt}$ **(b)** $\dfrac{dP}{dt} = \dfrac{dP}{dy} \cdot \dfrac{dy}{dt}$ **49. (a)** $\dfrac{200(100 - x^2)}{(100 + x^2)^2}$ **(b)** $\dfrac{200[100 - (4 + 2t)^2]}{[100 + (4 + 2t)^2]^2} \cdot (2)$

(c) Falling at the rate of $480 per week   **51. (a)** $.4 + .0002x$ **(b)** increasing at the rate of 25 thousand

persons per year   **(c)** rising at the rate of 14 ppm per year   **53.** $x^3 + 1$ **55.** 24

Exercises 3.3, page 277

1. $\dfrac{x}{y}$ **3.** $\dfrac{1 + 6x}{5y^4}$ **5.** $\dfrac{2x^3 - x}{2y^3 - y}$ **7.** $\dfrac{1 - 6x^2}{1 - 6y^2}$ **9.** $-\dfrac{y}{x}$ **11.** $-\dfrac{y + 2}{5x}$ **13.** $\dfrac{8 - 3xy^2}{2x^2y}$

15. $\dfrac{x^2(y^3 - 1)}{y^2(1 - x^3)}$ **17.** $-\dfrac{y^2 + 2xy}{x^2 + 2xy}$ **19.** $\frac{1}{2}$ **21.** $-\frac{8}{3}$ **23.** $-\frac{2}{15}$ **25.** $y - \frac{1}{2} = -\frac{1}{16}(x - 4)$,

$y + \frac{1}{2} = \frac{1}{16}(x - 4)$ **27. (a)** $\dfrac{2x - x^3 - xy^2}{2y + y^3 + x^2y}$ **(b)** 0   **29.** $-\frac{27}{16}$ **31.** $-\dfrac{x^3}{y^3}\dfrac{dx}{dt}$

33. $\dfrac{2x - y}{x}\dfrac{dx}{dt}$ **35.** $\dfrac{2x + 2y}{3y^2 - 2x}\dfrac{dx}{dt}$ **37.** $-\frac{15}{8}$ units per second   **39.** Rising at 3 thousand units per week

41. Increasing at $20 thousand per month   **43.** Decreasing at $\frac{1}{14}$ L per second   **45. (a)** $x^2 + y^2 = 100$

(b) $\dfrac{dy}{dt} = -4$, so the top of the ladder is falling at the rate of 4 ft/s.   **47.** $\dfrac{22}{\sqrt{5}}$ ft/s (or 9.84 ft/s)

Chapter 3: Supplementary Exercises, page 281

1. $(4x - 1) \cdot 4(3x + 1)^3(3) + (3x + 1)^4 \cdot 4$, or $4(3x + 1)^3(15x - 2)$ **3.** $x \cdot 3(x^5 - 1)^2 \cdot 5x^4 + (x^5 - 1)^3 \cdot 1$, or

$(x^5 - 1)^2(16x^5 - 1)$ **5.** $5(x^{1/2} - 1)^4 \cdot 2(x^{1/2} - 2)(\frac{1}{2}x^{-1/2}) + (x^{1/2} - 2)^2 \cdot 20(x^{1/2} - 1)^3(\frac{1}{2}x^{-1/2})$, or

$5x^{-1/2}(x^{1/2} - 1)^3(x^{1/2} - 2)(3x^{1/2} - 5)$

7. $3(x^2 - 1)^3 \cdot 5(x^2 + 1)^4(2x) + (x^2 + 1)^5 \cdot 9(x^2 - 1)^2(2x)$, or $12x(x^2 - 1)^2(x^2 + 1)^4(4x^2 - 1)$

9. $\dfrac{(x - 2) \cdot (2x - 6) - (x^2 - 6x) \cdot 1}{(x - 2)^2}$, or $\dfrac{x^2 - 4x + 12}{(x - 2)^2}$ **11.** $2\left(\dfrac{3 - x^2}{x^3}\right) \cdot \dfrac{x^3 \cdot (-2x) - (3 - x^2) \cdot 3x^2}{x^6}$, or

$\dfrac{2(3 - x^2)(x^2 - 9)}{x^7}$ **13.** $-\frac{1}{3}, 3, \frac{31}{27}$ **15.** $y + 32 = 176(x + 1)$ **17.** $x = 44$ m, $y = 22$ m

19. $\dfrac{dC}{dt} = \dfrac{dC}{dx} \cdot \dfrac{dx}{dt} = 40 \cdot 3 = 120$. Costs are rising $120 per day.   **21.** $0; -\frac{7}{2}$ **23.** $\frac{3}{2}; -\frac{7}{8}$ **25.** $1; -\frac{3}{2}$

27. $\dfrac{3x^2}{x^6 + 1}$ **29.** $\dfrac{2x}{(x^2 + 1)^2 + 1}$ **31.** $\frac{1}{2}\sqrt{1 - x}$ **33.** $\dfrac{3x^2}{2(x^3 + 1)}$

35. $\dfrac{5/x}{(5/x)^2 + 1} \cdot \left(-\dfrac{5}{x^2}\right)$, or $-\dfrac{25}{x(25 + x^2)}$ **37.** $\dfrac{x^{1/2}}{(1 + x^2)^{1/2}} \cdot \dfrac{1}{2}(x^{-1/2})$, or $\dfrac{1}{2\sqrt{1 + x^2}}$

39. (a) $\dfrac{dR}{dA}, \dfrac{dA}{dt}, \dfrac{dR}{dx}$, and $\dfrac{dx}{dA}$ **(b)** $\dfrac{dR}{dt} = \dfrac{dR}{dx}\dfrac{dx}{dA}\dfrac{dA}{dt}$ **41. (a)** $-y^{1/3}/x^{1/3}$ **(b)** 1   **43.** -3 **45.** $\frac{3}{5}$

47. (a) $\dfrac{dy}{dx} = \dfrac{15x^2}{2y}$ **(b)** $\frac{20}{3}$ thousand dollars per thousand unit increase in production   **(c)** $\dfrac{dy}{dt} = \dfrac{15x^2}{2y}\dfrac{dx}{dt}$

(d) 2 thousand dollars per week   **49.** Increasing at the rate of 2.5 units per unit time   **51.** 1.89 m²/year

CHAPTER 4

Exercises 4.1, page 291

1. 2^{2x}, $3^{(1/2)x}$, 3^{-2x} 3. 2^{2x}, 3^{3x}, 2^{-3x} 5. 2^{-4x}, 2^{9x}, 3^{-2x} 7. $2^{(1/2)x}$, $3^{(4/3)x}$ 9. 2^{-2x}, 3^x 11. 3^{2x}, 2^{6x}, 3^{-x} 13. 2^x, 3^x, 3^x 15. (a) 8   (b) $\frac{1}{8}$ (c) 5.66   (d) 17.15   (e) 1.15   (f) 1.87   (g) .18   (h) .07   17. 1   19. 2   21. -1 23. $\frac{1}{5}$ 25. $\frac{5}{2}$ 27. -1 29. 4   31. 2^h 33. $2^h - 1$ 35. $3^x + 1$ 37. $3^{5x} + 1$

Exercises 4.2, page 297

1. 1.1612, 1.105, 1.10   3. 1.005, 1.002, 1   5. $10e^{10x}$ 7. $4e^{4x}$ 9. $e^{2x} + e^{5x}$ 11. e^{1+x} 13. e^{2x} 15. 7.3891   17. .60653   19. $x = 4$ 21. $x = 4, -2$ 23. $xe^x + e^x$

25. $\dfrac{e^x}{(1 + e^x)^2}$ 27. $20e^x(1 + 5e^x)^3$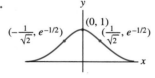

Exercises 4.3, page 304

1. $-e^{-x}$ 3. $5e^x$ 5. $2te^{t^2}$ 7. $\dfrac{e^x + e^{-x}}{2}$ 9. $-2(e^{-2x} + 1)$ 11. $3(e^x + e^{-x})^2(e^x - e^{-x})$

13. $-\frac{2}{3}e^{3-2x}$ 15. $3e^{3t}$ 17. $\left(3x^2 + 1 + \dfrac{1}{x^2}\right)e^{x^3+x-(1/x)}$ 19. $4(2x + 1 - e^{2x+1})^3(2 - 2e^{2x+1})$

21. $3x^2e^{x^2} + 2x^4e^{x^2}$ 23. $3xe^{x^3} - x^{-2}e^{x^3}$ 25. $-xe^{-x+2}$ 27. $\left(-\dfrac{1}{x^2} + \dfrac{1}{x} + 3\right)e^x$ 29. $\dfrac{2e^x}{(e^x + 1)^2}$

31. Max at $x = 1$ 33. Min at $x = \frac{11}{2}$ 35. Max at $x = 3$ 37. Min at $x = 1$; max at $x = 3$ 39. Max at $x = -6$; min at $x = -5$ 41. $.02e^{-2e^{-.01x}}e^{-.01x}$ 43. \$54,365.64 per year   45. $y = Ce^{-4x}$

47. $y = e^{-.5x}$ 51.   53. 1

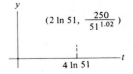

Exercises 4.4, page 310

1. -1 3. $-\ln 1.7$ 5. $e^{2.2}$ 7. 2   9. e 11. 1   13. $\frac{1}{2}\ln 5$ 15. $4 - e^{1/2}$

17. $\pm e^3$ 19. $\dfrac{\ln(.5)}{-.00012}$ 21. $\frac{3}{5}$ 23. $\dfrac{e}{2}$ 25. $3\ln\frac{9}{2}$ 27. $5\ln 6$ 29. $\frac{1}{5}\ln\frac{2}{5}$

31. $-\ln\frac{3}{2}$ 33. $(-\ln 3, 3 - 3\ln 3)$, minimum   35. $(\frac{1}{2}\ln\frac{3}{2}, \frac{1}{2})$, minimum   37. Max at $t = 2\ln 51$

39. 109.947

Exercises 4.5, page 314

1. $\dfrac{1}{x}$ **3.** $\dfrac{1}{x+5}$ **5.** $-\dfrac{\ln(x+1)}{x^2}+\dfrac{1}{x(x+1)}$ **7.** $\left(\dfrac{1}{x}+1\right)e^{\ln x+x}$ **9.** $\dfrac{1}{x}$ **11.** $\dfrac{2\ln x}{x}+\dfrac{1}{x}$

13. $\dfrac{1}{x}$ **15.** $\dfrac{\ln x-2}{(\ln x)^3}$ **17.** $2e^{2x}\ln x+\dfrac{e^{2x}}{x}$ **19.** $\dfrac{5e^{5x}}{e^{5x}+1}$ **21.** $2(\ln 4)t$ **23.** $\dfrac{6\ln t-3(\ln t)^2}{t^2}$

25. $y=1$ **27.** $\left(e^2,\dfrac{2}{e}\right)$, yes   **29.**

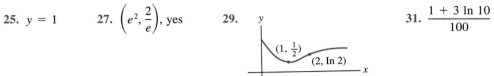

31. $\dfrac{1+3\ln 10}{100}$

33. $P(x)=300\ln(x+1)-2x$ and $P'(149)=0$. Since $P''(149)<0$, the graph of $P(x)$ is concave down at $x=149$.

So $P(x)$ has a relative maximum point there.   **35.** $\dfrac{1}{e}$

Exercises 4.6, page 319

1. $\ln 5x$ **3.** $\ln 3$ **5.** $\ln 2$ **7.** x^2 **9.** $\ln\dfrac{x^5z^3}{\sqrt{y}}$ **11.** $3\ln x$ **13.** $3\ln 3$ **15.** d

17. d   **19.** $\dfrac{1}{x+5}+\dfrac{2}{2x-1}-\dfrac{1}{4-x}$ **21.** $\dfrac{1}{x+1}+\dfrac{3}{3x-2}-\dfrac{1}{x+2}$ **23.** $\dfrac{1}{2x}-\dfrac{2x}{x^2+1}$

25. $(x+1)^3(4x-1)^2\left[\dfrac{3}{x+1}+\dfrac{8}{4x-1}\right]$ **27.** $(x-2)^3(x-3)^5(x+2)^{-7}\left[\dfrac{3}{x-2}+\dfrac{5}{x-3}-\dfrac{7}{x+2}\right]$

29. $x^x[1+\ln x]$ **31.** $e^x\sqrt{x^2-1}\left[1+\dfrac{x}{x^2-1}\right]$ **33.** $x^{\ln x}\cdot\dfrac{2\ln x}{x}$

35. $\dfrac{\sqrt{x-1}\,(x-2)}{x^2-3}\cdot\left[\dfrac{1}{2}\cdot\dfrac{1}{x-1}+\dfrac{1}{x-2}-\dfrac{2x}{x^2-3}\right]$ **37.** $y=cx^k$ **39.** $h=3,\ k=\ln 2$

Chapter 4: Supplementary Exercises, page 322

1. 81   **3.** $\frac{1}{25}$ **5.** 4   **7.** 9   **9.** e^{3x^2} **11.** e^{2x} **13.** $e^{11x}+7e^x$ **15.** $x=4$

17. $x=-5$ **19.** $70e^{7x}$ **21.** $e^{x^2}+2x^2e^{x^2}$ **23.** $e^x\cdot e^{e^x}=e^{x+e^x}$

25. $\dfrac{(2x-1)(e^{3x}+3)-3e^{3x}(x^2-x+5)}{(e^{3x}+3)^2}$ **27.** $y=Ce^{-t}$ **29.** $y=2e^{1.5t}$

31.

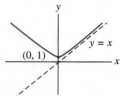

33.

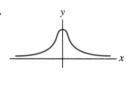

35. $\frac{2}{3}$ **37.** 1   **39.** $\sqrt{5}$

41. $\frac{1}{2}\ln 5$ **43.** $e^{5/2}$ **45.** $e,\dfrac{1}{e}$ **47.** $\dfrac{5}{5x-7}$ **49.** $\dfrac{2\ln x}{x}$ **51.** $\dfrac{6x^5+12x^3}{x^6+3x^4+1}$

53. $\frac{1}{x} + 1 - \frac{1}{2(1 + x)}$ **55.** $\frac{1}{x \ln x}$ **57.** $\ln x$ **59.** $\frac{e^x}{x} + e^x \ln x$

61. $(x^2 + 5)^6(x^3 + 7)^8(x^4 + 9)^{10}\left[\dfrac{12x}{x^2 + 5} + \dfrac{24x^2}{x^3 + 7} + \dfrac{40x^3}{x^4 + 9}\right]$ **63.** $10^x \ln 10$ **65.**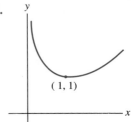

67.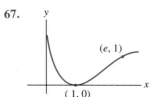

CHAPTER 5

Exercises 5.1, page 334

1. $P(t) = 400e^{.07t}$ **3. (a)** 5000   **(b)** $t = 6.9$ **5.** .017   **7. (a)** $P(t) = 5.4e^{.0263t}$ **(b)** 8.9 billion

(c) 2000   **9.** 71 min   **11.** a–F, b–D, c–A, d–G, e–H, f–C, g–B, h–E   **13. (a)** $P(t) = 30e^{-.08t}$

(b) 13.48 g   **15. (a)** $P(t) = 100e^{-.01t}$ **(b)** 74.082 g   **(c)** 69 y   **17. (a)** .13   **(b)** 7.7105 g

19. 58.275%   **21.** 8990 y   **23.** $f(t) = e^{-.081t}$ **25.** 184 y   **27.** 20,022 y   **29. (a)** $P(t) =$

$500e^{-.23t}$ **(b)** 10 mo   **33.** \$161,100   **35. (a)** 25 gm   **(b)** 3 years   **(c)** 15 gm/yr   **(d)** 6 yrs

Exercises 5.2, page 344

1. \$1210   **3.** $10,000(1.02)^{12}$ **5.** \$2316.37   **7.** \$616.83   **9.** 11.45%   **11.** 39% interest

compounded continuously   **13.** 8.9 yr   **15.** 1996   **17.** \$446.26   **19.** 7.25%

21. \$5488.10   **23.** 29%   **25.** a–B, b–D, c–G, d–A, e–F, f–E, g–H, h–C

Exercises 5.3, page 353

1. 20%, 4%   **3.** 30%, 30%   **5.** 60%, 300%   **7.** $-25\%, -10\%$ **9.** 12.5%   **11.** 5.8 yr

13. $p/(140 - p)$, elastic   **15.** $2p^2/(116 - p^2)$, inelastic   **17.** $p - 2$, elastic

19. (a) Inelastic   **(b)** raised   **21. (a)** Elastic   **(b)** increase   **23. (a)** 2   **(b)** yes

Exercises 5.4, page 367

1. (a) $f'(x) = 10e^{-2x} > 0, f(x)$ increasing, $f''(x) = -20e^{-2x} < 0, f(x)$ concave down   **(b)** As x becomes large,

$e^{-2x} = \dfrac{1}{e^{2x}}$ approaches 0   **(c)**

$$\begin{array}{l} y \\ \text{------} \; y = 5 \\ y = 5(1 - e^{-2x}) \\ \hline \qquad\qquad x \end{array}$$

3. $y' = 2e^{-x} = 2 - (2 - 2e^{-x}) = 2 - y$

5. $y' = 30e^{-10x} = 30 - (30 - 30e^{-10x}) = 30 - 10y = 10(3 - y), f(0) = 3(1 - 1) = 0$ **7.** 4.8 h

11. (a) 2500   **(b)** 1000 people/day   **(c)** day 12   **(d)** day 6 and day 13.5   **(e)** at time 9.78 days   **13. (a)** 2 cm

(b) 2 cm/wk   (c) week 5   (d) week 3 and week 10   (e) 6.4 weeks, 15 cm   15. (a) 400°

(b) decreasing at a rate of 100°/sec   (c) 16 sec   (d) 2 sec

Chapter 5: Supplementary Exercises, page 371

1. $29.92e^{-.2x}$ 3. $4$5488.12$ 5. .058   7. (a) $14.2e^{.018t}$ (b) 20.4 million   (c) 2011

9. (a) $36,693   (b) The alternative investment is superior by $3859.   11. 400%   13. 3%, decrease

15. Increase   17. $100(1 - e^{-.083t})$ 19. a–D, b–G, c–E, d–B, e–H, f–F, g–A, h–C

CHAPTER 6

Exercises 6.1, page 382

1. $\frac{1}{2}x^2 + C$ 3. $\frac{1}{3}e^{3x} + C$ 5. $3x + C$ 7. $-\frac{1}{4}$ 9. $\frac{2}{3}$

11. -2 13. $-\frac{5}{2}$ 15. $\frac{1}{2}$ 17. -1 19. 1 21. $\frac{1}{15}$

23. $\dfrac{x^3}{3} - \dfrac{x^2}{2} - x + C$ 25. $4\sqrt{x} - 2x^{3/2} + C$ 27. $4t + e^{-5t} + \dfrac{e^{2t}}{6} + C$ 29. $\frac{2}{5}t^{5/2} + C$ 31. C

33. $\dfrac{x^2}{2} + 3$ 35. $\frac{2}{3}x^{3/2} + x - \frac{28}{3}$ 37. $2 \ln x + 2$ 39. Testing all three functions reveals that (b) is the

only one that works.   41.

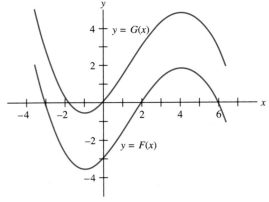

43. $\frac{1}{4}$ 45. (a) $-16t^2 + 96t + 256$ (b) 8 s   (c) 400 ft   47. $P(t) = 60t + t^2 - \frac{1}{12}t^3$

49. $20 - 25e^{-.4t}$ °C   51. $-95 + 1.3x + .03x^2 - .0006x^3$ 53. $5875(e^{.016t} - 1)$

55. $C(x) = 25x^2 + 1000x + 10,000$

Exercises 6.2, page 394

1. .5; .25, .75, 1.25, 1.75   3. .6; 1.3, 1.9, 2.5, 3.1, 3.7   5. 8.625   7. 15.12   9. .077278

11. 5.625; 4.5   13. 1.61321; error $= .04241$ 15. 1.08 L   17. 2800 ft   19. Increase in the

population (in millions) from 1910 to 1950; rate of cigarette consumption t years after 1985; 20 to 50   21. Tons of

soil eroded during a 5-day period   23. The first area is 5 times the second.   25. The area under the graph of

$f(x) + g(x)$ is the sum of the area under the graph of $f(x)$ and the area under the graph of $g(x)$.

Exercises 6.3, page 409

1. 0   **3.** 5   **5.** 30   **7.** $\frac{4}{3}(1 - e^{-3})$ **9.** 14   **11.** $\frac{1}{2}(1 - e^{-10})$ **13.** $\ln 2$ **15.** $1\frac{7}{9}$

17. $\frac{1}{5}$ **19.** $3\frac{3}{4}$ **21.** $\frac{115}{6} + \ln\frac{7}{9}$ **23.** 10   **25.** $2(e^{1/2} - 1)$ **27.** $6\frac{3}{5}$ **29.** 15

31. 9   **33.** 3   **35.** Positive   **37.** 49 trillion   **39. (a)** 30 ft   **(b)**

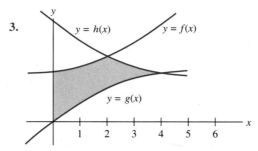

41. (a) \$1185.75   **(b)** The area under the marginal cost curve from $x = 2$ to $x = 8$.   **43.** The increase in profits resulting from increasing the production level from 44 to 48 units   **45. (a)** $368/15 \approx 24.5$ **(b)** The amount the temperature falls during the first 2 h   **47.** 2088 million m^3 **49.** 10

Exercises 6.4, page 420

1. $\displaystyle\int_1^2 f(x)\,dx + \int_3^4 -f(x)\,dx$ **3.**

5. $\frac{64}{3}$ **7.** $\frac{52}{3}$ **9.** 18   **11.** $\frac{32}{3}$ **13.** $\frac{32}{3}$ **15.** 72   **17.** 108   **19. (a)** $\frac{9}{2}$ **(b)** $\frac{19}{3}$ **(c)** $\frac{79}{6}$

21. $\frac{3}{2}$ **23.** $2 + 12\ln(\frac{3}{2})$ **25.** $\displaystyle\int_0^{20} (76.2e^{.03t} - 50 + 6.03e^{.09t})\,dt$ **27.** No; 20; the additional profit from using the original plan   **29. (a)** The distance between the two cars after 1 h   **(b)** after 2 h

Exercises 6.5, page 432

1. 3   **3.** 0   **5.** 8   **7.** 55°   **9.** ≈ 82 g   **11.** \$20   **13.** \$404.72   **15.** \$200

17. \$25   **19.** Intersection (100, 10), consumers' surplus = \$100, producers' surplus = \$250   **21.** \$3236.68

23. \$75,426   **25.** 13.35 y   **27. (b)** $1000e^{-.06t_1}\,\Delta t + 1000e^{-.06t_2}\,\Delta t + \cdots + 1000e^{-.06t_n}\,\Delta t$

(c) $f(t) = 1000e^{-.06t};\ 0 \le x \le 5$ **(d)** $\displaystyle\int_0^5 1000e^{-.06t}\,dt$ **(e)** \$4319.69   **29.** $\frac{4}{3}\pi r^3$ **31.** $\dfrac{31\pi}{5}$

33. 8π **35.** $\dfrac{\pi}{2}\left(1 - \dfrac{1}{e^{2r}}\right)$ **37.** $n = 4$, $b = 10$, $f(x) = x^3$ **39.** $n = 3$, $b = 7$, $f(x) = x + e^x$

41. The sum is approximated by $\displaystyle\int_0^3 (3 - x)^2\,dx = 9$.

Chapter 6: Supplementary Exercises, page 436

1. $-2e^{-x/2} + C$ **3.** $\frac{3}{5}x^5 - x^4 + C$ **5.** $-\frac{2}{3}(4 - x)^{3/2} + C$ **7.** $\frac{3}{4}$ **9.** $\frac{1}{3}\ln 4$ **11.** $\frac{5}{32}$

13. 8   **15.** $\frac{1}{3}(x - 5)^3 - 7$ **17. (a)** $2t^2 + C$ **(b)** Ae^{4t} **(c)** $\frac{1}{4}e^{4t} + C$ **19.** $.02x^2 + 150x + 500$

dollars   **21.** The total quantity of drug (in cubic centimeters) injected during the first 4 min   **23.** \$4708.71

25. .685714; .693147   **27.** \$433.33   **29.** 15   **31.** 11.48   **33. (a)** $f(t) = Q - \dfrac{Q}{A}t$ **(b)** $\dfrac{Q}{2}$

35. (a) The area under the curve $y = 1/(1 + t^2)$ from $t = 0$ to $t = 3$ **(b)** $1/(1 + x^2)$ **39.** $\frac{15}{4}$ **41.** True

43. 84,990 km³

CHAPTER 7
Exercises 7.1, page 450

1. $f(1, 0) = 1, f(0, 1) = 8, f(3, 2) = 25$ **3.** $f(0, 1) = 0, f(3, 12) = 18, f(a, b) = 3\sqrt{ab}$

5. $f(2, 3, 4) = -2, f(7, 46, 44) = \frac{7}{2}$ **11.** \$50. \$50 invested at 5% continuously compounded interest will yield

\$100 in 13.8 years   **13. (a)** \$18,750   **(b)** \$22,500; yes

15.
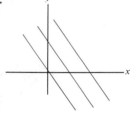

17.
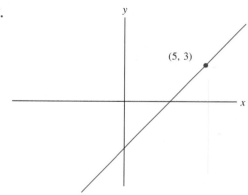

(5, 3)

19. $f(x, y) = y - 3x$ **21.** They correspond to the points having the same altitude above sea level.

23. d   **25.** c

Exercises 7.2, page 462

1. $5y, 5x$ **3.** $4xe^y, 2x^2e^y$ **5.** $-\dfrac{y^2}{x^2}, \dfrac{2y}{x}$ **7.** $4(2x - y + 5), -2(2x - y + 5)$

9. $(2xe^{3x} + 3x^2e^{3x})\ln y, x^2e^{3x}/y$ **11.** $\dfrac{2y}{(x + y)^2}, -\dfrac{2x}{(x + y)^2}$ **13.** $\dfrac{3\sqrt{K}}{2\sqrt{L}}$ **15.** $\dfrac{2xy}{z}, \dfrac{x^2}{z}, -\dfrac{1 + x^2y}{z^2}$

17. $ze^{yz}, xz^2e^{yz}, x(yz + 1)e^{yz}$ **19.** 1, 3   **21.** -12

23. $\dfrac{\partial f}{\partial x} = 3x^2y + 2y^2, \dfrac{\partial^2 f}{\partial x^2} = 6xy, \dfrac{\partial f}{\partial y} = x^3 + 4xy, \dfrac{\partial^2 f}{\partial y^2} = 4x, \dfrac{\partial^2 f}{\partial y\partial x} = \dfrac{\partial^2 f}{\partial x\partial y} = 3x^2 + 4y$ **25. (a)** Marginal pro-

ductivity of labor = 480; of capital = 40   **(b)** 240 fewer units produced   **27.** If the price of a bus ride increases

and the price of a train ticket remains constant, fewer people will ride the bus. An increase in train-ticket prices coupled

with constant bus fare should cause more people to ride the bus.   **29.** If the average price of videotapes increases and the average price of a VCR remains constant, people will purchase fewer VCRs. An increase in average VCR prices coupled with constant videotape prices should cause a decline in the number of videotapes purchased.

31. $\dfrac{\partial V}{\partial P}$ (20, 300) $= -.06$, $\dfrac{\partial V}{\partial T}$ (20, 300) $= .004$ **33.** $\dfrac{\partial f}{\partial r} > 0$, $\dfrac{\partial f}{\partial m} > 0$, $\dfrac{\partial f}{\partial p} < 0$

35. $\dfrac{\partial^2 f}{\partial x^2} = -\dfrac{45}{4} x^{-5/4} y^{1/4}$. Marginal productivity of labor is decreasing.

Exercises 7.3, page 473

1. $(-2, 1)$ **3.** $(26, 11)$ **5.** $(1, -3), (-1, -3)$ **7.** $(\sqrt{5}, 1), (\sqrt{5}, -1), (-\sqrt{5}, 1), (-\sqrt{5}, -1)$
9. $(\frac{1}{3}, \frac{4}{3})$ **11.** Relative minimum; neither relative maximum nor relative minimum   **13.** Relative maximum; neither relative maximum nor relative minimum; relative maximum   **15.** Neither relative maximum nor relative minimum   **17.** $(0, 0)$ min   **19.** $(-1, -4)$ max   **21.** $(0, -1)$ min   **23.** $(-1, 2)$ max; $(1, 2)$ neither max nor min   **25.** $(\frac{1}{4}, 2)$ min; $(\frac{1}{4}, -2)$ neither max nor min   **27.** $(\frac{1}{2}, \frac{1}{6}, \frac{1}{2})$ **29.** 14 in. $\times$ 14 in. $\times$ 28 in.
31. $x = 120, y = 80$

Exercises 7.4, page 485

1. 58 at $x = 6, y = 2, \lambda = 12$ **3.** 13 at $x = 8, y = -3, \lambda = 13$ **5.** $x = \frac{1}{2}, y = 2$ **7.** 5, 5
9. Base 10 in., height 5 in.   **11.** $F(x, y, \lambda) = 4xy + \lambda(1 - x^2 - y^2)$; $\sqrt{2} \times \sqrt{2}$ **13.** $F(x, y, \lambda) =$
$3x + 4y + \lambda(18,000 - 9x^2 - 4y^2)$; $x = 20, y = 60$ **15.** (a) $F(x, y, \lambda) = 96x + 162y + \lambda(3456 - 64x^{3/4}y^{1/4})$;
$x = 81, y = 16$ (b) $\lambda = 3$ **17.** $x = 12, y = 2, z = 4$ **19.** $x = 2, y = 3, z = 1$
21. $F(x, y, z, \lambda) = 3xy + 2xz + 2yz + \lambda(12 - xyz)$; $x = 2, y = 2, z = 3$
23. $F(x, y, z, \lambda) = xy + 2xz + 2yz + \lambda(32 - xyz)$; $x = y = 4, z = 2$

Exercises 7.5, page 495

1. $y = -2x + \frac{7}{3}$ **3.** $y = x + 2$ **5.** $y = .4x + .9$ **7.** $y = .7x - .6$ **9.** $y = .5875x + 1.85$
11. $\frac{2}{3}$ **13.** 46   **15.** (a) $y = -4.2x + 22$ (b) 8.6°C   **17.** (a) $y = -327.2x + 1591.4$ (b) 610

Exercises 7.6, page 503

1. $e^2 - 2e + 1$ **3.** $2 - e^{-2} - e^2$ **5.** $309\frac{3}{8}$ **7.** $\frac{5}{3}$ **9.** $\frac{38}{3}$ **11.** $e^{-5} + e^{-2} - e^{-3} - e^{-4}$
13. $9\frac{1}{3}$

Chapter 7: Supplementary Exercises, page 505

1. $2, \frac{5}{6}, 0$ **3.** ≈ 20. Ten dollars increases to 20 dollars in 11.5 y.   **5.** $6x + y, x + 10y$

7. $\dfrac{1}{y} e^{x/y}, -\dfrac{x}{y^2} e^{x/y}$ **9.** $3x^2, -z^2, -2yz$ **11.** 6, 1   **13.** $20x^3 - 12xy, 6y^2, -6x^2, -6x^2$

15. $-201, 5.5$. At the level $p = 25, t = 10,000$, an increase in price of \$1 will result in a loss in sales of approximately 201 calculators, and an increase in advertising of \$1 will result in the sale of approximately 5.5 additional calculators.

17. $(3, 2)$ **19.** $(0, 1), (-2, 1)$ **21.** Min at $(2, 3)$ **23.** Min at $(1, 4)$; neither max nor min at $(-1, 4)$

25. $20; x = 3, y = -1, \lambda = 8$ **27.** $x = \frac{1}{2}, y = \frac{3}{2}, z = 2$ **29.** $F(x, y, \lambda) = xy + \lambda(40 - 2x - y); x = 10,$
$y = 20$ **31.** $y = \frac{5}{2}x - \frac{5}{3}$ **33.** $y = -2x + 1$ **35.** 5160 **37.** 40

CHAPTER 8

Exercises 8.1, page 513

1. $\frac{\pi}{6}, \frac{2\pi}{3}, \frac{7\pi}{4}$ **3.** $\frac{5\pi}{2}, -\frac{7\pi}{6}, -\frac{\pi}{2}$ **5.** 4π **7.** $\frac{7\pi}{2}$ **9.** -3π **11.** $\frac{2\pi}{3}$

13.   **15.**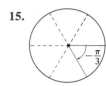

17.

Exercises 8.2, page 521

1. $\sin t = \frac{1}{2}, \cos t = \frac{\sqrt{3}}{2}$ **3.** $\sin t = \frac{2}{\sqrt{13}}, \cos t = \frac{3}{\sqrt{13}}$ **5.** $\sin t = \frac{1}{4}, \cos t = \frac{\sqrt{15}}{4}$ **7.** $\sin t = \frac{1}{\sqrt{5}},$

$\cos t = -\frac{2}{\sqrt{5}}$ **9.** $\sin t = \frac{\sqrt{2}}{2}, \cos t = -\frac{\sqrt{2}}{2}$ **11.** $\sin t = -.8, \cos t = -.6$ **13.** .4 rad

15. 3.59 **17.** 10.86 **19.** $b = 1.31, c = 2.73$ **21.** $\frac{\pi}{6}$ **23.** $\frac{3\pi}{4}$ **25.** $\frac{5\pi}{8}$ **27.** $\frac{\pi}{4}$

29. $\frac{\pi}{3}$ **31.** $-\frac{\pi}{6}$ **33.** $\frac{\pi}{4}$ **35.** Here $\cos t$ decreases from 1 to -1. **37.** $0, 0, 1, -1$

39. .2, .98, .98, −.2

Exercises 8.3, page 534

1. $4 \cos 4t$ **3.** $4 \cos t$ **5.** $-6 \sin 3t$ **7.** $1 - \pi \sin \pi t$ **9.** $-\cos(\pi - t)$

11. $-3 \cos^2 t \sin t$ **13.** $\frac{\cos(\sqrt{x} - 1)}{2\sqrt{x} - 1}$ **15.** $\frac{\cos(x - 1)}{2\sqrt{\sin (x - 1)}}$ **17.** $8(1 + \cos t)^7 \cdot (-\sin t)$

19. $-6x^2 \cos x^3 \sin x^3$ **21.** $e^x(\sin x + \cos x)$

23. $2 \cos(2x) \cos(3x) - 3 \sin(2x) \sin(3x)$ **25.** $\cos^{-2} t$ **27.** $-\frac{\sin t}{\cos t}$ **29.** $\frac{\cos(\ln t)}{t}$ **31.** -3

33. $y = 2$ **35.** $\frac{1}{2} \sin 2x + C$ **37.** $-\frac{\cos(4x + 1)}{4} + C$ **39.** (a) max $= 120$ at $0, \frac{\pi}{3}$; min $= 80$ at $\frac{\pi}{6}$,

$\frac{\pi}{2}$ **(b)** 57 **41.** Maximum of 73° occurs at $t = 208$ (July 28); minimum of 45° occurs at $t = 25$ (Jan. 25).

43. 0 **45.** (a) 69° (b) increasing 1.6°/wk (c) weeks 6 and 44 (d) weeks 28 and 48 (e) week 25, week 51

(f) week 12, week 38

Exercises 8.4, page 541

1. $\sec t = \dfrac{\text{hypotenuse}}{\text{adjacent}}$ **3.** $\tan t = \frac{5}{12}$, $\sec t = \frac{13}{12}$ **5.** $\tan t = -\frac{1}{2}$, $\sec t = -\dfrac{\sqrt{5}}{2}$ **7.** $\tan t = -1$,

$\sec t = -\sqrt{2}$ **9.** $\tan t = \frac{4}{3}$, $\sec t = -\frac{5}{3}$ **11.** 75 $\tan(.7) \approx 63$ ft **13.** $\tan t \sec t$ **15.** $-\csc^2 t$

17. $4\sec^2(4t)$ **19.** $-3\sec^2(\pi - x)$ **21.** $4(2x + 1)\sec^2(x^2 + x + 3)$ **23.** $\dfrac{\sec^2 \sqrt{x}}{2\sqrt{x}}$

25. $\tan x + x \sec^2 x$ **27.** $2 \tan x \sec^2 x$ **29.** $6[1 + \tan(2t)]^2 \sec^2(2t)$ **31.** $\sec t$

Chapter 8: Supplementary Exercises, page 544

1. $\dfrac{3\pi}{2}$ **3.** $-\dfrac{3\pi}{4}$ **5.** **7.** .8, .6, $\frac{4}{3}$ **9.** $-.8, -.6, \frac{4}{3}$ **11.** $\pm\dfrac{2\sqrt{6}}{5}$

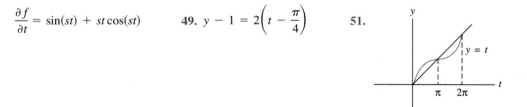

13. $\dfrac{\pi}{4}, \dfrac{5\pi}{4}, -\dfrac{3\pi}{4}, -\dfrac{7\pi}{4}$ **15.** Negative **17.** 16.3 ft **19.** 3 cos t **21.** $(\cos \sqrt{t}) \cdot \frac{1}{2} t^{-1/2}$

23. $x^3 \cos x + 3x^2 \sin x$ **25.** $-\dfrac{2 \sin(3x) \sin(2x) + 3 \cos(2x) \cos(3x)}{\sin^2(3x)}$ **27.** $-12 \cos^2(4x) \sin(4x)$

29. $[\sec^2(x^4 + x^2)](4x^3 + 2x)$ **31.** $\cos(\tan x) \sec^2 x$ **33.** $\sin x \sec^2 x + \sin x$ **35.** $\dfrac{\cos x}{\sin x}$

37. $3e^{3x} \sin^4 x + 4e^{3x} \sin^3 x \cos x$ **39.** $\dfrac{\tan(3t) \cos t - 3 \sin t \sec^2(3t)}{\tan^2(3t)}$ **41.** $e^{\tan t} \sec^2 t$

43. $2(\cos^2 t - \sin^2 t)$ **45.** $\dfrac{\partial f}{\partial s} = \cos s \cos(2t)$, $\dfrac{\partial f}{\partial t} = -2 \sin s \sin(2t)$ **47.** $\dfrac{\partial f}{\partial s} = t^2 \cos(st)$,

$\dfrac{\partial f}{\partial t} = \sin(st) + st \cos(st)$ **49.** $y - 1 = 2\left(t - \dfrac{\pi}{4}\right)$ **51.**

53. 4 **55.** $\dfrac{\pi^2}{2} - 2$ **57.** (a) $V'(t) = 8\pi \cos\left(160\ \pi t - \dfrac{\pi}{2}\right)$ (b) 8π L/min (c) 16 L/min

CHAPTER 9

Exercises 9.1, page 555

1. $\frac{1}{6}(x^2 + 4)^6 + C$ 3. $\frac{1}{4}(x^2 - 5x)^4 + C$ 5. $e^{(x^3 - 1)} + C$ 7. $-\frac{1}{3}(4 - x^2)^{3/2} + C$

9. $(2x + 1)^{1/2} + C$ 11. $\frac{1}{6}e^{3x^2} + C$ 13. $\frac{1}{2}(\ln 2x)^2 + C$ 15. $\frac{1}{5}\ln|x^5 + 1| + C$

17. $-\frac{1}{12}(x^4 + 4)^{-3} + C$ 19. $\frac{1}{4}(\ln x)^2 + C = (\ln \sqrt{x})^2 + C$ 21. $\frac{1}{3}\ln|x^3 - 3x^2 + 1| + C$

23. $-4e^{-x^2} + C$ 25. $\frac{1}{2}\ln|\ln x| + C$, or $\frac{1}{4}\ln|\ln x^2| + C$ 27. $-\frac{1}{10}(x^2 - 6x)^5 + C$

29. $-\frac{1}{2}(\ln x)^{-2} + C$ 31. $f(x) = (x^2 + 9)^{1/2} + 3$ 33. $2e^{\sqrt{x+5}} + C$ 35. $\frac{1}{2}\tan x^2 + C$

37. $\frac{1}{2}(\sin x)^2 + C$ 39. $2 \sin \sqrt{x} + C$ 41. $-\frac{1}{4}\cos^4 x + C$ 43. $-\frac{2}{15}(1 - \sin 5x)^{3/2} + C$

45. $-\cos(x^2 - x + 1) + C$ 47. $\frac{1}{2}(x^2 + 5)^2 + C = \frac{1}{2}x^4 + 5x^2 + \frac{25}{2} + C$; $\frac{1}{2}x^4 + 5x^2 + C$

Exercises 9.2, page 562

1. $\frac{1}{5}xe^{5x} - \frac{1}{25}e^{5x} + C$ 3. $\frac{x}{10}(2x + 1)^5 - \frac{1}{120}(2x + 1)^6 + C$ 5. $-xe^{-x} - e^{-x} + C$

7. $2x(x + 1)^{1/2} - \frac{4}{3}(x + 1)^{3/2} + C$ 9. $-\frac{1}{15}x(1 - 3x)^5 - \frac{1}{270}(1 - 3x)^6 + C$ 11. $-3xe^{-x} - 3e^{-x} + C$

13. $\frac{2}{3}x(x + 1)^{3/2} - \frac{4}{15}(x + 1)^{5/2} + C$ 15. $\frac{1}{3}x^{3/2}\ln x - \frac{2}{9}x^{3/2} + C$ 17. $x^2 \sin x + 2x \cos x - 2 \sin x + C$

19. $\frac{x^2}{2}\ln 5x - \frac{1}{4}x^2 + C$ 21. $4x \ln x - 4x + C$ 23. $-e^{-x}(x^2 + 2x + 2) + C$

25. $\frac{1}{5}x(x + 5)^5 - \frac{1}{30}(x + 5)^6 + C$ 27. $\frac{1}{10}(x^2 + 5)^5 + C$ 29. $3(3x + 1)e^{x/3} - 27e^{x/3} + C$

31. $\frac{1}{2}\tan(x^2 + 1) + C$ 33. $\frac{1}{2}xe^{2x} - \frac{1}{4}e^{2x} + \frac{1}{3}x^3 + C$ 35. $\frac{1}{2}e^{x^2} - x^2 + C$

37. $2x\sqrt{x + 9} - \frac{4}{3}(x + 9)^{3/2} + 38$

Exercises 9.3, page 569

1. $\frac{1}{18}$ 3. 0   5. $\frac{1}{2}$ 7. $\frac{64}{3}$ 9. $\frac{23}{6}$ 11. 9   13. $\frac{1}{3}(e^{27} - e)$ 15. 0   17. 4

19. $\frac{1}{\pi}$ 21. $\frac{\pi}{2}$ 23. $\frac{9\pi}{2}$ 25. $\frac{16}{3}$

Exercises 9.4, page 578

1. $\Delta x = .4$; 3, 3.4, 3.8, 4.2, 4.6, 5   3. $\Delta x = .5$; .25, .75, 1.25, 1.75   5. $\Delta x = .6$; 1.3, 1.9, 2.5, 3.1, 3.7

7. $(n = 2)$ 40, $(n = 4)$ 41, exact: $41\frac{1}{3}$ 9. .63107, exact: .63212   11. .09375, exact: .08333

13. 1.03741, exact: .8   15. $M = 72$, $T = 90$, $S = 78$, exact: 78   17. $M = 44.96248$, $T =$

72.19005, $S = 54.03834$, exact: 53.59815   19. $M = 573.41792$, $T = 612.10802$, $S = 586.31461$, exact:

586.26358   21. 3.24124   23. 1.61347   25. 25,750 ft^2 27. 2150 ft   29. .0018

Exercises 9.5, page 589

1. $164,840   3. $180,475   5. $63,142,300   7. 1,490,500

9. 1,399,600;

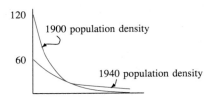

People moved away from the center of the city

11. 290,635

13. (a) $80\pi te^{-.5t}\Delta t$ thousand people   (b) $P'(t)$ (c) the number of people who live between 5 mi and $5 + \Delta t$ mi from the city center   (d) $P'(t) = 80\pi te^{-.5t}$ (e) $P(b) - P(a) = \int_a^b P'(t)\, dt = \int_a^b 80\pi te^{-.5t}\, dt$

Exercises 9.6, page 595

1. 0   **3.** No limit   **5.** $\frac{1}{4}$ **7.** 2   **9.** 5   **11.** 6   **13.** 2   **15.** $\frac{1}{2}$ **17.** 2

19. 1   **21.** Area under the graph from 1 to b is $\frac{5}{14}(14b + 18)^{1/5} - \frac{5}{7}$. This has no limit as $b \to \infty$.   **23.** $\frac{1}{2}$

25. Divergent   **27.** $\frac{1}{6}$ **29.** $\frac{2}{3}$ **31.** 1   **33.** $2e$ **35.** Divergent   **37.** $\frac{1}{2}$ **39.** 2

41. $\frac{1}{4}$ **43.** 2   **45.** $\frac{1}{6}$ **49.** \$50,000

Chapter 9: Supplementary Exercises, page 598

1. $-\frac{1}{6}\cos(3x^2) + C$ **3.** $-\frac{1}{36}(1 - 3x^2)^6 + C$ **5.** $\frac{1}{3}[\ln x]^3 + C$ **7.** $-\frac{1}{3}(4 - x^2)^{3/2} + C$

9. $-\frac{1}{3}e^{-x^3} + C$ **11.** $\frac{1}{3}x^2 \sin(3x) - \frac{2}{27}\sin(3x) + \frac{2}{9}x\cos(3x) + C$ **13.** $2(x \ln x - x) + C$

15. $\frac{2}{3}x(3x - 1)^{1/2} - \frac{4}{27}(3x - 1)^{3/2} + C$ **17.** $\frac{1}{4}\frac{x}{(1 - x)^4} - \frac{1}{12}\frac{1}{(1 - x)^3} + C$ **19.** $f(x) = x$,

$g(x) = e^{2x}$ **21.** $u = \sqrt{x + 1}$ **23.** $u = x^4 - x^2 + 4$ **25.** $f(x) = (3x - 1)^2$, $g(x) = e^{-x}$; then integrate by parts again.   **27.** $f(x) = 500 - 4x$, $g(x) = e^{-x/2}$ **29.** $f(x) = \ln(x + 2)$, $g(x) = \sqrt{x + 2}$

31. $u = x^2 + 6x$ **33.** $u = x^2 - 9$ **35.** $u = x^3 - 6x$ **37.** $\frac{3}{8}$ **39.** $1 - e^{-2}$

41. $\frac{3}{4}e^{-2} - \frac{5}{4}e^{-4}$ **43.** $M = 3.93782, T = 4.13839, S = 4.00468$ **45.** $M = 12.84089, T = 13.20137$,

$S = 12.96105$ **47.** $\frac{1}{3}e^6$ **49.** Divergent   **51.** $2^{7/4}$ **53.** $2e^{-1}$ **55.** \$137,668

57. (a) $M(t_1)\Delta t + \cdots + M(t_n)\Delta t \approx \int_0^2 M(t)\, dt$ (b) $M(t_1)e^{-.1t_1}\Delta t + \cdots + M(t_n)e^{-.1t_n}\Delta t \approx \int_0^2 M(t)e^{-.1t}\, dt$

CHAPTER 10
Exercises 10.1, page 605

5. $y = e^{t/2}$ **7.** Yes   **9.** $y = 5$ **11.** $f(0) = 4, f'(0) = 5$ **13.** $y' = k(C - y), k > 0$

Exercises 10.2, page 614

1. $y = \sqrt[3]{15t - \frac{3}{2}t^2 + C}$ **3.** $y = -\frac{1}{2}\ln(-t^2 + C)$ **5.** $y = (-\frac{2}{3}t^{3/2} + C)^{-1}$ or $y = 0$

7. $y = (e^{t^3} + C)^{1/3}$ **9.** $y = (\sqrt{t} + C)^2$ or $y = 0$ **11.** $y = 3 + Ae^{4t}$ **13.** $y = 2 - Ae^{-t}$

15. $y = \pm\sqrt{(\ln t)^2 + C}$ **17.** $y = \dfrac{1}{t - t \ln t + C} + 3$ or $y = 3$ **19.** $y = \dfrac{1}{2}\ln(2t^2 - 2t + e^6)$

21. $y = \sqrt[3]{3 \sin t - 3t \cos t + 8}$ **23.** $y = \frac{5}{3} - \frac{2}{3}e^{-3t}$ **25.** $y = -\sqrt{2t + 2 \ln t + 7}$

27. $y = \frac{2}{5} + \frac{3}{5}e^{(5/2)t^2}$ **29.** $y = (3x^{1/2} \ln x - 6x^{1/2} + 14)^{2/3}$ **31.** $y = A(p + 3)^{-1/2}, A > 0$

33. $y' = k(1 - y)$, where $y = p(t)$, $y(0) = 0$; $y = 1 - e^{-kt}$ **35.** $\dfrac{dV}{dt} = kV^{2/3}$, where $k < 0$. Solution:

$V = (3 - \frac{1}{8}t)^3$, $V = 0$ when $t = 24$ weeks.   **37.** $y = be^{Ce^{-at}}$, C any number

Exercises 10.3, page 622

1. 3   **3.** Increasing   **5.** $f(1) \approx -\frac{9}{4}$

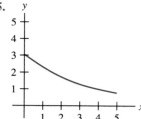

7. $f(2) \approx \frac{27}{8}$

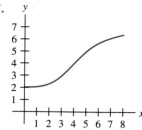

9. Euler's method yields $f(1) \approx .37011$; solution: $f(t) = 1/(\frac{1}{2}t^2 + t + 1)$; $f(1) = .4$; error $= .02989$

11. (a) $y' = k(1 - y)$, $y(0) = 0$, where $k > 0$ **(b)** $3k - 3k^2 + k^3$ **(c)** $y = 1 - e^{-kt}$, $y(3) = 1 - e^{-3k}$

(d) Euler's method gives .27100, exact solution

13.

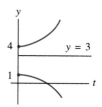

15.

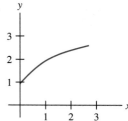

17.

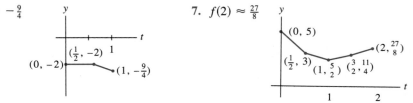

Exercises 10.4, page 632

1.

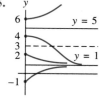

3.

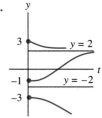

5.

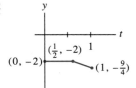

7.
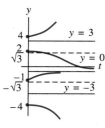

9.
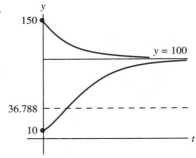

11.
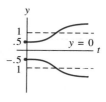

13.
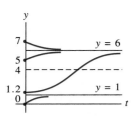

15.
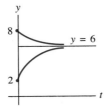

17.
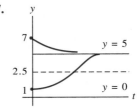

19.
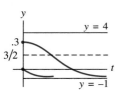

21.
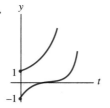

23.
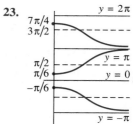

25.
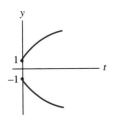

27.
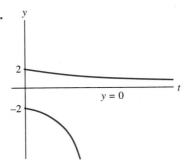

29.

31. $y' = ky(H - y), k > 0$
H = height at maturity
$y = f(t)$

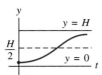

Exercises 10.5, page 646

1. $y' = k(100 - y), k > 0$

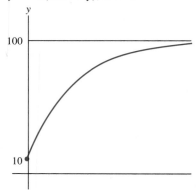

3. $y' = ky(M - y), k > 0$

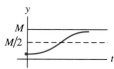

The reaction is proceeding fastest when $y = M/2$.

5. $y' = k(c - y), k > 0$

7. $y' = ky^2, k < 0$

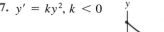

9. $y' = k(E - y), k > 0$

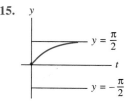

11. $y' = .055y - 300, y(0) = 5400$

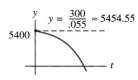

13. (a) $y' = .05y + 10,000, y(0) = 0$ (b) $y = 200,000(e^{.05t} - 1)$, $56,806

15. (a) $y' = .05 - .2y, y(0) = 6.25$ (b) $y' = .13 - .2y, y(0) = 6.25$

17. $y' = -.14y$ **19.**

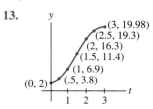

Chapter 10: Supplementary Exercises, page 649

1. $y = \sqrt[3]{3t^4 - 3t^3 + 6t + C}$ **3.** $y = Ate^{-3t}$ **5.** $y = \frac{3}{7}t + 3$

7. $y = \sqrt{4t^3 - t^2 + 49}$ **9.** $f(t) = 2t, f(2) = 4$ **11.** Decreasing

13.

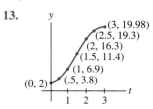

15.

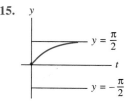

17.

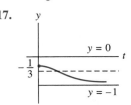

19.

21.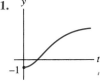

23.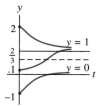

25. (a) $N' = .015N - 3000$ **(b)** There is a constant solution $N = 200,000$, but it is unstable. It is unlikely a city would have such a constant population. 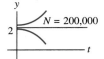   **27.** 13.863 y

CHAPTER 11
Exercises 11.1, page 660

1. $E(X) = \frac{4}{5}$, $\text{Var}(X) = .16$, standard deviation $= .4$

3. (a) $\text{Var}(X) = 1$ **(b)** $\text{Var}(X) = 4$ **(c)** $\text{Var}(X) = 16$

5. (a)

Accidents	0	1	2	3
Probability	.21	.5	.25	.04

(b) $E(X) = 1.12$

(c) Average of 1.12 accidents per week during year   **7. (a)** 25%   **(b)** $100c^2\%$

9. $E(X) = 90,000$. The grower should spend the $5000.

Exercises 11.2, page 672

7. $\frac{1}{4}$ **9.** $\frac{1}{15}$ **11.** 12   **13.**    **15.** $\frac{1}{12}$ **17.** $\frac{1}{2}$ **19.** $\frac{3}{4}$

21. $f(x) = \dfrac{1}{4\sqrt{x-1}}$ **23.** $\frac{1}{5}x - \frac{2}{5}$ **25. (a)** $\frac{19}{63}$ **(b)** $(x^3-1)/63$ **(c)** $\frac{19}{63}$ **31.** .586   **33.** 1.8

35. .6   **37. (b)** $1 - x^{-4}$ **(c)** $\frac{15}{16}, \frac{1}{16}$

Exercises 11.3, page 680

1. $E(X) = 4$, $\text{Var}(X) = 2$ **3.** $E(X) = 3$, $\text{Var}(X) = \frac{4}{3}$ **5.** $E(X) = \frac{5}{6}$, $\text{Var}(X) = \frac{5}{252}$ **7.** $E(X) = \frac{2}{5}$, $\text{Var}(X) = \frac{1}{25}$ **9. (a)** $F(x) = 10x^3 - 15x^4 + 6x^5$ **(b)** $\frac{53}{512}$ **(c)** $\frac{1}{2}$. On the average about half of the newspaper's space is devoted to advertising.   **(d)** $\frac{1}{28}$ **11. (a)** 2. The average useful life of the component is 200 h   **(b)** $\frac{1}{2}$

13. 8 min   **15. (a)** $F(x) = \frac{1}{6}x^2 - \frac{1}{54}x^3 - 1$ **(b)** $\frac{23}{27}$ **(c)** 412.5 worker-hours   **(d)** .5344   **17.** $E(X) = \frac{4}{3}$,

$\text{Var}(X) = \frac{2}{9}$ **19.** $3\sqrt{2}$ **21.** $3/\sqrt{2}$ hundred hours   **23.** $\frac{5}{6}$ min   **25.** *Hint:* Compute $\int_A^B xf(x)\, dx$
using integration by parts.

Exercises 11.4, page 692

1. $E(X) = \frac{1}{3}$, $\text{Var}(X) = \frac{1}{9}$ **3.** $E(X) = 5$, $\text{Var}(X) = 25$ **5.** $e^{-1} - e^{-2}$ **7.** $1 - e^{-2/3}$ **9.** e^{-3}

11. $1 - e^{-2}$ **13. (a)** $e^{-1/3}$ **(b)** $r(t) = e^{-t/72}$ **15.** $\mu = 4$, $\sigma = 1$ **17.** $\mu = 0$, $\sigma = 3$

23. (a) .4032   **(b)** .4013   **(c)** .8351   **(d)** .9772   **25. (a)** .4772   **(b)** .4772   **27.** .1587

29. The Capital Beltway   **31.** 1.22%   **33.** *Hint:* First show that $\Pr(a \le X \le b) = e^{-ka}(1 - e^{-kb})$.

Chapter 11: Supplementary Exercises, page 695

1. (a) .125, .2969   **(b)** $E(X) = 1.5$, $\text{Var}(X) = .15$ **3.** $1 - e^{A-x}$ **5.** $c_2 = \frac{1}{2}$, $c_4 = \frac{1}{4}$ **7. (a)** 5.01

(b) 100   **9. (a)** .96   **(b)** 1.8 thousand gallons   **(c)** $1 - x/2$ **11. (a)** $E(X) = 22.5$, $\text{Var}(X) = 2.0833$

(b) 21.5   **13. (a)** 1/300   **(b)** .375   **(c)** \$17,222   **15.** $2/\pi$ **17.** \$54.88   **19.** $k \approx .35$

21. .4938   **23.** 69.15%   **25.** $a \approx .25$ **27. (b)** .6826   **29. (a)** $\Pr\left(-\dfrac{1}{k} \le X \le \dfrac{3}{k}\right) =$

$\Pr\left(0 \le X \le \dfrac{3}{k}\right) \ge \dfrac{3}{4}$ **(b)** .95

CHAPTER 12
Exercises 12.1, page 709

1. $x - \frac{1}{6}x^3$ **3.** $5 + 10x + 10x^2 + \frac{20}{3}x^3$ **5.** $1 + 2x - 2x^2 + 4x^3$ **7.** $x + 3x^2 + \frac{9}{2}x^3$

9. $p_4(x) = 1 + x + \frac{1}{2}x^2 + \frac{1}{6}x^3 + \frac{1}{24}x^4$, $e^{.01} \approx p_4(.01) = 1.01005$

11.

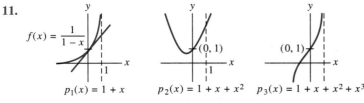

$p_1(x) = 1 + x$ $p_2(x) = 1 + x + x^2$ $p_3(x) = 1 + x + x^2 + x^3$

13. $p_n(x) = 1 + x + \dfrac{1}{2}x^2 + \dfrac{1}{3!}x^3 + \cdots + \dfrac{1}{n!}x^n$ **15.** $p_2(x) = x^2$, area $\approx .0417$

17. $1 + (x - 4) + (x - 4)^2 + (x - 4)^3$ **19.** $p_3(x) = -1 + \frac{1}{2}(x - \pi)^2$, $p_4(x) = -1 + \frac{1}{2}(x - \pi)^2 - \frac{1}{24}(x - \pi)^4$

21. 3.04958   **23.** $p_1(x) = 19 + 33(x - 2)$, $p_2(x) = 19 + 33(x - 2) + 24(x - 2)^2$, $p_3(x) =$

$19 + 33(x - 2) + 24(x - 2)^2 + 8(x - 2)^3$, $p_n(x) = 19 + 33(x - 2) + 24(x - 2)^2 + 8(x - 2)^3 + (x - 2)^4$, $n \ge 4$

25. $p_2(z) = \dfrac{1}{\sqrt{2\pi}} - \dfrac{1}{2\sqrt{2\pi}}z^2$ **27.** $f''(0) = -5$, $f'''(0) = 7$ **29. (a)** 1

(b) $|R_3(.12)| \le \dfrac{1}{4!}(.12)^4 = 8.64 \times 10^{-6}$ **31. (a)** $R_2(x) = \dfrac{f^{(3)}(c)}{3!}(x - 9)^3$, where c is between 9 and x

(b) $f^{(3)}(c) = \dfrac{3}{8}c^{-5/2} \le \dfrac{3}{8}9^{-5/2} = \dfrac{1}{648}$ **(c)** $|R_2(x)| \le \dfrac{1}{648} \cdot \dfrac{1}{3!}(.3)^3 = \dfrac{1}{144} \times 10^{-3}$

Exercises 12.2, page 719

1. Let $f(x) = x^2 - 5$, $x_0 = 2$; then $x_1 = 2.25$, $x_2 = 2.2361$, $x_3 = 2.23607$.   **3.** Let $f(x) = x^3 - 6$, $x_0 = 2$; then $x_1 = 1.8333$, $x_2 = 1.81726$, $x_3 = 1.81712$.   **5.** If $x_0 = 2$, then $x_3 = 2.79130$.   **7.** $x_3 = .63707$

9.

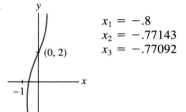

$x_1 = -.8$
$x_2 = -.77143$
$x_3 = -.77092$

11. .704   **13.** 8.21% per month   **15.** 1% per month

17. $x_1 \approx 3.5$, $x_2 \approx 3.2$ **19.** $-\frac{5}{4}$ **21.** $x_0 > 0$

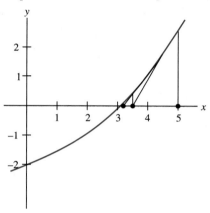

23. x_1 will be the exact root.   **25.** $x_0 = 1$, $x_1 = -2$, $x_2 = 4$, $x_3 = -8$

Exercises 12.3, page 730

1. $\frac{6}{5}$ **3.** $\frac{9}{10}$ **5.** 3   **7.** $\frac{25}{124}$ **9.** $\frac{21}{10}$ **11.** Divergent $(r = -\frac{8}{5})$ **13.** 25   **15.** $\frac{3}{11}$

17. $\frac{2}{9}$ **19.** $\frac{4007}{999}$ **21.** $.99\overline{9}\ldots = (.9)\dfrac{1}{1 - .1} = \dfrac{9}{10} \cdot \dfrac{10}{9} = 1$ **23.** \$190 billion

25. (a) $100 + 100(1.01)^{-1} + 100(1.01)^{-2} + \cdots = \displaystyle\sum_{k=0}^{\infty} 100(1.01)^{-k}$ (b) \$10,100   **27.** \$1,655,629

29. 20 mg   **31.** 5 mg   **33.** (a) 2.5   (b) yes; 3   **35.** 6   **37.** $\frac{1}{24}$ **39.** $\frac{15}{8}$

Exercises 12.4, page 740

1. $\dfrac{1}{3} - \dfrac{2}{9}x + \dfrac{2^2}{3^3}x^2 - \dfrac{2^3}{3^4}x^3 + \cdots$ **3.** $1 + \dfrac{1}{2}x - \dfrac{1}{2^2 \cdot 2!}x^2 + \dfrac{1 \cdot 3}{2^3 \cdot 3!}x^3 - \dfrac{1 \cdot 3 \cdot 5}{2^4 \cdot 4!}x^4 + \cdots$ **5.** $1 + 3x + 3^2x^2 +$

$3^3x^3 + \cdots$ **7.** $1 - x^2 + x^4 - x^6 + \cdots$ **9.** $1 - 2x + 3x^2 - 4x^3 + 5x^4 - \cdots$ **11.** $5 + \dfrac{5}{3}x +$

$\dfrac{5}{3^2 \cdot 2!}x^2 + \dfrac{5}{3^3 \cdot 3!}x^3 + \cdots$ **13.** $x - \dfrac{1}{2!}x^2 + \dfrac{1}{3!}x^3 - \dfrac{1}{4!}x^4 + \cdots$ **15.** $x - \frac{1}{2}x^2 + \frac{1}{3}x^3 - \frac{1}{4}x^4 + \cdots$

17. $1 - \dfrac{3^2}{2!}x^2 + \dfrac{3^4}{4!}x^4 - \dfrac{3^6}{6!}x^6 + \cdots$ **19.** $3x - \dfrac{3^3}{3!}x^3 + \dfrac{3^5}{5!}x^5 - \cdots$ **21.** $x + x^3 + \dfrac{1}{2!}x^5 + \dfrac{1}{3!}x^7 + \cdots$

23. $1 + \dfrac{1}{2!}x^2 + \dfrac{1}{4!}x^4 + \dfrac{1}{6!}x^6 + \cdots$ **25.** $1 + \dfrac{1}{2}x + \dfrac{1 \cdot 3}{2 \cdot 4}x^2 + \dfrac{1 \cdot 3 \cdot 5}{2 \cdot 4 \cdot 6}x^3$ **27.** $x - \dfrac{1}{2 \cdot 3}x^3 + \dfrac{1 \cdot 3}{2 \cdot 4 \cdot 5}x^5$

$- \dfrac{1 \cdot 3 \cdot 5}{2 \cdot 4 \cdot 6 \cdot 7}x^7 + \cdots$ **31.** 48   **33.** 0   **35.** $\left[x - \dfrac{1}{3}x^3 + \dfrac{1}{5 \cdot 2!}x^5 - \dfrac{1}{7 \cdot 3!}x^7 + \cdots \right] + C$

37. $\left[x - \dfrac{1}{4}x^4 + \dfrac{1}{7}x^7 - \dfrac{1}{10}x^{10} + \cdots \right] + C$ **39.** $1 - \dfrac{1}{3} + \dfrac{1}{5 \cdot 2!} - \dfrac{1}{7 \cdot 3!} + \cdots$

Exercises 12.5, page 750

1. $\frac{1}{16}$ **3.** $\frac{1}{8}$ **5.** $\frac{1}{12}$ **7.** $(\frac{1}{3})(\frac{2}{3})^n$ **9.** $\frac{9}{19}$ **11.** .08425   **13.** .44063   **15.** .95956

17. .99723   **19.** .01029   **21.** $\lambda = 2$ **23.** 461

Chapter 12: Supplementary Exercises, page 752

1. $x + \frac{3}{2}x^2$ **3.** $8 - 7x^2 + x^3$ **5.** $9 + 6(x - 3) + (x - 3)^2$ **7.** $p_2(x) = 2x^2, \frac{1}{12}$

9. (a) $3 + \frac{1}{6}(x - 9) - \frac{1}{216}(x - 9)^2$ (b) 2.949583   (c) 2.949576   **11.** $\frac{374}{105} \approx 3.5619$ **13.** $\frac{4}{7}$

15. $\frac{1}{7}$ **17.** 1   **19.** e^2 **21.** $\frac{9}{2}$ **23.** $1 - x^3 + x^6 - x^9 + x^{12} - \cdots$ **25.** $1 + 6x + 27x^2 +$

$108x^3 + 405x^4 + \cdots$ **27.** (a) $1 - \dfrac{2^2}{2!}x^2 + \dfrac{2^4}{4!}x^4 - \dfrac{2^6}{6!}x^6 + \cdots$ (b) $x^2 - \dfrac{2^3}{4!}x^4 + \dfrac{2^5}{6!}x^6 - \dfrac{2^7}{8!}x^8 + \cdots$

29. $1 + 2x + 2x^2 + 2x^3 + 2x^4 + \cdots$ **31.** (a) $p_5(x) = x^2$ (b) 0   (c) $\frac{13}{42} \approx .3095$ (exact value .3103)

33. (a) $2x + 4x^3 + 6x^5 + 8x^7 + \cdots$ (b) $2x/(1 - x^2)^2$ **35.** \$566,666,667   **37.** \$120,067

39. \$3,285,603   **41.** $\frac{7}{16}$ **43.** .19537   **45.** 4

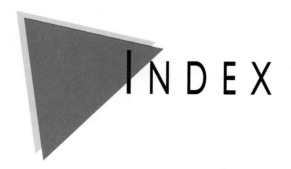

INDEX